SCHEDULES OF REINFORCEMENT

SCHEDULES OF REINFORCEMENT

by

C. B. FERSTER

Indiana University Medical Center

and

B. F. SKINNER

Harvard University

Copley Publishing Group
Acton, Massachusetts 01720

Printed in the United States of America

ISBN 0-87411-828-x

B. F. Skinner Reprint Series
edited by Julie S. Vargas.

For information about the series
please contact the B. F. Skinner Foundation,
P. O. Box 825, Cambridge, Massachusetts, 02238.

To
Members of the
"Pigeon Staff"

FOREWORD I

ONE OF B.F. SKINNER'S most important contributions to the experimental analysis of behavior was his discovery of schedules of reinforcement. In the 1956 paper, "A Case History in Scientific Method," Skinner described how his short supply of food pellets led him to ask why he should have to reinforce every bar press made by his rat subjects. His question, asked because of the purely pragmatic issue that he was running out of self-made pellets on a Saturday afternoon, led him to the discovery of the first "fixed interval" one minute (FI1) food delivery schedule of reinforcement. (It was fortuitous that Skinner did not use water as a reinforcer during this early experimental time since he would not have run short.) Until then he had used either a continuous reinforcement program, wherein every response was reinforced with a food pellet, or extinction, where no responses were reinforced. The number of bar presses was the variable of interest. The FI1 schedule arranged for the delivery of a pellet contingent upon the first response occurring after one minute since the last pellet delivery. Skinner reports that he was amazed that under this schedule the number of responses increased. Next, while attempting to maintain a stable rate of bar pressing by maintaining a constant level of hunger, he devised and scheduled "fixed ratio" reinforcement and ran extended-length sessions, some for 24 hours. These two types of intermittent reinforcement schedules produced very different patterns of responses. Skinner observed the results, and quickly realized the importance of differently scheduled consequences.

Scheduling the availability of consequences with regard to either time, or number of responses, or both, is the subject of this book. Operants are selected by the complex interaction of individual actions and the timing and frequency of historical and current consequences. These consequences, together with genetic influences thus determine operant behavior. Reinforcement schedules can induce adjunctive behavior (which is behavior maintained by a stimulus that gets its reinforcing value from some other ongoing schedule of reinforcement), and they direct the persistence and patterning of all behavior. Moreover, schedules contribute to a clarification of the topic of motivation in that they can account for patterns and cycles of behavior. The significance of schedules of reinforcement cannot be overestimated.

One concomitant development with the discovery of schedules was the evolution of the idea of considering operant behavior as a different form of "reflex" than that described by Pavlov. Instead of being elicited by an antecedent stimulus this kind of behavior operated upon the environment to produce consequences. Unlike respondent conditioning, with operants there was no need for individual trials. The behavior was free to occur at any time in that it was not dependent upon a prior stimulus.

For Skinner, this changed the dependent variable from latency or magnitude to responses per unit time. The total number of responses divided by the time over which the count was taken equals rate, and rate is the main dependent variable in the experimental analysis of behavior.

The discovery of schedules occurred around the years 1930 and 1931, or about 27 years before the publication of *Schedules of Reinforcement* (hereafter *Schedules*). *Schedules* was published the same year as *Verbal Behavior* in 1957, which was also a major contribution to our discipline, still not fully realized. The *Journal of the Experimental Analysis of Behavior* (*JEAB*), which was the first major professional journal in our discipline, did not appear for two more years. Next to Skinner's 1938 book *Behavior of Organisms, Schedules* was the most frequently cited reference source in the first 4 years of *JEAB*.

Ferster and Skinner exposed their research subjects to schedule parameters until the birds' performances were "stable," although the authors provide no objective definition of "stability" in *Schedules.* In a later book on research tactics, Murray Sidman noted that there is no final answer as to how to determine behavioral stability, but that one calls behavior stable when a behavior does not change over a sufficient period of time. The experimenter makes an informed judgement, based on his or her experience with subjects in a particular laboratory, as to whether a subject is showing transition or steady-state performance. Transition is the time when the subject is coming under the control of new schedule parameters, and stability is more representative of the final performance illustrating schedule control. Ferster & Skinner speak of transitions and display many records showing transitions from one schedule to another. The precise transition condition can be known only if the preceding performance is relatively stable and a final or terminal state of stability is regained. Orderliness of the performance is important if one is to judge the effects of some change in conditions, and experimental analysis and replication are paramount.

Ferster & Skinner determined stability by visual inspection; and no statistics were used in *Schedules.* They were able to employ this practice successfully because their experimental manipulations produced large behavioral changes that were easily seen. Often a change involved a beginning low rate of responding that then evolved into some high and typically patterned performance. A return to baseline was rarely reported in *Schedules,* but subsequent replications and research often included reinstatement of original conditions in a classic A-B-A single-subject reversal design.

Programming equipment used in *Schedules* is primitive by today's standards. It was either hand-made by Skinner, by graduate students on the pigeon lab staff, or constructed in the shop which was available to Psychology faculty at Harvard. Skinner typically made the first approximation of a needed piece of equipment from material on hand. The final models were more "professionally" constructed by Ralph Gerbrands, or by Rufus Grason and Steve Stadler, all of whom were mechanics in the shop and who later went on to found companies.

The prototype of the cumulative recorder currently in use was built by Gerbrands from Skinner's plans about 1951, and it allowed Ferster & Skinner to collect the data that are shown and described in *Schedules.* This recording device evolved from Skinner's earlier use of the physiologist's kymograph to record the occurrence of individual (usually muscle) events. The cumulative recorder is not totally unlike the invention of the microscope in that it, too, suddenly permitted the detailed inspection of ongoing life, in this case behavior and its relationship to environmental events. The cumulative record permits immediate inspection and estimation of rate, observation of changes in rate, as well as observation of the pattern of responding. This fact allows the experimenter to directly observe the effects of the independent variable and to develop statements of functional relationships as they occur with reference to a stable baseline. Rate has been the subject of extensive analysis, and keys called "rate meters" accompany cumulative records so as to allow the observer to visually gauge the rate of responding from the slope of the record. Since cumulative records show moment-to-moment changes in the rate and pattern of behavior as it occurs, as well as the operation of the food dispenser and other stimulus events, the records serve as discriminative stimuli to control the behavior of the scientist. Because the behavior and the contingencies are directly observed, such records reduce the temptation to make hypotheses about inner processes. By inspecting the record as it is being produced one can instantly determine many variables of influence during the session, and with surveillance video one can also watch the subject in real time. The experimental subject and the experimenter thus interact in a dynamic way impossible with group research, or work that requires extensive post hoc statistical analysis to detect effects.

Systematically scheduling reinforcement allows for an analysis of the conditions governing the probability that a given action will occur at a given time. These conditions specify the relation between independent and dependent variables. The history of contact with the schedule provides the behavior that is visible under current conditions.

Schedules was one of the first extensive presentations of single-subject behavior research to appear in the literature since Ebbinghaus's work 44 years earlier. The demonstration of such control and predictability had never been seen in behavior science before. In this regard *Schedules* was a major contributor to the formation of an independent science of behavior relations.

It was in 1976 that B.F. Skinner wrote "Farewell, My Lovely" as he observed in the *Journal of the Experimental Analysis of Behavior* the gradual decline in the use of cumulative records to report results. His choice of words for the title of that paper indicated not only the affection he had for the apparatus he invented but also the disappointment he felt for the decline in single-subject non-averaged behavior reports. During 1967, in four issues of *JEAB*, out of 55 papers there were 27 that contained cumulative records (49%); in 1977 there were 21 out of 69 (30%); in 1987 there were 12 out of 53 (22%); and in 1993 there were only two out of 42. (4%). We might predict that eventually there will be none.

The extensive research results presented in *Schedules,* which were collected over a three year period, are an incredible accomplishment, particularly considering the available manpower, the equipment constraints, and the many variations of schedules developed by Ferster and Skinner. The dedication of the book "To the Pigeon Staff" was intended to acknowledge the many graduate students who also contributed during the years of the project. There are more than 900 Figures in the book that illustrate original data, and 15 separate schedules are described and demonstrated. The results presented in *Schedules* illustrate over 70,000 recorded hours and ". . . approximately one-quarter billion responses."

Ferster & Skinner present the effects not only of the schedule itself on acquisition, transition, and steady-state performance but of many other independent variables as well. For example, drug treatment, brain ablations, added and novel stimuli, response contingent and non-contingent time out, body weight alterations, satiation, limited hold, pre-feeding, water deprivation, pre-aversive stimuli, and yoked subject conditions were used (p. 402). One dramatic example of the power of an independent variable affecting schedule controlled performance is the FI with added clock (p. 277). In this study a visual "clock" stimulus was co-related with the interreinforcement time during the fixed interval. When performance appeared to be in synchrony with the visual clock, the clock was suddenly reversed, that is, it ran backward. The result was that the FI behavior "scallop" also inverted. Such a demonstration illustrates how discriminative environmental variables, when paired with behavioral consequences, come to exert precise control over the actions of an organism.

The discovery and parametric manipulation of critical independent variables has been a milestone in every science. In biology there are many independent variables including genetic, environmental, and chemical ones. The discovery of schedules of response consequences as the critical variable in operant behavior was a major contribution to biobehavioral analysis.

The primary purpose of *Schedules* was ". . . to present a series of experiments designed to evaluate the extent to which the organism's own behavior enters into the determination of its subsequent behavior" (p. 3). A sometimes surprising finding for many experimenters, is how schedules of reinforcement generate comparable patterns of responding independent of the species producing the behavior. A schedule produces what the cumulative record portrays, a universal property of behavior, not that which is idiosyncratic to a particular organism. One obvious advantage that cumulative records provide is that they allow a direct comparison of these behavioral properties across species. As Skinner notes, once you have allowed for the different ways different species make contact with the environment, what remains of their behavior shows astonishingly similar properties. Homing pigeons were most frequently used by Ferster & Skinner, although White Carneaux pigeons were used, and some rat records are also shown. It is not possible to discern the species of the behaving organism by inspection of its cumulative record.

The commentary provided by Ferster & Skinner about the cumulative records shown in *Schedules* provides insight and observation that has yet to be fully exploited. It is not only a description of the records and what is happening but also an integration of the effects of behavior contacting the environment.

The results of accidents, mistakes, and sheer "luck" during the work are noted by Ferster & Skinner. As everyone who has conducted similar experiments has experienced, equipment breaks down, and when it does sometimes interesting things happen. A certain amount of serendipity also occurs in all research that begins with the question "I wonder what will happen when I do this?"

The book contains three introductory chapters in which the authors explain the schedules they studied and the way the records were obtained and prepared for presentation. Technical information about apparatus construction and operation, and methodology that was used in conducting the studies is presented in Chapter 3. Valuable information, some that is rarely available even today, describes how to maintain and train pigeons, as well as details of construction, purpose and use of programming instrumentation and apparatus. As Ferster & Skinner mention, it is important to plan ahead when collecting data in the form of cumulative records, especially from multiple chambers, because they accumulate quickly and arranging them for analysis can become a chore. The authors worked out a means of preparing records by trimming out the white space and "nesting" the remaining segments. In addition to the body of the book, a 140-item Glossary and an Index are included in the book. This may have been the first presentation of a glossary of operant analysis terms in the literature.

Since publication, no one has come close, or even attempted, such an extensive analysis of any variable in the experimental analysis of behavior. However, the results reported in *Schedules* have been systematically replicated literally hundreds of thousands of times in thousands of laboratories, classrooms, institutions, clinics and other sites around the world. The fact that intermittent schedules of reinforcement were first noted in 1938 yet no comprehensive treatment of them was forthcoming until 1957, (and only then by their own discoverer) illustrates how far ahead of his contemporaries Skinner was. His treatment of verbal behavior in a book with the same name, a schedule notation system, and the research reported with Robert Epstein in the Columban simulations are other examples of Skinner's versatility and creativity. We are still conducting parametric analyses and extensions of some of his seminal work.

Charles B. Ferster was a graduate student scheduled to finish his doctoral program at Columbia University with Fred S. Keller in June, 1950. Skinner asked for Keller's nomination of someone to work in his lab and Ferster came to Harvard as Skinner's full-time assistant on February 1, 1950.

According to Skinner, Ferster was first author of *Schedules* to underline his equal partnership because if Skinner, as a professor of psychology at the time, was first, then Ferster would be cast in the role of an assistant and Skinner reports that was not the case. Skinner always indicated that Ferster was an equal colleague in creating and conducting the work reported in *Schedules* and therefore he should receive equal credit. Authorship order was one way of insuring that credit.

Ferster's and Skinner's was possibly the most productive and creative five and one-half year collaborative effort we shall ever see in the experimental analysis of behavior. Fred Keller, Skinner's lifetime friend and colleague called it ". . . a landmark in the history of behavior science." Skinner himself wrote "It was the high point in my research history."

Schedules of Reinforcement, and a few other books, notably Keller & Schoenfeld's 1950 *Principles of Psychology,* the Holland and Skinner 1961 program and text *Analysis of Behavior,* and Sidman 's 1960 *Tactics of Scientific Research,* served as my own introduction to basic operant research. We read and used *Schedules* as a major reference source when I was a graduate student at Arizona State University in the early 1960s. Jack Michael or Iz Goldiamond did not assign us the book as a text in any particular course, but they suggested we learn as much from it as possible. And we did, because it was exciting reading and we learned as we collected data that resembled the curves in the book. We referred to it often, especially the early chapters on technique and instrumentation. Generating VI schedules could actually be fun. We used the well-known Fibonnacci numbers procedure (p. 335) to generate series of intervals, and some very long pieces of 16mm film to operate the equipment. We made particular use of the descriptions of the construction of equipment since there was still not much commercially available in those days and we frequently had to build our own apparatus from "state surplus" items. To this day I remain a great admirer of fine microswitches, a good soldering iron, and multimeters. The opportunity to scrounge and build equipment was a "hands-on" experience from which we learned about electrical components and programming, and also the mechanics of environment-behavior interaction. (We were engaged in serious and productive recycling decades before it became popular.) Mechner notation was our language and many evenings we tipped a few at Jack's while discussing a particular rat or pigeon subject's cumulative records with Bob Dylan on the "hi-fi " in the background. It was very common to have to break from these "seminars" in the middle of the night and go to the lab to check an ongoing experiment.

Schedules of Reinforcement has had a profound effect on the development of the experimental analysis of behavior and the understanding of both human and nonhuman behavior. It will remain a classic in the field.

Carl D. Cheney
Logan, Utah

March, 1997

FOREWORD II

I

Schedules of Reinforcement (*Schedules*) is an extraordinary monograph. It is an account of exciting scientific discoveries that were both important and original. The material was quite unfamiliar except to a small coterie who had been close to the work. A monograph of this magnitude is normally preceded by a series of technical papers in the scientific literature, describing reasonably coherent fragments of the work as it progresses, so that people in the field can have some familiarity (which often passes as understanding) with the new discoveries. But *Schedules* was not preceded by papers. It appeared full grown in 700 pages of almost entirely original material. To most psychologists even the nomenclature was unfamiliar, although some terms had been used before. In a word, it was an uncompromising challenge. Here it is, a mother lode of information on new discoveries: Go ahead and mine it. There are substantive written sections, mostly in the early chapters, but the bulk of *Schedules* is the 921 figures and their accompanying description. This atlas of figures contrasts sharply with the careful, analytical development in earlier books by Skinner (*Behavior of Organisms* and *Science and Human Behavior*) and even more with other books in psychology. An account of why *Schedules* is such a different book will give some perspective on the historical importance of the research and may help those approaching the book for the first time to understand it and appreciate its significance.

Ferster and Skinner discovered the incredible power of schedules of reinforcement to engender patterns of behavior. Their own behavior was so reinforced by the phenomena associated with schedule-controlled responding that, with the aid of automated equipment, they did research 24 hours a day, 7 days a week, year after year. Skinner had stable research funds from the Office of Naval Research that permitted Ferster and him to do uninterrupted research. Skinner did not publish any of this experimental work except for a report of a paper given at the 1951 Congress of Psychology in Sweden, and Ferster wrote one technical article about how to do research on operant behavior and published three experimental papers. Rather than stopping research to write reports, new experiments were planned on the basis of the results of those just conducted. Progress was evident from the capability to do experiments that were not possible or even conceivable earlier.

In the Festschrift volume for B.F. Skinner, Ferster gives a good description of the activities of the pigeon lab. He properly emphasizes the effort that went into technical developments and the availability of shop facilities to build equipment. Keys and feeders were tried and improved in a dozen iterations. The cumulative recording of responses, where each response causes a constant step movement of a pen perpendicular to the constant rate of movement of the paper, deserves special comment. Four different models of cumulative recorders were used, starting with one using a Ledex rotary switch as the main stepping mechanism and ending with a recorder built in the Psychological Laboratories by Ralph Gerbrands and later produced commercially by Gerbrands and Co. in a number of still more successively improved models. It was the cumulative recorder that permitted the recognition of the powerful effects of schedules. The information shown in a cumulative record is equally contained in a series of blips corresponding to the steps on a horizontal line of a polygraph, just as the information in most graphs can all be shown in a table of numbers. But the information *conveyed* to the observer by the cumulative record, as with a graph, is far greater. Changes in rate of responding, indicated by changes in slope, are more obvious in the cumulative record than in a polygraph. The cumulative record shows at a glance the pattern of changes in rate of responding in real time over periods of hours or longer. The characteristic properties of different schedules would not have been discovered without the cumulative recorder.

When at last Ferster and Skinner turned to writing an account of their research on schedule-controlled behavior, they described all of it rather than summarizing the main findings. Dealing with the cabinets filled with cumulative records from experiments over several years was a Herculean task that would have overwhelmed most people. Ferster and Skinner took to writing *Schedules* with boundless enthusiasm. Long before multiple schedule control had been discovered as an experimental phenomenon, it had been Skinner's practice to bring his professional activities under strong stimulus control by working without interruption in a particular place. The room with the cabinets of records was made the writing room. There were log books of the daily experiments, giving the details about schedules, parameter values, and the subjects that were studied each day. With these books it was possible to retrieve the records for all experiments. Ferster stopped doing any research (freeing about 10 independent experimental units for use by deserving graduate students), and for a long period neither Ferster nor Skinner came into the pigeon lab except for a look at the cumulative records of experiments after they had finished their daily stint of figure preparation.

Ferster's Festschrift description of the mechanics of preparing the figures captures the flavor of their joint activities. The general practice was for Ferster and Skinner together to look at cumulative records for each subject studied in a particular experiment and select records to be photographed. This selection was undoubtedly the most important intellectual activity involved in the creation of *Schedules* and its success is indicated by subsequent workers confirming the important characteristic features of schedule performances described in *Schedules*, but it is impossible for a reader now to assess how the selections were made or to appreciate the extraordinary talent required to understand the details of the records and to recognize the salient and replicable features. In Ferster's account of writing *Schedules*, he says "decisions about what to excerpt were made quickly, usually without much discussion, because we were both so familiar with the records." Because space limitations made it impossible to show photographs of all records as they were recorded without sacrificing details, they devised a method for collapsing the time scale by "telescoping" the pen tracings (pp. 26–27, also described by Ferster). Skinner loved making useful mechanical devices and also took pleasure in working with his hands, cutting out the pen tracings and pasting them on cardboard perfectly aligned with the coordinate scales showing representative slopes. Ferster later photographed the numbered figures in a part of the room equipped with a lighted stand and permanently mounted camera. After figures were mounted on cardboard, generally both Ferster and Skinner sat together and reviewed them, dictating descriptions of the figures, but sometimes Ferster alone dictated the descriptions. It is clear from reading the text that there was not much editing of the dictations, but Marilyn Ferster (later Gilbert) did do a final editing for consistency of usage. And in this way the accomplishments of their years of research were preserved for posterity.

Unfortunately, the importance of the work was not made obvious to the casual reader. The introductory sections of the book are helpful, but not enough explanatory material is presented to make parts of the book completely understandable to the uninitiated reader. The material in the introductory chapters explains the use of frequency of responding as an experimental datum, technical features about the experiments, the behavioral processes assumed to be important, and special features of fixed-ratio and fixed-interval schedules. Many figures show that responding can be differently controlled by different schedules hour after hour, day after day, without any broad conclusions about the importance and significance of these findings being made explicitly. Readers who understand the figures will certainly appreciate that Ferster and Skinner's studies were extraordinary, but even understanding the figures requires much work for the reader.

The summaries are mostly about particular individual experiments and there is little in *Schedules* to help a reader determine the optimum conditions for engendering definitive schedule performances characteristic of particular schedule conditions. Readers must work through the examples for themselves and undoubtedly some give up. For the most part, there is no indication of the chronological order of individual experiments. Technical advances led to an increased degree of control in later experiments, but these are reported together with the findings of earlier experiments. (In an intermediate design of a cumulative recorder, the displacement of the stepping pen indicating food presentations was horizontal rather than downward [p. 25], and in general, figures showing this feature are from experiments conducted before 1952.) If one leafs through the pages of any chapter there are clearly differences in the uniformity and reproducibility of performances under a particular type of schedule. Some of these differences in performances came from the continuing technical improvements in the designs of keys and feeders, others from differences in the past experience of subjects before exposure to the current condition or from the duration of exposure to current conditions, and, sometimes, from differences between subjects treated alike. (But often Ferster and Skinner did not use subjects with a common past experience, believing that a consistent finding established in subjects with diverse backgrounds showed greater generality than one established in subjects similarly treated.) The reader is helped by the chronological description of individual experiments. The figures that show the sequential development of behavioral performances toward a consistent pattern during continued exposure to unchanging conditions will generally be understandable to readers. The figures that show terminal performances may be misunderstood because in *Schedules* "terminal" means only the last day of exposure to that schedule and the figure may or may not be representative of the steady state under the particular schedule conditions. Many of the figures or sequences of figures show transitions following a schedule or schedule parameter change. Even after many sessions of steady-state responding, performances were generally immediately altered by changing the schedule contingencies. An important inference from such figures may be less evident: the features of schedules that are important in developing patterns of responding continue to operate in maintaining the patterns. It is not a matter of "learning" a pattern and then continuing to execute a "learned" pattern, but rather that the pattern of responding is maintained in steady state by the consistency of the schedule.

II

What does it mean to say Ferster and Skinner discovered the power of schedules of reinforcement? Fixed-interval (initially called periodic reconditioning) and fixed-ratio schedules had been conceived and studied by Skinner in the early 1930s and he had made insightful analyses of their features. In *Schedules*, the experiments on tandem schedules and differential reinforcement of rate follow from Skinner's earlier analysis of the effects under ratio and interval schedules of different probabilities of reinforcement by interresponse times of different durations. In the course of doing these and other experiments on chaining, it became increasingly clear that responding in any pigeon could be brought under discriminative stimulus control and reproducibly maintained for hours with suitable schedule parameters and past experience. Schedule histories, the sequential intertwining of responding and contingent consequences, are the primary determinants of current behavior. This basic fact had not been fully appreciated, even by Skinner, before this time. A dramatic way to show this new understanding is to describe the background for the first experiments on multiple schedules that evolved from studies on chained schedules.

It is now widely accepted that the behavior of an individual is generally under stimulus control and may differ under different circumstances, but there were no laboratory experiments to show this explicitly until the 1950s. The concept of multiple behavioral repertories under stimulus control was not part of any earlier psychological literature (consider how different *Science and Human Behavior* would have been without such a concept). In contrast, the chaining of sequential responses had been an established principle of behavior with experimental foundations from the time of Skinner's earliest work. It was a natural development for Ferster and Skinner to extend the concept of chaining by conducting systematic studies on chained schedules.

In a chained schedule, responding under a schedule in the presence of one stimulus produces a second stimulus, in the presence of which responding under another schedule is reinforced with food, water, etc. In studying two-component chained schedules where the initial and terminal components were different schedules, Ferster and Skinner observed instances in which the pattern of responding in each component was characteristic of the respective schedule. For example, in Fig. 841, segment A shows the performance in the initial component (a 2-min fixed interval schedule maintained by the onset of the stimulus for the terminal component) and segment B shows the terminal component (a 3-min variable-interval schedule maintained by food presentation). A reader who has worked through *Schedules* up to this figure will understand that responding in the two components is recorded separately and that following each mark on the response record in segment A the stimulus changes, recorder A stops and recorder B starts recording responses in the other stimulus condition. Following food presentations marked on the record in Segment B, responses in the initial component are again recorded in segment A. It was clear from cumulative records such as those shown in Fig. 841 that the performances in the two components were appropriate to the prevailing schedule condition. In a moment of insight, Ferster and Skinner realized that the performance in the initial component maintained by the stimulus change would also be maintained by food presentation. When this proved to be the case, multiple schedules became an experimental reality.

Under a multiple schedule, two or more independent component schedules, each with a distinctive discriminative stimulus, occur sequentially. Ferster had a favorite example of the power of schedule-controlled responding under multiple stimulus control, which is shown in *Schedules* in Figs. 640–642. A pigeon that was being studied under a multiple schedule with 5-min fixed-interval and 275 response fixed-ratio components alternating after each food presentation began to pause for long periods during the fixed-ratio component (strained ratio). In several instances, changing to the stimulus of the fixed-interval component resulted in immediate responding that increased to the terminal rate for the interval schedule. In Fig. 642, after a pause of about 80-min in the ratio component the schedules were changed. In the presence of the fixed-interval stimulus the pigeon responded appropriately to that schedule and made over 300 responses during the 5-min interval. The long pauses in the ratio component were caused by the number requirement of the fixed-ratio schedule, yet an even greater number of responses were made under the fixed-interval schedule condition. Everyone knows that people behave differently under different circumstances, for example with their friends, their parents or children. Ferster and Skinner showed that a repertory of different patterns of responding, each under discriminative stimulus control depending entirely on the schedule conditions, could be studied experimentally in laboratory animals.

The capability of studying responding under multiple schedule control completely changed what could be studied in behavioral experiments and the interpretative inferences that could be made. Prevailing psychological theory before the 1950s relied greatly on generalized states (drive reduction, anxiety, etc.), as explanations of behavior. Earlier work on schedule-controlled behavior had established that the pattern and output of responding varied with different schedules. With multiple schedules it was now apparent that discriminative stimuli associated with different schedule conditions could, at any time, control separate behavioral performances. Explanations of behavior in terms of generalized motivational states are untenable when an individual responds in different ways depending on the history of contingencies associated with the current stimulus conditions. The later findings, that the effects of drugs could differ and even be opposite in direction under different components of multiple schedules occurring during brief time periods, further established the biological significance of schedule-controlled responding under stimulus control.

Amazingly, most of the research presented in *Schedules* was conducted in only a four-year period from 1950 to 1953 and during the beginning part of this period there were continued modifications of apparatus, as described in the reminiscence by Ferster cited earlier. The pace of work generated great excitement in those familiar with it and clearly this fantastic research outpouring would have been slowed had Ferster and Skinner interrupted it by publishing research papers in the more conventional way. Yet the impact of the work was diminished by the limited analysis and interpretation of the results and elucidation of their significance in *Schedules.* Indeed, even Skinner's own writing after the 1950s did not as thoroughly incorporate these discoveries as one might have expected. When Skinner was actively involved in the conduct of research, his broader writings emphasized the sequential interplay between an individual's responding and the consequences of responding that characterize schedule-controlled activities. In later writings he gave a greater emphasis to contingencies than to the interplay of the behavior with contingencies. Probably this would not have happened if Ferster and Skinner had taken more time to analyze the important influence of exposure to prior schedule conditions in determining subsequent schedule performances.

In retrospect, it seems surprising that the concepts of multiple schedule control and schedule-controlled behavior were not appreciated earlier by individuals knowledgeable about operant behavior. While the significance of the work described in *Schedules* remains unfamiliar to most individuals interested in behavior, the technical advances that came from this work are evident everywhere behavioral research is conducted. Unfortunately there has been a decline in the use of the most important technical feature of the work, the cumulative recording of responses in real time, which Skinner considered to be his most important scientific contribution. At Indiana University and after he returned to Harvard University, Skinner had planned to apply the already developed techniques of operant behavior to the analysis of traditional psychological concepts, such as thinking, seeing, and attending. These plans were changed when schedule-appropriate behavior under discriminative stimulus control emerged as the primary determinant of an individual's behavior. *Schedules of Reinforcement* documents this important discovery in a highly original way.

W. H. Morse
P. B. Dews
Boston, Massachusetts

March, 1997

ACKNOWLEDGMENTS

MOST OF THE WORK reported in this book was carried out under contract to the Office of Naval Research under Contracts No. N5ori-07631 and N5ori-07656 with Harvard University, during the period between September 1, 1949 and June 30, 1955. We also wish to acknowledge generous help from the Milton and Robinson Funds of Harvard University, and from the Smith, Kline & French Foundation.

We are grateful for help and stimulating advice to Dr. W. H. Morse and Dr. R. J. Herrnstein, who worked closely with us during the later years of the period of the contract. Special contributions of Dr. P. B. Dews, Dr. Alfredo V. Lagmay, Dr. Clare Marshall, Dr. Merle J. Moskowitz, and Mr. George Victor are acknowledged in the text. Dr. Edward J. Green and Mr. Marvin Levine were associated with the work during the early part of the program.

Invaluable technical help was contributed by Mrs. Antoinette Papp, Mrs. Ada Hughes, and Mrs. Patricia M. Blough. Miss Dorothy Cohen and Mrs. Diana S. Larsen assisted in the preparation of the manuscript. We are especially grateful to Mrs. Marilyn Ferster for editing the manuscript and seeing it through the press.

C. B. FERSTER and B. F. SKINNER

Harvard University
Cambridge, Massachusetts

CONTENTS

Chapter One

• • •

INTRODUCTION

WHEN AN ORGANISM acts upon the environment in which it lives, it changes that environment in ways which often affect the organism itself. Some of these changes are what the layman calls rewards, or what are now generally referred to technically as reinforcers: when they follow behavior in this way, they increase the likelihood that the organism will behave in the same way again. Most events which function as reinforcers are related to biological processes important to the survival of the organism. Thus, food is reinforcing to a hungry organism. The capacity to be reinforced by food substances has presumably been acquired as part of the evolutionary development of the species. Because of the strengthening of behavior which follows, the behavior of the organism is "shaped up" so that it is maximally effective in any particular environment. The shaping process includes the differentiation of new forms of response, including the subtle refinements of form called skill. It also includes the development of appropriate stimulus control, so that a given response is generally emitted only upon an appropriate occasion. The traditional study of reinforcement in the field of learning has been almost exclusively concerned with the acquisition and retention of behavior in this sense.

Another important function of a reinforcement is to *maintain* an active repertoire of behavior. Once a response has been acquired and brought under appropriate stimulus control, the question remains whether it will continue to be emitted under appropriate stimulating conditions. In classical studies of learning this question is usually assigned to the field of motivation. The subject of an experiment acquires a motor skill or learns a list of nonsense syllables. The processes through which this is achieved are considered apart from the question of whether the subject will at any given moment actually oblige the experimenter by displaying his skill or testing his memory. But the maintenance of behavior in strength after it has been acquired is an equally important function of reinforcement. It is only indirectly related to the acquisition of the form of a response or of the control exerted by the stimulating environment.

The effect of reinforcement in maintaining behavior in the repertoire of an organism has been neglected partly because the contingencies of reinforcement actually studied have usually been of an all-or-nothing nature. An act has been reinforced, or it has not. For example, in the traditional study of "learning," "right" responses are always rewarded and "wrong" responses are always allowed to go unrewarded. But the most casual observation of the normal environment of an organism will show that these conditions are not typical. Behavior of a given form seldom has precisely the same effect upon the environment in two instances, and the kind of effect called a reinforcement is seldom inevitable. Most reinforcements, in other words, are intermittent.

In 1933 one of the present authors (Skinner, 1933) reported experiments in which reinforcements were intermittent, and later (Skinner, 1938) pointed out that they may be scheduled in many ways. Although only two types of schedules were reported in detail at that time, they clearly showed that subtle differences in scheduling might generate dramatic differences in behavior. The present book is an exhaustive extension of this earlier work. It is now clear that far from being a mere falling-short of the ideal of inevitable reinforcement, intermittent reinforcement constitutes an important condition of action. Many significant features of behavior can be explained only by reference to the properties of schedules. By the manipulation of schedules, a wide range of changes in behavior can be produced, most of which would previously have been attributed to motivational or emotional variables.

A schedule of reinforcement may be defined without reference to its effect upon behavior. Thus, a response may be reinforced on the basis of the time which has elapsed since the preceding reinforcement, or on the basis of the number of responses which have been emitted since the preceding reinforcement. A given schedule may be fixed, or it may vary either at random or according to some plan. These two possibilities yield four basic schedules: fixed-interval, variable-interval, fixed-ratio, and variable-ratio. But other possibilities exist, as well as many combinations of such schedules. The first step in our analysis of the field is therefore a purely formal statement of these possibilities, such as that given at the beginning of Chapter Two.

A second step is a description of the performances generated by such schedules. For each schedule in our logical classification, we present the typical performance under standard conditions of a representative organism. Such data may be collected without respect to any theoretical formulation of the results.

A more general analysis is also possible which answers the question of *why* a given schedule generates a given performance. It is in one sense a theoretical analysis; but it is not theoretical in the sense of speculating about corresponding events in some other universe of discourse. It simply reduces a large number of performances generated by a large number of schedules to a formulation in terms of certain common features. It does this by a closer analysis of the actual contingencies of reinforcement prevailing under any given schedule.

A schedule of reinforcement is represented by a certain arrangement of timers, counters, and relay circuits. The only contact between this system and the organism occurs at the moment of reinforcement. We can specify the stimuli then present in purely physical terms. These must include a description of the recent behavior of the organism itself. The extent to which features of the present or immediately past environment actually enter into the control of behavior is an experimental question. Under a given schedule of reinforcement, it can be shown that at the moment of reinforcement a given set of stimuli will usually prevail. A schedule is simply a convenient way of arranging this. Reinforcement occurs in the presence of such stimuli, and the future behavior of the organism is in part controlled by them or by similar stimuli according to a well-established principle of operant discrimination.

Some features of the current and recent behavior of the organism generated by schedules at the moment of reinforcement are common to more than one schedule. Therefore, we can simplify the empirical description of the results. The behavior of the organism under any schedule is expressed as a function of the conditions prevailing under the schedule, including the behavior of the organism itself. Some schedules lead to steady states, in which repeated reinforcements merely emphasize the control being exerted by current conditions. Under other schedules, reinforcement under one set of conditions generates a change in performance leading to a new condition at the time of reinforcement. The result may be a progressive change or an oscillation.

We deal with conditions at the moment of reinforcement in two ways: (1) inferentially, by comparing the effects of different schedules and particularly of schedules designed primarily to make such inferences most plausible; and (2) by direct manipulation or determination. The primary purpose of the present book is to present a series of experiments designed to evaluate the extent to which the organism's own behavior enters into the determination of its subsequent behavior. From a formulation of such results we should be able to predict the effect of any schedule.

Such a "theoretical" analysis is only one result, however, and possibly the least important. The experimental analysis of schedules now permits the experimenter to achieve a degree of control over the organism which is of an entirely new order. High levels of activity may be generated for long periods of time. Intermediate and low levels of activity may also be generated. Changes in level which have hitherto seemed capricious may be more readily understood. Through an application of scheduling, extremely complicated examples of behavior can be set up, and behavior can be brought under subtle and complex stimulus control. As a result, complex processes can be studied in the lower organism at a new level of rigor.

The technological use of schedules of reinforcement is rapidly expanding. In research in psychophysics, problem solving, and motor skills, lower organisms may now be used as conveniently as human subjects and with many advantages arising from the greater possibility of many types of control. Performances generated by particular schedules have proved to be useful in the study of motivation (e.g., in the analysis of ingestive and sexual behavior), of emotion (e.g., in the study of "anxiety"), of punishment, of escape and avoidance behavior, and of the effects of drugs. Techniques involving schedules have been adapted to a wide range of species. Surprisingly similar performances, particularly under complex schedules, have been demonstrated in organisms as diverse as the pigeon, mouse, rat, dog, cat, and monkey. At the human level the analysis of schedules has proved useful in the study of psychotic behavior and in the design of educational techniques for normal human subjects in the classroom. Other applications to the problem of the control of human behavior, as in law and penology, religion, industry, and commerce, offer considerable promise (Skinner, 1953).

Chapter Two

• • •

PLAN OF ANALYSIS

SCHEDULES OF REINFORCEMENT STUDIED

TWO NONINTERMITTENT SCHEDULES of reinforcement are:

Continuous reinforcement (crf), in which every response emitted is reinforced; and

Extinction (ext), in which no responses are reinforced.

Schedules of intermittent reinforcement analyzed here include the following:

A. *Fixed-ratio* (FR), in which a response is reinforced upon completion of a fixed number of responses counted from the preceding reinforcement. (The word "ratio" refers to the ratio of responses to reinforcements.) A given ratio is indicated by the addition of a number to the letters FR. Thus, in FR 100 the one-hundredth response after the preceding reinforcement is reinforced.

B. *Variable-ratio* (VR), similar to fixed-ratio except that reinforcements are scheduled according to a random series of ratios having a given mean and lying between arbitrary values. The mean may be noted by a number, as in VR 100.

C. *Fixed-interval* (FI), in which the first response occurring after a given interval of time measured from the preceding reinforcement is reinforced. A given interval is indicated by the addition of a number to the letters FI. (Unless otherwise noted, the number always indicates *minutes.*) Thus, in FI 5 the first response which occurs 5 minutes or more after the preceding reinforcement is reinforced.

D. *Variable-interval* (VI), similar to fixed-interval except that reinforcements are scheduled according to a random series of intervals having a given mean and lying between arbitrary values. The average interval of reinforcement (in minutes) is indicated by the addition of a number to the letters VI, e.g. VI 3.

The three main schedules involving both numbers of responses and intervals of time are:

E. *Alternative* (alt),[1] in which reinforcement is programmed by either a ratio or an interval schedule, whichever is satisfied first. Thus, in alt FI 5 FR 300 the first response is reinforced: (1) after a period of 5 minutes provided 300 responses have not been made; or (2) upon completion of 300 responses provided 5 minutes has not elapsed.

[1] This schedule was first used by W. H Morse and R. J. Herrnstein.

F. *Conjunctive* (conj),[1] in which reinforcement occurs when both a ratio and an interval schedule have been satisfied. For example, a response is reinforced when at least 5 minutes have elapsed since the preceding reinforcement *and* after at least 300 responses.

G. *Interlocking* (interlock),[2] in which the organism is reinforced upon completion of a number of responses; but this number changes during the interval which follows the previous reinforcement. For example, the number may be set at 300 immediately after reinforcement, but it is reduced linearly, reaching 1 after 10 minutes. If the organism responds very rapidly, it will have to emit nearly 300 responses for reinforcement. If it responds at an intermediate rate, it will be reinforced after a smaller number—say, 150. If it does not respond at all during 10 minutes, the first response thereafter will be reinforced. Many different cases are possible, depending upon the way in which the number changes with time.

On two important schedules a single reinforcement is received after two conditions have been satisfied in tandem order:

H. *Tandem* (tand), in which a single reinforcement is programmed by two schedules, the second of which begins when the first has been completed, with no correlated change in stimuli. In tand FI 10 FR 5, for example, a reinforcement occurs upon completion of 5 responses, counted only after a response after expiration of the 10-minute fixed interval. In tand FR 300 FI 5 a response is reinforced after the lapse of 5 seconds, timed from the completion of 300 responses. It is often important to specify which of the two schedules is the more substantial part of the schedule. This may be done by italicizing the important member. For example, tand *FR*FI describes the case tand FR 200 FI 5 sec while tand FR*FI* describes the case FR 1 FI 10. In the first the ratio of 200 is the more important part of the schedule. In the second the 10-minute fixed interval is the major part, except that the timing begins only after the execution of a single response. (The FR 1 is by no means trivial; for example, it has a marked effect in opposing the development of a pause after reinforcement.)

I. *Chained* (chain), similar to tandem schedules except that a conspicuous change in stimuli occurs upon completion of the first component of the schedule. The second stimulus eventually controls the performance appropriate to the second schedule, and, as a conditioned reinforcer, reinforces a response to the first stimulus. This schedule is usually studied when both contributing schedules are substantial.

The values in any schedule may be systematically changed as the experiment progresses in terms of the performances generated. We have, then, another schedule:

J. *Adjusting* (adj), in which the value of the interval or ratio is changed in some systematic way after reinforcement as a function of the immediately preceding performance. (In an interlocking schedule the change in value of the ratio occurs

[1] *Ibid.*

[2] Some early unpublished experiments on this schedule were done by Norman Guttman.

between reinforcements.) For example, a fixed ratio is increased or decreased by a small amount after each reinforcement, depending upon whether the time from the preceding reinforcement to the first response is less than or greater than an arbitrary value.

A complex program may be composed of two or more of these schedules arranged in any given order. Depending upon the presence or absence of correlated stimuli, we may distinguish:

K. *Multiple* (mult), in which reinforcement is programmed by two or more schedules alternating usually at random. Each schedule is accompanied by a different stimulus, which is present as long as the schedule is in force. For example, in mult FI 5 FR 100 the key is sometimes red (when reinforcement occurs after an interval of 5 minutes) and sometimes green (when reinforcement occurs after 100 responses). The key colors and their corresponding schedules may occur either at random or according to a given program in any determined proportions.

L. *Mixed* (mix), similar to multiple except that no stimuli are correlated with the schedules. For example, mix FI 5 FR 50 represents a schedule in which a reinforcement sometimes occurs after an interval of 5 minutes and sometimes after the completion of 50 responses. These possibilities occur either at random or according to a given program in any determined proportion.

A small block of reinforcements on one schedule may be introduced into a background of another schedule. Such cases are referred to as:

M. *Interpolated* (interpol). For example, a block of 10 reinforcements on a fixed ratio of 50 is inserted into a 6-hour period of reinforcement on FI 10 without change of stimulus.

THE BASIC DATUM

Most of the following experiments are concerned with pigeons and with the behavior of pecking a key on the wall of an experimental chamber. In some of the experiments rats were studied, and the response then consisted of depressing a small horizontal bar projecting from the wall of the experimental chamber. Such responses are not wholly arbitrary. They are chosen because they can be easily executed, and because they can be repeated quickly and over long periods of time without fatigue. In such a bird as the pigeon, pecking has a certain genetic unity; it is a characteristic bit of behavior which appears with well-defined topography. Its features may nevertheless be modified by differential reinforcement; and in more detailed analyses, such particular features as speed and direction must sometimes be specified.

Our basic datum is the rate at which such a response is emitted by a freely moving organism. This is recorded in a cumulative curve showing the number of responses plotted against time. The curve permits immediate inspection of rate and change in rate. Such a datum is closely associated with the notion of probability of action.

Among the independent variables which modify this rate or probability are some which are not primarily at issue in a study of intermittent reinforcement. Levels of deprivation, for example, although occasionally varied in a systematic fashion, are generally held constant. The principal variables to be considered are those arising from schedules of reinforcement and from the momentary stimulus conditions which they generate.

PROCESSES ASSUMED

Reinforcement and extinction

In most of the following experiments the organism is maintained at less than its normal body-weight, and food is used as reinforcement. The process of operant conditioning describes the fact that any behavior immediately followed by the presentation of food tends to occur more frequently thereafter. A particular contingency between the response of pecking the key and the presentation of food is arranged by certain reinforcing circuits; but any contingency, even when not specified by the apparatus, is assumed to be effective. In particular, any behavior which immediately precedes a specified response is subject to operant conditioning. Concurrent behavior, such as the assumption of a particular posture at the time of pecking the key, is also likely to be reinforced.

Whenever behavior which has been reinforced in this way is not followed by a reinforcement, it tends to occur with a reduced frequency thereafter. This change is the process of extinction.

Control exerted by stimuli present at the time of reinforcement

The effect of reinforcement is maximally felt when precisely the same conditions prevail. Thus, if a response is reinforced in the presence of stimulus *A*, any increase in frequency will be maximal in the presence of stimulus *A;* but a smaller increase in frequency may nevertheless occur in the presence of stimulus *B*, which differs from *A* in some particular. A comparable process is assumed for extinction. Thus, a given frequency of responding might prevail under stimuli *A*, *B*, and *C* because of earlier conditioning. If the response now goes unreinforced in the presence of *B*, the decrease in frequency is maximal in the presence of *B*, but less in *A* and *C*.[1]

In the process of stimulus *discrimination* a response is reinforced in the presence of one stimulus and extinguished in the presence of a second stimulus. This intensifies the degree of stimulus control, the frequency being maximal in the presence of the one stimulus and possibly zero in the other. Whether the two procedures of (1) reinforcement in one stimulus and (2) extinction in another are always necessary in establishing a differential stimulus control is not easy to decide. The experimental chamber is usually only part of the daily environment of the organism; and unless

[1] Excellent demonstrations of these gradients of stimulus control are afforded by recent experiments by Norman Guttman (1956).

the experiment is run continuously, it may be argued that some instances of the selected response are likely to go unreinforced outside the chamber. The effect of novel stimuli must also be considered in an evaluation of stimulus control.

A discriminative stimulus functioning as a reinforcer

A stimulus which has acquired the stimulus control described in the preceding section may also be a reinforcer; that is, it may be used to condition further behavior. Such a stimulus is an example of a conditioned reinforcer. The reinforcing stimuli which are made contingent upon behavior in the present experiment are all conditioned. The stimuli accompanying the presentation of food are made conditioned reinforcers through an explicit process described in Chapter Three. Unconditioned stimuli almost invariably follow. The conditioned reinforcers, such as a light or buzzer, are useful because they can be made contingent on a particular form of a response.

A discriminative stimulus as a reinforcer assumes special importance in the *chaining of behavior* when a response produces a stimulus which is the discriminative stimulus for another response. This condition is explicitly arranged in some of the experiments which follow, and is indirectly arranged by many schedules of reinforcement, as will be shown in later chapters.

Differentiation of the form of response

When separate instances of a given response can be shown to differ with respect to some property, reinforcement may be contingent upon one value of the property alone. Thus, the response of pecking the key may be reinforced when the beak passes from right to left across the key, but not when it passes from left to right. If the original responses to the key include instances of both types, differentiation is achieved by reinforcing responses of one form and extinguishing responses of the other form. The stereotyped posture adopted by the organism in executing a response often results from accidental differential reinforcement of this sort.

Differentiation of form is particularly important in extremes of frequency. With a very high rate of responding the extent of the excursion of the head is greatly reduced. Special postures may arise, and movements of the beak alone may be emphasized. Under such circumstances the original response of "pecking the key" can no longer be identified. The *speed* with which a response is executed may also be differentially reinforced. The actual speed of movement of the head may be differentially reinforced if it can be recorded in a form permitting the apparatus to distinguish between two speeds. Speed is indirectly differentiated when the contingencies are expressed in terms of times elapsing between successive responses, or between a given stimulus and the first following response. The differentiation of low rates of responding usually produces the chaining of mediating behavior, such as pacing in the experimental chamber, which does not directly affect the reinforcing circuits. Both cases illustrate the present point: an apparatus which differentially reinforces *rate* of responding may function through contingencies between the *form* of response and the reinforcement.

CONDITIONS PREVAILING UNDER A SCHEDULE OF REINFORCEMENT

When a more-or-less stable performance has been well-established under a given schedule, the organism is being reinforced under certain stimulus conditions. The experimenter arranges some of these, such as the details of the experimental chamber and special stimuli added for explicit purposes. But among the physical events occurring in the experimental chamber are the activities of the organism itself. These enter into the contingencies and must be specified as part of the animal's environment.

(Note: We are not interested here in identifying the specific end-organs involved. Just as we specify the color of the light on the key as a stimulus without investigating the mode of operation upon the organism, so we define certain typical features of the organism's own behavior without determining how the organism "receives" them. The procedure for finding whether the organism is sensitive to given kinds of stimuli is the same in both cases: we determine a sensitivity to color by demonstrating a differential reaction to colored stimuli, and we demonstrate a sensitivity to some aspect of the organism's behavior by demonstrating a differential reaction to that behavior.)

In certain cases the topography of behavior can serve as a discriminative stimulus controlling other behavior. In general, however, the important properties of such stimuli are precisely those of our dependent variable, namely, the rate. A given rate of responding may be both a dependent variable (a description of the bird's behavior at the moment of reinforcement) and an independent variable (a stimulus upon which reinforcement is contingent). This distinction may be the source of considerable confusion. We deal with a response both as an activity of the organism and as part of a series of events affecting the organism as a stimulus.

It can be shown that the organism reacts not only to the immediate rate at which it is responding at the moment of reinforcement, but also to recent changes in that rate. The ways in which reinforcements are scheduled lead us to emphasize certain features of this complex variable: we find it convenient to point to the *number of responses* which the organism has emitted since the preceding reinforcement or some other marking point. This number will not be a useful datum if rate of responding varies widely; but where the rate is reasonably constant, it may be a useful aspect of the stimulus situation. The *time* which elapses from the preceding reinforcement, or from some other marking point, may also be a useful aspect, although again it will be significant only if rate of responding is held constant or is changed according to some specified pattern.

The effect of a schedule in maintaining a set of conditions at reinforcement

When an organism is first reinforced on a given schedule, the actual conditions prevailing at reinforcement depend upon the behavior which the organism brings to the experiment. When a different schedule has previously been in effect, certain fairly predictable relations may prevail between the rate of responding and reinforcement. Sometimes, however, the current performance is quickly disrupted, and uniform contingencies do not arise for some time. When a schedule is based upon number of responses, the behavior may be "lost"; that is, the preceding schedules have not provided for a sufficient number of responses to produce the reinforcements needed for further maintenance of the behavior. This occurs, for example, when an organism is placed on a fairly large fixed ratio after a history of only continuous reinforcement.

As a schedule remains in effect, however, the behavior which it generates begins to establish fairly consistent contingencies at the moment of reinforcement. Early reinforcements lead to a performance which, in combination with the reinforcing circuit, causes reinforcements to occur in a specified relation to the behavior of the organism. The three characteristic states are:

Steady state. Under some conditions the performance generated by a given schedule produces just those conditions at the moment of reinforcement required to maintain that performance. The performance under the schedule may show no appreciable change for hundreds of hours. Such a schedule is a small fixed-ratio.

Oscillation. When a schedule generates a temporary steady state, the uniformity of the behavior may produce contingencies of reinforcement not specified in the schedule. These disrupt the performance; and during the period of disruption, the schedule again begins to reconstruct the steady state. Under reinforcement on a small fixed interval, for example, there is a temporary stage when the rate is fairly constant and when the number of responses at reinforcement, counted from the preceding reinforcement, also approaches constancy. This condition is not specified by the scheduling apparatus, and it has the effect of introducing features appropriate to ratio schedules. These features may disrupt uniform performance under the fixed interval and thus destroy the constancy of the number at reinforcement. The performance later returns to that appropriate to the fixed interval alone, whereupon another cycle begins.

Slow drift. Probably no performance obtained under a schedule remains unchanged indefinitely. A fixed interval of moderate size may produce a performance which is fairly stable for hundreds of hours; yet, slight changes may occur (for example, in the pause before the first response after reinforcement) which are traceable to the very large number of instances permitting a more sensitive temporal discrimination. A variable-interval schedule which produces a constant rate of responding for many hours may eventually be "learned" by the organism if the arrangement of intervals is not entirely random, but this effect is very subtle and may not be felt for hundreds of hours. These slight changes need not disrupt the performance appropriate to the schedule, and a return to an earlier state may never occur.

ANALYTICAL PROCEDURES

In order to achieve a generalized account of the effects of schedules, the stimuli which enter into reinforcing contingencies must be identified and evaluated. This can be done in several ways.

Inference from performance

Assumptions about conditions at the time of reinforcement may be checked by an analysis of performances appropriate to various schedules.

Direct control of contingencies

1. *Differential rate reinforcement.* A schedule which *tends* to arrange a particular contingency at reinforcement may be made to do so with additional controlling circuits. For example, a device which arranges a schedule under which reinforcements *usually* occur at a given rate of responding may be extended with an additional device which withholds reinforcement until *precisely* that rate is achieved. This may often be done without appreciable changes in other characteristics of the schedule.

3. *Supplementary correlated stimuli.* The importance of the organism's own behavior as a stimulus may be demonstrated by the introduction of additional stimuli entering into the same contingencies. For example, if the effect of a schedule seems to depend upon the differential stimulus control exerted by the behavior of pausing for different lengths of time, a clearly discriminable external stimulus may be introduced which varies with the length of pause. Supplementary stimuli have also been added which vary with the following: (1) the number of responses since the preceding reinforcement ("added counter"); (2) the time since the preceding reinforcement ("added clock"); or (3) the momentary rate of responding ("added speedometer"). If the added stimuli have a large effect upon the performance, the stimuli supplied by the behavior itself probably have not been particularly important or have different characteristics from the added stimulus.

3. *Time out.* A period of "time out" is injected into an experiment by the arrangement of any condition under which the bird does not respond. (See Chapter Three, *Technical Procedures.*) The assumption is that stimuli arising from the bird's own behavior just before such a time out will no longer be effective. For example, if the behavior of the organism during a fixed interval is suspected of being affected by the behavior in the preceding interval, a time out may be used to eliminate these stimuli.

NATURE OF THE FOLLOWING EXPERIMENTS

A reasonably satisfactory formulation of schedules of reinforcement emerged slowly as the experiments to be described here were conducted. Not all the experiments are crucial tests of the formulation, however; many belong to an earlier stage in the research where we were mainly interested in collecting representative performances under representative schedules. Since these early experiments yielded information which is of value in itself, a fairly full report is given. Points leading to and supporting the present formulation are emphasized.

The experiments also show different levels of technical development. The art of programming and recording the performances generated by schedules has greatly improved during this research. Since important characteristics of the performance were often impossible to predict, instrumentation was frequently inadequate. Nevertheless, we have reported those results which have appeared to be important for other reasons. For example, a schedule now recognized to be most effective when a period of time out follows each reinforcement may have been studied before the importance of the time out was discovered. Although the performance on the schedule in the absence of time out is less significant for our present analysis, it is nevertheless worth recording and, of course, ultimately demands explanation.

Chapter Three

• • •

TECHNICAL PROCEDURES

APPARATUS

The experimental box

A TYPICAL EXPERIMENTAL BOX is made from a picnic icebox approximately 30 by 12 by 15 inches. A removable panel divides the box into two compartments. The bird's compartment (Fig. 1) is 12 by 12 by 15 inches. A small blower, mounted on an outside wall, exhausts air from the box through a filter, which collects the feather dust

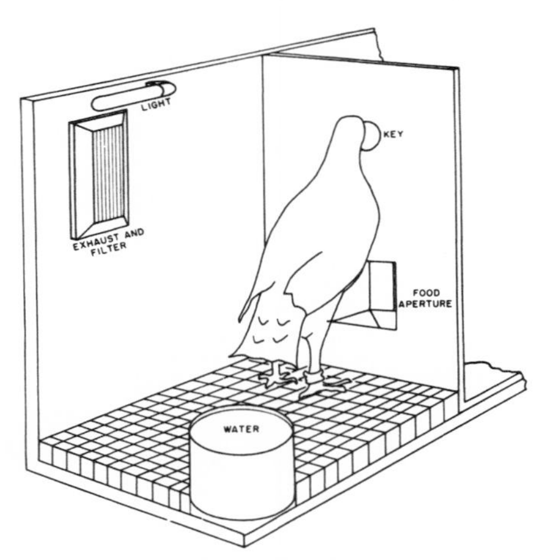

Fig. 1. Experimental chamber

which would otherwise blow about the room. The pigeon stands on a heavy No. 5 wire mesh floor, under which paper towels are spread. The mesh gives the bird a clean, dry surface on which it can stand and walk more easily than on a smooth floor. Fresh water is supplied in the cup at the rear of the cage. A hood above the cup is sometimes used to prevent pollution of the water by feces. The compartment is illuminated by a light near the ceiling. The magazine, the key, the lights, and whatever

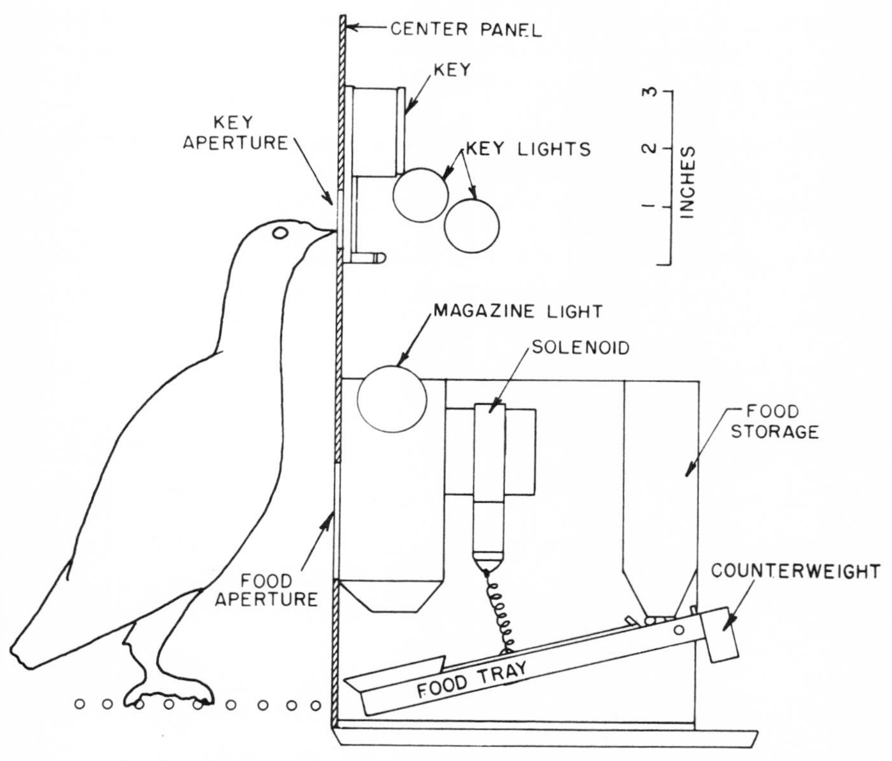

Fig. 2. Side view of experimental chamber and associated apparatus

electrical equipment and wiring are most conveniently placed in the box rather than with the rest of the programming equipment are mounted on the opposite side of the panel (Fig. 2). The key and the magazine are described below. A cable (not shown) enters the box through the wall and ends at a connector on the panel. Connections to the blower and the light in the bird's compartment lead to a second connector. When these leads are disconnected, the panel with all the working parts of the apparatus can be removed for repairing or for cleaning the box.

Sound insulation. Such a box attenuates outside noises roughly from 20 to 30 db. A masking white noise is provided in the box from an earphone (not shown) driven by a white-noise generator. Because of this insulation and masking, normal noises occurring in the laboratory do not produce effects upon behavior observable against existing baselines. Care is taken, however, to keep critical relays in the programming circuits as quiet as possible. Such relays are lightweight and shock-mounted. Under these conditions, no discriminations based on relay operation have been observed even after many hundreds of hours on a given schedule. The recorder can be heard inside the box, but this sound always accompanies the normal noise of the key and probably helps to maintain a stable form of the pecking response. Indeed, it is sometimes advantageous to emphasize the auditory feedback from a response by the addition of other stimuli generated by the operation of the key.

Four to eight such apparatuses are operated concurrently in a single experimental room. Each apparatus has its separate programming equipment, so that any auditory stimuli from the activity of one bird has no relevance to the schedule in another apparatus.

Key. Figure 3 shows the key used in most of these experiments. Its position in the apparatus is indicated in Fig. 2. The armature consists of a piece of frosted Plexiglas 2 by 3 by 1/16 inches. An axle passes through two Plexiglas blocks cemented to oppo-

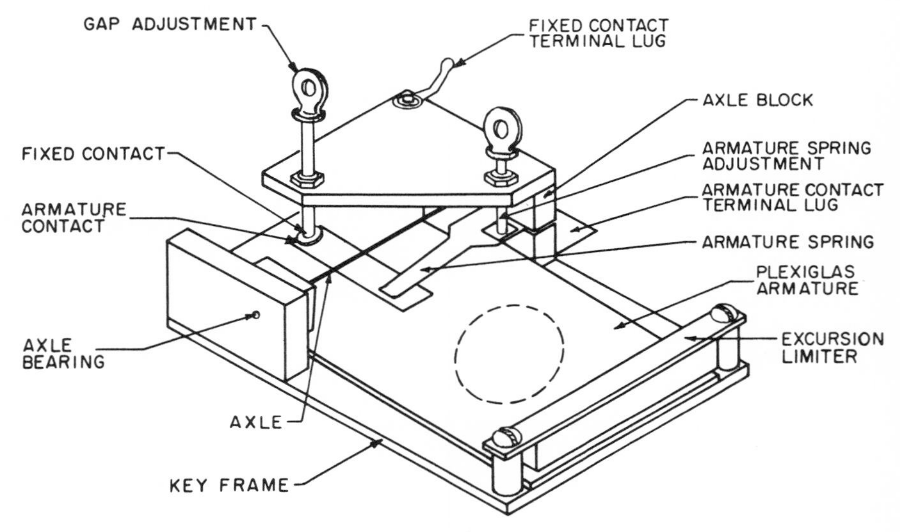

Fig. 3. The key

site edges. A silver contact is riveted on the Plexiglas above the axle and is in contact with a strip of phosphor bronze which runs halfway down the center of the armature, where it is riveted. An electrical connection to the moving contact is made from this strip through the spring which supplies the armature tension. The armature is mounted in a frame consisting of a $\frac{1}{8}$-inch piece of aluminum $2\frac{1}{2}$ by 3 inches. Two pillow blocks on the side of the frame provide bearings for the armature axle. The bird pecks through a 1-inch hole in this frame. A piece of brass fastened to one of the pillow blocks overhangs the upper part of the key. A thumbscrew passes through a hole tapped in the brass directly above the contact on the armature, and a small silver contact soldered to the end of the screw provides the fixed contact for the key. Pecking the Plexiglas armature *opens* this switch. Tension on the armature is adjusted by turning a second screw through the overhang which presses a phosphor bronze spring.

A sponge-rubber block on the bottom of the key limits the excursion. Six million responses per month over periods of from 1 to 2 years have been recorded from such a key with only a rare failure.

The height of the aperture which exposes the surface of the key to the pigeon is important. The rate of responding depends in part upon the energy required; if the key is too high or too low, the pigeon must stretch its neck excessively or assume awkward postures, so that a lowered rate results. The average homing pigeon operates most conveniently a key centered about 7½ inches from the floor; larger pigeons, such as White Carneaux, work best on a key approximately 8½ inches from the floor.

The Plexiglas armature has a low mass, so that high natural frequencies are possible. It also has a certain elasticity, which permits many thousands of responses to be made each day without injury to the beak or undue fatigue. A natural frequency sufficient to follow all rates of responding can be achieved with such a key by the placement of a limit on the excursion of the armature and by the proper adjustment of the tension. The advantage of a break contact is felt here. A very light touch on the key, moving the armature through only a very short distance, is sufficient to open the contact and thus to operate the controlling circuit. The tension is a compromise between too stiff a key and too slow an operation: if the spring is too tense, some responses may not break the circuit; on the other hand, if the spring is too light, the contact may not close fast enough after each response and the frequency of operation will be reduced. The key is designed to give the pigeon a mechanical advantage. In the lever system composed of the axle, contact, and the exposed area of the armature, the contact may be held closed with a substantial force yet opened by a fairly light touch.

The response which the pigeon makes in pecking such a key has the following desirable properties:

1. It uses an effector system (the beak, the head, and the neck of the bird) which occupies an important place in the normal activity of the organism.
2. The effector system has a relatively small mass and is capable of high rates of activity.
3. Successive responses to the key resemble each other closely. The topography of examples of this class of responses can be fairly narrowly specified.
4. Responses can be reliably recorded and reinforced through their action upon the key.

The part of the Plexiglas armature exposed through the aperture is illuminated from behind by 6-watt "Christmas tree" lamps, each with a 300-ohm resistor in series in order to prolong its life. When an experiment does not require discriminative stimuli, the key is usually illuminated with a white light. In establishing discriminations, several key colors may be programmed by mounting different lamps behind the key and changing the illumination through the controlling circuits. In critical experiments, two lamps of each color may be provided in case one burns out. In nearly all the experiments to be reported the stimuli are presented in this way. The fact that the bird pecks directly at the surface carrying the stimulus color is important; a weaker control follows when changes in color occur, for example, in the general illumination of the experimental box.

No effort has been made to specify such stimuli precisely. In any given experiment the only important consideration is that stimuli are grossly different. When we say, for example, that the bird is reinforced when the key is red and not reinforced when the key is blue, it should be understood that the two colors also undoubtedly differ in saturation, intensity, and possibly the distribution of light on the key.

In some instances (e.g., during a time out), all lights in the box are turned off. The bird is not in total darkness, however, since some light enters through the edges of the lid and the blower and cable holes.

Food magazine. Two types of food magazine have commonly been used. One type gives free access to food for a fixed period of time, while the other delivers a fixed amount of food. The latter has several disadvantages. Food may be left in the magazine and may be found during a later part of the experimental session when a reinforcement is not scheduled. Such reinforcements break down the stimulus control of the magazine-approach behavior, and lead to responses ("looking for food") which compete with the behavior generated by the independent variables in the experiment. With such a type of magazine, moreover, the amount of time the animal spends in eating varies from reinforcement to reinforcement. Recorded pauses following reinforcement are therefore difficult to interpret.

These objections are avoided by using a magazine which presents a supply of food for a fixed period of time. All remaining food, as well as the container in which the food is presented, is withdrawn at the end of the period. Although some cleaning or grooming behavior may follow, the total time occupied by eating usually assumes a constant value. This type of magazine also helps to bring responses to the magazine under complete stimulus control. Approach to the magazine in the absence of the normal stimulus ("spontaneous looking for food") quickly disappears during magazine training. The amount of food actually eaten from reinforcement to reinforcement varies only slightly after magazine training has taken place. A magazine which presents food for a fixed time is more successful for a pigeon than for such an organism as a rat, since the pigeon has no way of picking up more food than it can immediately eat. With the rat, powdered or liquid food generally must be used, to avoid small-scale hoarding operations.

The food magazine used in most of the following experiments is shown in Fig. 2. Grain is stored in the hopper (to the right of the drawing), from which it feeds into the shallow pivoted tray. A solenoid draws the tray into horizontal position, and the bird reaches the grain through a $\frac{3}{8}$-inch-square aperture in the hood at the front of the magazine. Two white, 6-watt lamps are wired in parallel with the solenoid, so that the grain is well-lighted during the operation of the magazine. The second lamp is present in case one burns out during an experiment. The pivoted tray is occasionally cleaned out to remove inedible debris which slowly accumulates in it. When the magazine is not operating, the grain is not only out of reach but unlighted and scarcely visible to the bird.

The actual operation of such a magazine generates the all-important conditioned reinforcing stimuli which supply a precise contingency between behavior and reinforcement. In this case the stimuli are both auditory and visual. The operation of the tray and the solenoid can easily be heard. The magazine lights illuminate the grain and usually throw a white light on the key. Such stimuli are usually amplified by arranging that all key lights and the house light go off simultaneously with the operation of the magazine.

In all experiments reported here, reinforcement consists of free access to grain for from 3.5 to 4 seconds. Under good conditions of accessibility of the grain and uniform size of individual grains, the amount consumed per reinforcement is approximately 0.25 gram. A daily ration of 60 reinforcements therefore equals 15 grams. This exposure to grain, together with the resulting amount eaten, was determined empirically, as an amount which sustained performance under the types of schedules considered here. In order to reinforce frequently during a daily session, however, the amount is kept close to the minimum which will produce this result. Unfortunately, it is not always easy to detect the fact that a reinforcement is too small. The resulting low rates and irregularity do not provide an immediate indicator. In work with a new organism and a new food, the amount of reinforcement that should be used is probably impossible to determine in advance. A rule of thumb for a preliminary trial value, however, is 0.5 gram per 1000 grams body-weight.

Automatic programming

Work on schedules of reinforcement is all but impossible without automatic programming of the contingencies of reinforcement. A large number of events need to be arranged during long experimental sessions. More than a million responses per month are often recorded from a single subject, with daily experimental sessions ranging up to 15 hours. In such cases manual programming is unthinkable. Moreover, the reaction time and variability of the human operator introduce errors in programming which defeat the purpose of such research. The required level of precision can be achieved only through the use of mechanical or electrical devices.

With automatic programming a single experimenter can run a large number of experiments concurrently and almost continuously. Long experimental sessions may extend through the night without the presence of laboratory personnel. Relatively unskilled personnel can start and stop particular experiments, while the experimenter's time is free for the design of new equipment and experiments and the treatment and evaluation of results.

Typical circuit. In the present experiment, programming of contingencies is carried out largely by various types of relays and electric and electronic timing and counting devices. Figure 4 shows a circuit containing many commonly occurring elements. This circuit is designed to program and record a fixed-interval schedule of reinforcement. Circuit elements which are drawn to the right of the vertical broken line are customarily wired into the pigeon's compartment. These control the magazine timing cycle, the lights, and the key, which usually do not change from experiment to experiment. The circuit elements to the left are wired outside and are connected to the bird's compartment by a cable. This part of the circuit is changed frequently and must be flexible.

In Fig. 4, S 2 is the normally closed switch on the key at which the bird pecks.

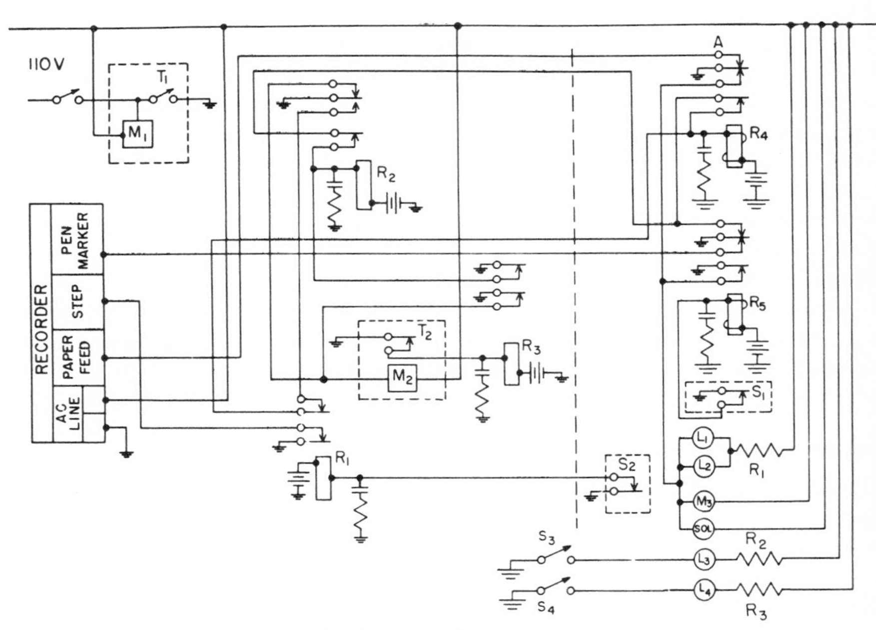

Fig. 4. Diagram of a typical circuit

Every time the bird pecks, S 2 opens and Relay 1 releases. The first pole on Relay 1 operates the recorder each time the bird pecks.

The magazine cycle is programmed by Relays 4 and 5. A pulse to the coil of Relay 4 locks up Relay 4 through its own coil, its first pair of contacts, and the normally closed contact of Relay 5. While Relay 4 is locked up, the normally open contact of the second pole supplies power to: L 1 and L 2, the lights above the magazine which illuminate the grain; M 3, the timing motor by which the duration of the magazine cycle is determined; and the solenoid which draws up the tray so that the bird can reach the grain. The end of the magazine cycle is determined by a cam on the shaft of the motor, M 3, which operates S 1, a microswitch. The microswitch operates Relay 5. Since Relay 4 is locked up by the fact that it draws its power through the normally closed contact on the second pole of Relay 5, the operation of Relay 5 causes Relay 4 to release. The first pole of Relay 5, however, is connected to the light, motor, and magazine solenoid, which continue to operate until the motor drives the cam beyond the point where microswitch S 1 releases. At this point a single pulse long enough to close Relay 4 can produce a second magazine cycle. The normally open contact on the second pole of Relay 5 is connected to the pen marker on the recorder, which makes a diagonal mark on the record whenever reinforcement occurs. The normally closed contact on the second pole of Relay 4 is connected to the paper feed on the recorder so that the paper does not feed during the operation of the magazine. Box light L 3 and key light L 4 are powered continuously through switches S 3 and S 4. R 1, R 2, and R 3 are 300-ohm, 5-watt resistors, which keep the voltage across the lamp filament low in order to prolong the life of the lamp. T 2 is a timing device consisting of a motor which actuates a switch once every 10 minutes. The switch on T 2 operates Relay 3. The second pole on Relay 3 operates Relay 2, which locks up through its first pole and through the normally closed contact of Relay 5. When Relay 2 is locked up, the normally open contact of the second pole is grounded and supplies a ground connection to the second pole of Relay 1, the key relay. The other contact on the second pole of Relay 1 is connected to the magazine; thus, whenever Relay 2 is locked up and Relay 1 (the key) operates, a pulse reaches Relay 4 which initiates the magazine cycle. The motor on T 2 is powered by a normally closed contact of the second pole of Relay 2 and by the normally open contact of the first pole of Relay 3. The conditions of operation of the motor are: (1) Relay 2, the reinforcement lock-up, is not operated; and (2) Relay 3, the relay operated by the switch on T 2, is operated. Thus, the motor on T 2 is so wired that the timer can never stop with the switch closed; but once a reinforcement has been set up (Relay 2 locked up), the motor will run only long enough for the switch on T 2 to open again. This circuit programs a fixed-interval schedule in which the interval of reinforcement is defined from the reinforcement. If one reinforcement is delayed because of a low rate of responding, the following interval is not shortened.

T 1 is a 12-hour programmer of the type used in kitchen appliances. It starts the experiment and stops it after a preset period of time.

Circuit elements. Wherever possible, d-c relays are used. These are usually of higher quality, have more dependable action, and are smaller than a-c relays. Each relay has a capacitor with a resistor in series connected between the ground and the coil for spark suppression. Spark suppression materially increases the life of the contacts and is especially important where relays must operate many millions of times. (A rough rule for designing spark suppression for coil is to begin with a small capacitor [0.01 microfarad, 400 volts] and to vary the size of the resistor by a potentiometer while operating the coil in a slightly darkened room. That value of resistance is chosen which gives the least amount of spark when the coil is operated or released. The size of the capacitor is then increased to a larger value and the process repeated. The most efficient spark suppression occurs when further increases in the size of the capacitor do not reduce the amount of sparking when the relay is operated or released.)

The keying relay operates more often than any other element in the circuit except the switch on the key, and is therefore more subject to wear. A good-quality, fast-acting d-c relay is used. Because the relay acts directly through the contacts of the key, it is chosen for low current drain, and spark suppression is carefully arranged to prevent pitting of the key contacts. The relay handles rates as high as 15 responses per second. (A slow-acting or slow-releasing relay can produce an erroneous record of the bird's actual behavior, especially in schedules involving differential reinforcement at high rates.)

The keying relay must deliver a pulse sufficiently long to operate the recording and controlling circuits. In most of the present experiments a special type of relay was used which guaranteed a pulse of 0.040 second provided the key contact was broken for 0.005 second or more. The relay has a two-sectioned coil on a common core, and is operated by one coil and sustained by the induced current in the other. The 0.040-second pulse guarantees the operation of the recorder, and particularly of the magazine lock-up relay, and of any counting circuits currently in effect. Since the releasing time of the relay is usually shorter than the operating time, a normally closed key contact is generally advantageous. The keying relay also should control the circuit on release.

It is particularly important that the sound of the operation of the relay which sets up a reinforcement should not affect the organism. The insulation and masking noises already described ordinarily insulate the bird from the controlling circuits. However, the additional precaution is taken of using a lightweight, high-impedance, sensitive relay which is shock-mounted and makes very little noise relative to the clicks of standard relays. (This is the relay which is closed by the timing circuit so that the next response will be reinforced. Poor sound insulation or a noisy lock-up relay would permit the bird to form a discrimination and to respond only after a reinforcement had been set up.)

Complex circuits. The circuit shown in Fig. 4 is basic to nearly all simple schedules of reinforcement. Changes from one schedule to another are made at position T 2. For example, when the schedule is arranged by a clock, T 2 is a timer measuring fixed or variable intervals. But when the schedule of reinforcement is based upon the number of responses, T 2 is a counter, counting a fixed or variable number of responses. Although a counter at T 2 could operate the magazine upon completion of a preset number of responses, this practice could introduce a slight delay between response and reinforcement. It is better to set the counter for one less than the number of responses in the ratio and to let it lock up Relay 2, in which case the next response is reinforced.

Complex schedules are arranged by stepping switches or selector switches which change scheduling devices and relevant stimuli. Figure 5 is a block circuit-diagram of a program of alternating fixed-interval and fixed-ratio schedules under stimulus control (mult FIFR).

Flexible circuit-building equipment. In research of this sort the controlling circuits are

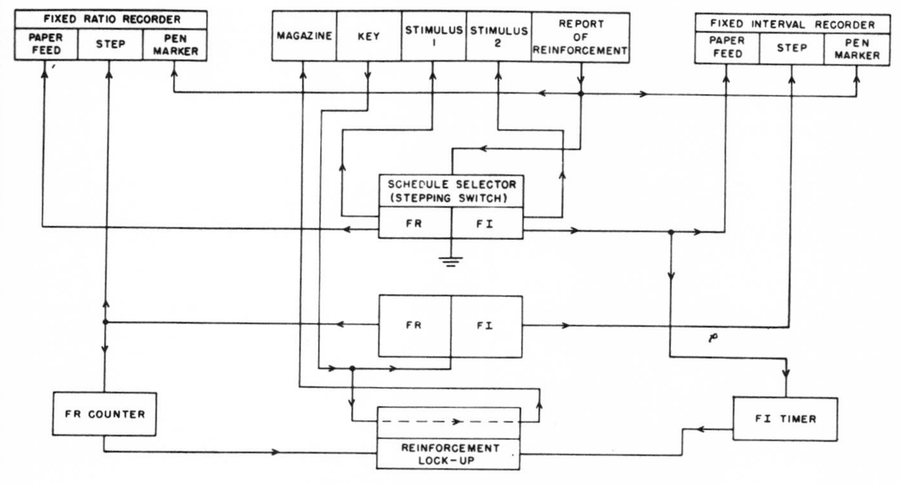

Fig. 5. Block diagram of a circuit for mult FIFR

frequently changed, and changes must often be made quickly. The relays and timing and counting devices are therefore attached to panels and prewired to a pattern of snap leads. The panels can be clipped into place on brass rods which carry alternating or direct current. Units are wired together with snap leads in a "functional wiring diagram." Changes in a circuit may be made very quickly, and a given apparatus redistributed for further use after it has been discontinued. A circuit such as that shown in Fig. 4, for example, can be assembled and ready for use in less than 10 minutes.

TREATMENT OF DATA

Recording

Cumulative curves. A graph showing the number of responses on the ordinate against time on the abscissa has proved to be the most convenient representation of the behavior observed in this research. Fortunately, such a "cumulative" record may be made directly at the time of the experiment. The record is raw data, but it also permits a direct inspection of rate and changes in rate not possible when the behavior is observed directly. Figure 6 shows how a cumulative record is taken. Each time the bird responds, the pen moves one step across the paper. At the same time, the paper feeds continuously. If the bird does not respond at all, a horizontal line is drawn in the direction of the paper feed. The faster the bird pecks, the steeper the line. Although the rate of responding is directly proportional to the slope of the curve, at slopes above 80 degrees small differences in angle represent very large differences in rate; and although these can be measured accurately, they cannot be eval-

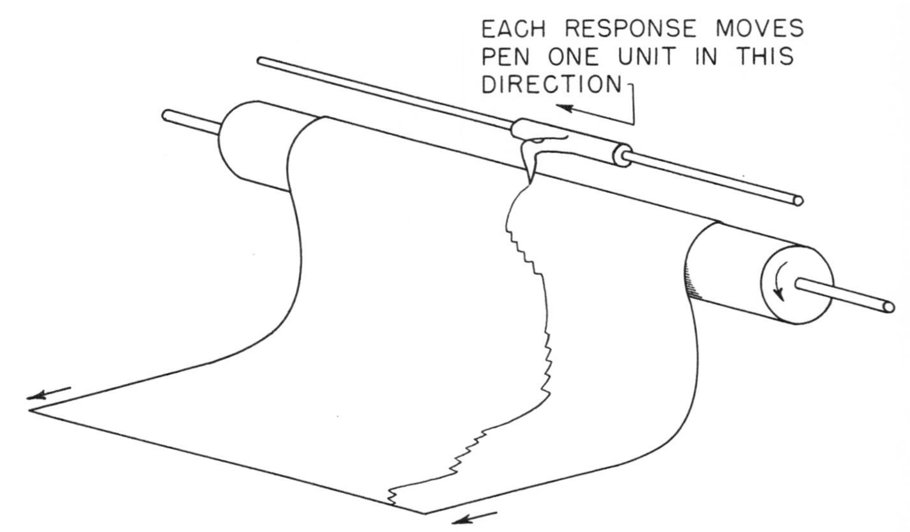

Fig. 6. Diagram of a cumulative recorder

uated easily by inspection. The speed of the paper movement and the scale of the pen movement are chosen with respect to rates to be observed. These depend in turn upon the organism used, the response used, and the schedule of reinforcement.

By proportionately increasing both the rate of the paper feed and the distance the pen travels per response, such a graph can be magnified or reduced without changing the slope. The values to be used are determined by a compromise between the precision of a large-scale record and the compactness and convenience of a small-scale record. Occasionally, two recorders are used: one to provide measurements and easy inspection of details, and the other to provide a compact summary of the whole session.

In the standard recorder the pen returns to its starting point after it crosses the paper. Since the paper itself is essentially endless, very long experimental sessions can be continuously recorded.

In multiple schedules two or more recorders may be used, only one of which operates at any given time. Thus, the behaviors appropriate to several conditions may be automatically separated and cumulated for study.

Cumulative recorders. Several types of cumulative recorders were used in the research described here. Improvements made during the research have been incorporated in a standard cumulative recorder manufactured by Ralph Gerbrands, 96 Ronald Road, Arlington, Massachusetts. Figure 7 shows one version of a cumulative recorder. The pen is stepped across the paper by the actuating mechanism of a telephone-type stepping switch capable of 250,000,000 operations. The stepping switch is linked to the pen-driving mechanism with a worm and gear. The maximum speed of opera-

Fig. 7. Photograph of a cumulative recorder

tion is achieved by operating the coil for the first instant of operation at a voltage higher than that for which it is rated. After the armature of the stepping switch begins to draw in, an interrupter spring opens; and from then on the coil draws its power through a voltage-dropping resistor. The pen is mounted on a carriage which is free to slide across the paper. It is moved by a pin soldered to a continuous chain driven by the stepping switch. When the carriage reaches the top of the paper, a cam disengages it from the pin, and a small weight returns the carriage to the bottom of the paper, where it is engaged by the next pin on the chain. In resetting, the pen draws a vertical line which is useful in the measurement of slopes or when the figures are cut for reproduction. The pen may be moved with respect to the pen carriage by a small relay coil. When the relay is operated briefly, the pen draws a small pip, which may be used to indicate the occurrence of a reinforcement or any other event. In one type of recorder used, this pip moved horizontally to the left of the record. In a later model, it moved to the right and downward. (See Fig. 8.) The pip drawn

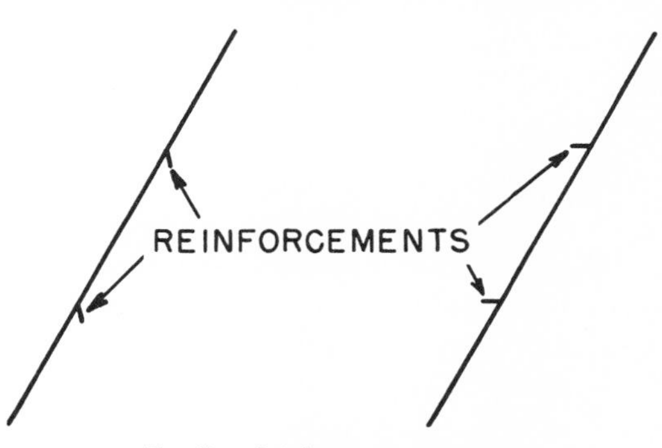

Fig. 8. Reinforcement marks

by this model is preferable, since it cannot coincide with any portion of the recorded behavior.

Another type of recorder used in some of these experiments can be reset to zero from the controlling circuit (and not, as in the above case, only when the pen reaches the top of the paper). If the pen is reset at each reinforcement, the records representing the performances between successive reinforcements are collected in a form convenient for inspection and measurement.[1] Both the stepping mechanism and the paper feed may be adapted for use in timing certain schedules. Unless the operation of the recorder is quite reliable, however, there is a certain danger in this practice. For example, in using the stepping mechanism to schedule a ratio, the possibility that the stepper is not reporting the behavior accurately may not be detected in the recorded performance.

In all experiments the paper feed of the recorder is stopped during the operation of the magazine. Any pause after reinforcement is thus not confused with the time when the bird is eating. The recorder is also stopped during periods of time out. The stepping mechanism remains connected, and any responding occurring during this time is recorded as a vertical line that is easily distinguishable from other features. Whenever the recorder is stopped for a substantial period, additional ink from the pen produces a darkened spot which is useful in reading the records.

Processing of data

Construction of graphs. The records for a given bird during an experiment are pieced together and accordion-folded for storage. The length of fold represents 1 hour of behavior. The reports of the experiments reproduced here are representative portions of such records. Because important features are usually not related to a zero-point in either time or number, standard coordinates would not be very useful. Instead, we indicate the scale of each record by attaching a small set of coordinates containing some representative slopes. Since full coordinates need not be indicated, the records may be compressed by the removal of unused parts of the paper. Figure 9A shows a cumulative curve as recorded. In Fig. 9B coordinates have been added and the record "telescoped" by cutting out diagonal strips. In cutting and pasting the correct angle of each portion of the record with respect to the vertical is carefully maintained. If a vertical line is drawn when the recorder resets, parts are left at each end to indicate continuity.

Where the actual rate of responding at a given point is important, careful measurements are made with a special type of protractor and are reported in the text. A rough evaluation of the absolute rate may be made from the coordinates supplied. In general, we are more concerned with changes in rate than with absolute values.

Description of rates. In the description of records of this sort, the word "rate" can have several interpretations. Occasionally, responses occur with a relatively constant inter-response time, so that uniform segments such as those in Fig. 10A are pro-

[1] The latest model supplied by Gerbrands has this resetting feature.

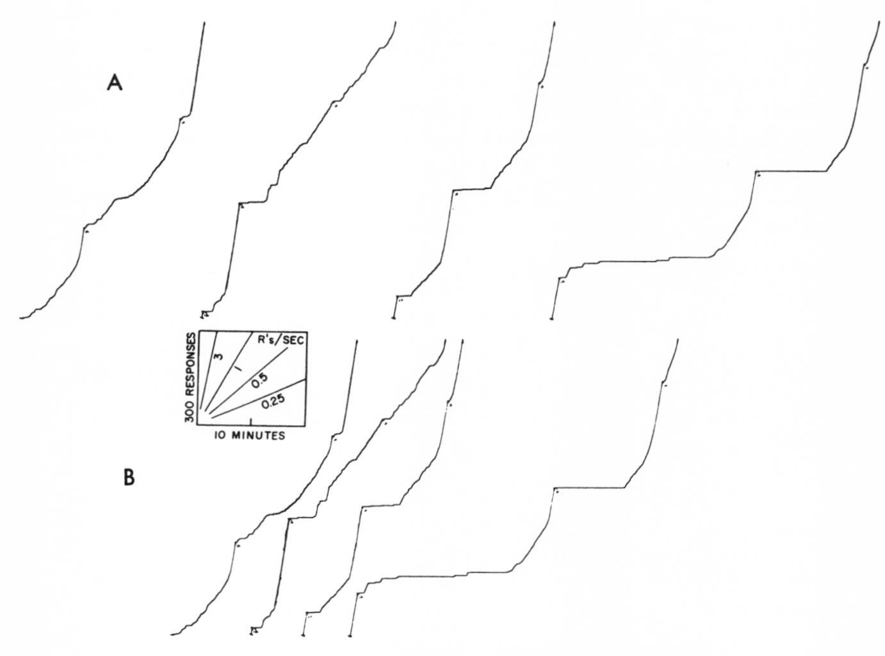

Fig. 9. Preparation of a telescoped figure

duced. Although the actual record is stepwise, a single constant slope may be identified; and when records are greatly reduced, as in most of those reproduced in this book, the record has the character of a straight line. Inter-response times may vary, however, even though we may still distinguish a mean rate as in Fig. 10B. In many experiments we must distinguish between a local rate, sustained for short periods of time, and the over-all mean rate. Thus, Fig. 10D indicates short bursts of responding at a constant local rate alternating with periods of no responding. Both of these combine to establish a single, intermediate over-all rate.

These distinctions may also be made when the rate changes in a reasonably uniform fashion. Thus, Fig. 10C is a smooth curve resulting from the continuous lengthening of inter-response times; to the right are two cases of roughly the same curve produced by much less uniform changes in inter-response times. Sometimes, considerable local changes may occur, although the record tends to approximate a constant over-all rate as at D. At E a fairly uniform local rate is maintained despite two local deviations, one caused by a temporary increase in rate followed by a compensatory decrease, and the other by a decrease in rate followed by a compensatory increase. The latter deviation is more common, but both occur. At F a smooth curve is evident in spite of a marked local deviation. When longer portions of an experiment are considered, it may be necessary to distinguish between three "rates." Thus, under certain conditions of fixed-ratio reinforcement, the ratio may be run off at (1) a constant high rate. But successive ratios separated by pauses produce (2) a well-marked over-all rate of intermediate value. Segments at this over-all rate may alternate with prolonged pauses, to yield (3) a total rate for the experiment of a still lower over-all value.

Recording special features. Additional recording instruments are used when some

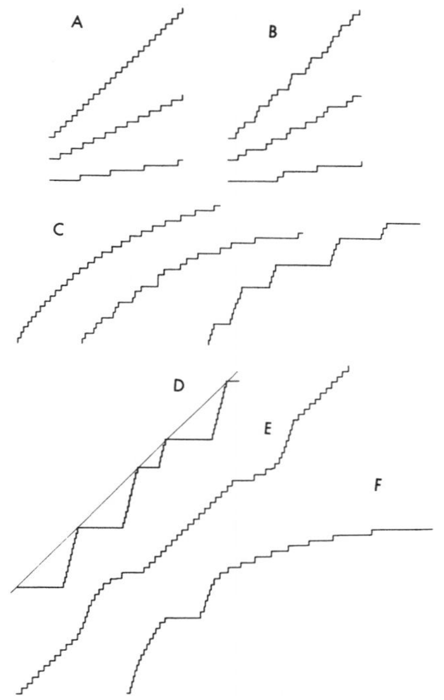

Fig. 10. Cumulative curves showing varieties of rate change

special aspect of the dependent variable is of particular interest. Recording all behavior in detail at all times would be uneconomical. The recording selected in any given experiment is determined by exploratory work. For example, in a study of the fixed-interval schedule of reinforcement, the curvature shown by the performance during an interval may appear to need careful analysis. Measurements taken from the cumulative records may not be precise enough or convenient. A polygraph can then be added to the recording equipment to provide more accurate measures, or a set of counters can be arranged to count responses during specified fractions of the interval.

With the latter procedure, measuring single cases or computing the mean result for an experiment would be unnecessary. Again, if a study of a fixed-ratio schedule suggests that the time which elapses before the first response is of special importance, a polygraph or other type of recorder can be arranged to provide a record of just this part of the bird's performance.

EXPERIMENTAL PROCEDURES

Care of pigeons

In most of these experiments, we used White Carneaux males, 1 to 2 years old at the start of the experiment, from the stock of the Palmetto Pigeon Plant, Sumter, South Carolina. These pigeons have the advantage of being uniform in body structure and in their metabolic reactions to schedules of feeding. A few homing pigeons of miscellaneous origin were also used. We have not discovered any species differences, except size, relevant to the variables we have explored.

As soon as a pigeon is received in the laboratory, an identifying band is put on the right leg. The first flight feathers are cut off along the line formed by the tips of the second row of feathers, and the long tail feathers are clipped along the line formed by the second set of tail feathers. The toe nails are inspected; and if they are unduly long, they are clipped with small wire-cutters. The birds are housed in individual cages, about 1 cubic foot in size. Fresh water and a standard health grit are available at all times in cups at the front of the cage. The floor is covered with a heavy grade of wrapping paper, which is changed weekly.

The food is a special mixture of 40% vetch, 50% Kaffir corn, and 10% hemp seed. The grains of this mixture are all approximately the same size. If a standard pigeon mixture containing whole corn as well as very small kernels is used, the pigeons will hunt for large grains or preferred grains, and unpreferred grains quickly accumulate in the magazine. To lessen changes in deprivation or reinforcing effect, the same grain mixture is fed both in the living cages and in the food magazine. The grain is kept in uncovered metal containers.

Control of body-weight

Percent body-weight is a useful method of controlling deprivation. Two pigeons of the same percent body-weight may not show the same rate of responding on comparable schedules, but they will differ less than when deprivation level is maintained in terms of a schedule of feeding or the amount fed per day.

The ad lib body-weight is first obtained by giving the bird continuous access to food for 2 or 3 days, or until the weight shows no appreciable increase. (This practice is reliable only if mature birds are used.) All food is then removed for 2 days; and beginning with the third day, 5 grams is given daily until the weight of the bird falls to 80% of its ad lib weight. Further changes may be advisable. If the bird continues to show emotional behavior in the experimental box and does not eat readily when magazine training is begun, the weight may be reduced below 80% temporarily. Rates of responding later observed in the experiments are used as the basis for a final adjustment of the percent body-weight used. Whenever possible, the adjustment is upward, since the birds remain healthier at higher weights. The body-weight can often be raised to 85% or even 90%, and the birds may still show a satisfactory reinforcing effect of food and a high stability of the behavior so reinforced. Sometimes, a return to practically 100% of body-weight is possible after the bird has worked in the experimental apparatus for some time. The actual percentage should not be taken too seriously, since some progressive change in body-weight may occur with aging.

Birds are sometimes matched in deprivation level as measured by the rate of responding under a variable-interval schedule. Each bird is run for a session of fixed length each day. A criterion rate of responding is selected to which each bird is to be adjusted. This is expressed as a number of responses per session. The weight to which each bird is fed after the session is adjusted upwards or downwards in terms of the difference between the number of responses emitted on that day and the criterion. A scale of adjustment is chosen as the number-of-grams added or subtracted from the weight for each 1000 responses shown in that difference. When the two birds are to be matched at a value which is otherwise unimportant, the average between the two rates shown by the birds before adjustment may be selected as a criterion.

A protocol card is prepared for each bird, showing its leg-band number, description, and source of supply. Daily entries of body-weight are made on the card, and a running curve of the body-weights may be plotted. In the final experimental procedure, the bird is weighed immediately before an experiment and again immediately after the session. Supplementary food is then given if necessary to bring the bird's weight to a chosen value. A bird whose experimental weight is set at 425 grams, for example, may be found to weigh 412 grams before a given session. During the experiment, it ingests some food and is found to weigh 422 grams when taken from the box. Three grams is therefore given to bring its weight up to the level of 425 grams. On the following day, the bird is again weighed before the session, and the same procedure is followed, all weights being recorded. Such measurements are subject to fluctuation due to the drinking of water, the eating of grit, and defecation. When a daily routine has been well-established, however, the times at which the pigeon drinks, eats grit, and defecates become fairly stable, provided water and grit are always available. If a bird is heavier than its designated weight, it is not used or fed that day.

Conditioning the pecking response

Adapting the pigeon to the apparatus. The pigeon may be adapted to the experimental box concurrently with adjustment of the body-weight. The bird is placed in the apparatus for several hours with its daily ration. Adaptation of emotional responses appears to occur more quickly when the pigeon eats its daily ration in its new environment. The process of emotional adaptation continues during magazine training.

Magazine training. The pigeon is conditioned to respond to the presentation of food in three stages. In stage 1 the bird has reached a suitable percentage of its ad lib weight and is placed in the box with the magazine in position for eating. It is allowed to eat for 1 or 2 minutes from a magazine held in place so that it has continuous access to grain. If a bird does not eat within 10 minutes, it is returned to the living cage without food until the following day. This procedure is used until the bird eats readily from the open magazine. In stage 2 the magazine is operated repeatedly until the bird begins to eat during the brief presentation of food. The first response of the bird to the operation of the magazine is usually an emotional pattern which must adapt out before the bird will eat. The bird usually comes to stand near the magazine because of the accidental contingencies favorable to the reinforcement of such behavior, and a third stage is necessary to break up this behavior. The magazine is opened (by hand) only when the bird has moved about the box. This stage is continued until the bird will turn and move toward the food magazine quickly from any position in the box. The strong conditioned reinforcing stimuli generated by the operation of the magazine are important here. The "superstitious" conditioning of stereotyped responses must be avoided at this stage. If the magazine is repeatedly presented by a clock device, an irregular schedule should be used; but stage 3 is best accomplished by an experimenter who can observe the bird's behavior.

The pigeon is ready for final conditioning when it turns promptly, approaches the magazine, and eats readily as soon as the magazine is presented, but does not approach the magazine in the absence of magazine stimuli.

Differentiating the peck. A pigeon can be conditioned in several ways to peck the key:

1. If the key is adjusted so that a very light contact will operate the magazine (a voice key can be adapted for this purpose during the initial stages of training), and if the key is lighted brightly from behind, a well-adapted pigeon will usually peck at the key during a 1-hour experimental session. A small spot on the key will increase the probability of such a response. If the bird has already been magazine-trained, conditioning almost always occurs at the first peck. Some birds may require several sessions to produce the response. The topography is that of a light exploratory peck, but it shifts under continuous reinforcement to a more vigorous and stable form.

2. Where the process of acquisition is not important, a small grain of corn can be attached to the key with Scotch tape. It is lighted from the pigeon's side so that it can be seen in its normal color. The pigeon will quickly peck at the grain, and the magazine will open. When conditioning to the key has taken place, the grain can then be removed, either at once, or by a progressive reduction in size until it disappears altogether. The resulting response has the topography of the type of peck with which a grain is picked off a surface.

3. The response of pecking the key can be "shaped up" by hand. The experimenter holds a switch which operates the food magazine. Any response made by the bird in the direction of the key is reinforced by the presentation of food. Later, only responses which bring the bird's head close to the key and, still later, only responses in which the head moves toward the key are reinforced. Eventually, an exploratory peck occurs and is, of course, immediately reinforced. If the experimenter can be seen by the bird, he must be below the bird's horizon, since a bird is relatively undisturbed by objects below it, but adapts only with difficulty to a person appearing above. Although this procedure is a useful demonstration of the technique of the shaping of a response through operant reinforcement, it is likely to produce a poor topography. The experimenter must be quite skilled in anticipating and reinforcing the correct movement. For example, the bird will often be reinforced when it moves its head past the key. A pendulum-like oscillation near the key often results. This movement may persist as superstitious behavior for a long time, and it requires so much time for execution that high rates are impossible.

4. A fourth method is useful especially when time in the experimental box is at a premium. A "punchboard" is constructed of a block of plywood, 8 by 8 by $\frac{3}{4}$ inches, in which several $\frac{3}{4}$-inch holes have been drilled to a depth of about $\frac{1}{4}$ inch. A sheet of aluminum, also 8 by 8 inches, is drilled with $\frac{3}{4}$-inch holes in the same places. The board is placed flat on a table; the holes are filled with grain; a sheet of tissue-paper, 8 by 8 inches, is placed over the board; and the aluminum template is fastened in place. The result is several exposed disks of paper, $\frac{3}{4}$ inch in diameter, behind each of which is a small pocket of grain. A few of these exposed segments are cut lightly with a knife blade so that some of the grain may be seen. A board so prepared is placed in the *living cage* of the bird. Most pigeons quickly learn to eat the grain behind the paper and to peck through the uncut disks. Slightly tougher papers are used on successive days. Since the white disks closely resemble the lighted key on the experimental box, well-adapted birds will then peck at the key almost immediately when confronted with it for the first time in the experimental box. If they have been well magazine-trained, conditioning usually takes place instantly. The topography of the resulting response resembles the behavior of puncturing and tearing a disk of paper.

In most of the following experiments, the process of acquisition is unimportant, except as it bears upon the topography of the resulting behavior. In most of the experiments reported here, the birds were conditioned with the punchboard technique. The rest were shaped by hand.

Immediately after conditioning, the response is continuously reinforced. At least three sessions of continuous reinforcement, containing 60 reinforcements each, are arranged.

SPECIAL TECHNIQUES

Differential reinforcement of rates

The differential reinforcement of low rates (drl). Drl is conveniently arranged with a timer which is reset by each response. If the timer reaches a given setting before a response occurs, it sets up a reinforcement. The device can be added to other scheduling circuits; for example, the reinforcement thus set up is effective only if a variable-interval timer has also set up a reinforcement.

The differential reinforcement of high rates (drh). Drh is more difficult. If we reinforce a response when the inter-response time falls below a small fraction of a second, the characteristics of the key and the keying relay become crucial. Any chattering of the contacts will produce an unscheduled reinforcement. Even with reliable contacts, this method tends to reinforce a topography of response which results in vibration of the key. It is better to take several responses into account in measuring the rate. This appears to be justified by the probability that high rates which serve as stimuli in the contingencies of reinforcement generated by schedules are rates sustained for a number of responses. A great many different possibilities can be investigated, depending upon the number of responses considered and the extent to which more remote responses are permitted to contribute to the result. We have studied only a few arbitrary cases in which electrical and mechanical devices have provided the crucial contingencies.

In some cases differential reinforcement of high rates is achieved by a condenser which is charged a given amount by each response but is constantly discharging through a high resistance. When the rate is above a given value, the charge reaches a point at which the condenser energizes another circuit, operating the food magazine. Figure 11 shows a schematic design of a similar arrangement with a mechanical "overtake." An arm (A) driven clockwise by a timing motor carries a contact (B). A second arm (C) pivoting about the same center is driven by a ratchet operated by the keying relay. A third arm (E), rigidly fastened to A, carries a break contact (F). If the bird is not responding, arm E overtakes arm C, and the break contact at F disconnects the timing motor. The whole system is then motionless. As soon as the bird begins to respond, arm C is driven toward A by the ratchet, the contact at F is closed, and the arms E and A begin to move. If the rate of responding is sufficiently high and sufficiently sustained, arm C will overtake arm A, and contacts D and B will operate the magazine. The angle between E and A may be changed from experiment to experiment so that the gap is altered through which arm C moves. This angle partially determines the number of responses which must be executed at a given rate in order to overtake arm A beginning at rest. Other arbitrary features are, of course, the speed of movement of A and the distance which C moves per response. Some actual values and the corresponding performances are described in Chapter Ten.

Added stimuli

The design of an external stimulus which varies with the stimuli automatically entering into the reinforcing contingencies under various schedules raises many problems, most of which we have not attempted to solve here. The stimulus most frequently used is a small spot of light projected on the key which the bird pecks. Two strips of opaque tape are applied to the back of the key, leaving a lighted horizontal slit 1/16 inch wide extending across the key. Two vertical shutters behind the key determine the length of the slit illuminated. These shutters meet in the middle of the key; and as they separate, the spot of light lengthens. In general, we have used a spot no smaller than 1/16 inch square; the maximum length is the width of the key, or 1 inch. When the shutters are driven with cams cut in particular ways, the rate of growth may have any character desired. If the cams are driven by a clock, the stimulus represents

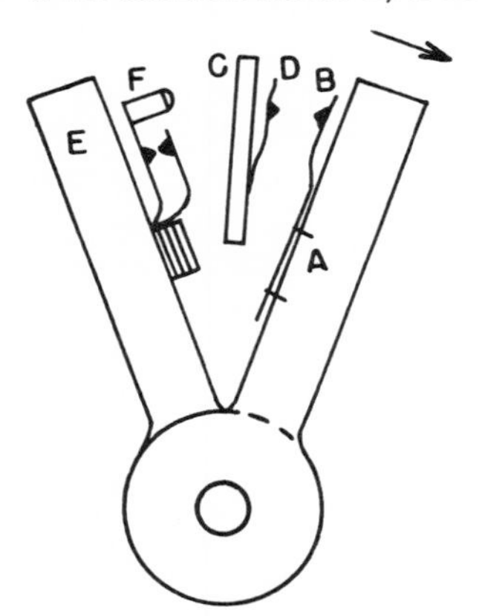

Fig. 11. A device for differentially reinforcing high rates

"added time"; when driven by a counter, "added number"; and when changed by the rate of responding, "added speedometer."

Time out

"Time out" (abbreviated as TO) is any period of time during which the organism is prevented from emitting the behavior under observation. A TO can be arranged in several ways.

1. The animal is removed from the apparatus. This is difficult to accomplish automatically and may generate unnecessary emotional behavior.

2. The key or other manipulandum is removed. This must be done quickly and in such a way that the organism is not injured if it is responding at the time. Some clearly discriminable stimulus correlated with the presence or absence of the key is advisable. Otherwise, the organism will continue to approach the place where the

manipulandum normally appears; or if this extinguishes, it will fail to approach when the manipulandum is in place. But if a conspicuous stimulus is to be arranged, procedure 4 is preferable.

3. For most birds and some mammals, all light in the apparatus may simply be turned off. A pigeon may be conditioned to respond in total darkness, but it normally does not do so. Such a "blackout" is the simplest method to use with the pigeon, since it is easy to arrange electrically and is effective immediately.

4. A stimulus previously correlated with no reinforcement is introduced. This method requires some preparation but has many advantages, particularly in working with animals such as rats, which continue to respond in the dark. For example, a discrimination is established between silence, in which responses are reinforced, and the sound of a buzzer, in which no response is ever reinforced. When discrimination training is well-advanced, the organism immediately stops responding when the buzzer sounds. It has been found experimentally, with organisms which do not respond in the dark, that such a stimulus has the same effect as turning out the lights.

A time out may have other effects than those for which it is primarily used:

A. Under some circumstances it functions as an aversive stimulus, which may be used to generate avoidance or escape behavior, to develop a conditioned "anxiety" suppression, or as a punishment. Time out has been used, for example, to improve performance under a complex discrimination where some responses are followed by reinforcement and others by time out. As Herrnstein (1955) has shown, a time out is aversive only when the schedule in which it occurs is positively reinforcing. When the schedule itself is aversive, the time out may serve as a positive reinforcement.

B. As a novel stimulus a TO may have an emotional effect in the form of a pause and/or a reduced rate. This eventually adapts out.

C. During a TO the bird may move about the box; and at the end of the TO it may be farther from the key than usual. A slight pause will then be introduced as the bird returns to the key.

D. During time away from the key, competing behavior—drinking, cleaning, etc. —may develop; once in progress, it may not terminate immediately upon cessation of the TO. The result may be longer pauses than in C.

E. Neither C nor D should produce a depressed rate once the bird has begun to respond. However, when a TO is used against the background of a very high rate, returning from other positions in the box may leave the bird in an unfavorable posture for executing a high rate. A brief reduction in rate may then appear.

F. When darkness is used as a TO, the bird may "roost." Little is known of this effect; but it might produce a pause and perhaps even a lowered rate after the termination of TO. This result is not inevitable, however, because many instances have been observed of an immediate start at a maximal rate after fairly long TO's.

G. When a TO follows a reinforcement, it may give the bird time to complete such behavior as cleaning its beak or grooming. The TO will eliminate any slight pausing characteristically observed after reinforcement because of such behavior.

H. A TO could possibly have side effects from incidental conditioning if it is also used to start and stop experiments. (See below.)

Yoked boxes

In one of the following experiments a control technique is used which may have a more general application. For example, when a schedule is changed from an interval to a ratio, residual effects from the earlier schedule may have to be considered. Although these intervals are not specially scheduled, they still naturally elapse between successive reinforcements. Since such intervals may be quite irregular, it is difficult to do a control experiment with the same irregular interval schedule. This problem may be solved with "yoked boxes." Two standard experiments are run completely independently except that reinforcements are set up in both boxes by a single relay. If this relay is operated by an interval timer, the same contingencies prevail in both boxes. But if it is operated by a ratio schedule involving the bird in the first box, the performance of this bird will be responsible for the scheduling of reinforcements for the other bird. Except for an occasional reinforcement which is missed before another is set up, the same temporal program obtains in both boxes. Nevertheless, the first bird receives reinforcements on a ratio schedule, while the second is on a variable-interval schedule. Differences in the "end effect" of the two schedules may be compared while the temporal pattern of reinforcements remains constant.

Probe

A probe is some momentary change of conditions against a background of a stable or slowly changing performance used only once or so rarely and unpredictably that it cannot begin to function as a discriminative stimulus. One effective probe is the time out noted above. Others include: (a) an unscheduled reinforcement of a response; (b) a free presentation of the food magazine when no response is made; and (c) a brief introduction of a discriminative stimulus which, because of other contingencies or other conditions of deprivation or aversive stimulation, controls a different rate of responding. These probes momentarily change some of the variables operating in the stable performance being studied. Thus, they may reduce the effectiveness of recent behavior (as with a time out), produce stimuli generated by responding at another rate, introduce stimuli generated by a reinforcement at some other time in a schedule, etc.

Length of experimental session

The length of a daily session may be expressed either as a period of time or as a number of reinforcements. Where the schedule involves the number of responses per reinforcement, the session is usually terminated on completion of a given number of reinforcements. Where the reinforcements are scheduled in terms of time, the experiment is usually terminated by a clock. The limiting factor is the amount of food the animal ingests during the experiment. This amount must be equal to or less than that needed to bring the animal back to its standard weight if experimentation is to continue daily. Suitable lengths of experimental sessions can be chosen by observing the body-weight of the bird from day to day. The most efficient experimental session is one in which the bird receives its whole daily ration during the experiment. Because this ration varies from bird to bird, experimental sessions are not always of the same length in a given experiment. In general, the experimental session is made as long as possible, since under these circumstances a process is least likely to be affected by incidental changes in the current history of the organism. The periods used in the following research vary from a few minutes to 15 hours.

Starting and stopping an experimental session

In experiments which are automatically controlled, the organism usually spends some time in the experimental box both before and after the actual experimental session. It must be prevented from responding during these periods. The same solutions are available as with the time out. In experiments with pigeons the experimental box is dark both before and after the experimental session.

Mistakes are reduced if some such routine as the following is observed. The apparatus is first tested to ensure that the proper circuits are in effect and operating, that the key has the proper sensitivity, that the magazine is full of grain, that water is available, and that all lights are operating. The animal is then placed in the box with time-out conditions in effect. The animal's number and the experimental procedures are entered on the cumulative record and in a permanent-records book. The recorder pen is cleaned and filled, and the paper supply checked. The experiment proper is started a few minutes later, either by hand or by a timing device. It is discontinued at the end of a pre-arranged time or number of reinforcements, and time-out conditions prevail until the organism is removed from the box. When a single experiment of only a few hours' duration is run during the night, these time-out intervals before and after the experiment may last for hours. The bird is fed up to weight immediately after being removed from the apparatus and returned to its living cage.

Procedural irregularities and mistakes

In general, an effort is made to experiment on a given organism day after day without interruption for the duration of the experiment. Time spent in the living cage usually can be shown to have some effect upon the results, and missing a day from an over-all schedule may be an important disturbance. This happens when for some reason the weight of the bird exceeds the set weight. (The food magazine may jam in an open position, the bird may be fed by mistake, and so on.) Occasionally, also, the apparatus fails. The organism may undergo extinction or may be reinforced on an unscheduled program. Although these difficulties are serious, they need not result in the total loss of an experiment. The prevailing performance can usually be recaptured with a day or two on the proper procedure. No important change in procedure is ever made until a stable performance has again been observed.

Length of exposure to schedules

The effect of a schedule often depends upon momentary accidental contingencies resulting from the interaction of the schedule and the existing performance of the organism. We do not expect two organisms to reach either a stable or an oscillating state at the same rate. The experiments are therefore not designed in advance. The pigeon is placed on a given schedule, and its performance is watched from day to day until it is stable or until several cycles have revealed the nature of any oscillatory results. Especially in the later experiments, generous exposures to a schedule were arranged.

Change of conditions

So far as possible, a change in conditions is made in the middle of an experimental period in order to detect first effects independently of the special variations produced by the beginning of an experimental session.

Reversibility

With some exceptions, the conditions to be described here are reversible; that is, the performance characteristic of one schedule may be recovered after the organism has been exposed to other schedules or conditions. This usually makes it possible to establish a baseline, introduce a variable, and return again to the baseline in such a way that the organism serves as its own control.

Number of subjects

Because of the very large number of hours of behavior recorded with this technique, and because we may use the reversibility of processes to provide control curves, the use of large groups of organisms was not necessary. Two pigeons are usually studied on a given schedule, but in some cases as many as four or five are used. If differences in performance are conspicuous and cannot be immediately explained in terms of differences in the behavior which the organisms bring to the schedules, the experimental conditions are further analyzed, and the experiment may be repeated with other pigeons. But this is rarely done. A given experiment is often repeated many times in later, more complex experiments. For example, the total number of pigeons studied on fixed-ratio schedules in this research is of the order of fifty, although no particular experiment involves more than three.

The research as a whole covers approximately 70,000 recorded hours, during which the experimental organisms emitted approximately one-quarter of a billion responses.

Chapter Four

• • •

FIXED RATIO

INTRODUCTION

IN A FIXED-RATIO SCHEDULE of reinforcement, every *n*th response produces a reinforcing stimulus. On a cumulative plot of responses as a function of time, reinforcement occurs when the curve crosses a line parallel with the base of the graph. The height of this line is the number of responses in the "ratio." The time to reinforcement varies, depending on how rapidly the required responses are emitted. At least five relevant conditions prevailing at reinforcement can be specified. These are manipulable conditions which may be used in the analysis of fixed-ratio schedules.

CONTINGENCIES RESULTING FROM FIXED-RATIO REINFORCEMENT

Frequency of reinforcement

The higher the rate at which the responses required for reinforcement are emitted, the shorter the time to reinforcement and, hence, the higher the frequency of reinforcement. The high rates of responding which occur under small fixed-ratio schedules, for example, may simply be due to a relatively high frequency of reinforcement. Experiments described in Chapter Eight isolate the contribution of frequency of reinforcement to behavior under a fixed ratio.

The reinforcement as a discriminative stimulus

A response is never reinforced just after a previous reinforcement. The operation of the magazine and the ingestion of food is therefore a stimulus which bears a constant relation to (the next) reinforcement. Like any other stimulus, its effect may extend beyond its termination. One common effect is a low rate of responding just after reinforcement.

Later effects of "time since reinforcement"

A response cannot be reinforced within a shorter period of time than that required to count out the ratio at the highest possible rate. But since the actual rates vary, there is only a rough correlation between time since reinforcement and the probability that a response will be reinforced. Control from this source is much less precise than in a fixed-interval schedule. However, time since reinforcement and the number of responses since reinforcement vary together when the rate is constant, as in most fixed-ratio performances. Hence, some allowance must be made for such a factor.

Number of responses since reinforcement

Evidence presented in Chapter Eleven shows that the number of responses emitted since reinforcement or some other arbitrary event is a discriminable stimulus. It is an important factor in a fixed-ratio schedule because the number is constant at reinforcement. There are two principal effects:

(A) *Number as a discriminative stimulus.* Probability of reinforcement is maximal when the number of responses emitted equals the ratio. The probability of response is therefore eventually maximal at that point. Smaller numbers of emitted responses exert proportionately less control. In particular, a count of zero responses just after reinforcement is the occasion upon which a response is never reinforced. Zero and *N* (the ratio) are thus end points on a stimulus continuum which exerts a gradient of control because of its correlation with probability of reinforcement.

(B) *Number of responses since reinforcement as a conditioned reinforcer.* As a discriminative stimulus, the number of responses emitted at reinforcement is also a reinforcing stimulus. More generally, at any point during a fixed ratio, a response may be reinforced because it increases the number and advances this stimulus toward the reinforced end of the continuum.

A chain of responses develops as follows:

$$S^D_{N-1} \cdot R \rightarrow S^D_{Mag} \qquad (1)$$

Here, the next to the last response is a discriminative stimulus in the presence of which a response opens the magazine. Then, N-1 responses since reinforcement would reinforce any behavior which produces it, as in:

$$S^D_{N-2} \cdot R \rightarrow S^D_{N-1} \cdot R \rightarrow S^D_{Mag} \qquad (2)$$

In (2) the second from the last response in the fixed ratio is the occasion on which a response produces the occasion for the last response, which is in turn followed by the appearance of the magazine. The chain can extend backwards to the preceding reinforcement.

$$S^D_{reinf} \cdot R \rightarrow S^D_1 \cdot R \rightarrow S^D_2 \cdots\cdots S^D_{N-1} \cdot R \rightarrow S^D_{Mag} \qquad (3)$$

Here, the reinforcement is the occasion on which the first response produces a count of one, which is the occasion on which a second response produces a count of two, and so on.

The preceding reinforcement, then, may have 2 opposed effects: (1) It may control a low or zero rate because a response is never reinforced when the bird has eaten recently or when only a small number of responses has been emitted. (2) It may control a substantial rate as the occasion on which responses are reinforced because they conspicuously increase the count or remove a zero count. (A discriminable increase in number might be a function of the number of responses already emitted. A psychophysics of the number of responses is possible here, just as with any stimulus dimension. Experiments bearing on these points are reported in Chapter Eleven, particularly in the sections on mix FRFR, mix FIFR, and mix FR ext.) The two types of control are usually present simultaneously. However, the first function can be separated from the second under conditions in which the conditioned-reinforcing properties of number are slight or lacking.

The importance of number of responses as a conditioned reinforcement in fixed-ratio behavior can be settled more generally by an analysis of how reinforcing properties of a conditioned reinforcer are established, maintained, and changed (Chapter Twelve).

Differential reinforcement of high rates

In interval schedules the probability that a response will produce a reinforcement increases with time after any unreinforced response has been made. After a long pause, the probability of reinforcement is high (the clock having run continuously during the pause). In a fixed ratio, however, the probability that a response will produce a reinforcement is independent of the time since the last response, the reinforcement being determined by a counter which operates only as a function of the bird's behavior. If the elapsed time between responses differs—that is, if responses are grouped—a reinforced response is more likely to follow a short inter-response time on a ratio schedule and a long inter-response time on an interval schedule (Skinner, 1938, p. 274).

The relative importance of these five properties of the reinforcing contingencies generated by fixed-ratio schedules will be evaluated here in experiments involving (1) a time out after reinforcement, (2) a time-out probe, (3) the number of reinforcements in the ratio, (4) drugs, (5) level of food deprivation, (6) novel stimuli, (7) brain damage, and (8) "added counter."

TECHNICAL PROBLEMS

Fixed-ratio schedules generate high rates of responding; in the pigeon they often exceed 10 responses per second. The key, the recorder, and the device which determines the ratio must be designed to follow these rates. Differences between the responses of these elements to high rates may lead to poor counting of the ratio not detected in the records, incomplete records, and so on. The conditioned reinforcers (such as the noises which accompany the operation of the magazine) must also be presented very quickly. Even good-quality, telephone-type stepping switches, although otherwise satisfactory for programming fixed-ratio schedules, are usually back-acting; that is, they advance at the end of a pulse rather than at the beginning. If the magazine is arranged to be operated by a pulse from the stepper, the slight delay may have the effect of reinforcing some later stage in the act of pecking the key—for example, the movement of withdrawing the head. If this contingency remains in force for a long time, it will alter the topography and possibly affect the rate. The solution noted in Chapter Three is to permit the next to the last response in the ratio to close another relay, so that the final response to the key operates the magazine directly.

THE TRANSITION FROM CONTINUOUS REINFORCEMENT TO SHORT FIXED RATIOS

A fixed-ratio schedule will take control directly after continuous reinforcement if the ratio is not too large. Under the new schedule a short period at a high but declining rate usually occurs, representing extinction of the effect of the previous continuous reinforcement. This is followed by a period of low rate with subsequent acceleration. The rate either continues to increase until the terminal rate of the fixed ratio is reached, or it enters a further slight decline before such an increase. Figure 12 is a stylized

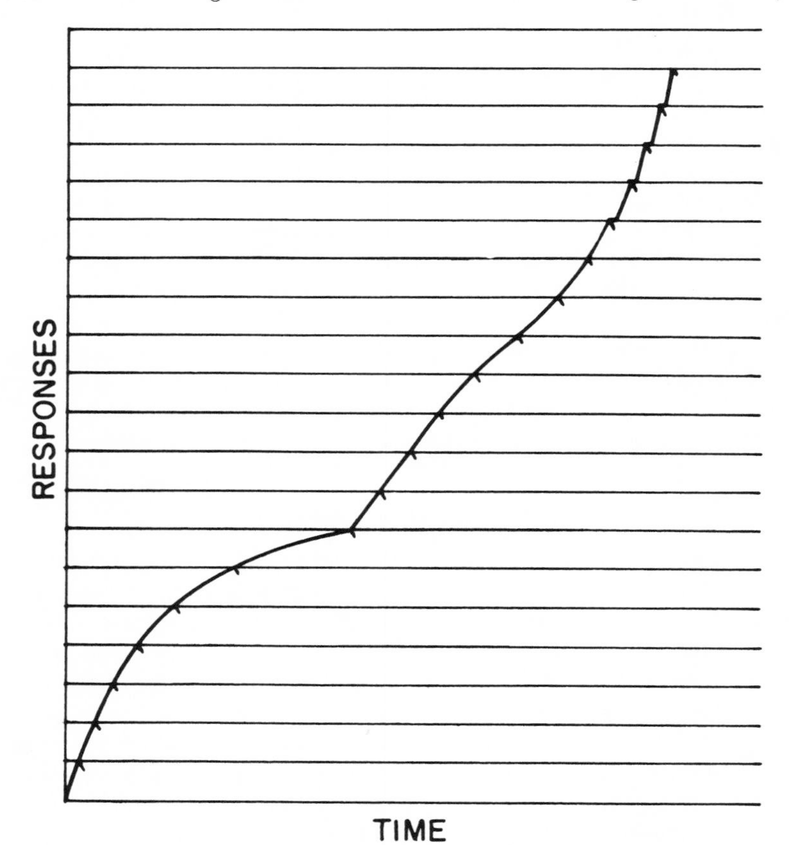

Fig. 12. Stylized plot of the transition from crf to FR

graph illustrating the main features of the transition. Occasionally, a third slight inflection occurs. The special contingencies imposed by a fixed ratio begin to operate during the first decline in rate; in particular, the pauses during the low rate are not likely to be followed by reinforcement as in interval schedules. Reinforcements tend to occur after responses which are members of small groups.

Figure 13 shows the transition from crf to FR 22 where a performance very close to the terminal performance of the schedule is reached during the 1st session. The over-all rate begins at about 1.5 responses per second at *a* and declines slowly. The fact that some of the pauses in the region *b* follow reinforcements shows a possibly early development of a discrimination based on the reinforcement. A low rate occurs at the

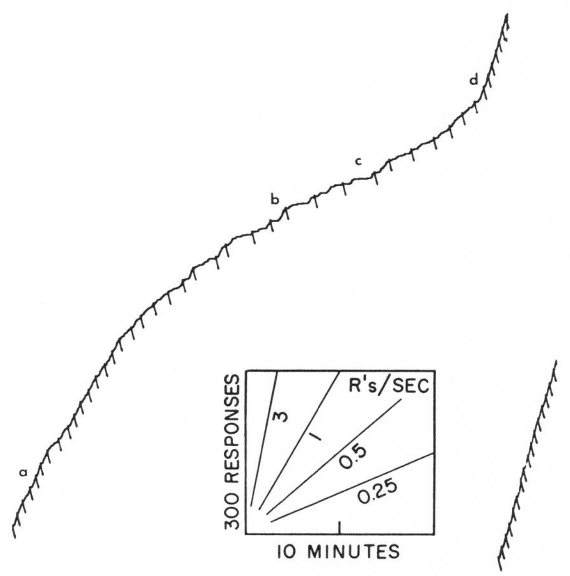

Fig. 13. Transition from crf to FR 22

inflection point at *c*, after which the rate increases continuously until a terminal rate is reached, beginning at *d*. The over-all rate is then approximately 2.5 responses per second, and pauses following reinforcement are almost wholly eliminated.

Figure 14A shows the transition to a larger ratio (crf to FR 40). A high rate at *a* appears after a slight "warm-up." The rate declines fairly smoothly until at *b* it is almost zero. The reinforcement following the pause at *b* follows a brief rapid run; this characteristic contrasts sharply with the long inter-response times just preceding it which were unreinforced. The result is the highest rate yet seen. A slight decline sets in quickly and continues through *c*. The rate increases again, and the bird continues for the remainder of the session at an over-all rate of approximately 3 responses per second, with no pauses following reinforcement. Toward the end of the session,

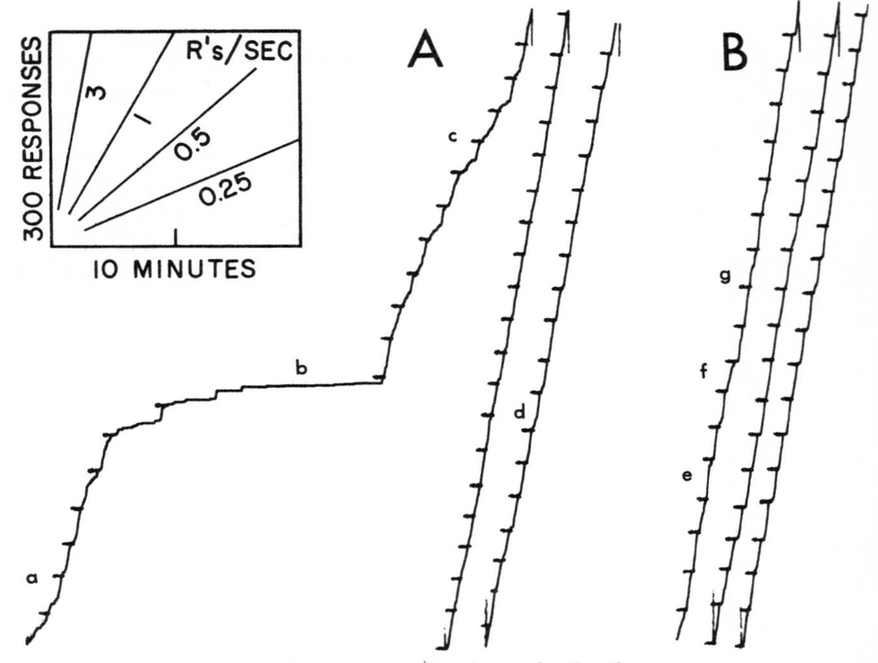

Fig. 14. Transition from crf to FR 40

some curvature tends to appear at the start of the ratio, as at *d*. In the 2nd daily session (Fig. 14B) a characteristic stable performance for this value of fixed ratio develops. Some irregularity occurs at the start of the session, where the rate falls after the maximum rate of the ratio is reached, as at *e, f,* and *g*. Thereafter, the performance is standard for a fixed ratio of this value—a pause after reinforcement of only a few seconds, possibly a brief period of acceleration, and a terminal rate of from 4 to 6 responses per second.

Figures 15 and 16 show two cases of the transition from crf to FR 45 in which the final performance develops more slowly. In Fig. 15 the characteristic high rate following continuous reinforcement at *a* declines to a lower rate at *b,* and a moderate rate of responding is resumed in the region *c*. A higher, fairly stable rate then prevails for some time (*d*). The "grain" of the record is rough, however, and this bird shows large changes in over-all rate from reinforcement to reinforcement. By the end of the session the bird is responding at a sustained rate during the last part of each ratio. However, the over-all rate does not exceed 1.5 responses per second, which is considerably below the normal rate on this size of ratio.

Figure 16 shows a second instance where the terminal performance develops slowly. As in Fig. 15, the initial portion of the curve is negatively accelerated, reaching its

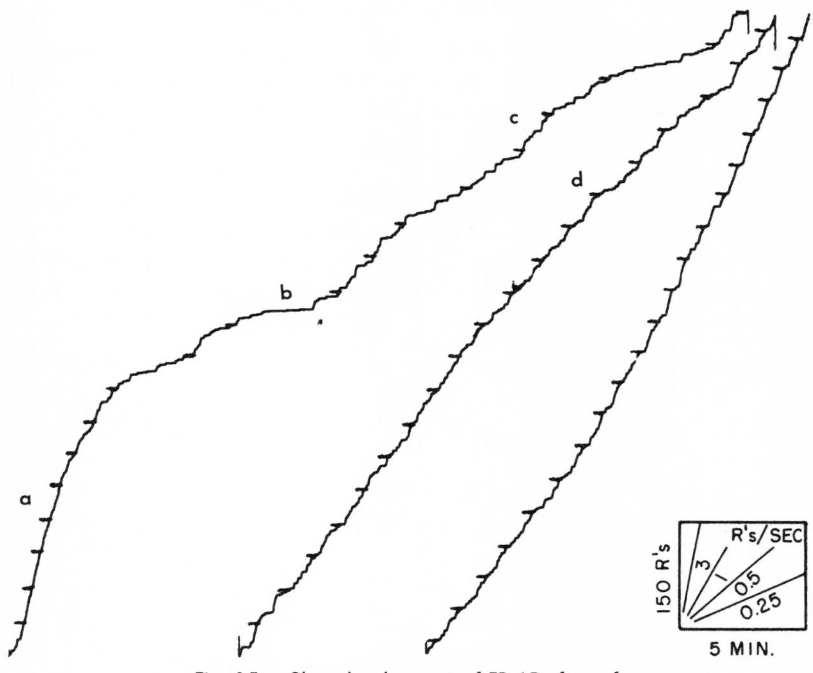

Fig. 15. Slow development of FR 45 after crf

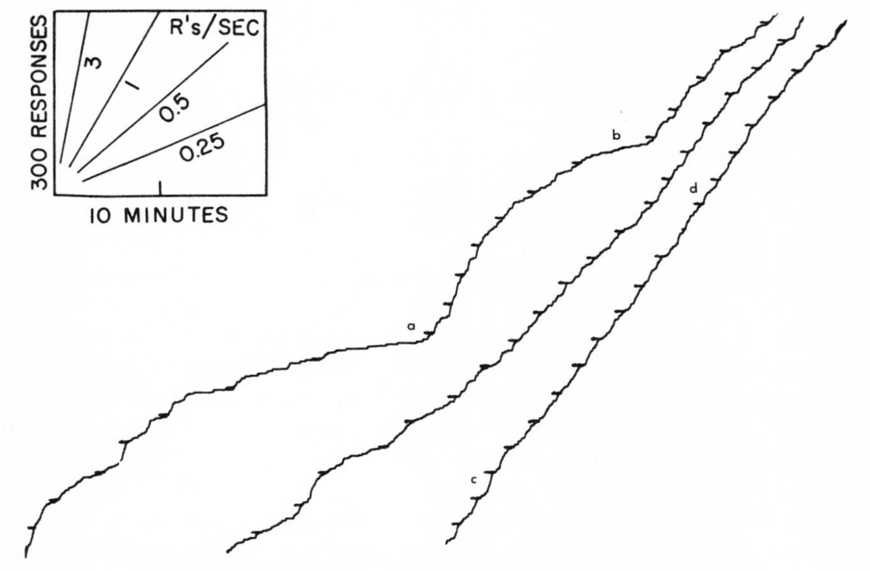

Fig. 16. Slow development of FR 45 after crf (second bird)

lowest rate at *a*. This negative acceleration is repeated and reaches an intermediate rate at *b*. Although the grain is rough, the over-all rate increases steadily for the remainder of the period. The over-all rate does not exceed 0.8 response per second by the end of the session. But some runs are at 2.5 and 3.5 responses per second, as at *c* and *d*, respectively.

Figure 17 shows a transition from crf to FR 50 in which the first negatively accelerated portion of the curve is considerably smaller than usual. The preceding crf has created only a slight tendency to respond in extinction. However, the rate declines as usual through the first 4 ratios to *b*. The rate is so low that the reinforcements at *a* and *b* are separated by about 15 minutes. The over-all rate increases during the next

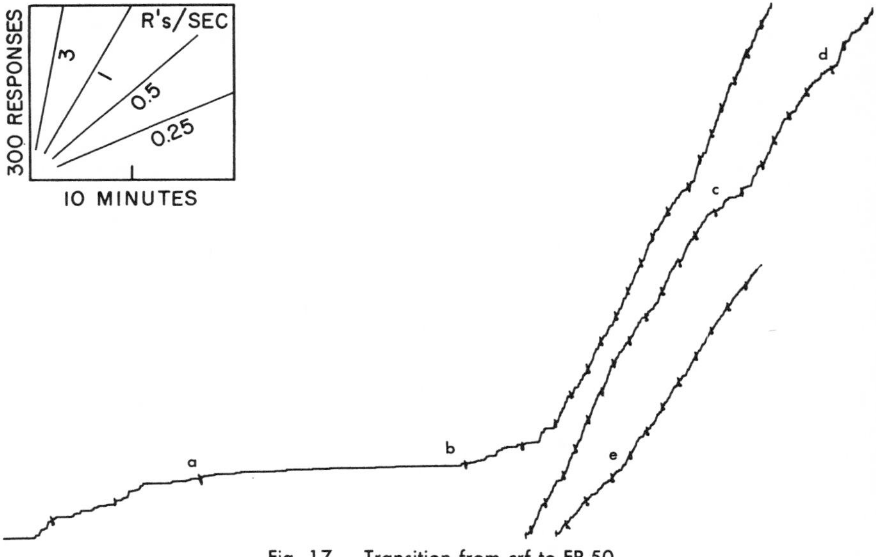

Fig. 17. Transition from crf to FR 50

3 ratios to approximately 1 response per second, which is maintained to the end of the session. The grain remains rough, and marked fluctuations in over-all rate occur, as at *c*, *d*, and *e*.

Figure 18 shows a transition from crf to FR 50 which does not conform to the usual pattern. An initial low rate accelerates slowly and continuously through the first 2 excursions of the figure. By the 3rd segment the rate has reached 1 response per second, which is maintained for the remainder of the session with some slight oscillation. The grain remains rough throughout the session. As in Fig. 15, 16, and 17, the fixed-ratio performance develops slowly.

Figure 19 shows examples of a rapid transition to a final FR 20 performance. In Record A the terminal rate is reached almost immediately, and no pauses are above 1 or 2 seconds. At *a*, however, 2 ratios are run off at a rate of only 1.3 responses per

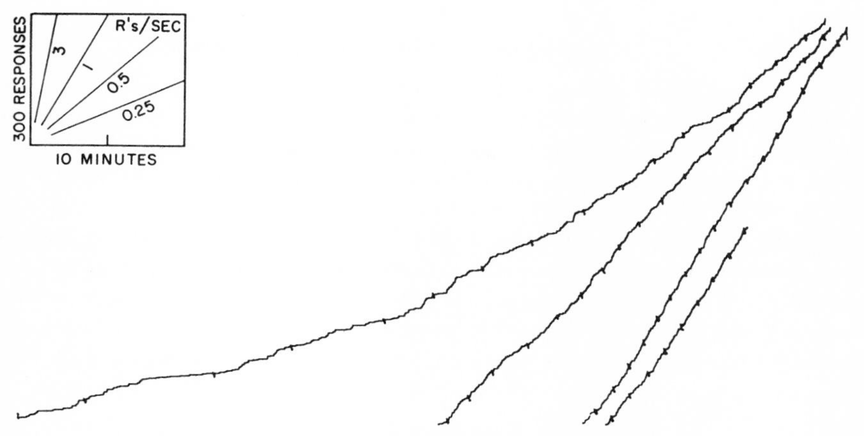

Fig. 18. Unusual transition from crf to FR 50

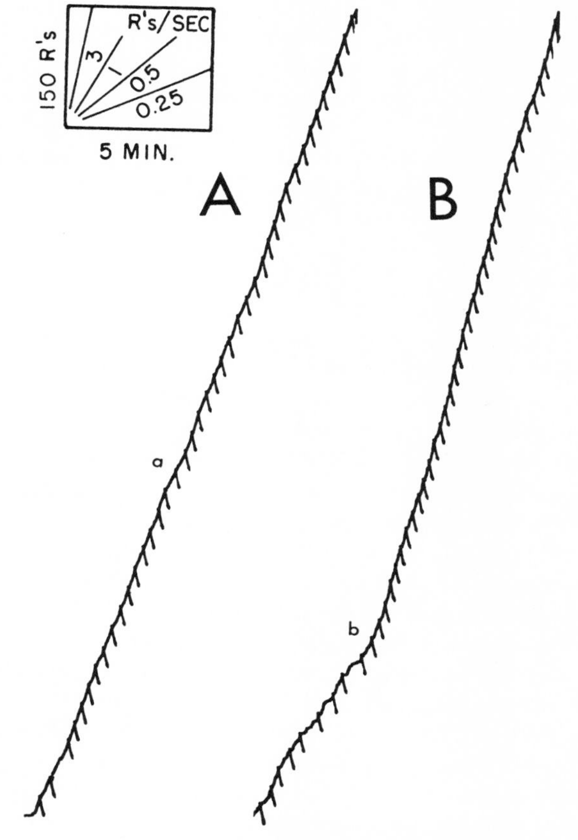

Fig. 19. Transition from crf to FR 20

second, instead of a prevailing rate of from 1.8 to 1.9 responses per second. The second bird, Record B, shows a trace of the typical pattern during the first 9 ratios, but a performance very near the final form begins abruptly at *b*. Variations in the early fixed-ratio curves after crf are to be expected, since the transition depends upon whether the preceding crf has generated not only the stable rate but the topography of response necessary to make the fixed-ratio contingencies effective. The actual contingencies are not precisely controlled. Nevertheless, we may say that fixed ratios of the order of 50 or 60 may be programmed directly following continuous reinforcement, assuming a standard level of deprivation represented by 80% body-weight. More drastic levels of deprivation would probably permit the sustaining of higher ratios. In any case, longer fixed ratios, as we shall see, can be set up by first establishing stable performances on shorter ratios.

FINAL PERFORMANCE UNDER SHORT FIXED RATIOS

Records A and B in Fig. 20 show final performances under FR 20 after 650 reinforcements. (Figure 19 shows the birds' transitions from crf.) The over-all rates are 2 and 2.5 responses per second, respectively, while the local rates (ignoring the slight pauses after reinforcement) are slightly above 3 per second in Record A and approximately 4 per second in Record B. Records C and D show the performances of 2 other birds on FR 31 after 1300 reinforcements. Here, the local rates are 3 per second for both birds, the over-all rate being 2.4 in Record C and 2.6 in Record D. The slightly lower rate in Record C is due to very slight pauses following reinforcements.

These curves represent the order of magnitude of rates to be expected under fixed ratios of this size as well as the order of consistency and uniformity. These performances are not likely to change with further exposure to the schedule.

With fixed ratios of 40 or more, the final performance occasionally develops much slower. Figure 21 shows the middle segments of 9 consecutive daily sessions on FR 45 after crf. (Here, the recorder resets after each reinforcement.) Rates of the order of 1.25 responses per second during the first 2 sessions increase to almost 4 responses per second by the 8th session (Record H). Pauses following reinforcement are common and may last up to 20 to 30 seconds. Acceleration frequently occurs in the early part of each ratio, and in Record H the rate falls abruptly to zero for approximately 30 seconds after the bird has begun to respond at the terminal rate. Such pauses and rate changes indicate that the schedule has not yet had its final effect, and that further exposure to the schedule will produce more uniform curves.

Figure 22 shows a typical performance after prolonged exposure to FR 60. The 2 records are for consecutive sessions occurring after more than 4000 reinforcements on this schedule. Occasionally, a pause follows the reinforcement, as at *a* and *b*, and the shift to the terminal rate is characteristically abrupt. In nearly all other cases, however, the bird begins pecking within a few seconds after reinforcement, and attains the terminal rate immediately.

This standard performance can be generated uniformly from subject to subject, al-

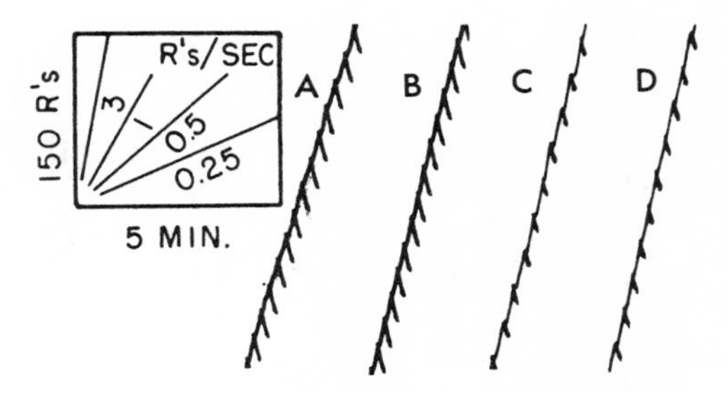

Fig. 20. Final performance on FR 20

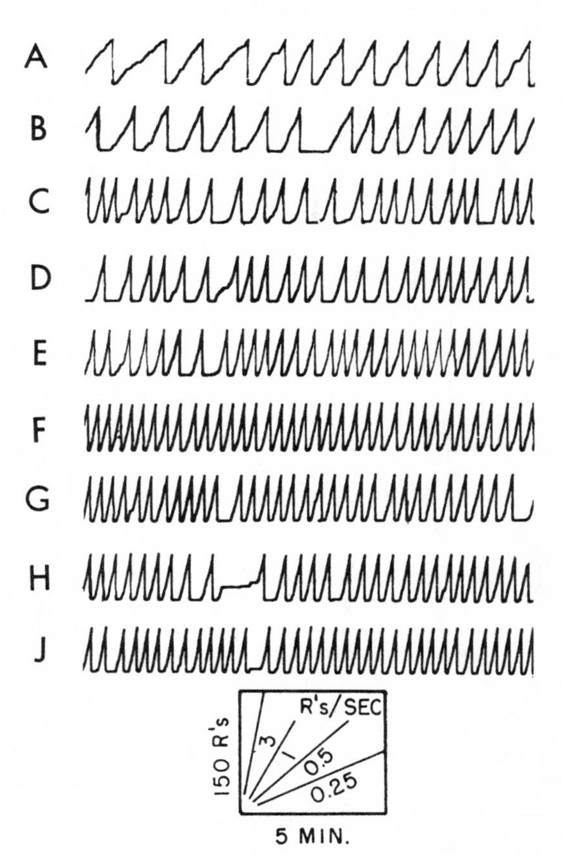

Fig. 21. Early development on FR 45

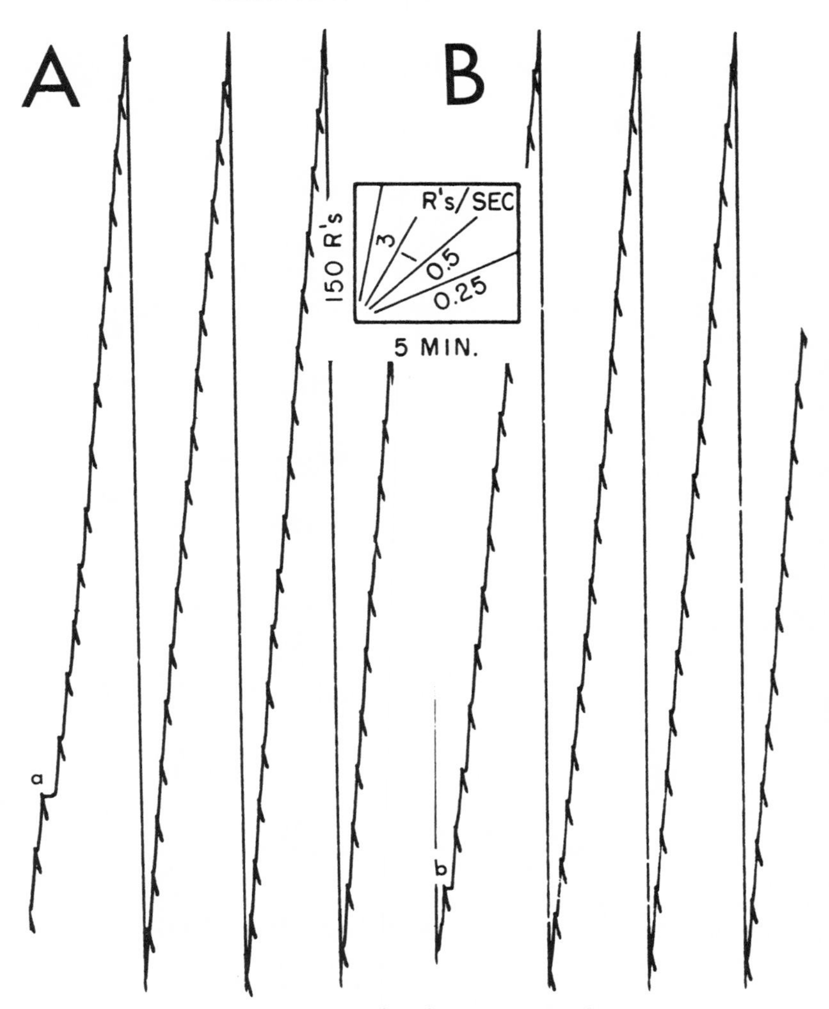

Fig. 22. Final performance on FR 60

though merely programming a fixed ratio of this size does not automatically guarantee such a result. A fixed-ratio schedule acts in concert with the particular behavior which the organism brings to the experiment, and appropriate manipulation may be required to achieve a performance similar to Fig. 22 in every case. If some other result is observed, the size of the ratio is dropped until no pauses occur after reinforcement and until the terminal rate is of the order of 3 responses per second. The ratio is then increased by 5 or 10 responses and held at the new value for several sessions in order to ensure that pauses or irregular features will not develop. If marked curvature, low terminal rates, or long pauses after reinforcement appear, the ratio is again decreased temporarily. (It is assumed of course that the amount of reinforcement is adequate, that the manipulandum and all circuit elements can follow high rates, and that the level of deprivation of the animals is satisfactory.)

INTERMEDIATE FIXED RATIOS

As the number of responses in the fixed ratio is increased, the pauses following reinforcement grow longer. Figure 23 shows a segment from a final performance on FR 120, after almost 6000 reinforcements on fixed ratios from 20 to 120. The pauses following reinforcement may be as long as 10 minutes (at *b*), although very short pauses still occur, as well as intermediate values (at *a*). Once the bird begins to re-

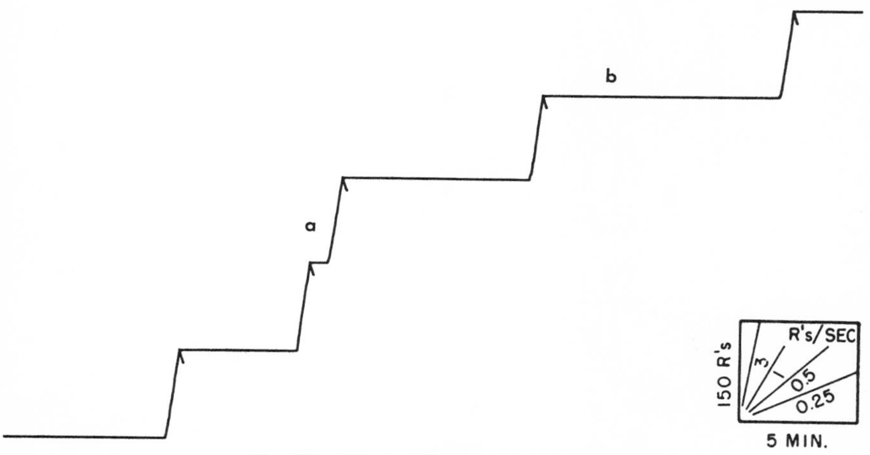

Fig. 23. Final performance on FR 120

spond, however, it reaches the terminal rate of the fixed ratio immediately and maintains it until the ratio is completed.

Figure 24A shows an example of a stable performance on FR 200 after a history of almost 4000 reinforcements on fixed ratios from 50 to 180. The bird pauses after each reinforcement for a period varying from a few seconds to more than a minute; the rate then shifts (usually abruptly) to 3.5 to 4 responses per second, which is maintained (though possibly with a slight decline) until reinforcement. A second bird (Record B) in the same experiment on FR 120 after a similar history shows somewhat longer pauses and an instantaneous rate change from zero to the terminal rate.

A well-marked, but rare, exception is the deviation called a "knee." An example is shown in Fig. 25 at *b*. After a long pause, the pigeon begins at nearly the terminal rate but decelerates smoothly to zero before returning to complete the ratio run. This bird showed slight negative curvature (especially at *d*) and irregular positive accelerations at *c* and *e*.

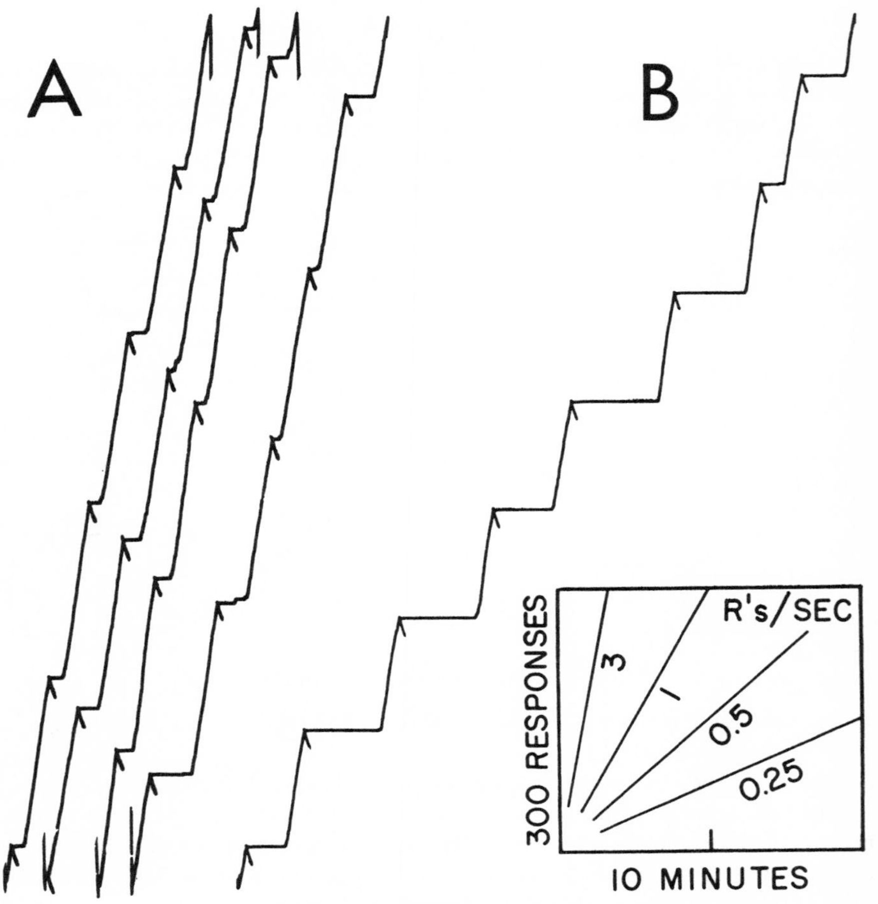

Fig. 24. Final performance on FR 200 and FR 120

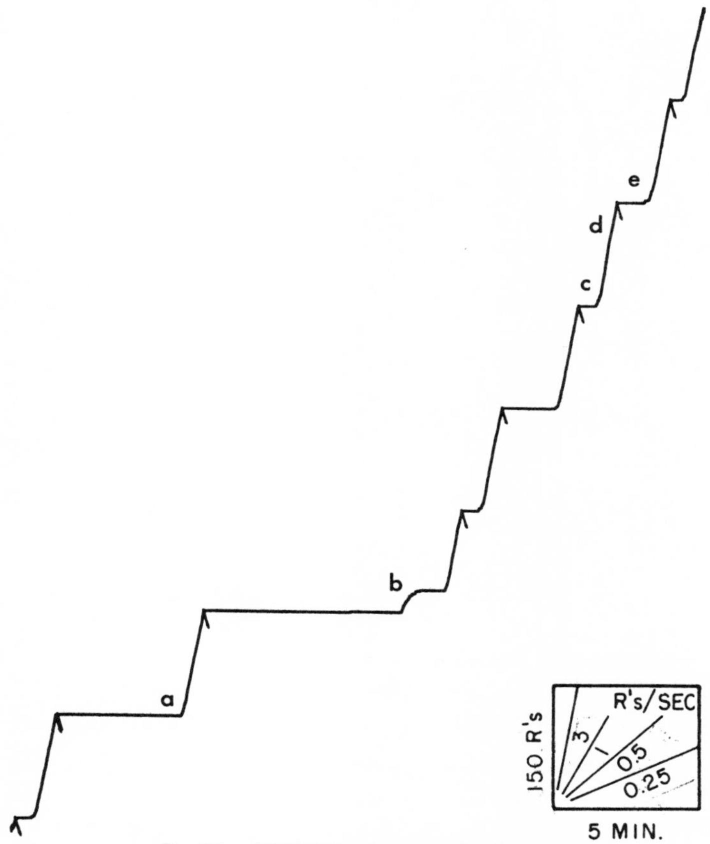

Fig. 25. FR 125 showing a marked knee

The over-all pattern for a daily session at a ratio which generates occasional long pauses is shown in Fig. 26, where the record has been pieced together to make the coordinates continuous in time and number. The bird had begun with a short fixed ratio after crf, and the ratio had been slowly increased over a long period of time to a size large enough to produce pauses following reinforcement in most cases. The average rate of responding is less than 1 response per second, but the actual running rate is about 5 responses per second. The pauses following reinforcement vary from practically zero (as at *a* and *b*) to over 10 minutes (as at *c*). Two other common features of the curve should be noticed. First, ratios beginning with short pauses generally tend to group. Second, a slight over-all curvature develops during the session which is fairly characteristic of performance under larger ratios. Even at the end of the session, where the over-all rate has fallen, responding occasionally begins with almost no pausing after reinforcement (as at *d*).

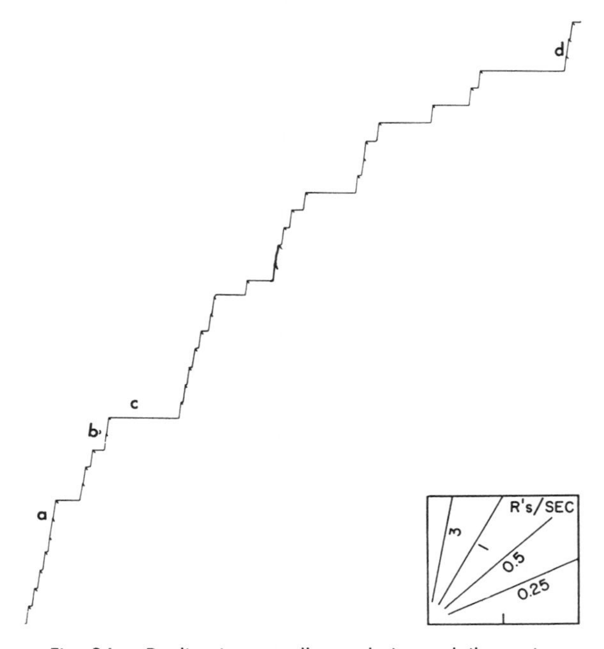

Fig. 26. Decline in over-all rate during a daily session

LARGE CHANGES IN THE FIXED RATIO

When a large fixed ratio which is producing pauses following reinforcement is reduced, pauses do not immediately disappear. Figure 27 shows a transition from FR 180 (Record A) to FR 65 (Record B). In the middle of the session on FR 180, after a history of over 6000 reinforcements on fixed ratios as high as 240, pauses range from only a few seconds (as at *b*) to almost 1 minute (as at *a*). At the beginning of the following session (Record B) the ratio was reduced to 65. The absence of a pause after the 1st reinforcement (at *c*) is characteristic of this bird at the start of a session, even under larger ratios which show considerable pausing elsewhere. The 2nd reinforcement (at *d*) shows a pause of slightly less than 1 minute. The next 2 reinforcements are followed by progressively shorter pauses until (at *e*) a characteristic performance on FR 65 has been established. This performance is maintained for the rest of the session.

Figure 28 shows a similar result. Record A is from the middle of the session on FR 185, after a history of over 7000 reinforcements on fixed ratios of from 40 to 170. (Note the examples of negative curvature in single segments at *a* and *b*.) When the ratio is dropped to 65 (Record B), the pauses following reinforcement do not disappear immediately, although the reduced size of the fixed ratio shortens them consider-

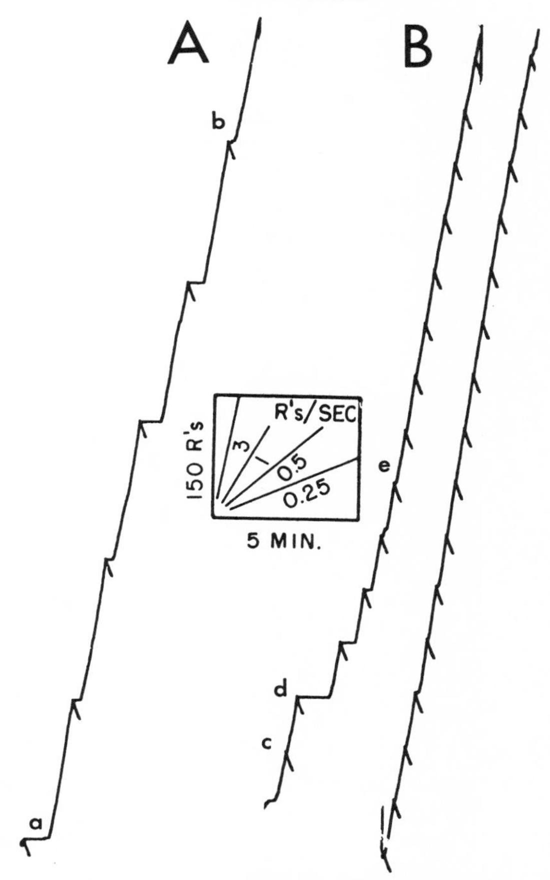

Fig. 27. Transition from FR 180 to FR 65

ably. The bird ultimately reached a stable performance on FR 65 similar to that in Fig. 27B.

Figure 29 illustrates a more extreme change in the size of the fixed ratio. The bird had had a history of tand *FR*FI before exposure to the fixed-ratio schedule. Record A shows 3 segments from a stable performance on FR 160, with only brief pauses after reinforcement. Record B shows the beginning of the following session, when the ratio was reduced to 35. The pauses following the reinforcement became progressively shorter and essentially disappeared by the 5th reinforcement. In the following session the size of the ratio was increased to 140; and Record C, containing 3 segments from the early part of the session, shows the complete absence of any pausing following reinforcement. Curvature and pausing again develop by the middle of the session (Record D).

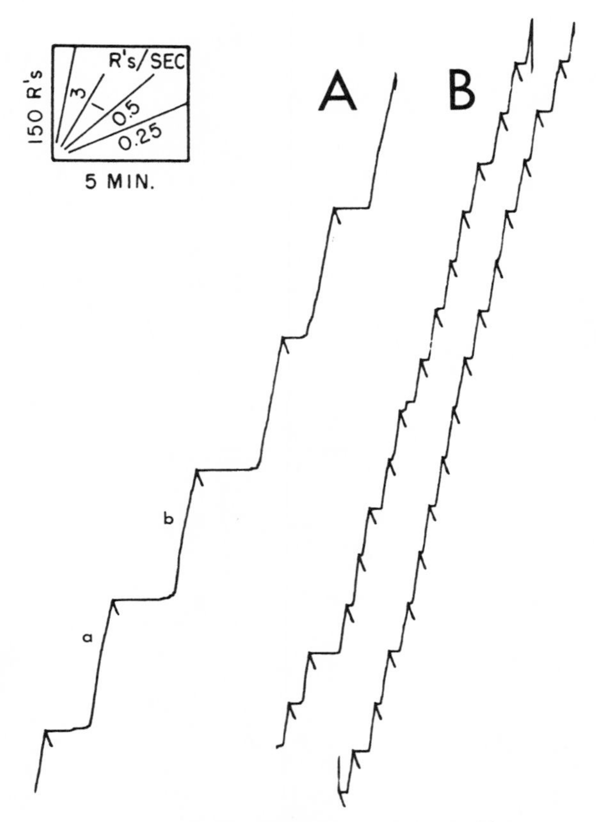

Fig. 28. Transition from FR 185 to FR 65

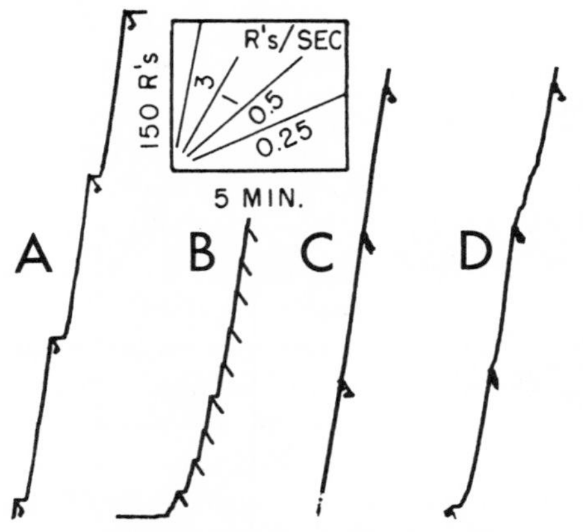

Fig. 29. Transition from FR 160 to FR 35

In another experiment the ratio was changed from 60 to 10 and then back to 70. Figure 30 shows the transition for 2 birds. Little effect is felt at Record A, because the higher ratio had not produced pauses. A marked effect is shown at Record B, however. Record B1 shows a segment from the final performance on FR 60, where pauses up to 60 seconds occur. When the ratio is reduced to 10 in the following session, a stable performance develops after a short period of acceleration. The return to a larger fixed ratio (B3 and B4) yields a result similar to that in Fig. 29. (Both birds show a lower rate under FR 10 than under FR 60. This decline is partly due to

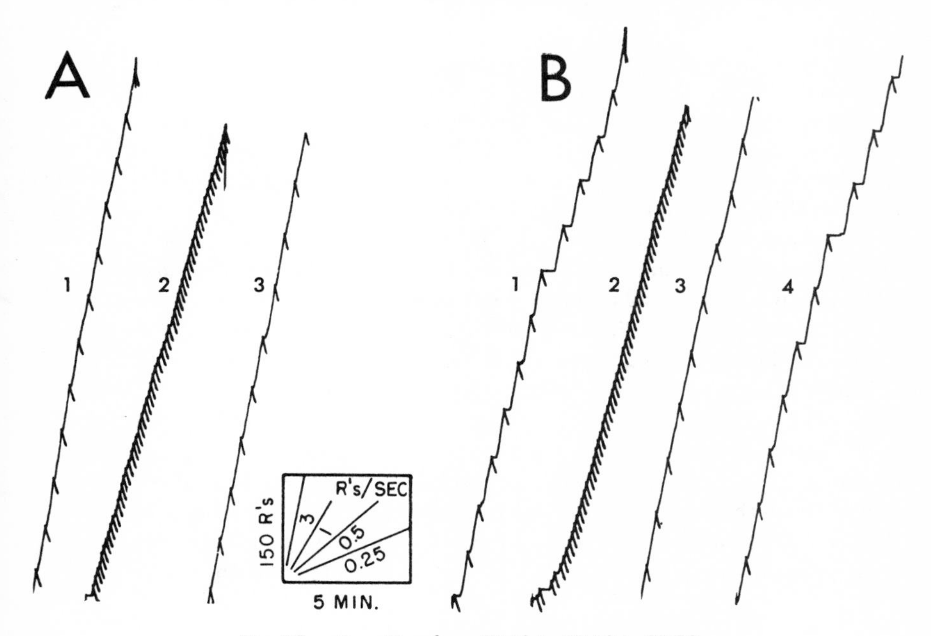

Fig. 30. Transition from FR 60 to FR 10 to FR 70

slight pausing inevitably involved in reinforcement, but the running rates during the ratios are also different. At very small ratios the running rate tends to be relatively low. (See Fig. 20.)

In summary, the pauses generated by large ratios do not disappear immediately when the ratio is reduced, nor do they return again immediately when the higher ratio is in effect following a period at a smaller ratio.

EXTINCTION AFTER FIXED RATIO

Extinction after fixed ratio shows the effects of the controlling contingencies arranged by that schedule. Most of the behavior occurs at the high running rate of the ratio. A decline in over-all rate develops because of increasingly longer periods of

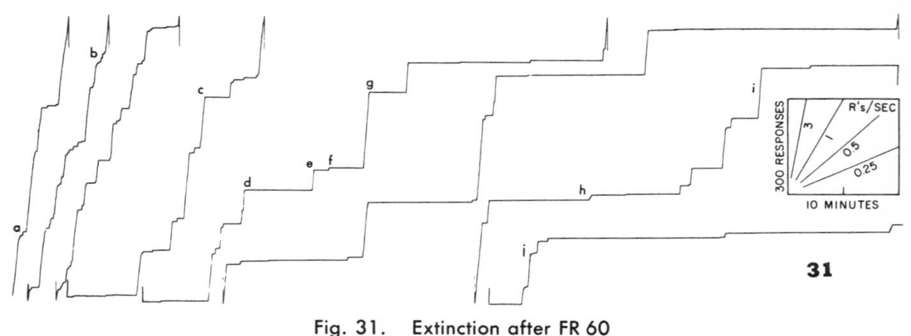

Fig. 31. Extinction after FR 60

no responding. Subjects vary considerably in the extent to which intermediate rates appear.

Figure 31 shows an extinction curve taken in a single session following 700 reinforcements at FR 60 (after crf). Nearly all the responding in this 3.5-hour session occurs at rates above 5 responses per second. Even at the end of the session (at *i,* for example), the rate is approximately 11 responses per second. Most of the transitions from a high rate to pausing occur abruptly—for example, at *c, d, e, f,* and *g.* Some examples of brief periods of responding at intermediate rates and with rough grain occur at the start of the session (as at *a* and *b*) and toward the end of the session (as at *h* and *j*).

Figure 32 is a curve showing roughly the same pattern but after a very different history. The bird had received 14,000 reinforcements on FR 60 with a total history of 35,000 reinforcements on various fixed ratios, including adjusting ratios. During the immediately preceding session only, the bird had been reinforced on FR 40 with "percentage reinforcement." (See below.) The over-all features of the curve resemble Fig. 31. The usual terminal rate continues for only a few hundred responses before negative acceleration sets in at *a.* The over-all rate then falls off, as a result of in-

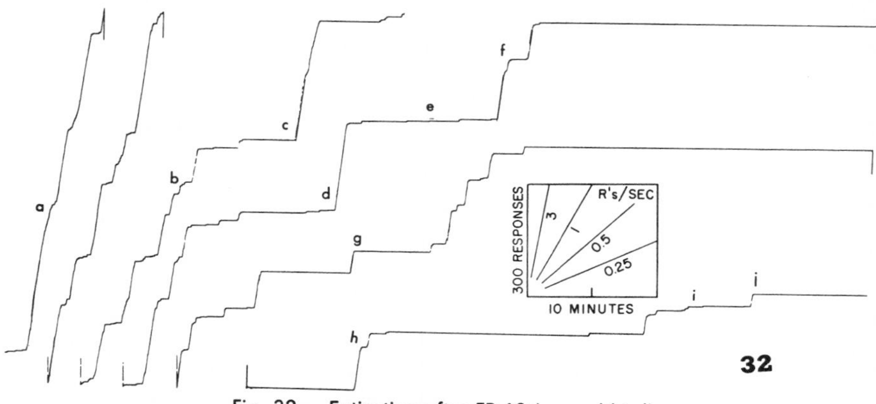

Fig. 32. Extinction after FR 60 (second bird)

creasingly longer pauses between bursts. While instances of abrupt shifts to zero rates and abrupt shifts from pausing to running at high rates occur (as at *c, d, g,* and *j*), frequent instances appear of slight negative curvature (as at *b, f,* and *h*) and desultory responding at low rates (as at *e* and *i*).

The bird whose extinction curve is shown in Fig. 33 had received more than 8000 reinforcements on various fixed-ratio schedules, with time out and added stimuli. The immediately preceding schedule was FR 170 for approximately 400 reinforcements. At the beginning of the extinction session, approximately 1000 responses occur at the usual ratio rate. A short pause appears at *a* and is followed by a brief acceleration to another run of 800 or 900 responses. A pause of 2 hours and 50 minutes occurs at the first break in the record. A 2nd run of approximately 650 responses at *b* is followed by another sustained pause of more than 1.5 hours. Except for the brief period

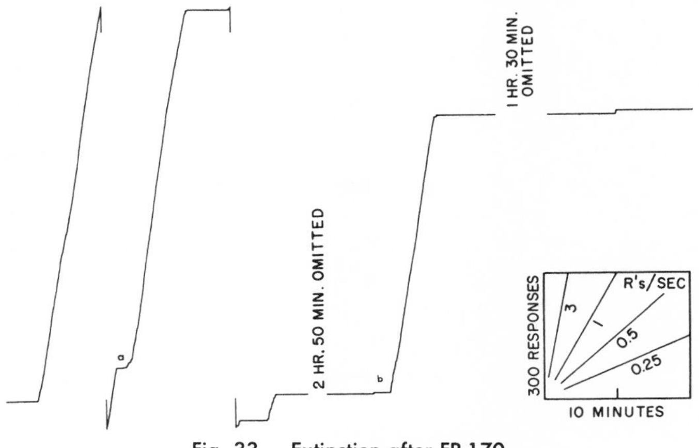

Fig. 33. Extinction after FR 170

of acceleration at *a,* the curve shows no intermediate rates of responding. The record shows the result of an especially well-developed contingency between high rate and reinforcement.

At the other extreme, Fig. 34 shows an extinction curve in which frequent and sustained intermediate rates occur in spite of a history of over 5000 reinforcements on fixed ratios of from 40 to 120. A segment of the performance on FR 120 immediately preceding extinction is illustrated. The terminal fixed-ratio rate shows some grain (as at *a*), and the running rate is 3 responses per second, which is minimal for a ratio of this size. When extinction is begun at *b,* a lower rate is quickly reached and the rest of the record shows considerable curvature and intermediate rates of responding. After a period of 2 hours and 35 minutes in which no responses occurred, the rate returns very close to that observed during the ratio performance (at *c*). Instead of an abrupt break to a zero rate, however, a period of negative acceleration (at *d*) follows. Another pause (at *e*) is again followed by negative acceleration. Even at interme-

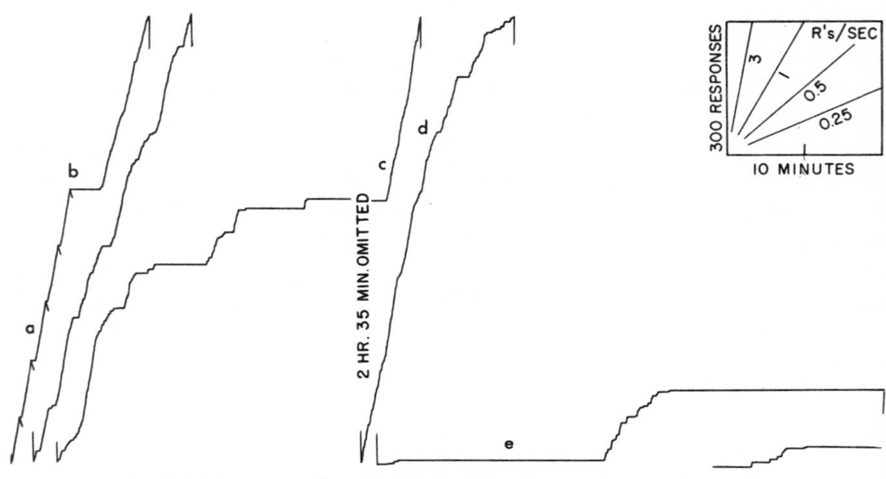

Fig. 34 Extinction after FR 120 showing sustained intermediate rates

diate rates, however, the fine grain of the record indicates that groups of responses are still occurring at the ratio rate. The marked and frequent curvature in this record is seldom observed after a fixed-ratio schedule. Even so, the long pauses and the return to the high rates 3 hours after extinction began are characteristic.

In general, a shorter history of reinforcement brings quicker extinction. Figure 35 shows a curve in which extinction is virtually complete after 20 minutes of rapid responding. The bird had a history of 360 reinforcements on FR 60 after crf. The final performance on FR 60 is shown at the beginning of the record. A single time-out

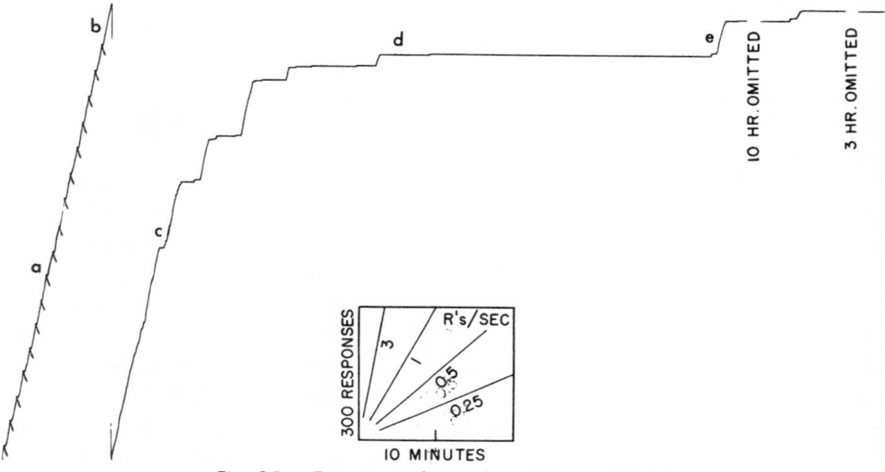

Fig. 35. Extinction after a short history of FR 60

probe (see below) occurs at *a*. When extinction is begun (at *b*), a run of over 500 responses is sustained until a short break occurs at *c*. Shorter runs and longer pauses then appear, until at *d* extinction is virtually complete. A short burst appears later at *e*, but further extinction during a 14-hour session produced less than 50 responses.

Figure 36 shows another extinction curve after a fixed-ratio schedule with a history similar to that of Fig. 35. A sample of the performance on FR 60, 350 reinforcements after continuous reinforcement, appears at the start of the figure, with the recorder resetting after each reinforcement. Extinction (beginning at *a*) shows a rough over-all decline composed of runs at the fixed-ratio rate and increasingly more frequent and longer pauses. There is no sustained intermediate rate, although some

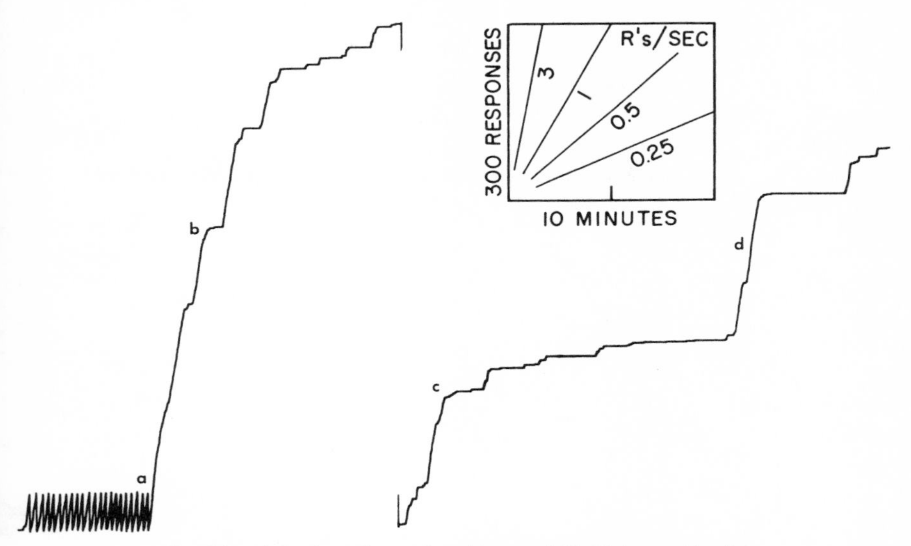

Fig. 36 Extinction after a short history of FR 60 (second bird)

curvature follows runs at high rates, as at *b* and *c*. Even near the end of the curve (at *d*), sustained runs of more than 150 responses occur at the fixed-ratio rate.

A ratio-type extinction curve requires a well-developed fixed-ratio performance. In Fig. 37 extinction curves were taken on 2 successive sessions before a stable fixed-ratio performance had developed. Record A begins with a brief performance on FR 60, 200 reinforcements after the bird was first exposed to this schedule after crf. The grain and the wide variations in rate show that a stable fixed-ratio performance has not yet been reached. Extinction proceeds at an intermediate rate. Some effect of FR has been felt: pauses as long as those at *a* and *b* are unlikely in the early stages of extinction after interval reinforcement. In the following daily session (Record B), reinforcement was reinstituted at FR 60, and another extinction curve then taken. Both the ratio performance and the extinction curve show an advance over the preced-

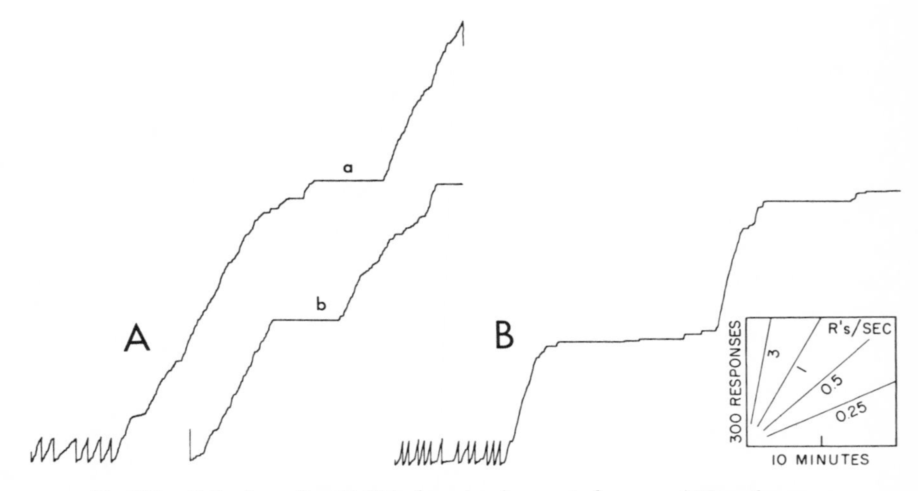

Fig. 37. Extinction after FR 60 before development of a normal FR performance

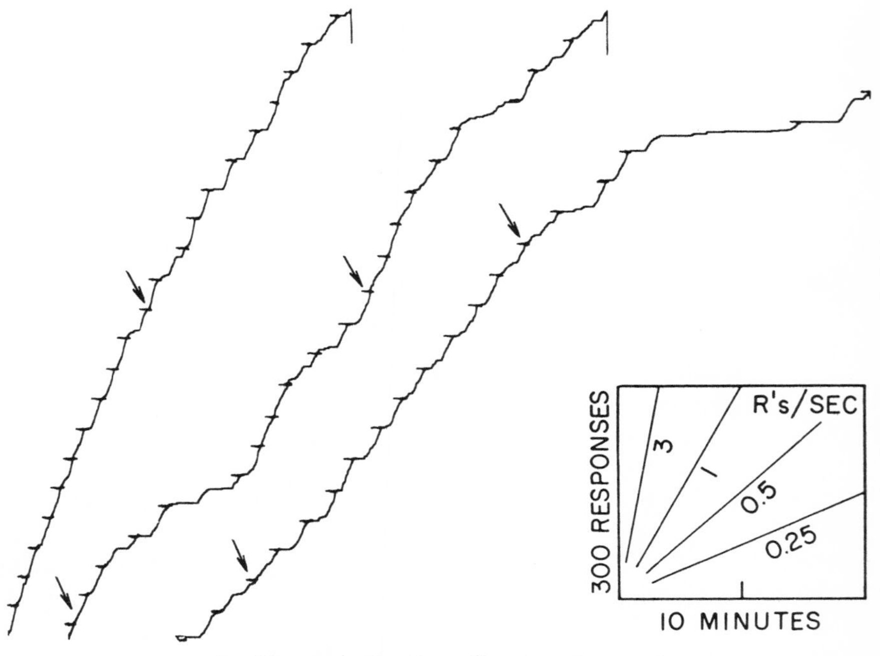

Fig. 38. Early FR with insufficient reinforcement

ing day. The extinction curve consists mainly of 2 sustained runs at rates higher than any occurring in the previous session. Because of the longer pauses, the over-all size of this curve is nevertheless not so great as that in Record A.

Summary

In extinction after fixed-ratio reinforcement, the organism responds predominantly at the fixed-ratio rate. An over-all decline in rate is due to increasingly longer pauses alternating between increasingly shorter runs. The curves are in some cases almost square, with most of the responding appearing very early in the session. Others, as in Fig. 31 and 32, show a fairly smooth decline in rate throughout the experimental session.

THE EFFECT OF INSUFFICIENT REINFORCEMENT ON FIXED RATIO

In an experiment on the effect of short time-out probes during the development of FR after crf, the magazine hopper became partially blocked. Eventually, the bird had access to only a few grains at each reinforcement. This condition was discovered at the end of the 5th session, when atypical curves were recorded. The blocking of the magazine had probably been progressive, beginning with normal operation on the first day of the experiment.

Figure 38 shows the 5th session on FR 40 after crf. (Arrows indicate 1-minute time-out probes, which will be discussed in a later section.) Before being affected by the gradually reduced reinforcement, the bird had reached a good fixed-ratio performance—no pausing after reinforcement, a high rate, and good grain. The reduced reinforcement produces a lower terminal rate of responding and an over-all decline in rate during the session. Intermediate rates and a rough grain appear. Pauses after reinforcement do not change greatly, however. The curve is therefore quite different from one in which too high a ratio has produced a similar decline in over-all rate.

Figure 39 shows the 3rd session for another bird in the same apparatus. Here, the reduced reinforcement produces sustained periods of responding at a low rate after reinforcement. Characteristic fixed-ratio performances occur at *a* and at *e* (following a probe). In general, however, the responding is somewhat scattered, beginning soon after each reinforcement. Intermediate rates appear at *b, c,* and *d.*

Figure 40 shows the 4th session of FR after crf for a third bird. Here, the fixed-ratio character of the behavior is maintained under reduced reinforcement despite a severe decline in over-all rate. Toward the end of the session, at *c* for example, 4 ratios are run off normally. A low over-all tendency to respond is indicated by the fact that only 8 more reinforcements occurred when the session was extended for 13 hours beyond the portion shown in the figure. Occasional disruptions of the fixed-ratio performance appear at *a, b,* and *d.*

The same kind of magazine failure occurred in another experiment on FR 20 after crf. Figure 41 shows the beginning portions of the 2nd, 3rd, 4th, 5th, and 6th sessions.

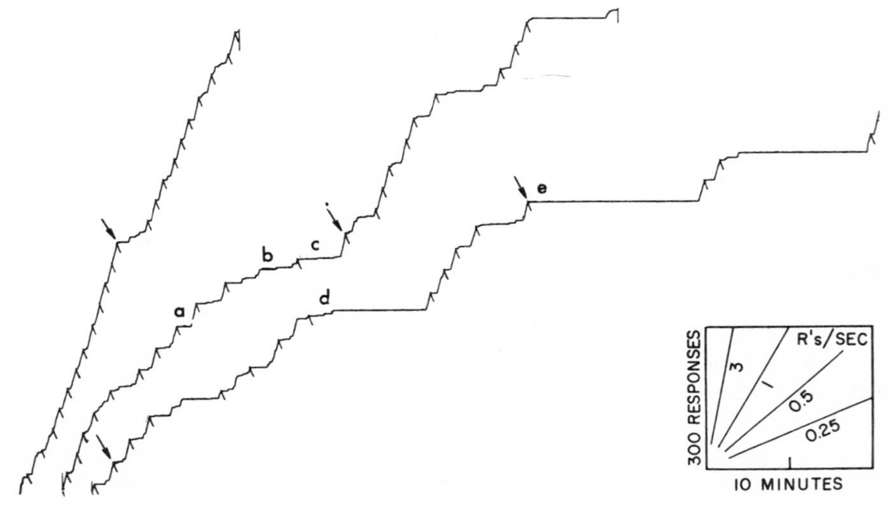

Fig. 39. Early FR with insufficient reinforcement (second bird)

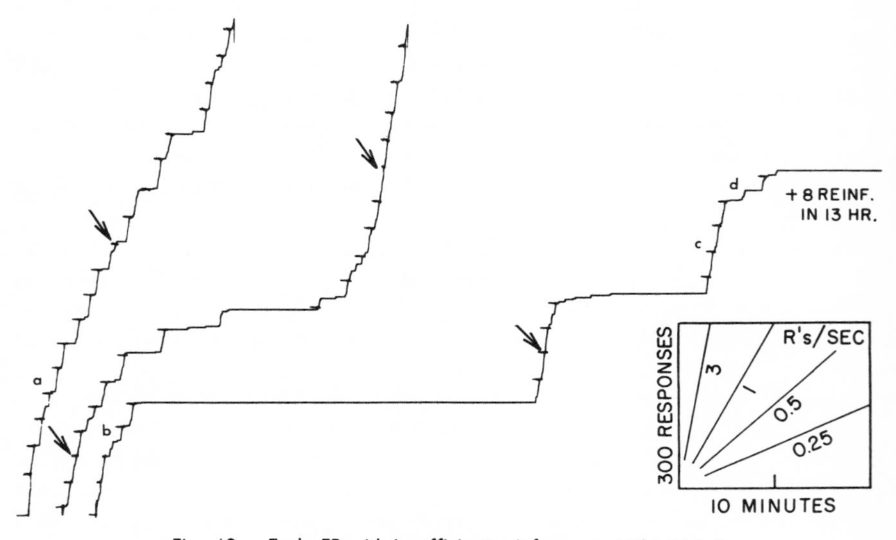

Fig. 40. Early FR with insufficient reinforcement (third bird)

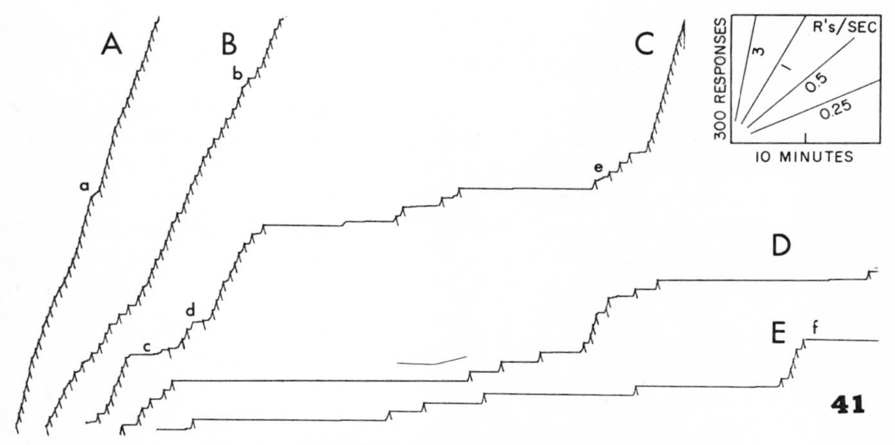

Fig. 41. Early FR 20 with insufficient reinforcement

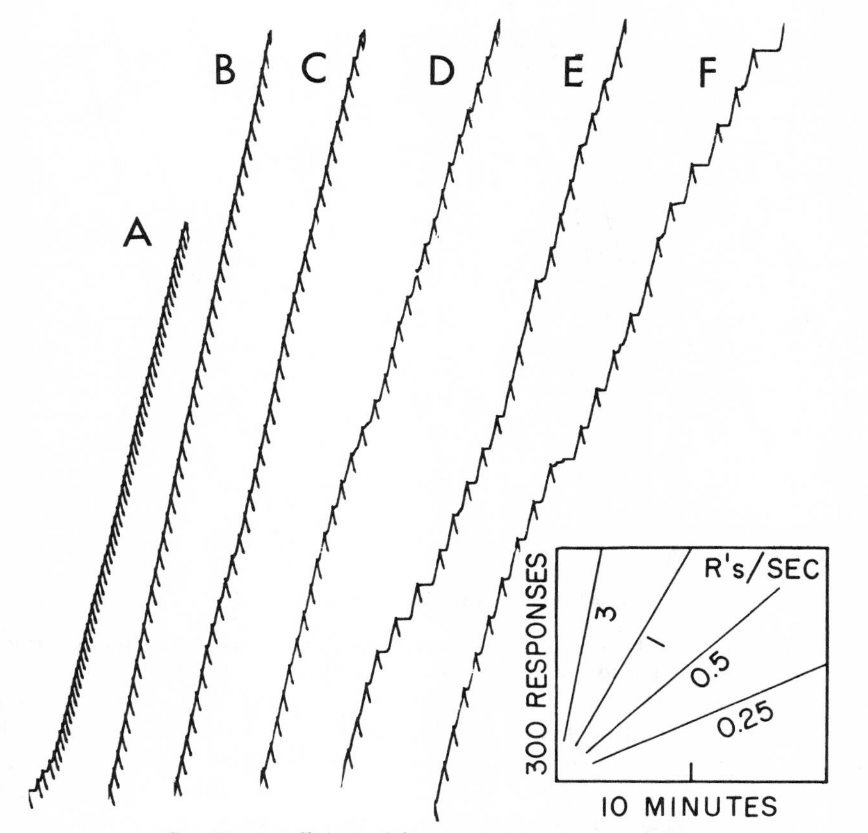

Fig. 42. Insufficient reinforcement during increase in FR

Under this shorter ratio the effect of the reduced reinforcement is felt almost entirely as a lengthening of the pause after reinforcement. Despite the reduced over-all tendency to respond, continued exposure to FR 20 through 5 sessions produces an increase in the actual running rate. Preceding *f* in Record E the bird pauses after reinforcement; but once it begins to respond, the rate is above 3 responses per second. Running rates of approximately 2 per second prevail in Record A. Occasional de-

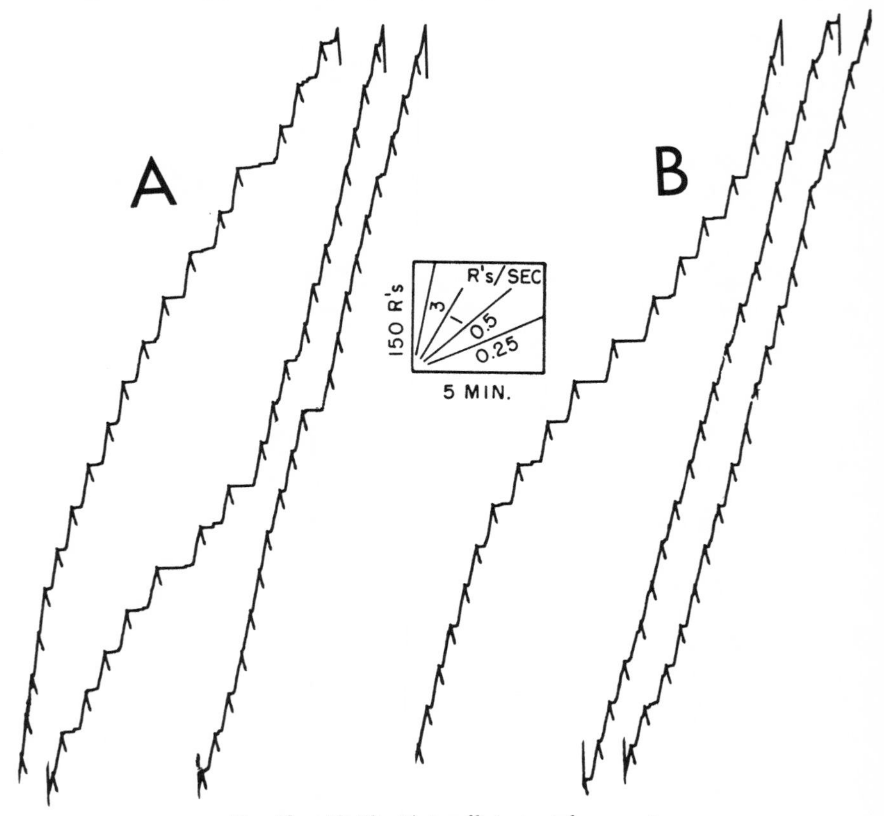

Fig. 43. FR 50 with insufficient reinforcement

partures from a characteristic fixed-ratio performance sometimes appear as intermediate rates (as at *a, b,* and *e*) and as broken segments (as at *c* and *d*).

The apparatus was repaired, and the ratio was reduced to 12 and then progressively increased over a period of 20 days to FR 50. During this period the magazine failed again. Figure 42 gives samples of the performance during this period. Figure 43 shows the marked inflections in the over-all rate in 2 later sessions, when the birds were still under inadequate reinforcement. Most of the changes in over-all rate occur as a result of pauses after reinforcement.

PERCENTAGE REINFORCEMENT

Another kind of intermittency can be introduced into any schedule of reinforcement by substituting some other event for a certain percentage of the reinforcements. This can be done in at least two ways:

1. The stimuli which have been present whenever food was available in the magazine can be presented for the same length of time, although the magazine hopper is either empty of grain or does not rise within reach. For the type of pellet dispenser commonly used with rats, a certain percentage of the pellet compartments is left empty; the magazine sounds but does not deliver a pellet. This method of percentage reinforcement is of interest as an instance of chaining. The reinforcing effect of the discriminative stimulus for approach to the magazine (say, the sound of the operation of the magazine) is studied as a function of the schedule according to which approach to the magazine is reinforced by food. This kind of percentage reinforcement introduces a type of complication into the study of fixed-ratio schedules. Since reinforcement is a stimulus which may in part control the subsequent rate of responding, a fixed-ratio performance after "reinforcement" with an empty magazine will begin under slightly novel circumstances.

2. If each reinforcement is normally followed by a time out, "reinforcements" with an empty magazine may also be followed by a time out, which may cover the period during which the preceding reinforcement remains an effective stimulus. The behavior which follows food plus time out and that which follows time out alone occur under nearly similar circumstances.

This method of percentage reinforcement also raises its own problems. On a ratio schedule the time out is produced by a response whenever it is substituted for the usual reinforcement and might have a punishing effect, which could reduce the rate.

In a first experiment on percentage reinforcement a 3-minute TO replaced the operation of the food magazine whenever a reinforcement was omitted. When a reinforcement occurred, it was not followed by a TO. Figure 44 shows the first effect. The points at which TOs were substituted for reinforcements are indicated by the dots at the left of the record. Reinforcements are marked as usual. Record A is a segment of the final performance on FR 40 after more than a year on fixed-ratio schedules. When TOs were substituted for 50% of the reinforcements in the following session, the over-all rate immediately increased (Record B). Record C shows the performance 7 sessions after Record B where the size of the fixed ratio has been reduced to 30 and a further increase takes place. (This increase in rate is an effect of time out per se rather than percentage reinforcement, as will be seen in a later section on time out.) Neither the omission of the reinforcement or the TO produces any disruptive effect at this stage. The procedure was then changed so that a 3-minute TO followed all ratios, but a percentage of these were preceded by the usual reinforcement. If the magazine operated, a 3-minute TO followed; if a reinforcement was omitted, a 3-minute TO followed the last response in the ratio.

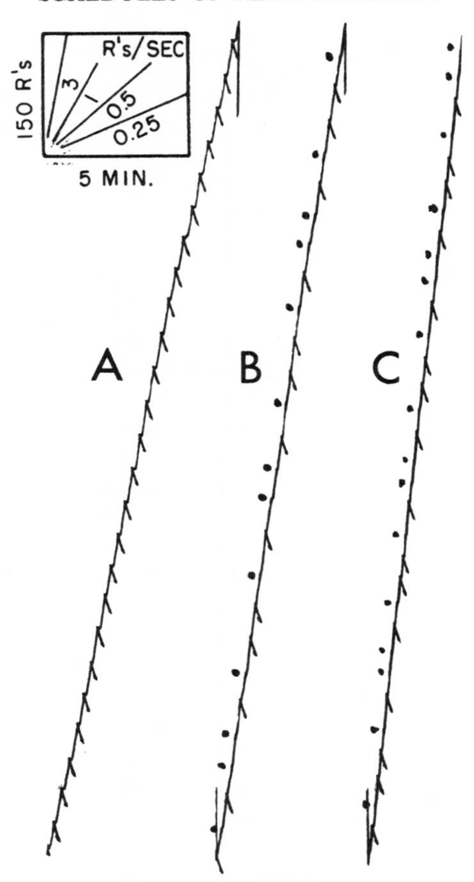

Fig. 44. First effect of percentage reinforcement

The 2 birds in the experiment reacted at different speeds to these conditions. One showed a lower tendency to respond while maintaining the general properties of the fixed-ratio performance. Figure 45 shows the 10th session on FR 30 for the first bird, in which 50% of the ratios are reinforced and then followed by a 3-minute TO, while 50% are simply followed by a 3-minute TO. Except for the size of the ratio and omitted reinforcement marks, this curve could be mistaken for one produced by too large a fixed ratio. The 3 longest pauses (at *a*, *b*, and *c*) follow omitted reinforcements. If this factor is significant, the length of pause must be influenced by the presence or absence of reinforcement which occurred 3 minutes earlier.

The second bird continued to respond at a high rate until the percentage of reinforcement was reduced below 50%. Figure 46 shows segments from the 20th to the

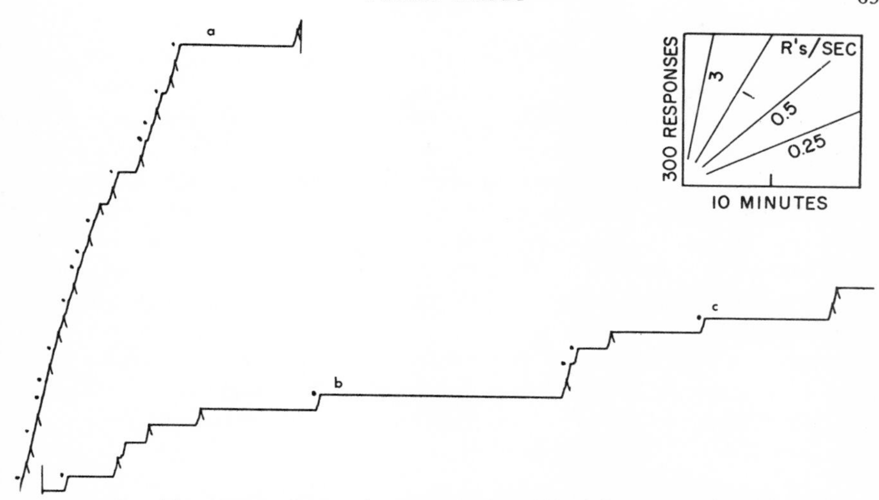

Fig. 45. Pauses after reinforcement under 50% reinforcement of FR 30

37th sessions of the experiment for this bird, in which the percentage of reinforcement omitted ranged from 50% to 85%. Record A shows the 20th session and Record B shows the 23rd session under FR 30, with 50% of the reinforcements omitted. It is similar to Fig. 44C, although some sign of disturbance is evident. Responses occur during TO (producing vertical segments, as at *b*), and the rate changes slightly within a ratio (as at *a*), so that the over-all record looks much less linear. Record C shows

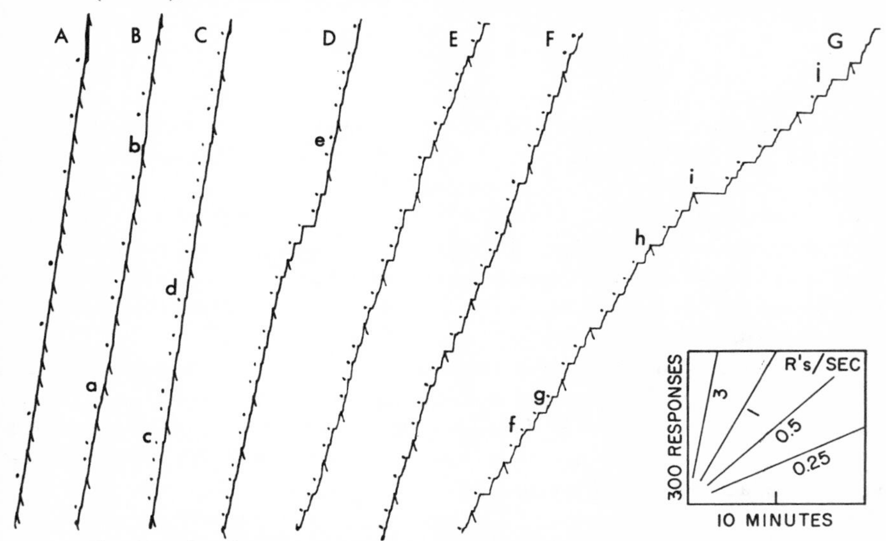

Fig. 46. Development of negative curvature under percentage reinforcement

the 4th session, where 75% of the reinforcements are omitted. The local rate changes of Record B have largely disappeared, but slight pauses tend to occur at the start of the ratio (as at *c* and *d*). Three sessions later, with 85% omitted (in Record D), the pauses become more pronounced under this same procedure, and changes in the local rate are beginning to appear (as at *e*). Records E, F, and G, at the same percentage, were taken 8, 13, and 16 sessions, respectively, after Record D. The fixed-ratio performance deteriorates progressively, with increased pauses following the reinforcement (as at *i* and *j*); and marked rate changes occur within a ratio (as at *f*, *g*, and *h*) as well as rough grain. Figure 47B illustrates in detail the performance in Record G. The punishing effect of the TO appears as a decline in rate just before reinforcement. Record A in the figure shows a later stage in the experiment in which the schedule is FR 70 with 50% reinforcement. The negative curvature has disappeared, but responding still occurs in bursts separated by slight pauses. The first bird (Fig. 45) eventually showed a similar effect of the punishment when the schedule was changed to FR 60 with 75% reinforcement.

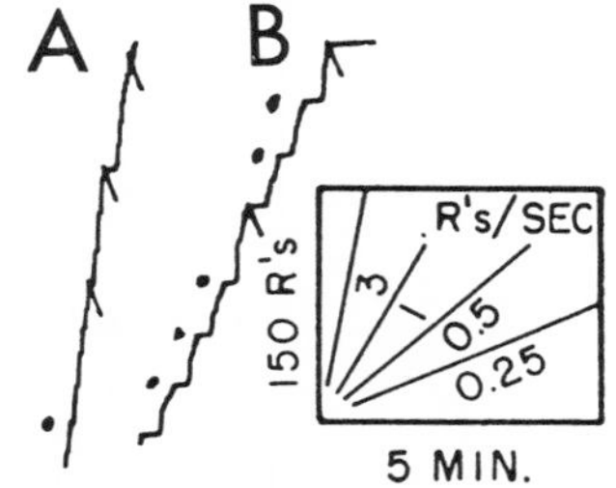

Fig. 47. Detail of negative curvature under percentage reinforcement

The reduction in rate towards the ends of the ratios in Fig. 47B is equivalent to the lowered rate observed during a pre-aversive stimulus, which is characteristically followed by a shock (Estes and Skinner, 1941). The number of responses already emitted constitutes a pre-aversive event which is functionally equivalent to the buzzer used in most such experiments, and the TO is equivalent to the shock. In more general terms, the removal of a positively reinforcing stimulus (the group of stimuli under which the key is pecked) is negatively reinforcing. By making a period of time out contingent upon a response, we punish a response by removing the "occasion" on which the bird is reinforced.

We have used TO in other experiments as a punishment in enhancing stimulus control. In some discriminative situations involving 2 or more responses (e.g., in matching experiments), the frequency of reinforcement is high enough to sustain behavior even in the absence of any stimulus control. A bird will usually behave without regard to available stimuli. By occasionally allowing an unreinforced response to produce a short time out, however, a much more precise stimulus control can be developed.

Summary

Little evidence of the effect of percentage reinforcement per se is forthcoming with these methods. The results are confounded by effects of the time out, including its effect as punishment. Before the punishing effects of the TO influence the performance, however, the percentage reinforcement increases the pause after reinforcement without disrupting the normal fixed-ratio pattern. This is similar to the effect of lessened deprivation.

THE EFFECT OF DEPRIVATION LEVEL ON FIXED-RATIO PERFORMANCE

The effect of deprivation on a stable fixed-ratio performance was studied in the following way. Two pigeons were maintained on FR 110 after more than 70 sessions on various values of fixed ratio with occasional TO probes after reinforcement. The schedule was then held at FR 110 for several months, during which the body-weights of the birds were varied over a wide range.[1] The principal effect was upon the pause after reinforcement. The local rates of responding show very little sensitivity to even wide ranges in deprivation. This finding confirms a published experiment by Sidman and Stebbins (1954).

The results are summarized in Fig. 48, which presents the average length of pause following reinforcement. Since each daily session showed a decline in over-all rate, pauses at the beginning and end of each session are presented. The top graphs show the mean pause for the last 2 excursions of the recorder pen, and the middle graphs show the mean pause for the first 2 excursions. When the over-all rate of responding was so low that 23 fixed-ratio segments did not occur during the 4-hour session allowed for the experiment, the value of the pause is indicated on the graph by *X*. Pauses for Bird 129 RP over 300 seconds and for Bird 130 RP over 200 seconds are indicated by an *X*. The weight changes in the experiment are shown in the graph at the bottom of the figures.

Bird 129 RP

During the first 40 days of this experiment, Bird 129 RP's weight increased from slightly over 400 grams to almost 550 grams. But this increase had little effect on the FR performance, except during the higher values at about 40 days and a single peak weight at 10 days. A slight increase in pausing is more marked toward the end of the session, as the top graph indicates. As the weight is increased further, beyond the 40th session, to 600 grams (approximately 100% of the ad lib weight for this bird), pauses increase to a maximum. In the last 10 days of the experiment, when the weight fell to 550 grams, the pauses became shorter at the beginning of the session but decreased only slightly during the last part. A differential effect is also present at the beginning and end of the session between 30 and 50 days.

Bird 130 RP

Bird 130 RP confirms the results of the first bird, although its weight changed in a more complicated way. Bird 130 RP was a smaller bird, whose lowest weight was 370

[1] See comments on change in body-weight in Chapter Six, p. 293.

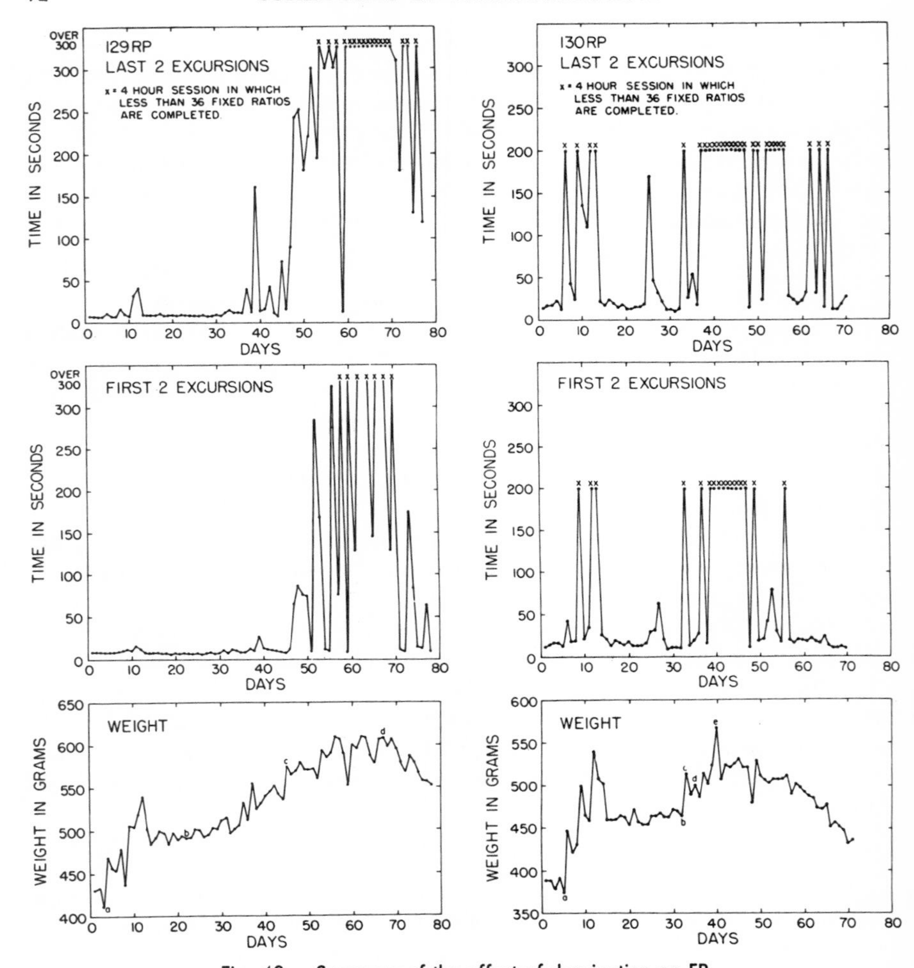

Fig. 48. Summary of the effect of deprivation on FR

grams compared with more than 400 grams for Bird 129 RP. During the first 15 sessions of the experiment, Bird 130 RP's weight increased rapidly to 540 grams, which was above 100% of the ad lib weight determined before the start of the experiment some months before. Corresponding with the increase in weight, the length of the pause following the reinforcement increases, too. Between the 15th and 30th day, the weight fell to about 460 grams, and the pause following reinforcement was correspondingly shorter. From the 35th to about the 50th day, the weight reached a maximum value, and both the top and middle curves show extreme pausing. The weight curve then fell slowly, and the length of the pause decreased correspondingly. Here, again, a given level of deprivation produces more pausing at the end of the session than at the beginning.

An ad lib weight determined more than 6 months before the experiment was begun is probably not very meaningful. However, a body-weight could be found at which very little pecking occurred on the fixed-ratio schedule, and this was used as a reference level. We will call the reference level the inactive weight. For 129 RP it was approximately 600 grams; and for 130 RP, 520 grams. Other weight levels in this experiment can be expressed as the percentage of the inactive weight. In the following section, figures will be presented which characterize the performance at 4 general ranges of body-weight in reference to each bird's inactive weight. These are sample daily sessions taken at *a, b, c,* and *d* in the weight curve for 129 RP, and at *a, b, c, d,* and *e* for Bird 130 RP in Fig. 48.

Figure 49A represents point *a* in Fig. 48 (left graph for Bird 129 RP). The bird

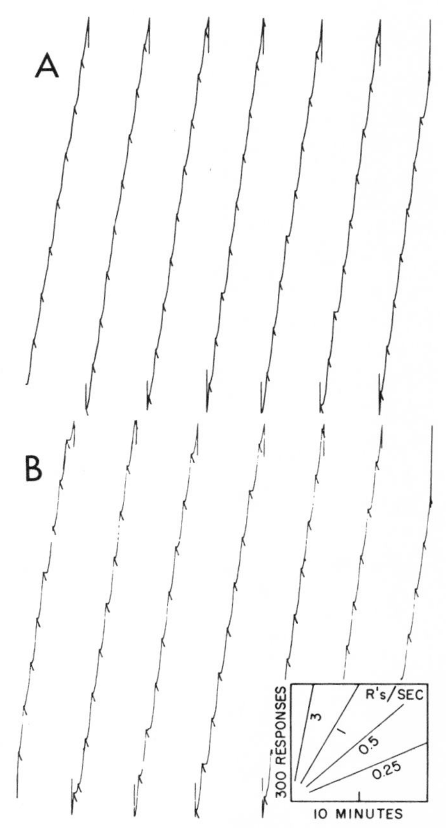

Fig. 49. FR at 68% and 82% of the inactive weight (129 RP)

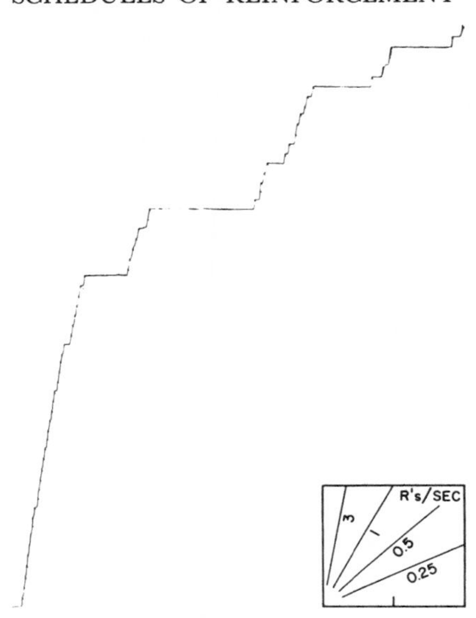

Fig. 50. Over-all rate changes under FR at 95% of the inactive weight (129 RP)

is at 68% of its inactive weight. This bird characteristically shows brief positive acceleration to a terminal rate of over 7 responses per second and an over-all rate of 4 responses per second. Pauses occur occasionally after reinforcement. Record B is for point *b* at 491 grams, or 82% of the inactive weight. At this weight level the fixed-ratio performance is stable, and both the over-all rate and the local rates are approximately the same as those in Record A.

Figure 50 shows a complete daily session for the still higher body-weight at *c*. This weight is 95% of the inactive weight, or 575 grams. At this weight level the over-all rate falls off during the period, mainly because of increasing occurrences of long pauses following reinforcement. (In order to show this, the curve has not been collapsed in the usual manner.) The ratios are still being executed at approximately the same running rate as in Fig. 49, although some positive acceleration occurs which cannot be seen at this reduction, and also occasional irregularities in rate. Figure 51 contains

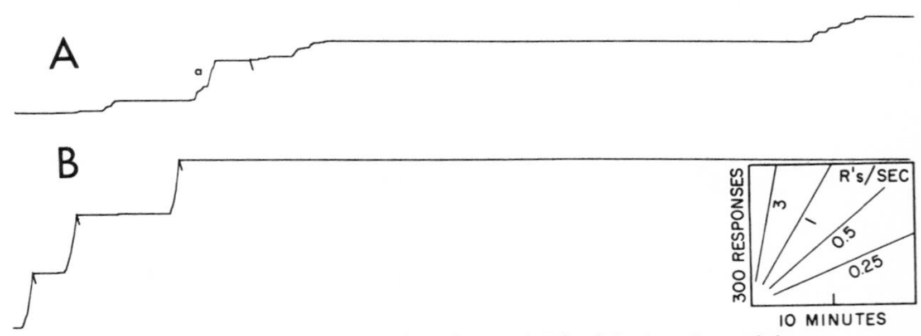

Fig. 51. Detail of a record on FR at 105% of the inactive weight

2 segments from point *d* which are both at 105% of the inactive weight, or 609 grams. This is essentially a free-feeding weight, and the curves represent at most a few hours of deprivation. Record A shows considerable disruption of the fixed-ratio performance and little evidence of the normal fixed-ratio rate at any time. Low or intermediate rates are usual, with only an occasional exception, as at *a*. Record B, a segment from the following session at the same ad lib feeding level, shows almost no disruption of the standard fixed-ratio performance. The pause following reinforcement is often

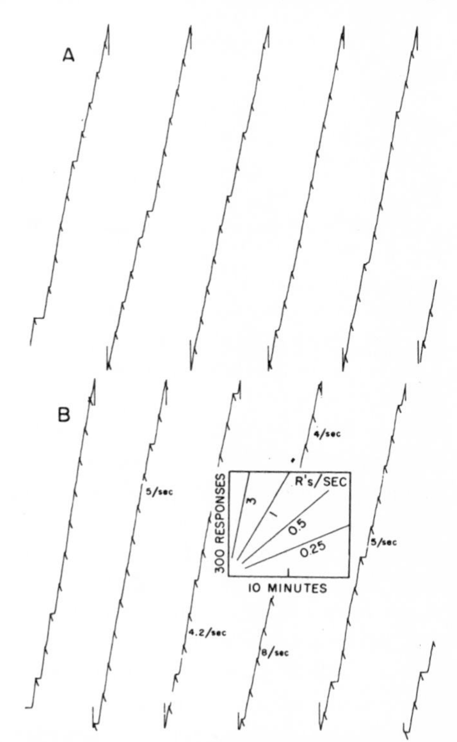

Fig. 52. FR at 71% and 88% of the inactive weight (130 RP)

extended (as, for example, the final pause shown in the record), and a brief acceleration from a near-terminal to a terminal rate of responding occurs. But once the bird begins to respond, the acceleration is rapid to a terminal rate of over 7 responses per second, as in Fig. 49.

Figure 52 shows 2 curves taken at *a* in the weight curve for Bird 130 RP at 71% of the bird's inactive weight, or 375 grams, and at *b* at 88% of this bird's inactive weight, or 465 grams. This bird showed generally lower rates than Bird 129 RP and larger pauses after reinforcement. No curvature is evident at the start of the ratio, however. In Record A, the session at the lowest body-weight, local rates are 3.5 to 4.5 responses

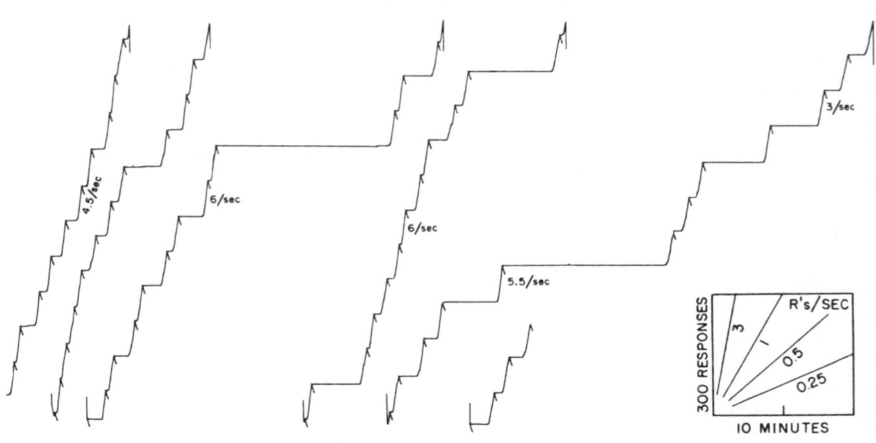

Fig. 53. FR at 97% of the inactive weight (130 RP)

per second, with an average over-all rate of approximately 3 responses per second. In Record B, at 88% of the inactive weight, instances of higher local rates than those in Record A occur, and the value of the local rates of responding varies widely. The over-all rate remains 3 per second, however, and the general form of the behavior does not change.

Figure 53 shows a complete daily session at an intermediate body-weight represented at *d* in the right curve in Fig. 48 at 97% of the inactive weight, or 502 grams. The performance here is similar to that in Fig. 50 at 95% of the inactive weight for Bird 129 RP. An over-all decline in rate during the period is due to a progressive increase in pauses following reinforcement. Local running rates are indicated in several cases, and are of the same order as those given in Fig. 52 at much lower body-weights.

Figure 54 shows samples from a daily session at 99% of the inactive weight, or 520 grams, for Bird 130 RP (*c* in the weight curve). The over-all rate is so low that only 10 fixed ratios are completed in the 4 hours allotted for the experimental session. Although pauses as long as 2 hours occur following reinforcements, the ratios are generally completed at rates comparable with those at lower body-weights if they are

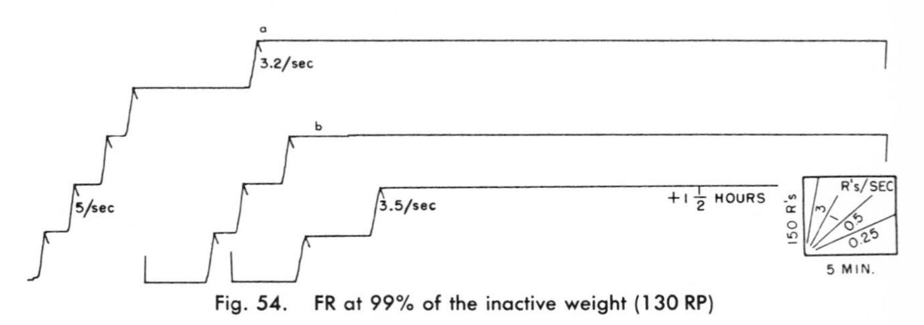

Fig. 54. FR at 99% of the inactive weight (130 RP)

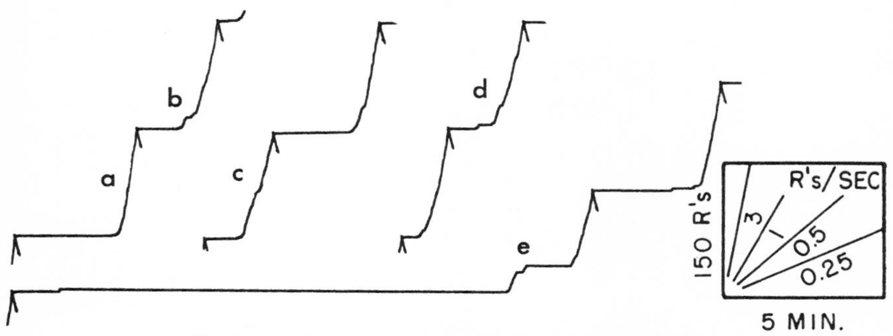

Fig. 55. FR at 112% of the inactive weight (130 RP)

started at all. At this body-weight, several responses occasionally occur after a pause without immediately leading into the terminal rate, as at *a* and *b*.

The performance under the highest weight recorded in this experiment is shown in Fig. 55 at 112% of the mean inactive weight, or 577 grams (*e* in the weight curve for 130 RP in Fig. 48). The segments shown in Fig. 55 show a range of local running rates from 4.5 responses per second at *a* to 2.5 at *c*. There is a somewhat higher distribution of running rates at lower body-weights in Fig. 52 and Fig. 54. The curves show frequent irregularities, including a well-marked knee at *e*, a smaller knee at *b*, and slight pauses after the terminal rate has been reached at *c* and *d*, where some grain is beginning to appear.

Summary

A fairly wide range exists between approximately 90% of the inactive weight and any lower weight at which the animal will remain healthy when performance under a fixed-ratio schedule is invariant. At higher weights, pauses after reinforcement become extreme, but responding occurs at the fixed-ratio rate whenever the bird responds at all. The characteristic performance of the individual ratio run is disturbed only at very high body-weights.

NOVEL STIMULI

Incidental stimuli in the experimental chamber often exercise considerable control over the behavior of an organism even though no explicit contingencies are arranged with respect to them. This control is seen when a novel stimulus is added or when a feature of the experimental chamber is altered. A characteristic effect, probably to be classified as emotional, is a general depression of the whole repertoire of the organism. The effect of a given stimulus depends upon the species and upon the stimuli to which the subject has already been exposed. A novel stimulus involving a change in the experimental chamber might conceivably function as a change in a discriminative stimulus, since the box is a situation where pecking is reinforced, compared with the home cage where it is not.

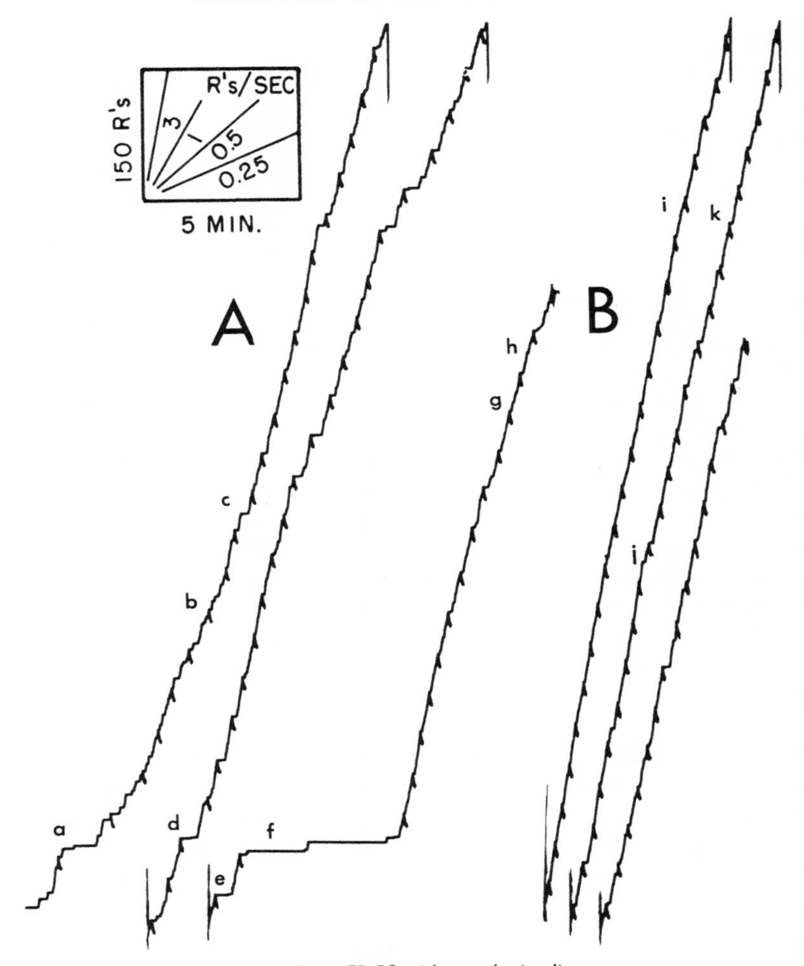

Fig. 56. FR 50 with novel stimuli

Some incidental effects of novel stimuli upon performance under various schedules have been observed. These were mainly the result of accident; a few of these will be reported.

Figure 56

A bird had received more than 5000 reinforcements on FR 50. By accident the key was changed to a novel color at the beginning of a session. The heretofore normal fixed-ratio performance was severely disrupted (Fig. 56A). Rough grain during the ratio, intermediate rates of responding (as at *b*), and pauses before the completion of a ratio (as at *a* and *c*) appear. Even toward the end of the session, some disturbance is still evident, in that the ratio is run off in small bursts of responses (as at *g* and *h*). The long pauses after reinforcement at *d*, *e*, and *f* are not typical of a fixed ratio of this size. Two sessions later, still with the new key-color, a normal fixed-ratio performance is recovered (Record B), although some sign of the disturbance still remains in the grain of the record at *i* and *k* and in the pause in the middle of the ratio at *j*.

Figure 57

Here, the effect of a novel stimulus is confounded with the passage of time since the bird was last run in the experiment.

After a history of approximately 15,000 reinforcements on an adjusting fixed ratio, followed by 6 months of no experimentation, 2 birds were run for the first time in a new apparatus, which differed in dimensions and other details from the earlier apparatus. Both birds were slow in responding to the key, and the first 2 responses of each were reinforced. The records in Fig. 57 begin immediately thereafter. Each bird failed to show its previous normal fixed-ratio behavior. In Record A, following the 1st reinforcement at *a*, the over-all rate declines continuously. This decline was produced by a reduction in running rate rather than by pauses (cf. *b* and *c*). Small deviations appear at *d* and *e*. A fairly standard performance under fixed-ratio reinforcement returns before the end of the session. The other bird (Record B) also shows a lower over-all rate, with a more marked negative acceleration during the ratios (as at *f*, *g*, *h*, and *i*). Even by the end of the session, Record B shows a low terminal running rate, considerable grain, and pauses that are too long for this size of ratio.

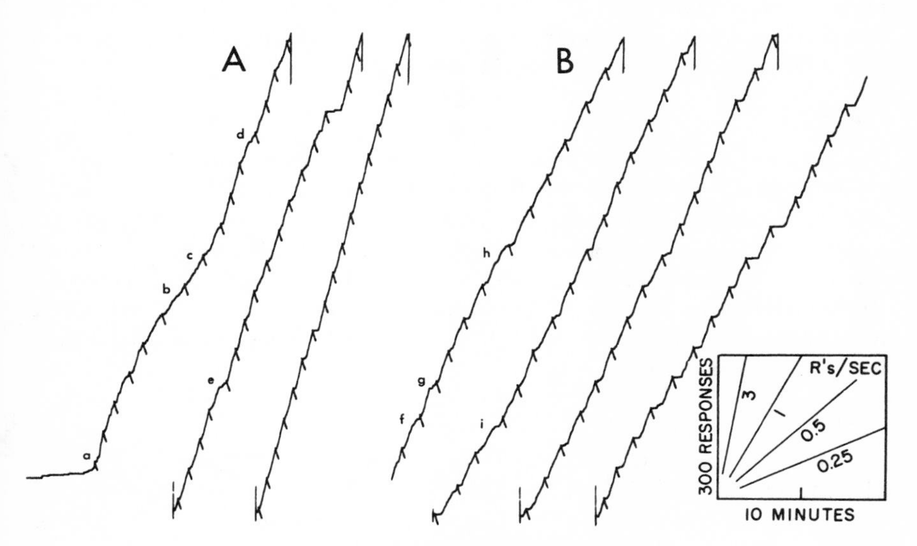

Fig. 57. FR 65 with novel stimuli

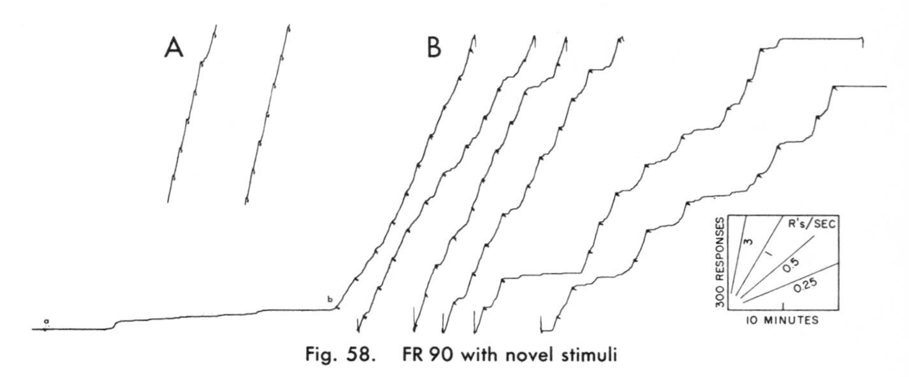

Fig. 58. FR 90 with novel stimuli

Figure 58

A long history of tand *FR*FI followed by about 1000 reinforcements on FR 100 produced the stable performance on FR 90 shown in Fig. 58A. The experiment was then transferred to a new experimental chamber with novel dimensions and details. Record B shows the 1st session complete. At *a* a reinforcement was set up so that a single response would open the magazine. A second reinforcement was set up at *b* after fewer than 90 responses. The program then remained FR 90 for the remainder of the session. The novel surroundings produced a much lower terminal running rate, marked curvature, irregular grain, and occasional negative curvature. This effect is probably partly a function of the previous tand *FR*FI, which has also produced extended periods of low or zero rate immediately following reinforcement and the slow and fairly smooth acceleration to a terminal rate in each ratio observed toward the end of the session shown. (See Fig. 541, tand *FR*FI.)

Figure 59

A history of more than 5000 reinforcements with FR varying between 40 and 110 yielded the stable performance on FR 90 seen at Record A. A slight negative acceleration (as at *a* and *b*) is characteristic. (Cf. Fig. 24B for the same bird.) The bird was then run with an "uncorrelated block counter," in which the key-color changed from white to green to yellow to blue, etc., every 10 responses. (See later section.) Record B shows the first effect of the changing key stimulus. Pauses after reinforcement are briefer, and slight pauses occur at many of the color changes; but these effects are not pronounced.

Figure 60

Another bird had developed a stable performance on FR 210 (shown in Fig. 24A). Here, again, the novel stimulus consisted of a change in key-color every 20 responses. Record A is the 1st segment under this novel stimulus; and thereafter every other excursion of the pen is shown. The novel stimulus produces a severe drop in both over-all and local rates. The rate at *b* is slightly over 1.5 responses per second, compared with a terminal rate of over 3 responses per second in Fig. 24. The rate increases progressively through the session, however; and by Record F the running rate has returned to nearly 3 responses per second. Some grain is still evident in the record, caused by slight pauses every 20 responses at the color changes. Note the unusual sudden changes at *a*, *c*, *d*, and *e* which occur only very rarely in a stable fixed-ratio performance.

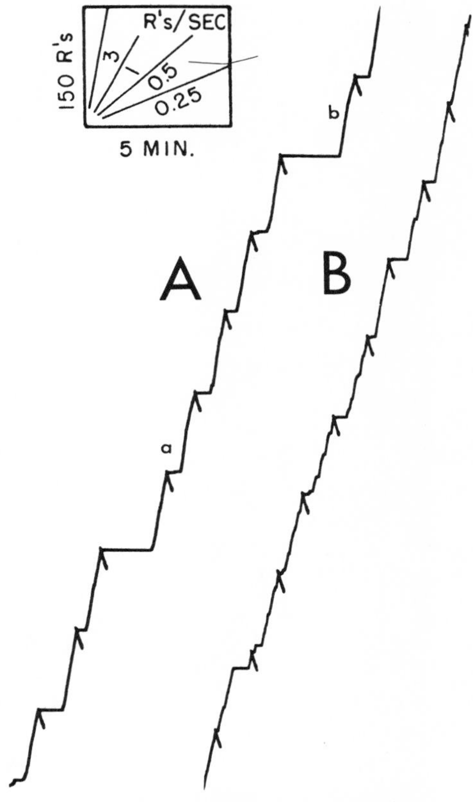

Fig. 59. FR 90 with novel stimuli

Figure 61

Record A shows a final performance on FR 105 following 54 sessions of FR after crf, when the fixed ratio was slowly increased. The key was blue. At the beginning of the following session the key-light had burned out. Record B shows the entire following session with no light behind the key. The adjustment to the new stimulus is continuous but slow. By the end of the session, the former stable fixed-ratio perform-

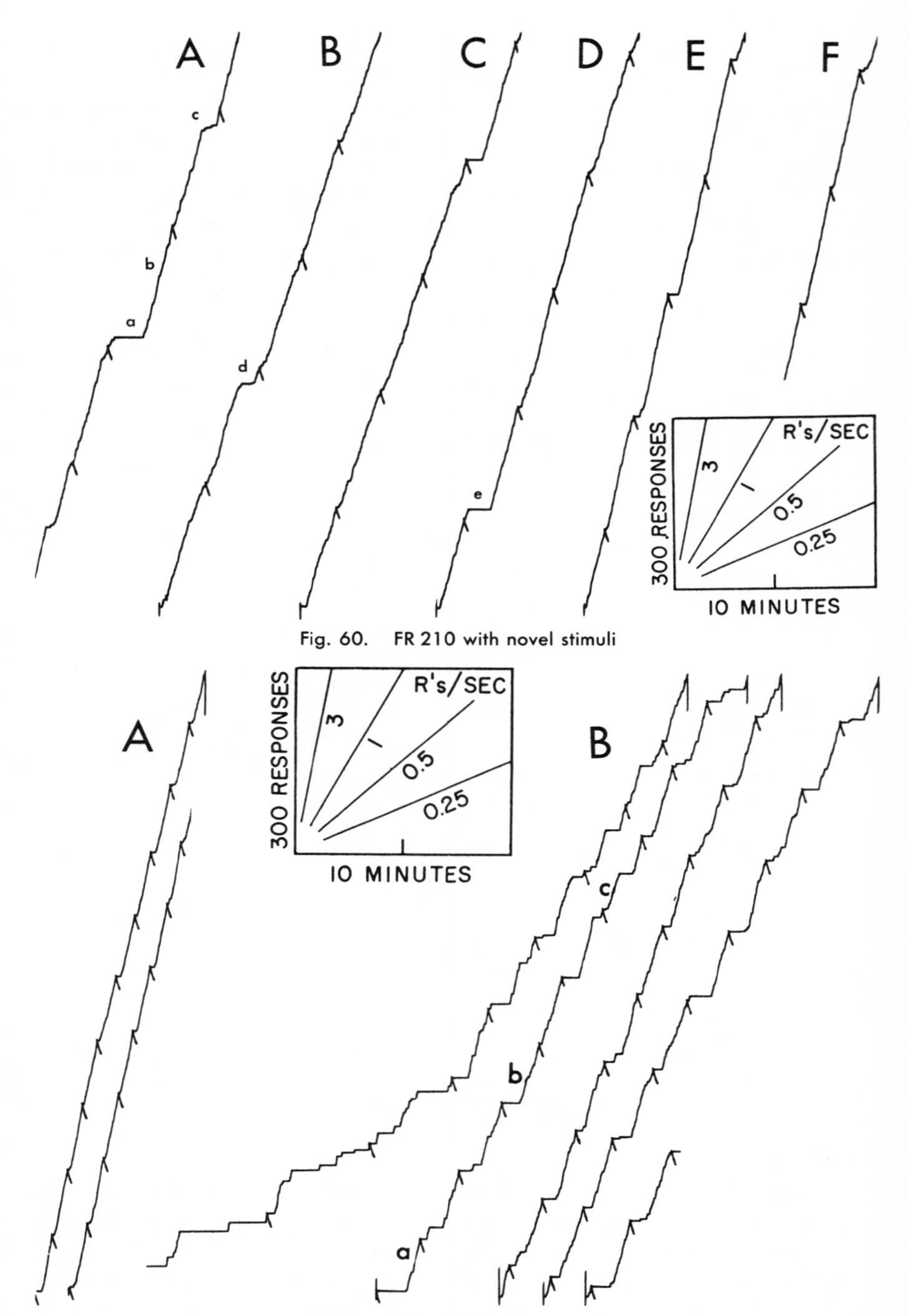

Fig. 60. FR 210 with novel stimuli

Fig. 61. FR 105 without a key light

ance has not yet been recovered. The normal pattern is especially disrupted during the first 8 ratios. Pecks occur in bursts, with negative curvature, very few sustained runs, and long pauses. The ratios at *a* and *b* show the beginning of a return to normal; but pauses still appear (as at *c*) after the terminal running rate has been reached. By the end of the session, marked pauses still occur after reinforcement, responses are in small bursts, and the rate varies widely within the fixed-ratio segment.

THE EFFECTS OF DRUGS ON FIXED RATIO

Some preliminary experiments have been undertaken to sample the changes which drugs produce in fixed-ratio behavior. The drugs were administered in collabora-

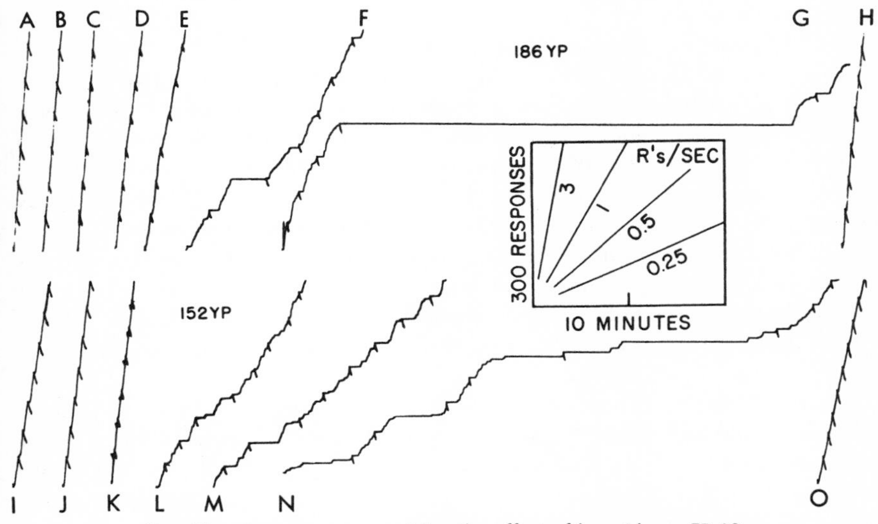

Fig. 62. Segments summarizing the effect of bromide on FR 60

tion with Dr. P. B. Dews of the Department of Pharmacology, Harvard Medical School.

Two birds were put on a sodium chloride-free diet, and desired bromide blood-levels were established by the feeding of sodium bromide. To reverse the condition, the birds were given free sodium chloride, and the excretion of bromide produced an exponential decline in the bromide level. Figure 62 shows segments from the 2 birds over a period of 2 months, in which the bromide blood-level was raised by the feeding of 100 milligrams of sodium bromide orally each day. Record A is the final performance for one bird under FR 60 before any bromide was given. Records B, C, D, and E are the 3rd, 10th, 12th, and 14th sessions under bromide, during which the bromide level increased progressively.

Between segments B and C (3rd and 10th days) the over-all rate increased; but this increase was not well-represented in the small segment shown in the figure. Just before Record D the bromide level reached 35 milliequivalents, and Record D shows the first depressive effect. Record E, 2 sessions later, shows a further decline in rate as well as marked shifts in local rate from one ratio to the next. In Record F, which is from the session following Record E, the bromide level had reached 47 milliequivalents. The fixed-ratio performance is severely disrupted. Segment G, for the following session, shows similar disturbances in augmented form. Marked negative acceleration during the ratio, rough grain, and long pauses are evident. At this high concentration the birds constantly staggered and fell easily if pushed. Bromide was then discontinued, and 100 milligrams of sodium chloride were administered orally each day. Record H shows the fixed-ratio performance almost 2 months later, after the blood level had returned to normal.

Record I shows the performance for the second bird before bromide. Three sessions later, in Record J, a total of 300 milligrams of bromide has produced an *increase* in both over-all and running rates; these increases are more easily seen in the samples. The high rate is maintained 7 sessions later (10 sessions after the first administration of bromide), with a blood level of 35 milliequivalents (Record K). At this time the bird was staggering badly under the effects of the bromide and lost its balance if lightly pushed. The 13th, 14th, and 15th sessions (Records L, M, and N) show a marked and progressive disruption, with long pauses, rough grain, intermediate rates, and negative acceleration during the ratio segments. The bromide blood-level has reached 51 milliequivalents at Record N. Approximately 1 month later, after daily feedings of 100 milligrams of sodium chloride, a normal fixed-ratio performance emerged (Record O).

Figure 63 illustrates further details of the progress of the bromide effect between Records K and L. On these 2 days an 8-minute TO was programmed following every 10th reinforcement. (See later section.) In Record A (11th session) a blood level of slightly over 35 milliequivalents of bromide has little effect on the fixed-ratio performance, although the bird is grossly staggering. The TOs after reinforcement, at the arrow, produce no effect except a very slight decline in rate, which can be seen if the curves are foreshortened (for example, at *a*). By the end of the session, rate changes are occurring (as at *b*) and pauses several seconds long are beginning to appear during the ratio (not shown at this reduction). On the following day shown (Record B), the bromide level reached 37 milliequivalents and produced a large fall in over-all rate as well as deviations from the normal running rate. The first 2 TOs (at *c* and *d*) have no effect, but the 3rd (at *e*) is followed by a fall in the rate extending through several reinforcements. The last 2 TOs (at *f* and *g*), where the fixed-ratio performance has deteriorated with wide fluctuations in rate, grain, and negative curvature, produce pauses of approximately 50 and 30 seconds, respectively.

Both birds show very strong resistance to any disruptive effect of bromide on FR compared with the effect on FI (Herrnstein and Morse, 1956) (Dews, 1955). Besides a general over-all decline in the rate, bromide produced a severe disruption of the

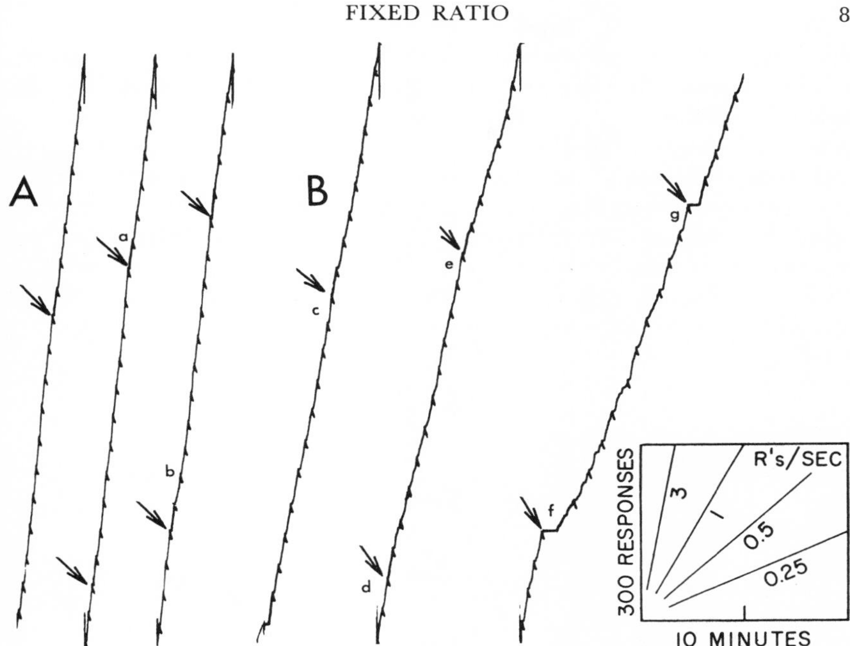

Fig. 63. FR 60 under 35 and 37 milliequivalents of bromide

normal fixed-ratio pattern. A more sensitive effect of central nervous system drugs is possible by modifications of the fixed-ratio schedule (Morse and Herrnstein, 1956).

FIXED-RATIO PERFORMANCE AFTER ABLATION OF BRAIN TISSUE

We are indebted to Dr. J. Lawrence Pool for suggesting some exploratory experiments on the effects of various brain lesions upon the performance of pigeons under various schedules of reinforcement. At the Columbia Presbyterian Medical Center in New York, Dr. Pool operated on 6 pigeons selected from various experiments in progress at the time. When he returned them to our laboratories, a rough examination of their performances under various schedules was undertaken. Four of these birds which had been on schedules involving fixed ratios were tested under fixed-ratio reinforcement after operation. Other results will be reported in appropriate places.

The circumstances of the experiment do not permit a precise comparison of performance before and after brain operation. The lesions were few in number and only roughly localized in post-mortem examination. Therefore, no attempt will be made to ascribe any particular behavioral effect to a specific lesion. The experiments will merely serve to supply examples of the kinds of disturbances which may be generated in intermittently reinforced behavior by ablation of cerebral tissue.

(It should be noted that the head is an important mechanical detail in these experiments. If, as the result of the operation, quick movements of the head produce painful stimulation, some disturbance in performance would be expected apart from any brain damage. Control experiments suggest that this difficulty is minimal. [See Chapter Ten.])

One bird (74G) had an extended history of FR, mult FIFI, and then FI 1. Parts of the cerebral hemispheres were then removed. Post-mortem sections showed a cyst in the right posterior aspect of the cerebrum and considerable atrophy in the left half of the brain. The performances were not normal and will be described in Chapter Five. The schedule was then changed to FR 60. Figure 64A shows the 10th session, where the schedule is still not producing a normal fixed-ratio performance. Pauses occur after reinforcement despite the small ratio, and most segments

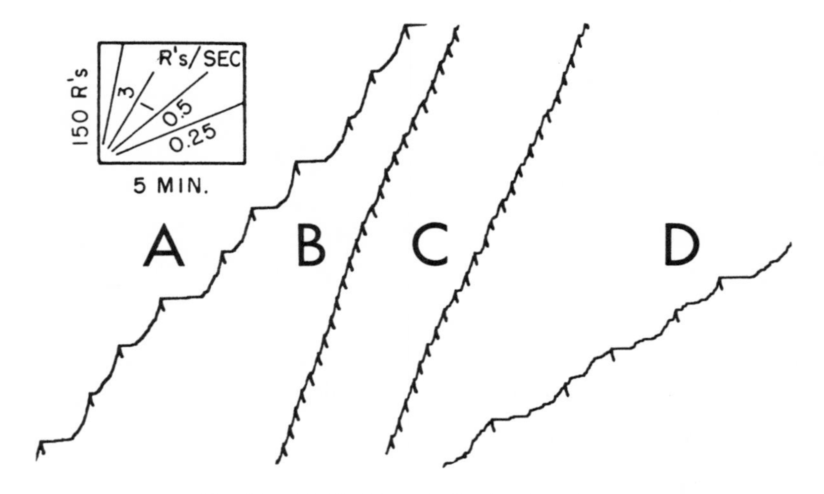

Fig. 64. FR 60, 20, and 40 after brain lesion

show considerable curvature. The ratio was reduced to 20 in the following session; Record B shows a segment of the performance. The smaller fixed ratio eliminated the pause after reinforcement; but the low rate, oscillation in over-all rate, and frequent acceleration after reinforcement are still not typical of a fixed ratio of this size. (Compare, for example, Fig. 20.) Record C shows a segment on the same FR 20, 9 sessions later. Here, the performance is even more disturbed, with pauses after reinforcement and breaks after responding has begun. The ratio was then increased to 40; Record D shows a segment after 5 sessions at this value. Further deterioration of performance beyond that of Record A is evident in the disappearance of the high terminal rate, even though the fixed ratio is only 40 here compared with 60 in Record A. The extreme deterioration in Figure 64D did not appear immediately. Figure 65 contains the complete 3rd session on FR 40, which occurred 2 sessions before Fig. 64D. The over-all curve is negatively accelerated, although some recovery can be observed toward the

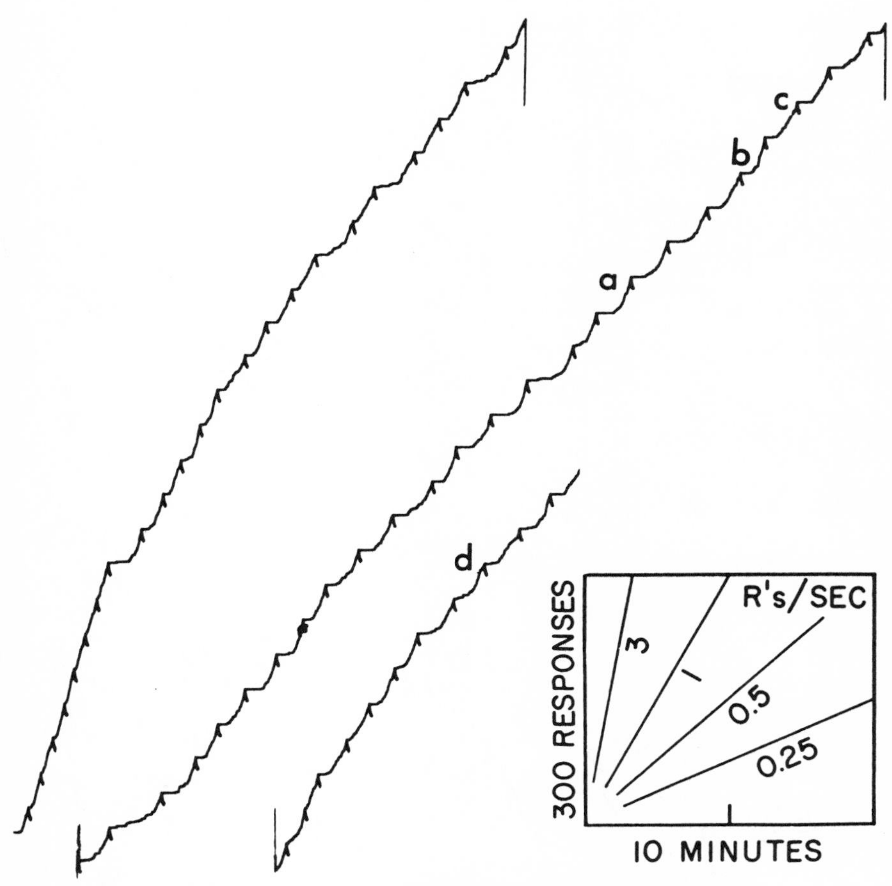

Fig. 65. FR 40: complete daily session after brain damage

end of the session. Parts of the performance resemble Record A for FR 60: a pause after the reinforcement is followed by smooth acceleration to a terminal rate. The terminal rate varies, however (cf. *b* with *c*). Substantial pauses or changes in rate after the terminal rate has been reached occur at *a* and *d*. Although this bird pecks the key at a fairly sustained rate, the fixed-ratio schedule does not produce a normal performance.

A second bird, whose post-mortem examination revealed damage in both hemispheres and the cerebellum, was reinforced on FR after pre-operative and post-operative performances on mult FIFR. The mult FIFR performance was severely disrupted by the operation. After 5 sessions on that schedule, the bird was given 1 session

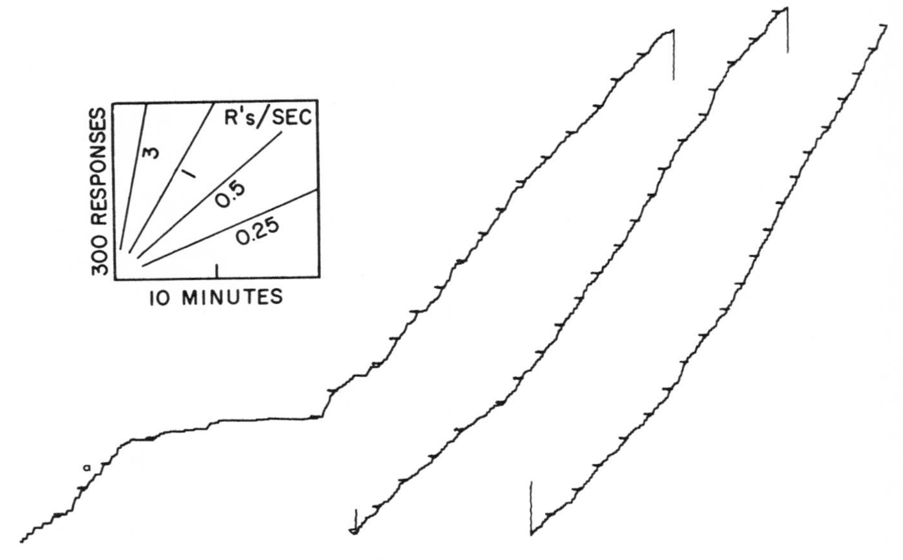

Fig. 66. crf to FR 40 after mult FRFI and brain lesion

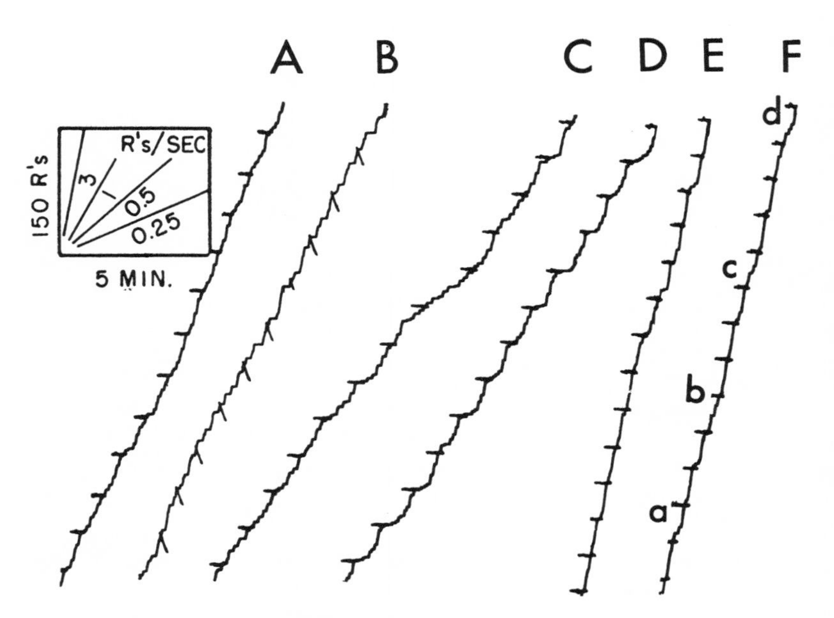

Fig. 67. Development on FR 40 after brain lesion

of crf and then placed on FR 40. Figure 66 illustrates the 1st session on FR 40 (37 days after the operation). The curve shows many features of a first exposure to FR after crf (cf. Fig. 16). But although the initial segment of the curve (at *a*) shows the negative acceleration generally seen in the transition crf to FR, the small bursts of responses followed by pauses are unusual. Further development of the performance on FR 40 is shown in Fig. 67, in segments chosen from the last parts of the next 6 sessions. The 2nd session on FR 40 (Record A) shows a slight increase in over-all rate. Short bursts of responses separated by pauses show in the grain. The 3rd and 4th sessions (Records B and C) show a progressive fall in over-all rate, and the rough grain continues. In the following session (Record D) the rough grain remains. The segments become more sharply scalloped; nevertheless, the over-all rate of responding increases. In the 6th session (Record E) the bird responds immediately after reinforcement at a higher terminal rate and with a smooth grain. In the 7th session (Record F) the ratio is being held fairly well; however, unlike a normal FR, breaks occur in the ratio as at *a, b, c,* and *d.* Figure 67 represents a very slow development for a fixed ratio of this size.

ADDED STIMULI

Added counter—Experiment I

An added stimulus is a stimulus some dimension of which is designed to remain proportional to some feature of a schedule of reinforcement or of the performance generated by a schedule. In a fixed-ratio schedule the most important feature is the number of responses since the reinforcement. Here, the added stimulus may be called an "added counter." If such a stimulus is clear-cut, it may acquire a more effective control than the bird's own behavior. We can then manipulate the "number of responses" in a way which is impossible when the bird's own behavior is the stimulus.

The first step in the analysis of an added counter is to demonstrate that it can achieve control. In the experiments described in this section, the key was opaque except for a small slit of light which could be made to grow as a function of the number of times the bird pecked the key. (See Chapter Three.) The maximum length of slit in this case was 3/4 inch. The slit returned to the minimum size, 1/16 by 1/10 inch, following reinforcement or during a short TO after reinforcement. As a modest effort to compensate for the fact that more responses were necessary to make the slit wider at the end of the ratio than earlier, the rate of growth was made roughly proportional to the size.

Development of FR with added counter

A long history of tand *FR*FI (Chapter Eight) was followed by 300 reinforcements on FR 190; during the last 100 reinforcements the slit which was to become the added counter was present but stationary at the smallest size. The first segment in Fig. 68A is the last performance for one bird with the slit small (the recorder resetting after each reinforcement). Beginning at the arrow, the slit grew as the bird pecked the key.

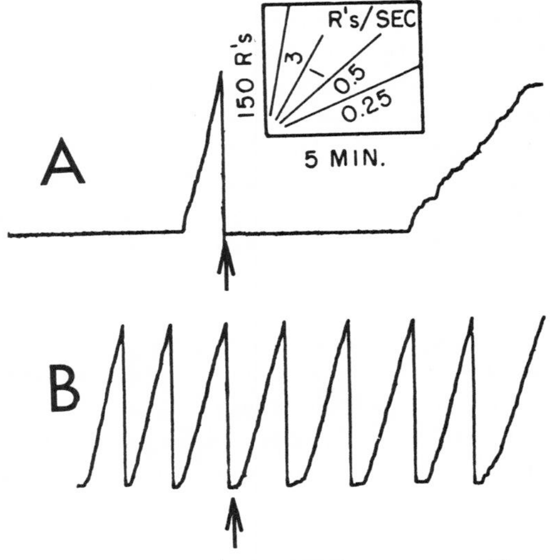

Fig. 68. First effect of a counter on FR 190

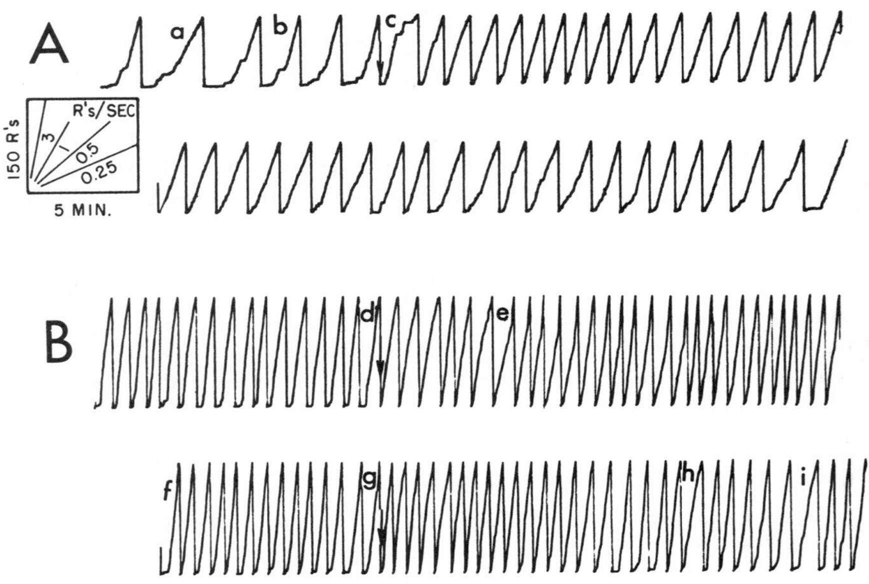

Fig. 69. Reversal of the counter

The novel stimulus disrupts the normal fixed-ratio pattern. (See section above on Novel Stimuli.) It also produces a rough grain with an intermediate rate of responding which falls to zero before the ratio is completed. The disruption produced by the (now) large slit is so severe that the food magazine was operated (*ad lib*) to restore the behavior. The second bird in the experiment showed less pausing after reinforcement both in the previous experiment on tand *FR*FI and in the later stages of this experiment. The first 3 segments of Fig. 68B are the last ratios during which the slit remains small throughout. At the arrow the slit begins to grow with the number of responses. Here, the only immediate effects of the novel stimulus are a slower acceleration to the also slightly lower terminal rate.

Reversing the added counter

Reversing the direction of growth of the slit easily shows the control acquired by the added counter as a result of the reinforcing contingencies in which it plays a part. Figure 69A shows the performance on FR plus counter for the bird in Fig. 68A after 90 reinforcements. A pause after reinforcement is followed by fairly slow acceleration to a terminal rate which varies rather widely (cf. *a* with *b*). At the arrow the direction of the added counter was reversed. Upon returning to the key after reinforcement, the bird finds the slit large. A high rate of responding follows immediately. The slit now shrinks with each response until it reaches its smallest size as the fixed ratio is completed. A rough negative acceleration of rate parallels this change. The curve at *c* under the reversed direction of growth is merely a 180-degree rotation of the earlier prevailing curve. Negative acceleration persists for several ratios, and the curves then pass through a linear phase with no pause after reinforcement. The combination of the added stimulus and the bird's behavior here has no differential effect on the rate. The new counter begins to take effect in the second line of Record A, however, and the segments begin to resemble those before the reversal.

Record B shows a similar result for the second bird. The curve at *d* just before the arrow is the last ratio during which the slit grows from small to large. At the arrow the slit was largest just after reinforcement but shrank as the bird completed the ratio. The bird begins to respond immediately in the presence of the large size of slit, and it shows only slight negative acceleration as the segment is completed. (This pattern confirms the less effective control shown in Fig. 68B for the same bird.) The next 2 segments show a lower-than-normal terminal ratio rate, but by *e* the curves have become positively accelerated, and by *f* (later in the same session) a slight pause reappears after reinforcement. Again, the reversal of the direction of growth at the 2nd arrow in Record B shows that the new counter has now taken control. The last segment before this 2nd reversal (*g*) shows a brief pause after reinforcement and a rapid acceleration to a high terminal rate which is maintained until reinforcement. In the 1st segment following the reversal the bird begins responding at a high rate immediately. However, this rate does not fall off as the slit reaches its largest size at the end of the ratio. Negative curvature is clear in the 2nd segment after reversal, and for some time thereafter. A linear phase then sets in; and by the end of the session, positively accelerated curves, with a pause following the reinforcement, appear again, although several instances of negative acceleration still occur, as at *h* and *i*.

The direction of growth of the slit was also reversed when the same 2 birds had stabilized at larger FRs with added counter. Records A and B in Fig. 70 show parts of 2 successive sessions on FR 380 after approximately 1900 reinforcements on various sizes of FR with added counter. Record A begins with the final performance on FR 380, with the slit growing from small to large.

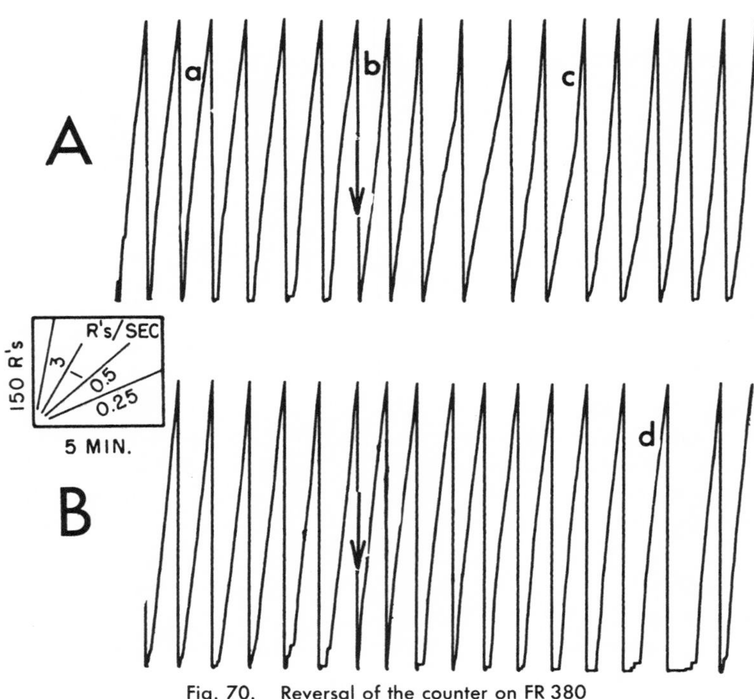

Fig. 70. Reversal of the counter on FR 380

The curves show marked negative acceleration with rates of approximately 10 responses per second for the first 100 responses of the fixed ratio, followed by an often abrupt shift to about 5 responses per second (as at *a*). This curvature occurs occasionally in FR without added counter. It suggests a nonlinearity of counter readings, possibly even when the bird's own behavior is the only stimulus influencing the contingencies. Pauses following reinforcement are either absent or only a few seconds long. At the arrow the direction of growth of the slit was reversed. During the 1st segment thereafter (*b*), the rate is constant at the value prevailing previously at the largest size. Some positive acceleration, which can be seen best when the curves are foreshortened, becomes marked by Segment *c*. Here, the rate continues at slightly over 3 responses per second until about 100 responses before the completion of the ratio, when it shifts abruptly to 8 responses per second. This positive acceleration, which is again almost a 180-degree inversion of some of the curves before the arrow, appears frequently during the remainder of the session, not all of which is shown in Record A. At the beginning of the following session the slit is still large just after reinforcement but shrinks during the ratio. Record B shows the middle part of the session. The performance before the arrow (where the direction of growth of the slit was again reversed to growing from small to large, as at the start of the session in Record A) shows a continuation of the performance at the end of the previous session, with the slit changing from large to small during the fixed-ratio segment. The rate does not vary between such wide extremes, however. Following reversal (at the arrow), the original performance returns immediately to that seen before the arrow in Record A, where the slit grew from small to large. In the segment at *d* the rate is approximately 10 responses per second for the first 100 responses, with a shift to slightly less than 6 per second during the rest of the ratio.

The other bird could not maintain so high an over-all rate, even with the help of an added counter. After approximately 1300 reinforcements at various sizes of ratio, long pauses continued to occur after reinforcement and were followed by slow acceleration to the terminal rate. The first two lines of Fig. 71A show a characteristic performance. One pause is more than 1.5 hours long, and all segments show extended curvature. At the arrow, during a pause, the size of the slit was changed from small to large and the direction of the growth reversed. (The slit now changes from large to small as the bird completes the fixed ratio.) The bird begins to respond within 5 seconds of the appearance of the large slit, although nearly 30 minutes had elapsed since the last reinforcement without a response. The decline in rate of responding as the spot becomes smaller again shows the powerful control exercised by the spot of light. The first 4 segments show marked negative acceleration throughout the whole ratio. Except for the absence of a long pause, which is prevented by the occurrence of reinforcement, these curves are again 180-degree rotations of the curves before the arrow. Continued reinforcement on the reversed counter produces a linear phase (at *a*), and then produces a pause after reinforcement, with subsequent positive curvature. Fifteen sessions later, ratios are being run off in blocks separated by long pauses. One such block is shown in Record B, where the pause was 3 hours long and the rate preceding the 1st segment was low. The 1st segment at *b* is run off at a constant rate, but the next segments show a progressive increase in curvature until another long pause follows the last segment in the record.

The long pauses after reinforcement at larger fixed ratios are clearly not due to a factor such as physical exhaustion or fatigue, but to the extremely unfavorable stimuli then present.

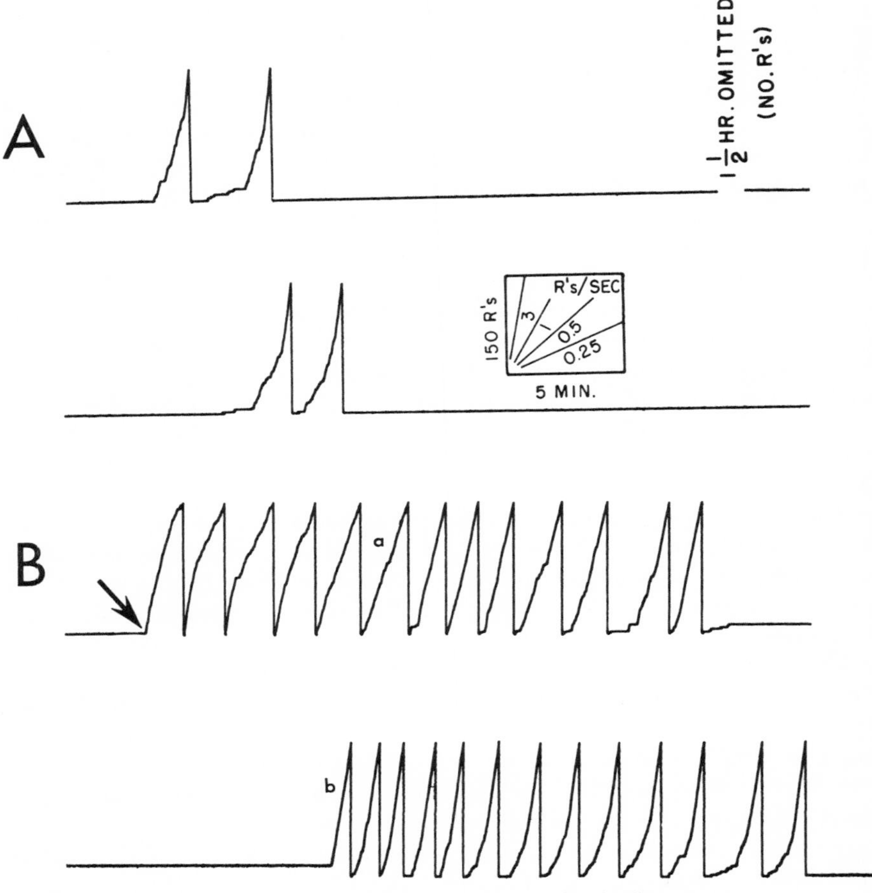

Fig. 71. Reversal of the counter on FR 200 showing long pauses

Stopping the added counter at its unoptimal reading

The beginning of Fig. 72 shows a final performance on FR 410 with added counter after approximately 2100 reinforcements on various fixed ratios with added counter. (This is the same bird as that in Fig. 70.) Considerable negative curvature occurs, with rates falling from 10 to 5 responses per second during the last 100 responses of the ratio. Following the reinforcement at the arrow, the slit became small as usual, but was allowed to remain small for the rest of the session. As a result the rate remained near 10 per second throughout each segment. At *a* the slit failed to reset to small for a few seconds. The large slit immediately after the reinforcement produced an immediate burst of responding at approximately 10 per second. The effect of the large

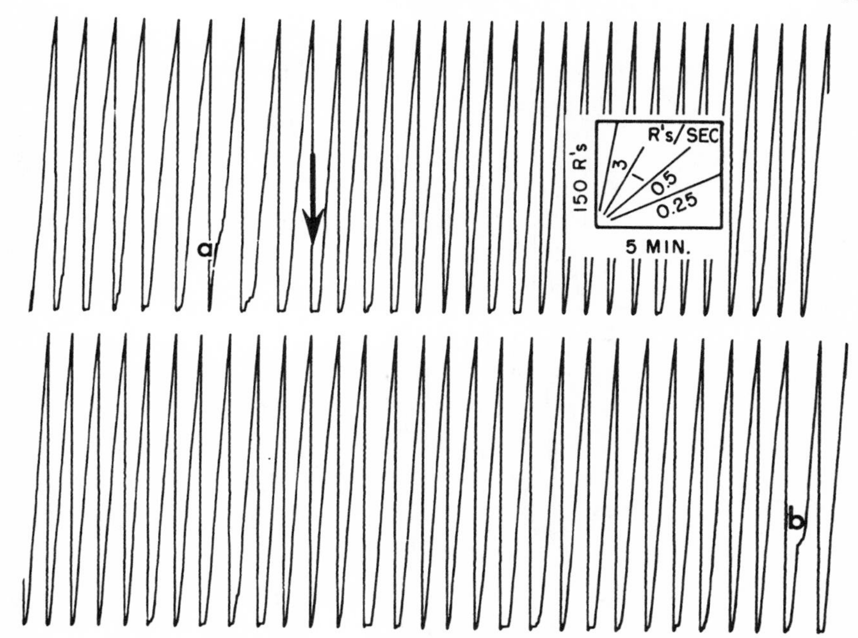

Fig. 72. Stopping the added counter at its unoptimal reading: FR 410

slit at *a* and the lower rate of responding at the end of the ratio segments are contradictory. Some interaction between the added counter and the bird's behavior must produce the lower rate at the end of the normal segment.

(An incidental effect is recorded at *b,* where the experimental apparatus was accidentally jarred. This jarring produced a lower rate, which then accelerated in roughly the same way as at the beginning of a segment.)

Increasing the size of the fixed ratio

The rate of growth of the slit has to be reduced to accommodate larger ratios if it is to reach its largest value upon completion of the ratio. Such a change is made for the first time in Fig. 73. The schedule at the beginning of the record was FR 190,

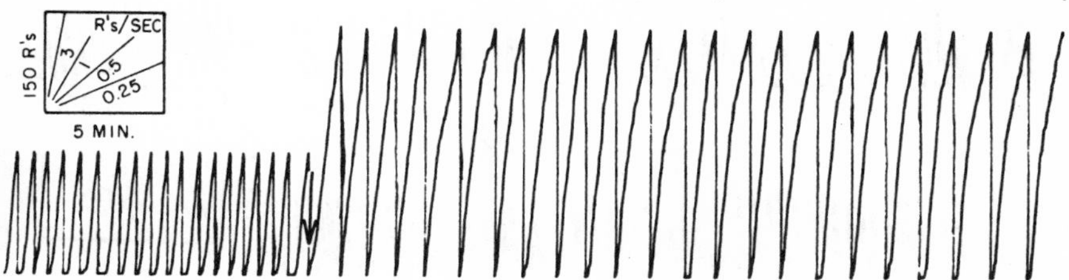

Fig. 73. Transition from FR 190 + counter to FR 360 + counter

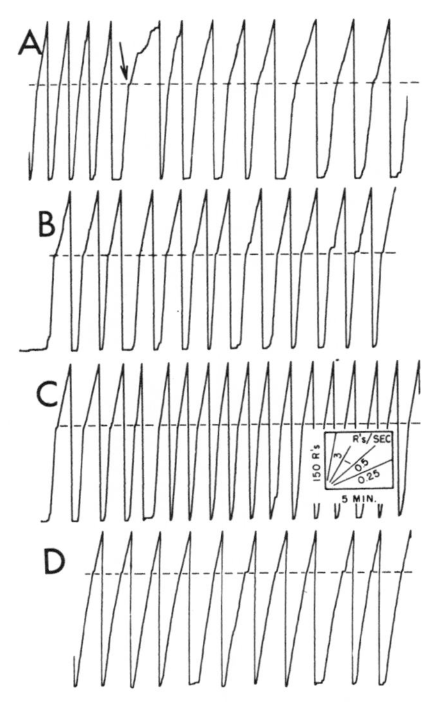

Fig. 74. Increased negative curvature on FR 410 from stopping the counter after 250 responses

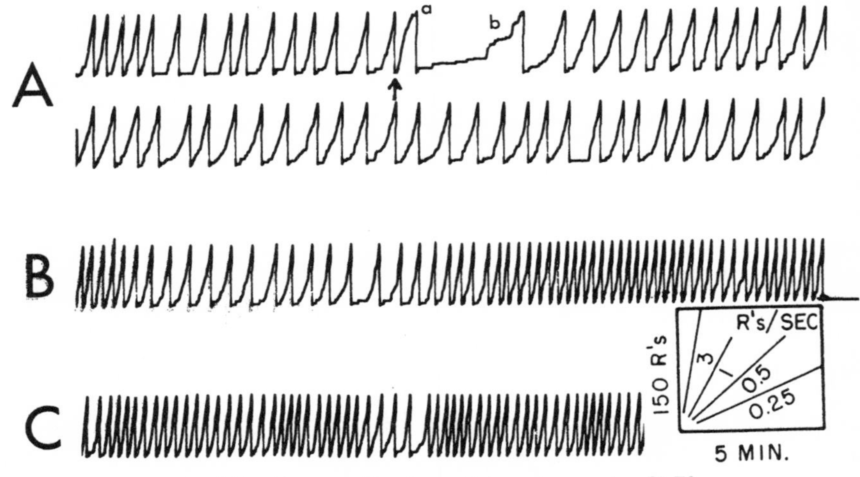

Fig. 75. First transition from slit large to counter: FR 70

with the slit growing to the full width. At the arrow the ratio is increased to 360, with the slit correspondingly reaching the same maximum size in 360 responses. While the method of recording does not permit a precise comparison of the 2 curves, no very gross differences are apparent. The pause after reinforcement appears less commonly in the larger ratio, and the negative curvature seen at FR 190 becomes clearer and more extensive at FR 360. A possible explanation for the reduced pause could be the novel stimuli from the bird's own "counter" at the new ratio.

More negative curvature in larger fixed ratios may occur because the necessarily smaller increase in the size of the slit per response is not so reinforcing. The correction for relative change in size was probably inadequate. The "increase in count," however, would be weakest as a reinforcer toward the end of the ratio. We checked on this effect in a further experiment. The slit grew to a maximum size in 250 responses; then, a blue light appeared on the key, the reinforcement occurring 160 responses later. If the decline in the rate of responding at the end of the fixed-ratio segment is due to inadequate reinforcement by changes in the counter reading, this procedure should produce more extended negative curvature, because only the bird's own counter operates during the blue light. Figure 74A shows the first effect of this procedure. At the arrow the blue light appears for the first time. The rate during the last 160 responses in the 1st fixed-ratio segment is approximately 2.5 responses per second compared with about 10 responses per second during the sustained run when the added counter operates. By the middle of the same session (Record B) the rate during the last 160 responses has increased to 3 responses per second, but the rate during the earlier sustained run with added counter has increased to over 12 responses per second. By the end of the session (Record C) the final rate has reached 4 per second. Several sessions later, in the segment shown in Record D, the blue light came on after 300 responses instead of 250. By this time negative curvature also developed during the early part of the ratio in which the added counter is operative.

This is essentially an experiment in chaining. A response is reinforced on FR 250 with added counter by a blue light. Under this stimulus a response is reinforced on FR 50.

Added counter—Experiment II

Development of FR with added counter. The 1st portion of Fig. 75A shows the performance (700 reinforcements after crf) on FR 70, with the slit remaining large. At the arrow the slit became small for the first time; it grew from 1/16 inch to 3/8 inch, each response producing the same change in size of slit regardless of the size of the ratio. When the slit changes to small for the first time at the arrow, responding begins instantly at a very high rate despite the novel stimulus dimensions. It then tapers off, until at the 1st reinforcement, at *a*, it is very low. The 2nd reinforcement, at *b*, is reached at a much lower rate, with rough grain and some positive acceleration. Within a few reinforcements the performance shows little change from the fixed ratio without a counter. The added counter eventually produces a very high rate. Records B and C show the 2nd and 3rd sessions. Pauses after reinforcement largely disappear, and the running rate characteristically reaches values of the order of 9 or 10 responses per second. Although performances of this order can be generated without added counter, the rapidity with which the performance shifts from that of Record A indicates one effect of the counter. The high rate of responding in the presence of the novel stimulus may be related to some innate tendencies of birds to peck small objects rather than the effect of the novel stimulus per se.

A second bird shows less clearly the tendency for a small spot to evoke more rapid responding apart from its place in reinforcing contingencies. The bird received approximately 700 reinforcements on FR 110 with the slit continuously at the larger size. Records A and B of Fig. 76 are segments from the last 2 sessions under these condi-

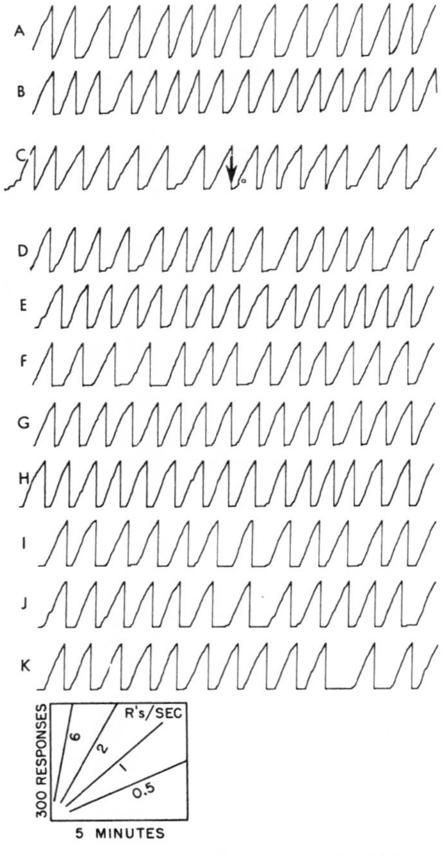

Fig. 76. Early performance on FR 110 + counter after FR 110 with slit at large

tions. Note the scale of this record. The paper was moving twice as fast as in preceding figures. During the reinforcement at the arrow in Record C, the slit was changed to small for the first time. This change has no immediate effect on the rate; but after approximately 15 seconds, there is a brief burst at 10 responses per second (at *a*). The rate then returns to the usual value. The next 4 ratio segments show the effect evident in Fig. 75. In the next daily session (Record D) the performance resembles that on FR alone. Records E through K show segments from the 2nd to 11th sessions following the addition of a counter. Note that most of the segments are S-shaped: they begin with a pause, and the rate is highest in the middle of the segment but declines before the end of the ratio. Eventually, a much higher over-all rate developed, presumably because of the added counter.

Two other birds received essentially the same treatment; but instead of being exposed to the largest size of the slit, they were exposed to a growing and resetting slit not correlated with the ratio. All sizes of slit had been equally correlated with reinforcement. Record A in Fig. 77 shows the final performance on FR 70 after approximately 900 reinforcements on FR 70 after crf, with the slit varying at random. The counter was added at the start of Record B. The pauses are reduced somewhat after reinforcement, and the over-all rate increases. Record A was recorded at standard

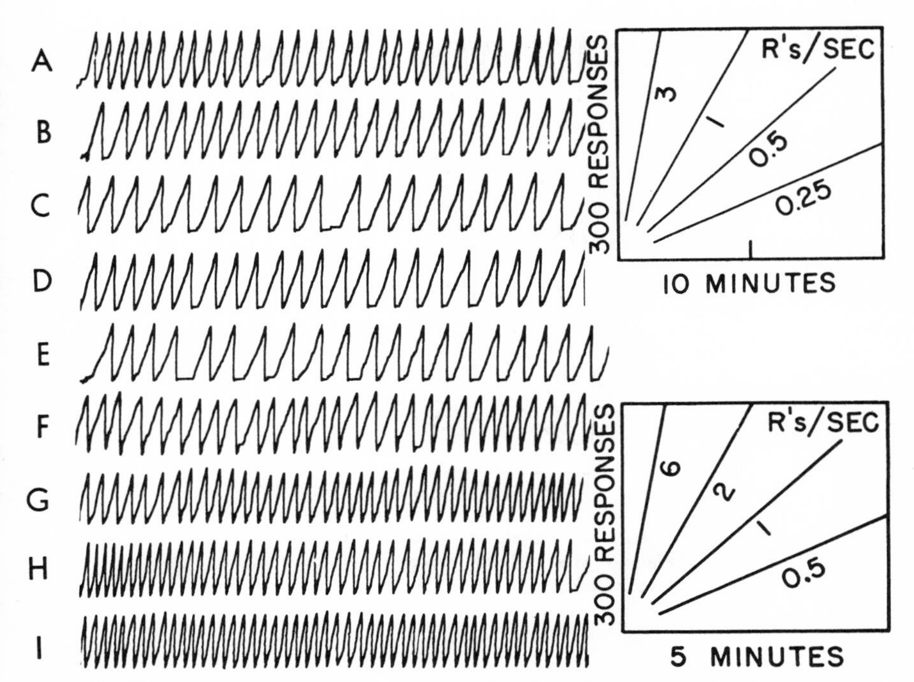

Fig. 77. Early development on FR + counter after FR with uncorrelated counter

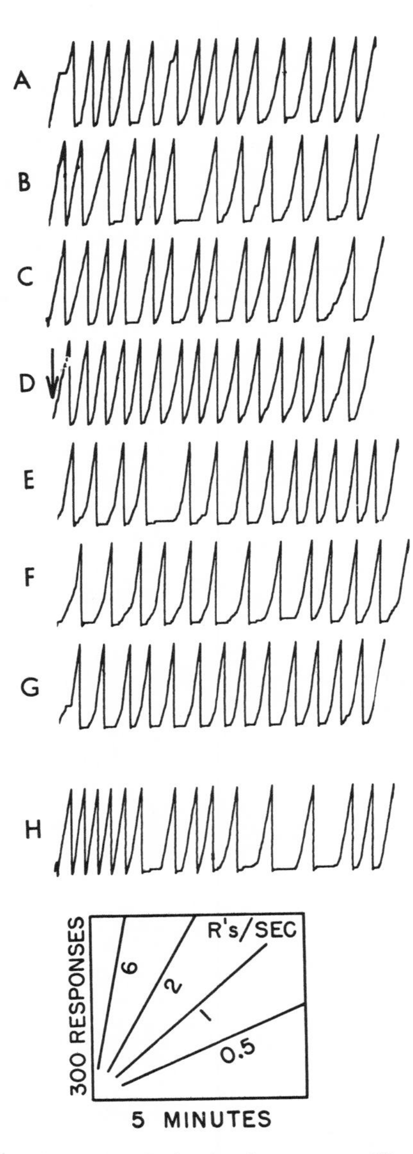

Fig. 78. Early development on FR + counter after FR with uncorrelated counter (second bird)

speed (top coordinate). The other records in the figure were made at double speed (lower coordinate). Records C through I show segments from 11 sessions, during which the influence of an added counter is felt in an increase in the rate and the elimination of curvature after reinforcement. Eventually, a final performance (Record I) shows average rates of the order of 6 per second and responding immediately after reinforcement.

The other bird previously exposed to all slit sizes was able to sustain a fixed ratio of 130. Records A, B, and C of Fig. 78 show segments from the 3 days before the addition of a counter, with the slit varying unsystematically during the session. At the start of Record D the slit was small following reinforcement and grew to the largest size as the ratio was counted out. Little or no immediate effect is apparent, nor any progressive change, through Records E, F, and G, which show segments from the 2nd, 3rd,

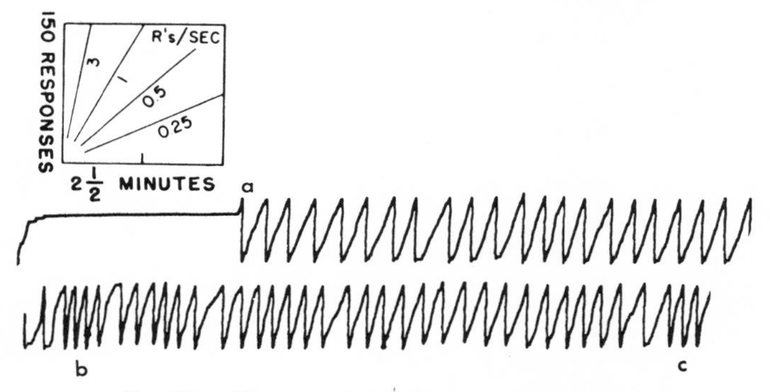

Fig. 79. Change in the scale of the added counter

and 4th sessions after the counter was added. The fact that the growing spot was acquiring control as a stimulus was shown at the beginning of the 5th session (Record H), where the slit was large after reinforcement and remained large for the rest of the session. The first 5 or 6 segments show the elimination of the pause after reinforcement. The performance then begins to resemble earlier records with or without the counter. (Since sustained rates of responding of approximately 7 responses per second occur before the counter is added, it is perhaps not surprising that the counter has no dramatic effect.)

Change in scale of added counter. In the experiments just described the counter reached a maximum size of $\frac{3}{8}$ inch. It was later changed to grow to $\frac{3}{4}$ inch at the completion of the same size of ratio. Figure 79 shows the session immediately following Fig. 77 I. The bird begins immediately; but as the slit grows to its novel size, the rate declines sharply. The 1st reinforcement (at *a*) occurs only after 4 minutes. The rate remains lower than normal, although before the end of the session (second line of the figure), instances appear (at *b* and *c*) where ratios are run off at rates comparable with those in Fig. 77.

With the slit reaching maximum size ($\frac{3}{4}$ inch), the ratio was increased to 125. Figure 80A shows the final performance. The curves are predominantly S-shaped, beginning immediately after reinforcement (or after a short pause) and accelerating quickly to rates of approximately 10 per second. These are maintained for varying numbers of responses; they finally give way to much lower rates of the order of 3 to 4 per second toward the end of the ratio. This is an unusual amount of curvature in a fixed ratio. Three sessions later, in Record B, the session begins on FR 110 with the slit large. At the arrow the counter is added, the slit again growing to $\frac{3}{4}$ inch at the end of the ratio. The over-all rate remains the same as before, but the curves become more nearly linear. In Record C, taken 4 sessions after Record B on the same schedule, the over-all rate has increased; rates of 10 per second in the middle part of the ratio give way to rates of 4 per second near the end. Occasionally, a low rate is maintained throughout the ratio, as at *a*. The highest *sustained* rates of responding are of the order of 8 responses per second, compared with over 10 responses per second in Fig. 77 on FR 60, where the counter was first added.

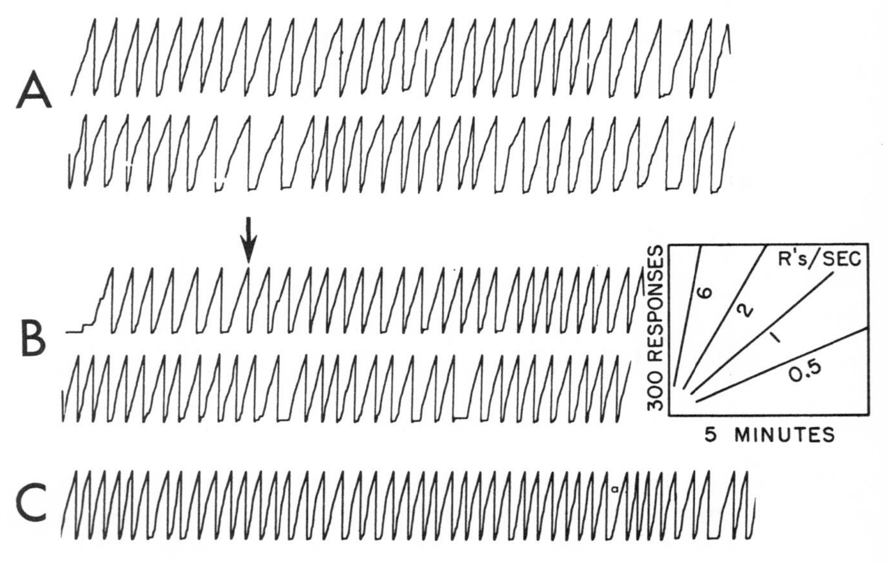

Fig. 80. Transition from FR 125 with large slit to FR 125 + counter

Transition from added counter to slit at optimal reading. Performances at large FRs with added counter were probed occasionally by letting the counter remain at the optimal size. Transitions in the reverse direction were also observed.

Two birds on FR 120 with added counter gave the curves at the left of Fig. 81. Beginning at the vertical dashed line, the slit remained large. In A the curves with the growing spot are S-shaped with marked negative curvature at the end of the seg-

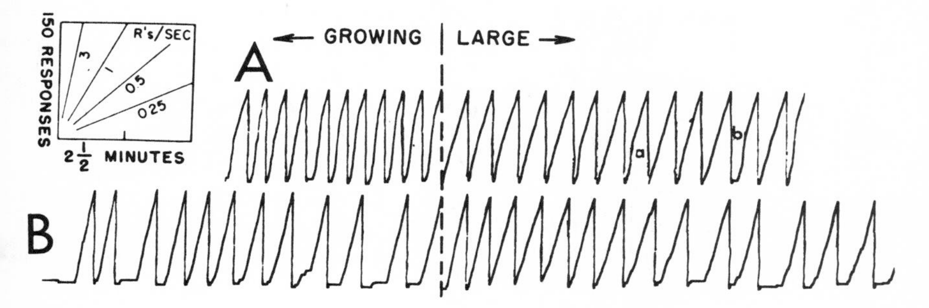

Fig. 81. Transition from FR 120 + counter to FR 120 with slit at large

ments. When the slit remains large, the slight pausing after reinforcement is temporarily dropped and the bird responds at a constant rate equal to that occurring at the end of the segment with the added counter. The old higher rate reappears in the middle of the fixed-ratio segment (as at *a* and *b*) as a normal fixed-ratio performance develops. In Record B (the second bird) the performance on the fixed ratio with the added counter is linear after a rough start. When the slit remained large, the running rate did not change; but the pause after the reinforcement disappeared for several ratios. Shortly after the transition in Fig. 81A, the ratio was increased to FR 220 with the slit remaining at large. Figure 82 shows segments from 10 consecutive sessions. (Records J and K show the beginning and end of the last session.) Instances of runs at very high rates early in the ratio appear, as at *a, b, c,* and *d* in Record A. These disappear in later sessions with the slit still at large, although S-shaped curves reappear toward the end of the session in Record K.

Figure 83 shows later transitions from FR with added counter to FR at optimal reading for 4 birds on FR 220. At this late stage in the experiment, after approximately 4400 reinforcements on various values of fixed ratio with and without added counter, a fairly high rate of responding prevails in the final performance on the fixed-ratio schedule with added counter (left of dashed line). When the slit remains large, the next segments begin at the terminal rate of responding, in the absence of the characteristic pause. In Record A the effect of the large slit lasts for about 10 segments, after which the performance becomes similar to that occurring with the counter. In Record B a short break appears near the beginning of the 3rd segment, and a pause appears after the 4th reinforcement. Toward the end of the record, the negative curvature becomes less frequent. In Record C pauses are eliminated only in the first 2 segments after the slit remains at large, although a group of 3 quick starts appears at *a*. The rest of the session shows a performance similar to that with the added counter. In Record D the optimal counter reading has the most prolonged effect. With the added counter, a pause of the order of 60 seconds consistently appeared after reinforcement. When the slit remains large, the pause is absent for the first 12 segments.

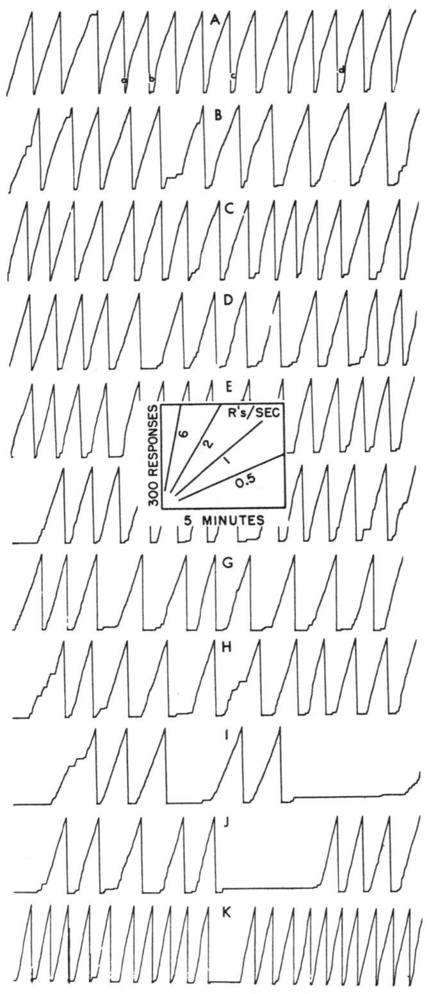

Fig. 82. Development on FR 220 with slit at large

Figure 84 shows transitions from a stable performance, with the slit stationary at large, to the added counter, where the slit grows from small to large. These are for the same bird but taken at early, intermediate, and late stages of the experiment, under 3 values of fixed ratio. Record A shows an early transition on FR 110. The amount of pausing after reinforcement is temporarily less. The pause increases, however, until by the end of the session it is of the same order of magnitude as before. Record B, taken about 1 month later under FR 220, shows a similar effect. When

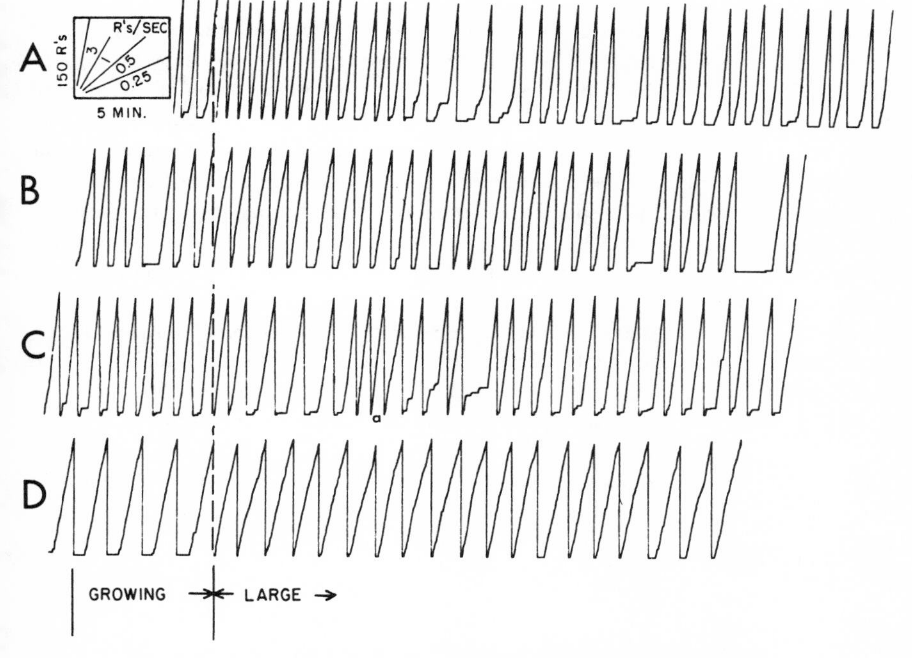

Fig. 83. Transition from FR 220 + counter to FR 220 with slit at large

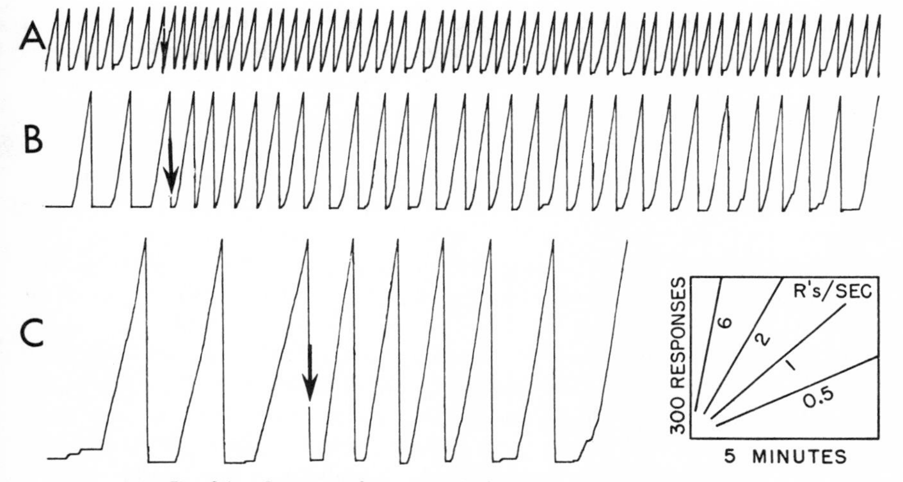

Fig. 84. Transition from FR with slit at large to FR + counter

the counter is added at the arrow, the pause disappears for a few reinforcements and then increases until the original performance is recovered. Record C shows a performance on FR 420 much later in the experiment. Before the addition of the counter, the bird pauses up to 1 minute or more, with a running rate of slightly less than 6 responses per second. The added counter, at the arrow, reduces the pause after the reinforcement for at least 4 segments, and increases the terminal rate from less than 5 responses per second to 8 responses per second. This rate change is not apparent in Records A and B.

Figure 85 shows 3 records, separated by 2 or 3 sessions where the slit remains continuously at large until the arrow, after which it grows as with an added counter. While the spot is large (left of the arrow), a pause characteristically follows reinforcement, and the curves are roughly linear with an occasional burst at a higher rate, as at *a, b,* and *c.*

Rate of growth of the added counter. An S-shaped counter is added in Record A of

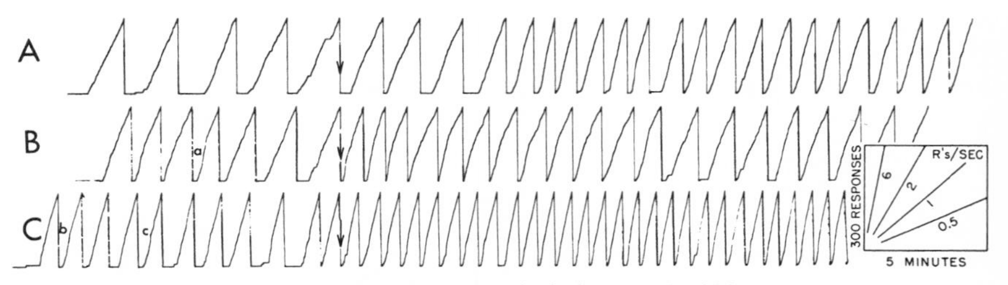

Fig. 85. Transition from FR 220 with slit large to FR 220 + counter

Fig. 85; that is, it grew most rapidly in the middle range. It has little immediate effect, although the running rate progressively increases and the pause after reinforcement progressively decreases. In Record B the counter grows most rapidly after reinforcement, and proportionally slower as the size of the slit increases. It immediately reduces the pause following the reinforcement and produces negatively accelerated curves. This pattern is temporary, however; and by the end of the period the negative curvature disappears, leaving a terminal rate similar to that without the added counter. In Record C the ratio is being held with short pauses. The S-shaped counter immediately reduces the pause after reinforcement. An increase in the running rate early in the ratio gives a sigmoid appearance, which is more obvious if the curves are foreshortened. The rate at the end of the ratio is not increased.

An added counter may help in maintaining a high ratio. Figure 86 shows 3 changes to an added counter for a bird which had difficulty in maintaining FR 220. (A linear counter was used in Record A and an S-shaped counter in Record B. In Record C the slit grew maximally at the smallest size of slit and proportionally less as the size of the slit increased. The day-to-day variability in these performances is such that we cannot safely ascribe any effect to the "shape" of the added counter.) The advantage of a counter in holding the ratio is clear. The performances to the left

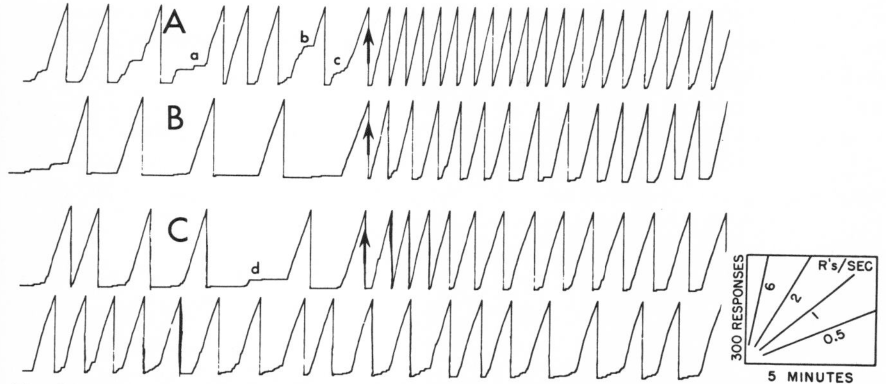

Fig. 86. Transition from FR 220 to FR 220 + counter reducing the pause after reinforcement

of the arrow (with the counter reading steadily maximum) show rough grain, long pauses, and occasional "knees," as at *a, b, c,* and *d.* When the counter is added, all of these characteristics disappear. As the session continues, the over-all rate tends to fall and the pause after reinforcement increases. This condition can best be seen in the longer Record C, when by the end of the session a pause or period of low rate consistently follows reinforcement. Even here, however, the form of the curve differs from that without counter. The rate of responding is high in the middle range of the segment, after which the rate shifts, sometimes abruptly, to a lower value for the remainder of the ratio.

The effect of added counter in making it possible to sustain a high ratio was felt more slowly in another bird, samples of whose records are given in Fig. 87. In this experiment the rate of growth of the slit was most rapid at the beginning and end of the ratio. (This counter was chosen in an effort to counteract 2 features of the usual segment: the reinforcing effect of the rapid growth at the beginning of the ratio should

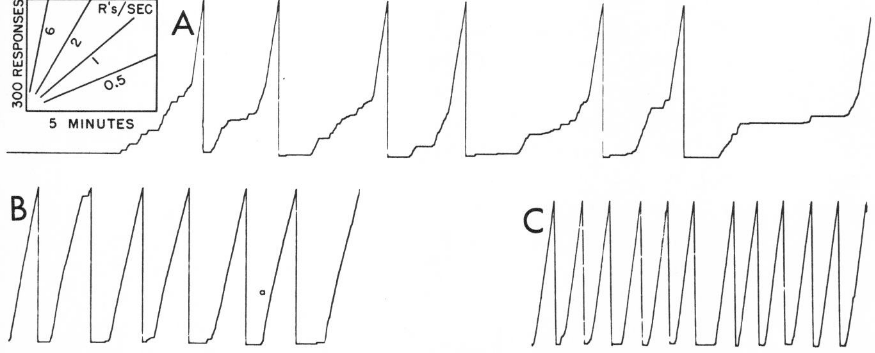

Fig. 87. Transition from FR 420 to FR 420 + counter reducing the pause after reinforcement

affect the stimulus-controlled pause, while that at the end should correct the frequently observed negative acceleration late in the ratio segment.) Figure 87A shows an early performance at FR 420 with this counter. There is considerable "straining" with ragged knees. In general, inverted Ss are suggested. At Record B, 5 sessions later, ratios are completed much more rapidly. The "shape" of the added counter fails to prevent some negative curvature (as at *a*). At C, 17 sessions (and 200 reinforcements) after A, the ratios are run off at an almost constant rate, and pauses after reinforcement have become fairly short. (Note that records are still at the faster paper speed.)

A second bird in this experiment showed rates of responding which corresponded much more closely to the rates of growth of the added counter. Figure 88 was taken

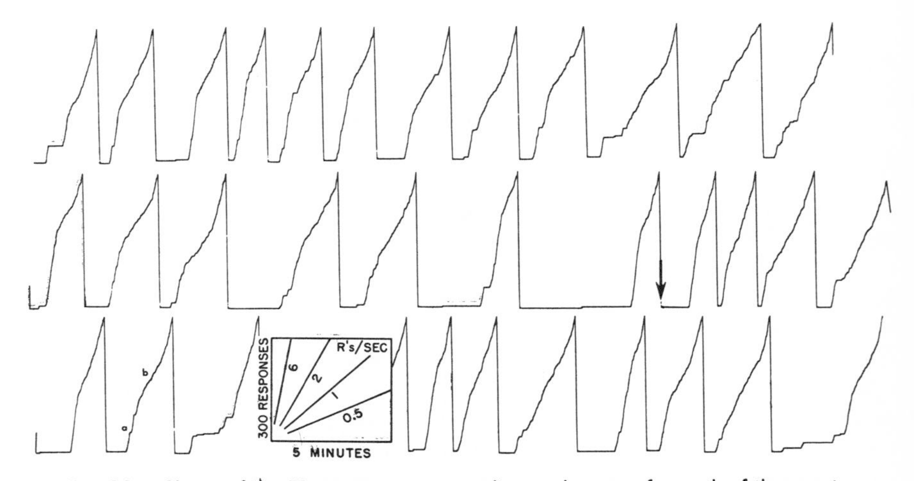

Fig. 88. Shape of the FR segment corresponding to the rate of growth of the counter

at a stage of development comparable with that about halfway between Records B and C in Fig. 87. The early, rapid growth of the slit has not eliminated the pause at the start of each segment, but the rate begins high. A relatively smooth deceleration to an intermediate rate is evident, and an acceleration to a high terminal rate of about the same value as the starting rate. The second ratio at the bottom line is a good example of this effect. After a pause of less than 1 minute, the rate accelerates briefly to 7.5 responses per second at *a*. A continuous decline in rate leads to an intermediate rate of slightly less than 2 responses per second halfway through the ratio (*b*). This rate is followed by a smooth acceleration to about 6 responses per second at the end of the ratio. The high rates occur when the slit on the key changes by the largest amount per response, and the lowest rate where the slit changes least.

At the arrow an injection of 0.5 cubic centimeter of saline was given intramuscularly. The apparatus was stopped, and the bird was taken out of the box, injected, and replaced. The brief interval required for the injection does not show in the record. The form of the segments following the arrow shows little, if any, effect of this disturbance.

The effect of sodium pentobarbital on FR with added counter. The bird shown in Fig. 88 had been injected with 5 milligrams of sodium pentobarbital 2 sessions before Fig. 88 in a similar procedure. Figure 89 shows the result. All segments before the pentobarbital injection show the inverted S (for example, at *a, b,* and *c*). Immediately after injection (at the arrow) a pause of about 1 minute occurs, as after the controlled saline injection in Fig. 88. The bird begins responding again; but with the onset of the drug less than 1 minute later, the rate abruptly drops to zero (*d*). No responses occur for the next 1.5 hours, although the bird is awake and moving about in the

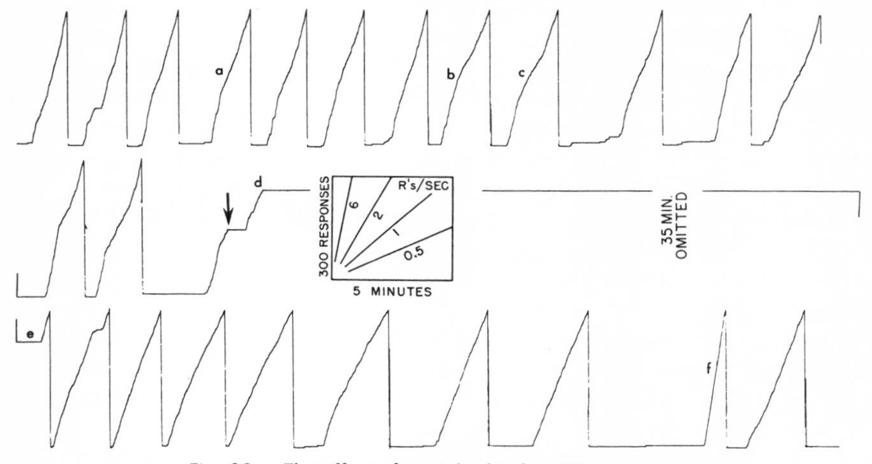

Fig. 89. The effect of pentobarbital on FR + counter

experimental apparatus. When the bird resumes responding abruptly at *e,* it completes the ratio and begins the next within a few seconds. During the remaining ratios of the session, the control previously exercised by the growth of the slit is almost completely lost. The first 4 ratios after injection show very short pauses after reinforcement and a sustained intermediate rate similar to the earlier rate in the middle of the ratio. At *f* an unusually long pause is followed by an abrupt shift to a very high rate, which is maintained until reinforcement.

This effect probably corresponds to the excitatory phase of the pentobarbital which is commonly observed in clinical work and in fixed-interval schedules with this drug, and may also show some loss of the complicated stimulus control by the slit.

Block counter

In the preceding experiments the slit on the key which serves as a counter increases in size *continuously* as the bird pecks the key. Another type of added stimulus is possible in which the counter advances in discrete steps. For example, the color of the key could change from red to orange after 30 responses, to green after another 30 responses, and to blue after another 30 responses, where a response to the key is reinforced 30 responses later (at FR 120). Such a "block" counter corresponds to many schedules of reinforcement in education where a certain amount of work is required to achieve a level of proficiency which is then the occasion when the next level of difficulty can be attacked. Such a counter is essentially a case of chaining (Chapter Fourteen). Each color is an occasion on which a response is reinforced by the next color in the sequence on FR 30. The final color is an occasion on which a response is reinforced on FR 30 by the magazine.

Another case of the block counter is possible where the color changes are not correlated consistently with the reinforcement. This is called an "uncorrelated block counter." The sequence of colors occurs in a constant pattern, but the cycle has no relation to the appearance of the magazine. A change in the counter reading might be reinforcing because it shows progress toward the reinforcement even though it does not indicate the distance still to be covered. If the number of responses required to produce a change in color is a factor of the fixed ratio, one of the color changes will occur simultaneously with reinforcement, and the first response after reinforcement will always occur at a new reading of the block counter. In the experiment to be described here, the number of responses required to change the block counter was not a factor of the fixed ratio. No relation exists between any given stimulus value and reinforcement.

Uncorrelated block counter. The stable performance on FR 120 with a constant blue key-light was first established, and the uncorrelated block counter was then added.

Figure 90 shows a badly strained performance on FR 120, with pauses after rein-

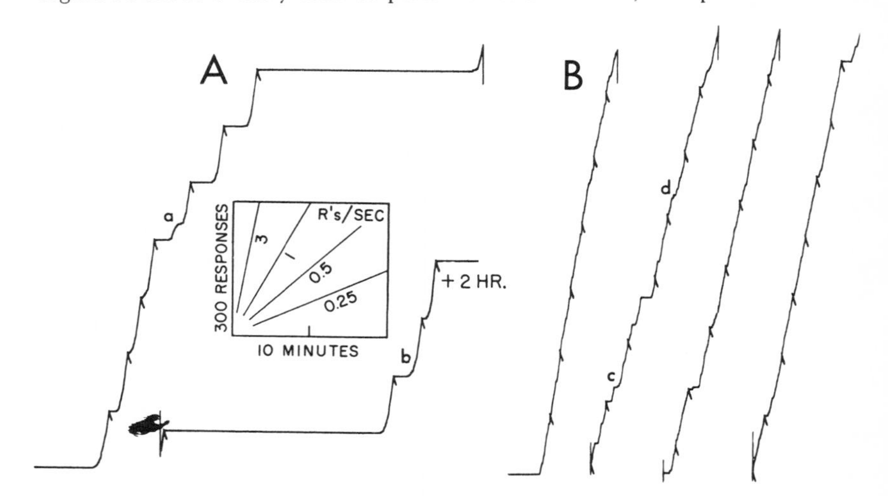

Fig. 90. Transitions from FR 120 to FR 120 + uncorrelated block counter

forcement as long as 2 hours, occasional "knees" (as at *a*), and curvature preceding the terminal run (as at *b*). The key-color was blue. On the following day on a schedule of FR 140, a block counter was added consisting of a key-color change every 10 responses in a sequence of yellow, blue, white, green, yellow, etc. Record B shows the resulting reduction in the amount of pausing following the reinforcement. The overall rate of responding is sharply increased; the rate difference is due to the elimination of the pause after reinforcement rather than to an increase in the local rate. Slight disturbances are evident as the key-color changes, and especially as they become more marked in the 2nd excursion of the pen. Rough grain appears at *c* and *d*, just after color changes.

The effect of an uncorrelated block counter in eliminating the pause is not necessarily due to the reinforcing effect of a counter. Pauses appear to result from a very unfavorable stimulus just after reinforcement. The present counter temporarily breaks up any contingencies between key-color and nonreinforcement.

Figure 91A shows a stable performance on FR 90 (steady blue key) for a second

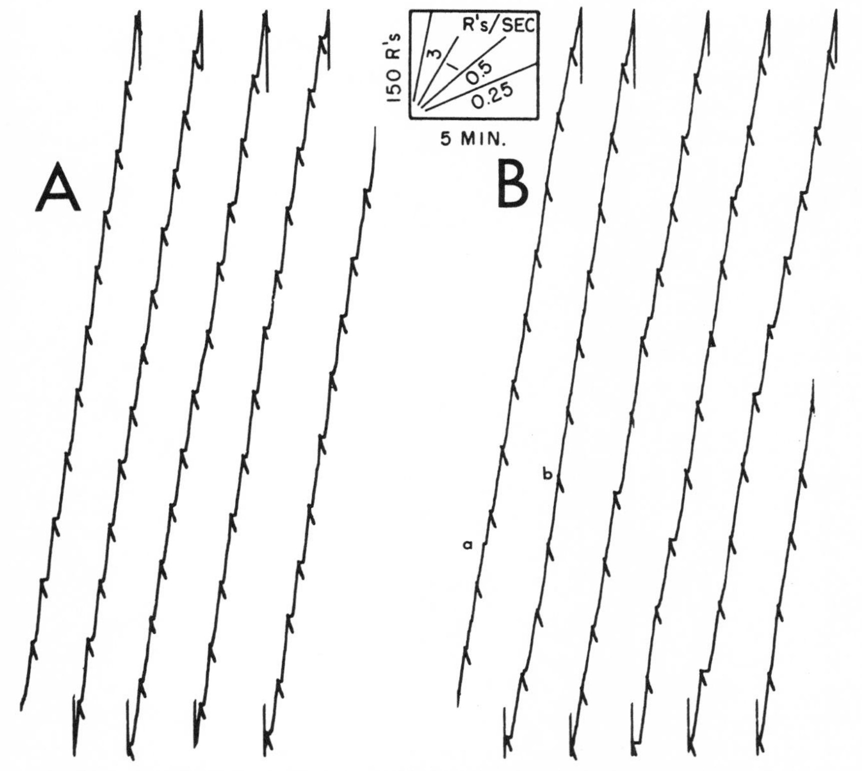

Fig. 91. Transitions from FR 90 to FR 90 + uncorrelated block counter (second bird)

bird in the experiment, in which a consistent pause after reinforcement is followed by an abrupt assumption of the terminal rate. The uncorrelated block counter added at the beginning of the following session (Record B) produces a few brief pauses at color changes (as at *a*) and a temporary reduction in pauses after reinforcement (as at *b*). The effect of the block counter disappears by the 2nd session, however. Pauses after reinforcement increase, and the grain and terminal rate of the ratio return to normal. No further effect of the uncorrelated block counter was observed in 5 further sessions. Reinstating the steady blue key had no effect. Presumably, the uncorrelated color changes now have no discriminative power.

Correlated block counter. The fixed ratio was increased to 120 for both birds in the preceding experiment, the key remaining blue. A block counter was then added; a yellow key after reinforcement changed to blue after 35 responses, to white after 70 responses, and to green after 105 responses, the magazine operating at the 125th response. The schedule may be regarded as 4 chained fixed ratios.

Figure 92 shows the early development with a correlated block counter for the bird

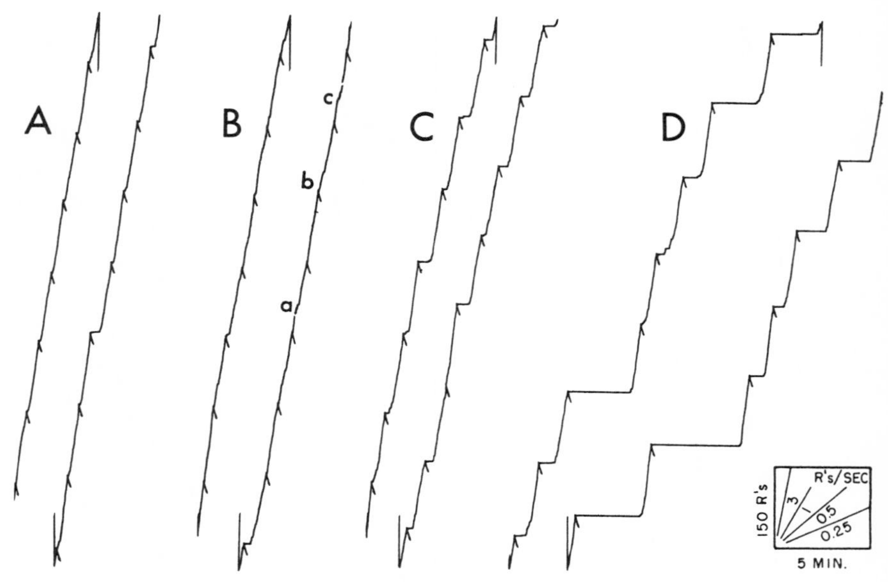

Fig. 92. Early development of correlated block counter

in Fig. 91. The correlated counter causes no immediate effect because of the history of the uncorrelated block counter. The 1st session (Record A) produces no change in performance. The 2nd session (Record B) begins to show some effect of the counter. Color changes show in the fine grain of the record (as at *a, b,* and *c*). Records C and D are segments from the beginning and end of the 3rd session. The correlation between readings on the block counter and the schedule of reinforcement now produces a marked increase in the pause after reinforcement. (There is now a "least favorable" key-color.) The running rate increases to approximately 6 responses per second at the end of Record D compared with 4 per second in Record A, even though the over-all rate falls.

This result is confirmed by the second bird in the experiment. Figure 93 gives segments from the first 3 sessions with the correlated block counter. The block counter

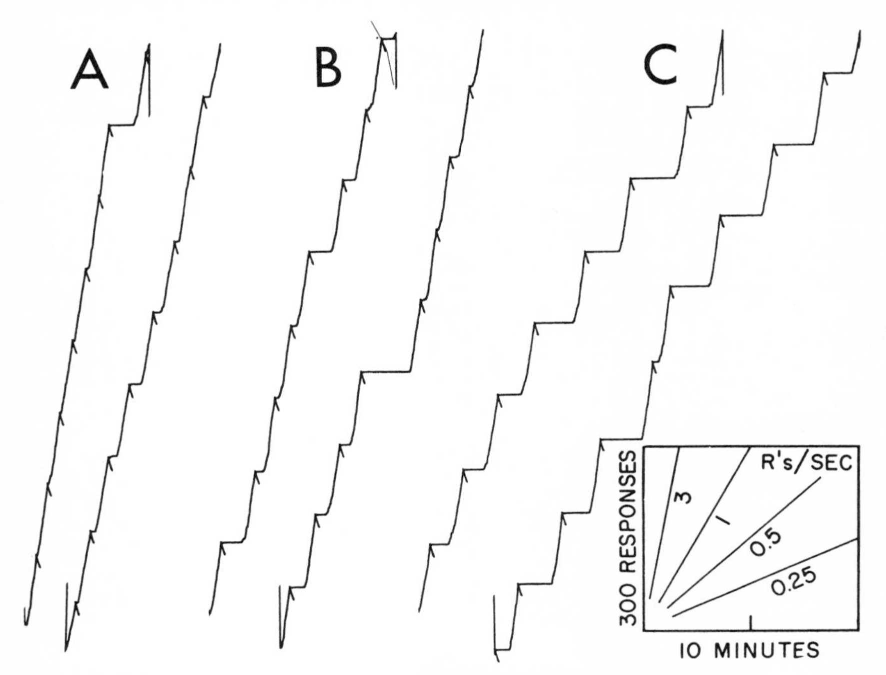

Fig. 93. Early development of correlated block counter (second bird)

(Record A) causes no immediate effect, because of the history of the uncorrelated block counter. In Records B and C, segments from the 2nd and 3rd sessions, there is a progressive increase in the pause after reinforcement, as well as the compensating increase in the running rate. The increase in the terminal rate is not sufficient to maintain the over-all rate, however.

In later sessions the pausing produced by the correlated block counter tends to disappear during the session, returning again on the following day. Figure 94A shows a sample for the bird in the preceding figure, taken after 720 reinforcements with the block counter. Not only do the pauses become short as the session proceeds, but the terminal rate increases. This pattern is the reverse of the usual order of events in a fixed ratio of substantial size, where the session customarily begins with very short pauses and a high over-all rate, which then declines during the session. Upon return-

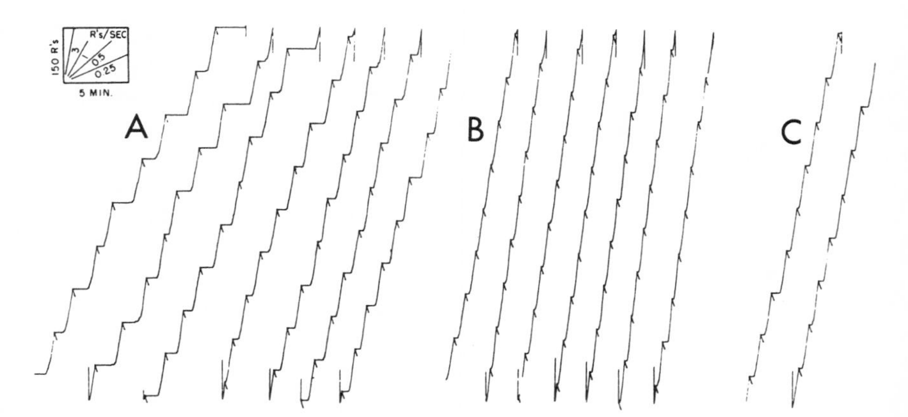

Fig. 94. Transition from FR + correlated block counter to FR

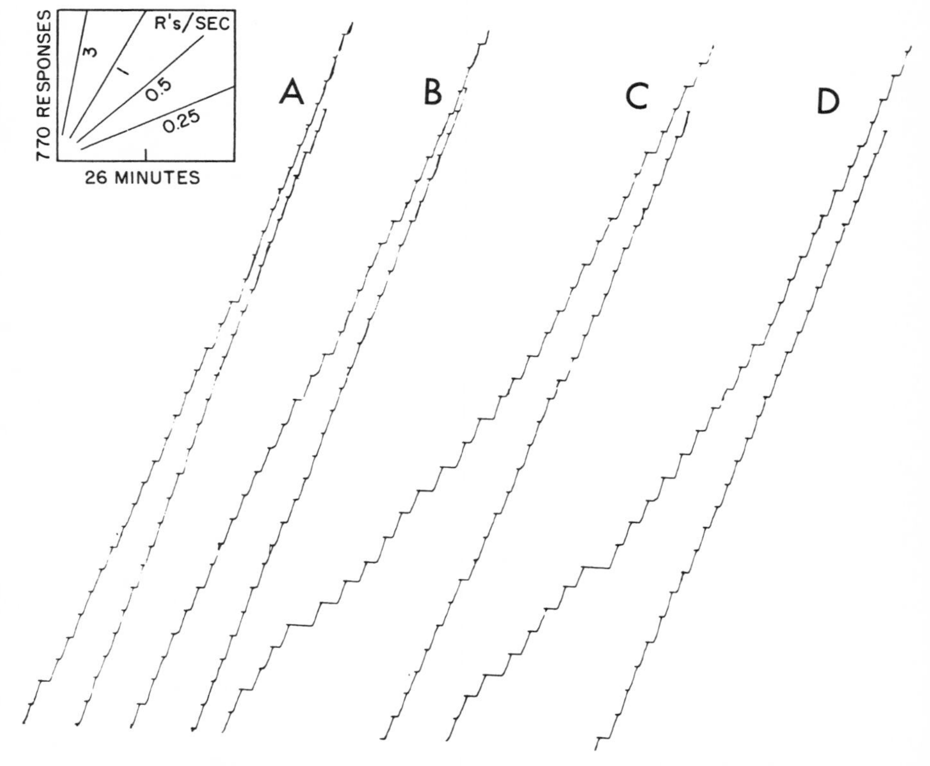

Fig. 95. Daily sessions showing progressive decrease in the pause after reinforcement

ing to the steady blue key, the birds show an immediate reduction in the pauses after reinforcement and some increase in running rate. In Fig. 94B the key is blue for the first time following the history of the block counter just discussed. The running rate is of the order of 6 responses per second compared with from 4 to 5 responses per second in Record A. This reduction is temporary, however; pausing after reinforcement and a lower running rate return. Record C shows a segment from the 6th session with the steady blue key.

One of these birds showed a continuing tendency to reduce the length of pause after reinforcement throughout the session. Figure 95 presents 4 complete sessions taken approximately 3 months after the performance shown in Fig. 94, during which the bird had been on FR 110 with a steady blue key. The other bird returned to a very stable performance under the same circumstances, with a slight pause after reinforcement and curvature at the start of the fixed-ratio segment. The last 8 sessions of this procedure for this bird are shown complete in Fig. 96, indicating very uniform performance from session to session. (Note that in Fig. 96 a faster paper speed is used to reduce the over-all slopes of the curve and to amplify local features.)

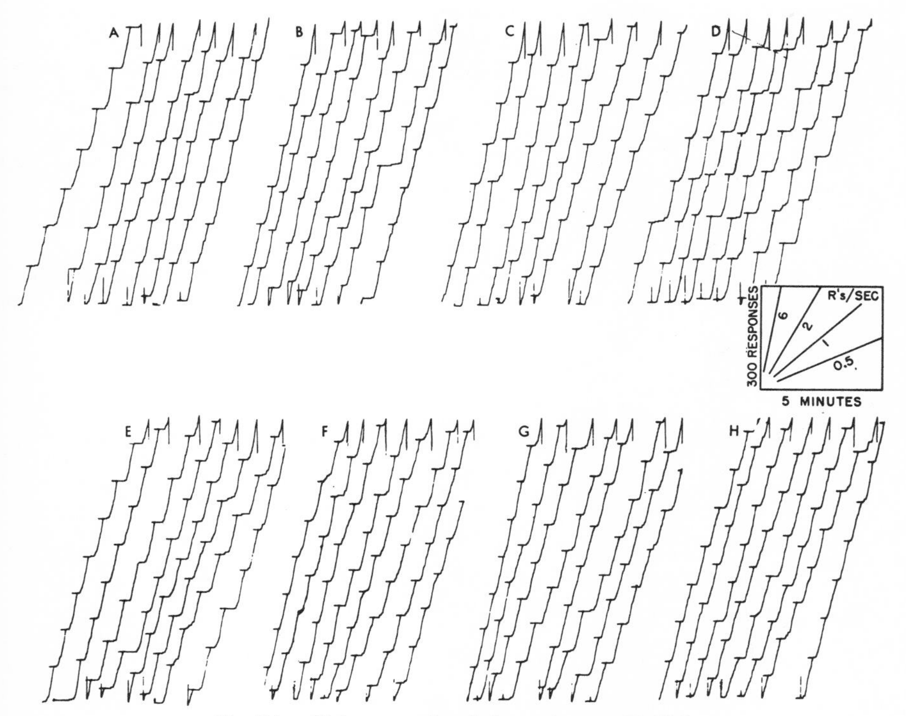

Fig. 96. Eight successive daily sessions on FR 110

Summary

The correlated block counter tends to accentuate the pause after reinforcement, while at the same time it increases the terminal rate of responding. In Fig. 96, in particular, the pause becomes remarkably uniform. Even where the size of the pause changes progressively through the session (Fig. 95), any local portion shows an unusual degree of uniformity. This characteristic may be explained as the effect of the first key-color in the ratio and its unfavorable position in the reinforcing contingencies. The block counter differs from the continuous counter in that it shows less stimulus induction between extreme readings.

TIME OUT IN THE ANALYSIS OF FIXED RATIO

A time out (TO) as an analytical tool has been described in Chapter Three. Its relevance to the study of fixed-ratio schedules is obvious. When a TO of sufficient duration occurs after reinforcement, it reduces the stimulus control from the preceding ratio performance, particularly the influence of the number of responses and rate as well as the reinforcement itself, which can be shown to be an effective stimulus for at least 30 seconds after it has occurred. When a TO is used as a probe during a fixed ratio, it interrupts current controlling stimuli in the ratio behavior. If a TO is programmed regularly, it will acquire controlling properties from its relation to reinforcing contingencies. When used as an occasional probe, however, this is unlikely. Time out was carried out in all of the experiments with the pigeons by turning out all the lights in the experimental chamber. The TO has been applied to the problem of FR in a series of experiments.

Time out after reinforcement

Early development after crf. Figure 97A shows a sample of a performance on FR 50 during the 2nd session after crf. The over-all rate is approximately 3 responses per

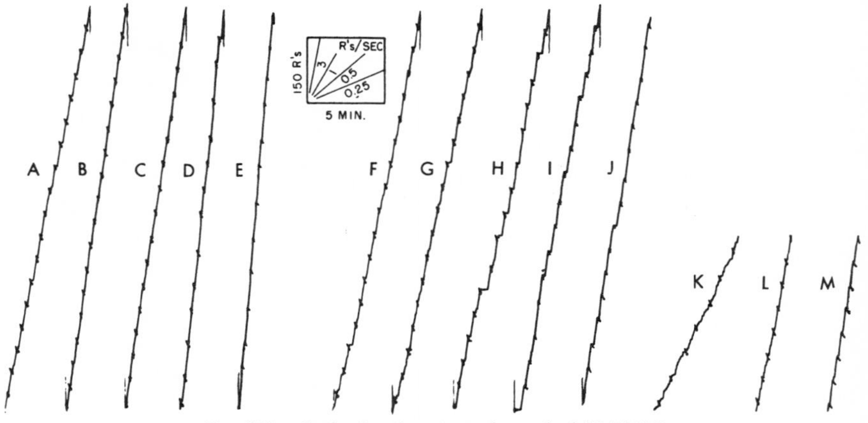

Fig. 97. Early development after crf of FR 50 TO

second; slight pauses follow reinforcement, but responding is otherwise at a single high rate. On the following day (Record B) a 1-minute TO followed every reinforcement. (The recorder was stopped during the TO.) The pause after reinforcement is immediately eliminated, and the terminal rate is slightly increased. The over-all rate is now 4 per second, a 33% increase over Record A. The next session (Record C) shows approximately the same rate; but in the following sessions (Records D and E) the over-all rate increases to almost 8 responses per second. (The counter used to program the ratio does not follow the rate accurately.)

A second bird showed the same result, though not so clearly (Records F, G, H, I, and J). Pauses still occur at the start of the fixed-ratio segment in spite of the TO, but the over-all rate nevertheless increases from slightly over 2.5 responses per second (Record F) to almost 4 per second (Record J). A similar result for a third bird is shown in Records K, L, and M, in segments from the 2nd, 3rd, and 4th sessions after crf, where the addition of a TO 1 after reinforcement in Record M eliminates the rough grain.

Time out after blocks of reinforcements. In another experiment, performances under FR and FRTO were compared within a single experimental session. Blocks of 10 ratios in which reinforcement was followed by a 3-minute TO alternated with blocks of 10 ratios without TO. A stable performance developed after several sessions in birds with a history (following crf) of 42 sessions during which the value of FR was increased from 20 to 60. Figures 98A and 98B show complete sessions for 2 birds with alternate blocks of TO. The TOs are indicated by small dots to the left of the reinforcements they followed.

The TO after reinforcement consistently produces a higher rate of responding. This may best be seen if the graph is foreshortened. The bird in Record A shows a higher rate of responding under TO. A similar increase in over-all rate in Record B is due partly to the elimination of the low rate after reinforcement.

At larger fixed ratios, TO after reinforcement may increase the pause at the start of the fixed ratio rather than decrease it as in Fig. 98. The running rate, however, may still remain higher following TO. Record A of Fig. 99 shows a performance on FR 85, 11 sessions after the performance shown in Fig. 98A. The over-all rate is lower for those segments with TO after reinforcement, but instances may be found where for brief periods the running rate is higher than in any segment without TO. (Note that over-all and local rates of responding are unusually low for a fixed-ratio performance after so much exposure to these schedules.) The second bird (Record B) showed a considerably higher over-all rate of responding and frequent pauses following the reinforcement, where there was no time out. The effect of TO here is both to increase the local rate of responding and decrease the amount of pausing at the start of the fixed ratio.

Further exposure to FR 110 TO after reinforcement produces pauses after the reinforcement, both with and without TO. Figure 100 shows a complete session on FR 110, 2 sessions after the performance shown in Fig. 99A. Pausing is most ex-

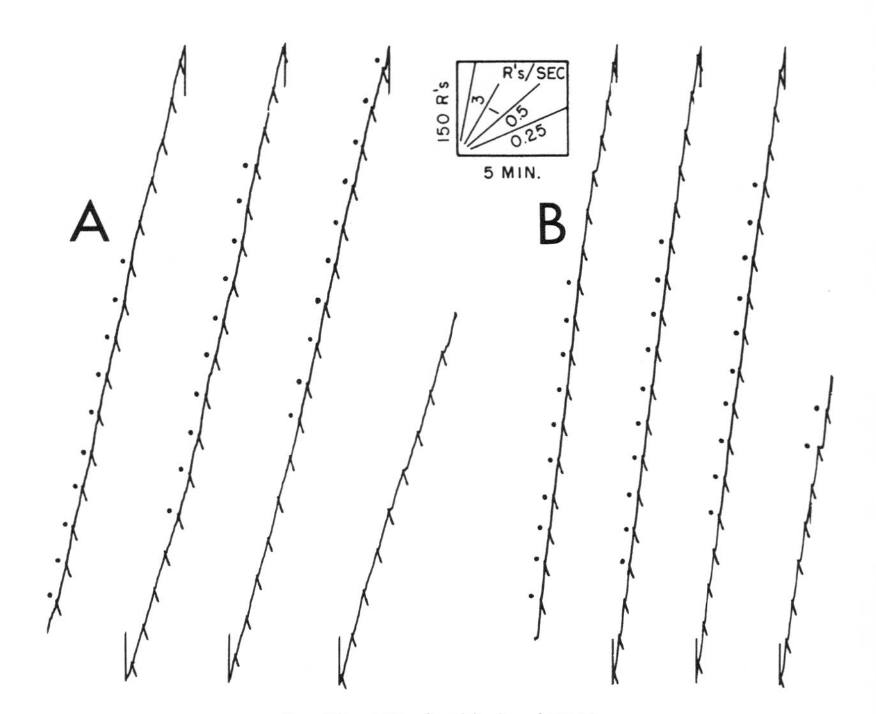

Fig. 98. TO after blocks of FR 60

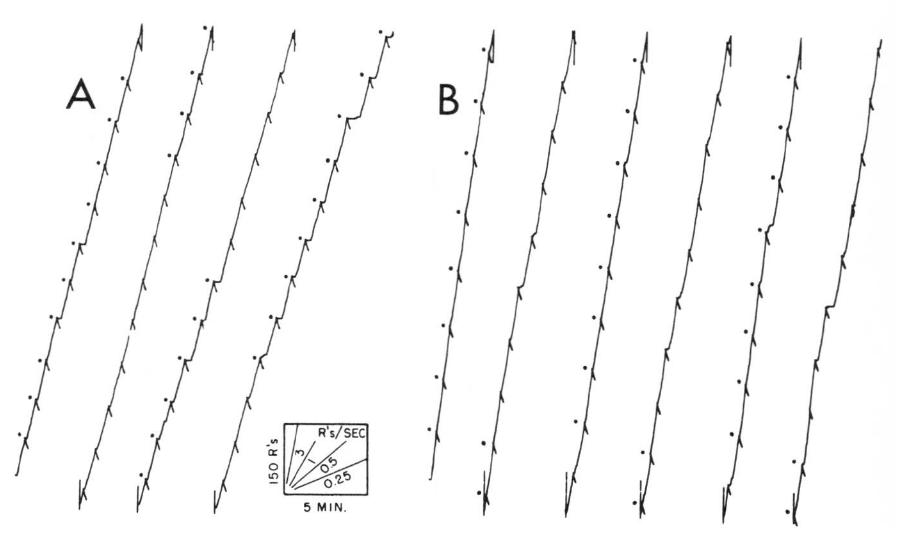

Fig. 99. TO after blocks of FR 85

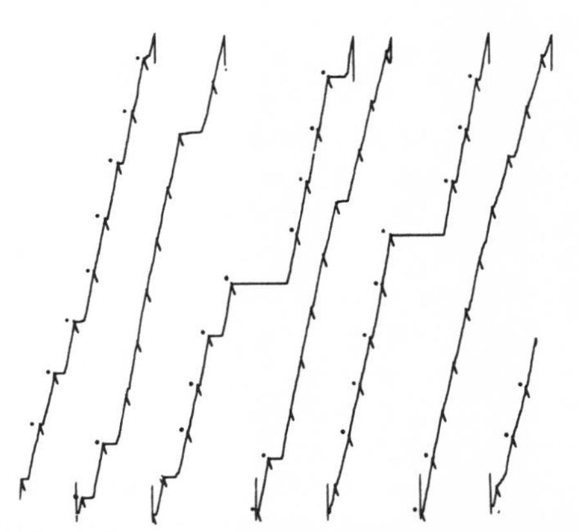

Fig. 100. TO after blocks of FR 110 showing marked pausing

treme when TOs follow reinforcements, and no pauses occur in the first few segments in each block of FR without TO. The running rate remains higher in segments preceded by TO.

Performances with and without TO were studied further by the scheduling of complete experimental sessions under each condition. Figure 101B shows the 5th session

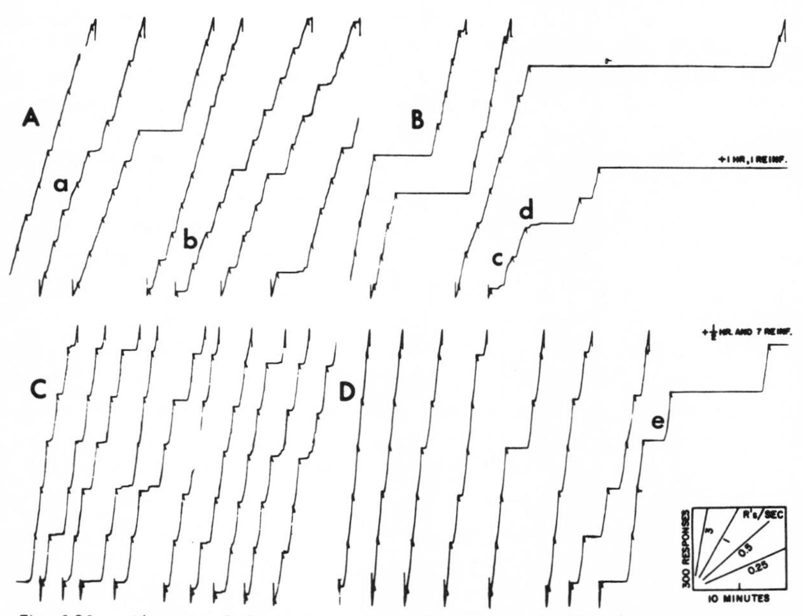

Fig. 101. Alternate daily sessions on FR with and without TO after reinforcement

with a 3-minute TO after every reinforcement. (The same bird was used as in Fig. 100.) The over-all rate of responding declines progressively during the session, mainly because of the increasing pauses after reinforcement. The session continued for more than 1 hour beyond the performance shown in the record, during which only 1 additional reinforcement was received. Marked negative curvature appears at *c*, and an intermediate rate at *d*. In the following session, which is shown in its entirety in Record A, TO was withheld. Negative curvature (as at *a* and *b*) appears more frequently than in Record B. Some pausing occurs after nearly every reinforcement, but never for more than a few minutes.

The other bird, which showed a generally higher over-all rate of responding, yielded similar results. Figure 101D shows the complete 8th session on FR 170, in which a 3-minute TO followed every reinforcement. The result is a high over-all rate, which declines during the session because of an increasing tendency to pause after reinforcement. A 30-minute period at the end of the session is not shown, in which only 7 ratios were completed. Some positive acceleration during the early part of a ratio becomes pronounced toward the end of the session (as at *a*). When TO was withheld in the following session (Record C), pausing develops early in the session; but it continues at approximately the same level without progressive increase, as in Record B. The terminal rate of responding is approximately 6 responses per second in Record A and 8 per second in Record B.

Where TO after reinforcement produces marked pausing on a short ratio, eliminating the TO often reduces the pause dramatically. Figure 102 shows this effect. Record A is the final part of a session showing marked pausing with a TO after all reinforcements. Just after the start of the following session in Record B, the elimination of the TO at the arrow eliminated the pause except in the 1st segment, when it was a few seconds long.

The effect of an extraneous stimulus on an FR performance with time out after blocks of reinforcements. In 1 experimental session in which TOs occurred in alternate blocks of 10 reinforcements (see Fig. 97, 98, and 99), a loud buzzer was sounded continuously. This extraneous (and possibly aversive) stimulus disrupted the performance during the segments preceded by TO, but it had little effect upon the performance when TO was omitted. Complete sessions for the 2 birds are shown in Fig. 103. Both records are from the 8th session on the alternate TO procedure (2 sessions after Fig. 98). The over-all rate falls by a very large factor when TOs follow reinforcements, but it is very largely unaffected in the absence of TO. The only deviations from a normal fixed-ratio performance occur at *c* and *d*, where the ratios begin in a normal manner but show a marked decline in rate before the end. This is the same kind of deviation produced after TO. Segments tend to begin with a high rate. The rate then sharply declines to near zero until reinforcement (as at *a*). Segments which begin with a pause or low rate may show an S-shaped curve, as at *b*. In general, the performance during each fixed-ratio segment suggests an extinction curve, but the over-all pattern is extremely ragged and the grain rough. Record B also shows a very gross difference in

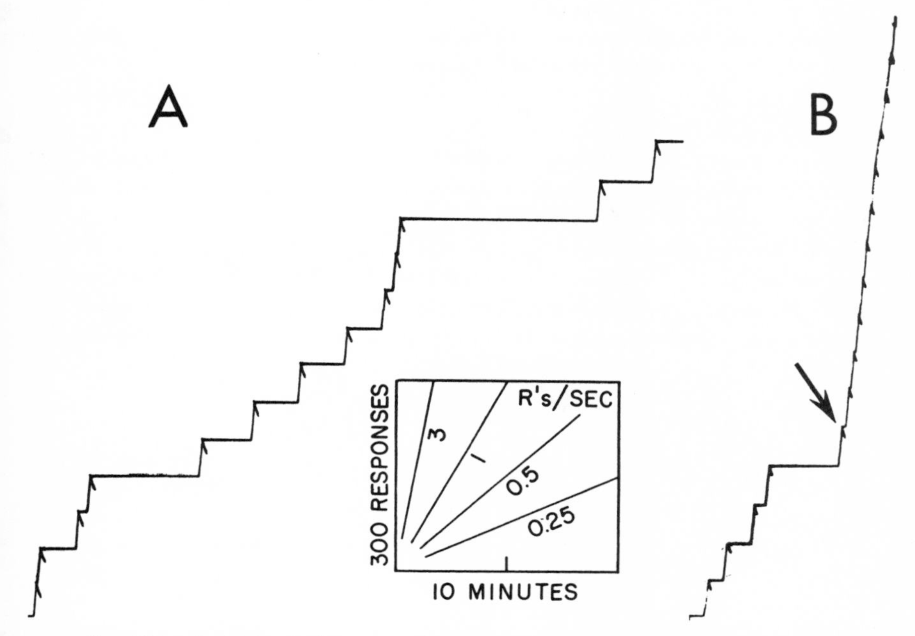

Fig. 102. Reduction in the pause after reinforcement by the removal of the TO

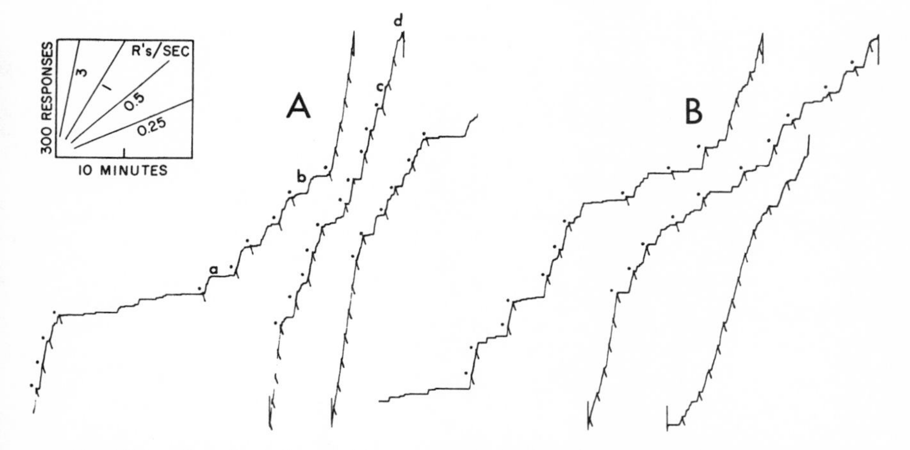

Fig. 103. The effect of a loud buzzer on FR with TO after blocks of reinforcements

the over-all rate between fixed-ratio performances with and without TO after reinforcement, although the character of the disturbance is not so uniform.

Time-out probes

Occasional probes. The effect of occasional TO probes on fixed-ratio behavior is not well enough understood for them to be classified. The following figures, however, present some examples of the kinds of effects.

Figure 104 shows the effect of three 8-minute TO probes on a well-developed performance on FR 60. This bird had a history of 33 sessions on fixed ratios of from 20 to 60. Probes had been used in the preceding session. The TO at *b* followed reinforcement; those at *a* and *c* occurred a few responses after the start of the fixed-ratio seg-

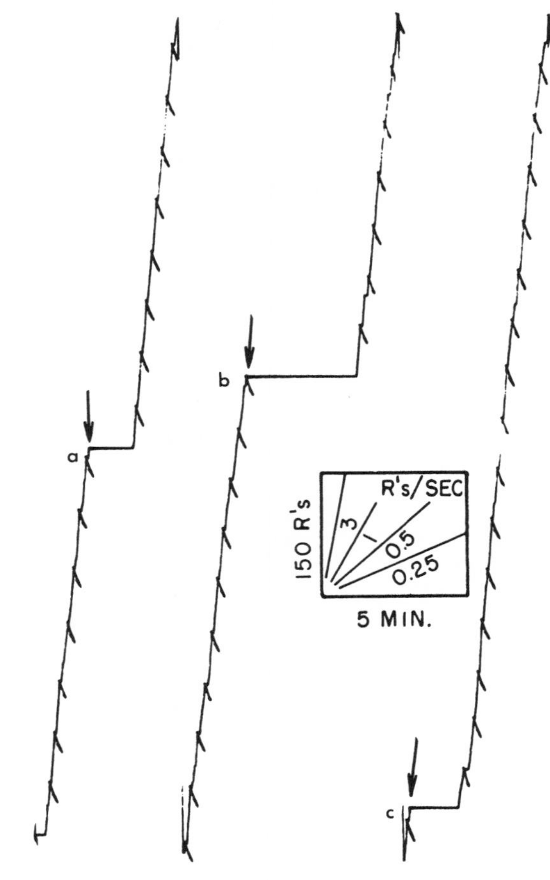

Fig. 104. TO 8 probe of FR 60 producing pausing

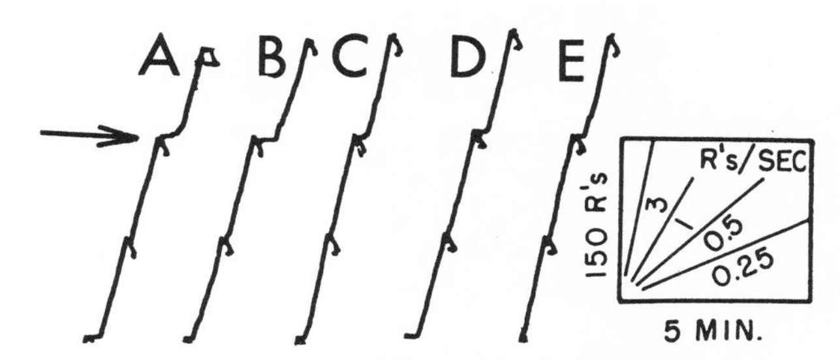

Fig. 105. TO 2 probe of FR 90 producing pausing

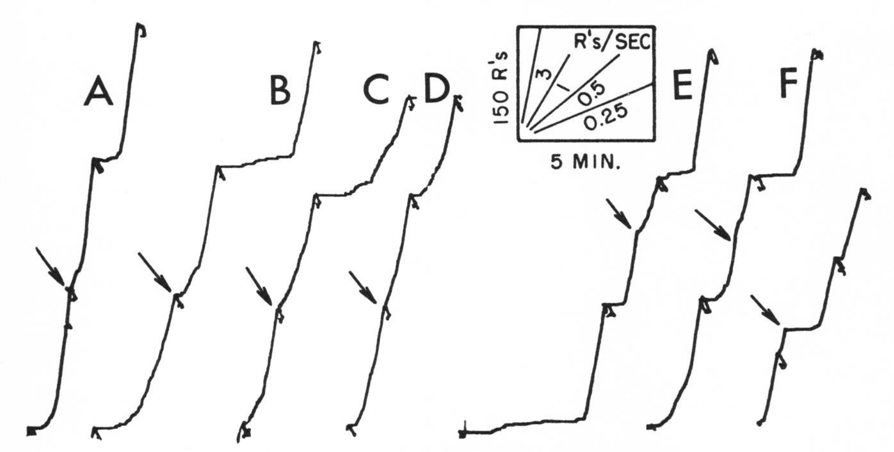

Fig. 106. TO 2 probe of FR 160 reducing pausing

ment. The fixed ratio is being held well, with no pause after reinforcement and good grain. The TO produces a pause in each case, varying from 1.5 minutes at *a* to 4 minutes at *b*. After the pause, responding begins at the terminal fixed-ratio rate, and the segment is run off in a normal manner.

Figure 105 shows 5 segments from the 2nd session on FR 90 after 190 hours on tand *FR*FI, and shows a similar effect of a 2-minute TO after reinforcement. The TO followed every 10th reinforcement. Two ratios preceding the TO and one following it are shown for all TOs in the session. In the 1st segment the TO is followed by a 30-second period at a low rate, with acceleration to the terminal rate. The other TOs produce a pause of from 15 to 30 seconds.

Figure 106 shows the behavior of a second bird in the early stages of FR 160 following tand *FR*FI. Records A, B, C, and D are sample segments from several sessions in which 2-minute TO probes followed occasional reinforcements. In Records A, C, and D the TO eliminates the pause completely, while in Record B it shortens it. In Records E, F, and G the TOs occur during fixed-ratio segments. In Records

E and G the result is a drop to a lower rate followed by acceleration to the terminal rate. In Record F a short burst at the terminal rate immediately follows the TO, but the rate falls to an intermediate level for a brief period.

Figure 107 shows 2 successive sessions on FR 40, beginning 12 sessions after continuous reinforcement, in which a 2-minute TO followed every 11th reinforcement. The only effect in Record A is a slight lowering of the rate after the 3rd and 5th TOs, at *a* and *b*. In the following session (Record B) all TOs except the 2nd have a marked effect. At *c* a pause occurs near the end of the ratio; at *d* the rate drops from 3.75

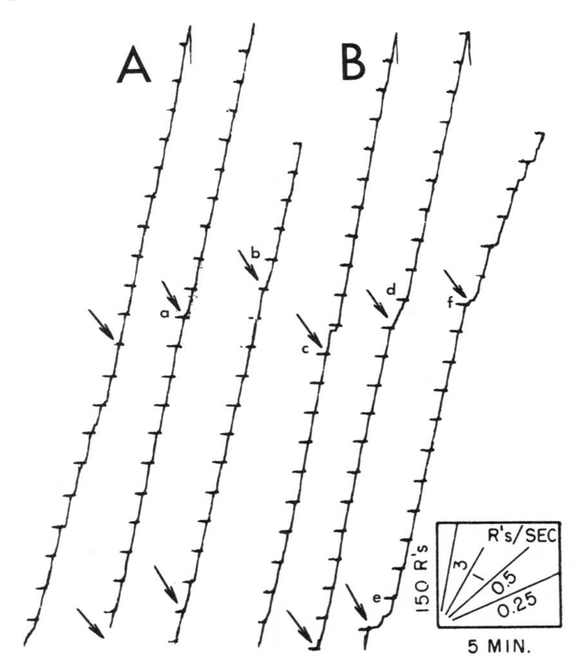

Fig. 107. TO 2 probe of well-developed FR 40

responses per second to approximately 2 per second for the whole ratio; at *e* a cusp follows a substantial pause and acceleration after TO; and at *f* a slight pause appears. The performance then deteriorates.

Figure 108 shows the complete 8th session on FR 50 after crf, in which every 11th reinforcement is followed by a 2-minute TO (indicated by vertical dashed lines above reinforcements). The considerable pausing after reinforcement and the low terminal rate show that this is not a final performance for this size of fixed ratio. It is nevertheless fairly stable at this stage. In every case except at *d* the TO reduces the pause after reinforcement, and in most cases (as, for example, at *a*, *b*, and *c*) eliminates it.

Figure 109 shows 2 fixed-ratio segments before and 3 segments after each TO in a

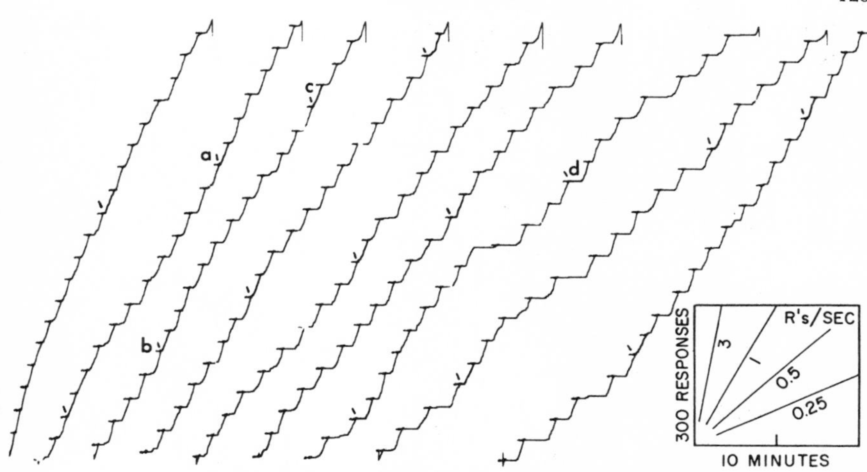

Fig. 108. Effect of TO 2 on FR 50 showing pausing

later series with the bird in Fig. 108. The character of the fixed-ratio behavior is now quite stable and is fairly typical, with either a pause of several seconds followed by a very slight and rapid acceleration to the terminal rate, or an immediate start after reinforcement with immediate or very rapid assumption of the terminal rate. Almost without exception, an 8-minute TO eliminated the pause following reinforcement. It often reduced the amount of pausing after the *next* reinforcement as well. The terminal rate remained unchanged.

Continuous sample of time-out probes after reinforcement and during the fixed ratio. A series of 25 sessions with 2 pigeons was arranged as follows: an 8-minute TO followed the 5th reinforcement of the session; a 30-second TO, the 25th response after the 13th reinforcement; and an 8-minute TO, 25 responses before the 20th reinforcement of the session. This arrangement was an attempt to compare the effect of duration of TO both after reinforcement and during the ratio.

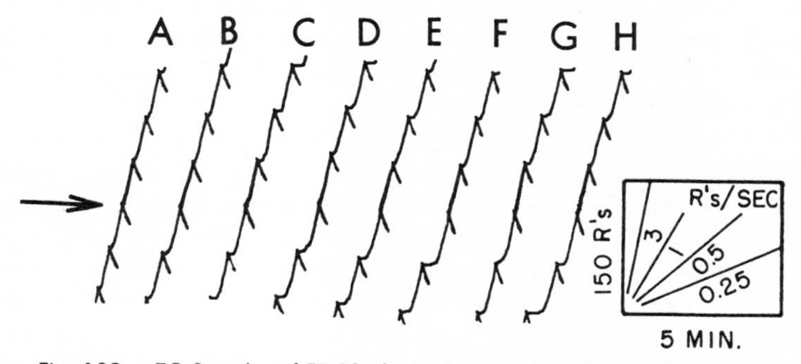

Fig. 109. TO 8 probe of FR 50 eliminating pausing after reinforcement

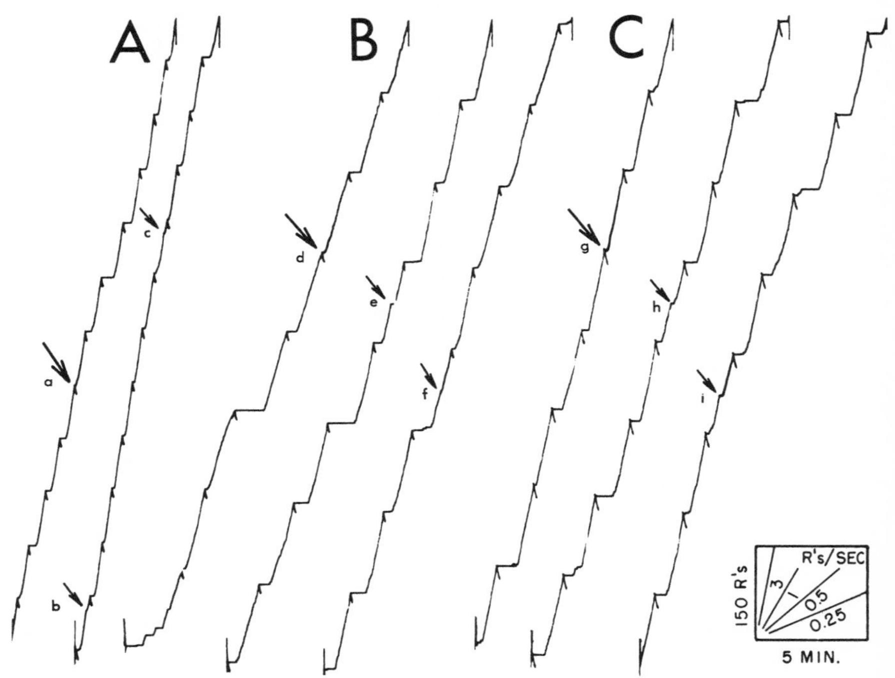

Fig. 110. Continuous sample of TO after reinforcement and during the ratio

Figure 110 shows the performance of one bird in the 11th, 16th, and 25th sessions under this procedure on FRs of 85, 125, and 125, respectively. This bird had been on a schedule similar to that in Fig. 107, 108, and 109, in which a short FR was probed by a TO after every 11th reinforcement for 22 sessions. The 8-minute TO following the 5th reinforcement of each session eliminates or reduces the pause after reinforcement (*a, d,* and *g*) at the large arrows. The 30-second TO occurring 25 responses before the 12th reinforcement produces a slight cusp at *b,* and a more pronounced pause and acceleration at *e* and *h.* At *c, f,* and *i* the 8-minute TO, 25 responses after the 20th reinforcement, produces a slight shift in rate at *c,* has almost no effect at *f,* and produces a pause and abrupt shift to a higher rate at *i.*

Figure 111 shows the second bird on this procedure. Two excursions of the recorder are shown for the 6th session, and segments from 3 other adjacent sessions were added. In Record A the 8-minute TO after the 5th reinforcement eliminates the pause after reinforcement which occurs at this size of fixed ratio. There is no effect on the subsequent ratio. At Record B the 30-second TO occurring 25 responses before the 13th reinforcement produces a slight cusp and a decline to a lower rate, similar to the effect occurring in Fig. 110 at *c, e,* and *h.* Record C shows the 8-minute TO occurring 25 responses before the 20th reinforcement. A cusp and lower terminal rate of re-

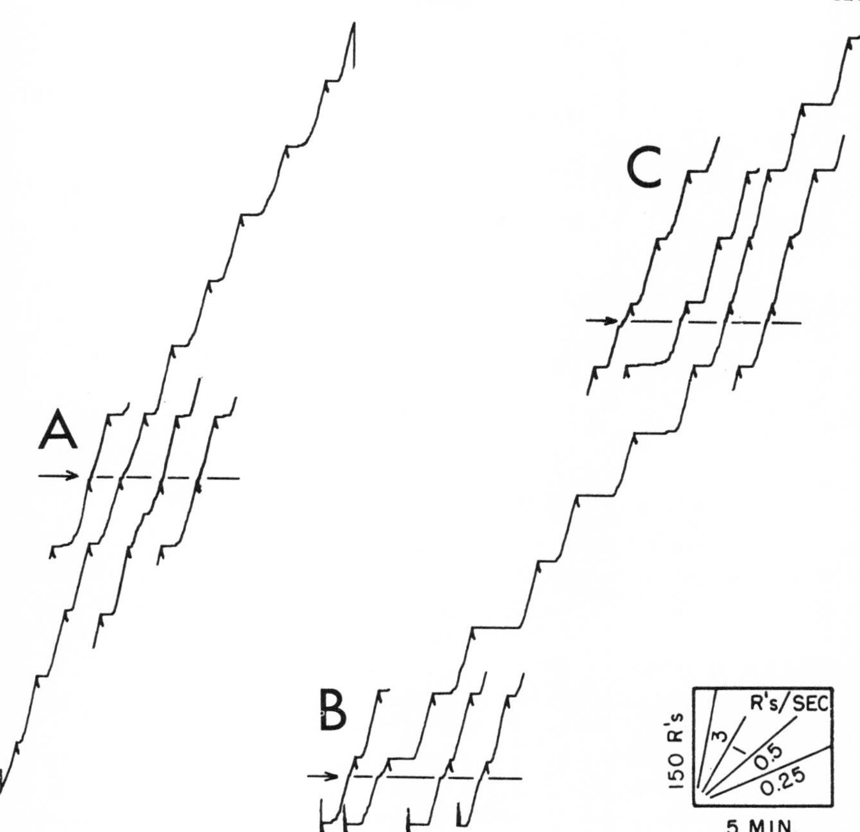

Fig. 111. Continuous sample of TO after reinforcement and during the ratio (second bird)

sponding occur in 2 cases, while only a slight shift to a lower rate occurs in the others. Figure 112 shows a complete series of TOs on the same procedure (for the same bird as that in Fig. 111) 13 sessions after the size of the ratio was increased to 120. The fixed-ratio segments immediately preceding and following the TOs are presented for the following session also. In Record A the 8-minute TO following the 5th reinforcement still characteristically reduces or eliminates the pause, as at *b*; but occasional instances occur in which there is little if any effect, as at *a*. In Record B the 30-second TO during the fixed ratio produces a slight cusp at *f* or a more marked effect at *g*. The 8-minute TO in Record C, however, frequently has no effect, as at *h*, or produces a pause, as at *k*. The large pause at *k* is from a session during which there was great difficulty in maintaining a performance on this size of fixed ratio and in which long pauses occurred following reinforcement. At this stage in the schedule, moreover, intermediate rates occur (for example, at *i* and *j*), as well as instances of smooth posi-

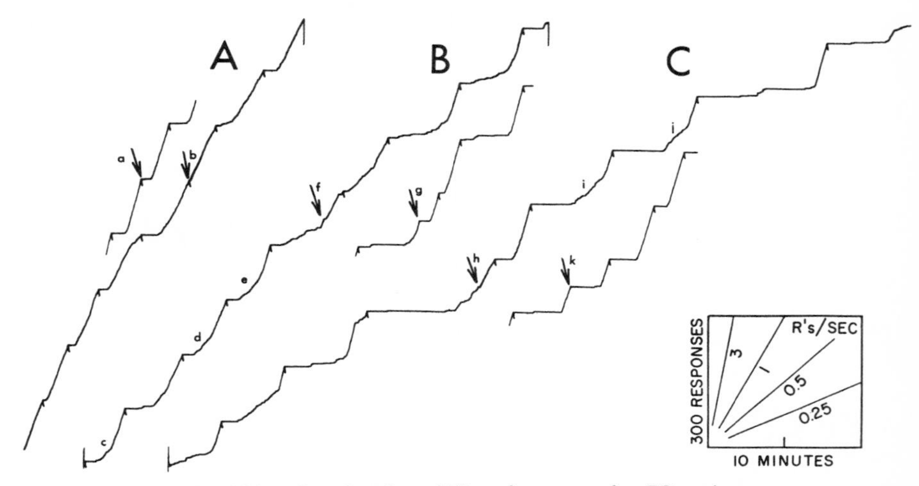

Fig. 112. Deterioration of FR performance after TO probes

tive acceleration (at *c, d,* and *e*). This deterioration in the fixed-ratio performance is probably due to the punishing effects of TO, similar to the effect in percentage reinforcement with TO.

An 8-minute TO after reinforcement eliminated or considerably reduced the pause in all of the cases studied. The 30-second TO occurring before the end of the 13th fixed ratio almost always produced a cusp and decline to a lower rate; but an 8-minute TO in a similar position later in the session frequently had little effect.

Summary

Under some conditions a TO after reinforcement will consistently eliminate the pause characteristically observed in a ratio performance. This effect will obviously not be observed if the performance shows no pausing; it is likely to be obscured by pauses due to other causes which may resemble the characteristic ratio pause.

Under some conditions, however, a TO *produces* pauses in a fixed-ratio performance. Introduced at times other than after reinforcement, it may produce some slight pausing and scalloping before returning to the normal high rate of responding. Some collateral effects of TO which may explain these are discussed in Chapter Three.

When the TO reduces a characteristic pause, the effect is clear and unambiguous. Failure to get this result, however, does not mean very much in our present state of knowledge.

The effect of time out on FR 875

Chapter Eight shows that after a history of tand FI 45 FR, in which the tandem ratio is continuously increased from 10 to 400, 2 pigeons were able to respond in a sustained fashion on FR 875. Early performances on FR 875 with TO 15 after rein-

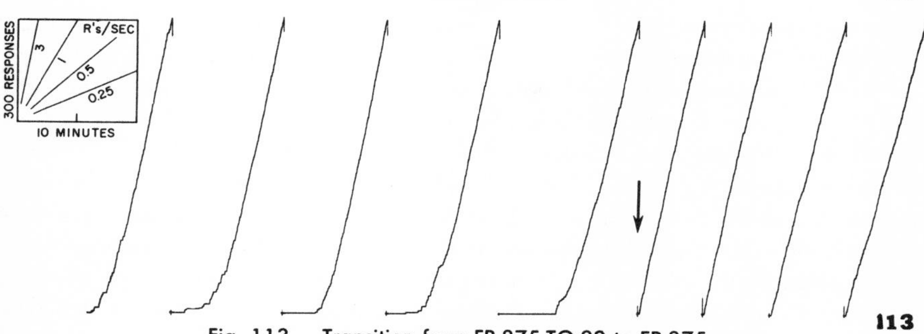

Fig. 113. Transition from FR 875 TO 20 to FR 875

forcement are described in Chapter Eight in Fig. 522. The beginning of Fig. 113 shows a stable performance in one bird after a total of 80 reinforcements on FR 875. (The recorder resets during reinforcement.) A terminal rate of 3 responses per second is reached after an acceleration extending roughly over the first 200 responses. Measurement of the pause before the 1st response in the last 51 fixed-ratio segments with TO after each reinforcement showed 21 cases of a 1st response within 2 minutes, 11 cases of a response within 4 minutes, and 17 cases in which the 1st response occurred more than 4 minutes after the termination of the TO. The time required to complete the entire fixed-ratio segment follows the same pattern. Out of the 51 cases, 20 ratios were run off within 10 minutes; 16, within 15 minutes, and 10, between 15 and 25 minutes. At the arrow in Fig. 113 the TO was discontinued. The bird begins to respond at the terminal rate immediately after reinforcement and maintains that rate until the next reinforcement. Twenty-five further reinforcements were received in the remainder of the session not shown in the figure. In each case, responding began almost immediately after reinforcement and continued at a constant rate until the next reinforcement. At a later stage, long pauses with prolonged acceleration to the terminal rate occasionally occur. Figure 114 shows a segment from the following ses-

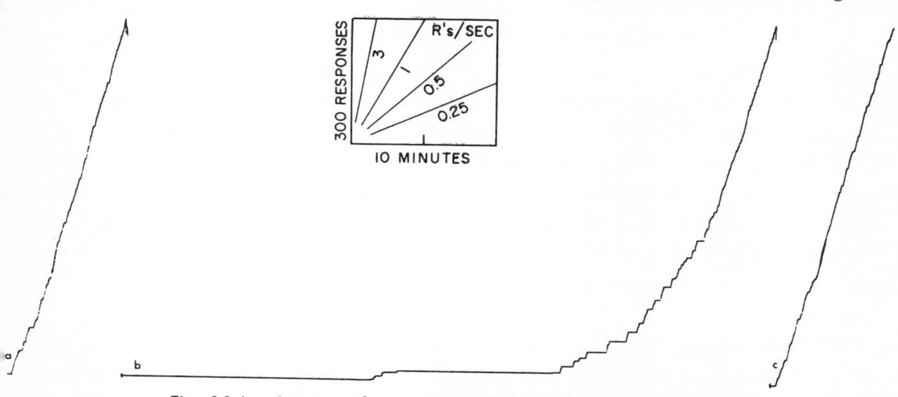

Fig. 114. Later performance on FR 875 after FR 875 TO 20

sion, 75 reinforcements after the last reinforcement shown in Fig. 113. A long pause appears at *b*; but frequent instances still occur in neighboring segments where the bird assumes the terminal rate immediately after reinforcement and maintains it throughout the ratio, as at *a* and *c*. Of 170 reinforcements recorded under FR 875 without TO for this bird, only 13 were followed by pauses and acceleration, as at *b*.

The final performance for the second bird under FR 875 with a TO 15 after reinforcement is also shown in Fig. 523, discussed in Chapter Eight. The pause after reinforcement was measured for the last 35 fixed ratios with TO after reinforcement. In 32 instances it was more than 20 minutes long, and in the remaining 3 cases was over 10 minutes. The first omission of TO occurred at the beginning of the session shown in Fig. 115. The first fixed-ratio segment has been omitted from the record because

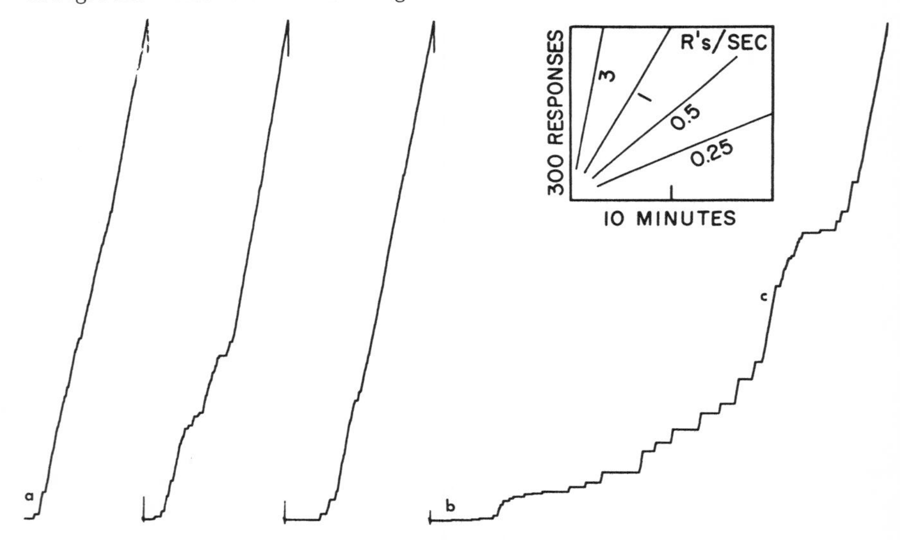

Fig. 115. Later performance on FR 875 after FR 875 TO 20 (second bird)

the start of the session functions as a TO. It begins after the usual pause and slow acceleration to the terminal rate. After the first reinforcement not followed by TO, a pause of less than 1 minute (at *a*) occurs. Responding begins almost at once at the terminal rate and continues until the next reinforcement. The 4th segment, beginning at *b*, shows intermediate rates of responding and a marked knee at *c*. The terminal rate is sustained for only the last 200 responses of the ratio. During the rest of the session (not shown) the pausing increases progressively, although very few instances occur where it reaches the order of magnitude observed with TO after reinforcement.

The remainder of this performance for this bird on FR 875 without TO was characterized by groups of ratios run off with little or no pause, and very rapid assumption of the terminal rate interspersed with instances where pauses as long as 6 hours occurred, followed by prolonged acceleration. When 15-minute TOs were again introduced after reinforcement, the performance returned to normal as in Fig. 523 (Chapter Eight).

Summary of time out on FR 875

Both birds confirm the effect of the time out. With the time out after every reinforcement, consistent pausing and a period of acceleration to the terminal rate occur. Without the time out one bird shows groups of ratios run off with no pause or acceleration, interspersed with long pauses and slow, rough grain extending through most of the ratio. The other bird shows a similar effect of the removal of the time out, except that instead of the groups of ratios being run off with no pause following

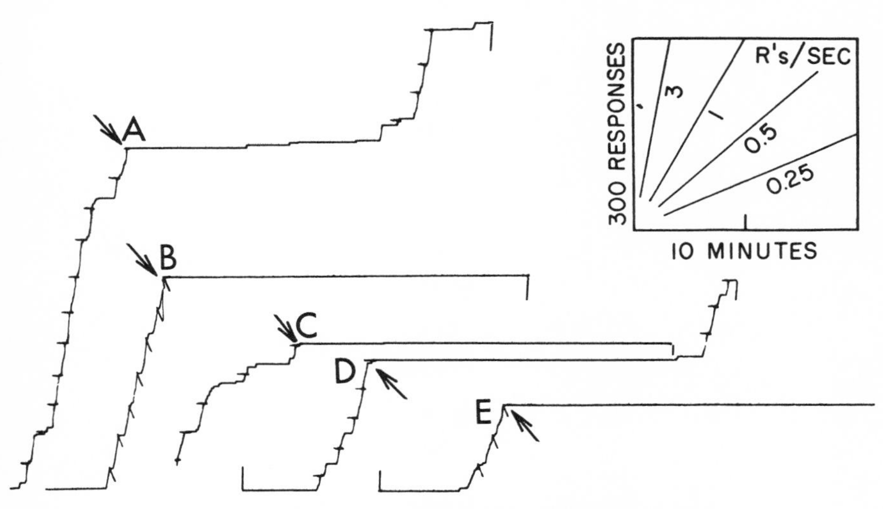

Fig. 116. Effect of TO on FR with insufficient reinforcement

the reinforcement, the amount of pausing is considerably less than with time out. Also, the long pauses and slow accelerations to the terminal rates, which are interspersed in the more rapidly run-off ratios, follow the longer pauses (as long as 6 hours). The omission of the time out in the large fixed ratio is similar to the effect in large fixed-interval schedules. Here, the carryover from the preceding interval produces immediate responding after the reinforcement instead of the appropriate pause.

The effect of time out on short fixed ratios with an insufficient amount of reinforcement

Figure 116 shows instances in which TO probes of an *early* fixed-ratio performance were followed by long pauses. Records A and C are for one bird, 5 sessions after crf; and Records B, D, and E are for another bird, 3 sessions after crf. An undeveloped fixed-ratio performance and the insufficient reinforcement are suggested by the intermediate rates and pauses that occur even after the bird begins to respond at the terminal rate.

A similar result has already been described in Fig. 39, where the first and last TO produced a long pause. Other TO probes in Fig. 38 and 40 had, if anything, the opposite effect.

Chapter Five

• • •

FIXED INTERVAL[1]

INTRODUCTION

IN A FIXED-INTERVAL SCHEDULE of reinforcement (FI) the first response after a designated interval of time is followed by a reinforcing stimulus. It is programmed by a timer which starts from zero after each reinforcement (or from the start of the session) and closes a circuit ("sets up" a reinforcement) at the end of a designated time. The first response following this period operates the magazine.

Since the reinforced response may occur some time after the reinforcement is set up, we have the option of timing the next interval from the reinforcement or from the end of the preceding interval. By timing from the end of the previous interval, we maintain the designated fixed interval as an average about which the actual intervals will be distributed. If the interval is timed from the last reinforcement, none of the intervals will be less than the designated fixed interval. Some will be larger, however, and the average interval will exceed the designated interval. In practice, there is little difference between these procedures, since fixed-interval schedules normally generate a substantial rate of responding at the time of reinforcement, and the reinforced response usually occurs within a second or two of the designated fixed interval. In all the experiments reported here the fixed interval is timed from the reinforced response, so that no instance occurs of a response being reinforced after less than the designated fixed interval.

CONTINGENCIES RESULTING FROM FIXED-INTERVAL REINFORCEMENT

Differential reinforcement of low rates

In Chapter Four we pointed out that fixed-ratio schedules increase the number of instances where a reinforced response is immediately preceded by another response or group of responses rather than by a pause. This indirect contingency results in a differential reinforcement of high rates. A fixed-interval schedule has the opposite effect of differentially reinforcing responses following pauses—that is, lower rates of responding. In the fixed-interval schedule, reinforcement is programmed by elapsed time;

[1] This is the "periodic reinforcement" of *The Behavior of Organisms*.

therefore, the longer the interval since the last response, the more likely the next response is to be reinforced. This contingency operates only when responses occur in groups. In both fixed-ratio and fixed-interval schedules the indirect contingency would not be felt if the rate were constant. And in both schedules, as we shall see, the stable performance shows a nearly constant rate in the region of the reinforced response; thus, this indirect contingency can be appealed to only during the development of such a performance or in transitions from one schedule to another.

Rate of responding at the moment of reinforcement

As will be seen later, a fixed-interval schedule normally generates a stable state in which a pause follows each reinforcement, after which the rate accelerates to a terminal (usually moderate) value. In the stable state a given rate of responding is correlated with reinforcement. Moreover, another (low) rate is correlated with nonreinforcement at and near the beginning of the interval. Although reinforcement does not always occur whenever the rate reaches its terminal value, that rate of responding nearly always accompanies reinforcement. Occasional instances where the bird does not respond throughout the interval and is reinforced for a single response weaken this correlation. The extent to which this contingency is a controlling factor in the fixed-interval performance depends on the stage of development of the performance.

Number of responses emitted in the fixed interval

In a quite stable performance, in which the rate is low or zero after reinforcement but increases in a standard fashion as the interval elapses, the number of responses emitted at the reinforcement should be fairly constant. Since this condition depends in a sensitive way upon the "triggering" of the acceleration to a higher rate, the number of responses at reinforcement will usually vary considerably. If either the number of responses emitted since reinforcement or the rate of responding at reinforcement is an important factor in fixed-interval contingencies, we must allow for the chaining of responses. Thus, if a given number of responses characteristically precedes reinforcement, any responding which brings the actual number closer to this value should be reinforcing. Note that any such effect should produce an oscillating state, since the increase in rate due to chaining should upset the number contingencies responsible for chaining. A similar argument may be made for rate.

Time since the preceding reinforcement

This is constant in a fixed-interval schedule, but it probably acts through mediating behavior. The responding discussed in the preceding paragraph is the principal example of such mediating behavior.

The reinforcement as an occasion for nonreinforcement

The stimuli associated with the presentation of a reinforcer and with the appropriate consummatory behavior (eating, cleaning, etc.) enter into the fixed-interval contingencies in an important way. Because they constitute an occasion upon which a response is never reinforced, a low rate quickly develops immediately after reinforcement. The duration of this control is in part a function of the temporal properties of the stimuli. Residual stimuli—from food in the mouth, swallowing, etc.—may extend past the moment of reinforcement. Other behavior may be set in motion (e.g., washing for the rat) which may also control a low rate of responding because of its relation to nonreinforcement. Very roughly speaking, the effect of reinforcement as a stimulus of this sort appears to last about 30 seconds for the pigeon. The effect is to start the new interval with a period of zero or a very low rate of responding.

TRANSITION FROM CONTINUOUS REINFORCEMENT TO A FIXED-INTERVAL SCHEDULE

Introduction

When a bird which has previously been reinforced on continuous reinforcement (crf) is placed directly on a fixed-interval (FI) schedule, a characteristic performance is usually observed. Figure 117 illustrates the important features.

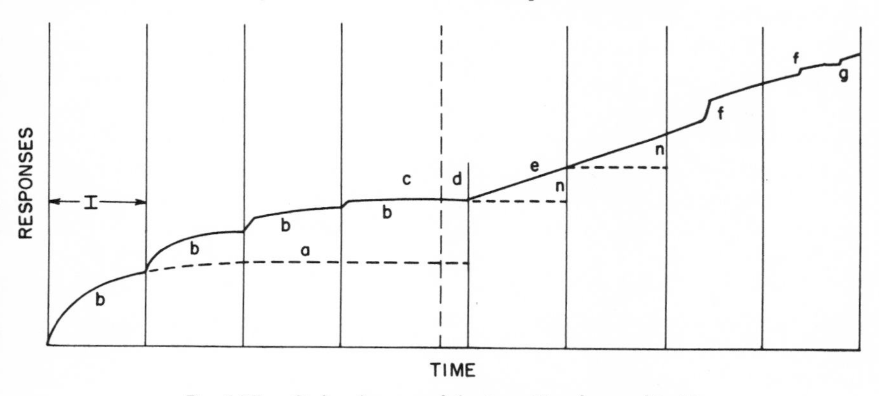

Fig. 117. Stylized curve of the transition from crf to FI

1. The preceding crf produces a negatively accelerated "extinction curve," suggested by the dashed curve at *a*. These curves are quite variable with respect to the number of preceding reinforcements, as unpublished experiments by W. H. Morse have shown. When the preceding crf has a substantial effect, however, the first part of the fixed-interval performance is negatively accelerated. The rate usually reaches a low level of responding, as at *c*, which is considerably below the rate which will eventually be maintained by the fixed-interval schedule of reinforcement.

2. Reinforcements are now being received on schedule at each vertical line. (Reinforcements are usually delayed at this stage because the bird is not responding when a reinforcement becomes available, as in the interval at *c*. The next interval, at *e*, is timed from the reinforcement *d*.)

Each such reinforcement is generally followed by an increase in rate, and the interval is usually marked by a small negatively accelerated segment (*b*, *b* . .). The larger negative acceleration attributed to extinction is combined with these smaller curves. The curvature is not often very uniform because of the rough grain usually prevailing at this stage.

3. A fairly uniform rate of responding emerges during an interval and from interval to interval, as at *e*. This constant rate seems to develop regardless of the size of the interval, and is presumably due to the special probability of reinforcement at low rates arising from the contingencies up to this point.

4. Because of this uniformity in rate, the number of responses at reinforcement becomes fairly constant. This condition appears to produce the fourth characteristic—occasional brief runs at higher rates, as at *f*. These are not the compensatory high rates which follow pauses (*g*). The rates are often as high as in early ratio performances. (See Chapter Four.) No instances of rates this high have been observed up to this point, nor, of course, have such rates ever been reinforced. The brief runs appear, therefore, to be due to the automatic reinforcement resulting from progress toward the number of responses characteristically prevailing at reinforcement. Since such a run destroys the constancy of this number, the situation is unstable.

When a pigeon is placed directly on FI after crf, the performance will depend in part upon the amount of extinction resulting from the crf, the distribution of inter-response times or the tendency for pecks to occur in bursts (the "grain" of the record), and the accidental contingencies resulting from the interaction of this performance with the schedule of reinforcement. Seventeen pigeons, after varying amounts of crf, were placed directly on FI schedules ranging from 1 to 45 minutes. Eleven of these transitions will be described to show particular features.

Transitions

FI 1 and FI 2. Figure 118 shows a performance under FI 1 after crf which begins with only a small extinction curve (*a*). (The 4 lines of the figure were recorded continuously.) The lowest rate of responding occurs at *b*. Then the rate becomes fairly constant in the region *c*. A short burst of responding at a higher rate occurs at *d* and much later at *e* (cf. Fig. 117*f*). The over-all rate during the session remains low, and the grain of the record is rough. There is already some tendency toward a lower rate just after the reinforcement (e.g., at *f*, *g*, *h*, *i*, and *j*).

Figure 119A shows the 1st session of a transition, crf to FI 1, which begins with a much higher initial rate of responding. The first 3 intervals average about 100 responses per interval. The rate then falls. At *a* and *b* the reinforced responses occur after pauses. Unfortunately, the curve is too short to show other features; but in the following session the over-all rate continued to decline, and it reached a stage similar to the last segment of Fig. 118.

Record B in Fig. 119 shows a similar transition to FI 2 after crf. The over-all negative curvature at the beginning of the session shows the effect of the previous con-

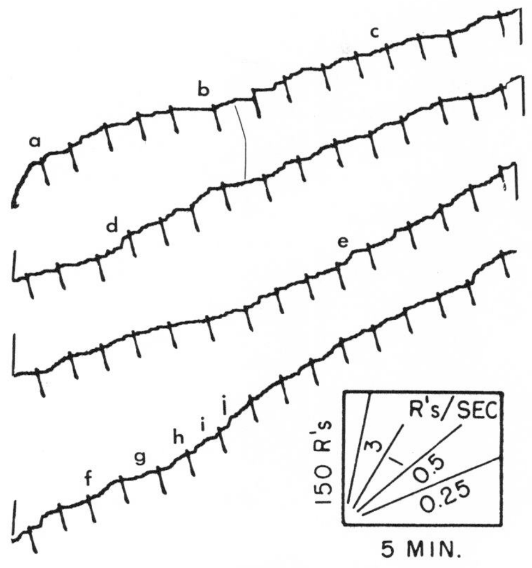

Fig. 118. Transition from crf to FI 1

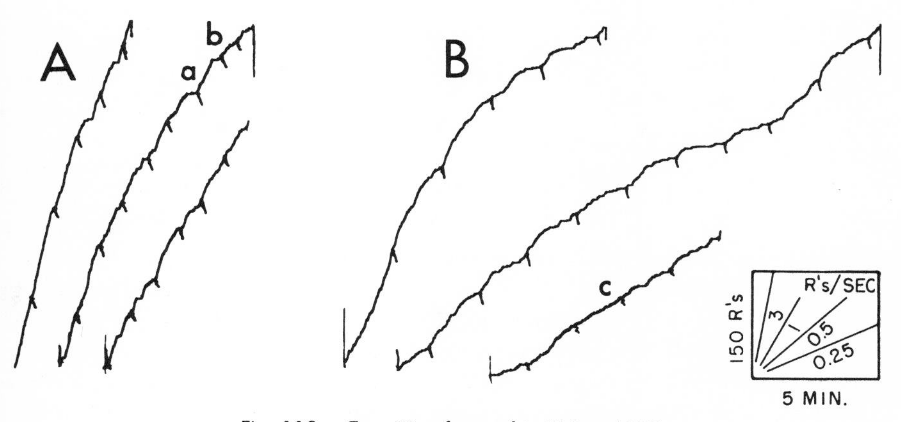

Fig. 119. Transition from crf to FI 1 and FI 2

tinuous reinforcement. The negative curvature within each interval is the result of the intermittent reinforcement. The linear phase of the typical performance just begins to appear at *c*.

FI 4. Figure 120 shows an exceptionally irregular transition from crf to FI 4. This bird started slowly under crf. The present session begins with a pause of more than 1 minute at *a*. The 1st interval contains only a single burst of responses, and the 1st reinforcement (at *b*) occurs after a pause of more than 2 minutes. It is followed by a small, negatively accelerated curve. Before the next reinforcement at *c*, however, the bird begins to peck rapidly, and the next interval shows a large, negatively accelerated curve. The reinforcement (at *d*) is received after a pause of about 2 minutes, and 4 intervals thereafter a prolonged pause precedes the reinforced response at *e*. Responding begins immediately after reinforcement at the highest sustained rate yet seen, and a large, negatively accelerated curve extends through the interval to the next reinforcement at *f*. Thereafter, a more uniform rate of responding emerges which yields a stable over-all rate before the end of the session, between *g* and *h*. Responding during this period is characterized by rough grain, which is due to short bursts of responding separated by pauses.

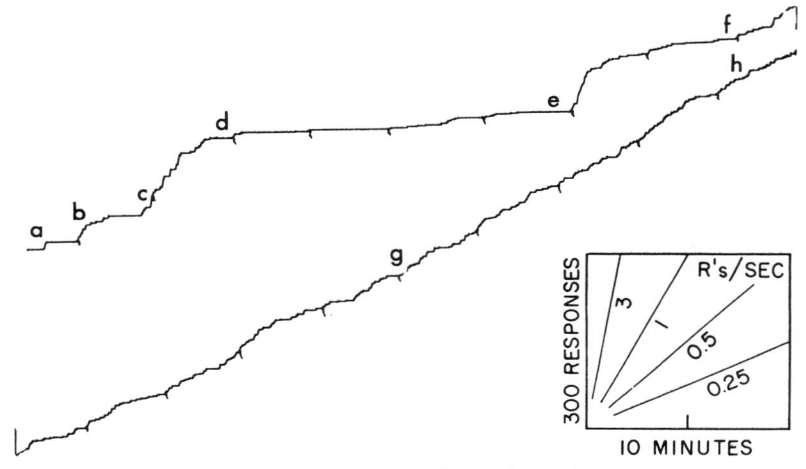

Fig. 120. Transition from crf to FI 4

Figure 121 shows a transition from crf to FI 5, with a rapid development of a high over-all rate of responding. The negatively accelerated curve due to the preceding crf extends through the first 2 segments. The first 10 segments show negative curvature within each interval. As the session progresses, the rate immediately after reinforcement falls, and beginning at *a* the over-all curve is nearly linear. The mean rate has increased to about 0.6 response per second, compared with 0.5 response per second at the start of the session. As the over-all performance becomes more linear toward the middle of the session, instances of brief periods of rapid responding occur at *b*, *d*, and *i*, and more sustained high rates at *c* and *e*. Many short pauses are followed by compensatory increases in rate, which give the impression of a nick or bite in a smooth over-all curve, as at *f*, *g*, *h*, *i*, and *j*. Nicks also appear in the other direction, at *b* and *k*, where a brief moment of rapid responding is followed by a pause. The grain of the record remains rough throughout this 1st session.

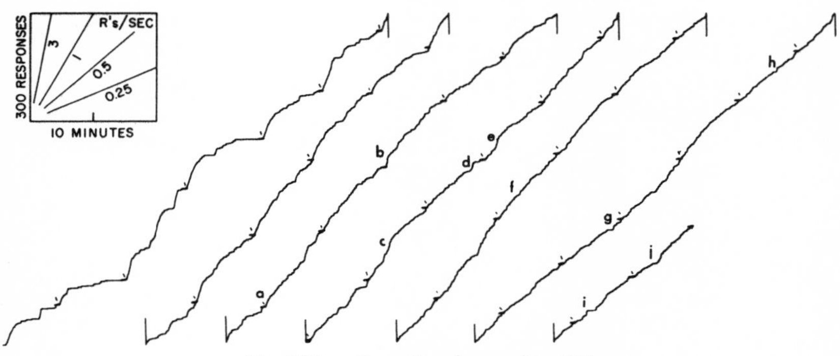

Fig. 121. Transition from crf to FI 5

FI 8. Figure 122 shows a transition from crf to FI 8. The first part of the curve, to the reinforcement at *a*, is negatively accelerated as a whole; and each interval shows a similar acceleration. The rate of responding immediately after the reinforcement becomes lower and sustained for fewer numbers of responses. The rate is roughly linear for the remainder of the session, with the over-all rate increasing gradually. Bursts of responding at rates which have never been reinforced occur at *b*, *c*, and *d*. By the end of the session, a pause of the order of 10 or 15 seconds appears after each reinforcement, and the linear performance gives way to marked oscillations, as at *e*.

FI 17. In the transition from crf to larger fixed intervals, extinction has a greater

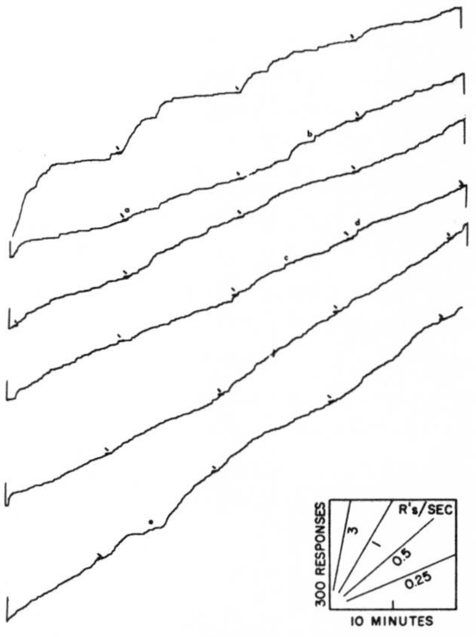

Fig. 122. Transition from crf to FI 8

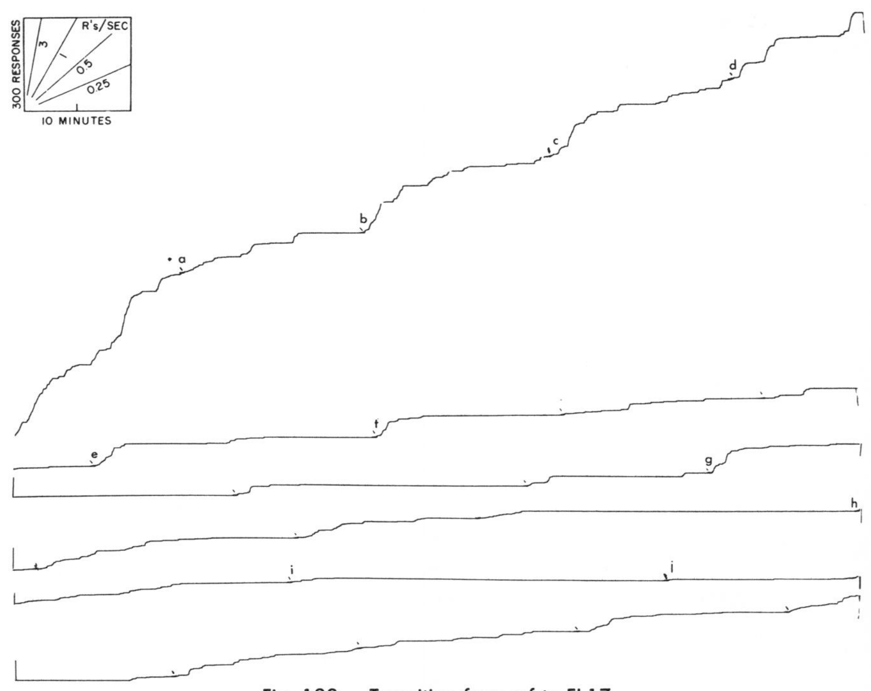

Fig. 123. Transition from crf to FI 17

opportunity to occur. Very low over-all rates of responding may prevail before the final performance on the schedule is developed. Figure 123 shows a transition from crf to FI 17. The top line comprises a large extinction curve consisting of over 500 responses, mainly due to the previous crf. The bird is still responding at an intermediate rate at the 1st reinforcement, at *a*. In the 2nd interval the rate reaches zero, and the 2nd reinforcement, at *b*, occurs after a 6-minute pause. The reinforcement reinstates a high rate, which then declines through the next interval. The same performance is repeated in the next 4 intervals, beginning at *c*, *d*, *e*, and *f*, when the size and rate of the initial run, as well as the total number of responses in the interval, decrease. The rate is very low for the rest of the session except following the reinforcement at *g*. Reinforcements at *h*, *i*, and *j* occur long after the designated interval has elapsed. Toward the end of the session a more sustained rate of responding develops (last line). This low rate was maintained through the 2nd session. The over-all rate does not increase further until the end of the 2nd session, after 13 hours of exposure to the schedule.

Another transition from crf to FI 17 is shown in Fig. 124, where extinction from the preceding crf generates a smaller curve. Except for the run at *a*, the first 6 intervals show fairly smooth negative acceleration, and reinforcements occur after pauses. A

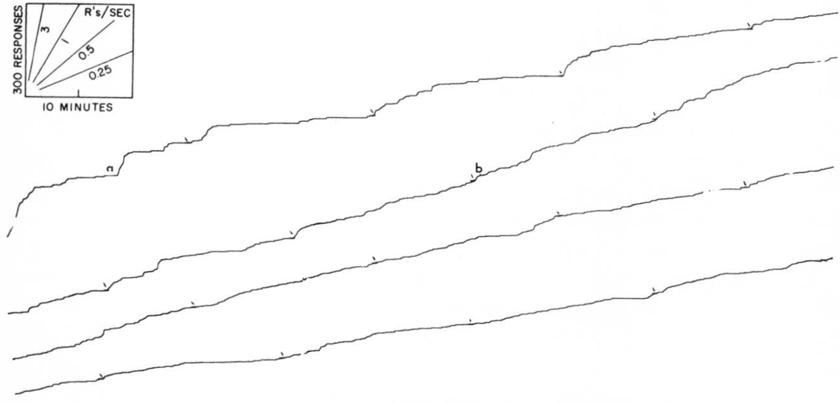

Fig. 124. Transition from crf to FI 17

substantial rate of responding is reached more quickly than in Fig. 123. Although the grain of the record is rough, with responses occurring in bursts separated by pauses, the over-all effect, beginning in the region of *b*, is a fairly uniform rate of responding.

FI 45. Two birds were put on FI 45 immediately following crf. The extinction of the preceding crf did not produce any substantial responding, and both birds showed a very low rate for many hours. One bird showed no effect of the fixed-interval schedule after 5 hours. During this period only 6 reinforcements and 400 responses occurred. Most of the reinforced responses occurred long after the designated fixed interval had elapsed. The 7th reinforcement led to a period of fairly active responding (at *a* in Fig. 125) followed by a negatively accelerated segment suggesting an extinction curve. This segment is followed by a fairly uniform rate, although some rough grain is present. Figure 143 shows subsequent sessions in which the over-all rate rises slowly. The effects of the early reinforcement were not so marked for the

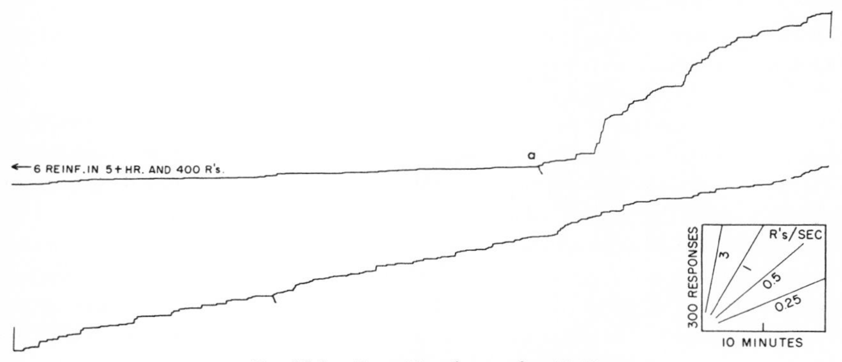

Fig. 125. Transition from crf to FI 45

second bird in the experiment. This bird maintained a considerably lower over-all rate, and took a longer period of time to reach a constant rate of responding.

Summary

For very short fixed intervals, the 1st reinforcement occurs so soon that the rate due to the preceding crf does not decline greatly. All reinforcements tend to occur while the rate is high. For long fixed intervals, however, reinforcements usually occur after the rate of responding has fallen nearly to zero.

With exposure to the fixed-interval schedule the negatively accelerated fixed-interval segments give way to a linear performance with rough grain. This condition occurs because the high rate of responding early in the segment is correlated with nonreinforcement, while the low rate toward the end constitutes a favorable stimulus. The negatively accelerated fixed-interval segments generate low rates, and the relation between the low rate and the reinforcement is the most important difference between transitions from crf to fixed-ratio and fixed-interval schedules.

In the early exposure to a fixed-ratio schedule, reinforcements occur when the local rate of responding is high. Very few instances of reinforcement are observed when the local rate is low.

THE DEVELOPMENT OF FI AFTER crf

FI 1 and FI 2.5

The transition from crf to FI 1 shown in Fig. 118 was followed by the performance in Fig. 126, for 3 sessions on FI 1. The 2nd session, Record A, begins with a constant over-all rate of the order of 0.4 response per second, which gradually increased until it reached 0.7 response per second at the end of the session (Record B). The relatively constant rate (yielding a relatively constant number of responses at reinforcement) in the early intervals of Record A produces a burst of 20 responses at more than 3 responses per second at *a*. The whole curve shifts upward (that is, no compensatory drop in rate occurs). A slight tendency to pause after reinforcement appears at *b* and *c* and becomes quite marked in Record B. An increase in terminal rate accompanies the development of this pause. The terminal rates at *d* and *e* are about 1.25 and 1.5 responses per second, respectively. The over-all rate remains the same because of the longer pauses after reinforcement. In Record C, the 4th session following crf, the interval segments are consistently scalloped; the terminal rate reaches 1.5 to 1.7 responses per second, and the period of pausing and acceleration extends over half of the interval. The over-all rate of responding remains about the same.

A transition to FI 2.5 from crf is shown in Fig. 127,[1] which presents 16 segments at the ends of the first 6 sessions. Any effect of the previous crf has already disappeared in Record A. The rate is roughly constant, and a slight pause begins to follow reinforcement. The pauses increase in the 2nd and 3rd sessions (Records B and

[1] We are indebted to George Victor for help in this experiment.

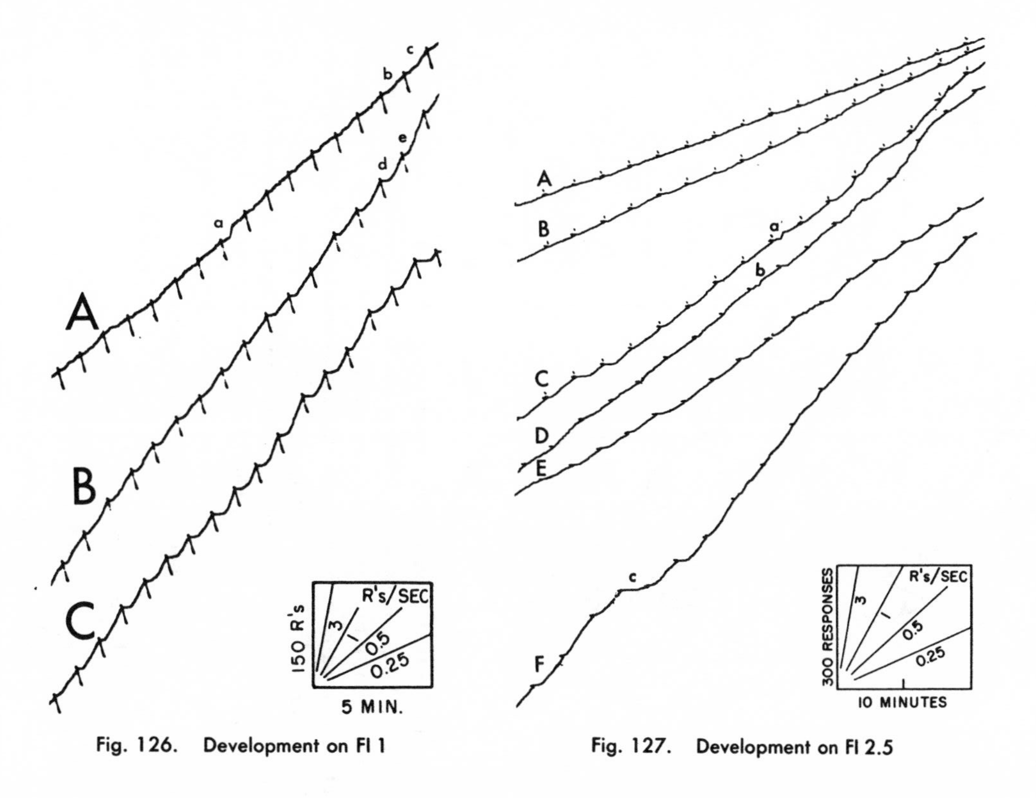

Fig. 126. Development on FI 1

Fig. 127. Development on FI 2.5

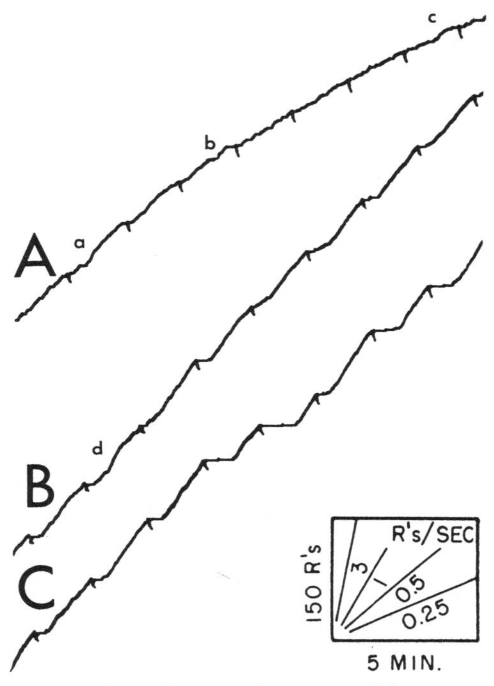

Fig. 128. Development on FI 2

C); but in the 4th session a linear performance returns temporarily. Brief periods of high rates of responding occur at *a* and *b*. Record E shows consistent pausing and curvature; but a brief return to a constant rate occurs on the following day. Except for Record E the over-all rate increases consistently throughout 6 sessions. A "second-order effect" (see below) appears at *c*.

The further development of FI 2, the transition to which was shown in Fig. 119B, is indicated by the segments in Fig. 128, from the 2nd, 4th, and 6th sessions following crf. Record A shows a fairly constant rate, with brief increases at *a*, *b*, and *c*. Pausing or slower responding after reinforcement is clear by Record B; a brief period of an exceptionally high rate appears at *d*. By Record C, pausing is marked, with the usual concomitant higher terminal rate.

Accidental contingency on FI 1

Figure 129 shows an example of the effect of an accidental contingency in the early stages of FI. The bird developed a sequence of responses in which it pecked the panel at the side of the key before pecking the key itself. Record A is the 13th session on FI 1 after crf. A lower rate is still resulting from the accidental chaining of the response to the panel. Such responses are most common near the end of the intervals,

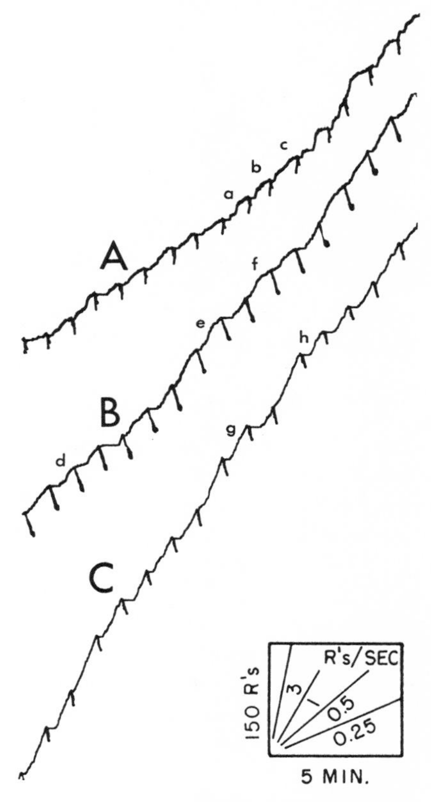

Fig. 129. Accidental contingency on FI 1

and they produce small S-shaped curves (as at *a, b,* and *c*). Traces are still evident in Record B for the 23rd session (cf. *d, e,* and *f*), and a slight roughness of grain and a tendency to retard before reinforcement (as at *g* and *h*) can still be seen by the 56th session (Record C).

FI 5

Figure 130 shows further development of FI 5 (after the transition in Fig. 121). Record A reports the 40th to 45th reinforcements, with a somewhat reduced rate beginning to show at *a* and *b*. Record B, for the 75th to 80th reinforcements, and Record C, for the 83rd to 88th, show a further development of the pause, with a slight increase in terminal rate. In Record D, the 161st to 166th reinforcements, the pause after reinforcement is consistently of the order of 45 seconds. The terminal rate of responding is still not very high, and there is no extended acceleration. Record E, for the 193rd to 198th reinforcements, shows the same order of magnitude of pausing, with a higher

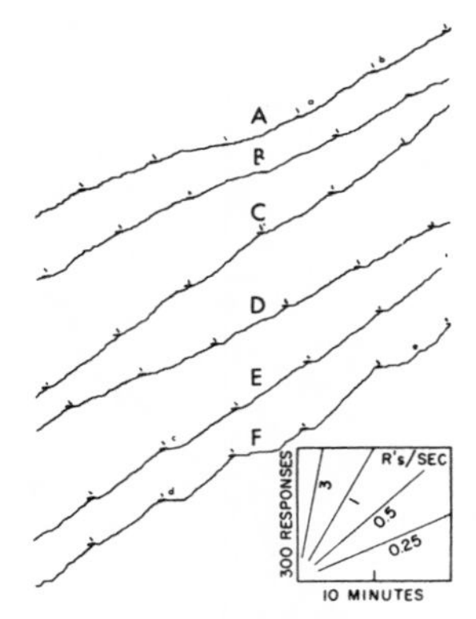

Fig. 130. Development on FI 5

terminal rate and less grain. Occasional instances occur where the lower rate of responding following the reinforcement is more extended (as at *c*). In the next session a much more marked pause and acceleration begin to show (Record F). The pause following reinforcement is now as long as 90 seconds (as at *d*), and the acceleration from the zero rate following the reinforcement to the terminal rate is occasionally quite extended (as at *e*).

Two birds previously used in an experiment on matching-to-sample, where one class of responses was reinforced (crf) and the other extinguished, were placed directly on FI 5. They had had no previous exposure to a fixed interval. Nevertheless, smooth accelerations throughout the interval developed. Figure 131 shows sample records of both birds. Record A follows 109 reinforcements on FI 5; Record B follows 44 rein-

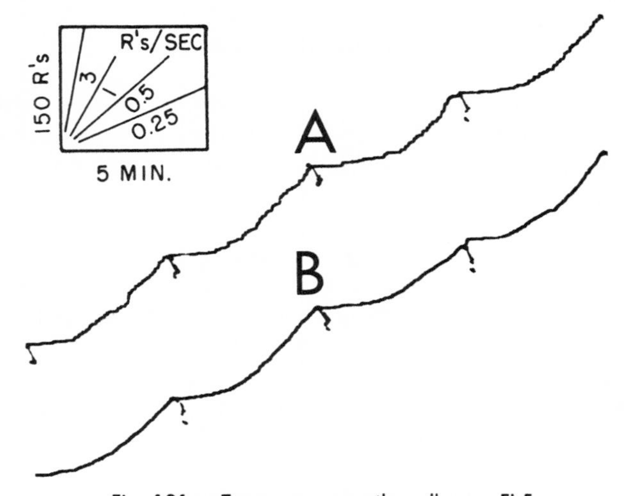

Fig. 131. Temporary smooth scallop on FI 5

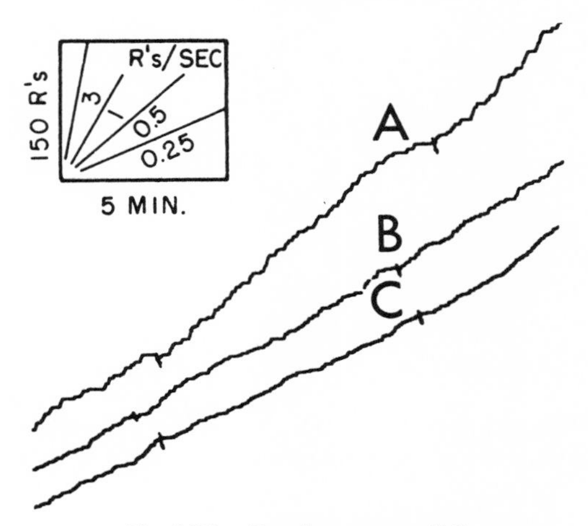

Fig. 132. Development on FI 8

forcements on FI 5. The curvature is much more marked than in Fig. 130F after more than 200 reinforcements. The performances shown in Fig. 131 were transient, however. The smooth scallops appeared 4 reinforcements before the curves shown in the graph, and they disappeared after a few other reinforcements. At this stage the over-all performance was linear except for pauses after reinforcement.

FI 8

Figure 132 shows the development of a linear performance on FI 8 after crf. The segments are from the first 3 sessions. Record A was taken 10 hours after the beginning of the 1st session on FI 8. The grain is very rough, the over-all rate low, and the curvature unrelated to the schedule. By Record C, after 26 hours on the schedule, the rate is constant except for slight pauses after reinforcement. The following session, shown complete in Fig. 133, is an excellent example of the instability of the contingencies generated by the FI schedule. A lower rate after reinforcement begins to appear at *a*, *b*, *c*, and *d*. The curvature at *e* is more marked, and acceleration extends

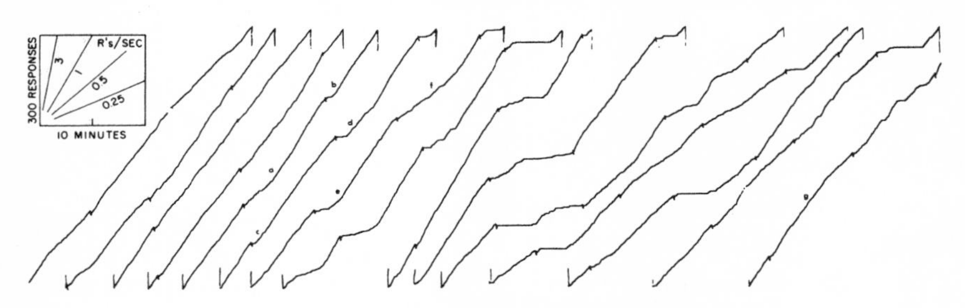

Fig. 133. FI 8: 23 hr after crf

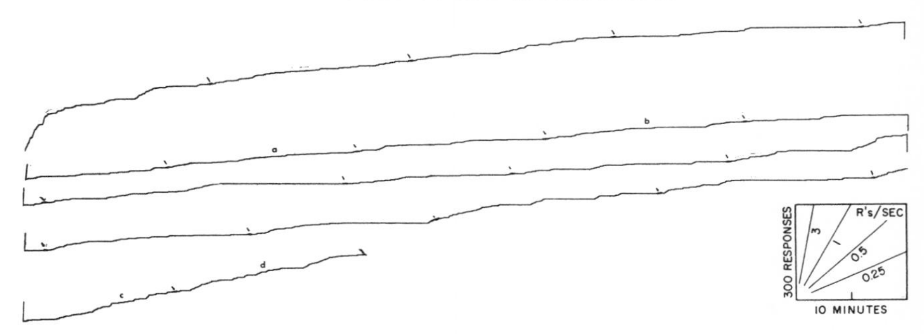

Fig. 134. FI 17: second session after crf

throughout the following interval (*f*). The scalloping is unstable, however, and by the end of the period the over-all performance is essentially linear (*g*).

FI 17

Further development of the performance on FI 17 begun in Fig. 123 is shown in Fig. 134 (2nd session complete) and Fig. 135 (segments at 15, 22, 25, 35, 37, and 48 hours after crf). Figure 134 begins with some recovery from the level of responding of Fig. 123 (probably because of special stimuli from the starting of the experiment). Before the end of the session a steady but low rate has been assumed. The rates at *a* and *b* are of the order of 0.1 response per second, and at *c* and *d* are of the order of 0.15 response per second.

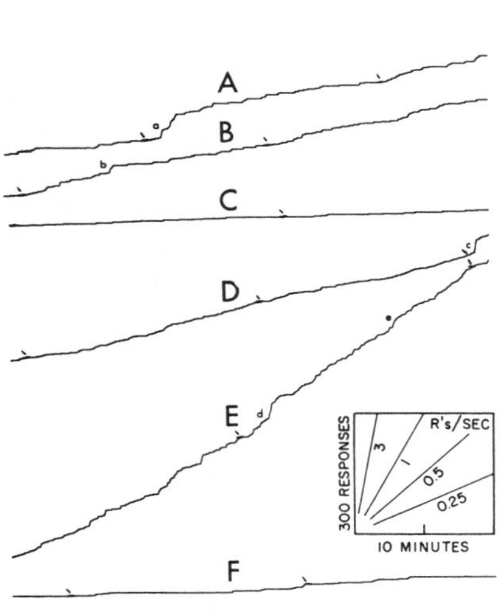

Fig. 135. Development on FI 17 to 48 hr after crf

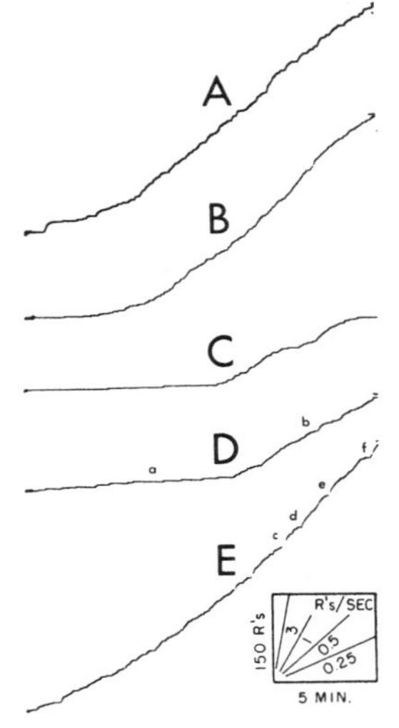

Fig. 136. Development on FI 17 to 72 hr after crf

In Fig. 135 a brief break-through occurs at *a* (Record A, 15 hours), and a smaller one at *b* (Record B, 22 hours). A very low rate then supervenes (Record C, 25 hours). At Record D (35 hours) a higher rate prevails, and short bursts at about 3 responses per second appear in Record E (*d, e*), 37 hours after crf. The over-all rate has now risen, but it falls to the very low value seen at Record F (48 hours).

A still further development of FI 17 for this bird is shown in Fig. 136, where single-interval segments have been selected. Record A at 52 hours shows smooth curvature, rough grain, and a slight negative acceleration toward the end of the interval. Record B at 60 hours shows a similar performance, in which the curvature is smoother and the grain less marked. In Record C at 61 hours the period of lower rate at the start of the interval is more extended, and a sharp decline in rate reduces even further the number of responses in the interval. Record D at 65 hours shows a frequently occurring performance, in which the interval begins with an extended period at a low

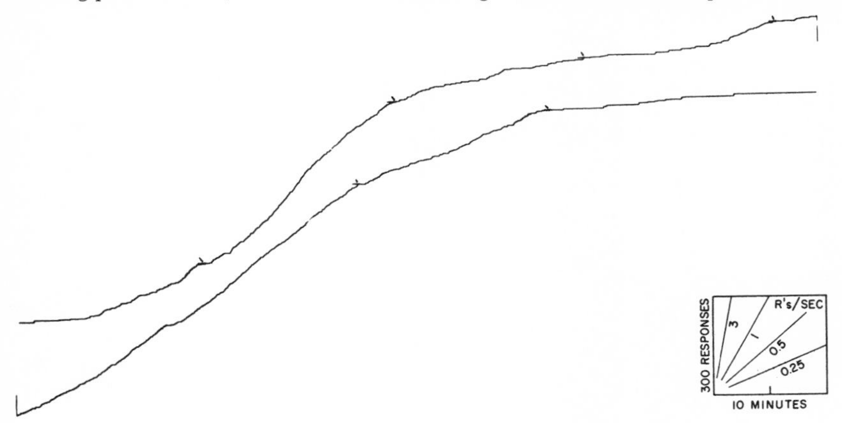

Fig. 137. FI 17: whole session 78 hr after crf

rate (*a*) that leads abruptly to an immediate terminal rate (*b*) maintained for the rest of the interval. Record E at 72 hours shows an instance in which the rate of responding accelerates throughout most of the interval. The over-all curve is smooth in spite of the 4 inflections at *c, d, e,* and *f*. In the sessions following Fig. 136 the over-all rate alternated between high and low values, with wave-like rate changes. The two lines in Fig. 137 form a continuous portion of the curve 78 hours after crf, and exemplify the fairly smooth oscillations in rate.

Another bird on FI 17 after crf (transition shown in Fig. 124) also maintained a low over-all rate in which accelerations related to the schedule appear and disappear. Figure 138 shows a complete session beginning 23 hours after crf. Note that curvature appears in spite of the low terminal rate.

Figure 139 shows features of the early development of FI 17 after crf in another bird. In Record A (6 hours after crf) there is still some tendency to respond rapidly just after

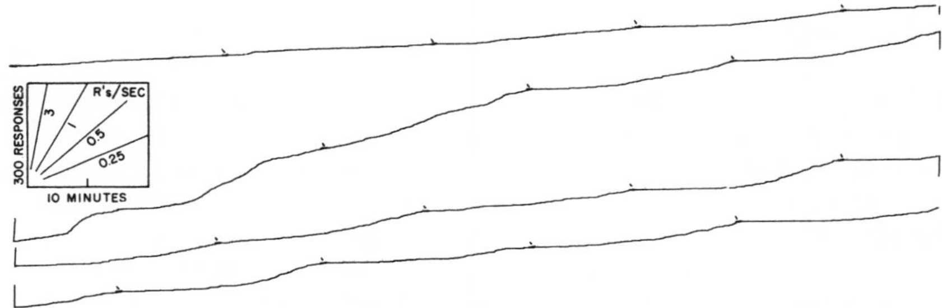

Fig. 138. FI 17: 23 hr after crf (second bird)

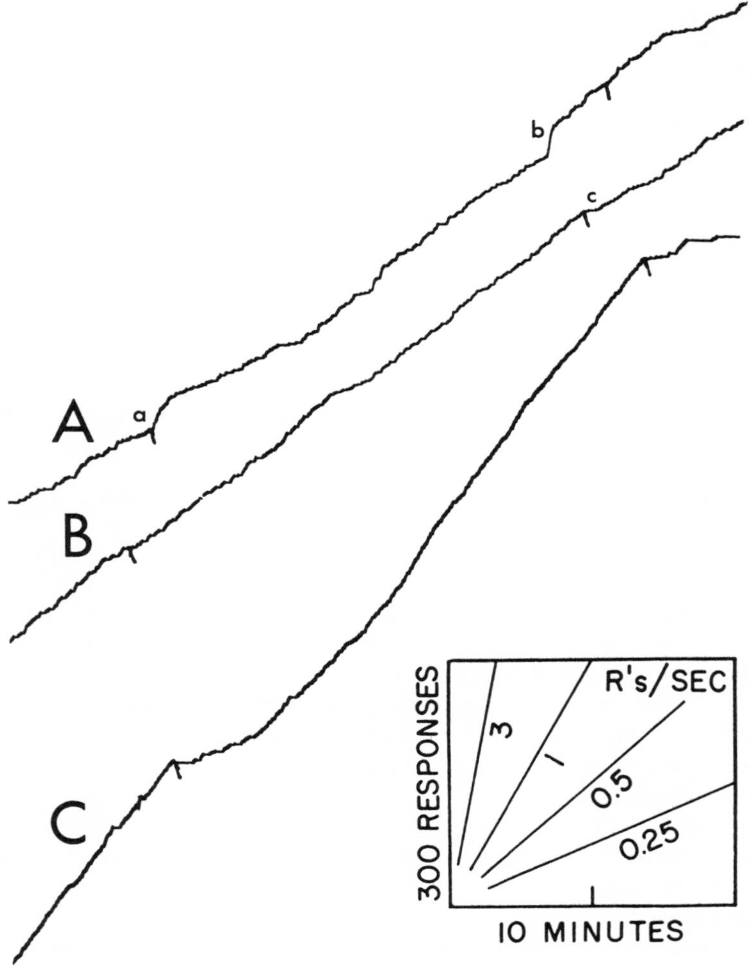

Fig. 139. Development on FI 17 (third bird)

reinforcement (*a*). A "ratio-like" break-through at a very high rate appears at *b*. In Record B, after 11 hours, the rate is fairly constant except for a slight pause following the reinforcement (as at *c*). In Record C, at 36 hours, a scallop has developed, the over-all and terminal rates have increased, and the grain is finer. After 72 hours on the schedule, oscillations such as those shown in Fig. 140 are evident. Figure 141 shows a complete 9-hour session for this bird after 87 hours of exposure to the schedule. The over-all rate declines progressively during the session, because of the lower rates of re-

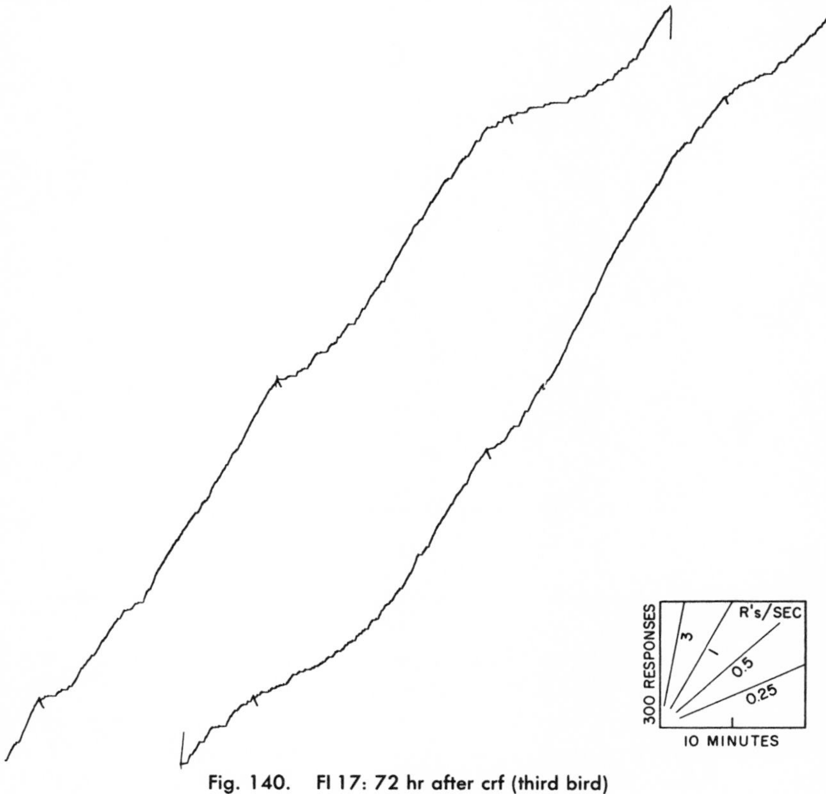

Fig. 140. FI 17: 72 hr after crf (third bird)

sponding appearing in the first parts of the intervals. Smooth curvature (as at *a, b,* and *d*) is infrequent, and there are only a few instances of pauses longer than a few seconds following reinforcement. Many intervals show inflection points as at *e* and *f*; and the grain may be extremely rough, with pauses and bursts of responding occurring even after the terminal rate is reached, as at *c* and *h*. The performance at 111 hours after crf (Fig. 142) continues the trend of Fig. 141. The over-all rate falls off even more sharply, and extended curvature develops earlier in the session.

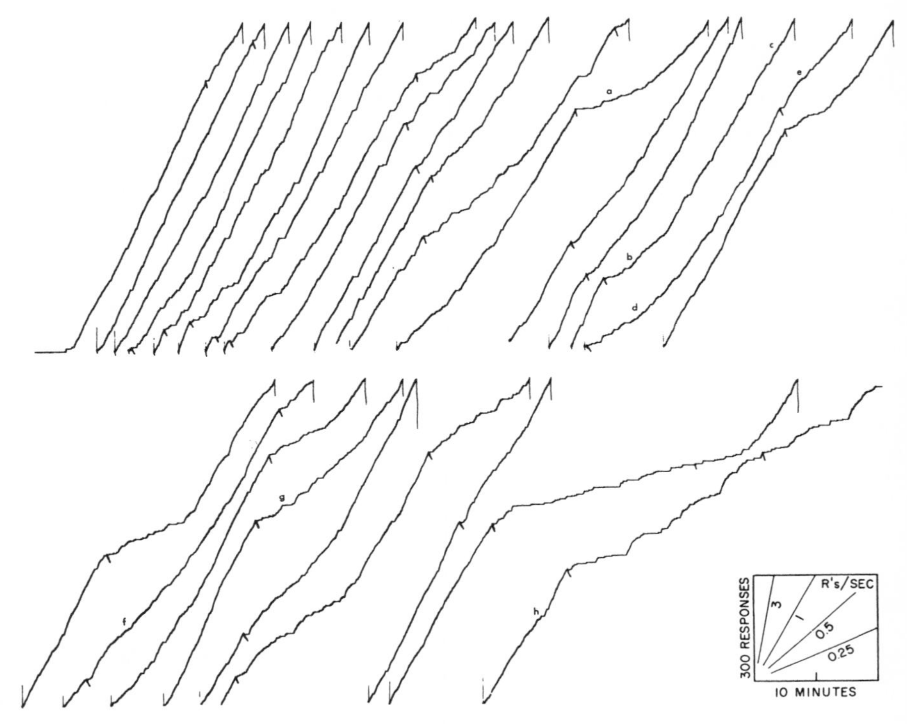

Fig. 141. FI 17: complete session after 87 hr (third bird)

FI 45

Figure 143 shows segments at 11, 14, 19, 26, and 34 hours on FI 45 following the transition from crf shown in Fig. 125.

In Record A a characteristic break-through occurs at *a* and grain is rough. Record B shows an unusually long pause at *b* and minor break-throughs at *c* and *d*. Record C shows a moderately high rate following the reinforcement at *e* which falls progressively during the remainder of the interval. The grain is marked by short runs of 10 to 20 responses. Record D shows the emergence of an almost linear perfomance, although occasional break-throughs occur, as at *f*. Record E is the final performance recorded for this bird before the schedule was changed to tand *FI*FR. (See Chapter Eight.)

A second bird on FI 45 showed a low rate in the 1st session following crf. Approximately 1000 responses occurred during a 15-hour session. Figure 144 shows the complete 2nd 15-hour session following crf. There are many instances of moderately high rates which tend to fall off smoothly as in extinction. Some of these occur following reinforcement (as at *a*, *b*, and *f*), while others occur during the interval (as at *c*, *d*, and *e*). Near the end of the record, at *g*, the bird assumes a rate of about 0.3 response

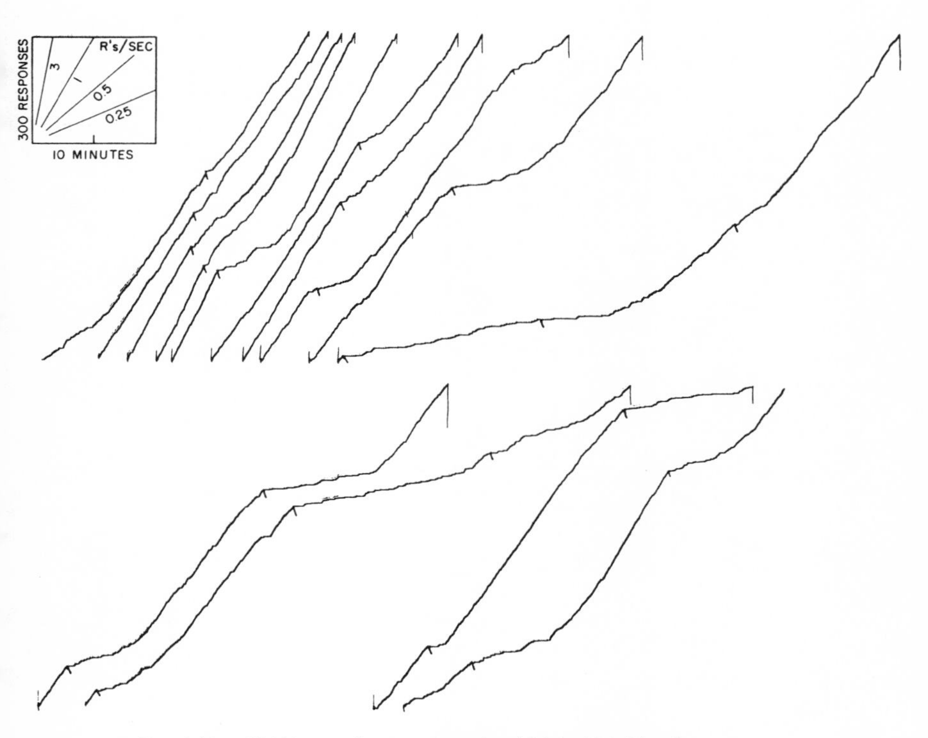

Fig. 142. FI 17: complete session after 111 hr (third bird)

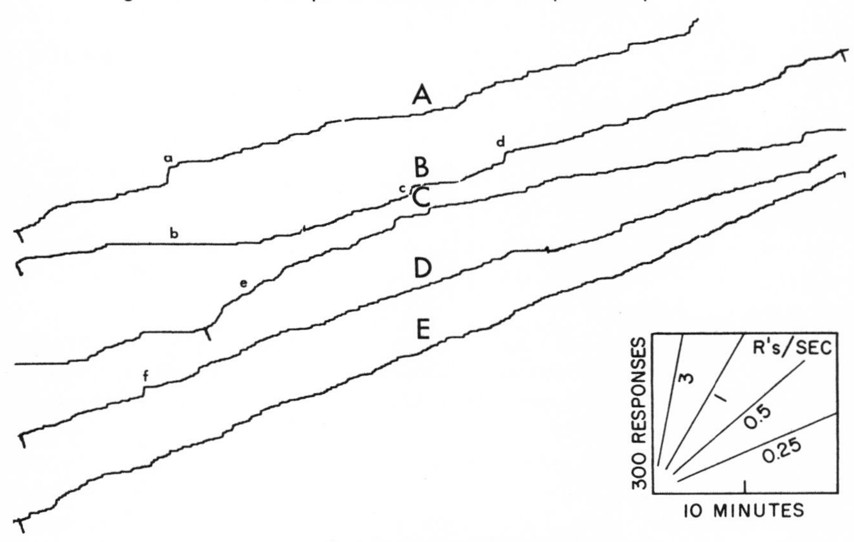

Fig. 143. Early development on FI 45 after crf

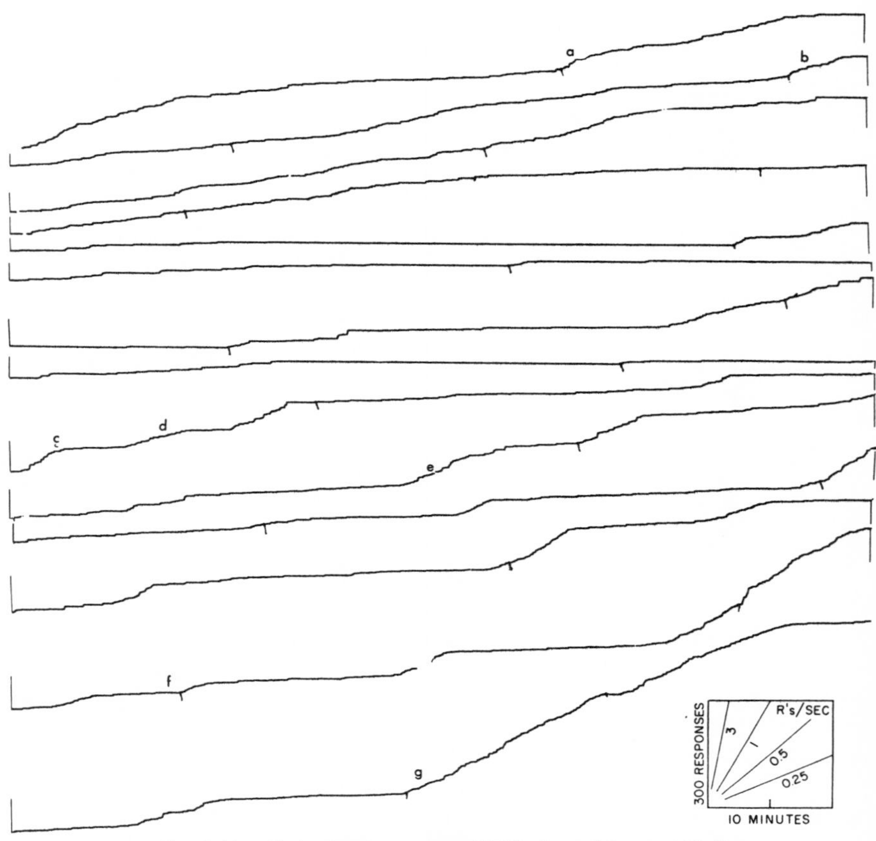

Fig. 144. Early development on FI 45 after crf (second bird)

per second immediately following reinforcement; it maintains this rate for more than 30 minutes, after which the rate falls off to nearly zero.

Transition from crf to FI 10 in the rat

In experiments with rats which are started and stopped automatically, some stimulus is made the occasion for nonreinforcement so that it will produce a zero rate of responding. (See Chapter Three.) This preliminary training gives the rat considerable exposure to extinction, which doubtless has some effect on the first exposure to a fixed-interval schedule. These curves also need to be qualified by the amount of reinforcement that was used. The reinforcement was increased from 0.05 gram to 0.1 gram later as a result of experience in this experiment.

Three rats were first reinforced continuously in the presence of one stimulus and extinguished in the presence of a second, until responding in the presence of the second

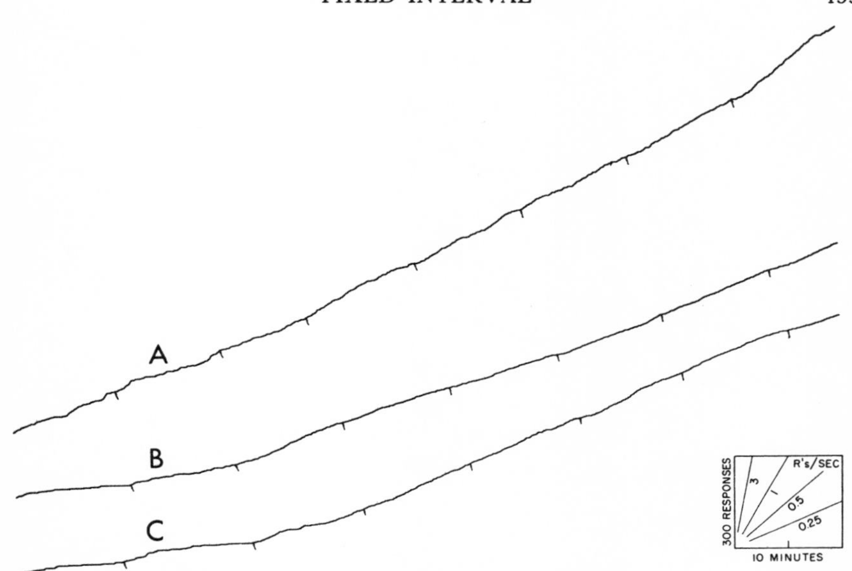

Fig. 145. Transition from crf to FI 10 in the rat

was essentially zero. They were then placed on FI 10 in the presence of the first stimulus, and the second stimulus was used to start and stop the experiment. Figure 145 shows the first 7 reinforcements for each rat on FI 10. The over-all rate of responding is low, with little curvature or local changes in rate. Except in the early intervals, very few pauses longer than a few seconds occur, and all of the reinforcements are received within a few seconds of the scheduled time. The over-all rate increases slightly through the whole first experimental session.

Figure 146 shows the complete 2nd session for the rat in Fig. 145B. A pause and lower rate follow the 4th reinforcement in the session (*a*). A 2nd scallop occurs at

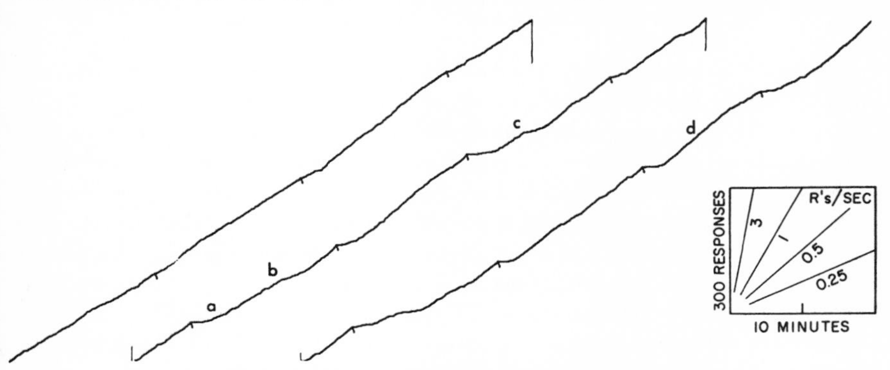

Fig. 146. Second session on FI 10 after crf (second rat)

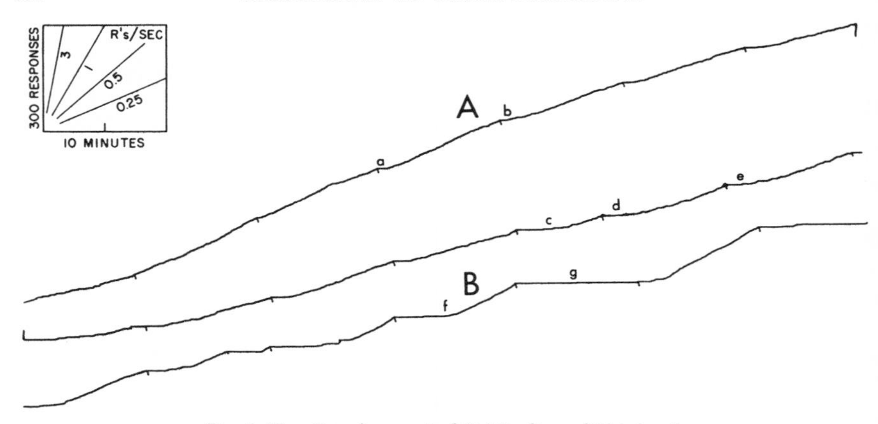

Fig. 147. Development of FI 10 after crf (third rat)

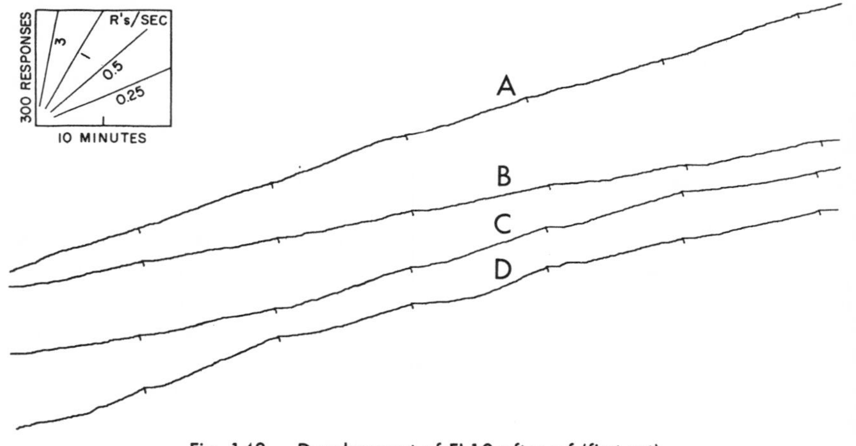

Fig. 148. Development of FI 10 after crf (first rat)

b, after an initial acceleration. Another example of this appears at *c.* At *d* the rate is slightly higher than heretofore and falls off before the end of the segment.

Figure 147 shows a continuation of the performance by the rat in Fig. 145C. Record A (top 2 curves, recorded continuously) shows the 9th to the 22nd reinforcements. Pauses after reinforcement begin to appear at *a, b . . .* and scalloping (with rather rough grain) appears at *c, d,* and *e,* possibly because of a too-short interval programmed accidentally at *c.* Marked acceleration to a higher terminal rate is evident at *f* in Curve B, which shows the 56th to 61st reinforcements. At *g* a pause after reinforcement extends through the interval.

Figure 148 shows segments from the session continuing Fig. 145A with features similar to Fig. 146 and 147.

THE EFFECT OF SUSTAINED FI AFTER crf

FI 1

Figure 149 shows the further development of the performance on FI 1 shown in Fig. 126. The over-all rate and degree of scalloping remain fairly stable over the 5 sessions represented by Segments A through E. A 15- to 30-second pause after reinforcement is followed by a smooth acceleration to a terminal rate, which is maintained until the next reinforcement. Frequently, however, the terminal rate is assumed immediately following the reinforcement and is maintained until the next reinforcement (at *a* through *g*). Two intervals together (3 in the case at *a*) thus form a single positively accelerated curve.[1] Record F, the 16th session, shows a lower over-all rate which is due mainly to a fall in the terminal rate. Frequently, the pause and low rate after reinforcement extend through the major part of the interval, as at *h, i,* and *k*. Instances still occur, as at *j*, where the rate of responding is high throughout the interval.

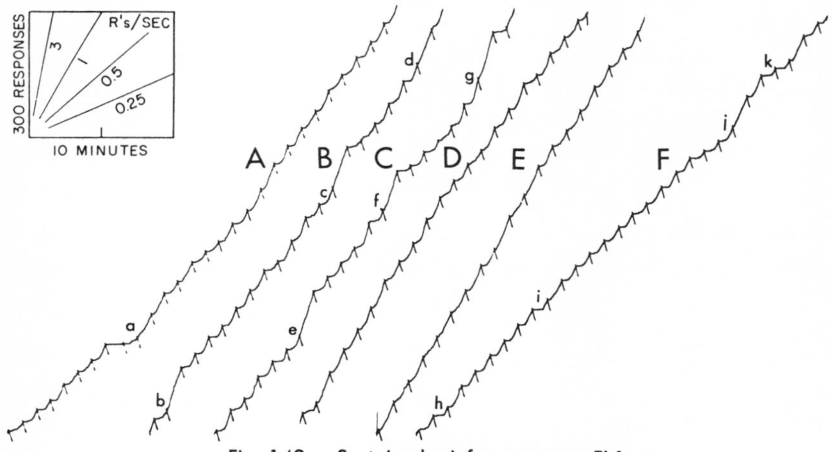

Fig. 149. Sustained reinforcement on FI 1

FI 2

Figure 150 shows further development of a performance on FI 2 (following Fig. 119B and 128). Record A shows the 22nd session; Record B, the 41st; and Record C, the 50th. In Record A a moderate over-all rate results from a consistent period of pausing and slow responding during the first half of the interval. Occasional instances of the terminal rate being maintained throughout the interval appear at *a* and *c*. A suggestion of a large second-order effect occurs in the 4 intervals following *b*. In Record B the terminal rate of responding has increased greatly, and on occasion the bird does not

[1] See *The Behavior of Organisms*, p. 123, on these "second-order effects."

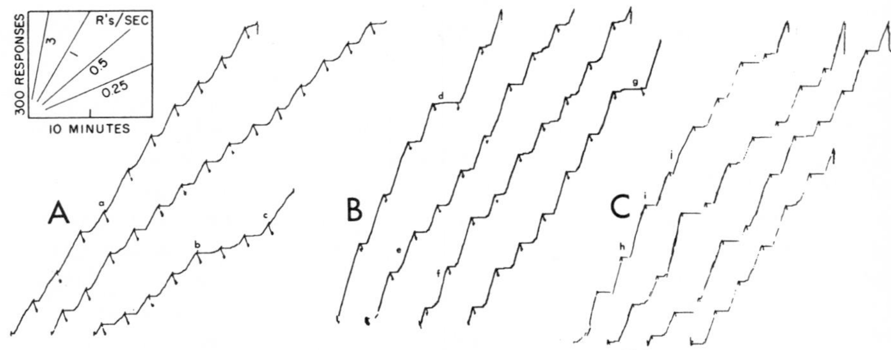

Fig. 150. Sustained reinforcement on FI 2

reach the terminal FI rate (as at *d* and *g*). These segments are characteristically followed by pauses that are shorter than usual. Pauses after reinforcement are otherwise of the same order of magnitude as before, and the increase in the over-all rate of responding is due to the higher terminal rates. Negative curvature shows slightly, as at *e* and *f*.

By the 50th session after crf (Record C) the pause after reinforcement is extended, and it usually gives way to an abrupt shift to a higher rate. Frequently, as at *h*, *i*, and *j*, the bird begins to respond at an unusually high rate, which declines smoothly to a more moderate value.

FI 4 or FI 5

After prolonged exposure to FI 4 or FI 5, a pause following the reinforcement is rarely absent, and most segments show a smooth acceleration to a moderately high ter-

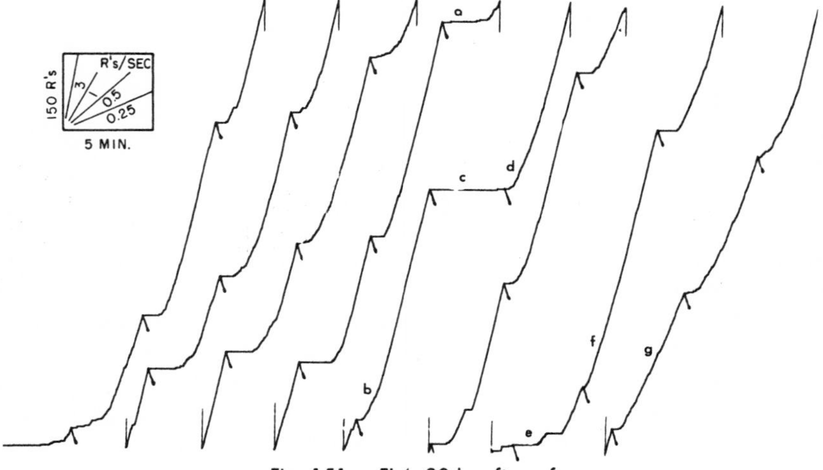

Fig. 151. FI 4: 38 hr after crf

minal rate. There is some grain and an occasional knee in the accelerated portion. Successive intervals are related to each other in the manner of a second-order effect: a slow development of the terminal rate in one interval, and hence a low number of responses during the interval, is generally followed by an interval in which the pause is shorter and the acceleration to the terminal rate more rapid.

Figure 151 shows a complete daily session on FI 4 (38 hours after crf) for the bird whose transition from crf was shown in Fig. 120. A pause of from 1 to 2 minutes is followed by smooth acceleration to a terminal rate of approximately 2.5 responses per second. Occasional instances of a lower sustained rate occur, as at *g*, where the rate is 1.25 responses per second. When very few responses occur during an interval, as at *a, c,*

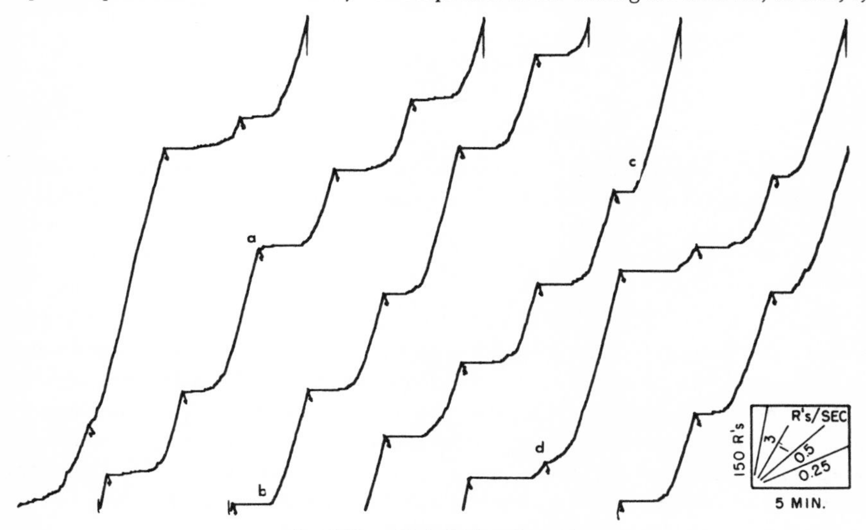

Fig. 152. FI 4: 66 hr after crf

and *e,* the next interval usually shows a shorter pause than usual (*b, d,* and *f*). Figure 152 shows a later performance for the same bird, 66 hours after crf. Here, the terminal rate has fallen slightly, and the pause and curvature after reinforcement have increased. Responding usually begins earlier in intervals which follow intervals containing small numbers of responses (*d*). The 3 following intervals comprise a good example of the second-order effect. Occasionally, the bird shifts abruptly to a rate only slightly less than the terminal rate during an FI segment, as at *b.* Note the exceptional case at *a,* where 4 or 5 responses occur immediately after reinforcement but the rate then falls to zero and the usual FI performance emerges.

A performance on FI 5, 66 hours after crf, is shown in Fig. 153, which shows the first 34 reinforcements of a session. The performance is similar to that in Fig. 152 except that the terminal rate is lower and the curvature more extended. Second-order effects are frequent, as at *a, b, d,* and *i.* Occasional negative curvature appears near

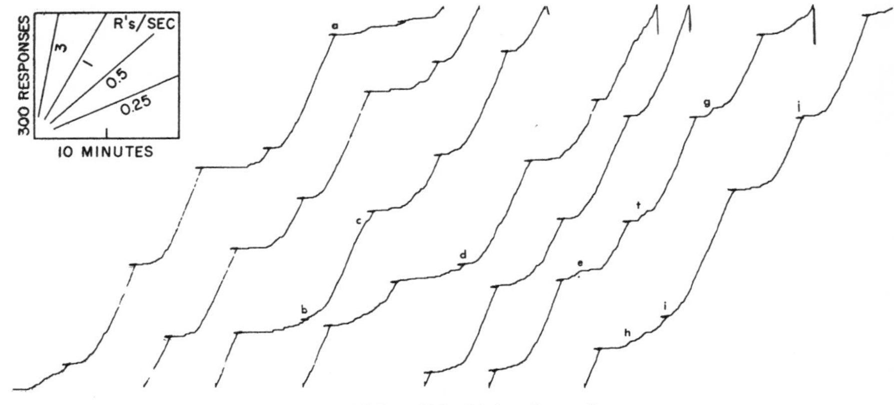

Fig. 153. FI 5: 66 hr after crf

the end of the interval, as at *c*. Irregular knees occur at *e*, *f*, *g*, and *h*. At *j* a few responses immediately after the reinforcement are followed by a pause and a typical FI performance.

FI 8

Late performances. Figure 154 shows a performance on FI 8 for a complete session beginning 186 hours after crf. A high over-all rate is maintained, but the performance in the interval is by no means stable. The terminal rate may be maintained throughout the interval except for a brief pause after reinforcement (as at *f*) or after a longer pause (as at *a*). Smooth curvature may lead to the terminal rate (as at *c*), or a marked knee may occur (as at *e*). Short runs at a higher rate may appear (as at *d*). The terminal rate may decline slightly before a reinforcement is received (as at *b*).

A wide variety of types of curves appeared during the 186 hours of exposure to the FI 8 schedule of the bird whose late performance is shown in Fig. 154. Figures 155 through 163 show examples of the various types of performances, which have been collected and grouped in terms of the shapes of the curves. These performances occurred at many stages of exposure to the schedule.

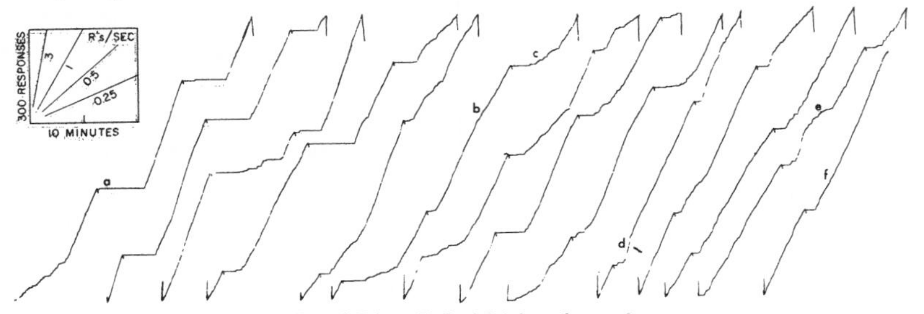

Fig. 154. FI 8: 186 hr after crf

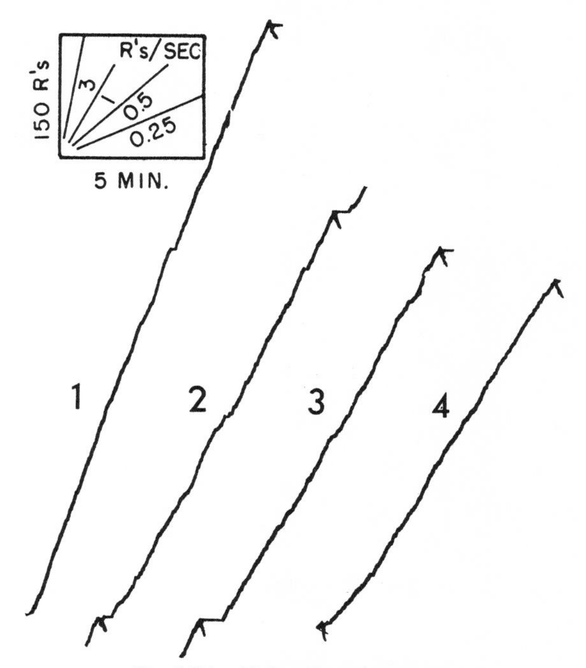

Fig. 155. FI 8: collected segments

The following is a rough account of the development leading up to the performance in Fig. 154. After the usual early features already described, the curve becomes essentially linear except for brief pauses presumably controlled by the preceding reinforcements. Figure 155 shows some of these linear segments, taken from various points in the whole experiment. By the 11th hour, marked deviations from a straight line begin to appear. The first good example of a smooth positive acceleration of the sort exemplified in Fig. 156 appears in the 28th hour, and several instances occur in the 2 or 3 hours following. These then become fairly common, alternating with straight segments (Fig. 155) or with slighter curves. In a temporary stage in the 38th hour, the grain is rough and a high sustained rate is not reached. Straight segments then predominate, and a more positive curvature emerges. Second-order effects are clear by the 52nd hour. The over-all rate is now high, and the terminal rate increasing. Multiple runs, exemplified by Curve 1 in Fig. 162, begin to appear. The linear segments of Fig. 155 return from the 70th to the 74th hours, followed by a more marked curvature and occasional intervals at a low over-all rate or with only a slight acceleration. In the 78th and 79th hours, almost all intervals show fairly smooth curvature, with only a slight suggestion of a second-order effect. By the 81st hour a much lower over-all rate has appeared, marked by rough grain and deviations of the sort presented in Fig. 159 and 160. Second-order effects are frequent, although a series of repeated

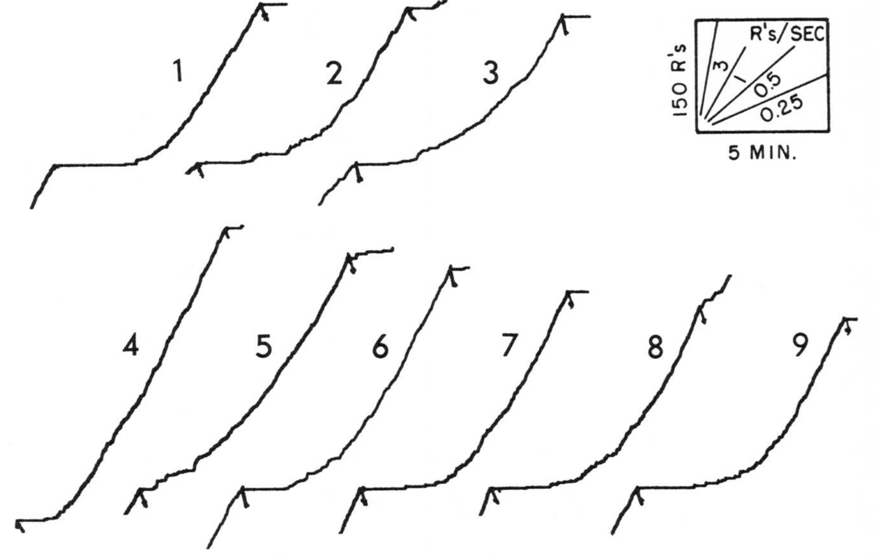

Fig. 156. FI 8: collected segments

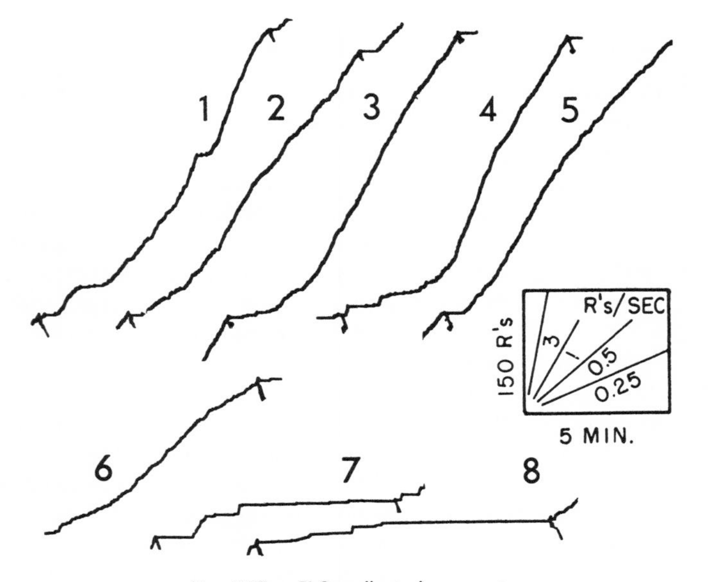

Fig. 157. FI 8: collected segments

smooth scallopings may extend through as many as 8 or 10 successive intervals. By the 100th hour, scallops are quite smooth, there are some second-order effects, and the grain is fair. Negative curvature appears occasionally, as exemplified in Fig. 157. Vertical displacement similar to Curves 6, 7, and 8 in Fig. 158 appears in the 110th hour. Between 110 and 130 hours the scallops are fairly uniform, marked by only minor deviations or second-order effects. The over-all rate remains moderate (0.5 response per second). At the 142nd hour a very good series of successive scallops free of second-order effects appears. The later performance is well-represented by Fig. 154.

Types of rate changes under sustained exposure to FI 8. Figure 155 shows linear segments which are especially common early in the history of such a schedule. (Cf. Fig. 151 for similar segments on a shorter interval.) Figure 156 exemplifies smooth accelerations extending throughout the segment. The curvature ranges all the way from arcs of circles, Curve 3, to a sharp acceleration immediately after reinforcement, as in Curve 4, where a further increase in rate is only very slight. We take these curves to be typical of basic processes in FI. The following figures note various kinds of deviations which may be imposed upon them.

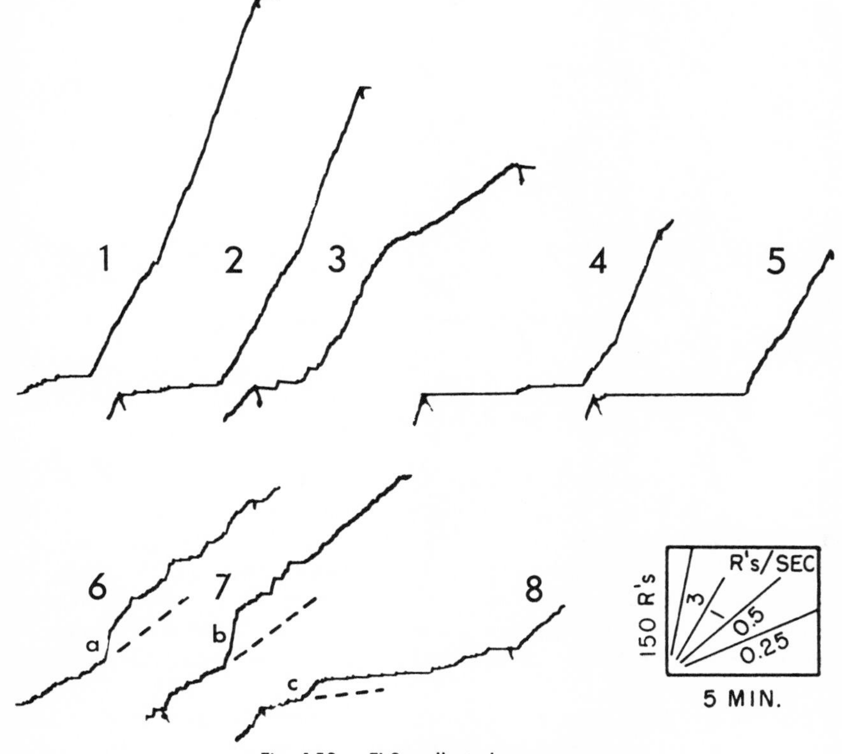

Fig. 158. FI 8: collected segments

One type of deviation consists of an early acceleration to a high rate, which is not, however, maintained. Examples of such negative acceleration at different over-all rates are shown in Fig. 157. In Curves 1 through 5, Fig. 158 shows various abrupt changes in rate; and in Curves 6 through 8, displacements due to a brief run for which no compensatory decline occurs.

A common deviation is a convex "bump" or "knee." In Fig. 159, Curve 1 shows only a very slight example; in Curves 2 and 3 the effect is more pronounced. Minor variations and evidence of more than one inflection are apparent in Curves 4 and 5.

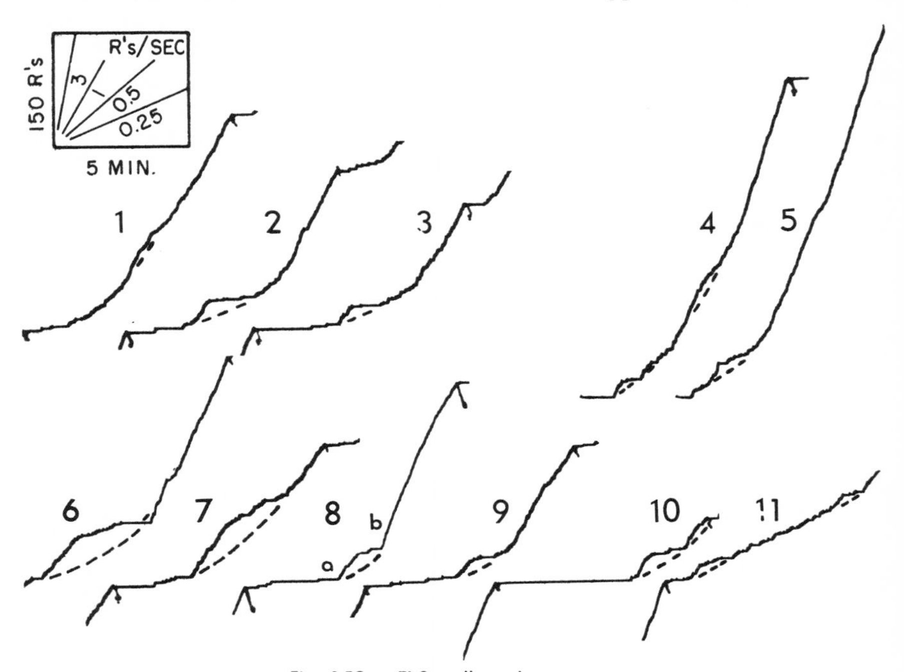

Fig. 159. FI 8: collected segments

In Curve 6 a fairly high rate falls off to zero; the rate then shifts abruptly to the final running rate. In Curve 7 an abrupt shift occurs early, and the rate falls off slightly and again recovers in a smooth curve. In Curve 8 the major changes are abrupt, as at *a* and *b*, but the negative curvature in the knee is fairly smooth. Curve 9 resembles Curve 8 except for the less abrupt changes. Curve 10 presents a variation on Curve 9, in which the knee occurs very late and little time remains for the maintenance of a terminal rate. Curve 11 shows a similar effect at a very low over-all rate throughout the interval.

A different type of deviation is a concave rather than convex deflection of the curve. Curve 1 in Fig. 160 seems to be a fairly smooth, positively accelerated curve, except that a pause appears early in the curve; this pause is followed by a rapid compensatory run which brings the curve roughly back to the extrapolation of the earlier portion. Curve 2 shows another example. Whether Curve 3 belongs in Fig. 159 or Fig. 160 is difficult to say. Curve 4 shows a nick which is followed by a continuing decline and the recovery of the terminal rate. The possibility of repeated deviations of both sorts is suggested in Curves 5, 6, 7, 8, 9, and 10.

The kind of deviation exemplified in Fig. 160 frequently occurs just after reinforcement. (See Fig. 161.) It suggests that competing behavior prevents the bird from returning to the key, and that the delay is made up in a compensatory run.

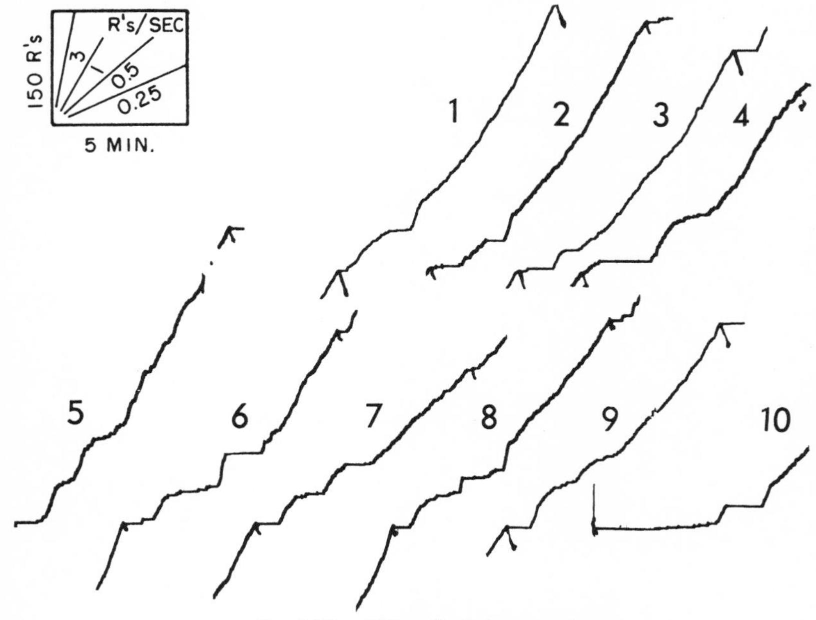

Fig. 160. FI 8: collected segments

Some of the most irregular segments encountered in such an experiment are illustrated in Fig. 162. In general, the rough course of the record may be guessed at in spite of the deviations.

Some instances of second-order effects from these experiments are shown in Fig. 163. In Curve A the resumption of responding at *d* almost immediately after reinforcement is related to the only moderate pausing at *c* and the low rate during the whole interval at *b*. The over-all effect is a positive acceleration extending through 3 intervals. In this particular case some negative acceleration sets in at *e*. One interpretation of this instance is that the reinforcement at *a*, occurring after a fairly sustained run and with a high count, accentuates the nonoptimal character of the pause after reinforcement. The bird responds only late and irregularly during the interval. But the re-

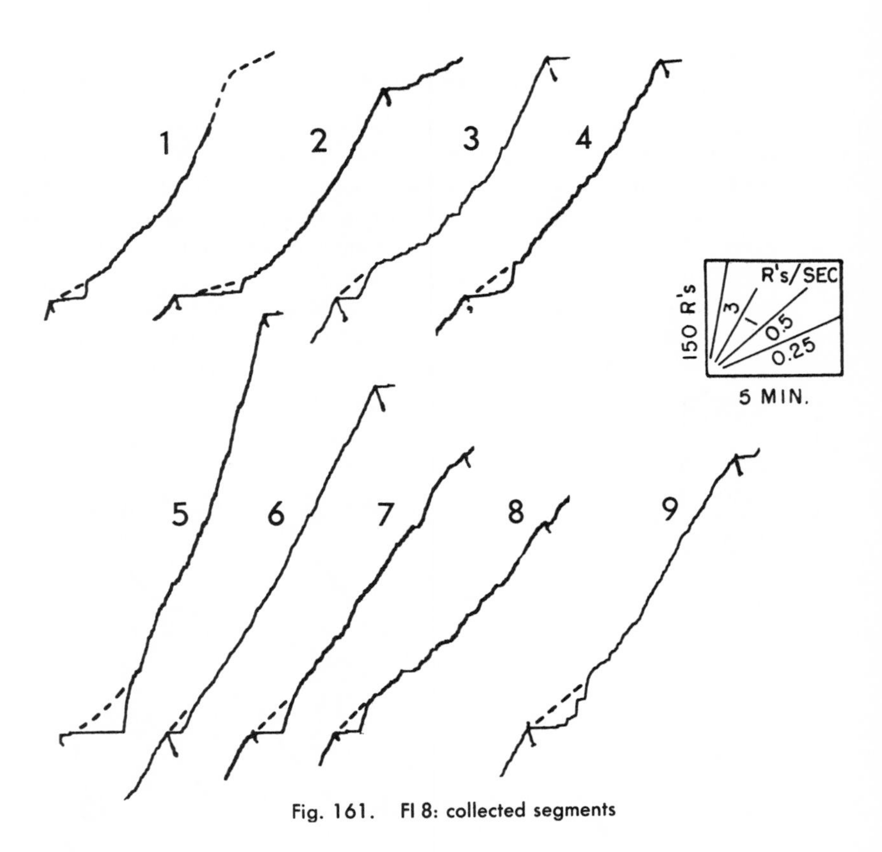

Fig. 161. FI 8: collected segments

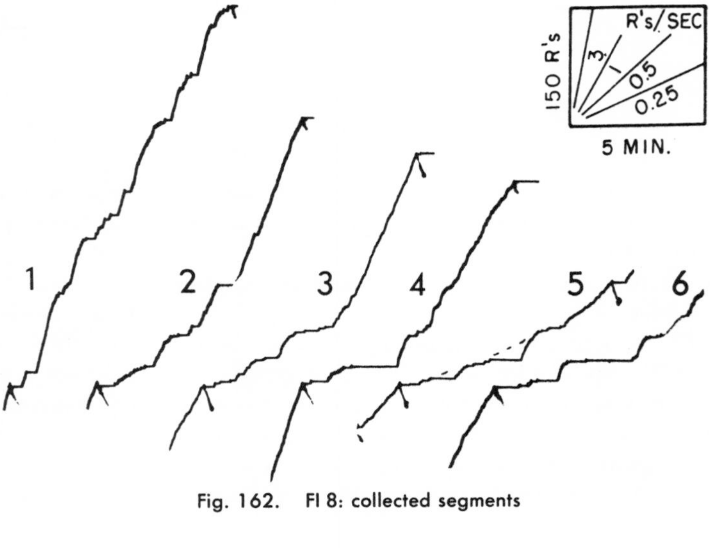

Fig. 162. FI 8: collected segments

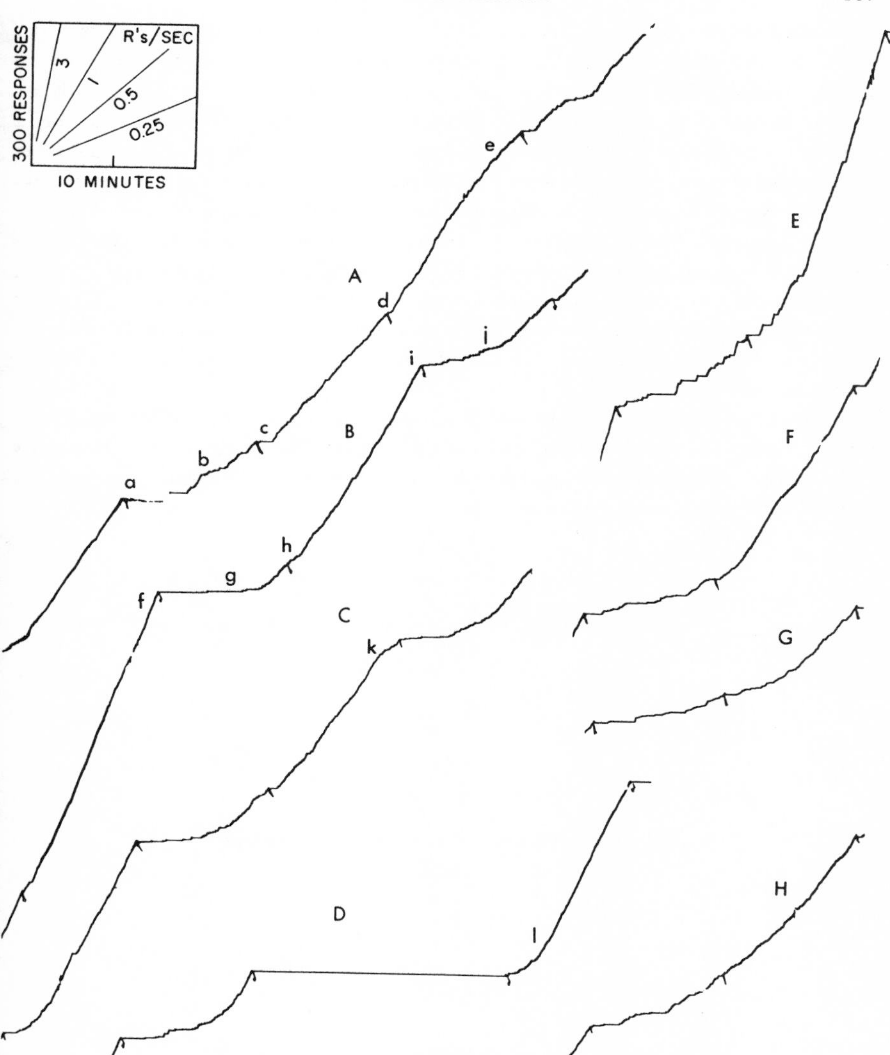

Fig. 163. FI 8: second-order effects

inforcement at *c* occurs after an irregular *low* rate and at a *low* count. It has an immediate effect in that it makes the situation just after reinforcement more favorable. Hence, responding begins fairly soon. The rate is still below maximal, so that the effect carries through the next reinforcement (at *d*). Both the count and rate become excessive, however, and negative curvature then develops at *e*.

Applying this interpretation to Record B, we could say that the very high count at reinforcement *f* produces the low rate at *g*; but the reinforcement at a low rate and low count at *h* reverses the effect. Therefore, responding begins early in the following interval and develops a high count and rate before reinforcement at *i*. This development in turn reduces the rate at *j*. Here, the second-order curve consists of only 2 intervals. In Record C the fact that a second-order effect during only 2 intervals shows some negative curvature at the very end suggests that the count at *k* has become excessive. Record D shows an example of an exceptionally long pause after reinforcement. The postponed reinforcement is received with only 1 response. The effect is to reinstate responding almost immediately, because the pause following reinforcement resembles the pause after which reinforcement has just been received. The bird quickly assumes the terminal rate at *l* and continues throughout the subsequent interval. The curves at Records E, F, G, and H show other examples of second-order effects at various over-all rates.

A second bird on FI 8 showed fewer deviations from a smoothly scalloped performance. Figure 164 shows the performance at 50 hours (Record A) and at 100 hours (Record B). Marked pauses and extended curvature tend to alternate with intervals in which the rate is constant throughout.

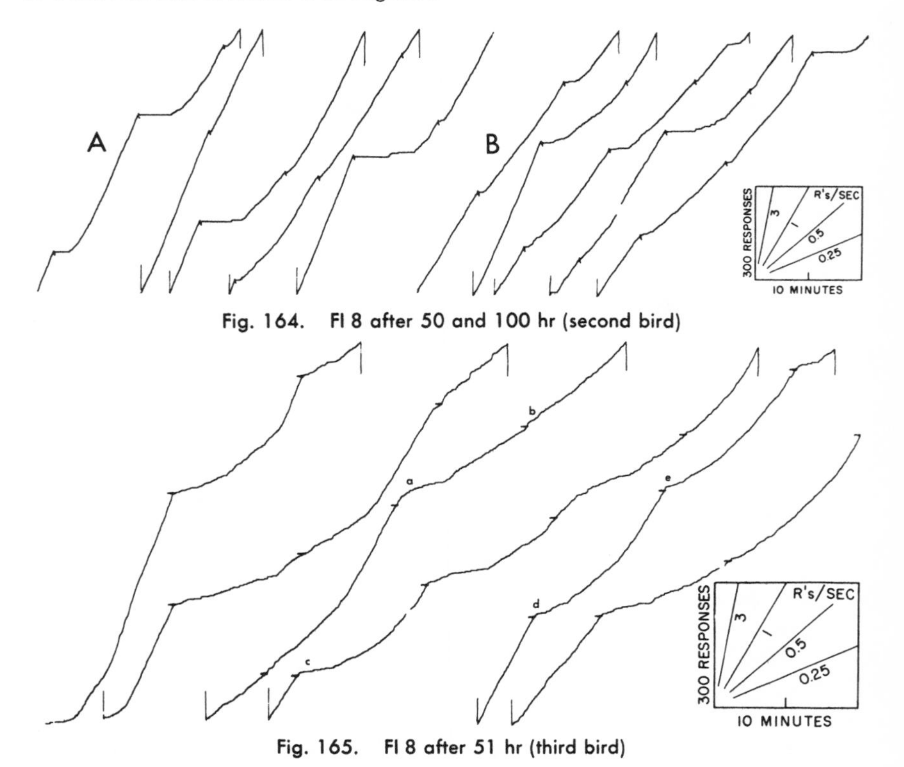

Fig. 164. FI 8 after 50 and 100 hr (second bird)

Fig. 165. FI 8 after 51 hr (third bird)

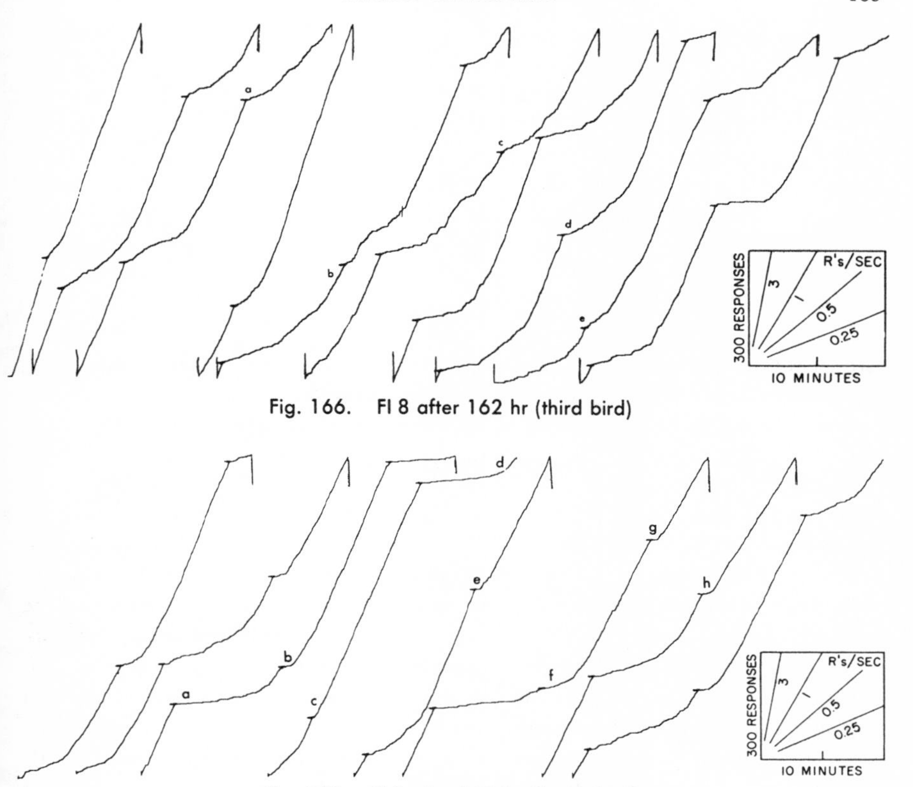

Fig. 166. FI 8 after 162 hr (third bird)

Fig. 167. FI 8 after 100 hr (fourth bird)

Figure 165 shows an intermediate performance on FI 8, 51 hours after the transition from crf shown in Fig. 122. The over-all rate is lower than in the preceding cases. The distinguishing feature is a species of "running-through." The terminal rate continues for from 5 to 20 responses after the reinforcement and then declines, often smoothly, to either a normal scallop (as at *c, d,* and *e*) or to a lower rate of responding, which may or may not be sustained for the remainder of the interval (as at *a* and *b*).

Figure 166, 105 hours later, still shows instances of responding just after reinforcement at the terminal-interval rate. When a pause follows reinforcement, it still does not exceed more than a few seconds, as at *a* and *d.* Rough second-order effects may be seen at *b, c,* and *e.* The final rate is considerably higher than before.

Another bird with the same history gave the performance shown in Fig. 167, 100 hours after crf. A high terminal rate develops early in the experiment, and running-through does not occur. Most of the interval segments are smooth curves with varying speeds of acceleration. At *c,* however, a pause of only a few seconds is followed by an abrupt shift to the terminal rate. At *a* and *d,* acceleration continues

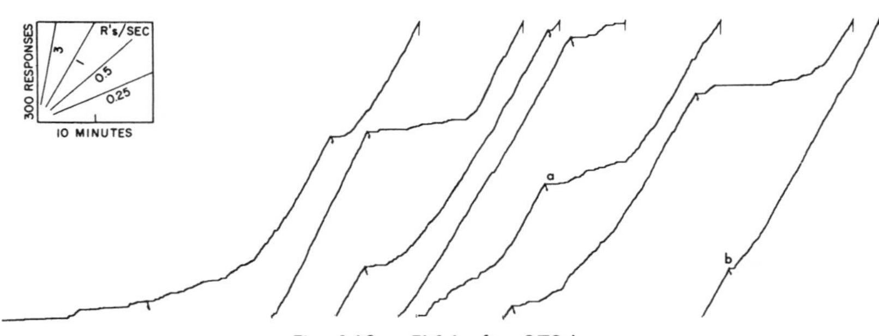

Fig. 168. FI 16 after 272 hr

throughout the whole interval. Where the bird begins to respond early in the interval, the uniform length of pause, as at *b, c, e,* and *g,* suggests a separate control of a low rate by the reinforcement. Second-order effects are evident at *b, f,* and *h.*

FI 16 or FI 17

Figures 168 and 169 show 2 successive daily sessions after 272 and 280 hours, respectively, on FI 16. These records show the principal features of a fixed-interval performance at this stage. In Fig. 168 the relatively low over-all rate is due to a very pronounced scalloping, where a low rate often extends through more than half the interval. Following the reinforcement, there is a consistent brief pause from approximately 30 seconds at *b* to over 1 minute at the start of the interval at *a.* Instances still occur where the terminal rate is reached soon after the reinforcement, but responding never begins immediately after reinforcement as in a run-through. The performance in the following session, in Fig. 169, shows a much higher over-all rate, which is due to an increase in the terminal rate from about 1 per second in the previous session to about 1.5 per second in this session. The curvature becomes more marked during the period and shows considerable uniformity. (Compare the last 6 intervals of the session.)

The schedule was maintained for a total of 338 hours. The last 2 sessions are shown in Fig. 170. Here, instances occur where responding begins immediately fol-

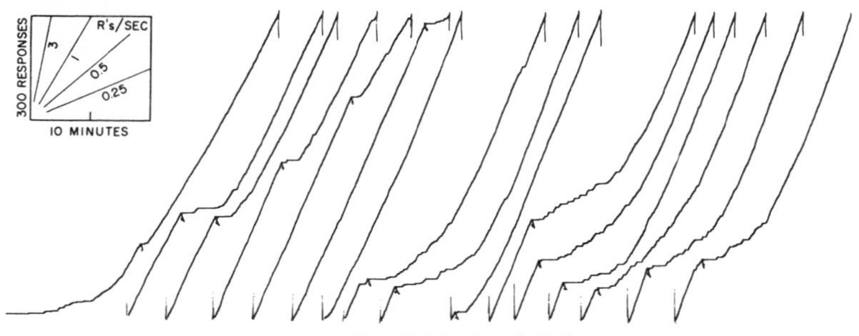

Fig. 169. FI 16 after 280 hr

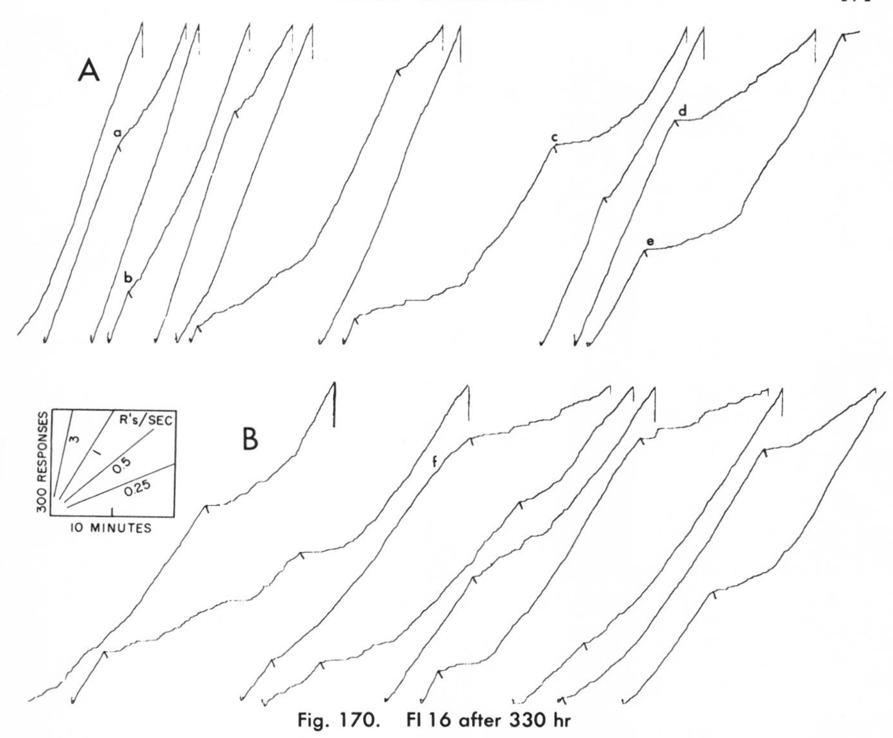

Fig. 170. FI 16 after 330 hr

lowing reinforcement, as at *a, b,* and *c*; and the pause following reinforcement varies from only a few seconds, as at *e,* to almost 2 minutes at *d.* Slight negative curvature may appear, as at *f.* The rate changes within the interval are now less reproducible from interval to interval than those in the earlier performance.

Figure 171 shows a late performance on FI 17 (100 hours after crf). (Early stages for this bird are represented by Fig. 136 and 137.) The curves at *a* and *d* show continuous acceleration through the interval. Other intervals show a more rapid acceleration to a terminal rate (as at *b* and *g*). The rate often declines toward the end of the interval (as at *c* and *e*). A short (reinforcement-controlled) pause occurs at the beginning of each interval. At *h* the terminal rate is not reached, and only a few responses occur during the interval. The bird characteristically showed a higher over-all rate of responding early in the session. The FI 17 reinforcement schedule was continued for a total of 334 hours.

A still further development is shown in Fig. 172, which contains the first 16 intervals of a session beginning after 277 hours of exposure to the schedule. The segments occurred in the order shown. At the end of each interval the recorder reset to the bottom of the graph; reinforcements are indicated by the reset line. The performance at this stage of development is radically changed. No instances occur here of a smooth

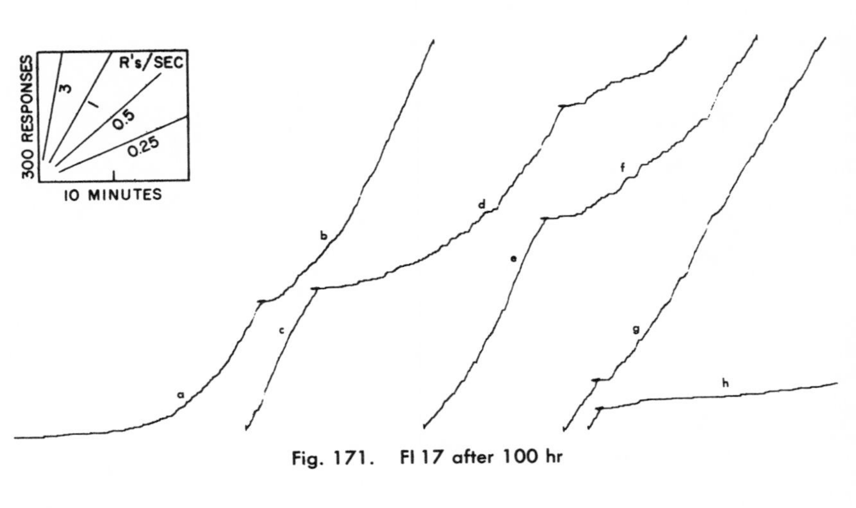

Fig. 171. FI 17 after 100 hr

Fig. 172. FI 17 after 277 hr

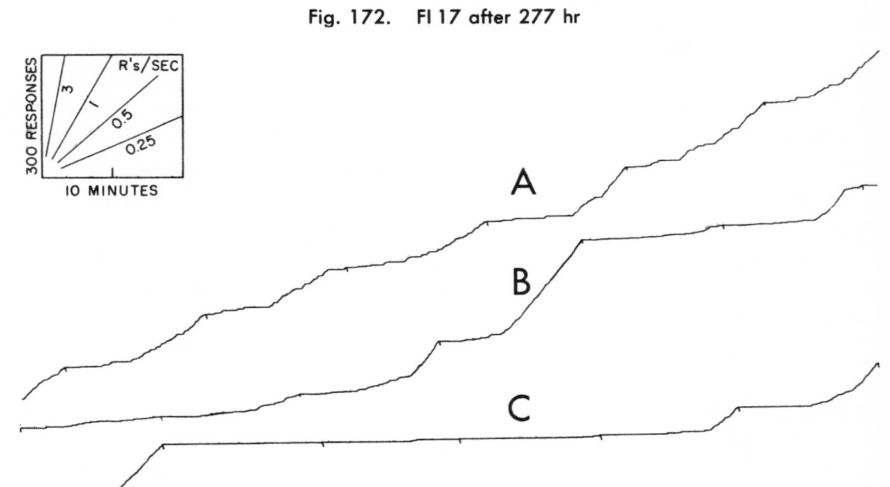

Fig. 173. Sustained FI 10 in the rat

acceleration during the interval. The characteristic pattern is a rapid assumption of the terminal rate after a pause of the order of 20 to 30 seconds after reinforcement. Extreme negative curvature appears (as at *a* and *b*). The number of responses per interval at reinforcement, as well as rate, varies widely. Such a performance interacting with the scheduling device produces no consistent set of contingencies.

The effect of a sustained FI 10 in rats

Figure 173 shows later performances for the rats responsible for Fig. 147 and 148. The low over-all rates are due mainly to the amount of reinforcement (0.05-gram pellet), which was later determined to be too small. The first rat, Record A (50 hours after crf), has a consistent pause following reinforcement, and curvature extends throughout most of the interval despite a rough grain. In Record B, for the second rat (62 hours after crf), the terminal rate is higher and the grain smoother, but many intervals show a very slow acceleration to a low rate. In Record C, for the third rat (140 hours after crf), the over-all rate is extremely low. Reinforcements are frequently postponed and are achieved with only 1 or 2 responses per interval. When responding begins, however, a fairly typical fixed-interval performance may be produced, as in the last interval shown.

These rats were then used on a mult FI avoidance schedule. (See Chapter Ten.) As a control condition in this experiment, the schedule of reinforcement under both stimuli was FI 10. At this stage of the experiment, we discovered that the reinforcement had been inadequate; and we increased the amount to 0.1 gram by loading the magazine with 2 pellets instead of the 1 which had been delivered previously. Under these conditions all 3 rats showed similar performances. Figure 174 shows a

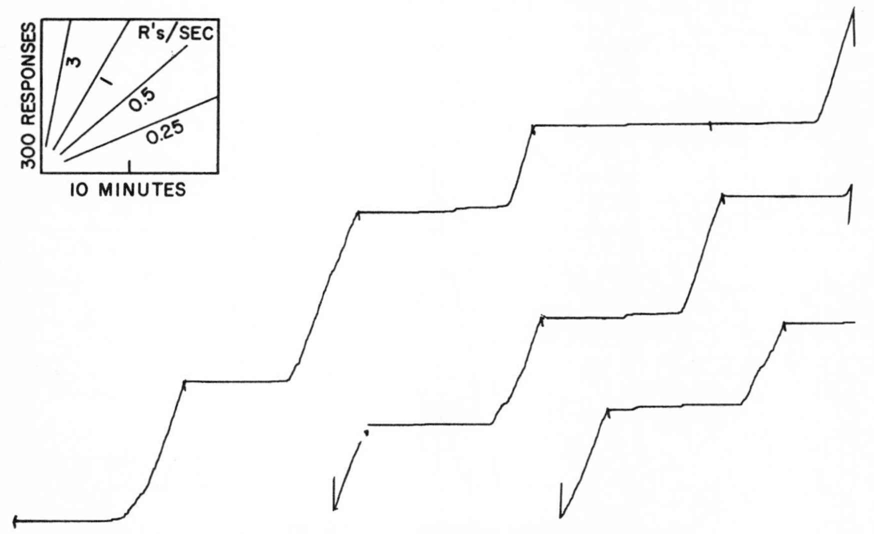

Fig. 174. FI 10 in the rat with larger amount of reinforcement

daily session after 94 hours on FI 10 alone after the mult FI 10 avoidance schedule. The performance from interval to interval is fairly reproducible. Occasionally, only a few responses may occur during an interval, as at *a*; but the usual performance is a period of 5 to 8 minutes when the rate is either zero or near zero followed by a quick acceleration to a terminal rate of from 1.5 to 2 responses per second.

TRANSITION FROM ONE VALUE OF FI TO ANOTHER

In the transitions from one value of FI to another the actual contingencies on the second value are the product of the new schedule and the performance generated by the old. Eventually, a fairly stable set of contingencies will arise under the new schedule. In an early experiment on this point the value of FI was advanced rapidly from 5 to 10, 15 to 25, and finally to 35. The performance was not allowed to stabilize at each value. The actual contingencies of reinforcement therefore varied widely. The result was an irregular performance, many parts of which suggest a variable-interval schedule (Chapter Six) or that special case of variable interval called a two-

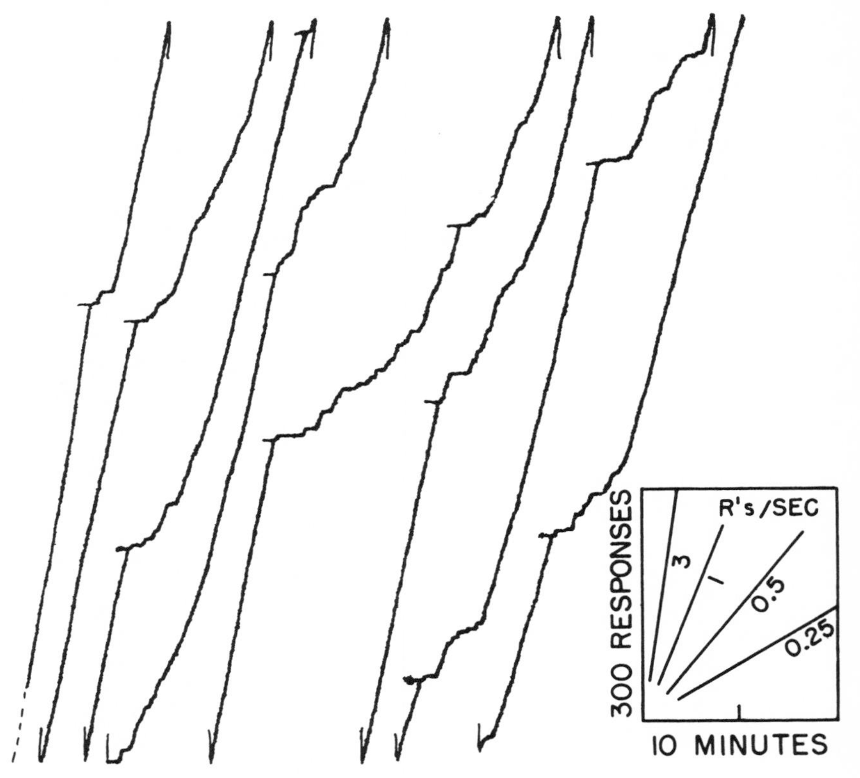

Fig. 175. Third session on FI 10 after FI 5

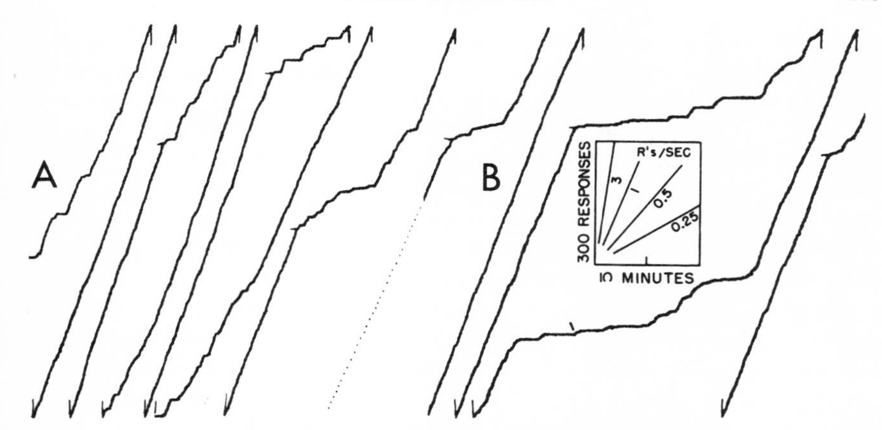

Fig. 176. Third and fourth sessions on FI 35 after FI 25

valued schedule. A terminal rate which is often maintained for many hundreds of responses develops quickly, though it may be preceded by quite irregular fluctuations in rate. The terminal rate often falls off fairly smoothly; this condition suggests extinction after a variable-interval schedule. Two samples for 1 bird are shown in Fig. 175 (the 3rd session on FI 10 after FI 5) and in Fig. 176 (two sessions on FI 35 after FI 25). Note the stable terminal rate, the occasional interval curvature, the knees and other irregularities early in each interval, and the running-through after many reinforcements.

FI 5 to FI 30

Figure 177 shows one transition from FI 5 to FI 10. The record begins with a fairly stable performance reached after 55 hours on FI 5. Running-through is especially marked. After a short period of slow responding, a high terminal rate is maintained throughout most of the interval. At the arrow the interval is increased to 10 minutes. The break in the first 10-minute interval at *a* is possibly due to a slight

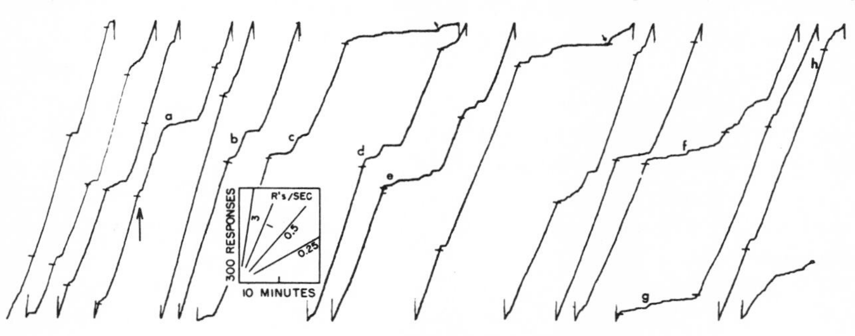

Fig. 177. Transition from FI 5 to FI 10

disturbance while the controlling circuit was changed. In the second 10-minute interval the terminal rate is held. During the remainder of the session the terminal rate declines somewhat, and the period of slow responding early in the interval is extended. At *e*, *f*, and *g* a low rate is maintained for half or more of the interval. At the end of the session, 3 intervals ending at *h* are run off with almost no pause following the reinforcement. Marked negative acceleration then occurs in the next interval. Small bursts of responses immediately after reinforcement continue to appear, varying from approximately 5 responses in interval *e* to approximately 30 responses at *d*. Examples of knees are evident at *b* and *c*.

Figure 178 gives the transition to FI 20 after 122 hours on FI 10. The 1st interval shows a scallop common under FI 10, and the terminal rate is maintained throughout the new 20-minute interval. This produces an exceptionally prolonged lower rate after reinforcement at *a*, but the number of responses emitted in the 2nd interval is still

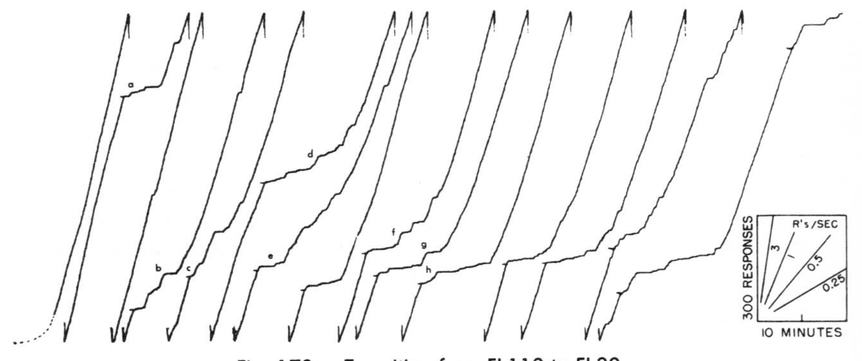

Fig. 178. Transition from FI 110 to FI 20

above 1000. Scalloping becomes progressively more extended as the session proceeds. At *d* the initial period of lower rate of responding extends over the major part of the interval. This is true for most of the rest of the session. Pauses sometimes follow reinforcements (as at *a*, *c*, and *h*), although they last only a few seconds and are probably based entirely on the effects of the recent reinforcement. Well-marked knees appear (as at *h*), and multiple knees (as at *b* and *f*). Many cases occur of fairly smooth acceleration to the terminal rate (as at *d* and *e*), though the curvature is marred by rough grain. As in the previous transition, the first effect is an increase in the number of responses emitted during the interval as the high terminal rate from the preceding schedule is maintained throughout the larger interval. A more characteristic FI performance emerges as the terminal rate of responding falls and the extent of the scallop becomes greater.

Figure 179 shows the final performance on FI 20 after 300 hours of exposure to the schedule. By this time the session begins with a pause and a fairly smooth and rapid acceleration to a terminal rate (as at *a*). The rest of the session is characterized by

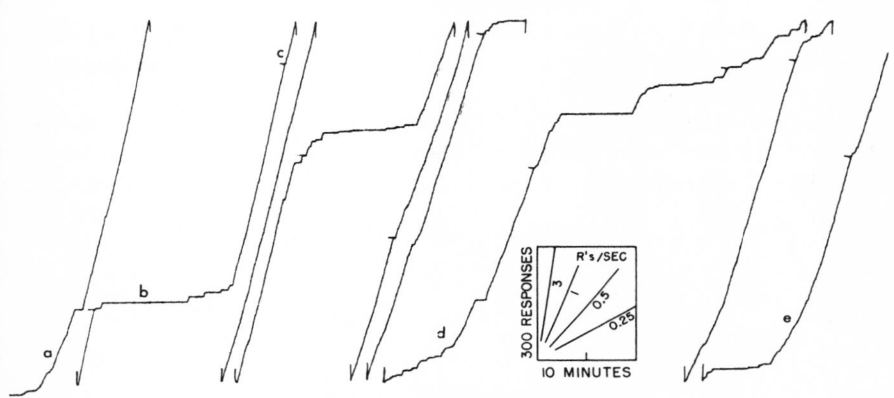

Fig. 179. FI 20 after 300 hr

a rough kind of alternation between performances (as at *b, d,* and *e*) where slow responding extends over more than half the interval and is followed by a high terminal rate, and a performance (as at *c*) where the rate is assumed immediately after reinforcement. A slight pause follows reinforcement, but generally some running-through occurs even when the terminal rate is not immediately assumed.

The transition from FI 20 to FI 30 was lost through a recorder failure, but the performance after 64 hours of exposure to FI 30 is shown in Fig. 180. The reinforcement at *a* is followed by a few responses at the terminal rate. After a short pause, the terminal rate continues without much change until the next reinforcement, at *b*, giving more than 2000 responses in this interval. The next interval shows running-through, a marked period at a low rate, and a shift to the terminal rate which continues with some grain (at *c*) until the next reinforcement. The following interval shows an early start at the terminal rate at *d*, a decline marked by rough grain at *e*, but a quick resumption of the terminal rate. The rate falls slightly just before reinforcement at *f*. The interval starting at *f* is low throughout. The following interval be-

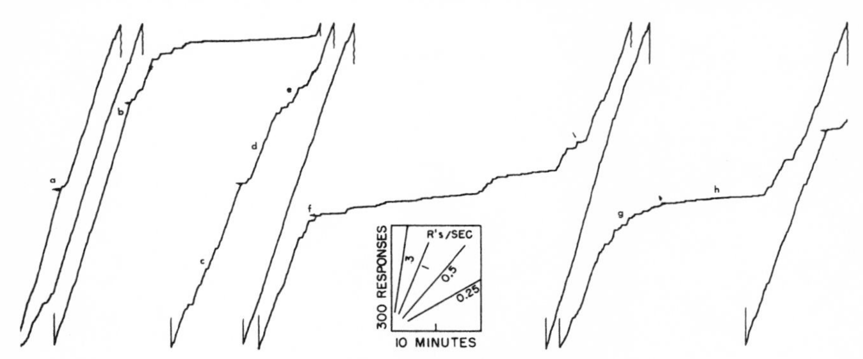

Fig. 180. FI 30 after 64 hr

gins at the terminal rate but leads to marked negative acceleration (at *g*) before the next reinforcement. The next interval begins with a low rate at *h*, followed by an acceleration with rough grain to the terminal rate.

Figure 181 shows an early performance on FI 45, recorded 29 hours after the change from FI 30. Except for the 1st interval, scalloping is not well-marked; and the grain is rough, even in long stretches at the terminal rate. Later, however, strong

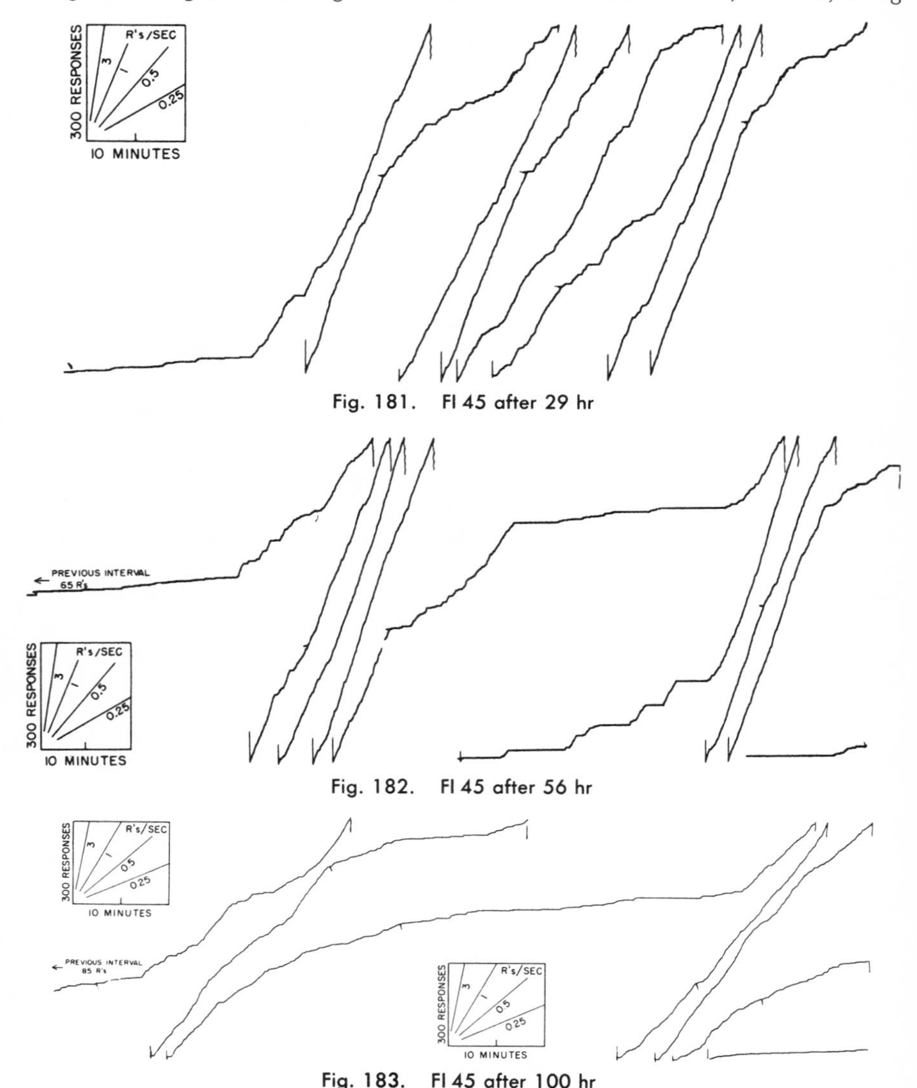

Fig. 181. FI 45 after 29 hr

Fig. 182. FI 45 after 56 hr

Fig. 183. FI 45 after 100 hr

scalloping develops (Fig. 182). Many intervals now contain very few responses. This decline in over-all rate at FI 45 continues. A session at 100 hours after the change from FI 30 is shown in Fig. 183.

A second bird showed a much slower transition from FI 10 to FI 20. Successive stages are shown in Fig. 184, 185, 186, and 187. The transition (not shown) is similar to those described: the terminal rate holds through the 1st extended interval, and the 2nd interval begins with a slight pause and a fairly abrupt shift to the terminal rate,

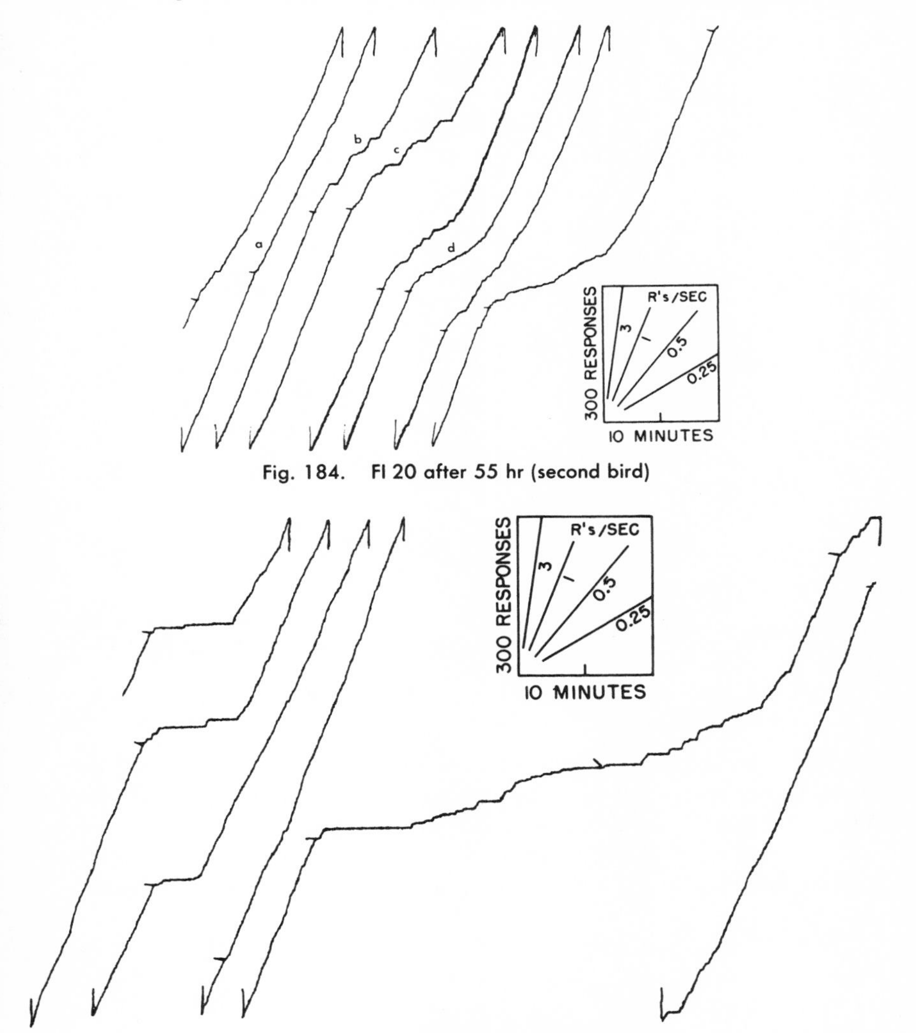

Fig. 184. FI 20 after 55 hr (second bird)

Fig. 185. FI 20 after 61 hr (second bird)

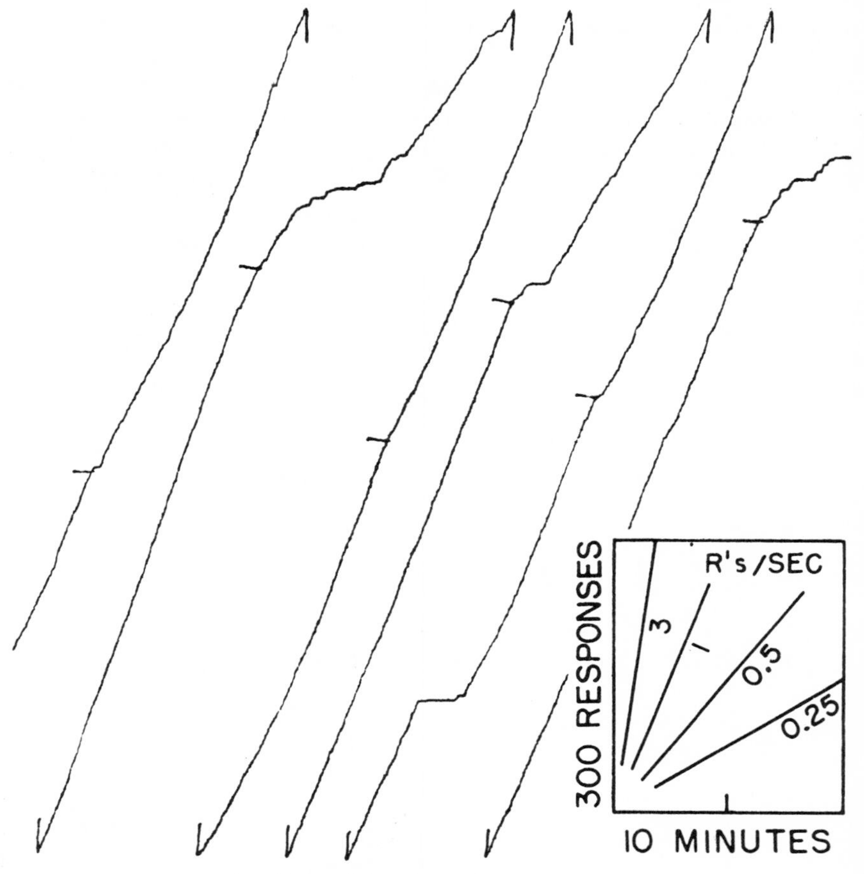

Fig. 186. FI 20 after 110 hr (second bird)

which falls off just before the end of the interval. Subsequent intervals showed substantial pauses, occasional knees, and rough grain. Figure 184 shows a portion of the session after 55 hours on FI 20. There is consistent running-through of the terminal rate. At *a* the terminal rate is then held throughout the next interval. In general, however, as at *b, c,* and *d,* the rate declines and then accelerates to the terminal value again.

Figure 185, at 61 hours after crf, shows a change in the interval curvature and a long second-order acceleration. Figure 186, at 110 hours after crf, shows a return to the earlier type of performance. By 220 hours after crf (Fig. 187), many intervals begin with deep scalloping (as at *a, d,* and *e*), others with marked knees and a briefer post-

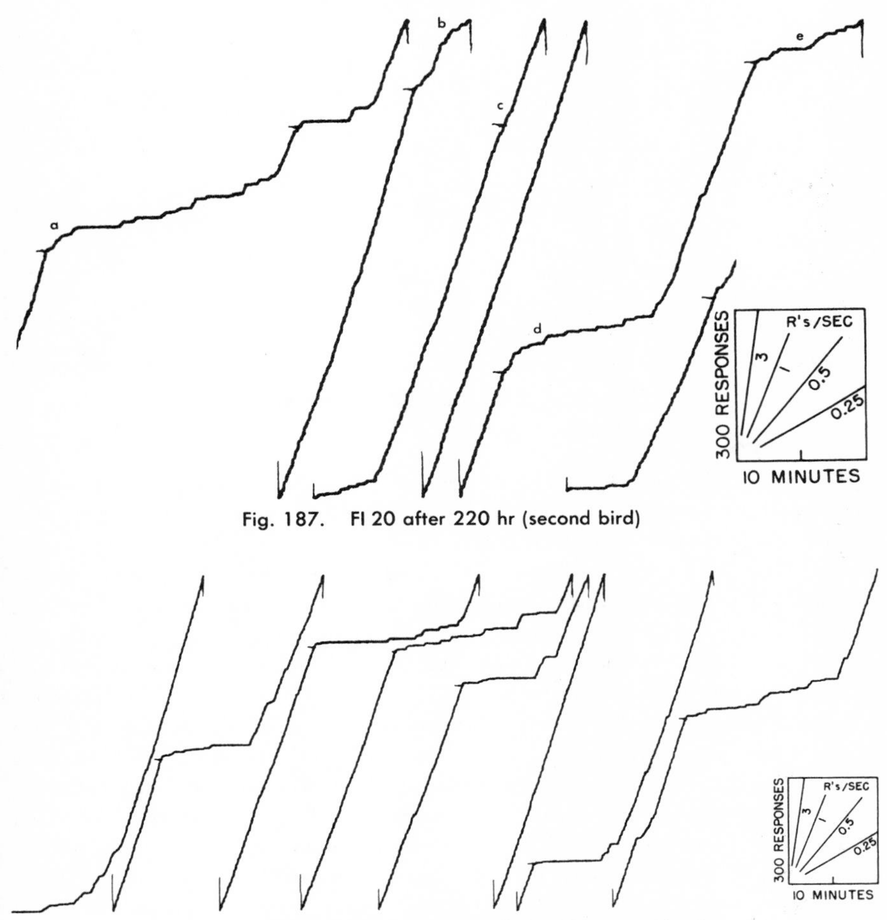

Fig. 187. FI 20 after 220 hr (second bird)

Fig. 188. First session on FI 30 after FI 20 (second bird)

ponement of the terminal rate (as at *b*), while some show the terminal rate throughout (as at *c*). No pausing occurs after the reinforcement. (The fine grain of this and the following record is an artifact due to excessive pen pressure above a fluted paper-feed cylinder.)

Figure 188 shows the transition to FI 30. The terminal rate, once assumed, usually continues until reinforcement; but the strains imposed by the longer interval reduce running-through to a few responses. Later, however, with further reinforcement on the longer interval, running-through increases, and whole intervals are sustained at the terminal rate, as in Fig. 189, taken 15 hours after the beginning of FI 30. Figure 190 shows the performance after 27 hours on FI 30.

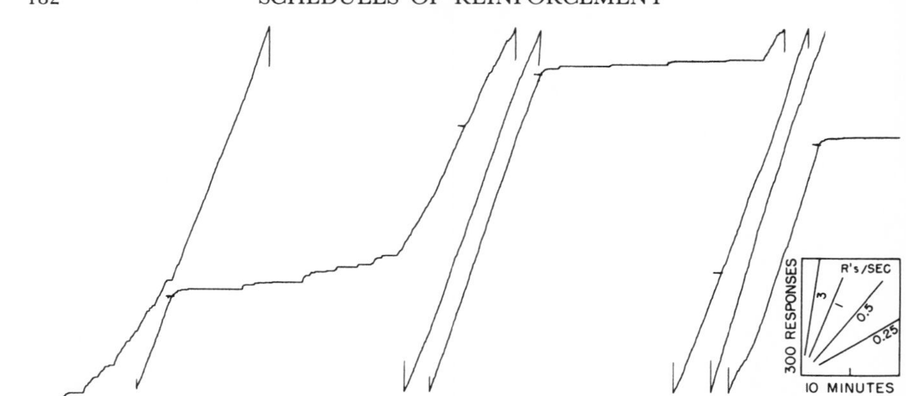

Fig. 189. FI 30 after 15 hr (second bird)

FI 30 to FI 45

The transition from FI 30 to FI 45 is shown in Fig. 191. The terminal rate assumed at *a* is held during the new interval except for the 2 pauses at *b* and *c*. The pause at *b* seems to follow a period of higher-than-normal rate. The rest of the session shows a few good scallops appropriate to the new interval, with some running-through. The rate is low at *d* and the grain rough, possibly because of the very large number of responses emitted per interval. A later pattern is exemplified by Fig. 192. The 1st interval shows a rough but deep scallop; but the 2nd is executed throughout at the terminal rate (from *a* to *b*). As a result the following interval shows a low rate and rough grain. A fairly smoothly accelerated curve begins at *c*, with reinforcement at the terminal rate at *d*. The next interval shows a low rate, negative curvature, and rough grain. Figure 193 shows a later performance 57 hours after the transition to FI 45. The fixed-interval segments from 1 session have been assembled in order of the number of responses. The numbers indicate the order in the session. At this stage in the development of FI 45 the performances from segment to segment are quite diverse. Only 3 segments, 2, 6, and 8, show a lower rate of responding early

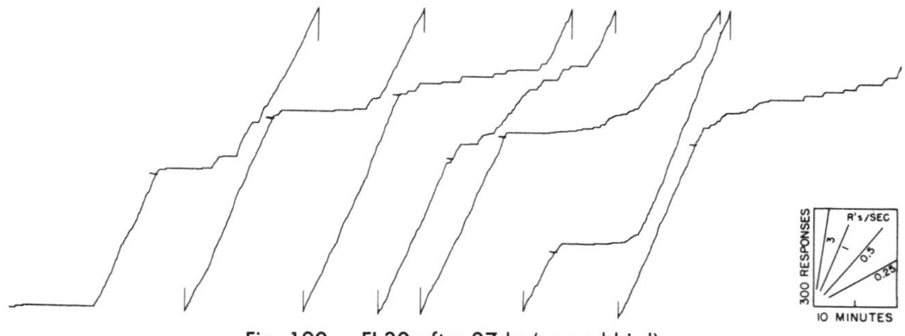

Fig. 190. FI 30 after 27 hr (second bird)

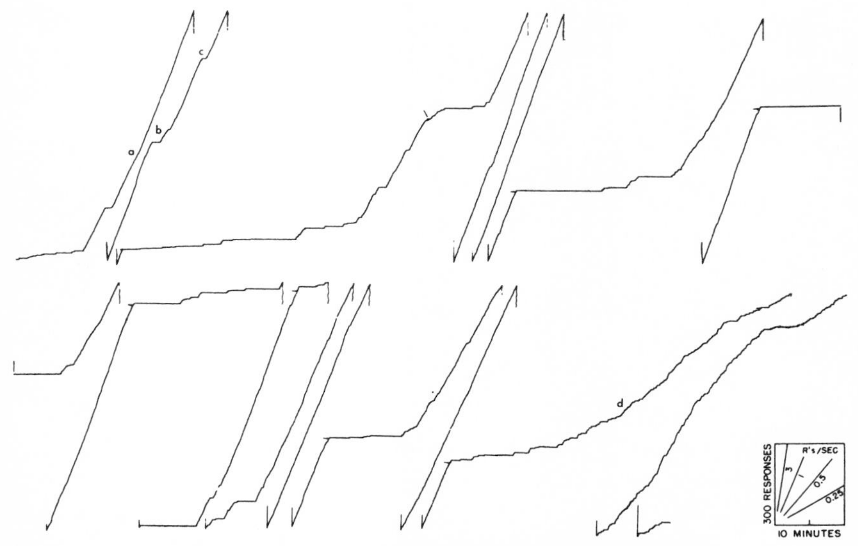

Fig. 191. First session on FI 45 after FI 30 (second bird)

in the interval and a later acceleration to a higher rate. Two segments, 3 and 9, show sustained responding immediately after the reinforcement, but the rate is not held.

The last recorded performance on FI 45 is shown in Fig. 194 in order of the number of responses. By this time the effect of the fixed-interval reinforcement is more clearly evident. Most of the curves show responding immediately after the reinforcement, although the segments at *a, b,* and *c* begin with substantial pauses. The maximal rate of responding has dropped to about 1.25 responses per second, compared with slightly less than 2 per second in Fig. 192.

At one point in the experiment on FI 45 just described the bird was briefly extin-

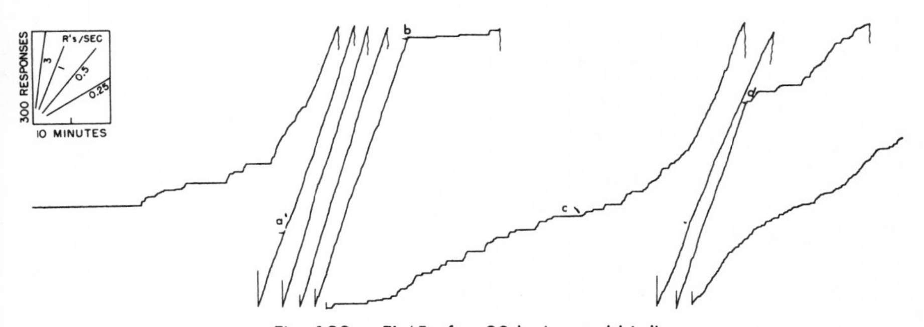

Fig. 192. FI 45 after 30 hr (second bird)

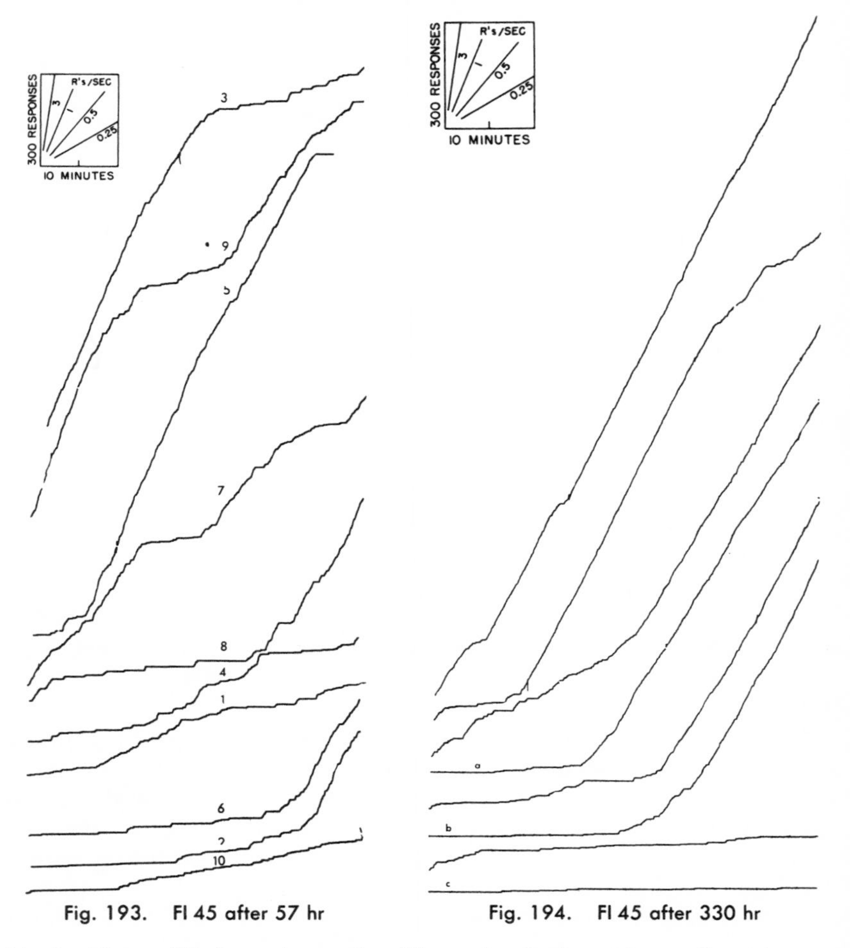

Fig. 193. FI 45 after 57 hr

Fig. 194. FI 45 after 330 hr

guished. Figure 195 shows the result. The session begins with a scallop (*a*) appropriate to the prevailing intervals of reinforcement; and the terminal rate is held with only a slight decline in rate until *b*, 97 minutes after the start of the experiment. At this point the rate drops to zero. The resulting pause initiates another extended scallop, reaching a somewhat lower terminal rate marked by greater irregularity of grain, which is sustained until *c*, 67 minutes after point *b*, where the rate falls to zero again. A 3rd slow acceleration shows a small knee at *d*. A somewhat lower sustained rate then declines gradually as the grain grows rougher, at *e*.

Figure 196 shows a fairly stable performance on FI 39 after 240 hours, arranged by the number of responses, for a third pigeon with a history of more than 700 hours on various variable-interval schedules. The over-all rate of responding is relatively low. Many intervals frequently do not show the terminal rate. Some segments are

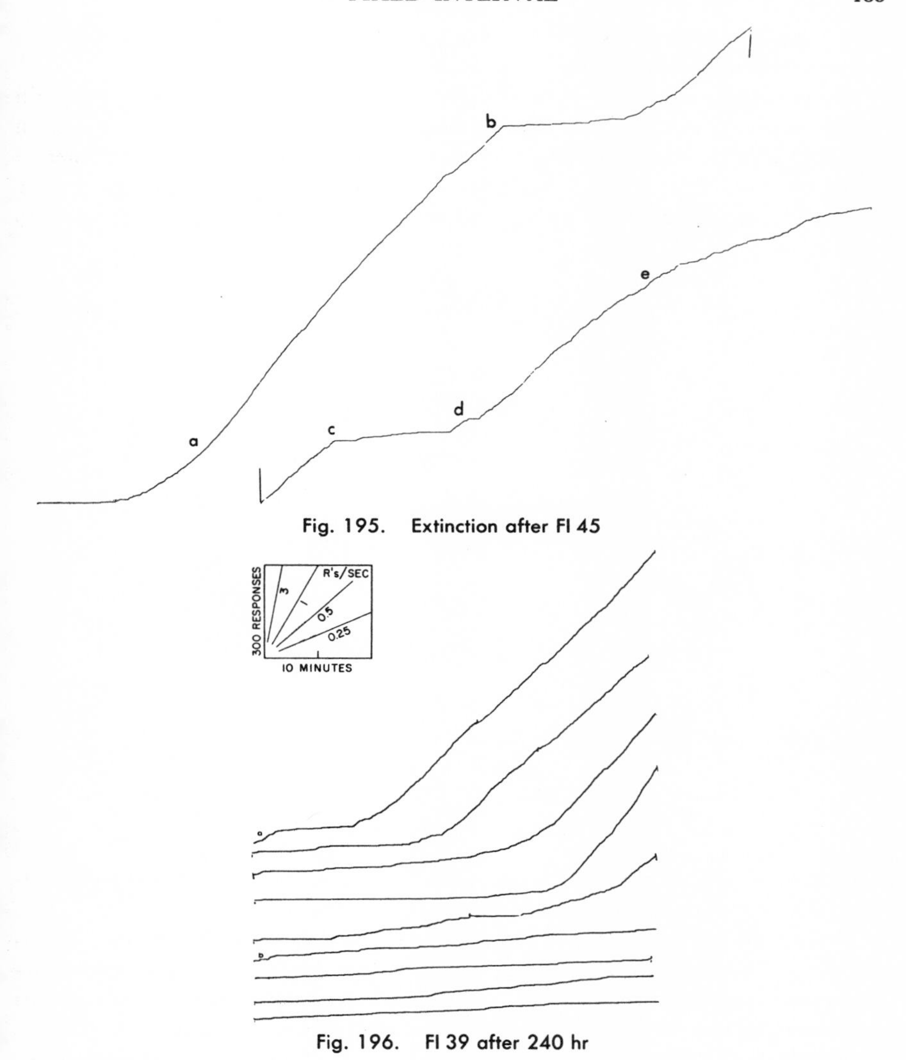

Fig. 195. Extinction after FI 45

Fig. 196. FI 39 after 240 hr

roughly scalloped, but a few scattered responses usually occur shortly after reinforcement.

TIME OUT

The effect of time out on FI 45

Time out after reinforcement. Figure 194 showed a fairly stable performance under FI 45. In the following session a 20-minute TO occurred after each reinforcement.

The complete 1st session is shown in Fig. 197, where the segments are arranged in terms of the number of responses. The order of occurrence is indicated.

The principal effect of the TO is to prevent running-through. All segments begin with pauses. The over-all rate of responding also shows a decline. The 2nd session with TO after reinforcement (Fig. 198) shows a better adjustment to fixed-interval reinforcement. Each segment begins with a pause (ranging from approximately 1 minute at *a* to 6 minutes at *c* and *d*); and all except the last show an acceleration to

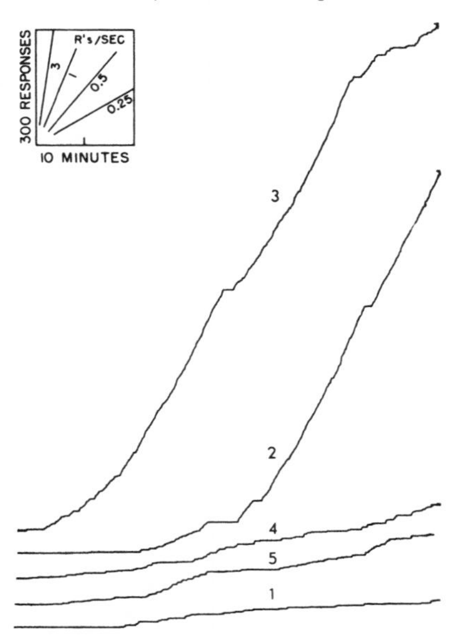

Fig. 197. First TO on FI 45

a higher rate. In the segment marked *b* the terminal rate is reached before the middle of the interval, and negative curvature appears before the end.

Time out after alternate reinforcements. We made a further effort to determine the effect of a TO by arranging a 20-minute TO after every other reinforcement, the intervening cases serving for comparison. Curves for 2 sessions were divided into segments and separated according to the location of the TO. Segments preceded by TO are assembled in Fig. 199; those not preceded by TO, in Fig. 200. They are arranged in order of the number of responses emitted in the interval. The preceding TO produces a substantial pause at the start of all segments. The curves are for the most part unevenly scalloped, however, and the acceleration is marked by rough grain, with responses occurring in bursts of 15 to 30. The alternate segments which were not preceded by TO (Fig. 200) show a much higher over-all rate of responding. Some responding occurs immediately after reinforcement, although at *a* and *b* this is followed by a zero rate before the acceleration to a terminal rate. The terminal rate is some-

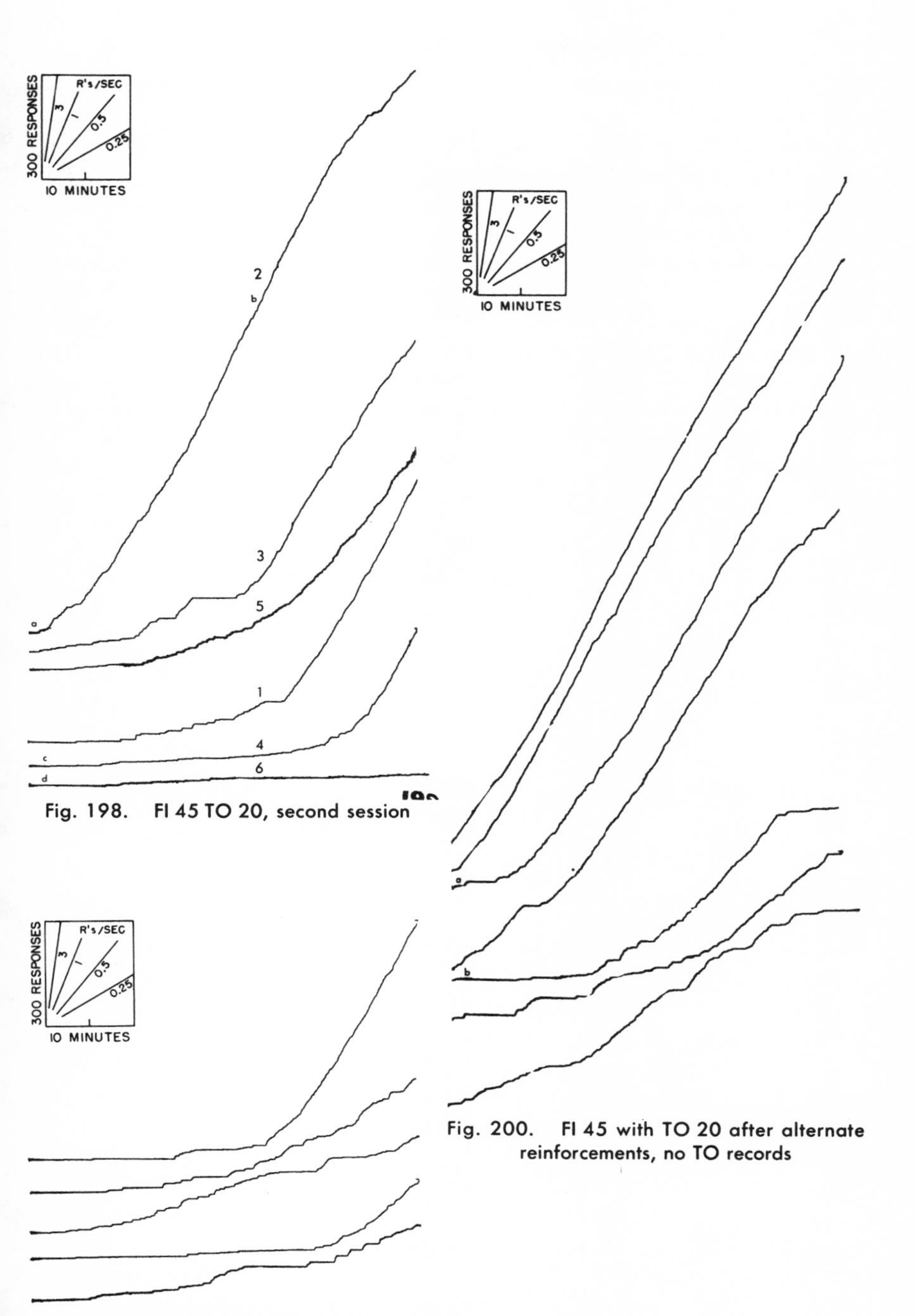

Fig. 198. FI 45 TO 20, second session

Fig. 199. FI 45 with TO 20 after alternate reinforcements, TO records

Fig. 200. FI 45 with TO 20 after alternate reinforcements, no TO records

times reached immediately (as in the 1st segment, which later shows a slight negative acceleration).

A complete daily session 50 hours after the alternating TO procedure was begun is shown in Fig. 201, in which the segments appear in their actual order. The TOs of 20 minutes are marked by arrows. The bird starts slowly at *a*, the beginning of the session, and the 1st reinforcement is received at *b*, after only scattered responding. No time out follows this reinforcement, and the following interval also begins at a low rate, at *c*. This interval is completed with an extended and fairly smooth scallop to the 2nd reinforcement, at *d*, which is followed by a 20-minute TO. Although the rate of responding just before the reinforcement at *c* was high, the following interval begins with a pause and ends with a fairly typical acceleration. The 4th interval, at *e*, is not preceded by a TO, and the high rate prevailing just before the reinforcement continues throughout the whole interval until the reinforcement, at *f*. A 20-minute TO follows at *g*, and the following interval contains only a few responses. A reinforcement becomes available at *h*, but is not received until approximately 5 minutes later.

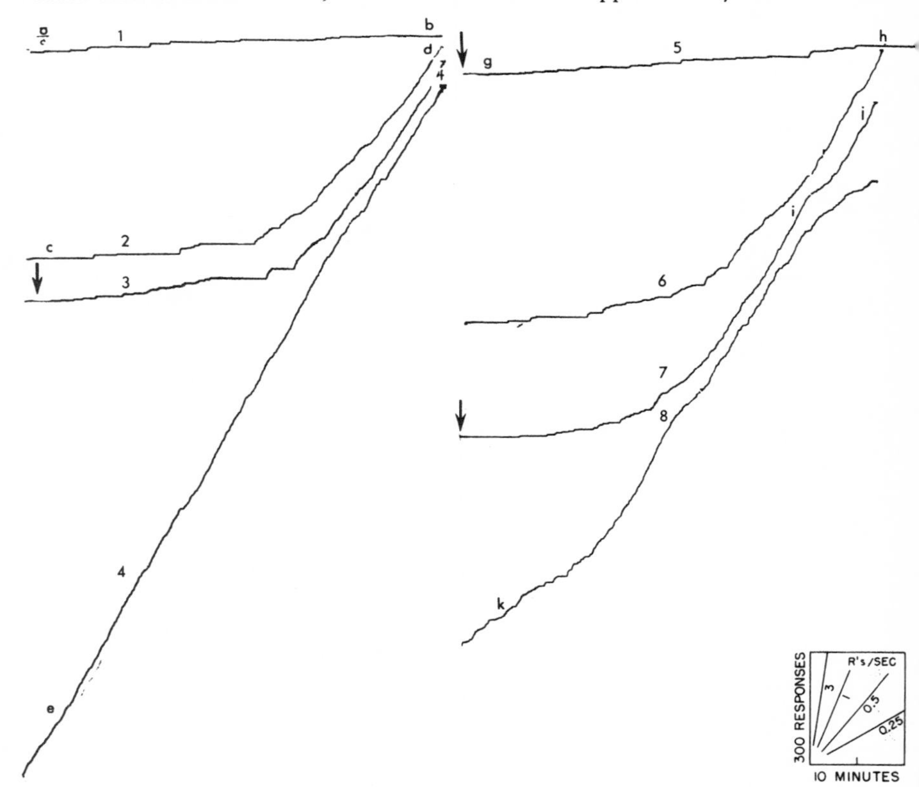

Fig. 201. FI 45 with TO 20 after alternate reinforcements after 50 hr

Although the next interval is not preceded by a TO, it begins with a pause and the rate accelerates to a high value. The next interval, following a TO, begins with a long pause and shows a typical FI acceleration. A rate somewhat higher than usual is reached at *i*; and this rate is followed by a drop to the normal terminal rate at *j*. No time out follows reinforcement at *j*, and the subsequent interval begins at a substantial rate of responding, at *k*.

In summary, this figure confirms the general effect of the time out seen in a comparison of Fig. 199 and 200. The 2 instances when an interval beginning without TO shows a pause and a normal acceleration to a terminal rate are preceded by intervals in which only a few responses occurred.

The 1st intervals of the 2 sessions are omitted from Fig. 199 and 200 because they have no relevance to the problem of the effect of the preceding interval. Actually, the 1st segment could be thought of as following a 15-hour period of TO. The 1st intervals from 3 sessions each for 2 birds have been collected in Fig. 202 and arranged by the number of responses. They all resemble intervals after TO, with either a consistent pause and acceleration following the start of the interval or a very low rate continuing throughout the interval.

In these experiments a TO after reinforcement results in a pause at the beginning of the next interval which may or may not be followed by an acceleration to a terminal rate. When TO is omitted after reinforcement, some responding usually occurs immediately (running-through). The over-all rate is highest when TO is omitted.

The difference is presumably attributable to the loss of the effect of the preceding terminal rate during the TO. A strict alternation of TO and no TO encourages a difference. The higher rate following no TO also follows an interval with relatively few responses, because this interval in turn was preceded by a TO.

Effective duration of time out. In order to determine the minimal effective duration of TO in an interval performance, 0-, 10-, 20-, and 30-minute TOs were programmed to alternate after reinforcements. This procedure was in force for 9 sessions. Analysis of the results showed no differential effect between TOs of 10, 20, or 30 minutes, either upon the magnitude of the pause following reinforcement or on the amount of time elapsing before the terminal rate is reached. Figures 203 through 207 reproduce all intervals except for segments following no TO and 8 anomalous segments. They have been arranged roughly in order of the number of responses emitted in the interval and collected in terms of similarity of curvature. They provide an example of the performance under a large fixed interval with a TO after reinforcement, and illustrate the kinds of curvature which are produced by this type of schedule. Figure 207 shows the anomalous segments. The principal feature is the negative curvature which sets in before the end of the interval.

Interspersed among the performances shown in Fig. 203 through 207 are segments preceded by no TO after reinforcement. These show a large number of responses per interval, as in the previous experiment.

The minimal effective TO was then sought among smaller values. A full session

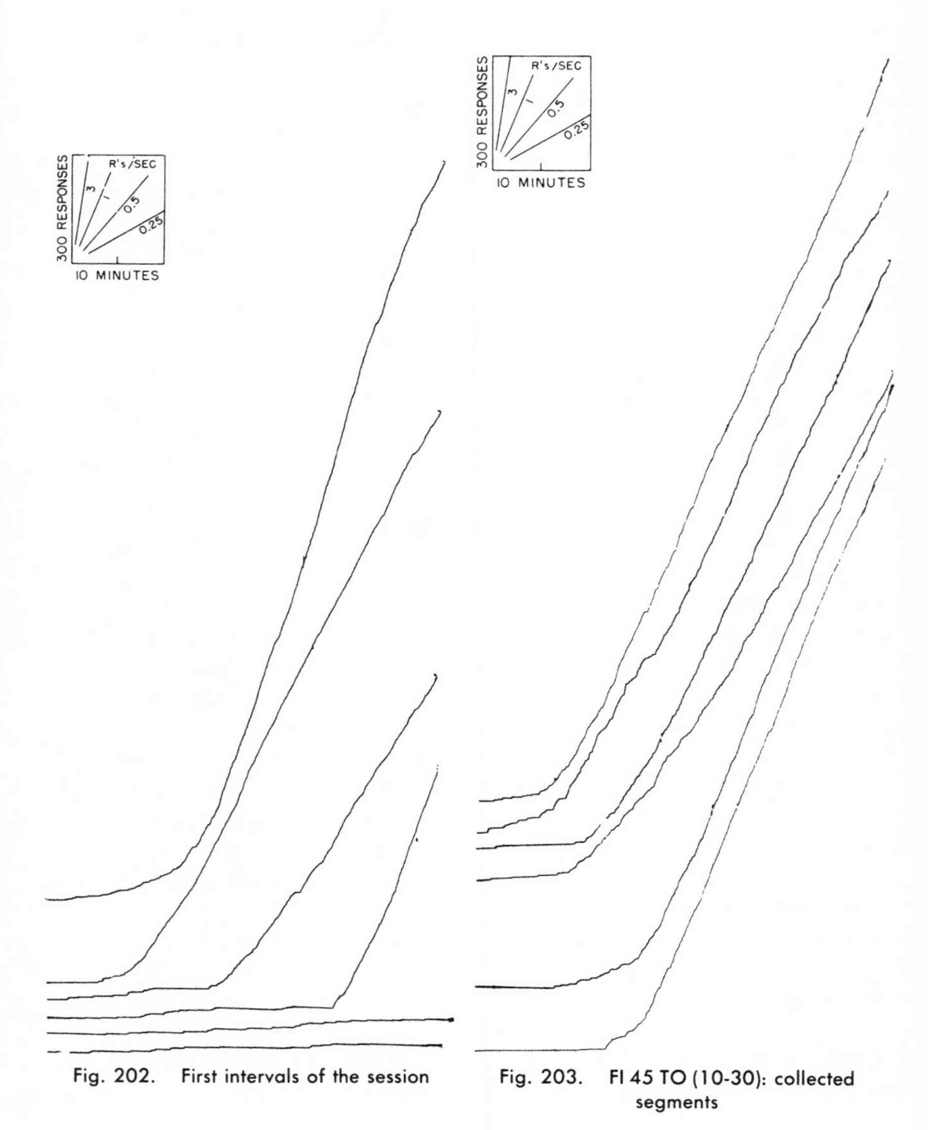

Fig. 202. First intervals of the session

Fig. 203. FI 45 TO (10-30): collected segments

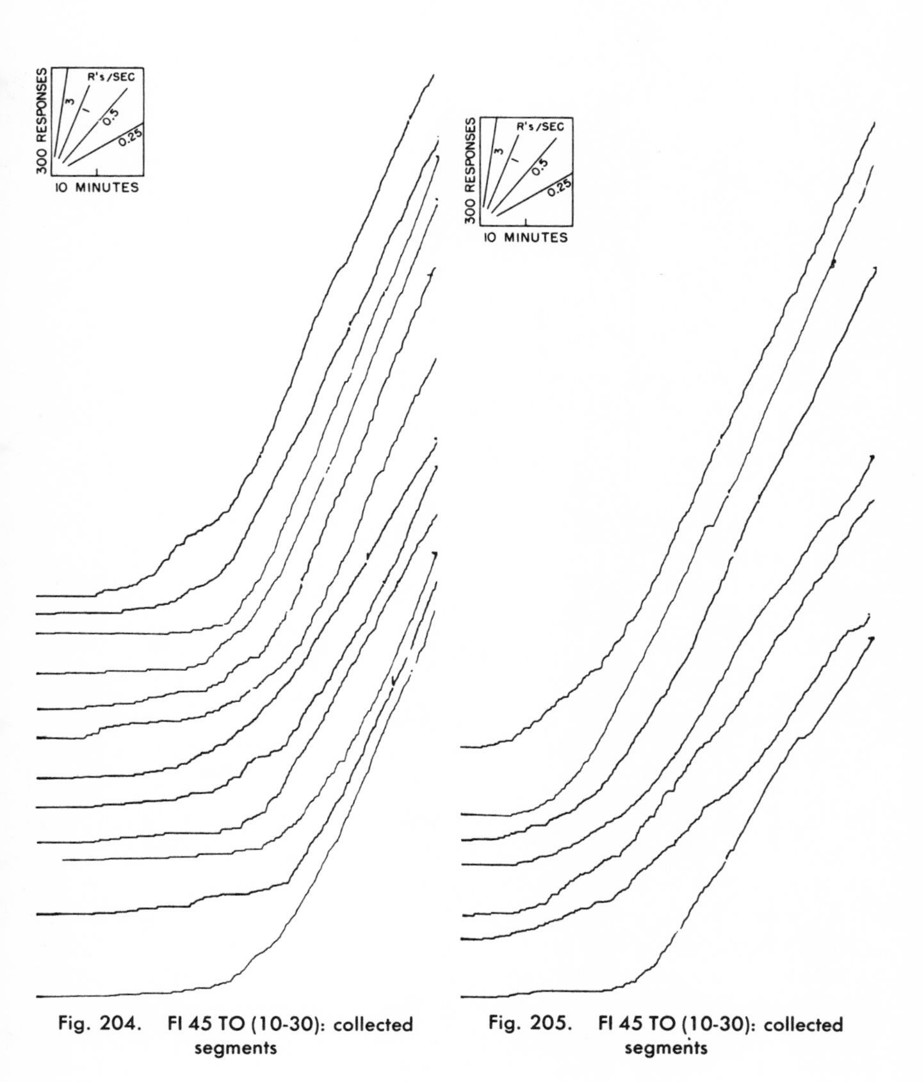

Fig. 204. FI 45 TO (10-30): collected segments

Fig. 205. FI 45 TO (10-30): collected segments

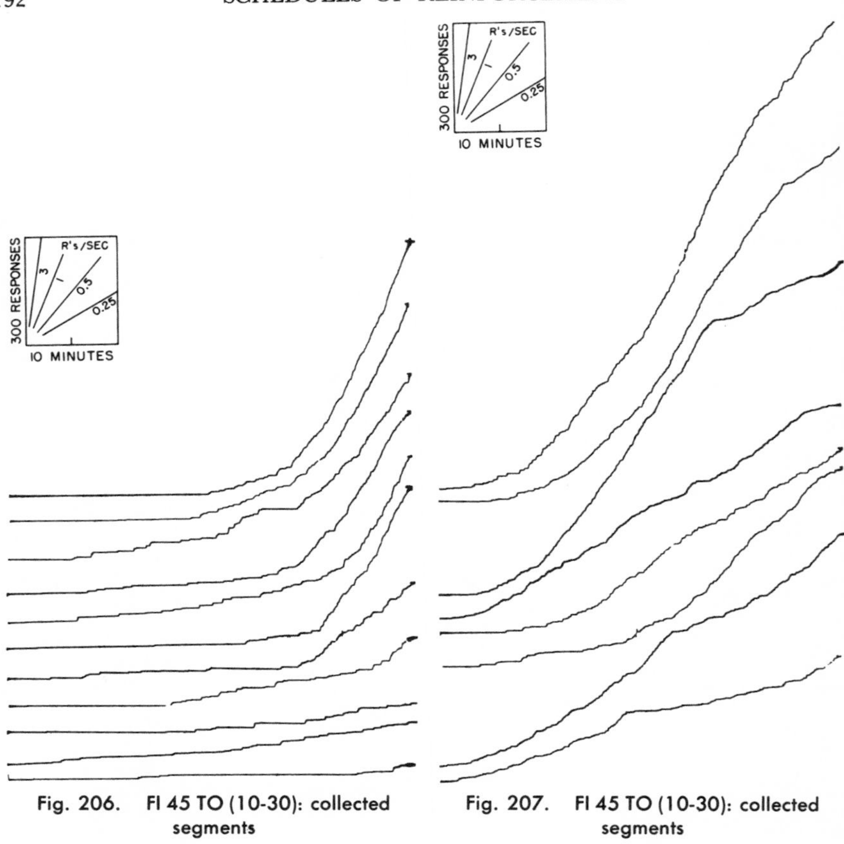

Fig. 206. FI 45 TO (10-30): collected segments

Fig. 207. FI 45 TO (10-30): collected segments

each was devoted to TOs of 0, 1, 3, 7, and 10 minutes. The effect was largely on the performance immediately following the reinforcement; but this condition, in turn, affected the curvature and general orderliness of the FI 45 performance. In order to compare the effects of these different lengths of TO, portions of the curves in the region immediately preceding and following reinforcement have been collected for 2 complete sessions on each value in Fig. 208. Column A shows all instances of reinforcement which were not followed by TO. The top 8 instances show a rapid assumption of the terminal fixed-interval rate, with some responding immediately after reinforcement in 5 out of the 8 cases. In the next 5 instances, substantial pauses either follow the reinforcement or a small amount of running-through occurs. The bottom 3 show a low rate both before and after reinforcement. Column B for a 1-minute TO resembles Column A fairly closely. The curves are irregular; and, in general, some responding after reinforcement occurs which may continue at the terminal rate or lead to pauses with a return to the terminal rate later. The effect of a 3-minute TO in Column C is marked. Responding occurs immediately after TO in only 3 instances.

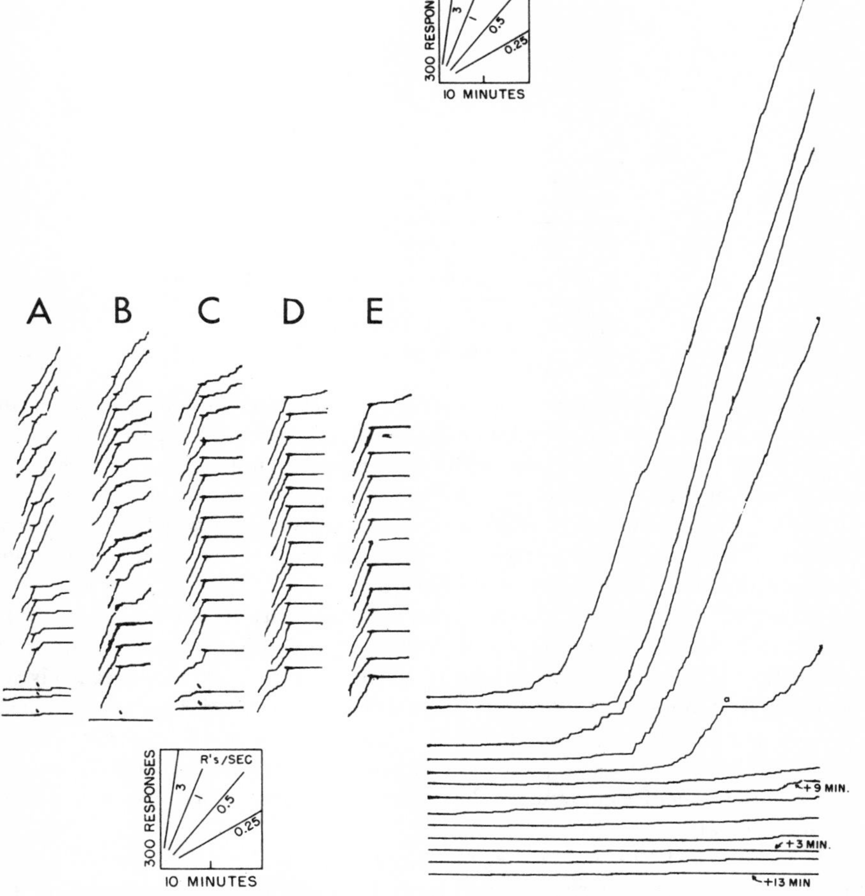

Fig. 208. FI 45 TO (0-10)

Fig. 209. FI 45 TO 20 after 23 hr (second bird)

The TOs of 7 and 10 minutes (Columns D and E) always yield a complete cessation of responding immediately following reinforcement.

The effective duration of TO for preventing carryover from the preceding interval is in the region of 3 to 7 minutes. Longer TOs should be used to minimize the interaction between successive intervals, whenever the TO is not too costly in experimental time.

Second bird. For the second bird the immediate effect of TO on FI 45 was lost because of apparatus difficulties; but a performance after 23 hours of reinforcement with TO after reinforcement is shown in Fig. 209, where the segments have been arranged in terms of the number of responses in the interval. The over-all rate of responding fell

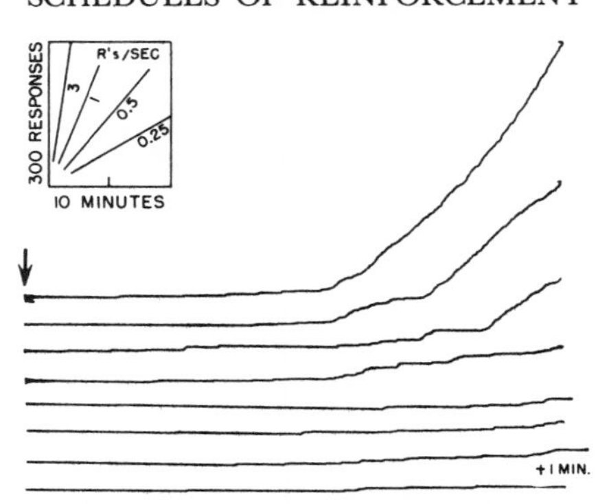

Fig. 210. FI 45 with TO 20 after alternate reinforcements, TO record

below that which prevailed previously without TO after reinforcement, as, for example, in Fig. 183. But where a substantial level of responding does occur, it follows the pattern of a period of acceleration with rough grain to a terminal rate, which is generally held until the reinforcement. A break followed by a 2nd acceleration is apparent at *a*. The over-all rate of responding is low because of the very frequent appearance of intervals in which the terminal rate is not reached even though responding begins within the first 15 minutes of the interval.

When TOs are arranged after alternate reinforcements, this bird confirms the result already noted. Segments from a single long session 55 hours after exposure to alternating TOs have been separated into 2 figures. Figure 210 shows segments preceded by a 20-minute TO. Long pauses and delayed acceleration are usual. Figure 211 shows segments not preceded by TOs. Segments 1, 4, and 5 show some FI scalloping with fairly consistent positive acceleration. Segments 2, 6, and 7, however, show marked negative curvature. The mean over-all rate is, of course, much higher than in Fig. 210.

In Fig. 212, 213, 214, and 215, segments from 3 complete daily sessions on FI 45 TO have been arranged in terms of the type of curvature. Within each figure the segments are arranged roughly in order of the number of responses. The first 3 segments from these sessions are reproduced separately in Fig. 216.

Extinction was carried out after 490 hours on FI 45 TO. The previous history was 620 hours on FI 5 to FI 45 after crf. The curve is slightly displaced in the coordinate frame of Fig. 217. The performance begins with the characteristic scallop (at *a*), and the terminal rate is maintained with only a slight decline for about 7000 responses. A sharp break occurs at *b* to essentially a zero rate, persisting for almost 2 hours. A 2nd interval scallop then follows at *c*, and the terminal rate is sustained for almost 2000 responses before it falls off with rough grain to another period of low rate at *d*. This period is again followed by an interval scallop and eventually by a more abrupt break to a zero rate of responding. A few brief bursts near the terminal rate

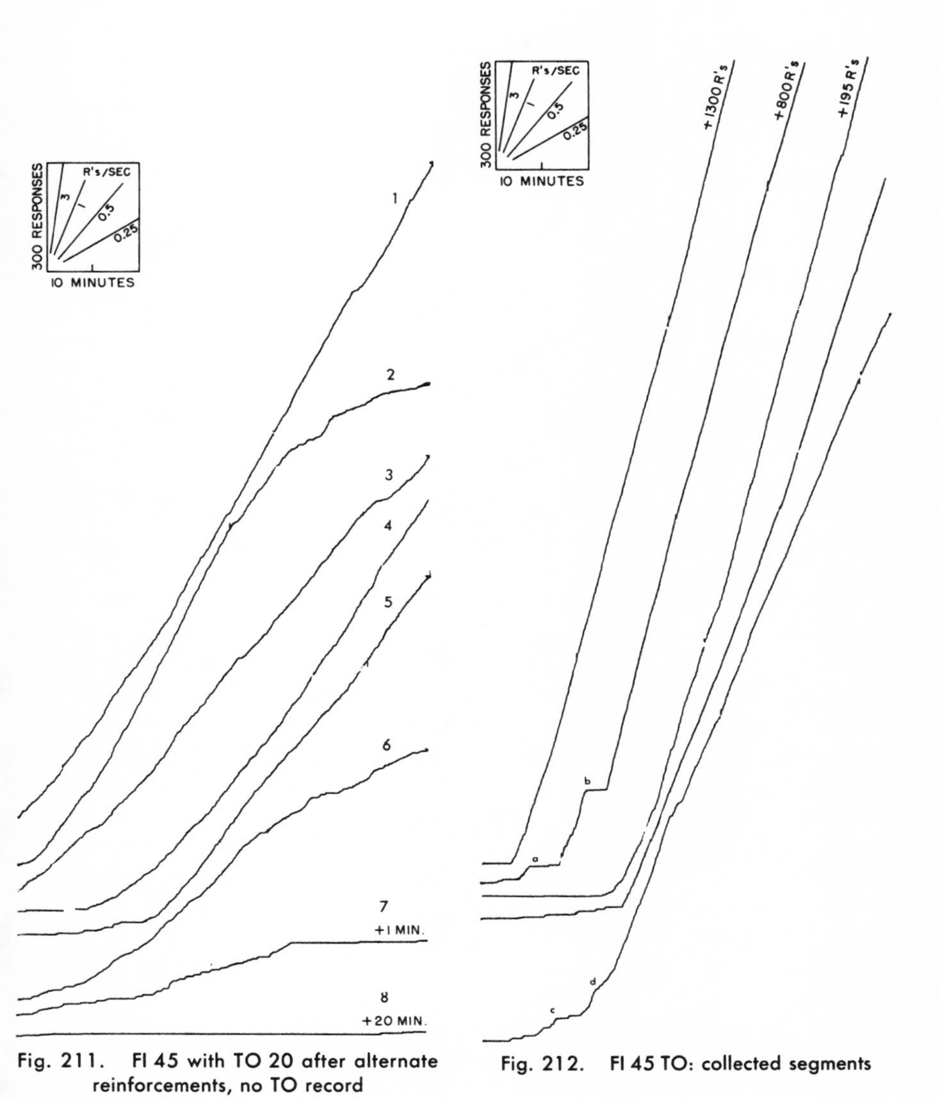

Fig. 211. FI 45 with TO 20 after alternate reinforcements, no TO record

Fig. 212. FI 45 TO: collected segments

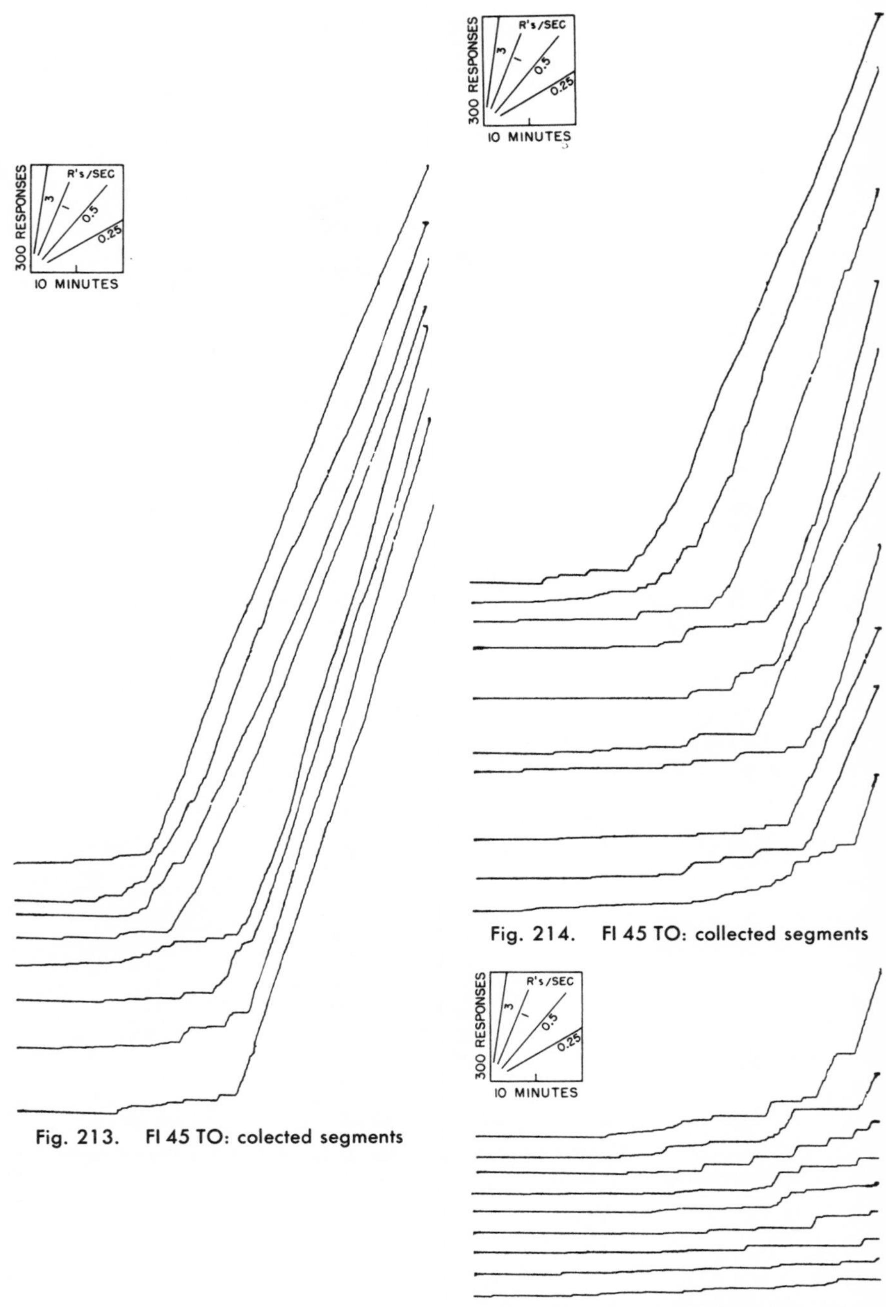

Fig. 213. FI 45 TO: colected segments

Fig. 214. FI 45 TO: collected segments

Fig. 215. FI 45 TO: collected segments

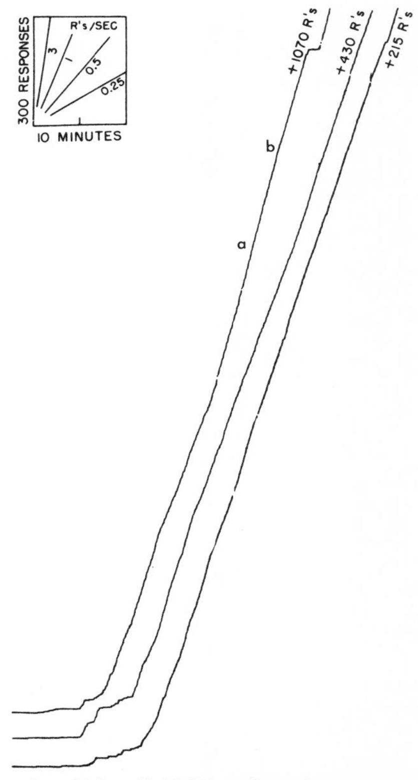

Fig. 216. FI 45 TO: collected segments

occur at *f*, but the curve ends at essentially a zero rate at *g*. The most obvious interpretation is that the interval performance follows whenever the bird has been pausing for any length of time. Cessation of responding at *b*, *d*, and *f* suggests some kind of exhaustion, which is probably the essential property of the process of extinction, over and above changes in rate due to self-generated stimulus changes.

The effect of time out on other values of FI

FI 1. Figure 129C describes a fairly stable performance on FI 1, 56 sessions following crf. Reinforcements are usually followed by a pause of 10 to 30 seconds, with later acceleration to a terminal rate. Occasionally, this bird began responding immediately after reinforcement, continuing at a constant rate through the interval. In the session immediately following Fig. 129C a 5-minute TO followed each reinforcement. The first effect is shown in Fig. 218A. The vertical lines following reinforcements show some responding during the 5-minute TO. Responding during the 1-minute interval is roughly constant. The 3rd pen excursion of the session is shown

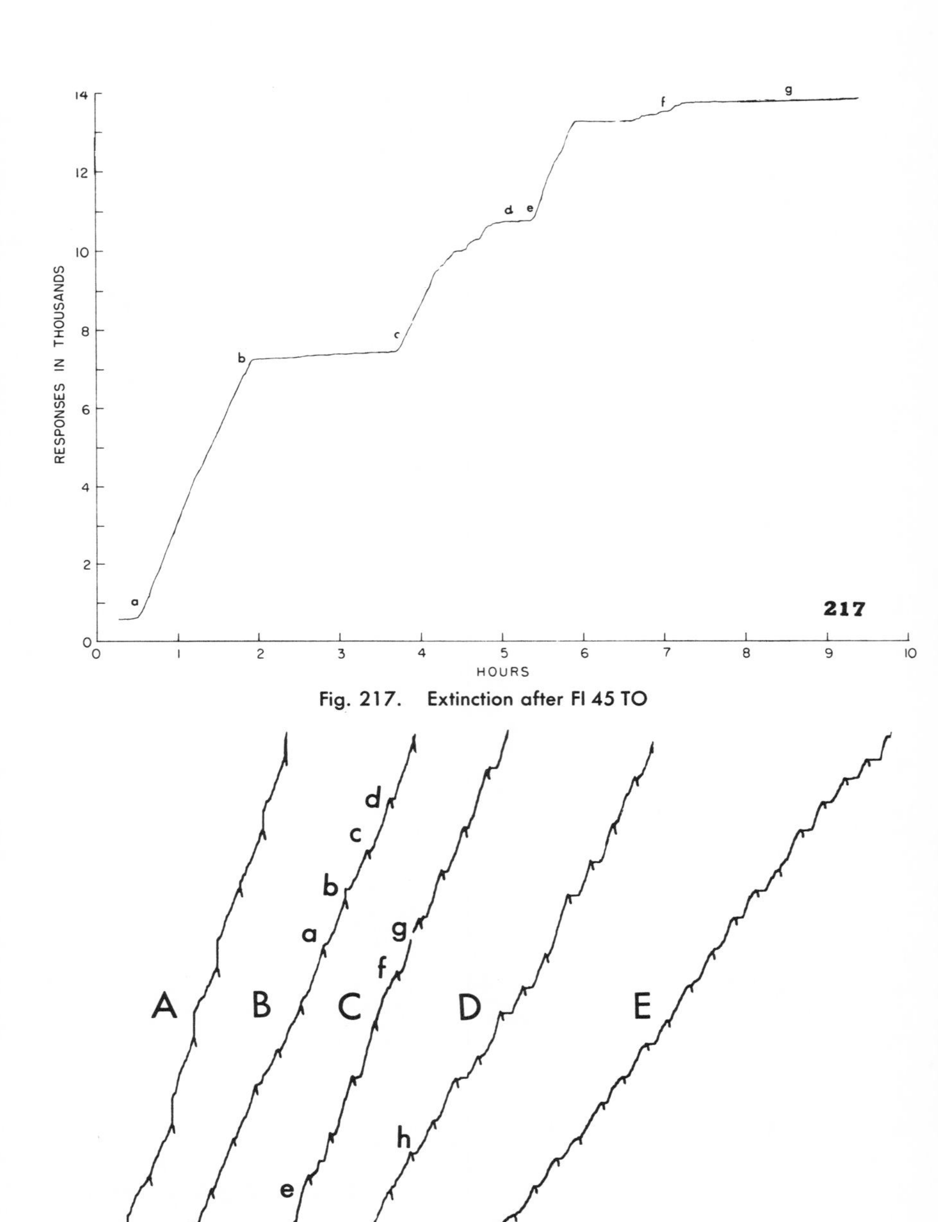

Fig. 217. Extinction after FI 45 TO

Fig. 218. The effect of TO 5 after reinforcement: FI 1

in Record B; responding during TO is now only slight, and pauses are beginning to appear after reinforcement (at *a, b, c,* and *d*). A lower rate during the earlier part of the interval develops slowly and irregularly with continued exposure to the schedule. Record C is a segment from the 9th session on FI 1 TO 5; it shows more pausing at the start of the interval, though still less than in Fig. 129. A pause or lower rate at the start of the interval is frequently lacking. The terminal rate is higher than without TO, and negative curvature frequently appears after the highest rate is reached (as at *e, f,* and *g*). The TOs did not occur in the 11th session (Record D). The immediate effect is the complete elimination of the pause following reinforcement; but the pause returns after the 5th reinforcement (*h*), and the rest of Record D shows a fairly

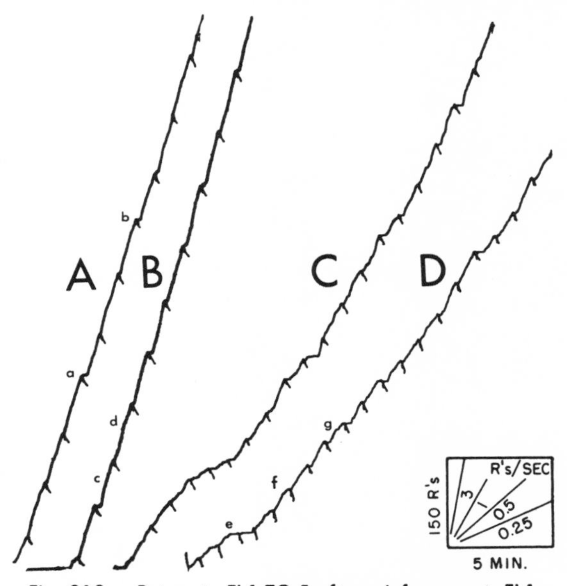

Fig. 219. Return to FI 1 TO 5 after reinforcement: FI 1

normal FI performance. Four sessions later (Record E) the negative curvature observed in Fig. 129 appears. This negative curvature probably represents a return to the superstitious behavior that was seen earlier in the experiment. A 5-minute TO followed reinforcement in the next session, and the result is shown in Fig. 219A. The TO produces a marked increase in over-all rate and eliminates the pause after reinforcement, except at *a* and *b*. After 4 sessions of further exposure to FI 1 TO 5 (Record B), the pause after the reinforcement has increased slightly, and occasional instances of negative curvature (as at *c* and *d*) occur. The over-all rate remains high. When the TO was removed from the program at the start of the following session, Record C was roughly similar to Fig. 218D. The over-all rate declines through the first few segments. The grain of the record is rough, and the earlier terminal rate is seldom reached or sustained. The last recorded performance on FI 1 alone, 4 sessions later, is shown in Record D, where a pause and lower rate during the early part of the interval are more marked and the over-all rate lower (cf. Fig. 218E). The performance is irregular, however; and negative curvature often occurs after the terminal FI rate is reached (as at *e*, *f*, and *g*).

In summary, a TO after reinforcement on FI 1 first reduces the pause and lower rate at the start of the interval and increases the terminal and over-all rates of responding. With further exposure to the schedule, some pausing at the start of the interval returns, although it is not so marked as without TO. When a TO is removed, the pause after reinforcement is temporarily eliminated. In this bird, rough grain and a failure to hold a terminal rate characterized the final performance under FI 1 with-

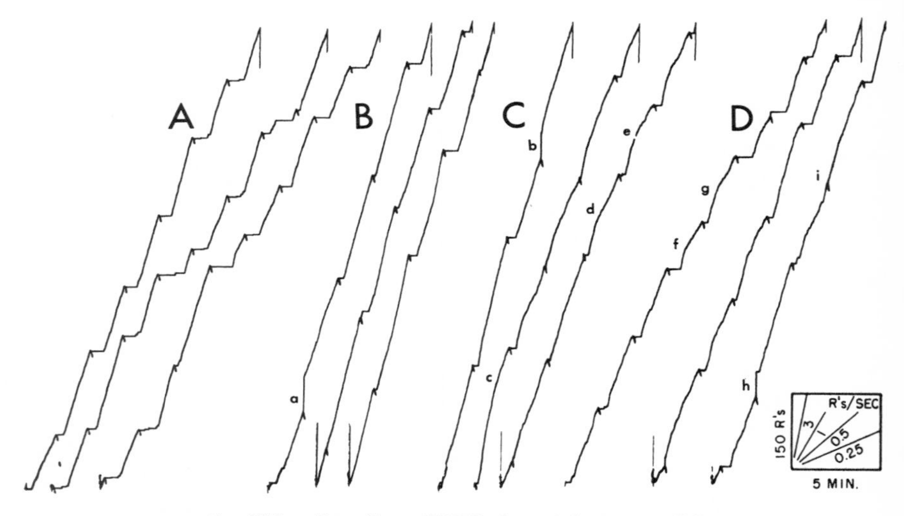

Fig. 220. The effect of TO 5 after reinforcement: FI 2

out TO; but this condition is probably due to the factors which caused the negative acceleration in Fig. 129A.

FI 2. Figure 220A shows a fairly stable performance on FI 2 after 1300 reinforcements. (For earlier stages, see Fig. 119 and 150.) The pause after reinforcement varies considerably; but no instance occurs where the reinforcement is not followed by some pause. Occasionally, a high rate develops in the middle of the interval and is followed by slight negative acceleration to a lower rate. Figure 220B shows the session following Record A, in which a 5-minute TO followed each reinforcement for the first time. Some responding during TO is evident at *a*. Pausing is reduced markedly and the terminal rate increased. Where a pause occurs, the acceleration to the higher rate is more abrupt than in Record A, where a few responses occur in most segments before the terminal rate is struck. The 5th session on FI 2 TO 5 (Record C) shows marked negative curvature (as at *c*, *d*, and *e*). Pauses after reinforcement are either absent or much reduced. (Responding during the TO occurs at *b*.) In Record D, after 10 further sessions with TO, strong negative curvature appears (as at *f* and *g*). No pause takes place at *i*, and responding occurs during the TO at *h*. Pauses and scalloping are considerably less than before the introduction of TO.

In the session following Fig. 220D the TO was removed. The performance (Fig. 221A) shows very rough grain and irregular rate changes. Slow responding appears at *a*, *b*, and *c*, and marked negative curvature is still evident. Record B shows the beginning of the next session, also without TO. Some further progress is made toward recovering the original FI 2 performance of Fig. 220A. This session continued for a

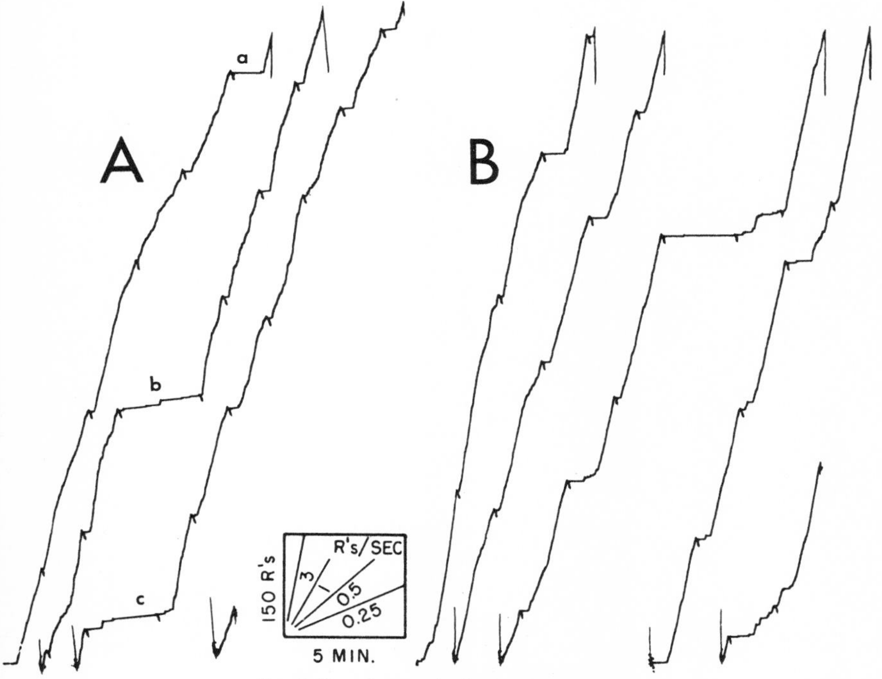

Fig. 221. Removal of TO 5 from FI 2

total of 528 reinforcements, generating a very long satiation curve. The remainder of the session will be presented in Fig. 380 and 381 later, under the section on satiation. The bird took several days to return to the designated running weight. Figure 222A shows the 1st performance following the satiation curve. A substantial period of slow responding follows most reinforcements. Toward the end of the session, however, marked negative curvature appears later in the interval, along with an over-all lower rate. When the TO was added, at the start of the following session (Record B), the

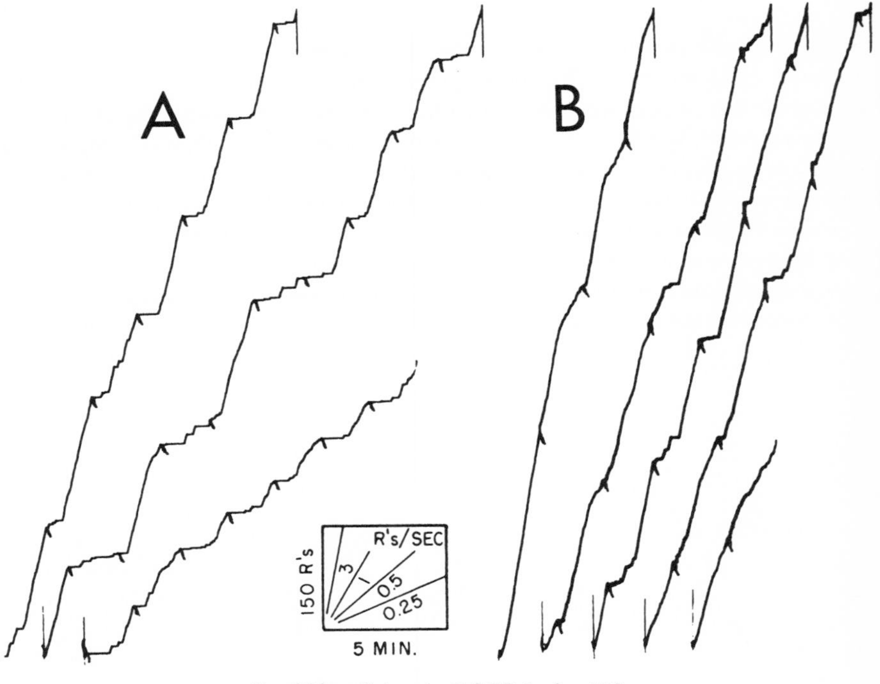

Fig. 222. Return to FI 2 TO 5 after FI 2

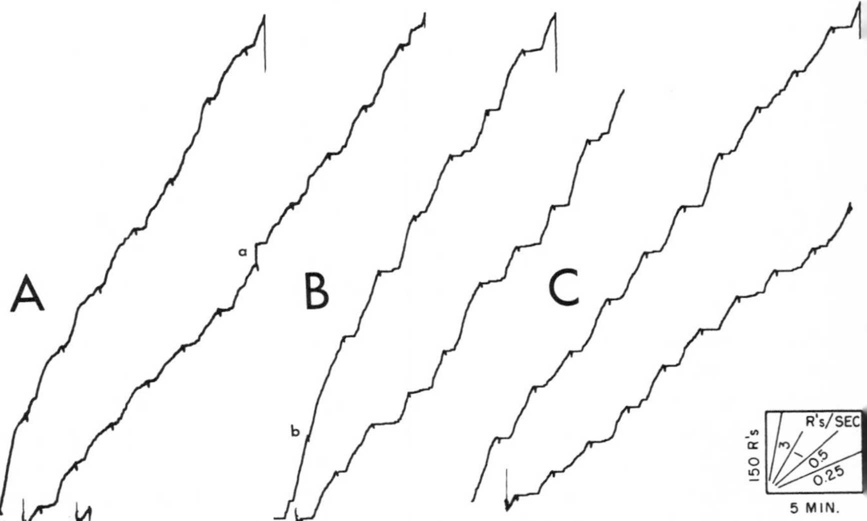

Fig. 223. Second return to FI 2 TO 5 after FI 2

pause after the reinforcement immediately disappeared and the terminal rate increased. As the session progresses, a short pause following the reinforcement reappears. When a pause occurs, the shift to the terminal rate is more abrupt than in Record A.

The FI 2 TO 5 was continued for 145 reinforcements, and the final performance is shown in Fig. 223A. A pause or period of slow responding at the start of the interval develops slowly as the session progresses. Toward the end of the session the segment is usually S-shaped. The grain is rough, and there are marked local fluctuations in rate. Responding during time out occurs occasionally (as at *a*). In the following session (Record B) the TO after reinforcement was removed for the second time. A pause following the 1st reinforcement at *b* is absent, but the 2nd reinforcement is followed by a substantial pause, and the remainder of the period shows marked scalloping. The terminal rate is not held, however, and most intervals show negative curvature. This is a more rapid transition than the 1st return to no TO. Record C shows a later performance, 190 reinforcements after Record B. Here, the over-all rate of responding has declined mainly because of a greater tendency for lower rates to occur during the 1st part of the interval. The negative curvature during the latter part of the interval is marked and the grain rough.

Summary. The effect of the time out on FI 2 was similar to the results reported for FI 1. The TO after reinforcement increases the over-all rate of responding by increasing the terminal rate of responding during the interval, and eliminating or shortening the pause after reinforcement. Unlike the performance on FI 1, the final performance on FI 2 TO showed marked negative curvature, which then persisted even when the TO was removed. At this size FI the introduction of a TO does not produce a good, scalloped interval performance. When the TO is removed, the FI 2 performance continues to show irregularities beyond those which occur in FI 2 after crf.

FI 4. A 5-minute TO after reinforcement was added to the fairly stable performance on FI 4 shown in Fig. 152, the last recorded segments of which are shown in Fig. 224A. In the following session, Record B, the TO eliminates the pause and scallop at *a* and *b*; but slower responding appears after the reinforcement in the 4th interval (*c*), with a smooth acceleration to a terminal rate. The curvature increases progressively, and a pause appears after the reinforcement at *d*. By the 10th reinforcement (at *e*) the pause and the scallop resemble those in Record A. Figure 225A shows the last performance after 10 hours of FI 4 with TO after reinforcement. The performance here is somewhat more irregular than in Fig. 152 and 224, possibly because of some aversive effect of the TO. The pause after reinforcement and the terminal rate (cf. *a* and *b*) vary considerably. The bird may begin to respond soon after reinforcement and strike the terminal rate early (as in *c* and *d*). When TO after reinforcement was removed on the following session, the performance (Record B) showed considerable disruption. The bird responds soon after reinforcement, although a few pauses still occur (as at *e* and *h*). The grain is rough, and marked rate changes occur after the terminal rate is reached (as at *g*). Slow responding may extend through the interval (*f*), and second-order effects appear (as at *i*).

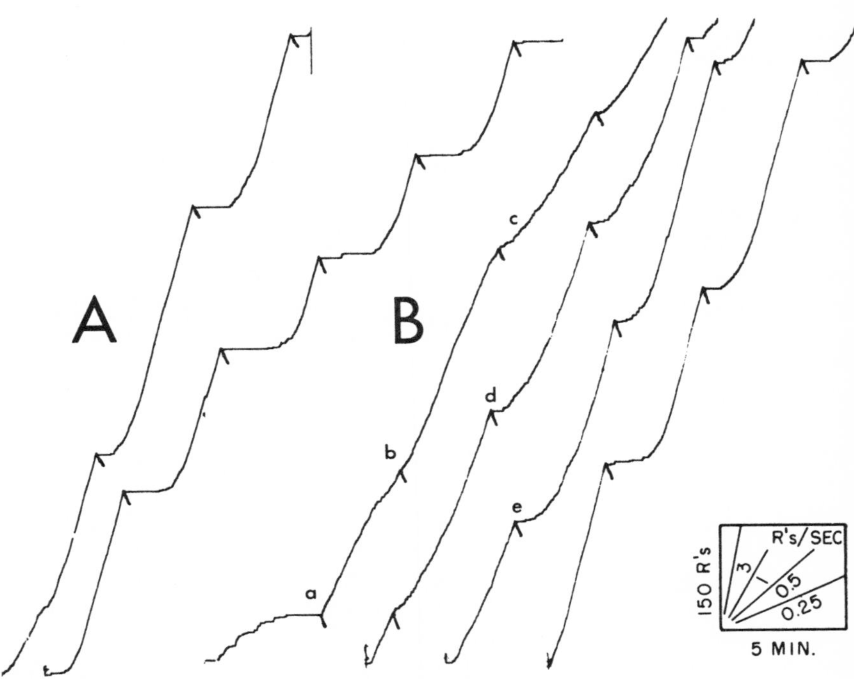

Fig. 224. The effect of TO 5 after reinforcement: FI 4

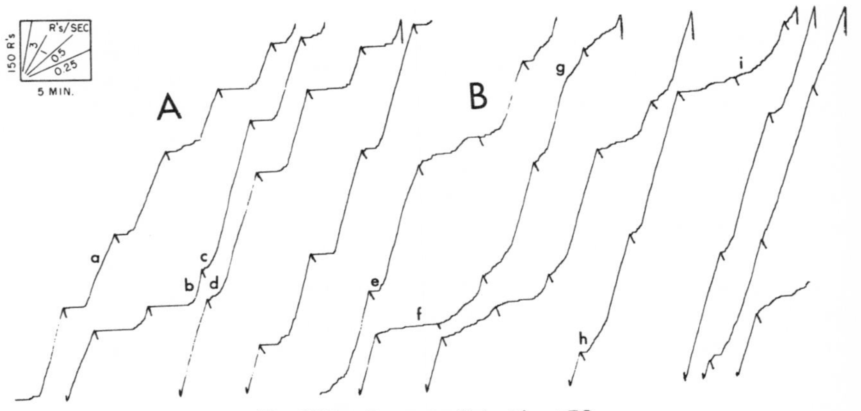

Fig. 225. Return to FI 4 without TO

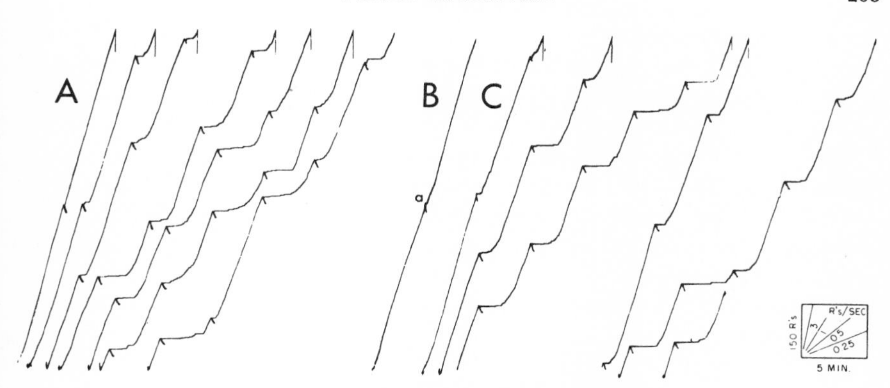

Fig. 226. Return to FI 4 TO 5 after FI 4

Time out was again added to the reinforcement. The first performances were lost through trouble with the apparatus. Figure 226A, however, shows the 3rd session of FI 4 without TO. The session begins with a high rate sustained throughout the interval. Pauses and acceleration in the early part of the interval develop progressively. The last part of the session shows the earlier standard performance. When TO was added to the reinforcement on the following day, the responding was immediately resumed after the 1st reinforcement. The experiment was stopped; but it was continued the next day (Record C), giving a performance similar to Record A. Figure 227 shows a later performance on FI 4 TO 5. The over-all rate of responding is somewhat higher than normal; and several instances occur in which the bird starts soon after

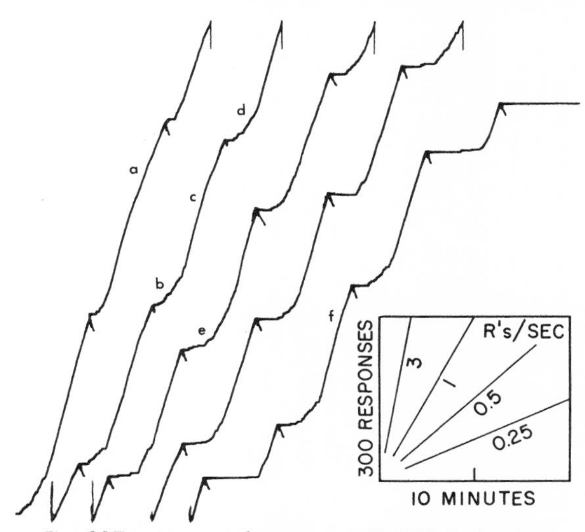

Fig. 227. Later performance on FI 4 TO 5 after FI 4

the TO after reinforcement, with subsequent marked curvature (*b, d,* and *e*). Negative curvature is occurring in many of the intervals (as at *a, c,* and *f*).

A complete experimental session 7 hours after a second return to FI 4 only is shown in Fig. 228A. The terminal rate is lower than in Fig. 227, with a resulting lower over-all rate. Marked negative curvature occurs (as at *a, b,* and *c*), and the grain is much rougher than with TO. This performance gives way to that shown in Record B, 2 hours later, which resembles the final FI 4 performance shown earlier in Fig. 152. A remnant of the previous negative acceleration is apparent in the short runs at higher rates occurring at *d* and *e*.

FI 8. Figure 229A shows the first exposure to FI 8 TO 5. The immediately pre-

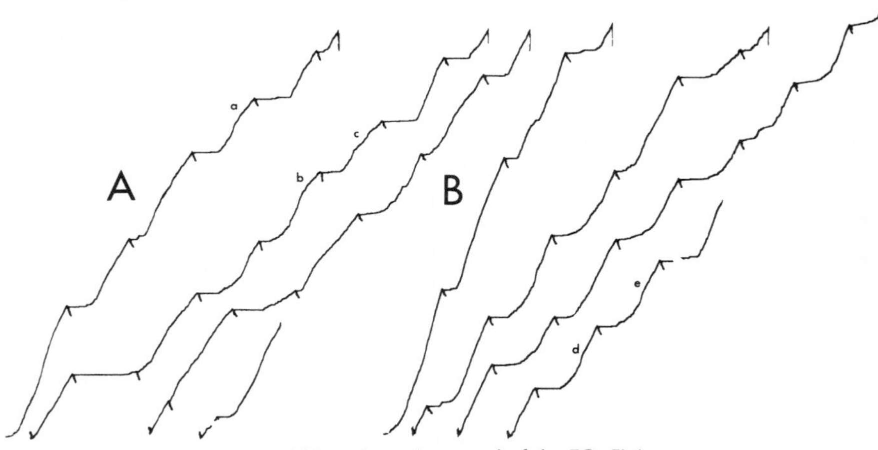

Fig. 228. Second removal of the TO: FI 4

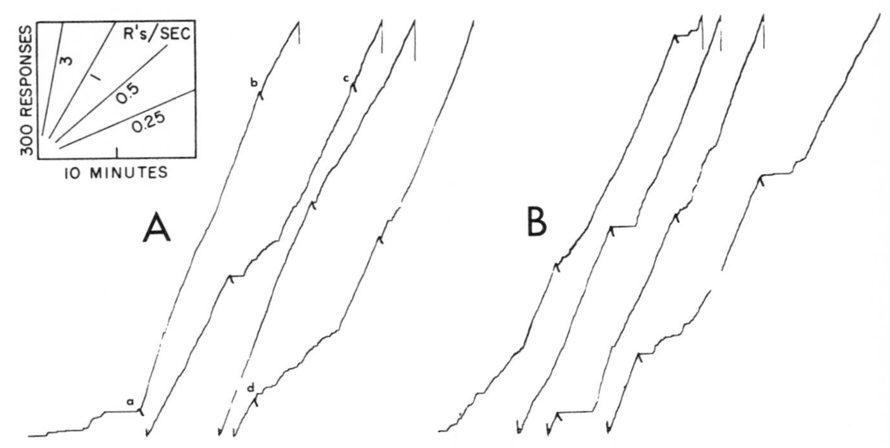

Fig. 229. The effect of TO 5 on FI 8

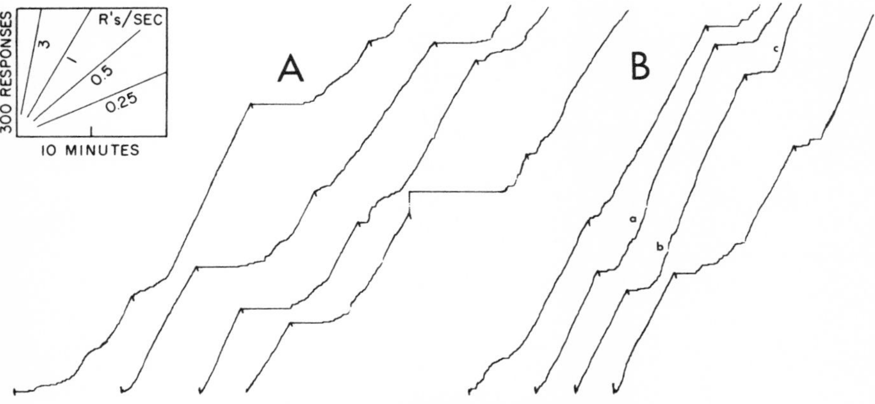

Fig. 230. FI 8 TO 5 after 3 sessions

ceding session on FI 8 (186 hours after crf) presented earlier (in Fig. 154) shows a very irregular performance, although every reinforcement was followed by at least some pausing. The first time out at *a* is followed by sustained responding throughout the FI segment; the remainder of the record shows a higher over-all rate, a reduction in the amount of scalloping, and instances of responding at the start of the interval (*a, b, c,* and *d*).

The pause and lower rate at the start of the interval become slightly more pronounced at the start of the 2nd session (Record B); and in Fig. 230A, at the start of the 4th session with TO, the pause and scallop at the start of the interval are of the same order as without TO. The abrupt shift from the pause to the terminal rate, however, has given way to fairly smooth curvature. By the start of the 6th session (Record B) the pause again becomes shorter and the shift to the terminal rate more abrupt. Responding occurs at the start of the 2nd interval in spite of the TO. Segments at *a, b,* and *c* show the highest rate of responding in the middle of the segment. The TO was then removed for 70 hours, and the return to FI 8 TO 5 (in Fig. 231A) resembles the first time the TO was used, although the transition is somewhat faster. A constant rate throughout the interval gives way to a pause and scallop that become very marked by the 7th segment. Nearly all of the segments show a pause with smooth and extended curvature 5 sessions later (Record B).

A 2nd return to FI 8 TO 5 after 165 hours without TO produces an even quicker development of a pause and smooth scallop (Fig. 232). By the end of the session, most of the intervals are showing long pauses and smoother and more extended curvature than heretofore.

Figure 164 showed a performance on FI 8 without TO in which a pause after reinforcement and a smoothly curved scallop were common. Another bird in the same experiment failed to develop similar segments. Figure 233A shows a performance after 170 hours on FI 8 after crf. The over-all rate is low, as it had been throughout

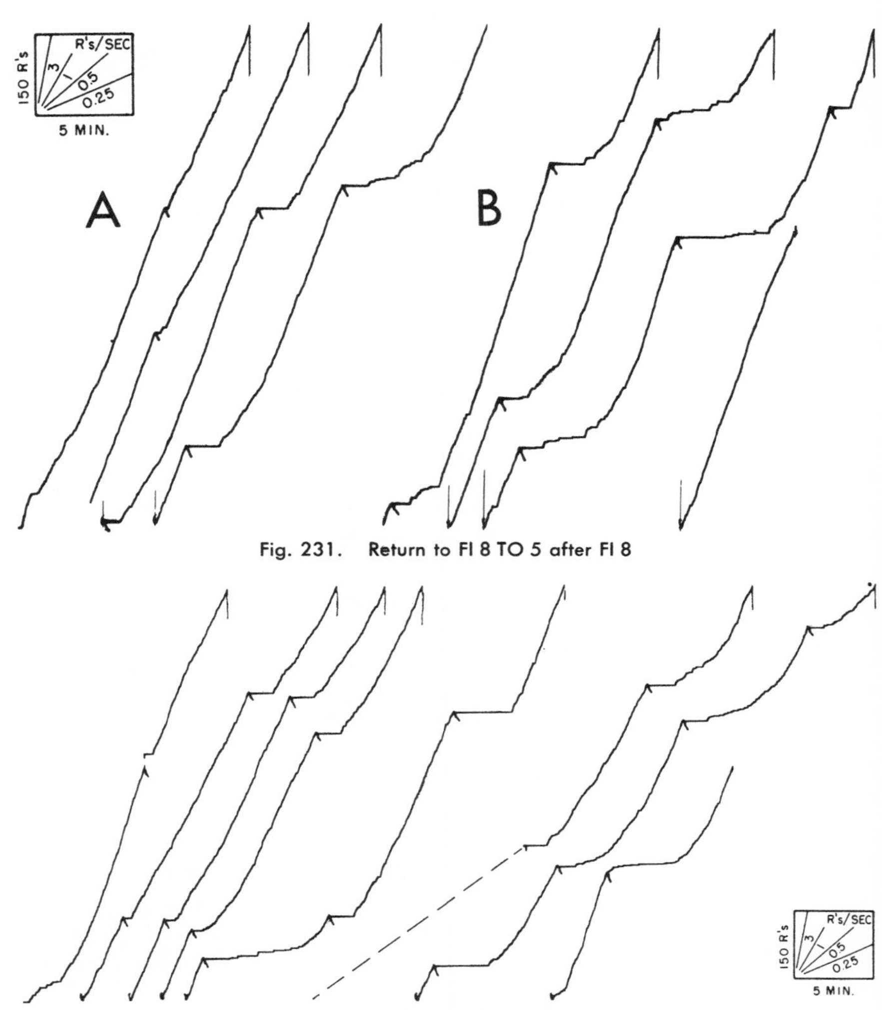

Fig. 231. Return to FI 8 TO 5 after FI 8

Fig. 232. Second return to FI 8 TO 5 after FI 8

the experiment, and the grain is rough. Slight pauses may or may not occur after reinforcement. Breaks occur elsewhere, and little or no scalloping appropriate to the schedule appears. A 5-minute TO after reinforcement was introduced at the start of the following session (Record B). The over-all rate of responding increases to 3100 responses per hour, compared with a mean rate of only 2300 responses per hour for a 10-hour sample before the introduction of TO. The scallop at *a* at the beginning of the session is characteristic of this bird and occurred without TO in earlier sessions.

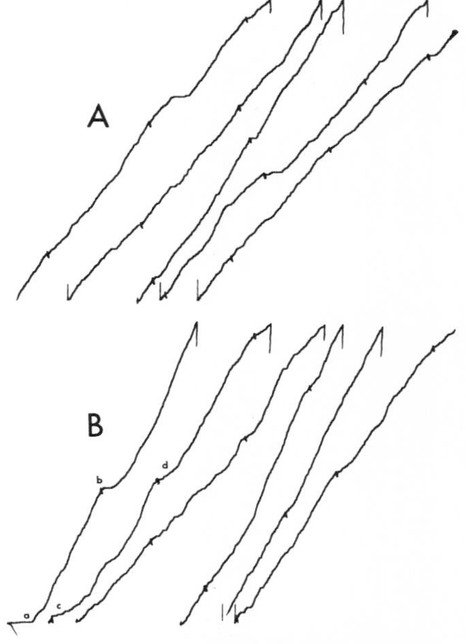

Fig. 233. The effect of TO 5 on FI 8 (second bird)

Up to this point the only effective TO was overnight in the home cage. The 1st TO (at *b*) reproduces a lower rate and smooth acceleration, which are repeated at *c* and *d*. Thereafter, bad grain develops, and the over-all pattern is not very different from Record A except for the over-all rate. Figure 234 shows the remainder of the 1st session with TO after reinforcement following immediately after Fig. 233. For some time the performance remains roughly linear, with occasional pauses or slower responding after reinforcement (as at *a* and *b*). Slight curvature begins to appear consistently at *c* and *d*, and the curvature becomes well-marked, at *e*, *f*, and *g*, before the end of the session. In a single experimental session of about 90 minutes, a 5-minute TO produces well-marked scallops, although these had not developed during 170 hours under FI 8 alone. The bird was apparently unable to develop a scallop or to respond appropriately to the fixed-interval schedule because of the carryover of stimuli from the preceding interval.

The subsequent history of this bird on FI 8 TO 5 is sampled in Fig. 235 through 238. These show an instability ranging all the way from a smooth performance conforming to the fixed-interval pattern of reinforcement to a performance showing high over-all rates with only a slight decline after reinforcement.

Figure 235 shows the entire 4th session after 12 hours of FI 8 TO 5. Part of this performance is not very different from that prevailing in the absence of TO (Fig. 233A).

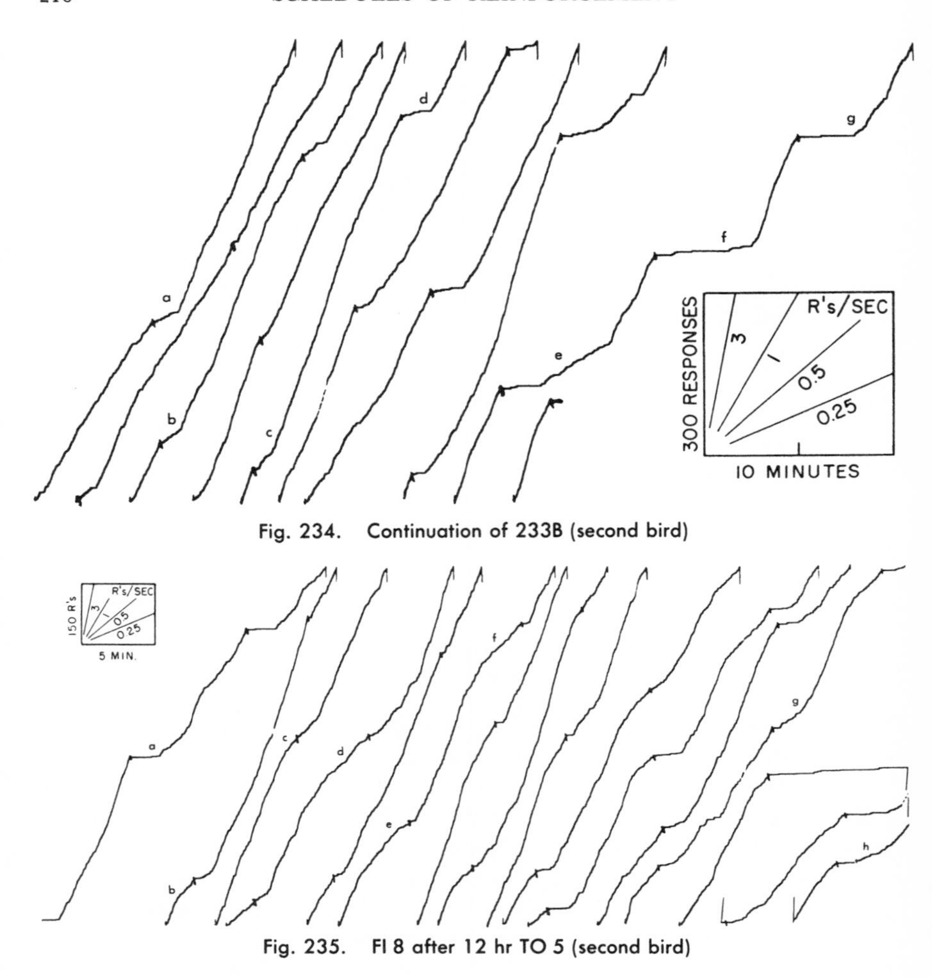

Fig. 234. Continuation of 233B (second bird)

Fig. 235. FI 8 after 12 hr TO 5 (second bird)

Occasional instances of substantial pauses or substantial reductions in rate after reinforcement appear, however. The many instances of negative curvature that occur (as at *b, c, d, e,* and *f*) suggest that the higher rate and earlier start due to the TO produce more responses than can be sustained on FI 8. Extinction therefore sets in during almost every interval; but most segments begin with some positive curvature appropriate to the fixed interval. This curvature is most conspicuous at *a, g,* and *h.* The grain is rough, and an occasional interval is run off at a very low rate.

A higher terminal rate later develops as in Fig. 236 after 32 hours of exposure to FI 8 TO. A longer pause after reinforcement prevents an excessive number of responses per interval, but the terminal rate is frequently broken by a decline before the

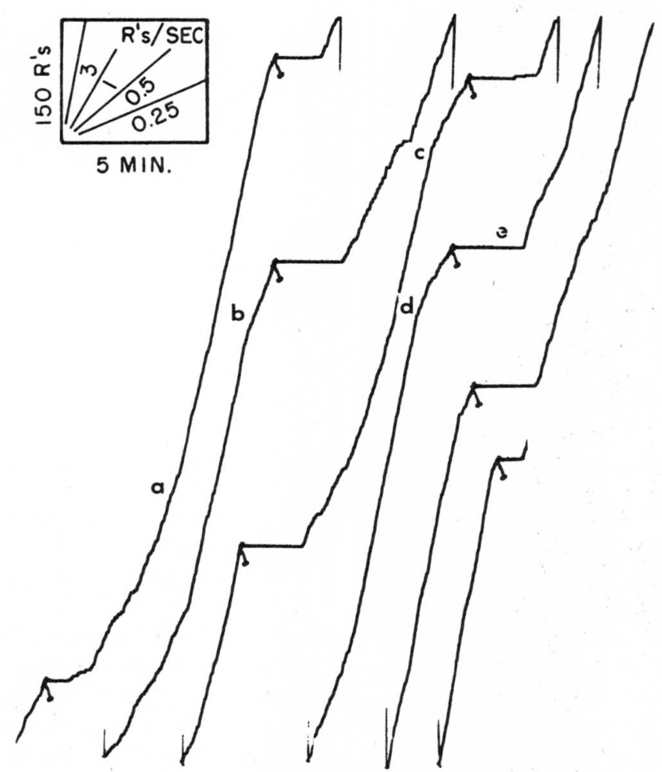

Fig. 236. FI 8 after 32 hr TO 5 (second bird)

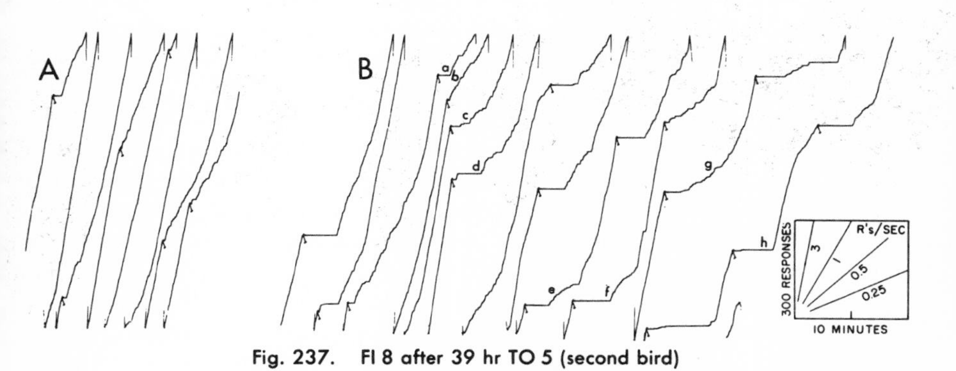

Fig. 237. FI 8 after 39 hr TO 5 (second bird)

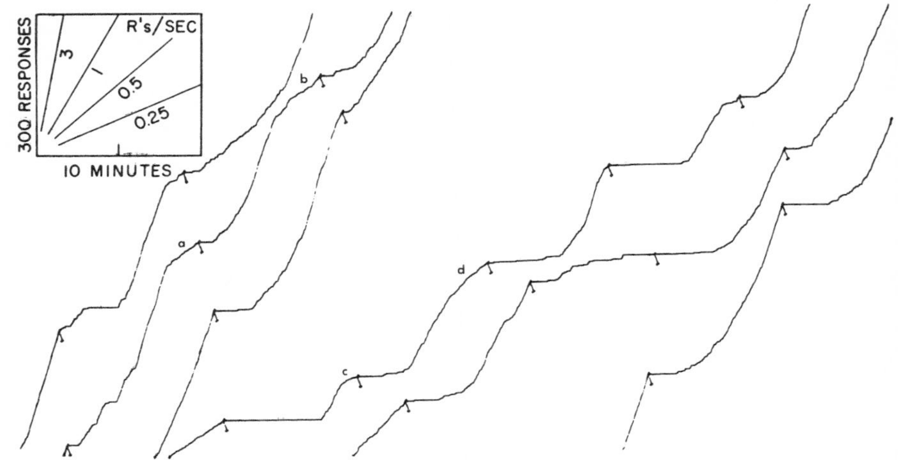

Fig. 238. FI 8 TO 5 showing negative curvature and marked pausing

interval is completed (as at *b* and *c*). Occasional instances occur of fairly smooth curvature to a terminal rate which is sustained almost to reinforcement (as at *a*). At *e*, after a long pause, the rate begins higher than usual and declines before the terminal rate is again reached and maintained.

Figure 237A shows a somewhat different performance after 39 hours of exposure to FI 8 TO 5. Pauses after reinforcement are sharply reduced; instead, intervals tend to begin with periods of slower responding, with rough grain. The terminal rate remains high until the reinforcement. This is a passing phase, however, and pauses again return by the end of the session shown at Record B. Occasionally, as best exemplified at *g*, there is a smooth interval scallop similar to the more regular types of performance on this schedule of reinforcement. At *a, b, c,* and *d* a series of intervals show a progressive deepening of the scallop. A long pause after reinforcement leading abruptly to a high rate followed by negative curvature is evident at *h*. Marked examples of knees occur at *e* and *f*.

The change from one to the other of these different performances in a single session must be due to very subtle conditions. Figure 238 shows a portion of a session after 4 hours; the over-all rate is considerably lower, because of marked pauses and scallops and failure to maintain the terminal rate of responding (*a, b, c,* and *d*).

Figure 239 shows a complete session toward the end of the experiment, beginning after 72 hours of exposure to FI 8 TO. Here, a variety of changes in rate occur in single segments: (1) negative curvature in *a, b, c, d,* etc.; (2) a series of intervals showing a gradual increase in the amount of scalloping, as at *e, f, g, h, i,* and *j*; and (3) fairly smooth acceleration to a terminal rate in spite of a rough starting grain, at *k* and *l*. The over-all rate shifts from a low value during the 1st hour of the experiment to a higher value in the middle part of the session.

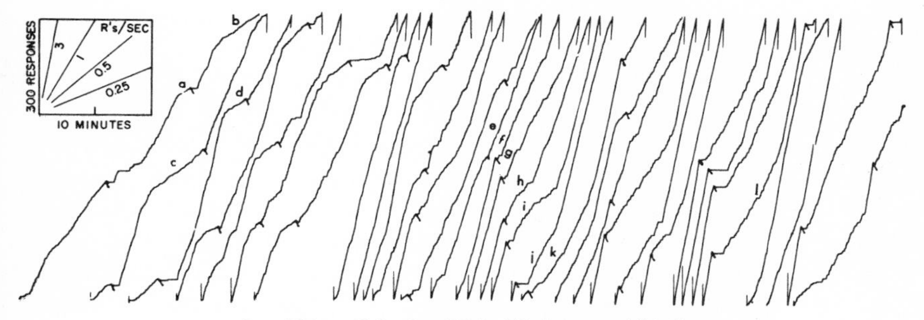

Fig. 239. FI 8 after 72 hr TO 5 (second bird)

Figure 240A shows the final performance for another bird on FI 8 (already represented in Fig. 164) for the 140th hour after crf. It is marked by reinforcements followed by smooth acceleration to a terminal rate (*a* and *c*), but also by occasional intervals in which the terminal rate is assumed immediately after reinforcement and maintained until reinforcement (at *d*). Some running-through appears at *b*. On the following day, Record B, a 5-minute TO followed each reinforcement. In general, scalloping is more marked. The 1st interval of the session at *e* is typical of the start of the session. The TO is effective at *f* even though the terminal rate is reached fairly quickly. However, no instances occur here, or in any subsequent session, of imme-

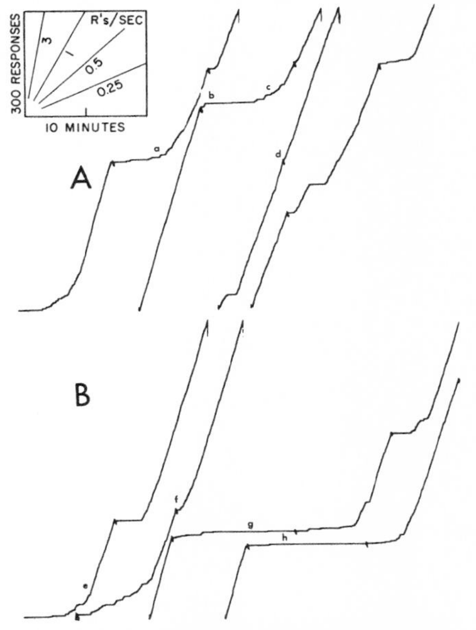

Fig. 240. The effect of TO 5 on FI 8
(third bird)

diate responding at the start of the interval. An early effect is to produce occasional intervals with almost no responding (as at *g* and *h*).

With further exposure the TO produces a more uniform performance from interval to interval. Figure 241 shows the complete record for the 2nd session with TO after reinforcement. The pauses following reinforcement are mostly brief, except for the segments at *b* and *i*. Many instances occur of fairly smooth curvature (as at *b, d, e,* and *j*), and also instances of fairly abrupt shifts to the terminal rate (as at *a* and *c*). Some intervals—for example, *f* and *g*—never show a terminal rate. A slight negative acceleration occurs at *h*. In general, however, the record shows a fairly narrow range of the number of responses emitted in intervals and only an occasional succession of intervals which could be regarded as a second-order effect. Another characteristic of FI 8 TO 5 exemplified by this record is an occasional set of intervals showing a progressively slower but smooth acceleration to the terminal rate.

In order to show sets of intervals in which the number of responses progressively declines, all intervals from a daily session have been arranged on a common baseline in Fig. 242 in the order of their occurrence. This performance occurred 16 hours after the first introduction of TO. The 2nd interval, at *a*, showed the largest number of responses per interval during the day and is the first of a set of 7 intervals showing a progressive decline. At the end of this series, reinforcements occur after a small number of responses; this occurrence appears to set up a new contingency, which leads to

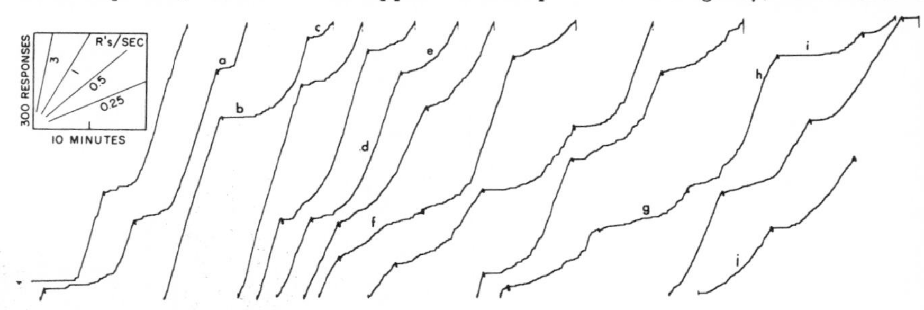

Fig. 241. Second session of TO 5 on FI 8 (third bird)

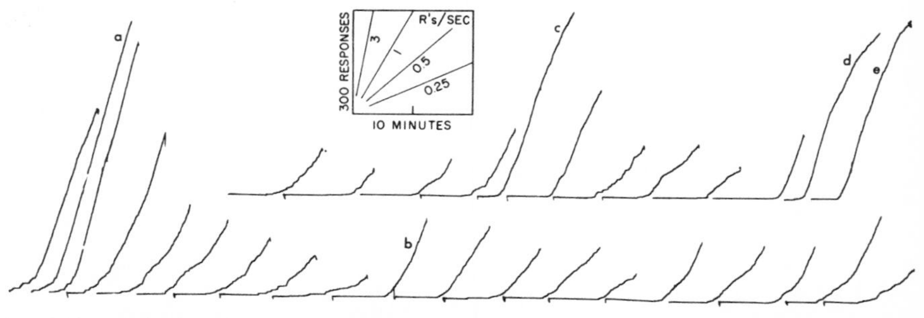

Fig. 242. FI 8 TO 5 showing a progressive decrease in the number of responses per interval

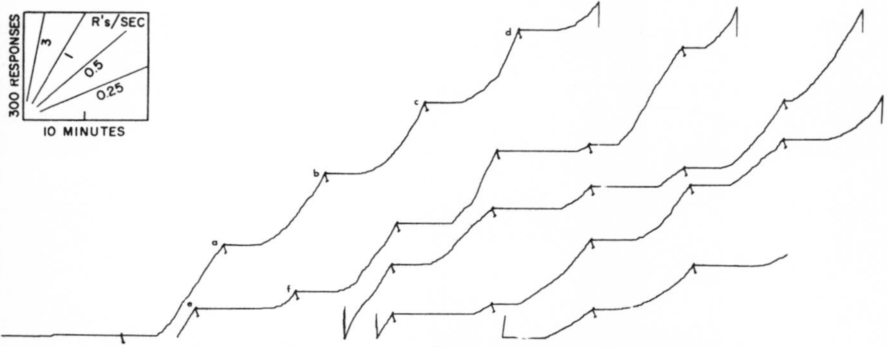

Fig. 243. Temporary stable condition on early FI 8 TO 5

a higher terminal rate, at *b*. The set of segments beginning at *b* shows a rough decline, as does another set of 5 intervals beginning at *c*. Toward the end of the session, marked negative curvature has begun to appear (at *d* and *e*) in intervals in which a high rate is reached early in the segment. The high rate presumably occurs because increasing the number of responses is now reinforcing.

These orderly changes show that even with a 5-minute TO after reinforcement, the performance on FI 8 is not stable. Contingencies arise under one level of performance which produce a change. Specifically, in this case, if the number of responses from interval to interval becomes stable and small, number becomes a discriminative stimulus which reinforces early responding in the interval. This in turn increases the rate and disrupts the stable FI performance. This instability may well be a property of the middle range of FI values in which number is of the magnitude required to show a sensitive control.

Occasionally, a fairly stable condition may be maintained for some time. Figure 243 begins after 34 hours of exposure to FI 8 with TO after reinforcement. The number of responses emitted per segment does not change greatly during the intervals ending at *a, b, c, d,* and *e*. However, some progressive reduction occurs, and the small count at *f* sets off an oscillating disturbance. In order to show the kinds of curvature which occur under FI 8 TO 5, a complete daily session after 50 hours on this schedule is given in Fig. 244, where the segments have been arranged in separate groups roughly according to curvature. Within each group the curves are arranged in order of the number of responses in the interval. The number of responses per interval here ranges between about 40 and 400. The set of segments in Group A contains 17 intervals, which seem appropriately classified as examples of simple positive acceleration. At the lower rates of responding there is some "sticking of the trigger." One curve (at *a*) shows a continuous acceleration throughout the whole segment. The rest start at zero and then accelerate to a constant terminal rate before the end of the interval. The terminal rate is low in the lower half of these curves.

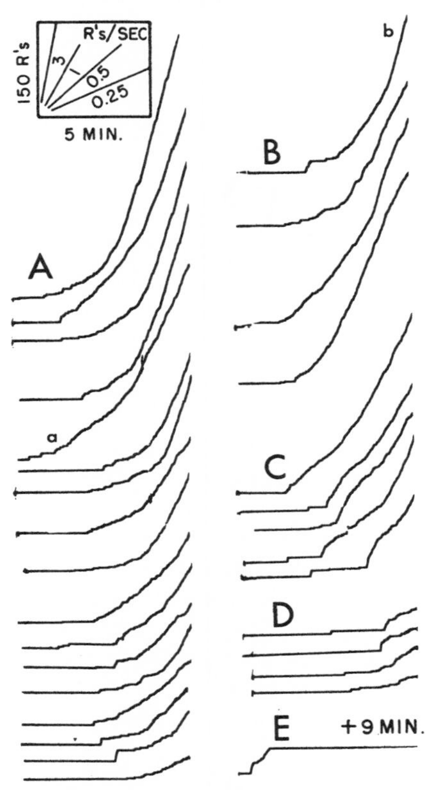

Fig. 244. Types of curvature under FI 8 TO 5

Group B shows 4 segments with negative acceleration. The decline is only just setting in at *b*, but it is more clearly marked in the other cases. A group of 5 intervals at C show a considerable delay in starting, and then a shift to a rate which appears to be a compensation for the delay. Another group of 4 intervals showing delay in starting, followed by a shift to a high rate briefly, and then negative curvature is at D. The number of responses in these segments is very small, however; and it is not clear whether these records could possibly form part of the series in Group A. One interval (at the start of the session) showed early responding followed by a very long pause extending 9 minutes beyond the end of the figure before a response occurred and was reinforced. (See Record E.)

FI 16. The effect of a 5-minute TO on FI 16 was tested on the performance already described in Fig. 170. The TO was introduced in the session following Record B in that figure. It reduced or eliminated pauses at the beginning of the interval, and the quicker acceleration to the terminal rate increased the over-all rate. Figure 245 shows later samples of the performance. In Record A, the start of the 4th session, the

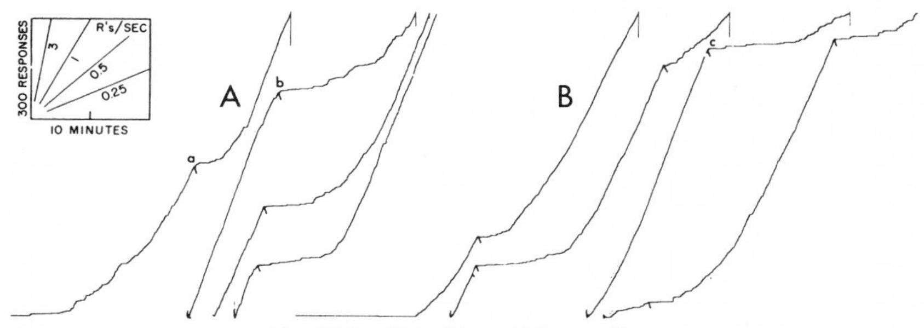

Fig. 245. The effect of TO 5 on FI 16

rate accelerates fairly smoothly to a terminal rate which is maintained until reinforcement. Pauses after reinforcement are absent at *a* and *b,* however, and the previous terminal rate is maintained for a few responses. Record B, the end of the 6th session, shows a similar performance, where there is a slight run-through at *c.* At this stage the rate following the reinforcement tends to be somewhat lower than in Fig. 170. In Fig. 246, the 7th session after the introduction of TO, the intervals arranged on a common baseline in the order in which they occurred in the experiment show a progressive increase in the amount of curvature over the first 9 intervals. By the 9th interval, however, the curvature is so extended that a terminal rate is not reached. After reinforcement at this low rate, the segments begin to oscillate between high and low numbers of responses.

When a TO no longer followed reinforcement, the performance was badly disrupted. Figure 247 shows the 1st session. The terminal rate is assumed quickly after the 1st reinforcement; and although the local rate varies considerably, a large number of responses is emitted in the interval. Nevertheless, the terminal rate is immediately assumed. Marked negative curvature then appears at *a.* Thereafter, the rate oscillates slowly and has little relation to the schedule of reinforcement.

When the FI 16 schedule was continued for a total of 35 hours, the performance sta-

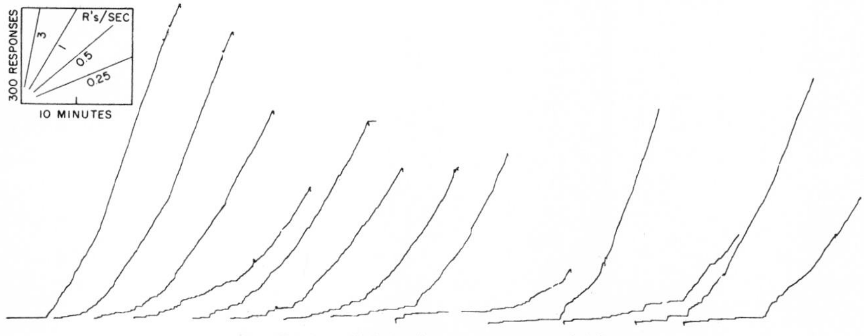

Fig. 246. FI 16 after 6 sessions of TO 5

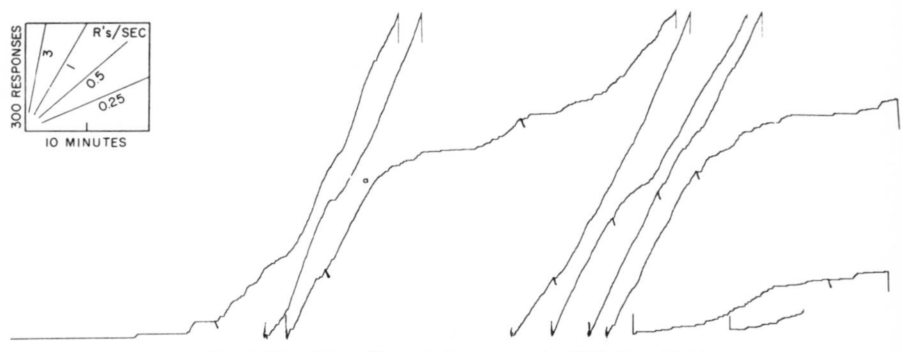

Fig. 247. The effect of the removal of TO 5 on FI 16

bilized and a normal FI performance (similar to that in Fig. 170) emerged. The TO was then added to the reinforcement for the 2nd time; the result is shown in Fig. 248. Record A shows the start of the 1st session, and Record B shows the end of a session 25 hours later. Note the progressive increase in the extent of the scallops at *a, b,* and *c.* A lower rate of responding and pauses after the reinforcement develop more rapidly than during the 1st introduction of the TO. The final performance (Record B) shows a more extended curvature and a more consistent interval-to-interval performance than under FI 16 without TO.

When TO was again removed from the program, considerably less disruption resulted than when the change was first made (Fig. 249). The performance is nevertheless irregular, with running-through at *a* and *b,* and many intervals showing a sustained rate of responding throughout (*c* and *d*).

After a further 71 hours of FI 16 without TO (Fig. 250), a fairly good fixed-interval performance develops. The over-all effect resembles the performance with TO after reinforcement, except that TO after reinforcement has a greater depth of scalloping. Note that in Fig. 250 the scalloping is more marked and the performance more con-

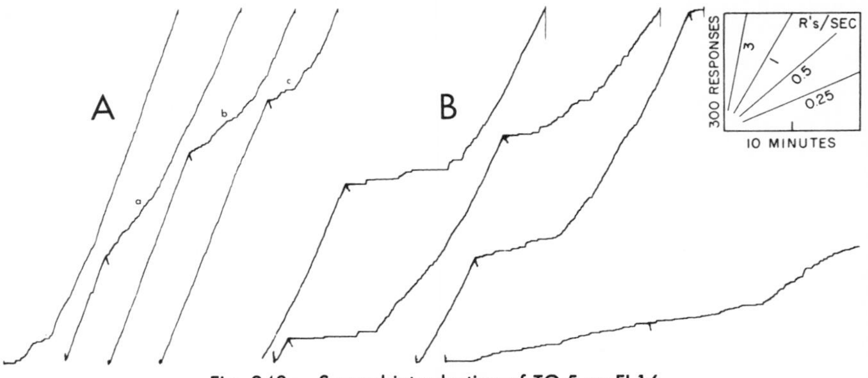

Fig. 248. Second introduction of TO 5 on FI 16

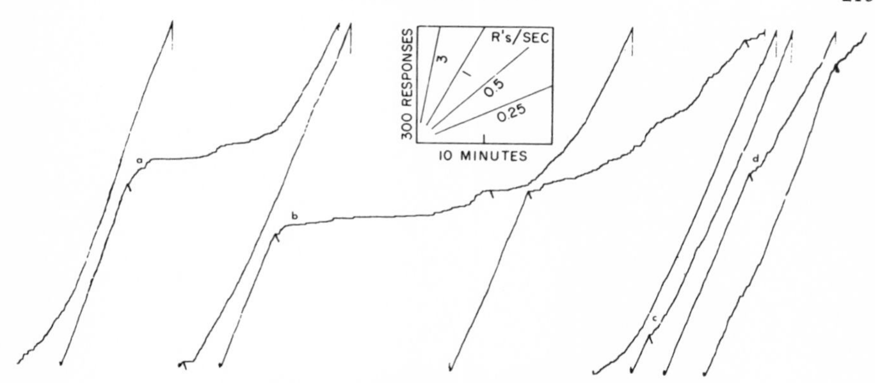

Fig. 249. Second removal of TO 5 from FI 16

sistent than in Fig. 170, which was the "final" performance after more than 300 hours of exposure to FI 16. The intervening exposure to FI 16 TO has resulted in a more marked and stable interval curvature without the aid of TO.

The most regular fixed-interval performance shown by this bird with TO after reinforcement is reproduced in Eig. 251, in which the intervals are represented in the order of their occurrence but on a common baseline. Except for the last 2 intervals, the number of responses from interval to interval varies within a narrow range. The grain is rough, and the curvature varies considerably (cf. *b* and *c*). The pause after reinforcement varies from zero at *a* to several minutes at *d*. The terminal rate is fairly uniform and the number of responses per interval fairly constant.

Time-out probes in FI 1

A regularly scheduled TO, as in the experiments just described, can enter into the reinforcing contingencies as an important stimulus. An occasional TO is relatively free of this complication. It can be used as a "probe," as described in Chapter Three.

In a fairly stable performance on FI 1 (similar to that in Fig. 149) a 30-second TO

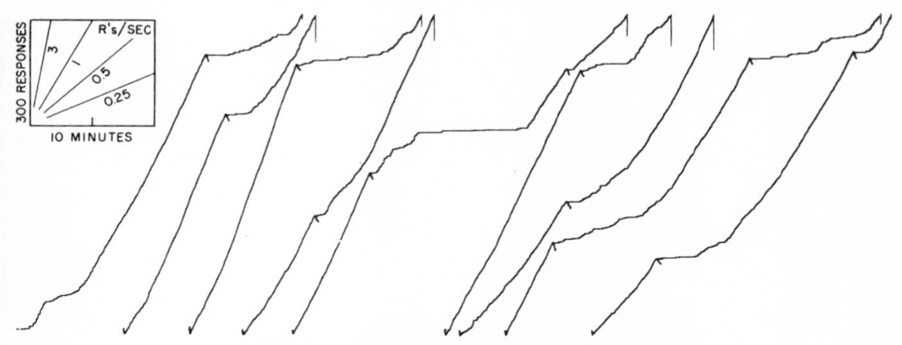

Fig. 250. FI 16 showing scallops without TO

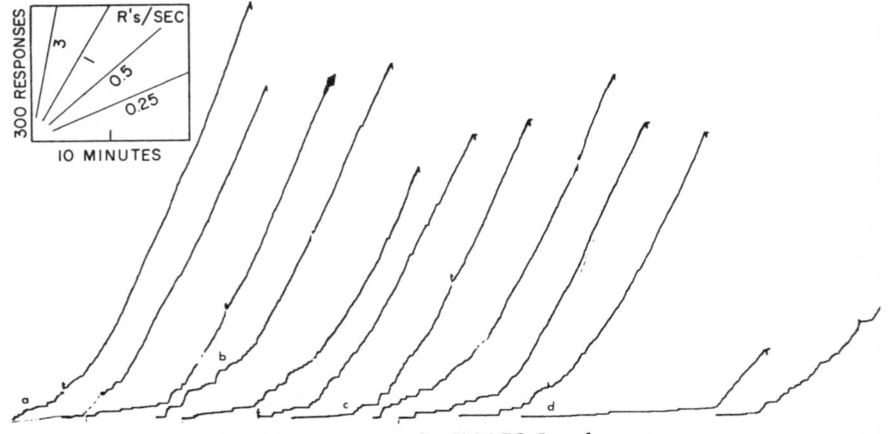

Fig. 251. Most regular FI 16 TO 5 performance

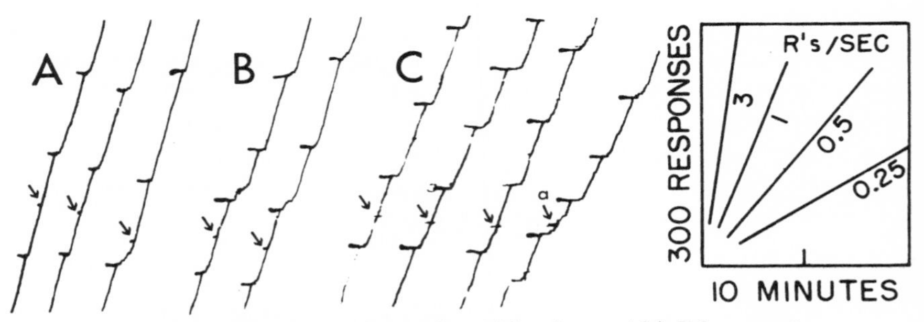

Fig. 252. TO 30sec probes in the middle of every 10th FI 1 segment

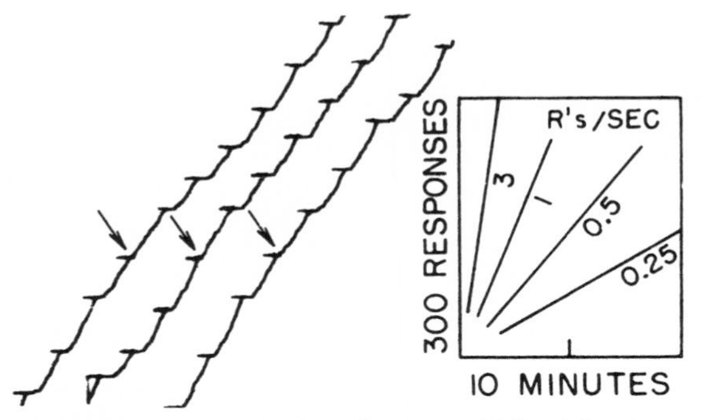

Fig. 253. TO 30sec probes after every 10th reinforcement

probe was programmed after every 10th reinforcement. The 9 intervening segments provided a baseline for judging the effect of the TO and for reducing, if not preventing, the development of a stimulus function.

The FI 1 segment was first probed midway in the interval. The terminal rate of responding had usually been reached at this point. Figure 252 shows segments from 3 daily performances over a 5-day period. Two reinforcements following the TO and one preceding it are presented. Arrows mark the probes. No reinforcements occurred at these points, of course. Record A shows the 1st session in which the TO probe occurred. In no session does it have any effect. (The last TO at *a* occurs before the terminal rate is reached, and a somewhat lower rate follows.)

Fifteen sessions later, the TO probe was programmed immediately after every 10th reinforcement. It had the effect of eliminating the pause. Figure 253 shows 3 examples from a single-session probe typical of the performances of 2 pigeons.

In a later session with the same birds the probe occurred 45 seconds after the start of the interval. Figure 254 shows a representative performance approximately 10 sessions after Fig. 253. Here, there appears to be some effect on the *following interval segment.* The probe reduces the amount of pausing after the next reinforcement. Note the negative acceleration during the intervals after the probe. This suggests that the probe has the effect of an aversive stimulus, which has been demonstrated elsewhere. Measurements were made over the 10-day period in which the TO probes occurred 45 seconds after every 10th reinforcement. The numbers of responses in the interval segments preceding and following the probed interval were counted. In one bird the segment following the probed segment shows more responding in 21 cases out of 27, and the mean numbers of responses are in the ratio of 1.5 to 1. In a second bird the figures are 19 and 29 in a ratio of 1.3 to 1. A third bird, however, failed to confirm this result.

The effect of a TO probe in eliminating the pause at the start of the fixed interval seems to indicate that the normal pause and period of slow responding under FI 1 are

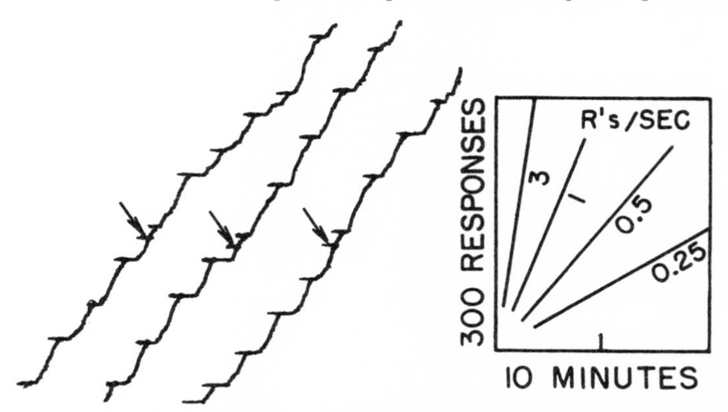

Fig. 254. TO 30sec probes 45 sec after the start of the FI 1 segment

due to a discrimination based on recent eating. The TO probe covers the time when eating is an effective stimulus. Since a 30-second TO eliminates the pause, we may conclude that the effective duration of the reinforcement as a controlling stimulus is 30 seconds or less.

Any effect on segments subsequent to the one in which a TO probe occurs, as in Fig. 254, would presumably be felt from a reduction in the stimulus carryover from responding in earlier segments from either the number of responses or the rate of responding. The order of magnitude of the effect of the TO after reinforcement, however, indicates that the main controlling factor in FI 1 is the preceding reinforcement. At a larger value, number and rate may be more effective.

Summary of time out

The effect of a TO depends on the value of the FI. At shorter FIs (up to 4 minutes) the TO after reinforcement increases the over-all rate by eliminating the pause and curvature. Continued reinforcement with the TO re-forms the scallop, although not so well as without the TO.

The final performances on FI 4 and FI 8 with and without TO were similar except that the curvature is somewhat smoother with TO.

On larger FIs (e.g., 16 minutes and over) the TO produces a stable performance with a pause and extended curvature at the start of the interval. In contrast, curves for performances without TO are very irregular, and pausing and smooth scalloping are either rare or absent. Instances still occur, however, where there is responding at the start of the interval in spite of the TO. Those become rare as the FI becomes larger (FI 45).

The transition from TO to no TO and vice versa temporarily eliminated the scallop and stable performance except for FI 45 to FI 45 TO. Each transition produced a temporary increase in the over-all rate when the TO was either added or removed.

Time-out effects upon FI schedules in the rat

The time out we have most frequently used with pigeons is a "blackout," programmed by turning out all the lights in the apparatus. An untrained pigeon will ordinarily not respond in the dark. Because a rat is nocturnal, its behavior cannot be controlled in the same way, i.e., *via* illumination of the experimental chamber. As noted in Chapter Three, a well-developed discrimination may be used instead.

In the following experiments the rat was first submitted to a procedure in which it was always reinforced in the presence of a light and never reinforced in its absence. Technically, this schedule is mult crf ext. In a given case lever presses occurring in the absence of the light are not reinforced for a 5-minute period. The light is then turned on, the response reinforced, etc. Eventually, responses only rarely occur in the absence of the light, which can then be used as a TO in the same manner as a blackout with pigeons.

Time out after reinforcement. Records A through I of Fig. 255 contain a representa-

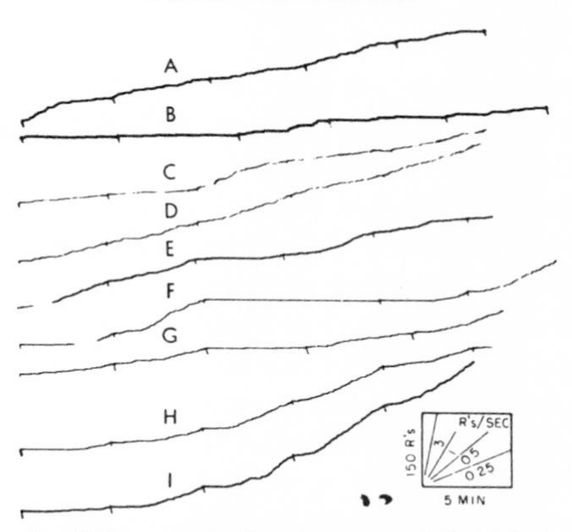

Fig. 255. Early development of FI 5 TO 5 in the rat

tive segment from each of the first 9 sessions on FI 5 TO 5 after a history of mult crf ext. The 1st session (Record A) shows a slow development of a fairly constant rate. Responding during the TO was near zero; it was not significantly higher than during the previous mult crf ext. By the 4th session, in Record D, a fairly sustained over-all rate develops. Some scalloping is clear in Record E, the 4th session; and by the 6th session, in Record F, a higher terminal rate is evident, although reinforcement is postponed in some intervals because the rat is not responding. The final performance in the 9th session following crf (Record I) shows a higher over-all rate of responding, rougher grain, and much less pausing after the reinforcement, even though the segments are consistently scalloped.

Figure 256 contains the entire daily session for this rat after 32 hours of exposure to FI 5 TO 5. A well-marked pause follows each reinforcement, and the rate accelerates fairly smoothly to a terminal value except at *a* and *b,* where pauses cover most of the interval. Two other rats showed similar performances.

Time-out probes. The importance of TO in the FI 5 performance shown in Fig.

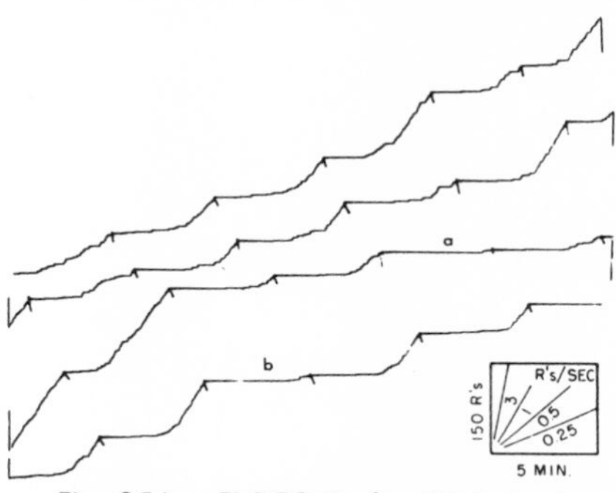

Fig. 256. FI 5 TO 5 after 32 hr (rat)

256 was probed by the omission of the TO after single reinforcements or small groups of reinforcements. All cases are presented in Fig. 257, which gives portions of 4 daily sessions. At the arrow, TOs after reinforcement were omitted. In Record A, after 31 hours of FI 5 TO 5 after reinforcement, the omission of TO produces some responding immediately after reinforcement, although this is followed by a zero rate and normal acceleration. In Record B, for a second rat, after 18 hours of FI 5 TO 5 after reinforcement, the over-all rate of responding is very low and the omission of the TO has no effect. However, the same rat showed a higher over-all rate after 6 hours of further exposure to the schedule (Record C), and the omission of the TO (at the arrow) produces a continuation of responding at the terminal rate in the preceding interval. In Record D, for the rat in Record A after 44 hours of FI 5 TO 5 reinforcement, the TO was omitted for 4 consecutive reinforcements, indicated by the arrows, at *a, b, c,* and *d.* Since the rate of responding was low preceding the first probe, the omission of the TO did not result in any increase in rate. A slightly higher rate of

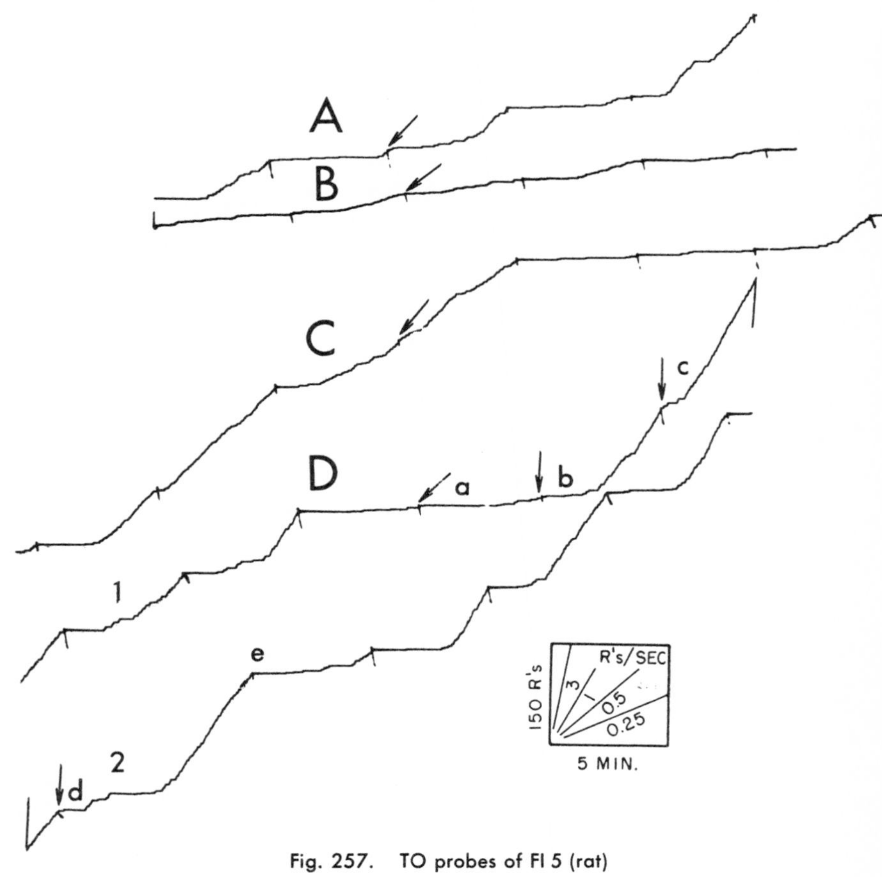

Fig. 257. TO probes of FI 5 (rat)

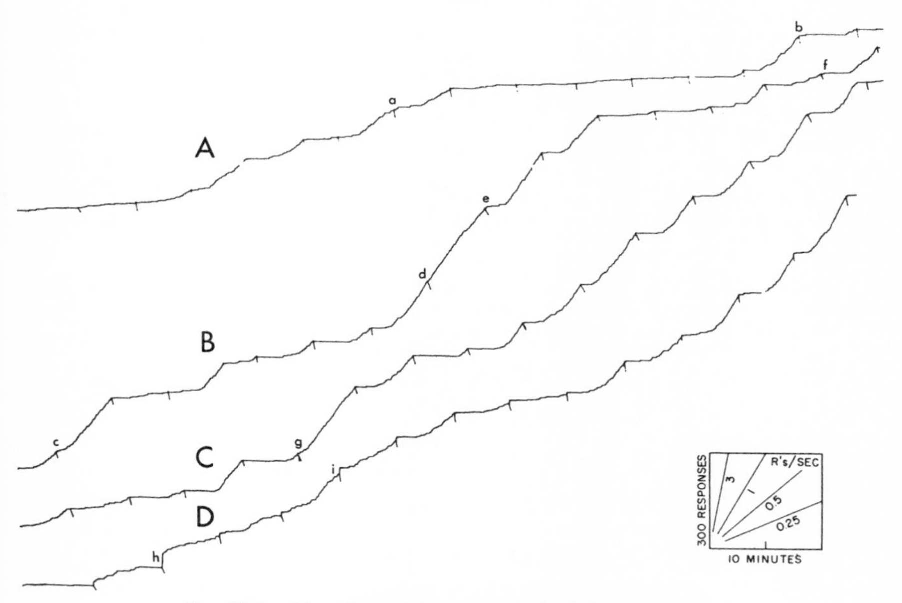

Fig. 258. The effect of the removal of TO 5 on FI 5 (rat)

responding occurs in the interval at *a*; and the 2nd reinforcement at *b,* not followed by a TO, shows some responding immediately following the reinforcement. The 3rd reinforcement with omitted TO (at *c*) is followed by a substantial rate, which, after some decline, leads quickly to a terminal rate. The 4th omission of the TO at *d* also produces a few responses immediately following the reinforcement, as well as a knee. The remainder of the session from *e,* with TO after reinforcement, shows consistent pausing and marked scalloping. (The experiment illustrated by Fig. 201 showed that the omission of TO produced a high rate after reinforcement only when the bird had reached the terminal rate of responding in the preceding interval.)

All TOs were then withdrawn while FI 5 continued in force. Figure 258 shows segments from the 1st, 2nd, and 4th sessions, all without TO, for 1 rat. In the 1st session (Record A) the over-all rate is low and no terminal rates comparable with those with TO after reinforcement (Fig. 256) are reached. Most intervals show pauses after reinforcement, although some running-through occurs at *a* and *b*. The over-all rate of responding increases in the 2nd session (Record B), and at *c, d, e,* and *f* no pausing occurs after reinforcement. In the segment beginning at *d* the terminal rate is maintained throughout. In the 3rd session (Record C) the performance is similar; most intervals show a pause after reinforcement with a normal acceleration to a terminal rate, although at *g* the responding begins immediately after reinforcement.

After 12 hours of exposure to FI 5 without TO, TO was again added. Record D shows some responding during the TO, particularly at *h* and *i*. The first 3 segments

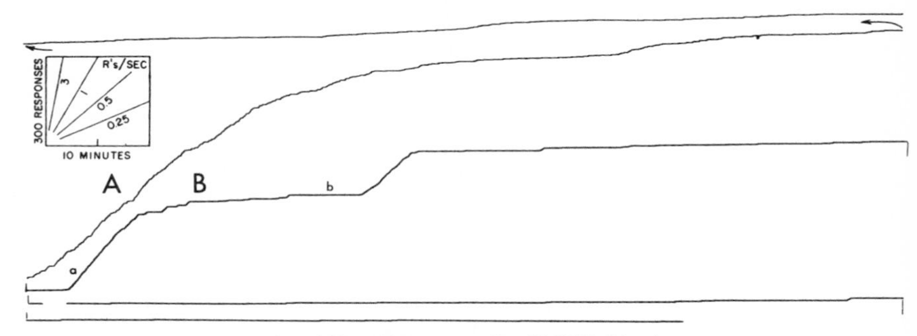

Fig. 259. Extinction after FI 5 TO 5 (rat)

show some negative acceleration and considerable disruption. Otherwise, however, a more orderly performance emerges.

Extinction after FI. Figure 259 shows 2 extinction curves after FI 5 TO 5 in the preceding experiment. The curve at A (completed by the segment above it) is for extinction after 29 hours of FI 5 TO 5. It shows an orderly decline in rate with few oscillations, suggesting extinction after VI. (See later chapters.) The curve at B, for another rat after 36 hours on FI 5 TO 5, shows an FI interval effect much better developed. When reinforcement is missing at *a,* 5 minutes after the start of the record, responding continues at the terminal rate and then declines by *b.* This pattern sets off a typical interval performance; a low rate prevails for the rest of the session.

Summary

The TO probes of the rats' FI 5 performance confirm the findings of the preceding pigeon experiments. Omitting an occasional TO produced a continuation of the terminal rate from the preceding interval.

The performance with TO after reinforcement showed consistent pauses and acceleration, contrasted with FI without TO where occasional instances occurred of responding at the start of the interval. Each time the TO after reinforcement was added or removed, however, the FI performance was disrupted.

The FI segments in the rat contain a relatively longer period of pausing and curvature than comparable performances with pigeons. This probably is a species difference.

The over-all and terminal rates of responding in this experiment reflect the inadequate amount of reinforcement (0.05 gram) used in the experiment.

TRANSITION FROM CRF TO FI TO 5

Performances under FI 1, 2, 4, 8, and 16 were recorded following crf with TO 5 after every reinforcement. One bird was exposed to each schedule for approximately 20 reinforcements per day every other day. The first 4 sessions will be described for each.

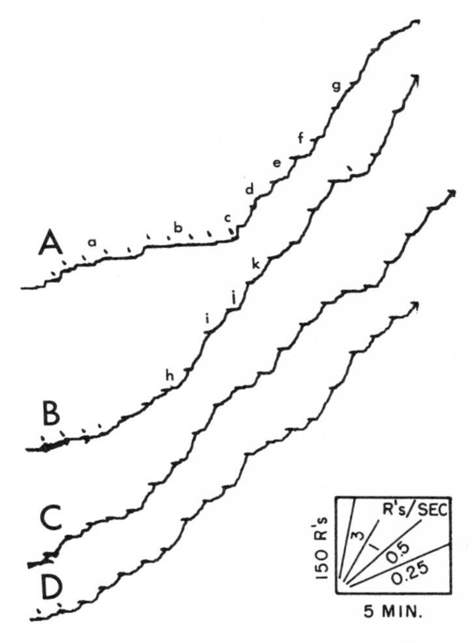

Fig. 260. Transition from crf to FI 1 TO 5

FI 1

The transition to FI 1 TO 5 shown in Fig. 260A is more like the transition to FR than FI, with a small extinction curve at *a* and a low rate of responding at *b*, followed by an abrupt shift to a higher over-all rate. Some responding during the TO occurred at *c* and *d*; and at *e* and *f* there are substantial periods of lower rates at the beginning of intervals. These are followed by a fairly linear performance at *g*. In the 2nd session, Record B, there is an over-all acceleration during the first 7 reinforcements. Well-marked scallops soon appear at *h, i, j,* and *k*. The over-all performance is irregular, however, and intervals occur, as in Record A, where little acceleration develops during the segment. The 3rd session following crf (Record C) shows a marked pause and a period of lower rate at the start of most intervals; this period of lower rate is even more pronounced during the 4th session (Record D).

This transition should be compared with Fig. 118 for FI 1 after crf without TO. The main difference is the rapid development of the pause after reinforcement and a high terminal rate of responding in Fig. 260. A comparable performance emerged after 160 reinforcements without TO. In the present experiment a well-marked scallop appears after only 40 reinforcements. Despite the slower development, however, Fig. 118 shows a greater uniformity from segment to segment.

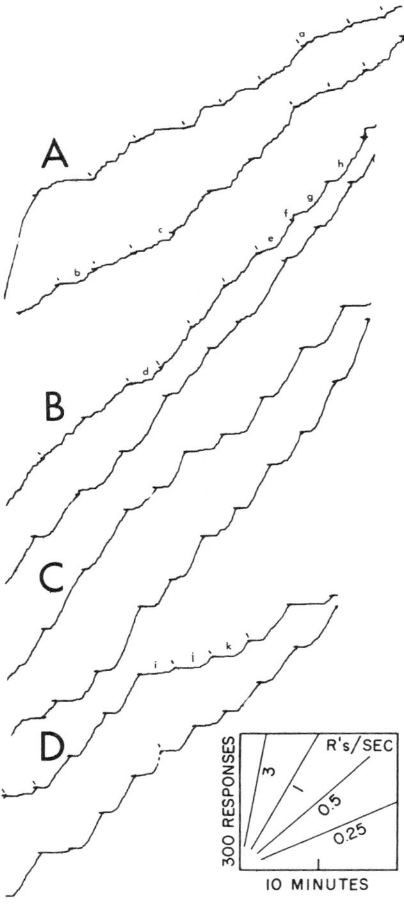

Fig. 261. Transition from crf to FI 2 TO 5

FI 2

Figure 261 shows the first 4 sessions of the transition crf to FI 2 TO 5. The overall form of the 1st session (Record A) is similar to the transition to FI 2 without TO, with each segment showing negative acceleration which becomes progressively less marked until a roughly linear phase is reached (*a* to *c*). Unlike the normal transition, however, the linear phase very quickly gives way to a rough FI scallop. One instance at *b* breaks into the linear phase. The 2nd session (Record B) begins with a roughly linear phase; but the 4th interval (*d*) shows a rough positive acceleration which becomes more marked at *e*, *g*, and *h*. Responding occurs during the TO at *f*; and for the remainder of the session, the start of each interval shows a marked pause and scallop. The same performance is sustained in Record C beginning 40 reinforcements after crf. By the 4th session (Record D) the scallops become even more marked, with the initial pause often extending through most of the interval, as at *j*, *k*, and *l*.

FI 4

Figure 262 contains the first 4 sessions on FI 4 TO 5 after reinforcement following crf. Only the first 6 segments of the 1st session are presented (Record A) because of a recorder failure during the remainder of the session. These segments are all roughly negatively accelerated, and the available portion of the over-all curve shows over-all negative curvature. The complete 2nd session is shown in Record B. The first 3 segments are negatively accelerated. But a very brief linear phase follows in the segment at *b*, after which there is a progressive development of positive acceleration at *c*, *d*, and *e*, which extends throughout most of the segment, although with very rough grain. Some responding during TO occurs at *a* and *b*. During the remainder of the session, the pause following the reinforcement becomes longer, the curvature becomes more abrupt, and the terminal rate increases and is sustained for a longer period of time (as at *f*, *g*, and *h*). The 3rd session following crf (Record C) shows over-all negative acceleration because of the progressive development of pausing and scalloping during the session. The 1st segment of the session shows a sustained high rate. The following 3 segments, at *i*, *j*, and *k*, show fairly linear performances with rough grain. These are followed at *l* and *m* by extended curvature throughout most of the interval, although with very little pausing following the reinforcement. After Segment *m*, the session shows a marked decline in the rate. The over-all rate falls to approximately 0.2 response per second. The performance continues to be markedly scalloped, with most of the responding occurring toward the end of the interval. As the number of responses in the interval becomes very low, the pause following the reinforcement becomes more marked, extending through most of the interval (as at *n*). The 4th session following crf shown in Record D begins at a much lower over-all

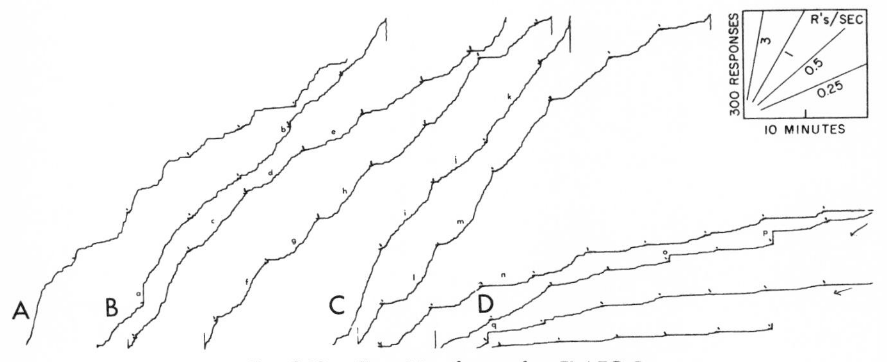

Fig. 262. Transition from crf to FI 4 TO 5

rate than the 3 previous sessions and shows a continuous decline, until at the end of the session only a few responses per interval segment occur. In spite of the extremely low rate of responding, however, the reinforcement is not being postponed beyond the designated fixed interval. Responding occurs during TO at *o, p,* and *q*. This transition should be compared with Fig. 120.

FI 8

The record of the 1st session was lost because of a recorder failure; however, Fig. 263 shows the complete 2nd session. The 1st interval shows slight negative acceleration and rough grain. The bird responds during the first TO at *a,* and the 2nd interval is roughly linear at a lower rate of responding. A few responses occur during the 2nd TO, at *b.* Well-marked pauses then begin to appear at the start of the intervals (as at *c* and *d*). The curvature becomes more uniform and extended. In the 3rd segment, at *e,* and in the 4th segment, at *g,* the bird pauses beyond the designated interval,

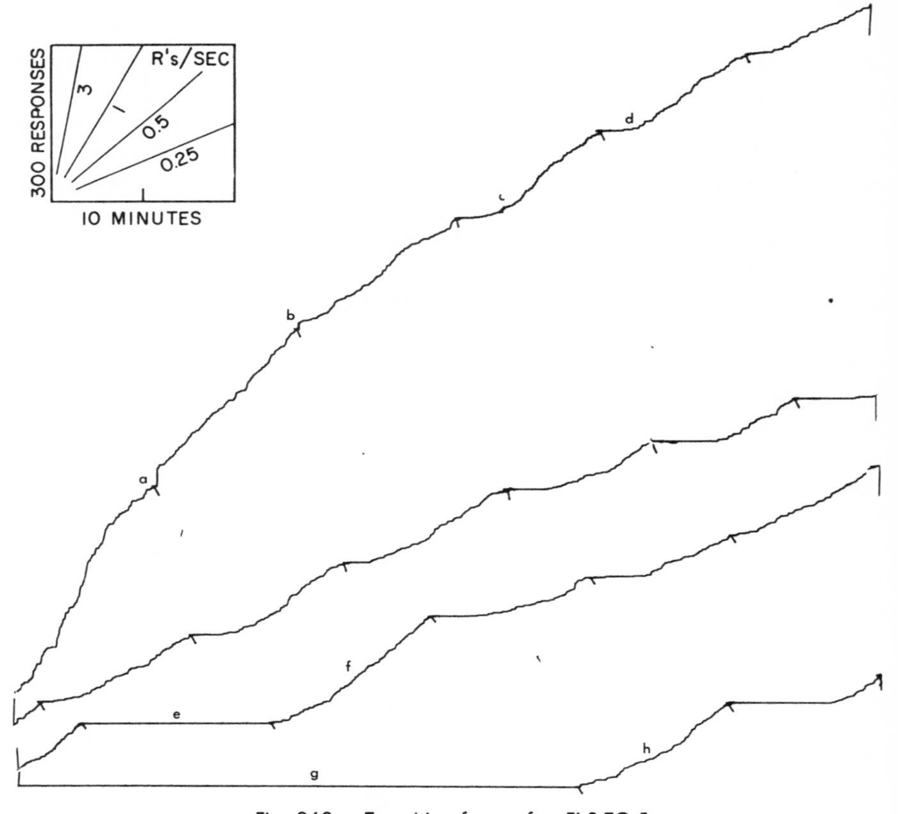

Fig. 263. Transition from crf to FI 8 TO 5

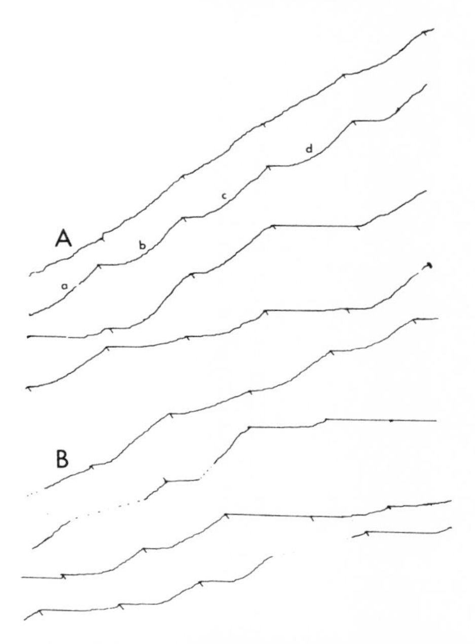

Fig. 264. Further development on FI 8 TO 5 after crf

and reinforcement is late. In both subsequent intervals (*f* and *h*) the curvature is slight, responding beginning immediately at nearly the terminal rate.

Figure 264 shows the 3rd and 4th sessions after crf. The 40th to 60th reinforcements are shown in Record A; the 60th to 80th, in Record B. The scallop is lost overnight. Record A begins with a roughly linear performance and shows only a slow increase in positive curvature, which becomes quite smooth by the segments at *a, b, c,* and *d.* The over-all rate of responding shows the continuous decline through the session seen in performances under FI 1, 2, and 4. Record B, the 4th session following crf, shows an even lower over-all rate of responding, because of increased pausing and failure to reach as high a terminal rate. Portions of Record B which were poorly recorded are indicated with dotted lines, which are not to be construed as the probable course of the curve. This transition should be compared with Fig. 132 and 133, showing the transition from crf to FI 8 without TO, where the performance even after many hours of exposure to the schedule did not show any good adjustment to the fixed-interval schedule.

FI 16

The effect of the first 20 reinforcements in the transition from crf was lost because of a recorder failure. Figure 265 shows the 2nd session, containing the 21st to 40th

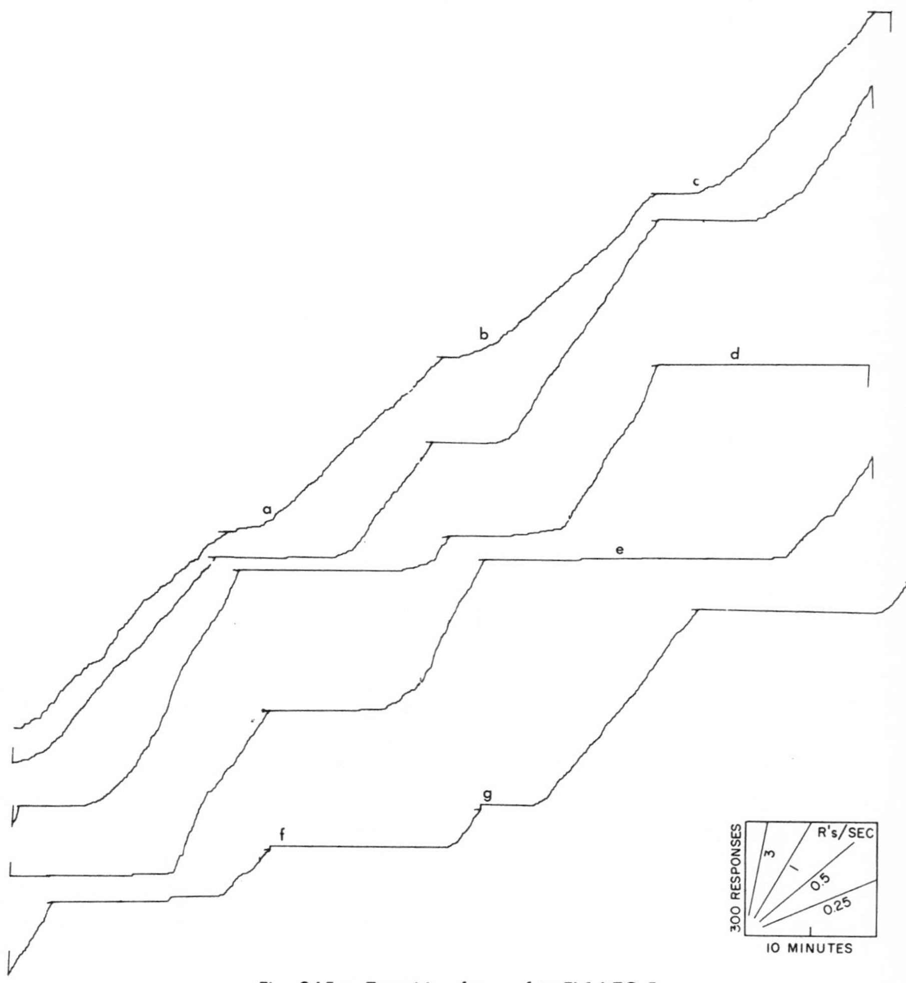

Fig. 265. Transition from crf to FI 16 TO 5

reinforcements. The 1st segment begins with a linear segment with rough grain. A marked pause and acceleration develop during the first 3 intervals, at *a, b,* and *c.* The amount of pausing after reinforcement increases progressively, and instances occur at *d* and *e* where the reinforcement is postponed beyond the designated interval. The terminal rate is not held well; and although the over-all rate is not markedly different during the early and latter parts of the interval, the nature of the responding during the latter part of the interval shows considerable variation. Responding occurs during TO at *f* and *g.*

Figure 266 shows the 3rd session after crf (the 41st to 60th reinforcements). The 1st interval, at *a,* shows an anomalous performance in which there is an extended pause through most of the interval. The 2nd interval, at *b,* is essentially linear. A pause and acceleration develop at *c.* A progressive increase in the pause at the start of the interval follows as in the previous session in Fig. 265. At *d* the reinforced response is the only response in the interval, and it is followed by a continuation of the near-zero rate. The reinforcement at *e* follows a long run at a high rate. In spite of the 5-minute TO the bird continues at a high rate for a few responses, and slows down only slightly before finishing the next segment at the terminal rate. The rapidity of the transition from crf to FI 16 with TO is evident from a comparison of Fig. 265 and 266 with Fig. 123 and 124, which show the transition from crf to FI 17 without TO.

Summary of crf to FI TO 5

All 5 birds studied here on FI 1 to FI 16 TO 5 after crf show a much more rapid production of a markedly scalloped FI performance than those without TO. In every case a stable performance with a marked pause following the reinforcement and an acceleration to a terminal rate occurred well within 80 reinforcements following crf. (Part of the rapidity with which the FI scallop develops here may possibly be due to the fact that the reinforcements are spaced out over several sessions.)

The difference between transitions from crf to FI with and without TO after reinforcement is greatest at longer fixed intervals. At FI 16, for example, a marked pause following the reinforcement and acceleration to a stable terminal rate occur between the 40th and 60th reinforcements after crf. This is an extremely rapid development of a close conformation to the fixed-interval pattern of reinforcement. The rapid transition is consistent with the view that the major effect of TO is to prevent the carryover of behavior as a stimulus from one interval to the next. This carryover is especially important at longer intervals of reinforcement, where schedules without TO result in unstable FI performances conforming only poorly to the temporal properties of the fixed-interval schedule.

TRANSITION FROM ONE FI TO ANOTHER WITH TO 5 AFTER REINFORCEMENT

Five schedules of reinforcement, FI 1, 2, 4, 8, and 16, plus TO 5, were rotated among 5 birds, so that each bird changed once from FI 2 to FI 1, FI 4 to FI 2, FI 8 to FI 4, FI 16 to FI 8, and FI 1 to FI 16, all with TO. The transitions were from a larger value of FI to a smaller, except for the change from FI 1 to FI 16. (Also, 1 bird was changed from FI 2 to FI 16.) Earlier performances for these birds at values of FI assumed after crf are shown in Fig. 129, 150, 152, and 170. Fairly stable performances with TO after reinforcement have already been described in Fig. 219, 223, 227, 232, and 251. In the present experiments a given value of reinforcement was maintained until a fairly stable condition developed. The transition to another value of FI was then made. The results provide information about performances on var-

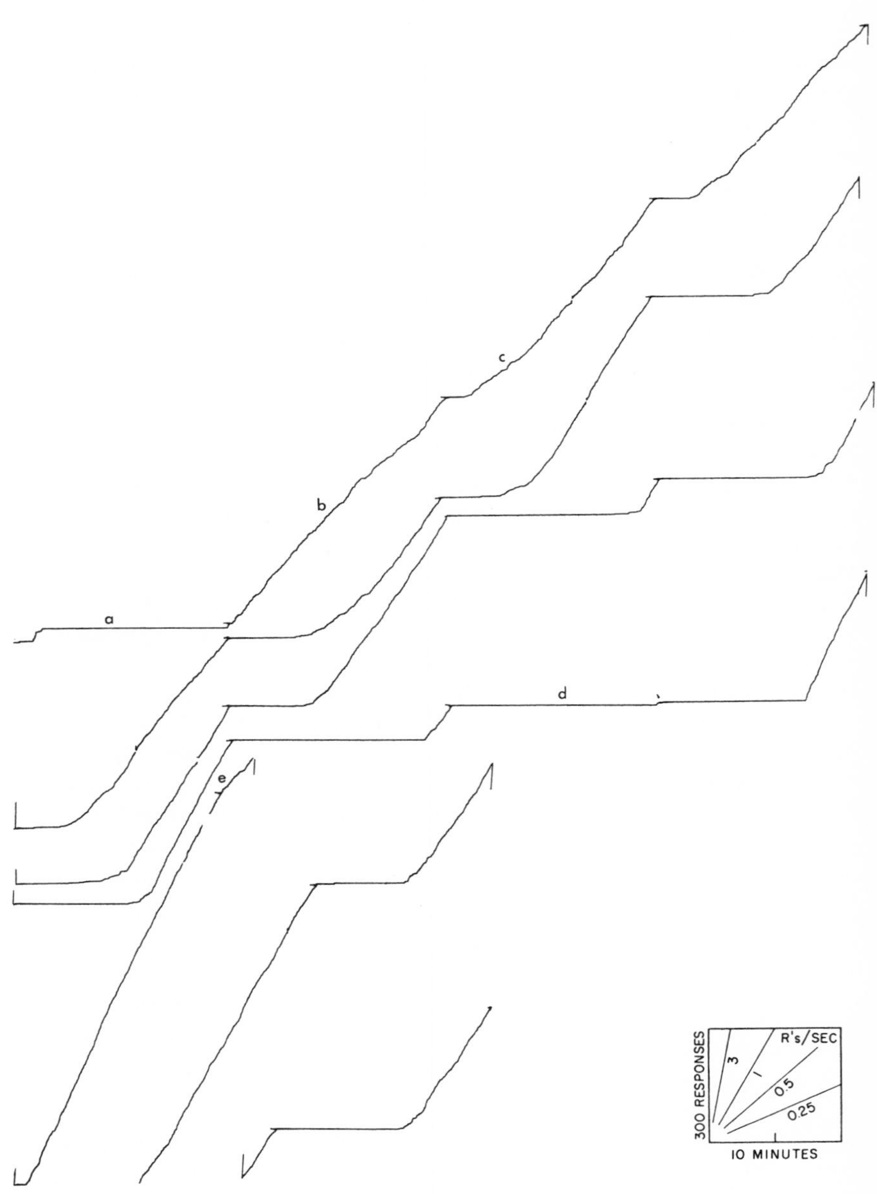

Fig. 266. Third session on FI 16 TO 5 after crf

ious FI schedules of reinforcement with TO, and show the extent to which we are now prepared to describe in detail the transitional states.

The 5 birds approached a given case of transition from different histories in this experiment. In the following reports each history is briefly indicated. Earlier experiments on FI 1 with and without TO are not indicated. All schedules included TOs.

FI 1 to FI 16

Bird 49Y. History: FI 2; FI 1 (240 reinforcements). Figure 267 shows the 1st session on FI 16 after FI 1. The previous FI reinforcement produces a high over-all rate of responding, which declines slowly throughout the session, as in extinction after variable-interval reinforcement. A high rate is maintained throughout the 1st FI 16 segment, which ends at the reinforcement at *a*. This high rate is the equivalent of

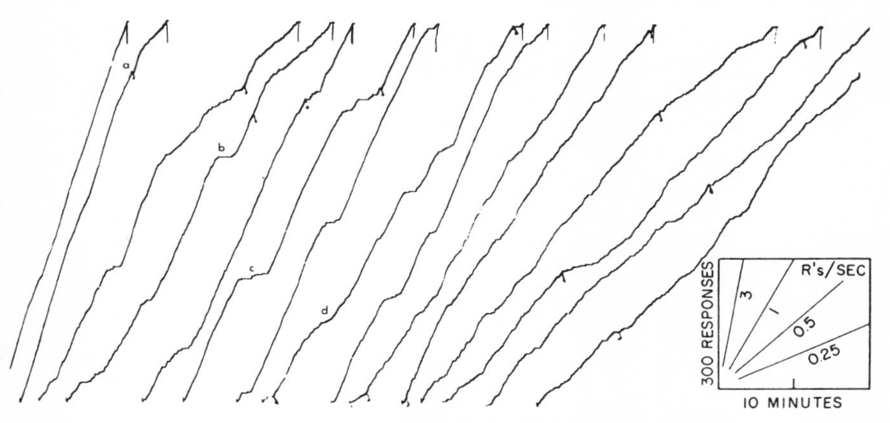

Fig. 267. First session on FI 16 TO 5 after FI 1 TO 5 (bird 49Y)

the first part of an extinction curve after FI 1. Thereafter, lower rates develop some time after reinforcement. Oscillations involving either no responding (as at *b*) or a lower rate of responding (as at *c* and *d*) are characteristic of the early transition.

The rate during the first part of the fixed-interval segment continues to fall during the 2nd session on FI 16; and by the end of that session the fixed-interval curvature is well-marked, as in Fig. 268A. The end of the 3rd session under FI 16 is shown in Record B. The period of slow responding during the first part of the interval has become more extended, as at *b, c,* and *d.* The terminal rate of responding has increased correspondingly. The high rate immediately following the reinforcement at *a,* breaking into a more appropriate scallop, is probably a result of the previous FI reinforcement.

Figure 269 shows segments from the 4th, 6th, and 9th sessions on FI 16. In Record A, the end of the 4th session, the over-all rate of responding is low. By the end of the 6th session, in Record B, the pause after reinforcement is longer and the terminal

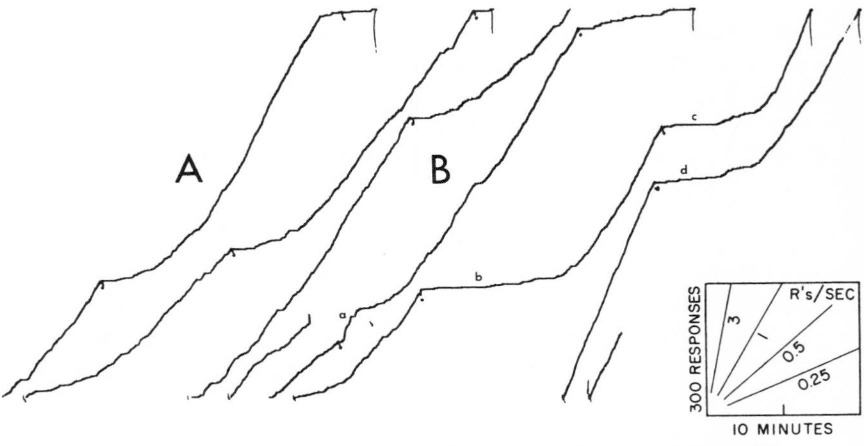

Fig. 268. Second and third sessions on FI 16 TO 5 after FI 1 TO 5 (bird 49Y)

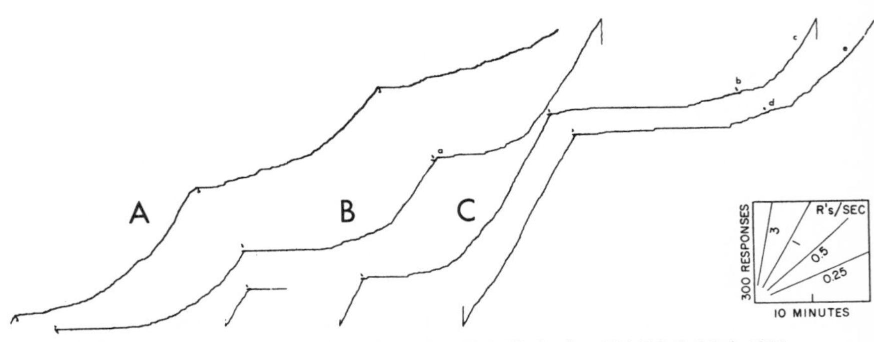

Fig. 269. Fourth to ninth sessions on FI 16 TO 5 after FI 1 TO 5 (bird 49Y)

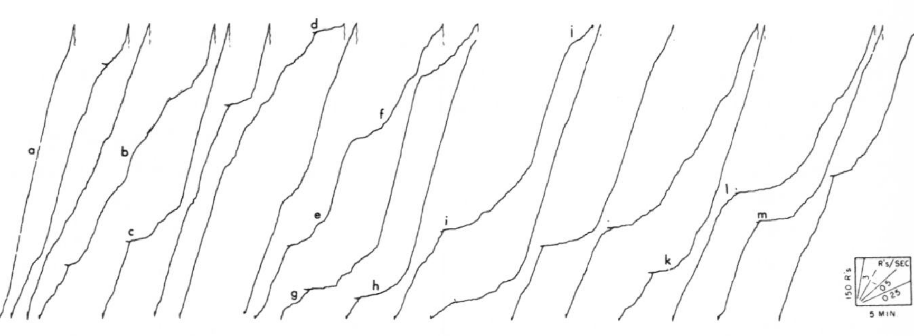

Fig. 270. First session on FI 16 TO 5 after FI 1 TO 5 (bird 52G)

rate higher. At *a* the terminal rate from the preceding interval continues beyond the reinforcement for approximately 10 responses, despite the 5-minute TO period separating the 2 segments.

Bird 52G. History: FI 4; FI 2; FI 1 (350 reinforcements). Figure 270 represents the transition. It is similar to Fig. 267 in that it shows a high over-all rate at the start of the session (*a*) which declines as the effect of the previous FI 1 reinforcement is extinguished under the new schedule. However, the bird showed a much more rapid development of a pause and a period of slower responding in the early part of the interval. The high rate of responding resulting from the previous FI 1 performance at *a* continued to appear throughout the remainder of the session, although the rate usually fell off before reinforcement, as at *g, j,* and *l.* In spite of these frequent reinforcements at lower rates of responding, the over-all rate and number of responses per interval remained high. Running-through occurs at *c, d, i,* and *m,* possibly because of the previous FI 1 reinforcement. The early appearance of a scallop in this bird might be due to the earlier history on FI 4. The oscillations appearing at *b, e,* and *f* and the type of curvature at *h* and *k* suggest previous performances on intervals of 4 minutes.

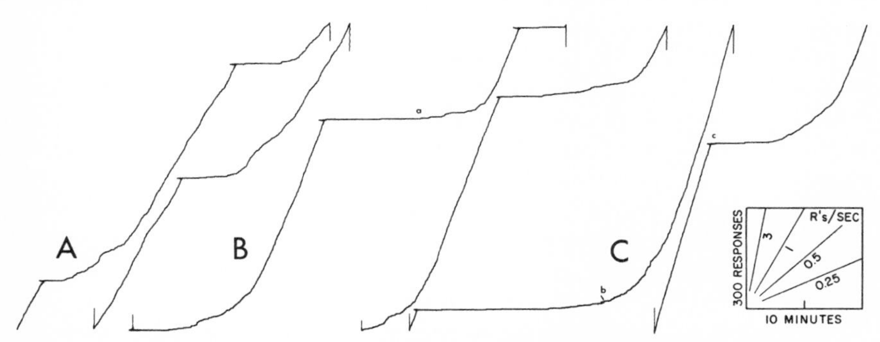

Fig. 271. Second to fifth sessions on FI 16 TO 5 after FI 1 TO 5 (bird 52G)

Figure 271 shows the subsequent performance on FI 16 for this bird. Record A, from the end of the 2nd session, shows a well-marked pause following the reinforcement and consistent curvature to a well-sustained terminal rate of responding. Record B, from the end of the 3rd session after 15 hours of exposure to FI 16, shows a much more extended pause and period of lower rate. At *a,* for example, the terminal rate is not reached until 13 minutes after the start of the interval. The terminal rate is now 1.5 responses per second, compared with approximately 1 in Record A. Record C describes a segment from the performance at the end of the 5th session after 25 hours of exposure to FI 16. The first 2 segments of the record show a second-order effect at *b*, occurring in spite of the TO after reinforcement. Again, the very low count at the reinforcement at *b* seems to be responsible for the early start of responding in the next segment. Fifteen hundred responses are emitted before the reinforcement at *c.* The

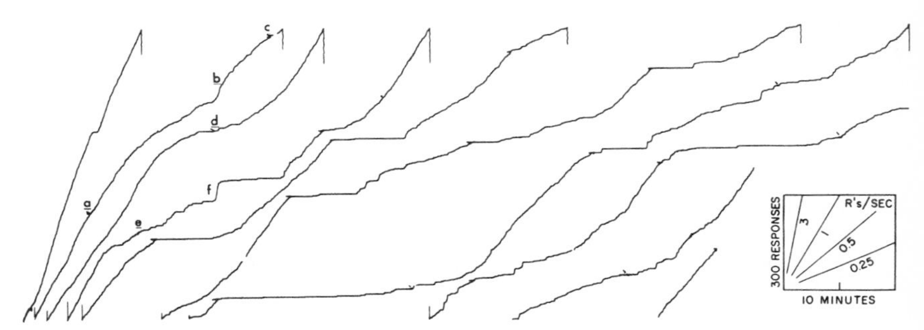

Fig. 272. First session on FI 16 TO 5 after FI 1 TO 5 (bird 89R)

next segment is followed by a pause of normal size and a very smooth acceleration to a terminal rate of about 2 responses per second.

Bird 89R. History: FI 8; FI 4; FI 2; FI 1 (120 reinforcements). The 1st session (Fig. 272) shows a decline in the over-all rate which is much more marked than in the figures already described. The rapid transition may be due to the fact that this bird had already undergone transitions to FI 8, FI 4, FI 2, and FI 1. The high rate from the preceding FI 1 reinforcement has fallen off by approximately one-third at the 1st reinforcement, at *a*. The 2nd, 3rd, and 4th fixed-interval segments, ending at *c, d,* and *e,* show marked negative curvature. Two bursts at high rates of responding at *b* and *f* may be reminiscent of earlier shorter intervals. A marked, irregular curvature characterizes the remainder of the session. The grain is extremely rough.

Figure 273 shows the further development of the FI 16 performance. Record A shows a segment 6 hours after the beginning of the 2nd session. The performance is still irregular, with rough grain and negative acceleration, but the over-all rate has increased considerably. Responding at intermediate rates is fairly continuous, unlike the performance in the previous session, where pauses occurred frequently at all points during the fixed-interval segments. A segment of the 3rd session after 15 hours of exposure to FI 16 (Record B) shows a very regular scallop in which the acceleration extends throughout most of the fixed interval. Note the slight running-through (*a* and *b*) in spite of the 5-minute TO. In Record C, from the 4th session after 20 hours of ex-

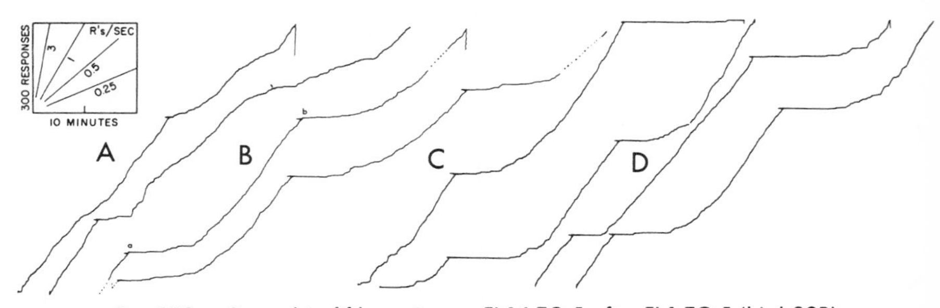

Fig. 273. Second to fifth session on FI 16 TO 5 after FI 1 TO 5 (bird 89R)

Fig. 274. End of sixth session on FI 16 TO 5 after FI 1 TO 5 (bird 89R)

posure to FI 16, the terminal rate is higher and is reached earlier in the interval. The final performance in Record D after 25 hours of FI 16 reinforcement shows a somewhat wider variation in the length of the pause and a period of lower rate of responding at the start of the fixed-interval segment.

Figure 274 shows a still later performance after 30 hours on FI 16.

Bird 98R. History: FI 16; FI 8; FI 2; FI 1 (120 reinforcements). The 1st session (Fig. 275) shows the same decline in the over-all rate as the other transitions from FI 1 to FI 16, but the transition is most rapid here. A pause follows the 1st reinforcement, at *a,* and the set of intervals beginning at *a* through *i* shows a smoothly developing curvature. Later intervals show a further increase in the extent of the scallop. Note that the bird had an earlier history of FI 16.

The start of the 2nd session on FI 16 in Fig. 276 shows some effect of shorter fixed intervals at the start of the session, but orderly scallops with a substantial pause at the start of the interval and a stable terminal rate develop soon thereafter.

Bird 46Y. History: FI 2 (31 reinforcements). In Fig. 277 the transition from FI 2 to FI 16 (made by accident) is similar to Fig. 267, except that the 1st segment on FI 16 does not show a sustained rate, and the decline to a lower rate throughout the session is less orderly. The record is very rough and marred by short rapid runs. But there is no consistent rate change correlated with any feature of the schedule of reinforce-

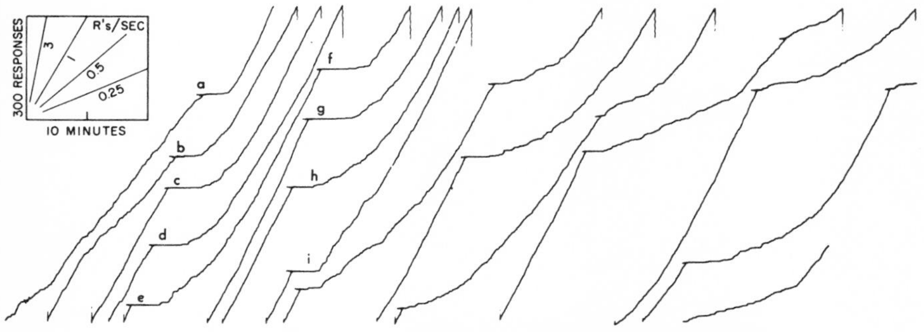

Fig. 275. First session on FI 16 TO 5 after FI 1 TO 5 (bird 98R)

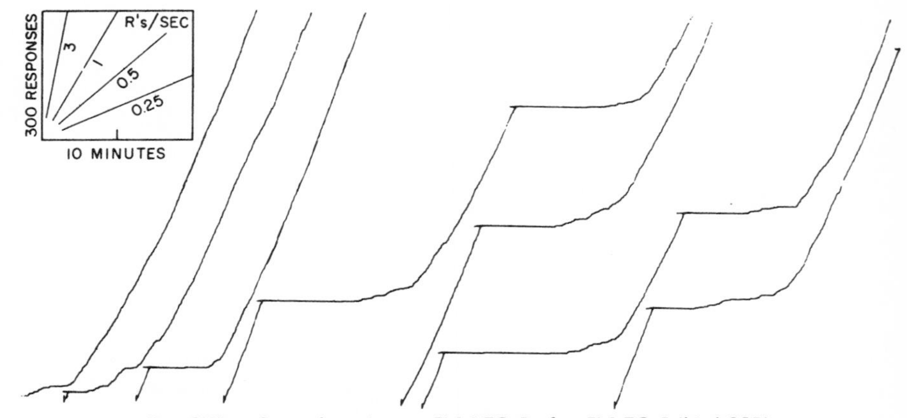

Fig. 276. Second session on FI 16 TO 5 after FI 1 TO 5 (bird 98R)

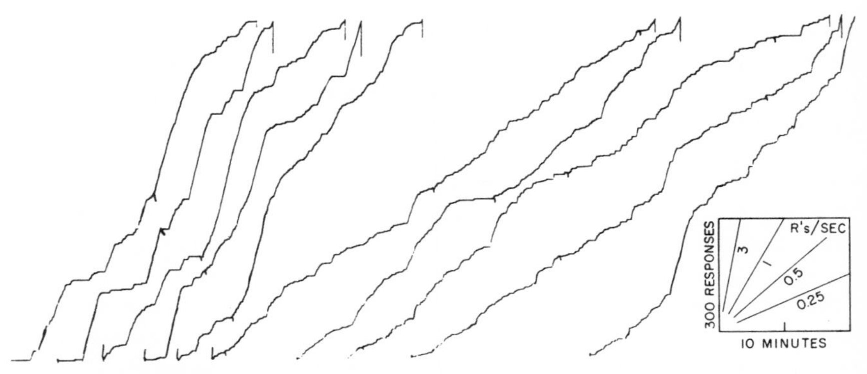

Fig. 277. First session on FI 16 TO 5 after FI 2 TO 5 (bird 46Y)

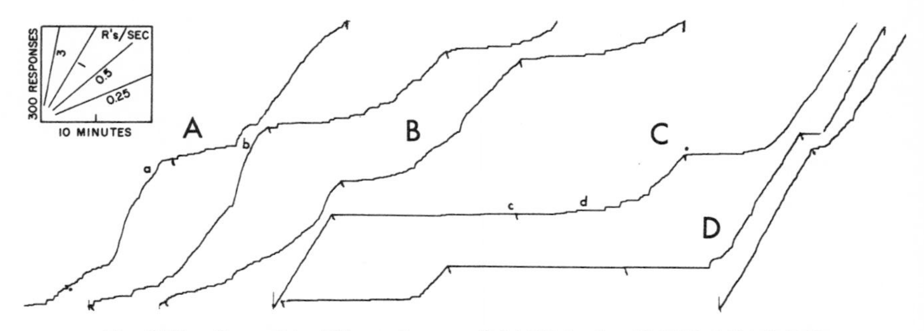

Fig. 278. Second to fifth sessions on FI 16 TO 5 after FI 2 TO 5 (bird 46Y)

ment. In the 2nd session, after 7 hours on FI 16 (Fig. 278A), a rough scallop has developed; but the performance is quite irregular, with negative curvature (as at *a* and *b*) and marked changes in performance from segment to segment. Record B, a segment from the 3rd session, after 15 hours of exposure to FI 16, shows less negative acceleration. In Record C, a segment from the 4th session, after 20 hours on FI 16, the pause following the reinforcement has become much more extended (as at *c*, where only a few responses occur during the whole 16-minute interval). The segment at *d* does not begin earlier, because of the reinforcement at a low rate in the previous segment. Record D is a segment from the 5th session after 23 hours of FI 16 reinforcement. Figure 279A shows the start of the 6th session after 49 hours of FI 16 reinforcement. A high terminal rate has developed and is maintained, with a concomitant increase in the pause following reinforcement. The performance is not stable, however; and the end of the session (Record B), after 53 hours of FI 16 reinforcement, shows negative curvature near the ends of intervals and wide changes in local rates uncorrelated with any features of the schedule.

FI 16 to FI 8

Bird 52G. History: FI 4; FI 2; FI 1; FI 16 (25 hours). Figure 280A, the 1st session on FI 8, begins at the low rate characteristic of the start of the FI 16 interval. The

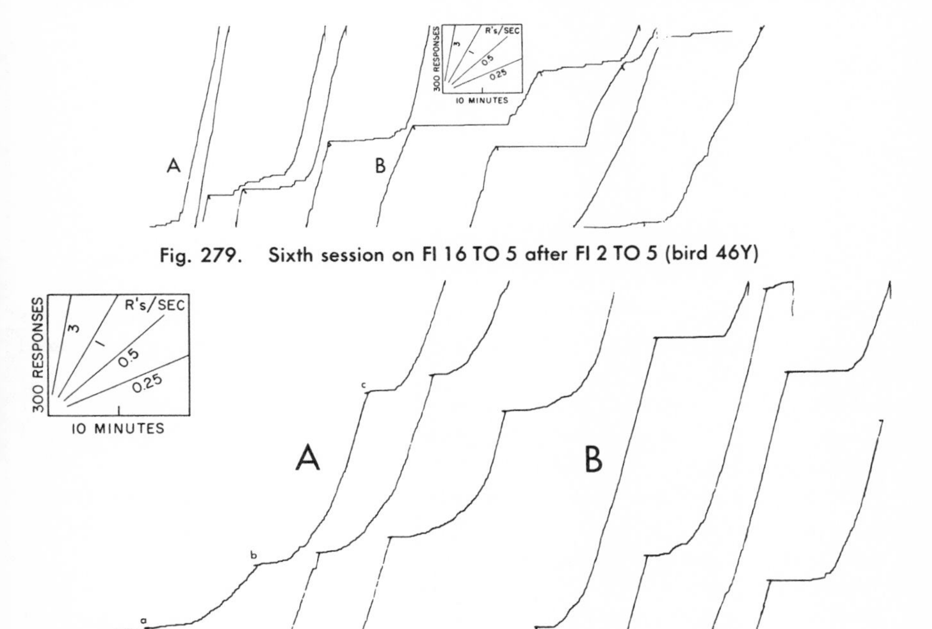

Fig. 279. Sixth session on FI 16 TO 5 after FI 2 TO 5 (bird 46Y)

Fig. 280. First and last sessions on FI 8 TO 5 after FI 16 TO 5 (bird 52G)

1st reinforcement, at *a*, occurs before the terminal rate is reached. Because of this early reinforcement the 2nd segment shows a more rapid acceleration to the terminal rate. The 2nd reinforcement, at *b*, is followed by an even more rapid acceleration to the terminal rate; and the 3rd reinforcement, at *c*, shows some running-through. The remaining segments of Record A then show a progressive deepening of the scallop, because of the development of a substantial pause at the beginning of the interval and a very smooth acceleration to a high terminal rate. Record B shows the final performance after 14 hours of exposure to FI 8. A short pause at the start of the interval is followed by a moderate period of acceleration to a high terminal rate, which is maintained for about 500 responses.

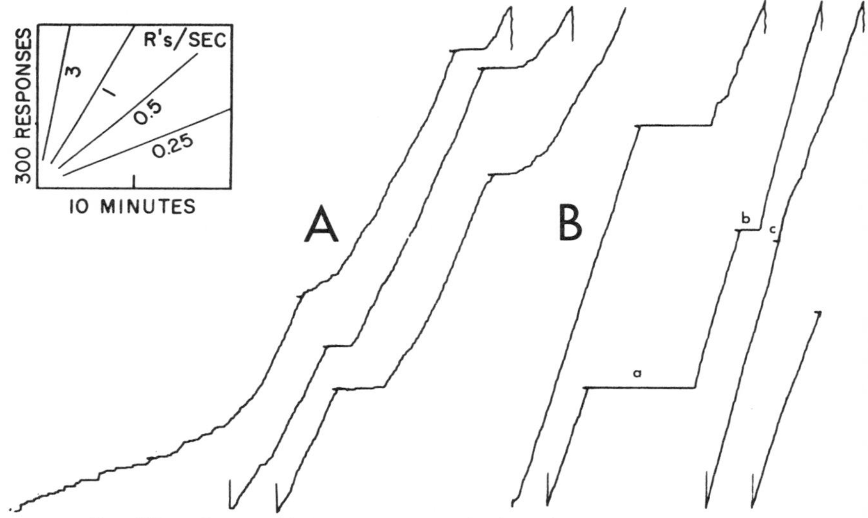

Fig. 281. First and last sessions on FI 8 TO 5 after FI 16 TO 5 (bird 49Y)

Bird 49Y. History: FI 2; FI 1; FI 16 (49 hours). The 1st session (Fig. 281A) shows roughly the same pattern as Fig. 280A, except that the smooth curvature is replaced here by a pause and an abrupt shift to the terminal rate. Record B shows a final performance after 17 hours. The pause after reinforcement varies from slightly over 1 minute at *b* to over 5 minutes at *a*. Responding begins abruptly at the terminal rate of responding after the pause, and most of the segments show slight negative curvature. At the start of the interval at *c* the bird begins responding at the terminal rate in the preceding interval and continues at a slightly lower rate for the rest of the segment. This performance is rarely observed with a 5-minute TO.

Bird 98R. History: FI 16. The 1st session (Fig. 282A) shows a rapid adjustment to the curvature appropriate to the FI 8 schedule. In Record B, after 30 hours of exposure to FI 8, each segment begins with a substantial pause and accelerates, often rather abruptly, to a terminal rate. An exceptional case of a knee occurs at *a*.

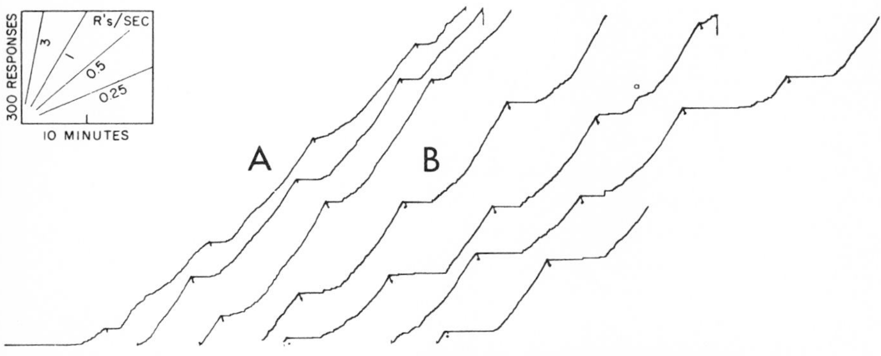

Fig. 282. First and last sessions on FI 8 TO 5 after FI 16 TO 5 (bird 98R)

Bird 89R. History: FI 8; FI 4; FI 2; FI 1; FI 16. This bird characteristically began each session with a high over-all rate with no fixed-interval scallop. Later in the session, a very stable and regular fixed-interval performance with marked scallops emerged. (Figure 273 is for this bird.) The 1st reinforcement on FI 8, at *a,* in Fig. 283A is received at a high rate of responding, as is the 2nd, at *b.* Thereafter, a lower rate develops progressively during the early part of the interval. Figure 283B shows the performance after a total of 8 hours of exposure to the FI 8 reinforcement schedule. A substantial pause follows each reinforcement, and the acceleration to a higher rate is irregular. The terminal rate varies from segment to segment, and instances of negative acceleration occur after the terminal rate has set in.

Bird 46Y. History: FI 2; FI 16 (55 hours). The previous performance on FI 16

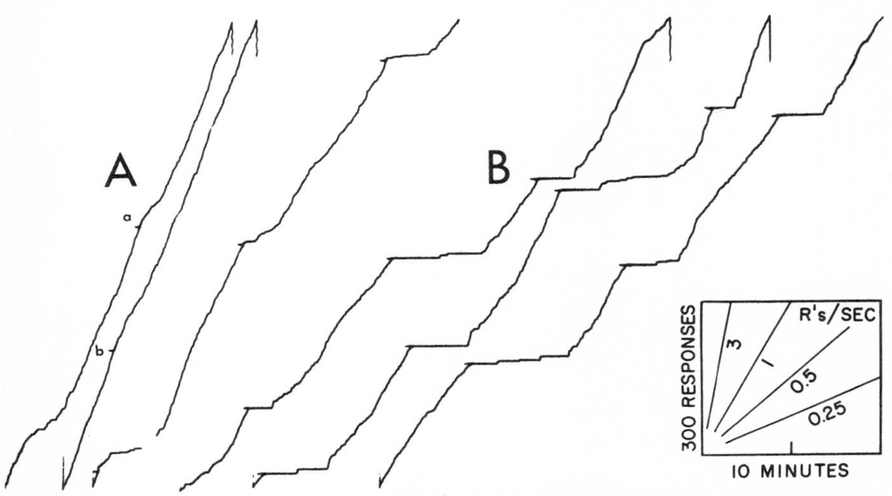

Fig. 283. First and last sessions on FI 8 TO 5 after FI 16 TO 5 (bird 89R)

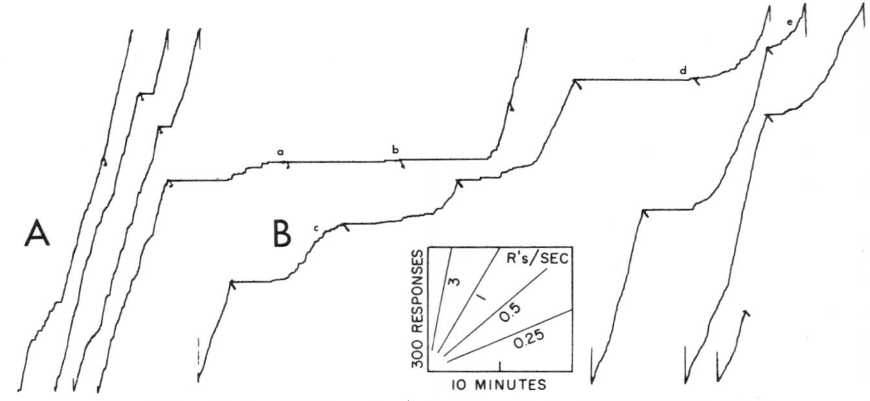

Fig. 284. First and last sessions on FI 8 TO 5 after FI 16 TO 5 (bird 46Y)

shown in Fig. 279 was not a stable performance on that schedule. The 1st reinforcements on FI 8 in Fig. 284A reflect the previous FI 16 performance in which the terminal rate was high. Responding is sustained almost continuously through the first 4 segments. The over-all rate then falls markedly during the next 3 segments, with a substantial pause following the reinforcement. The terminal rate is not reached at *a* and *b*. After 8 hours of reinforcement on FI 8, the curvature becomes appropriate to the new schedule of reinforcement. After 23 hours on FI 8, the performance is still irregular, as seen in Fig. 284B, which shows negative acceleration at *c*, failure to accelerate to the terminal rate of responding at *d*, and an early acceleration to the terminal rate at *e*. Figure 285 shows the following session complete. The high sustained rate of responding from the performance under the previous FI 16 reinforcement re-

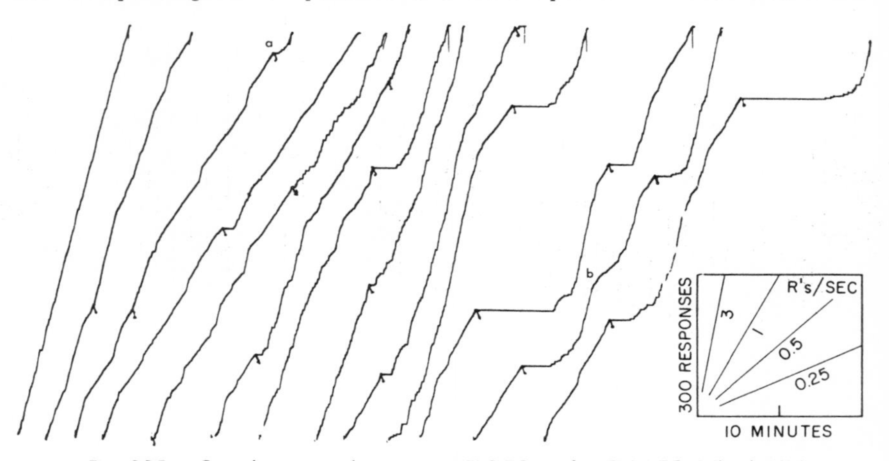

Fig. 285. Complete second session on FI 8 TO 5 after FI 16 TO 5 (bird 46Y)

turns at the beginning of the session but gives way to a lower over-all rate. Pausing appears after the 3rd reinforcement at *a* and develops progressively throughout the remainder of the session. The acceleration to the terminal rate of responding is generally premature, and in almost every case there is a marked negative curvature before the end of the interval.

FI 8 to FI 4

Bird 98R. History: FI 16; FI 8 (32 hours). This bird had developed a stable fixed-interval curve on FI 8. (See Fig. 282.) In the 1st session on FI 4 (Fig. 286) the

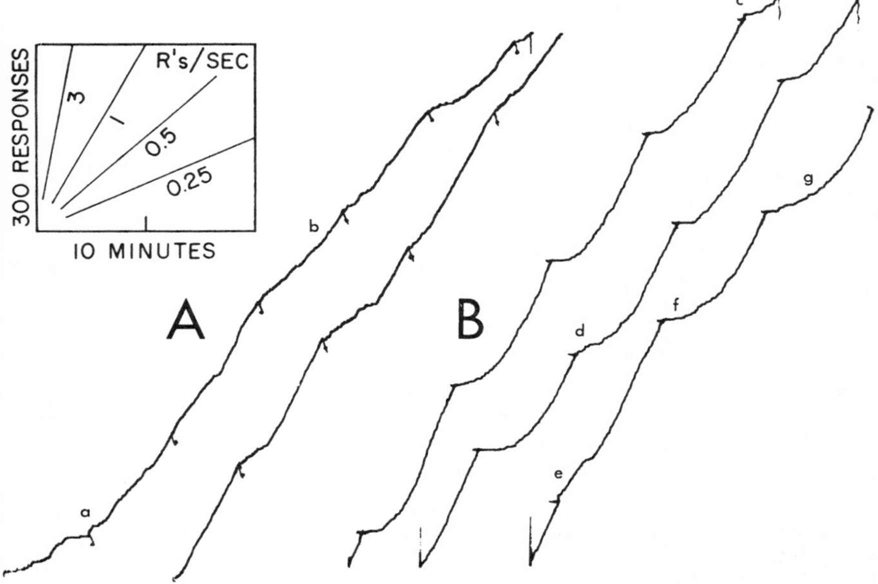

Fig. 286. First and last sessions on FI 4 TO 5 after FI 8 TO 5 (bird 98R)

1st reinforcement at *a* occurs before the terminal rate from FI 8 reinforcement has been reached. The early reinforcement produces immediate responding in the 2nd segment. Approximate scalloping begins at *b*. After 13 hours of reinforcement on FI 4, a performance approaching the final effect of the schedule is shown (Record B). Here, most intervals begin with a pause, and the curvature extends through most of the interval (cf. *f, g*). There is some responding at the start of the interval at *d* and responding during the time out at *c* and *e*.

Bird 52G. Figure 287 shows the transition (Record A) and later performance (Record B) on FI 8 TO 5 following the performance on FI 16 TO 5 shown in Fig. 280B. The early reinforcement at *a* is followed by a marked scallop before *b*. Strong negative curvature appears later (*c, d, e, f, g*).

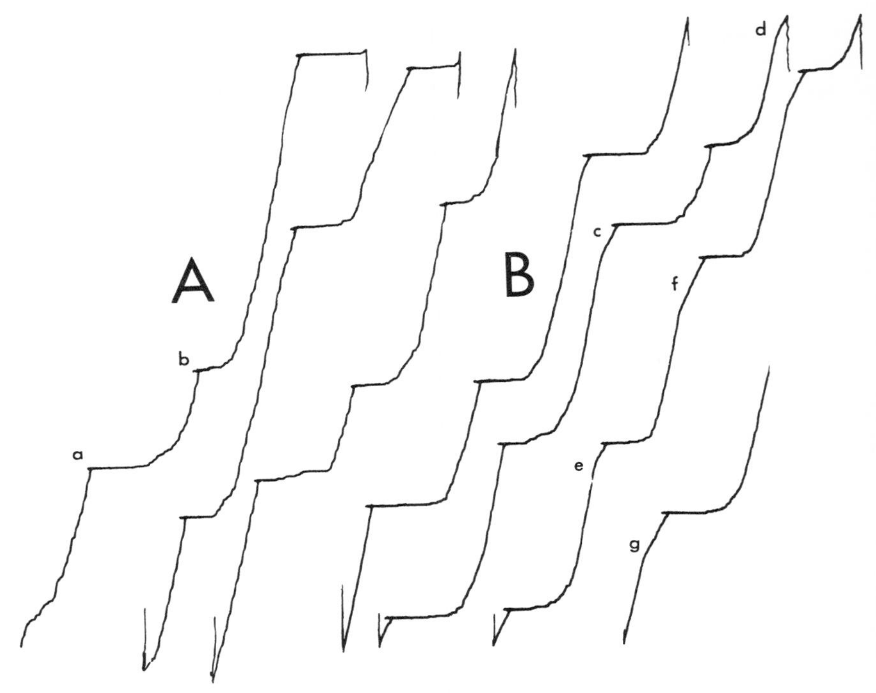

Fig. 287. First and last sessions on FI 4 TO 5 after FI 8 TO 5 (bird 52G)

Bird 49Y. History: FI 2; FI 1; FI 16; FI 8 (17 hours). Figure 280B showed a stable and smooth performance on FI 8. In the transition to FI 4 (Fig. 288A) the 1st reinforcement (at *a*) occurs after sustained responding during the first 4 minutes of the session. The first 5 segments show sustained responding at the terminal rate. A pause occurs at the start of the interval (*b*), however, with an abrupt shift to the terminal rate. The next segment (*c*) shows curvature appropriate to the FI 4 schedule. In the final performance in Record B, after 9 hours of exposure to FI 4, most segments begin with a substantial pause and show extended acceleration to a terminal rate. However, instances occur (as at *d* and *e*) where the pause at the beginning of the fixed-interval segment is brief and the acceleration to the terminal rate relatively abrupt.

Bird 46Y. History: FI 2; FI 16; FI 8 (14 hours). Although this bird did not show a very stable performance on FI 8 (Fig. 284B), an interval scallop appropriate to FI 4 reinforcement developed rapidly during the 1st session (Fig. 289A). Record B shows the final performance after 6 hours of reinforcement on FI 4.

Bird 89R. History: FI 8; FI 4; FI 2; FI 1; FI 16; FI 8. The 1st reinforcement on FI 4 (Fig. 290A) produces sustained responding throughout the next 4-minute interval in spite of the markedly scalloped performance which had previously been occurring on FI 8 (Fig. 283B). The linear performance then gives way to some positive curva-

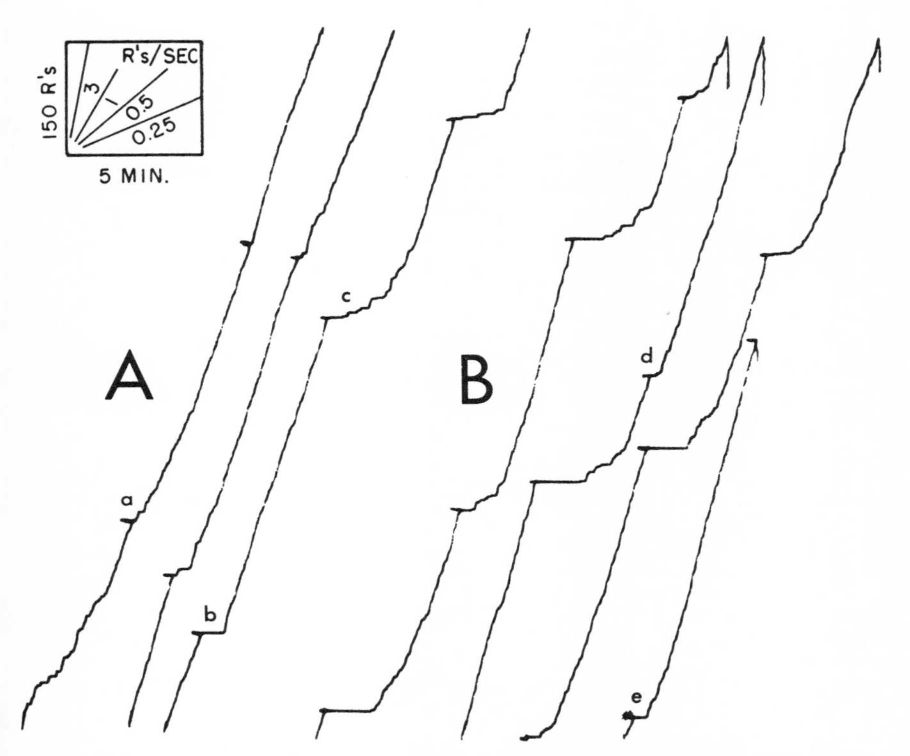

Fig. 288. First and last sessions on FI 4 TO 5 after FI 8 TO 5 (bird 49Y)

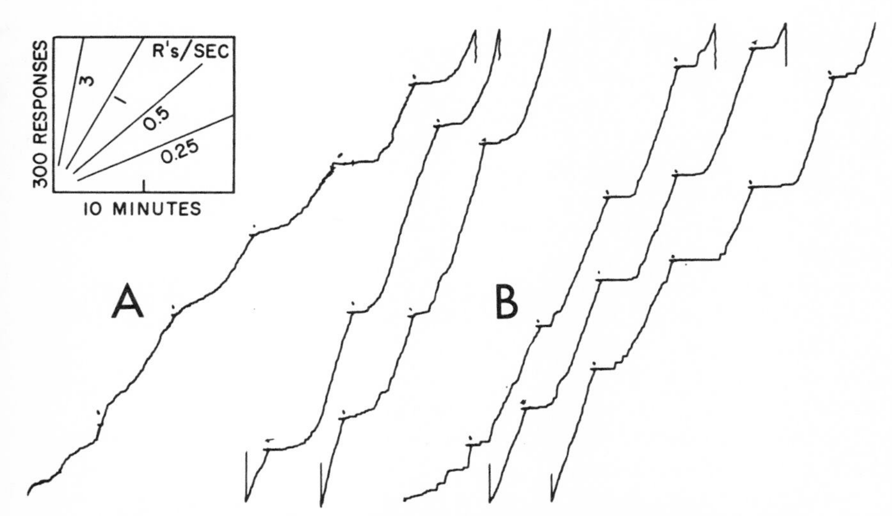

Fig. 289. First and last sessions on FI 4 TO 5 after FI 8 TO 5 (bird 46Y)

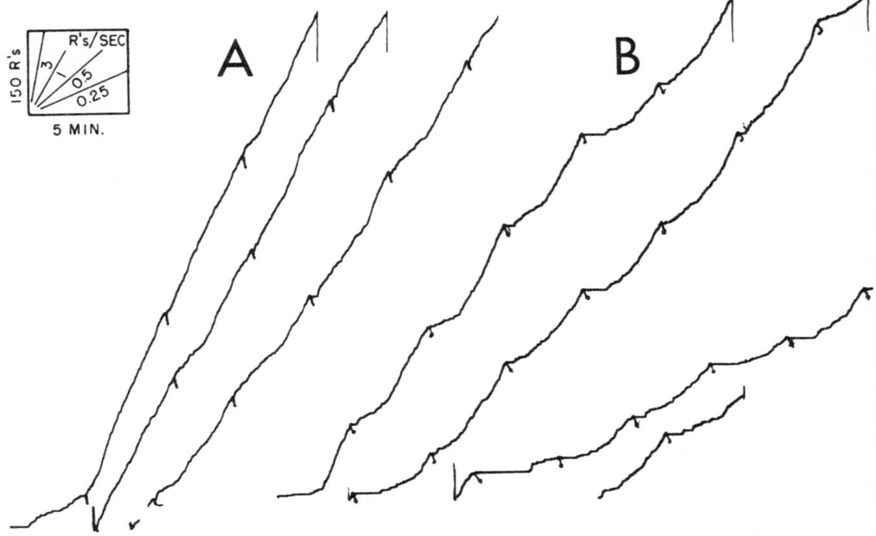

Fig. 290. First and last sessions on FI 4 TO 5 after FI 8 TO 5 (bird 89R)

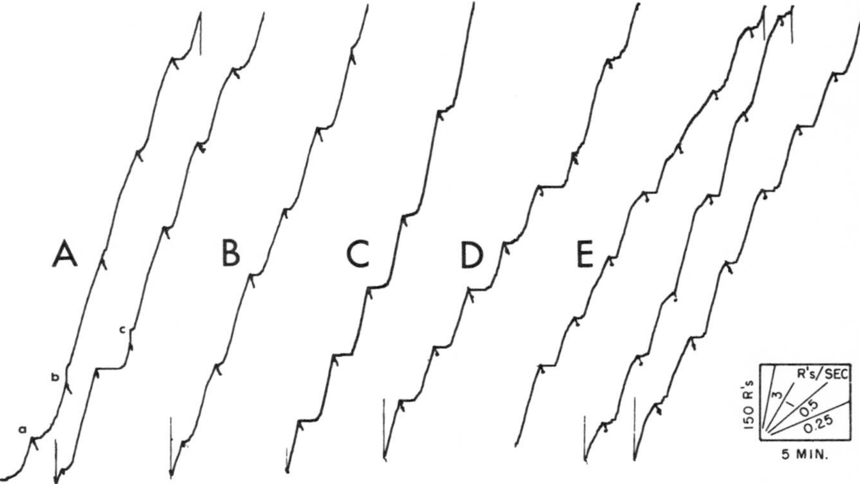

Fig. 291. First to twelfth sessions on FI 2 TO 5 after FI 4 TO 5 (bird 52G)

ture. The final performance, Record B, after 13 hours of reinforcement on FI 4, shows an over-all negatively accelerated curve as a result of: (1) the increase in the length of the pause during the session; (2) a decline in the terminal rate; and (3) an increase in the period during which acceleration occurs.

FI 4 to FI 2

Bird 52G. History: FI 4. The performance in Fig. 291A shows the first change in schedule for this bird after the original transition from crf to FI 4. The previous performance under FI 4 TO had been showing S-shaped curves. The 1st reinforcement on FI 2 (at *a*) occurs after the terminal rate of responding has been reached. Unlike the transitions from larger fixed intervals previously described, a similar pause and lower rate at the beginning of the interval appear in the second segment. Some responding occurs during the TO at *b* and *c*, and S-shaped curves soon appear. They disappear, however, by the end of the 2nd session (Record B). Record C shows a segment from the end of the 6th session of exposure to the FI 2 reinforcement, where the pause and period of lower rate following the reinforcement have become more extended and the terminal rate of responding has increased. The increase in the terminal rate more than compensates for the increase in the period of lower rate at the start of the interval. By the end of the 8th session (Record D), the terminal rate has fallen to a value of about the same order as in Record B, and the pause has increased, with a resulting decline in the over-all rate. After 11 sessions of reinforcement on FI 2, very marked negative curvature sets in. The entire 12th session is shown in Record E. Here, most of the intervals begin with pauses, followed by fairly abrupt accelerations to a high rate. Before the end of the interval, however, this rate falls off by a factor of 3, and the reinforcement occurs at a much lower rate. The S-shaped curves in Record E proved to be a special property of this interval of reinforcement at this stage of development of the performance; later performances for this bird on FI 16 and FI 8 in Fig. 271 and 280 showed normal, positively accelerated curves.

Bird 98R. History: FI 16; FI 8; FI 4. Figure 292 shows the first exposure to FI 2 after the final performance under FI 4 shown in Fig. 282. The middle part of the session was lost because of a recorder failure; hence, only the first 5 and last 3 segments of the session are shown. Some interval curvature develops. Record B shows the beginning of the 2nd session, during which a pause after reinforcement develops, together with an extended scallop throughout the interval. The over-all rate falls as the pause and scallop develop. Responding occurs during the TO at *a*. Record C shows the final performance after 5 sessions of FI 2 reinforcement. Responding occurs during the TO at *a, c, d,* and *e*. Both the over-all and terminal rates have fallen considerably below those of Record B.

Bird 89R. History: FI 8; FI 4. Figure 293A shows the complete 1st session on FI 2 after the final performance on FI 4 already shown in Fig. 290. The over-all rate declines slightly; but no pause develops after reinforcement or curvature within the interval, even though these were present on FI 4. By the 3rd session, a segment of which

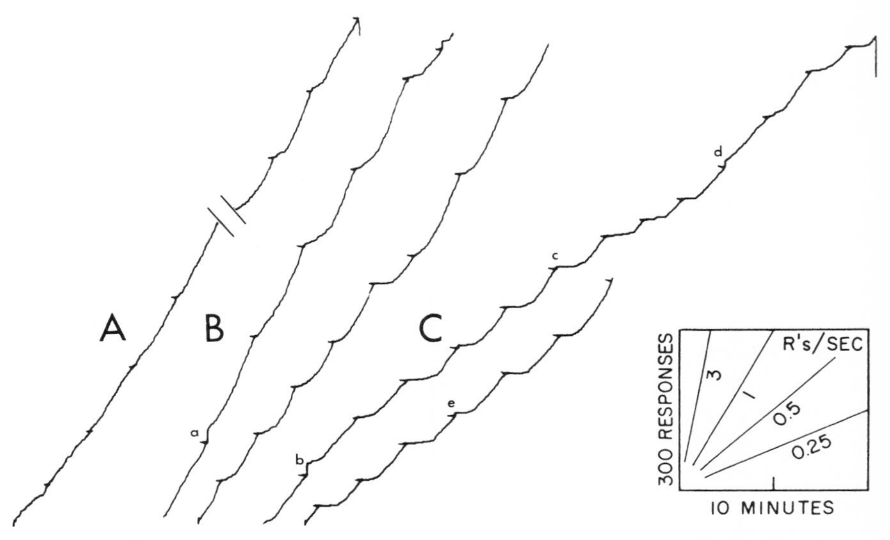

Fig. 292. First, second, and last sessions on FI 2 TO 5 after FI 4 TO 5 (bird 98R)

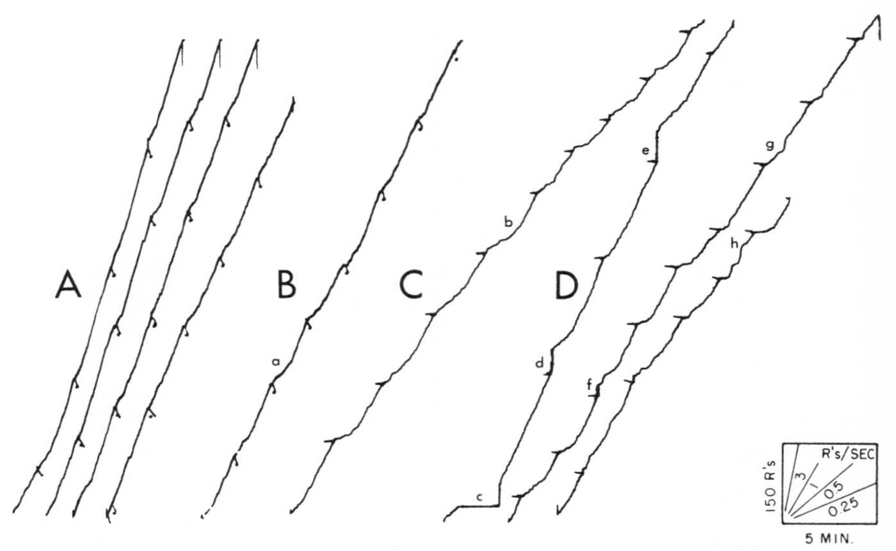

Fig. 293. First to ninth sessions on FI 2 TO 5 after FI 4 TO 5 (bird 89R)

is shown in Record B, slight positive curvature begins to appear (as at *a*). This curvature becomes more marked by the 7th session, represented by the segment in Record C. There is still no pause after reinforcement, and most of the intervals begin with a short burst of responding at the terminal rate of the preceding segment. The interval at *b* shows the greatest extent of the curvature at this stage. Record D shows the entire 9th and last session on FI 2. Considerable responding occurs during the TO at *d, e,* and *f*. The performance is generally irregular, with frequent break-throughs at high rates of responding (as at *g* and *h*). Although most of the intervals are roughly scalloped, no pause follows the reinforcement.

FI 2 to FI 1

Figure 294 shows the transitions for all 5 birds.

Previous performances can be identified through earlier figures as follows: A, Fig.

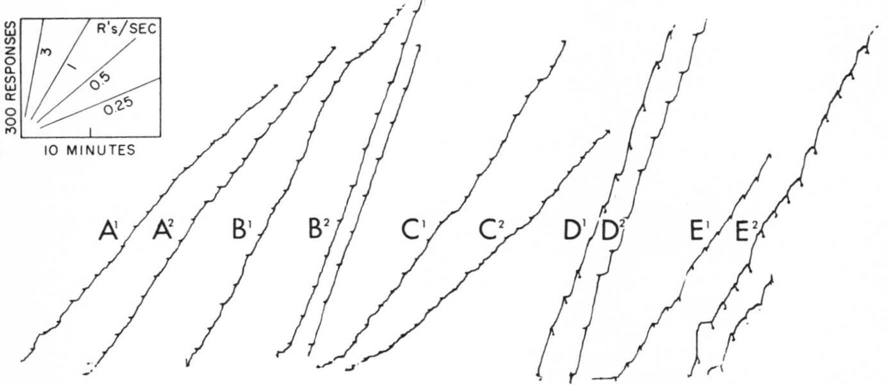

Fig. 294. First and last sessions on FI 1 TO 5 after FI 2 TO 5 (all birds)

292; B, Fig. 289; C, Fig. 293; D, Fig. 291; E, Fig. 288. Transitional (^{1}) and late (^{2}) performances are shown. As usual, FI 1 TO is very irregular. In B^2 and D^2, high terminal rates give a fairly smooth over-all curve. In A^2 and C^2, some slight stability is evident; E^2 is highly variable, with responding during TO and much negative acceleration.

Summary of fixed-interval transitions with time out after reinforcement

From small to larger fixed intervals. In the first exposure to a large fixed interval after reinforcement on a smaller one, the terminal rate from the shorter interval is maintained through most of the new larger interval. The large number of responses per interval thus produced cannot be maintained, and the terminal rate begins to fall, as pauses with later acceleration develop at the start of the interval. The over-all curve during the first several sessions shows negative acceleration. The intermediate rates and acceleration to the terminal rate become irregular. The transition is frequently prolonged, as many as 10 sessions being needed to develop the final performance on the larger interval. During the transition the curvature may extend through most of the interval. However, further exposure to the schedule of reinforcement produces a longer pause after reinforcement, a more restricted interval over which the curvature occurs, and a more sustained period of responding at the terminal rate.

From large to smaller fixed intervals. The transition from a larger interval of reinforcement to a smaller one is much faster. It is sometimes complete within 2 sessions. The scallop which had developed under longer fixed intervals gives way to a linear performance, often after a single reinforcement on the shorter interval. This performance is followed by the development of a scallop appropriate to the new fixed interval. An intermediate stage frequently occurs where the segment is uniformly accelerated throughout in a form similar to the arc of a circle.

In this experiment it was not practical to maintain the schedule of reinforcement at each interval until a final performance had developed. In some transitions to a fixed interval to which the bird had already been exposed, the previous performance was not fully recovered. Other experiments strongly suggest, however, that each schedule would produce a similar performance.

LARGE FIXED INTERVALS

Two birds from the experiment reported in Fig. 267 through 294 were exposed to a series of progressively increasing fixed intervals up to 134 minutes with time out. The data show the transitions on some fairly stable performances at large fixed intervals. Since the procedures were slightly different, the results will be described separately for the 2 birds.

Bird 89R

FI 8 to FI 64. History: FI (1–16) 1000 hours. In one experiment a pigeon which had had a history of FI reinforcement at values between 1 and 16 minutes ending with FI 8 TO 5 was exposed for the first time to FI 64 TO 5. The complete session is represented in Fig. 295. Each segment in the record is one 64-minute interval. In Segment 1 the bird accelerates to the former terminal rate, which continues far beyond the preceding interval, dips smoothly to a low over-all rate marked by occasional oscillations, and is reinforced at an intermediate value. Segment 2 begins at a lower rate, accelerates more slowly to a terminal rate, and breaks somewhat more sharply to a low rate at *a*. Reinforcement is received after some further acceleration (at *b*). Subsequent intervals show oscillations between a fairly high terminal rate and very low values, most changes being made fairly smoothly. The over-all rate declines. No curvature appropriate to the 64-minute interval appears. This bird failed to develop a good 64-minute performance, even after 107 hours on the schedule. Figure 296 shows the last session.

FI 29. Although the bird could not maintain a substantial performance on FI 64

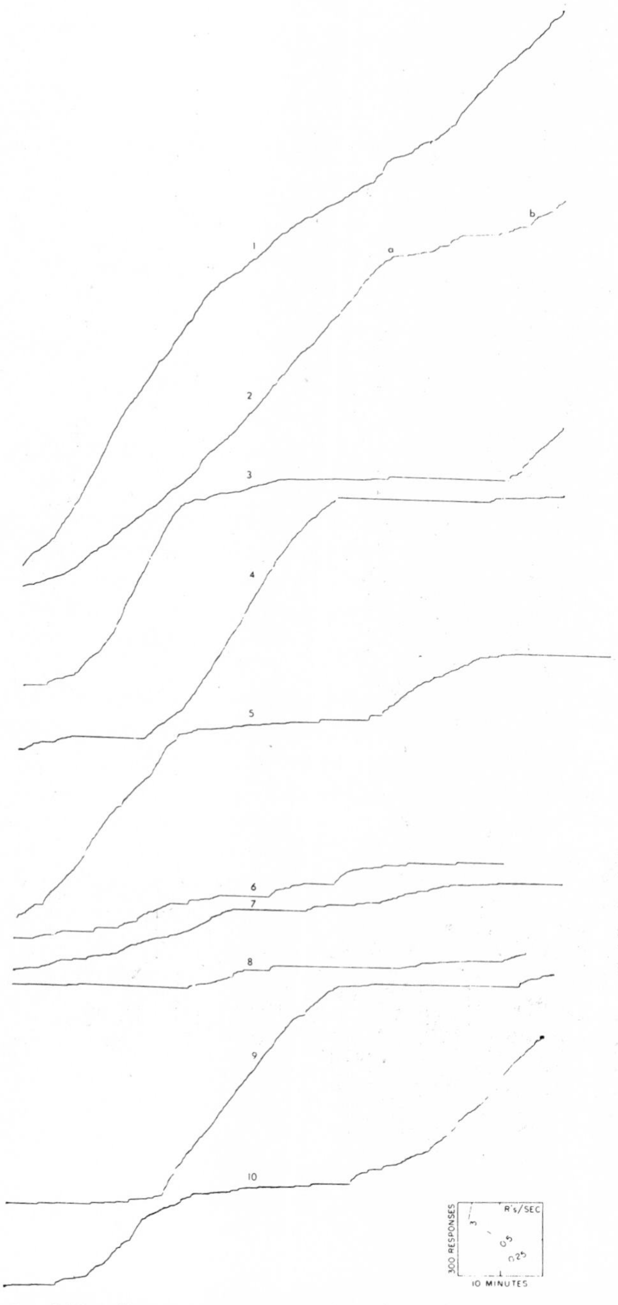

Fig. 295. First session on FI 64 TO 5 after FI 8 TO 5 (bird 89R)

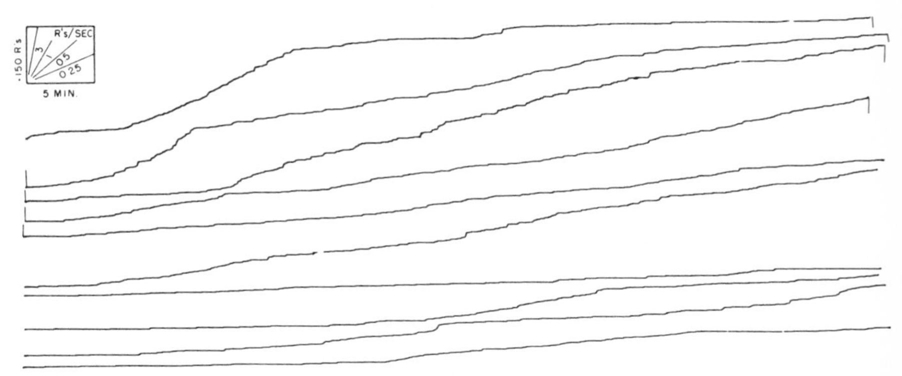

Fig. 296. FI 64 TO 5 after 107 hr showing failure to develop a scallop

at this stage, a change to FI 29 produced the fairly stable performances of Fig. 297 in the 3rd session. The continuous decline in the over-all rate has been characteristic of this bird. The session is characterized by extended curvature throughout the interval. Even toward the end of the session, in the segment at *a,* where the over-all rate is of the order of 0.3 response per second, the curve is fairly uniformly accelerated throughout.

A final performance during 2 successive sessions, beginning after 40 hours of reinforcement on FI 29, is shown in Fig. 298, where the fixed-interval segments have been arranged according to the number of responses occurring during the segments.

FI 77. When the interval was extended to 77 minutes, strong over-all negative curvature appeared during the session; but each new session began with a high over-all rate. The 5th session is shown in Fig. 299, where the order of occurrence of the segments is shown by the numbers near the beginnings of the segments. The 1st interval is normal, with about 4000 responses; Intervals 2 and 3 are almost empty; Interval 4 shows a normal curvature to about 800 responses; Interval 5 has a normal curvature to about 500 responses, Interval 6 to nearly 2000 responses, and Interval 7 to about

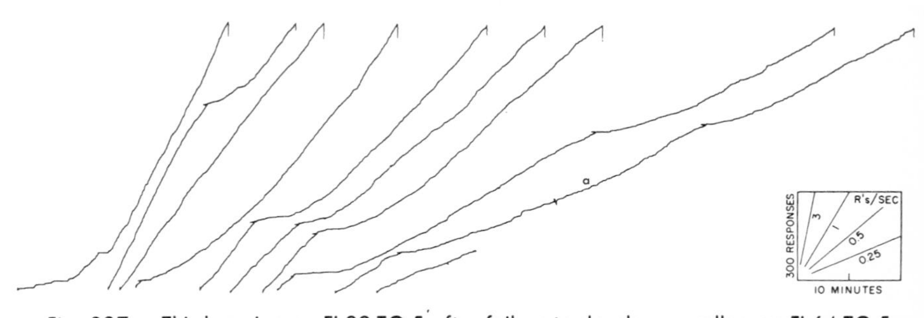

Fig. 297. Third session on FI 29 TO 5 after failure to develop a scallop on FI 64 TO 5

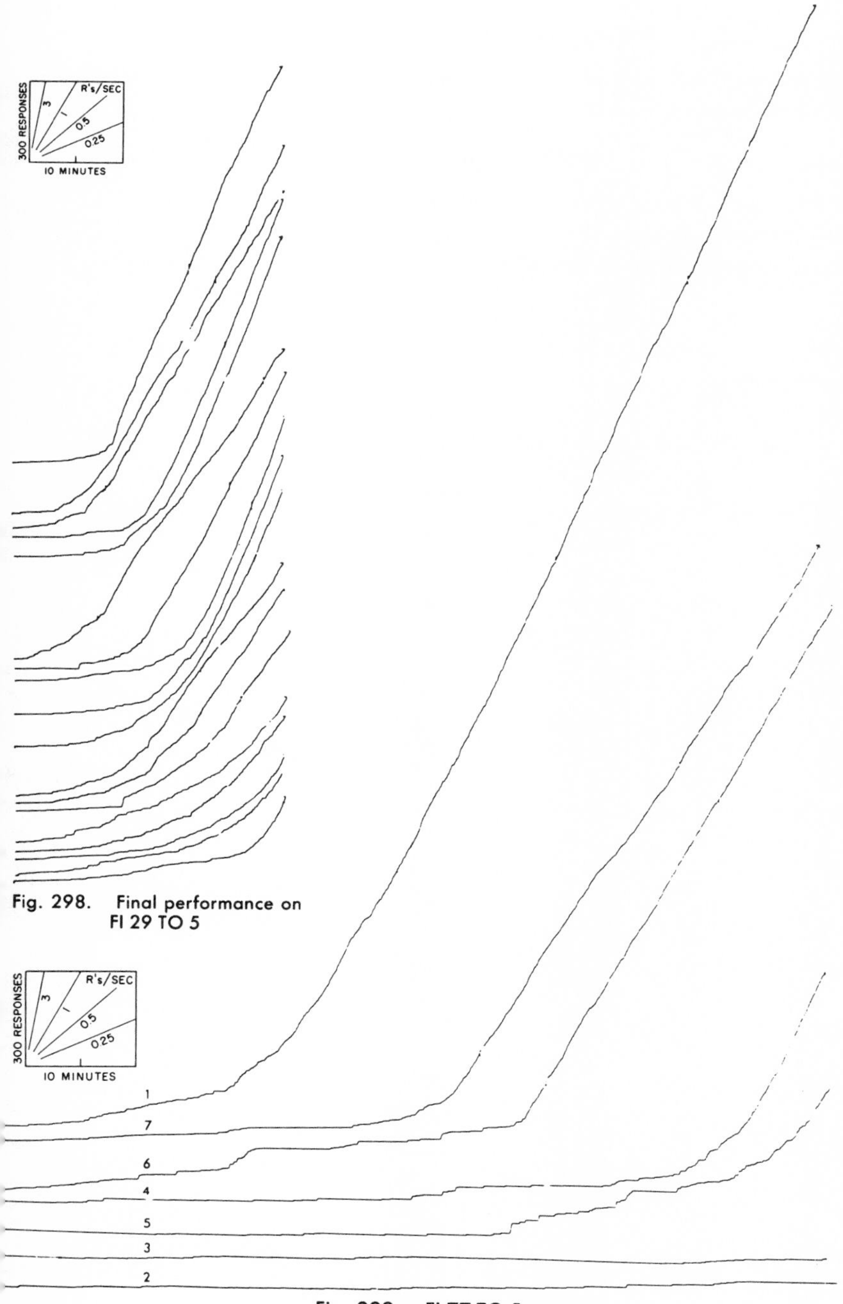

Fig. 298. Final performance on FI 29 TO 5

Fig. 299. FI 77 TO 5

2000 responses. By this time the interval pattern is well-established, although some intervals are practically empty.

Bird 98R

FI 29. Figure 300 shows the performance for the second bird on FI 29 TO 5, for 2 sessions beginning 30 hours after FI 16 TO 5, already described in Fig. 275. (Ten reinforcements occurred on FI 2 accidentally in the 21st hour.) The interval segments

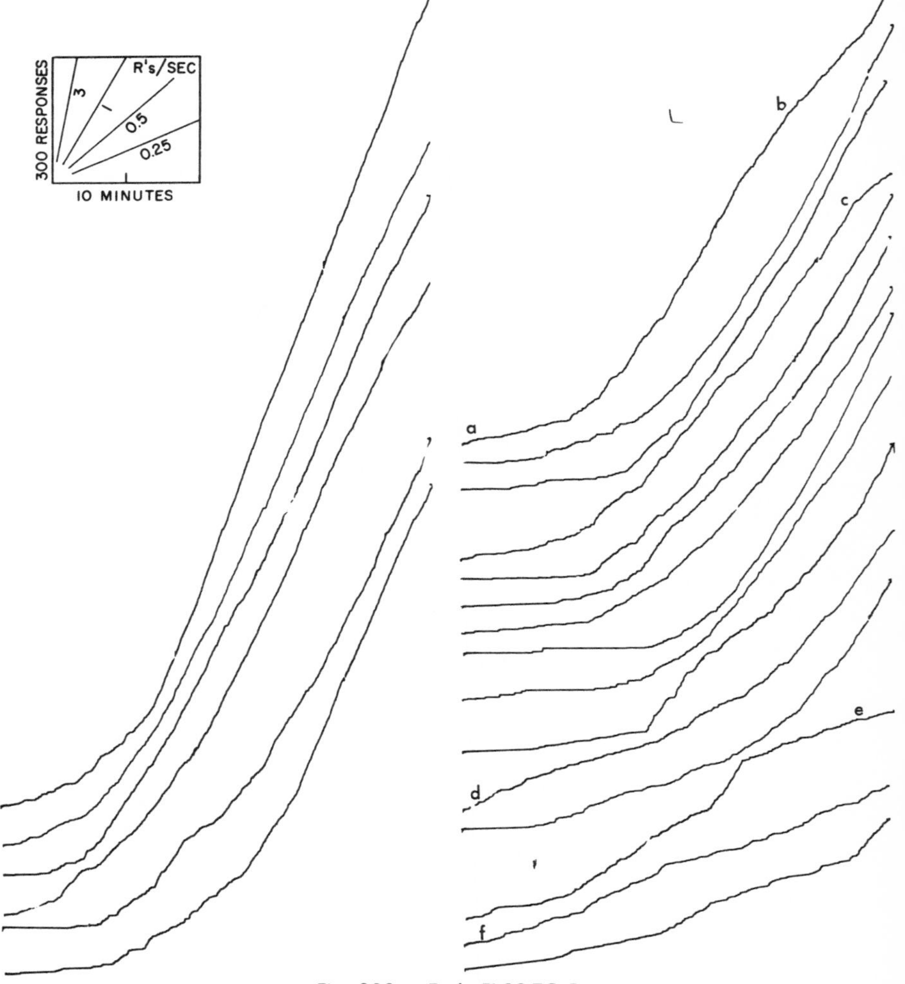

Fig. 300. Early FI 29 TO 5

have been assembled according to the numbers of responses. The performance is similar to that in Fig. 298. Most segments show a pause following the reinforcement and a moderately smooth and extended acceleration to a terminal rate which is maintained until the reinforcement. Occasional responding occurs at the start of the interval (as in *a*, *d*, and *f*) and negative curvature toward the end of the segment (as at *b*, *c*, and *e*). The following session is shown in Fig. 301 in the original order of segments. This performance after 39 hours of exposure to FI 29 TO 5 is similar to that in the previous figure except that no cases of negative curvature occur.

The curve resulting from the 10 accidental reinforcements on FI 2, in Fig. 302, shows an over-all shape almost identical with the curvature that had been occurring under FI 29 (cf. Fig. 300). The slow development of a higher rate of responding demonstrates the degree of control by the stimuli responsible for the low rate of responding during the early part of the FI 29 performance.

FI 64. After 54 hours of FI 29, the schedule was changed to FI 64 TO 5. All segments from the 1st session are shown in Fig. 303 and 304, where segments showing positive curvature and negative curvature have been separated. The positively accelerated curves in Fig. 303 show rough grain throughout; but the acceleration is usually continuous except for segments containing only a few responses. Note the absence of a pause following the reinforcement at *a* in spite of the 5-minute TO. The 2 marked negative accelerations occurring in Segments 1 and 4 in Fig. 304 are presumably the

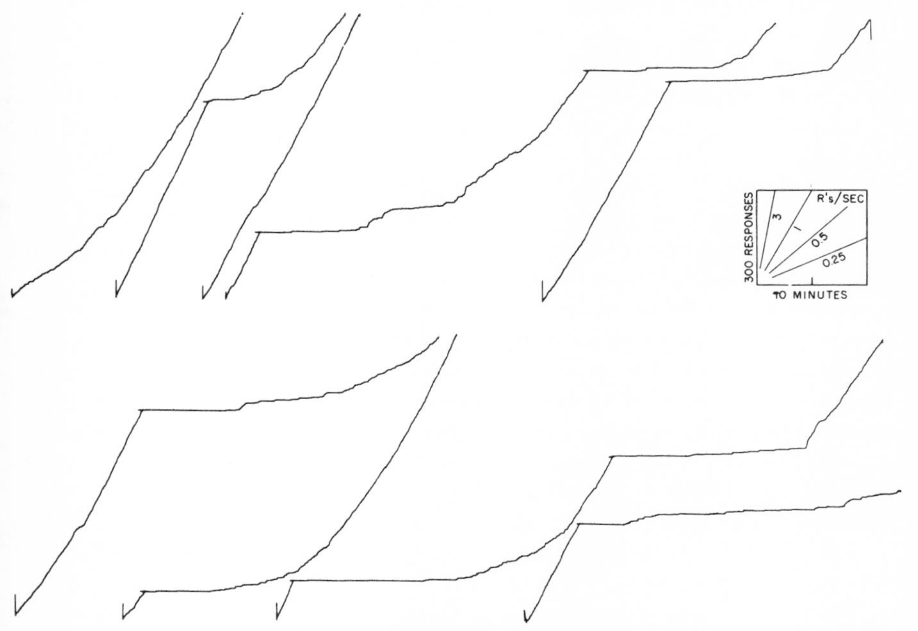

Fig. 301. FI 29 TO 5 after 39 hr

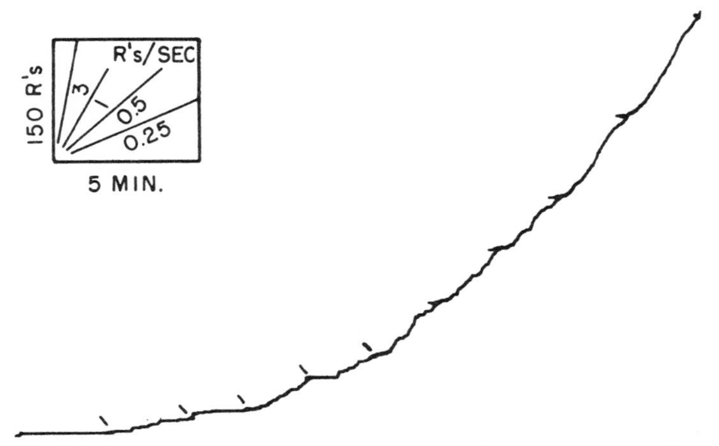

Fig. 302. Slow development of a high rate on FI 2 TO 5 after FI 29 TO 5

result of strain produced by the terminal rate generated by the preceding FI 29 performance.

Figure 305 shows the performance after 100 hours on FI 64 TO 5. The upper ends of the first 3 segments have been displaced to the left of the graph.

FI 77. The effect of increasing the interval of reinforcement from 64 to 77 minutes is some temporary negative curvature with occasional intervals at low rates. A fairly stable performance develops quickly. Figure 306 shows the performance during 2 sessions extending up to the 40th hour on FI 77. The segments have been rearranged in terms of the number of responses. In the first segments (A1 and B1) of the daily performance, responding begins early. All segments show some positive acceleration, although the segments at the bottom of the figure do not reach the same terminal rate. The relation between the number of responses in the segment and the time before the acceleration to the terminal rate begins is roughly linear.

FI 92. Figure 307 shows the transition from FI 77 TO 5 to FI 92 TO 5. All segments from the first 3 sessions on FI 92 have been arranged in order of the number of responses except for the 1st segments, which have been omitted. These 1st segments simply show a steady performance at a maximal rate throughout the intervals, which contain about 4500 responses each. The bird sustains a fairly high over-all rate of responding in spite of the infrequent reinforcement. The segments under FI 92 reinforcement are similar to those under FI 77, except that a larger number of responses occurs per fixed-interval segment and the terminal rate is somewhat delayed.

The other bird was also reinforced on FI 92 TO 5. The performance resembled that on transition from FI 64 to FI 77 (Fig. 299).

FI 111. The main effect of a transition from FI 92 TO 5 to FI 111 TO 5 is a gradual decline in the over-all rate, with further postponement of the acceleration to a ter-

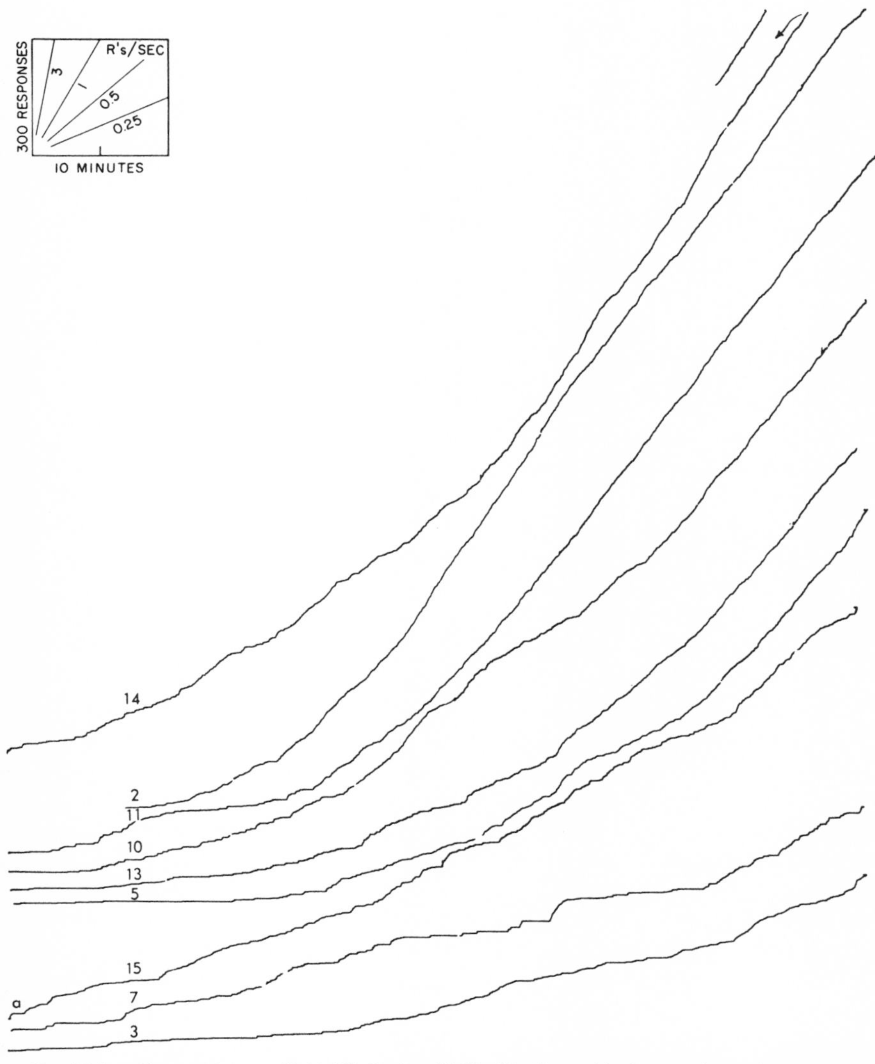

Fig. 303. First session on FI 64 TO 5 after FI 29 TO 5; positively accelerated segments

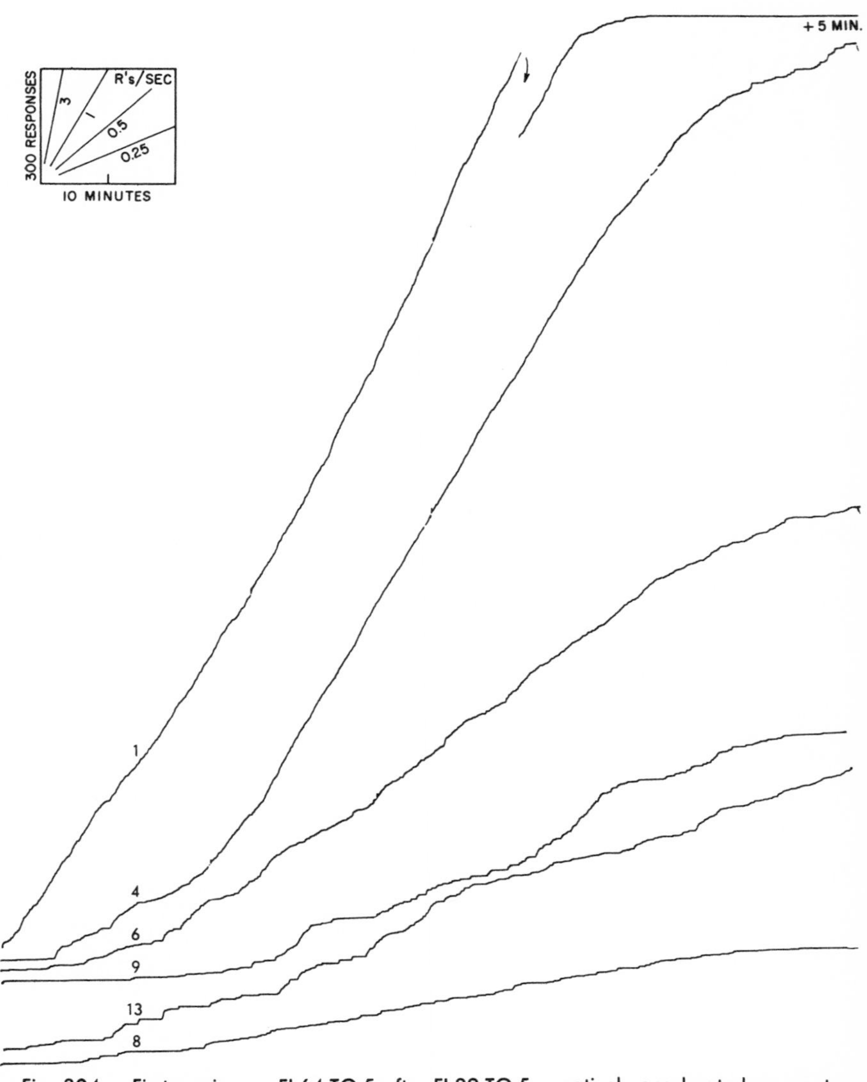

Fig. 304. First session on FI 64 TO 5 after FI 29 TO 5 negatively accelerated segments

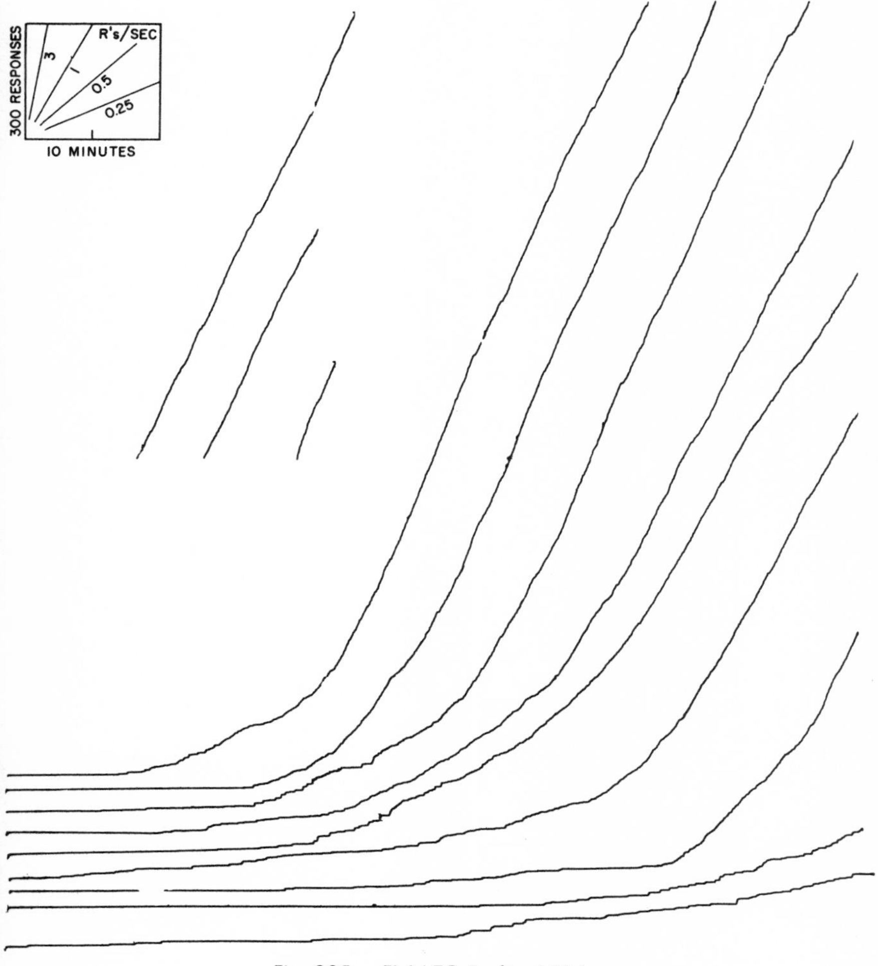

Fig. 305. FI 64 TO 5 after 100 hr

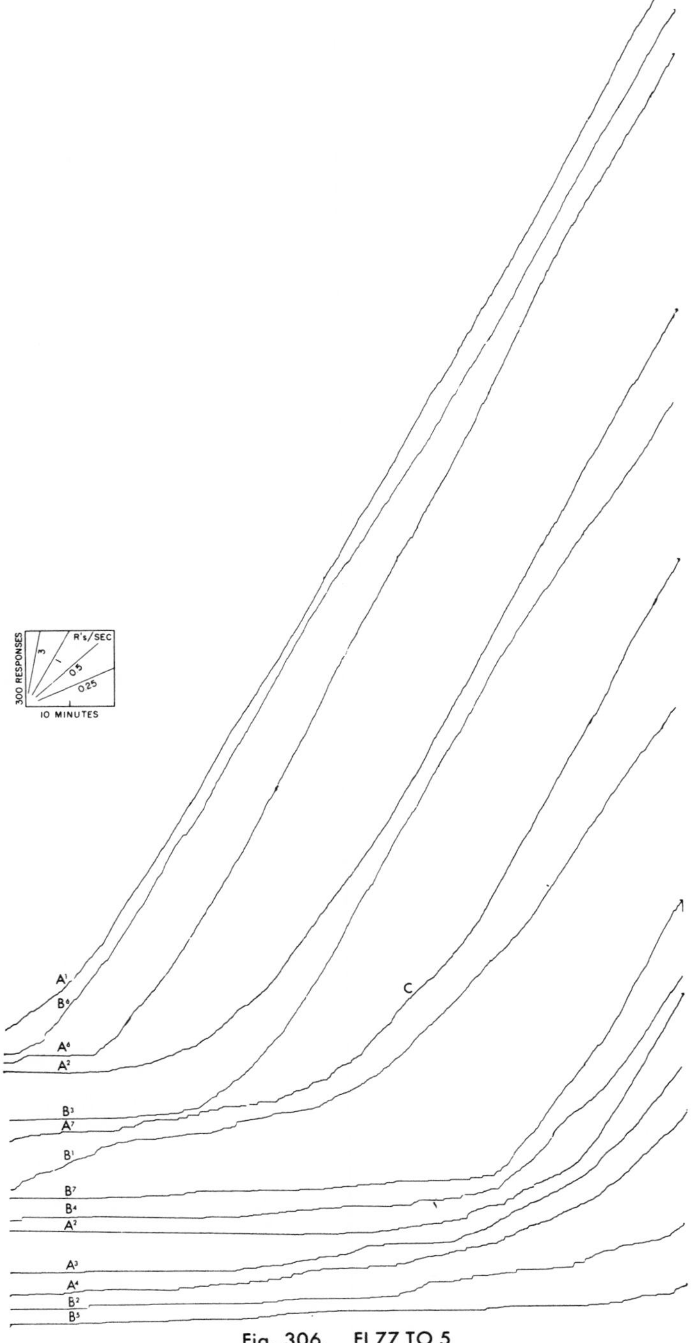

Fig. 306. FI 77 TO 5

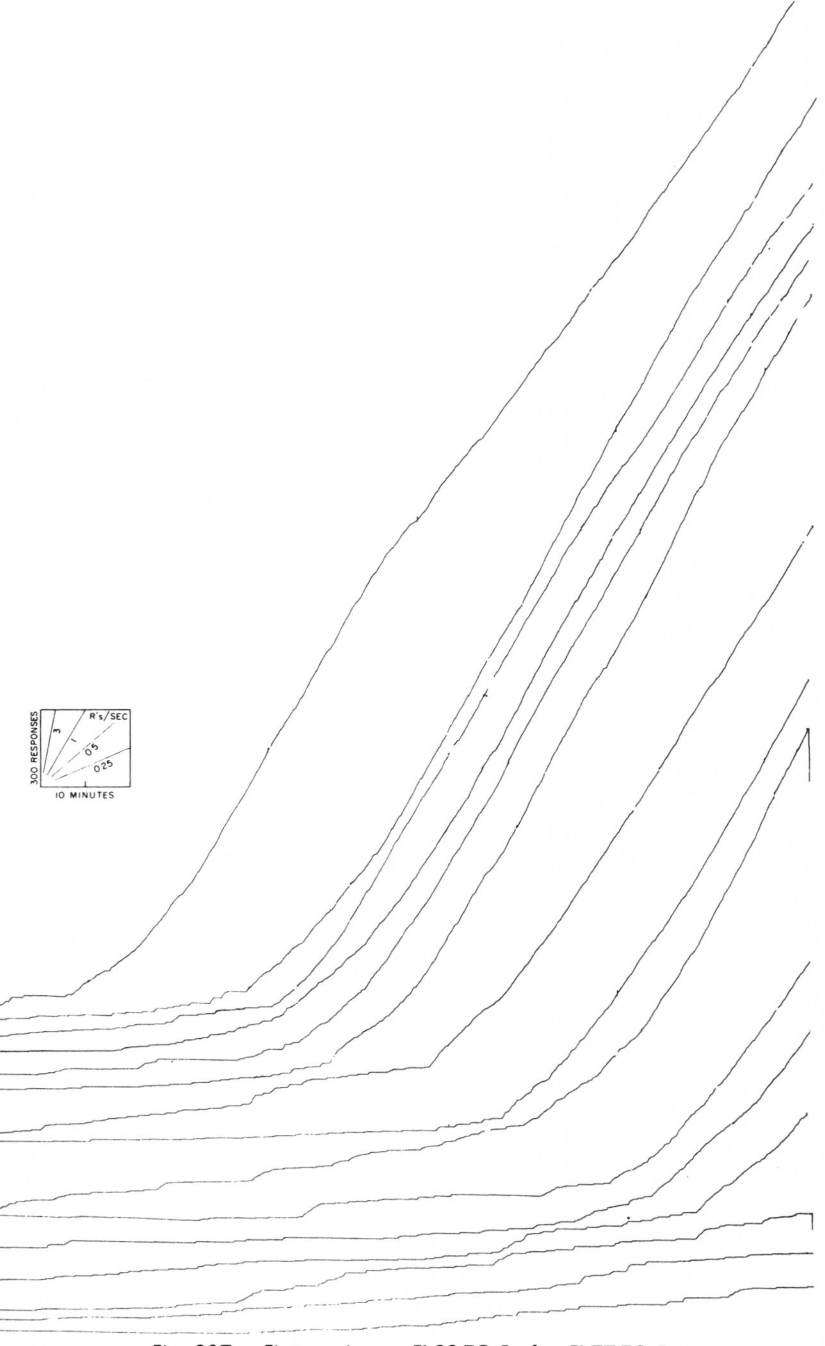

Fig. 307. First session on FI 92 TO 5 after FI 77 TO 5

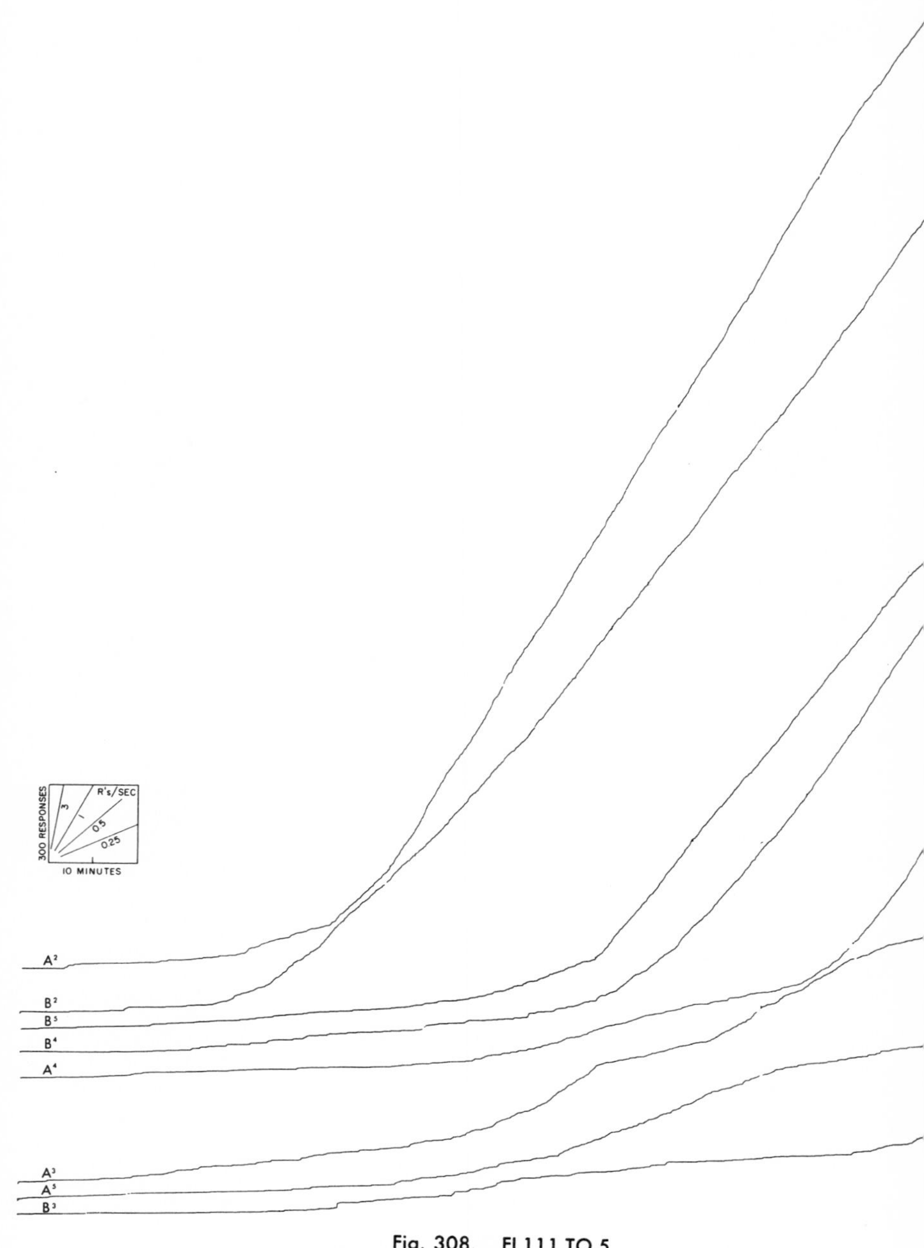

Fig. 308. FI 111 TO 5

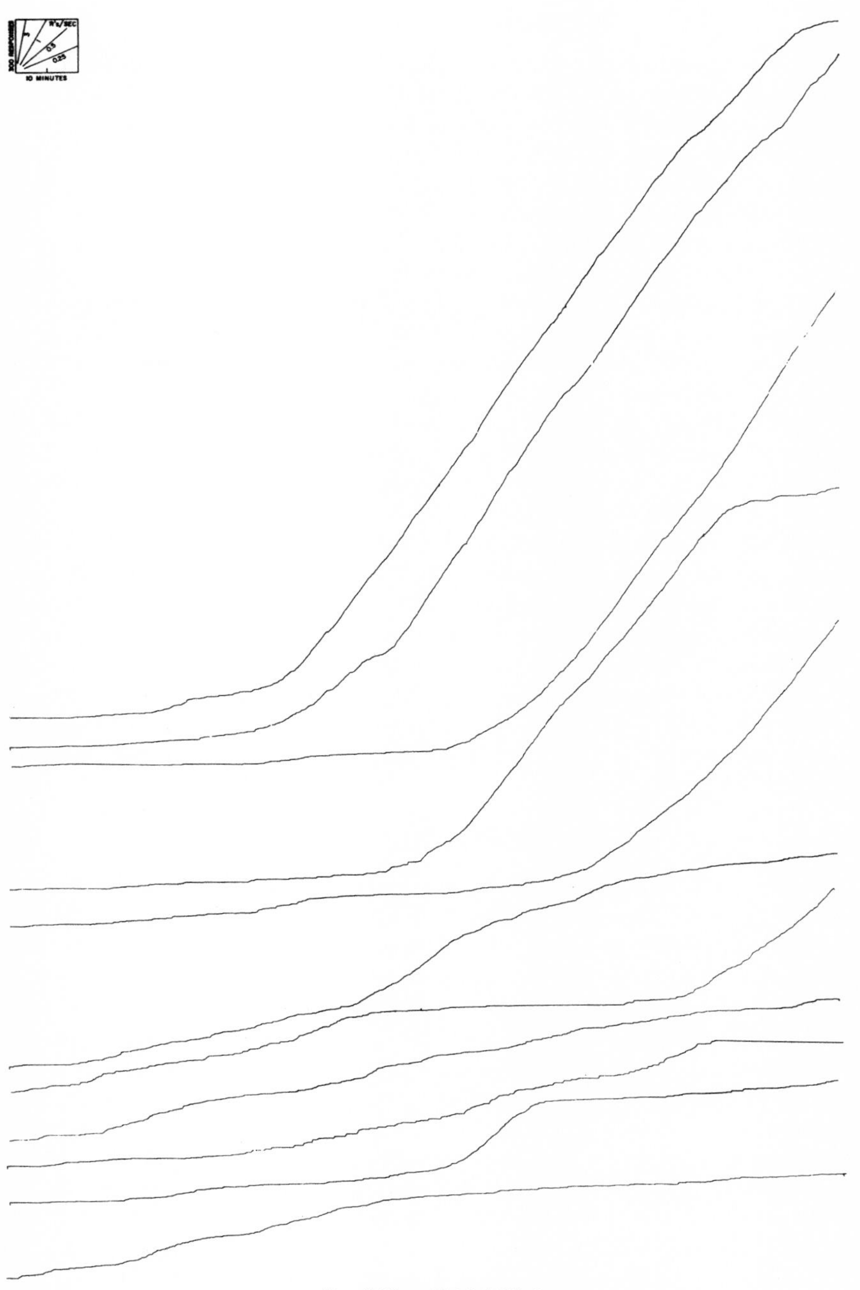

Fig. 309. FI 134 TO 5

minal rate of responding. Figure 308 shows the performance after 61 hours on FI 111. The figure contains all but the 1st segments from 2 successive daily sessions reassembled in terms of the number of responses emitted. The acceleration to the terminal rate is roughly appropriate to the large interval of reinforcement. The terminal rate is never reached very rapidly, although the first 2 curves show fairly quick starts, with almost 4000 responses per interval.

In the transition from FI 92 TO 5 to FI 111 TO 5 the other bird showed a low over-all rate with very irregular grain. A terminal rate reminiscent of shorter fixed intervals was occasionally held for approximately 1000 responses, but negative acceleration usually set in before reinforcement. During 72 hours of the schedule the performance consisted mostly of slow responding, with an occasional break-through to a higher rate.

FI 134. Figure 309 shows an early performance on FI 134 TO 5. Fixed-interval segments from 3 sessions beginning after 21 hours of exposure to the schedule have been arranged in order of the number of responses. The 1st segments (omitted to facilitate presentation of these records) showed: (1) a low rate of responding during the whole interval; (2) a fairly quick shift to an unsteady terminal rate, with the segment containing more than 4500 responses; and (3) a fairly normal interval curve containing about 1800 responses. In spite of the very infrequent reinforcement, the bird maintains a fairly substantial over-all level of responding. Some sign of strain, however, appears in the frequent negative curvature.

After a total exposure of 84 hours on this schedule, the second bird continued to show a very low over-all rate, with only occasional increases in rate in the form of the S-shaped curves of Fig. 296. The bird could not sustain a performance under FI 134 TO 5 at this level of deprivation.

FIXED INTERVAL PLUS ADDED CLOCK

To the extent that the bird's behavior is an event varying in time and correlated consistently with the FI schedule, it can be thought of as a clock by which the bird may modify its behavior with respect to reinforcement. The terminal rate at the end of the fixed interval represents the most optimal setting of the clock, and the curvature represents intermediate clock-settings. With a perfect clock an interval should contain only a single response. The bird's own behavior, however, is a relatively poor clock. The inherent instability of an FI performance, with its oscillations in rate, destroys the correlation between the number of responses and reinforcement, or the rate and reinforcement.

We have attempted to get some notion of the control exercised by the bird's own behavior by adding an external stimulus that varies uniformly in some dimension during the fixed interval. This stimulus is arranged by the experimenter, and can be consistently correlated with the schedule of reinforcement. In the following experiment, various kinds of fixed-interval performances were studied, with an added clock similar to the added counter of Chapter Four. A spot of light is projected on a translucent key through a slit, the length of which is determined by a circular wedge driven by a synchronous motor. As the motor turns the wedge, the slit grows from a small spot approximately 1/16 inch square to a slit 1/16 inch wide by 3/4 inch long, centered horizontally on the key. The reinforcement schedule was locked to the slit by the use of the same motor to program the reinforcement. By changing the speed and direction of the motor, and the design of the wedge, the added stimuli could be manipulated. A short TO (of a few seconds) followed each reinforcement as the clock was reset to zero. (The recorders used for these experiments were powered by a new type of synchronous motor; we discovered later that these motors could run at 2 speeds. Any time the recorder motor was stopped and started [as at reinforcement], it might shift speed. This variation in *recorded* time did not affect the bird, however, since the motor programming the slit and reinforcement schedule operated correctly. When interpreting the following records in fine detail, the length of the interval must be examined to discover at what speed the recorder was running. Since we will not be

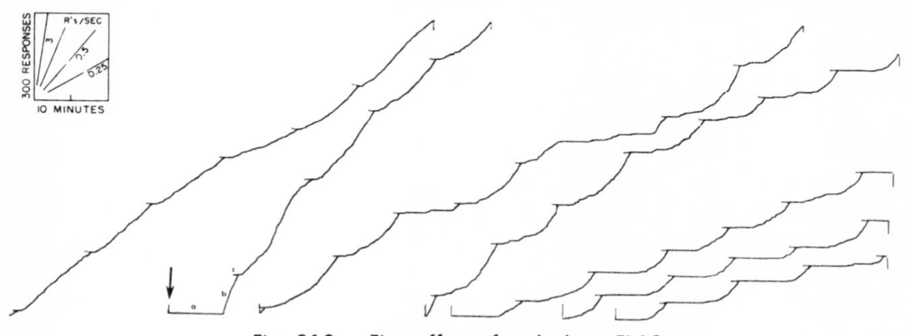

Fig. 310. First effect of a clock on FI 10

concerned with absolute rates but mainly with changes in rate during the intervals, this defect is not important for the conclusions which will be drawn from the experiment.)

Development of FI performance with clock

FI 10 + clock. Figure 310 shows the first effect of introducing a clock. The 1st excursion of the record is the final performance for this bird after 38 hours of exposure to FI 10, following 14 hours of FI 5 after crf. All 6 intervals show a brief pause following the reinforcement and a brief acceleration to a moderate terminal rate. During the 38 hours of exposure to FI 10 the key was illuminated through a 1/16-inch-wide slit, 3/4 inch long. At the beginning of the interval, indicated by the arrow, the slit on the key was at its smallest size; it grew linearly throughout the 10-minute interval to the maximum size at reinforcement. The small size was a novel stimulus and has the usual effect of suppressing responding for some time at *a*. During this time the slit is growing, however; and before the end of the interval the bird begins responding at a much higher rate than it had ever shown in its history. The remainder of the interval at *b* shows continuous negative acceleration; but the reinforcement at *c* occurs while the rate is considerably higher than normal. The next few intervals are run off at a high over-all rate. Each successive interval shows more marked curvature, however. The remainder of the session shows the development of a longer pause following the reinforcement, a rapid acceleration to the terminal rate, and a decline in the over-all rate.

Figure 311 shows segments from the 2nd and 3rd sessions with the clock. Records A and B show the 4th, 5th, 6th, 23rd, 24th, and 25th reinforcements of the 2nd session. These show a further development of the performance in Fig. 310. The pause extends over more than half the interval. The terminal rate, however, is now

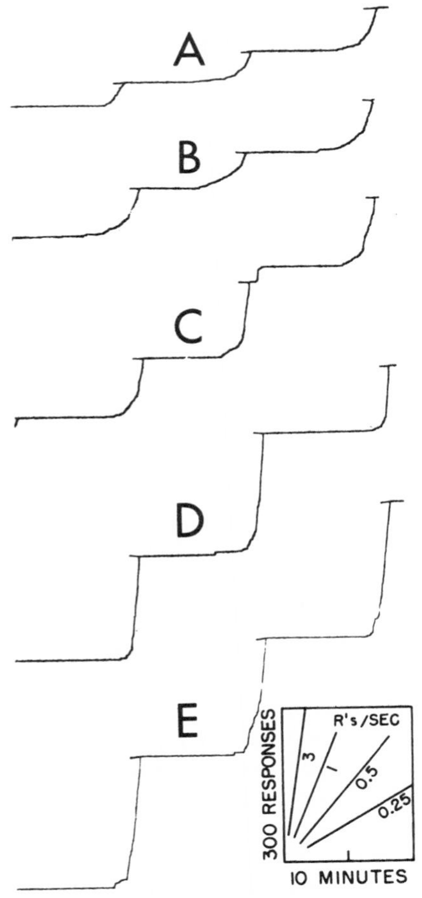

Fig. 311. Development of FI 10 + clock

higher than the intermediate rate occurring during the previous session. Records C and D show the 27th, 28th, 29th, 30th, 39th, and 40th intervals of the same session. The extent of the pause is about the same as before, but the acceleration to the terminal rate is more rapid and the terminal rate higher. The last segment in Record D, where the timing is precise, shows a rate of 10 responses per second during the last few seconds of the interval. Record E shows the 20th, 21st, and 22nd segments from the 3rd session following the introduction of the clock. The performance is similar to Record D, and represents a performance very close to the final performance under FI 10 + clock. Figure 312 shows a large segment from the final performance taken after 25 hours of exposure to the clock. The performance is uniform from segment to

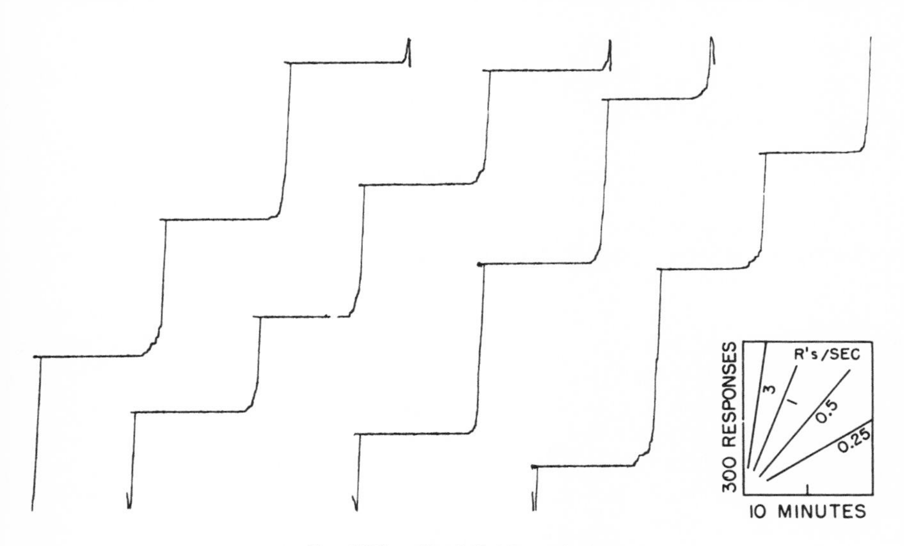

Fig. 312. Final FI 10 + clock

segment, with a pause of from 7 to 8.5 minutes, followed by a short period during which the rate accelerates to a terminal rate well above 10 responses per second. The variation in the number of responses from interval to interval is relatively small in view of the very high rate at which these responses are emitted. Variations of this order are caused by differences of a very few seconds between the times at which the terminal rate is reached. The terminal rate shows a 30- to 35-fold increase over the rate of 0.3 response per second prevailing before the clock was introduced. A second bird, 23Y, confirmed the general features of this transition.

The clock continues to control the rate even during extinction, when the slit on the key maintains its usual cycle. An instance is shown in Fig. 313. The usual procedure occurs at *a, b,* and *c.* Extinction begins at the arrow. The cycling spot of light produces scallops characteristic of the clock performance. The over-all rate even-

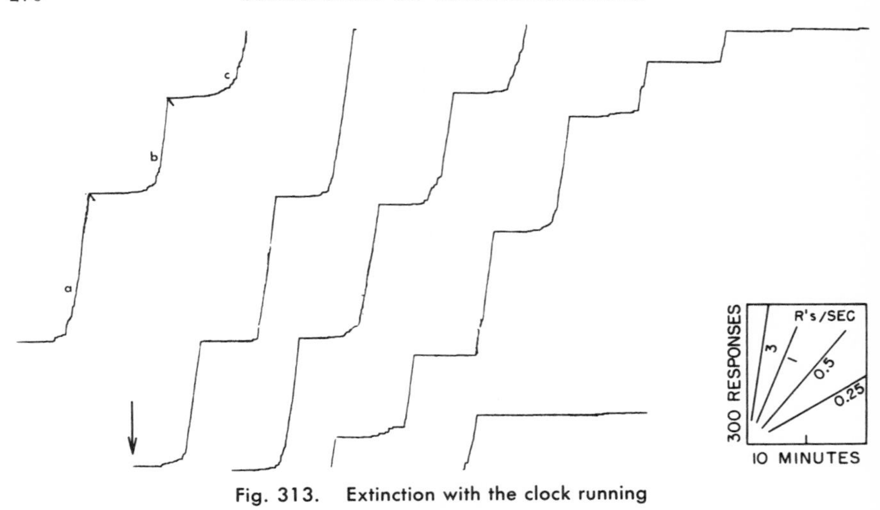

Fig. 313. Extinction with the clock running

tually declines as the number of responses emitted at the larger slit-values decreases. The stimulus control, however, does not become disturbed as a result of the extinction.

FI 30 + clock. After an intervening set of experiments with a clock (to be described later in the chapter), the bird responsible for the performance just described was placed on FI 30, with the slit on the key growing from 1/16 to 3/4 inch long during the interval. The performance after 5 sessions is shown in Fig. 314, where the last 7 intervals of the session have been arranged in the order of their occurrence from a common margin. At this value of FI the clock shows even better control than in Fig. 312 under FI 10. In the 1st segment of the figure, for example, the terminal rate is not reached until the slit is approximately 19/20 of its maximum size. The same order of magnitude of the slit at the start of the terminal rate appears in the remainder of the figure. Small knees at *a* and *b* probably reflect the previous exposure to another kind of clock (to be described). This schedule was continued for 5 sessions with no further change.

Development of FI 10 + clock after extended history of reinforcement on mix FIFI. Figure 769 in Chapter Eleven shows segments taken during the first 34 hours of exposure to FI 10 + clock after an extended history of FI, VI, and mix FIFI, and an immediately preceding history of mix FI 10 FI 2 with a reduced counter. Records A through D contain the last 2 segments of sessions after 8, 15, 24, and 29 hours of exposure to the clock, respectively; and Record E is a segment from the middle of a session after 34 hours of exposure to FI 10 + clock. Segments *a* and *b* resemble the terminal performance on FI 10 + clock in Fig. 312, but the terminal rate is more moderate. A performance for a bird having a similar history is shown in Fig. 769, Record A, in Chapter Eleven. This performance is after 100 hours of exposure to FI 10 + clock.

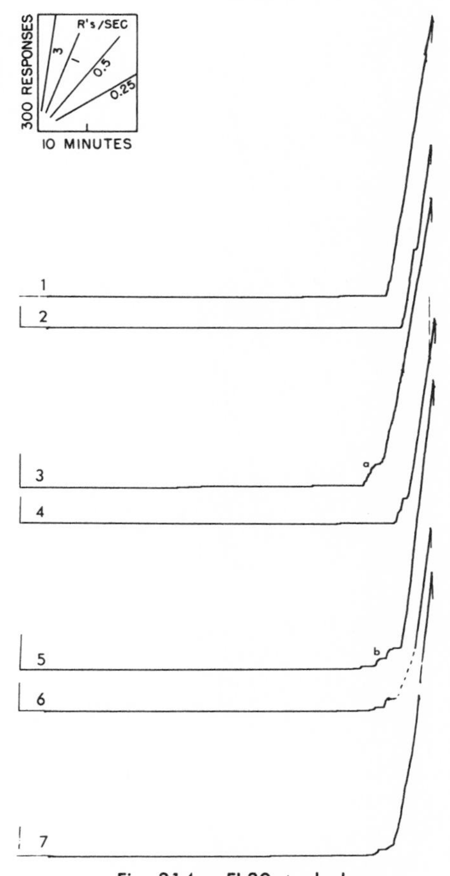

Fig. 314. FI 30 + clock

Both figures undoubtedly show an incomplete development under this schedule.

FI 1 to FI 6. A clock was added to the FI performances of 3 birds on the 2nd session following crf. Figure 315 shows the complete 1st session with clock for 1 bird. A typical FI performance develops very rapidly, with a consistent pause after reinforcement and a high terminal rate of responding. Although pauses after reinforcement develop without the aid of a clock, they develop more slowly (cf. Fig. 126 and 149), and the terminal rates do not reach the same order of magnitude as with the added clock. Figure 316 shows later performances after 6 sessions under FI 1 + clock for all 3 birds. Figure 316A shows little change beyond the performance at the end of the 1st session with clock as shown in Fig. 315. Record C shows a much lower over-all rate, although the order of consistency from interval to interval is approximately the same. The bird had been run by mistake on FI 10 briefly follow-

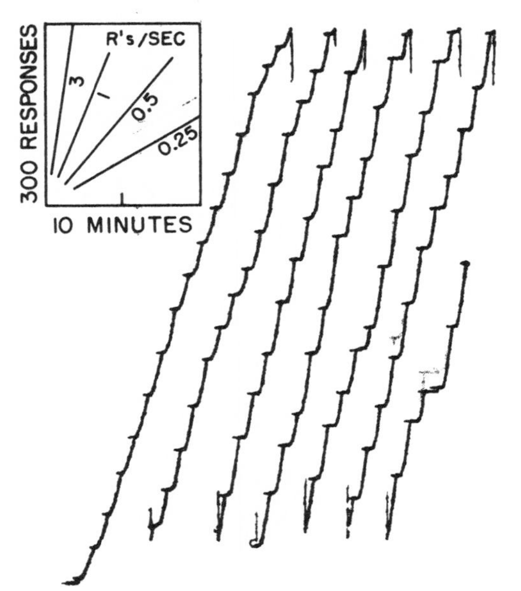

Fig. 315. First exposure to a clock on FI 1

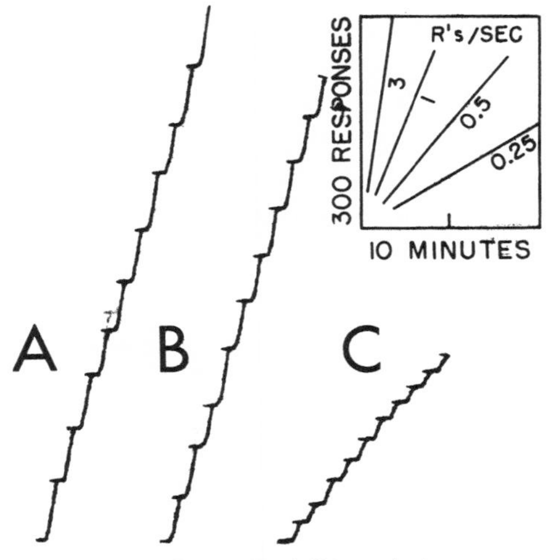

Fig. 316. Final FI 1 + clock

ing crf; and the lower rate of responding may be due to exposure to this schedule.

All 3 of these birds were then placed on FI 3 with the same clock, and final performances are shown in Fig. 317. The birds are assigned the same letters in Fig. 316 and 317. Record A was taken after 5 sessions; Records B and C, after 8 sessions. The clock grew to full size in 3 minutes, instead of in 1 minute as in the previous schedule. In the resulting performance the pause after reinforcement characteristically extends for more than half of the 3-minute interval. Most segments show a fairly abrupt shift to the terminal rate, although varying degrees of curvature are present. The terminal rates are approximately 3 per second in Record C, slightly less than 4 per second in Record A, and slightly more than 5 per second in Record B. While these do not approach the rate of more than 10 per second recorded under FI 10 + clock, they are higher than are usually recorded for this size of FI (cf. Fig. 150C for FI 2, 50 sessions after crf).

The session following Fig. 317 began with reinforcement on FI 3; but the interval was soon increased to 6 minutes, the clock growing from small to large in each interval. One performance is shown in Fig. 318. The effect (beginning at the arrow) is to enlarge the FI 3 curve roughly by a factor of 2. The interval at *c* is very close to a

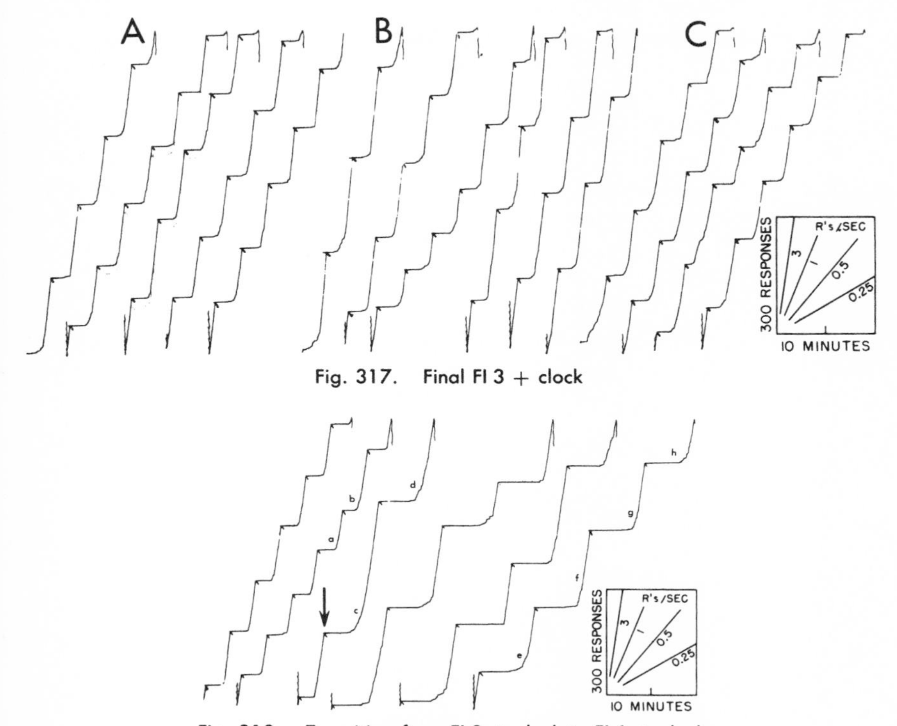

Fig. 317. Final FI 3 + clock

Fig. 318. Transition from FI 3 + clock to FI 6 + clock

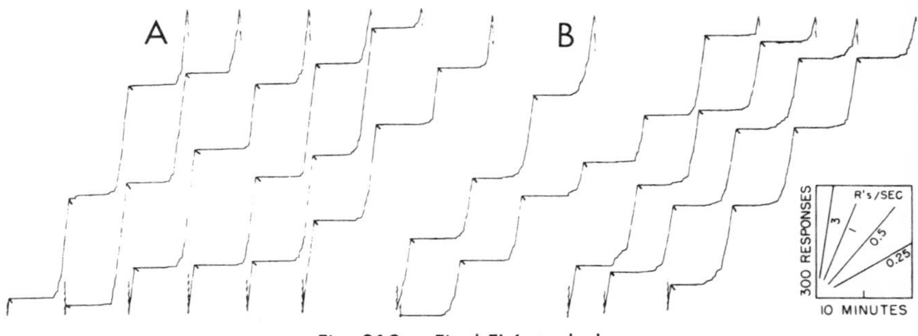

Fig. 319. Final FI 6 + clock

simple magnification of the curve at *b,* while *d* suggests magnification of *a.* With further exposure to FI 6 with clock, however, the number of responses in the interval becomes less as the bird pauses for a greater portion of the interval. Thus, in the curves at *e, f, g,* and *h* the numbers of responses in the interval are approximately only one and one-half times those under FI 3.

A performance under FI 6 + clock after 4 sessions is shown in Record A of Fig. 319, and for another bird after 5 sessions in Fig. 319B. (The third bird had been dropped from the experiment after only a brief exposure to FI 6 + clock; however, its early transition to this interval confirmed the performances of the other two.) The terminal rates in Fig. 319 are about the same as under FI 3, but the proportion of the interval spent during the pause remains greater under FI 6. This characteristic is apparent in the fact that the number of responses per interval does not reach twice the value occurring under FI 3.

Multiple clock

In order to compare the kinds of performances established by various speeds of clocks, a program was set up in which the bird was reinforced on any one of 4 intervals of reinforcement: 3, 5, 7.5, or 10 minutes. The slit on the key always began at the

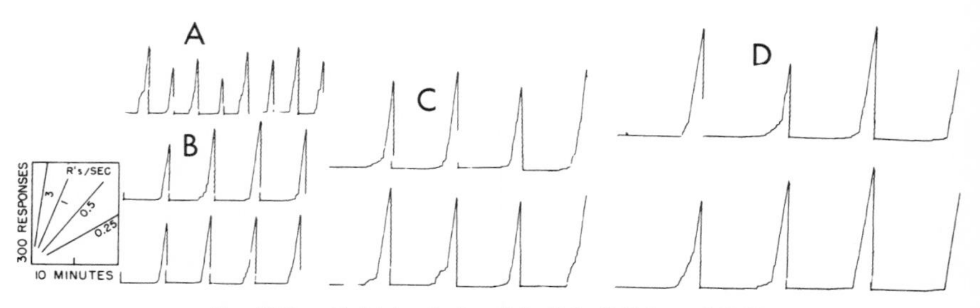

Fig. 320. Multiple clocks: FI 3, FI 5, FI 7.5, and FI 10

smallest size just after reinforcement and grew to maximum size just before reinforcement; thus, it grew most slowly on FI 10 and most rapidly on FI 3. Figure 320 shows performances for one bird under 4 fixed intervals alternating at random after 18 hours of reinforcement. The segments have been arranged in groups corresponding to the intervals of reinforcement. Segments of FI 3 are shown in Record A. Two instances occur of knees, which have been absent from the performance of this bird in the past; but the clock control is otherwise similar to the final performance on FI 3 + clock shown in Fig. 318. Record B shows the FI 5 segments, Record C the FI 7.5 segments, Record D the FI 10 segments. The acceleration just before the terminal rate is slightly more prolonged than at a single clock value, possibly because other factors contributing to the control vary from interval to interval in this experiment. The ratio of the number of responses in Record D to that in Record A is of the order of 3 to 8; therefore, the clock control must be greater at the longer intervals.

After 20 hours of reinforcement on 4 intervals (Fig. 320), the intervals were increased to FI 7, 13, 19, and 25. These occurred in random order, with the clock completing a full excursion during each interval, growing more slowly during the larger intervals. Figure 321 shows the first 45 intervals of a session 87 hours later. The records have been rearranged according to the interval of reinforcement. As in Fig. 320 the clock control is relatively better in Record A under FI 25 than in Record D under FI 7. The shift to the terminal rate of responding is usually rapid, although instances of knees similar to those in Fig. 320A occur frequently.

The procedure of alternating 4 different intervals under the appropriate clock control minimizes the importance of other stimuli related to the passage of time since the last reinforcement. The extent of such control was determined when the clock was removed by fixing the slit at large. The schedule then becomes a 4-ply mixed FI schedule. (See Chapter Eleven.) The first effect of the new procedure was a constant high rate of responding associated with the large slit. This rate falls, however, and Fig. 322 shows the performance after 32 hours of exposure to the 4-ply schedule for the bird giving the best curvature on this schedule. The record begins with the start of an experimental session; and the 1st interval is run off with marked acceleration at *a* to a high rate of responding comparable with those which were occurring with the clock. The next interval shows a more moderate terminal rate, maintained with some grain throughout the interval to the 2nd reinforcement (at *b*). Thereafter, each reinforcement is followed by a period of positive acceleration, which continues until terminated by the reinforcement. At *c*, several hundred responses suddenly appear at approximately 9 responses per second, appropriate to the fixed setting of the former clock. This rate then becomes more moderate; but before the appearance of the next reinforcement, the high rate again returns (at *d*) for a smaller number of responses. The other bird also showed these sudden shifts to the terminal rate characteristic of the clock performance. It did not show curvature related to reinforcement, however; and the over-all curve showed wide oscillations in rate.

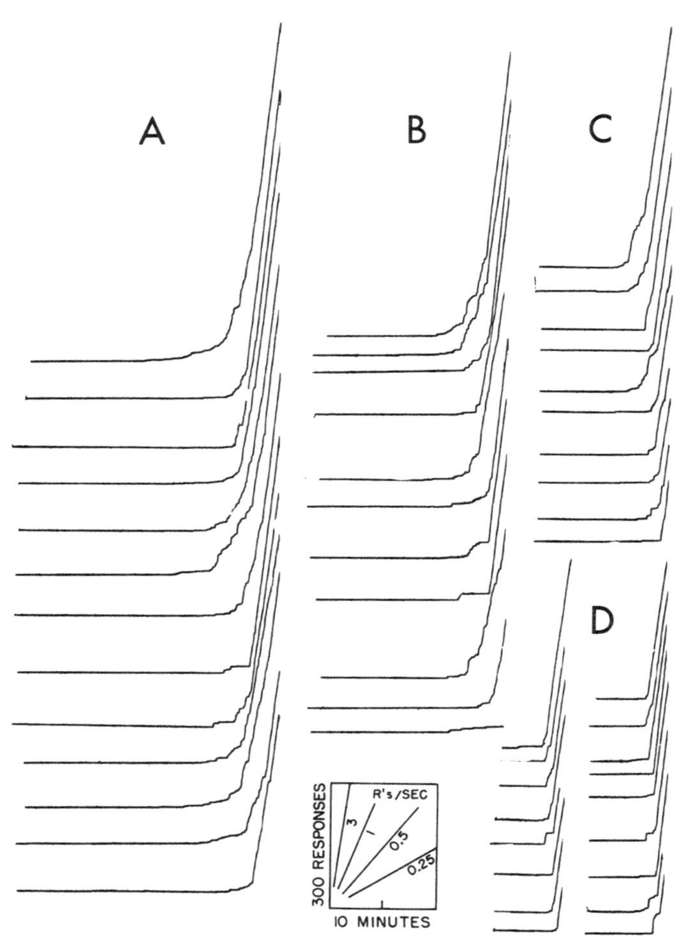

Fig. 321. Multiple clocks: FI 7, FI 13, FI 19, FI 25

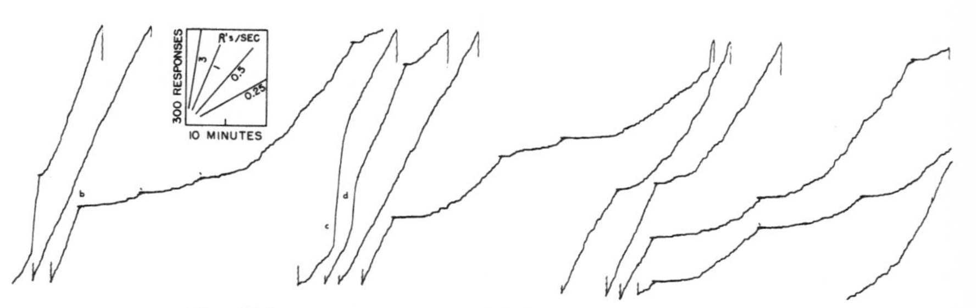

Fig. 322. Mix FI 7 FI 13 FI 19 FI 25 after multiple clocks

Reversing the clock

The extent of the control by the slit of light on the key is well-illustrated when the direction of movement of the clock is reversed. Figure 323A shows a final performance on FI 3 + clock in an experiment which has already been discussed. Record B follows Record A immediately, but the direction of growth of the slit was reversed at the arrow. The slit now remained large after reinforcement and shrank as the interval progressed until it reached the smallest size just before reinforcement. The large slit correlated with the end of the fixed interval now occurs immediately after the reinforcement, and the small size correlated with nonreinforcement now appears just

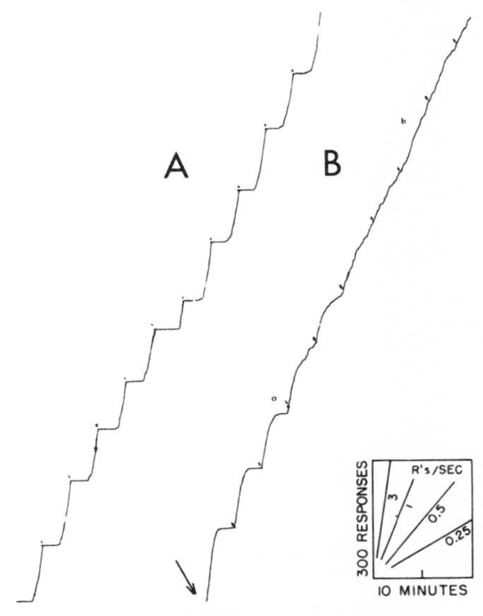

Fig. 323. Clock reversal on FI 3

before reinforcement. The first 2 segments are inverted images of the standard clock performance. The rate of responding is completely determined by the value of the slit on the key. Any stimuli from behavior or other stimuli also varying in time seem to exercise no control. The new contingencies of reinforcement, however, quickly alter this pattern. The bird reaches a stage in which the rate of responding is roughly constant (at *b*). Note that the new control arises first at the smallest size. Just before the slit reaches this size, at *a*, the bird begins to respond. The next 2 segments show a similar effect. The nature of the clock is arbitrary, of course; and although this experiment was not continued with the reversed direction of change, a normal clock performance would have developed.

The direction of growth of the clock was reversed twice for another bird on FI 30, whose performance has already been shown in Fig. 314. Figure 324 shows the first segment in which the clock was reversed. At the beginning of the interval, when the slit remains large instead of resetting to the smallest size, the bird begins responding appropriately, and shows a declining rate as the slit shrinks to the small size which had been present at the start of the interval and correlated with nonreinforcement. The curve is very close to the inversion of the standard FI 30 curve + clock shown in Fig. 314. The bird was not responding when the slit reached the small size, and it remained in the apparatus for nearly 2 hours without responding. The bird was removed without being reinforced on the small slit.

The bird was then placed on a normal FI 30 with added clock for 2 days. An extinction curve was then taken during which the clock cycled in the reverse direction

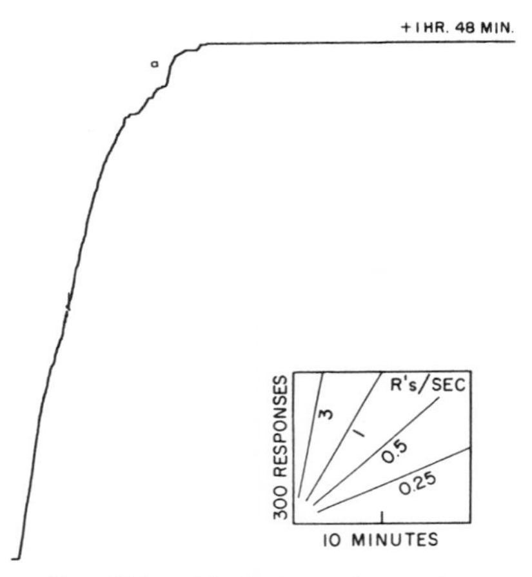

Fig. 324. Clock reversal on FI 30

continuously. There was no history of reinforcement with a reversed clock. The record (Fig. 325) begins with the clock at small, at *a*; the clock shifts abruptly to the large-size slit, at *b*. This shift institutes the high rate normally occurring when the slit is large; but as the slit shrinks during the next 30 minutes, the rate drops to zero, at *c*. This curve is similar to Fig. 324, where the direction of growth of the slit was also reversed for a single segment. The smallest value of the slit is reached at *d,* 30 minutes after *b*; and the clock abruptly changes to large, so that the high rate is again reinstated. Extinction is carried out for a total of 11 cycles, with continued control by the clock throughout. The duration of the responding at the high rate when the slit is large becomes less as extinction proceeds. The differential control remains in force. The result here closely resembles the performance of another bird in Fig. 313, except that the curves in the present figure are inverted because the clock has been reversed.

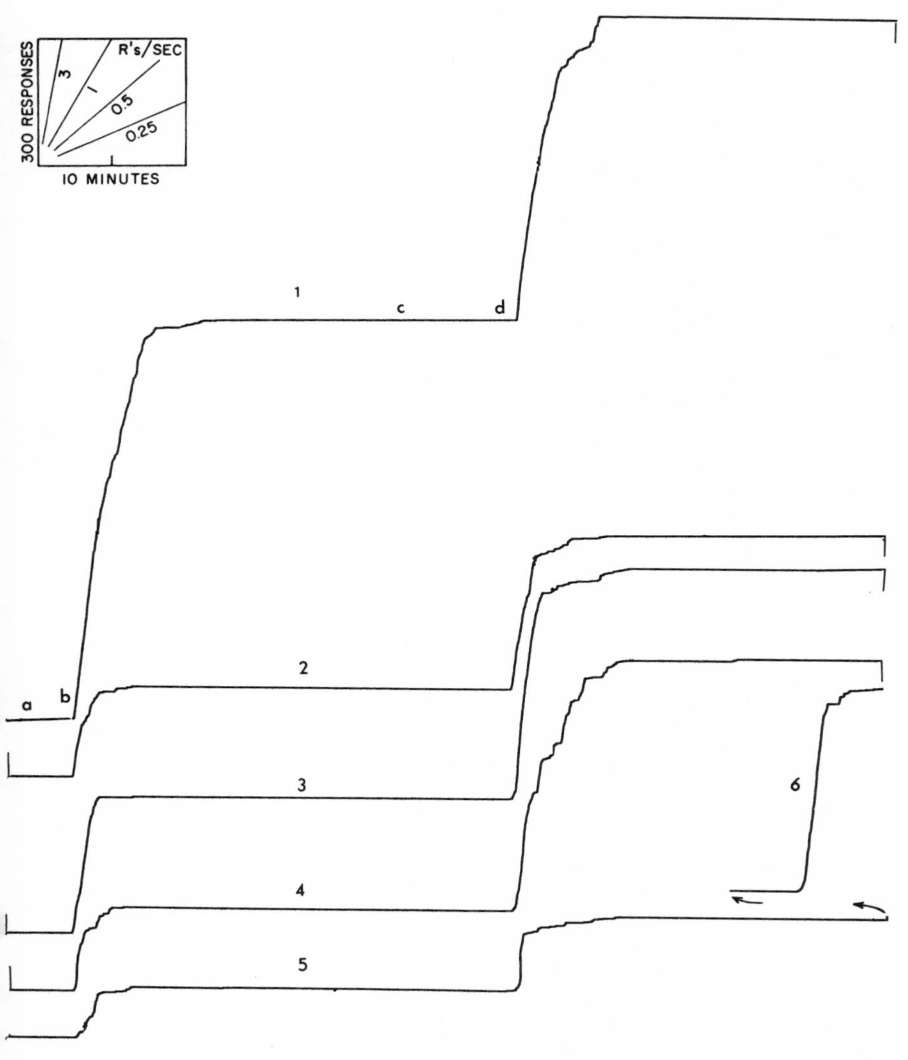

Fig. 325. Clock reversal during extinction

Transitions from clock to stationary slit

Transition from FI 10 + clock to FI 10 + clock at the optimal size. In the session following that in which the final performance on FI 10 + clock shown in Fig. 313 was recorded, the clock remained stationary at the optimal setting after 3 reinforcements on the former schedule. In Fig. 326 the first 3 intervals with clock are shown at *a, b,* and *c,* the slit remaining large as marked. The optimal size of slit controls the terminal rate throughout the first interval except for some irregularity and decline just before reinforcement, at *d.* The next reinforcement begins at a lower rate, which continues only for a few responses before a shift to a higher rate at *e*; this higher rate is maintained for several hundred responses before falling off to a lower value at *f.* Subsequent intervals show a gradually declining and slightly oscillating rate, with little or no curvature appropriate to the FI 10 schedule in spite of the continuing reinforcements.

Subsequently, FI 10 with clock was reinstated, and the transition from FI 10 + clock to FI 10 with the slit at the optimal size was repeated 12 times. Each daily session began with the program which was in force at the end of the preceding session; and during the session the procedure was changed from FI 10 + clock to FI 10 with the slit at the optimal size or vice versa. The bird begins to shift somewhat more read-

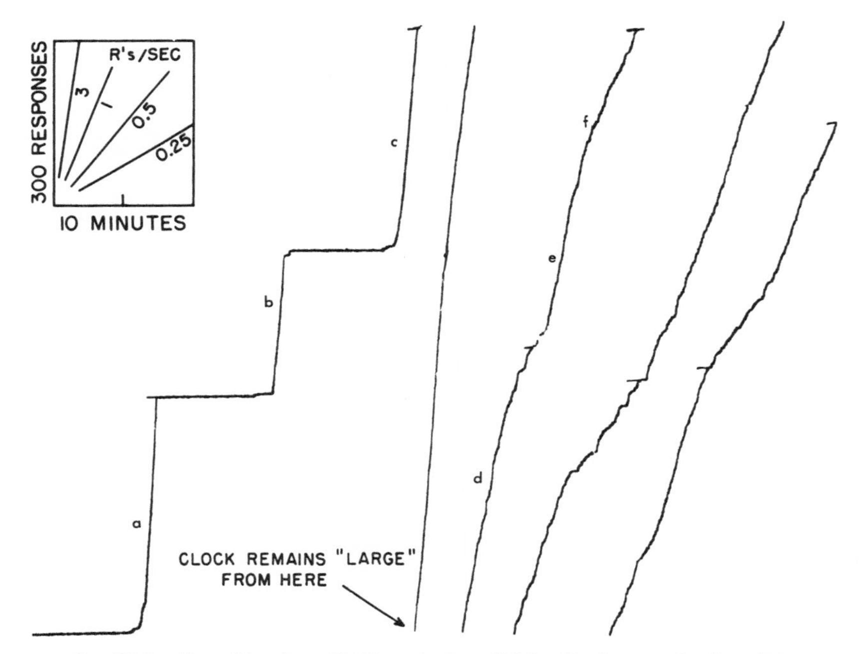

Fig. 326. Transition from FI 10 + clock to FI 10 with slit at optimal setting

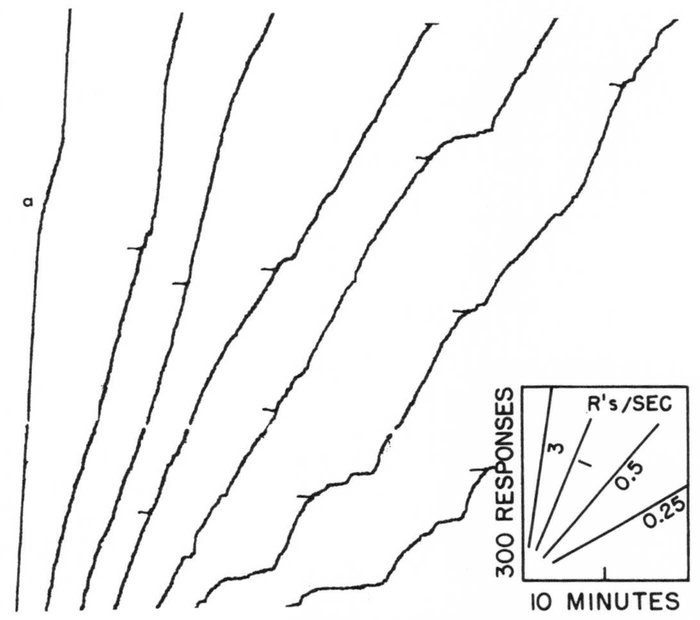

Fig. 327. Fourth transition to the optimal clock setting

ily and more abruptly to the performance characteristic of the schedule. Figure 327, for example, shows the 4th transition to FI 10, with the slit at the large value after previous exposure to FI 10 + clock. The preceding FI 10 performance with clock is not shown, but was standard. The terminal FI clock rate is maintained for approximately 600 responses to *a*; here, it drops to a lower rate for a short time before returning to a higher rate briefly and continuing to the 1st reinforcement at a slightly lower rate. This responding is of the same order of magnitude as in the 1st transition to the large slit in Fig. 326, but responding at the full, terminal FI clock rate is considerably reduced. The remainder of the session shows an irregular and declining over-all rate. Before the end of the session some tendency is evident toward a lower rate of responding during the first part of the interval.

Figure 328A shows the 6th transition, where the segments have been arranged without breaks to show the over-all shape of the curve. When the slit remains large after reinforcement at the start of the record, the usual terminal rate is maintained for only 500 responses (to *a*), where it falls to a lower value. The terminal rate recovers briefly at *b*, but the rate falls again and the reinforcement at *c* occurs at a lower rate. This reinforcement reinstates the terminal rate, which is again maintained for almost 500 responses before it shifts abruptly to an intermediate rate at *d*. The actual amount of time on the FI clock rate is about the same as for the 4th transition

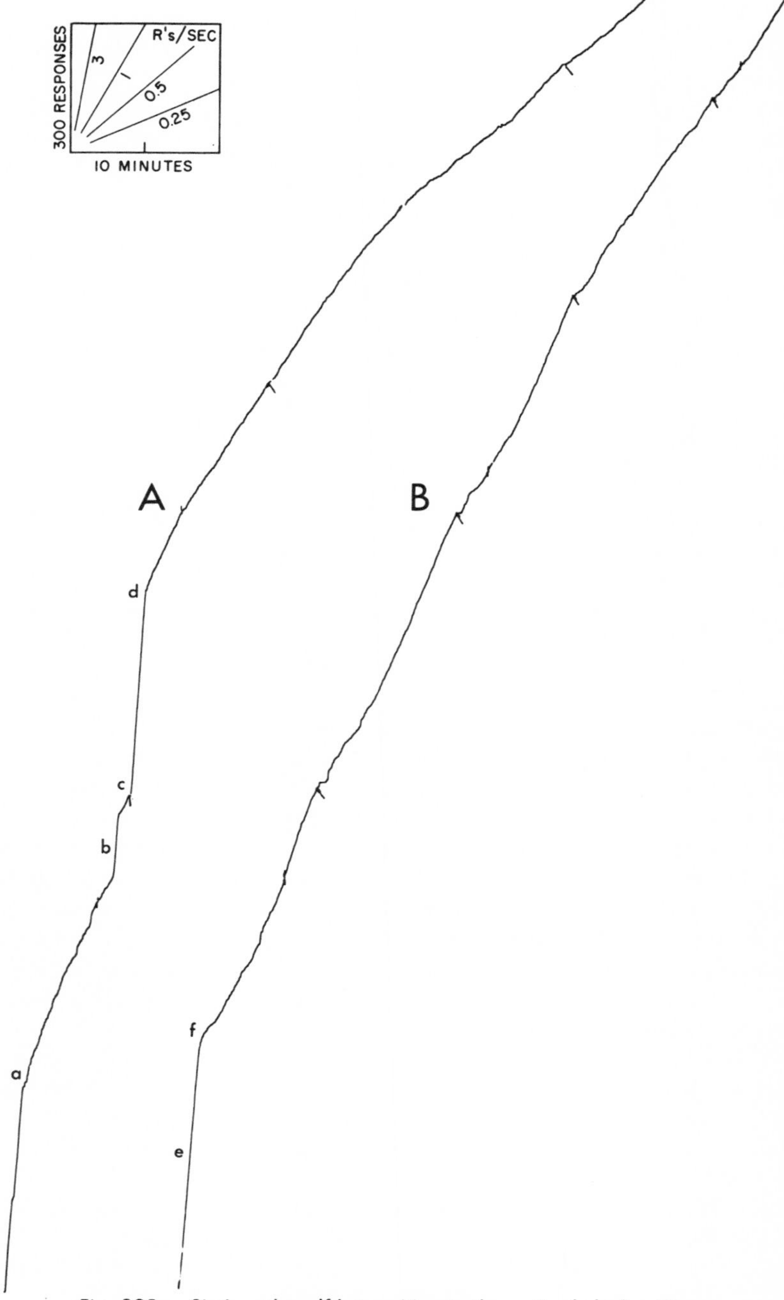

Fig. 328. Sixth and twelfth transition to the optimal clock setting

in Fig. 327. The other features of this record, however, show that the transition to the FI performance without benefit of clock occurs more rapidly.

Figure 328B shows the 12th transition from FI 10 + clock to FI 10 with the slit stationary at the optimal size. There is only an initial burst (at *e*) at the rate appropriate to the clock performance before the rate falls sharply (at *f*) to an intermediate value which is maintained thereafter. This rate is considerably higher than is usually maintained by an FI 10 performance without benefit of added stimuli. (The second bird in the experiment showed the same general pattern, but with greater irregularities in the absence of a clock. See Fig. 330.)

Transition from FI 10 to FI 10 + clock. The converse cases in this experiment, where the procedure was changed from FI 10 with the slit at large to FI 10 + clock, showed transitions similar to those in Fig. 329. (The uniform pause following the reinforce-

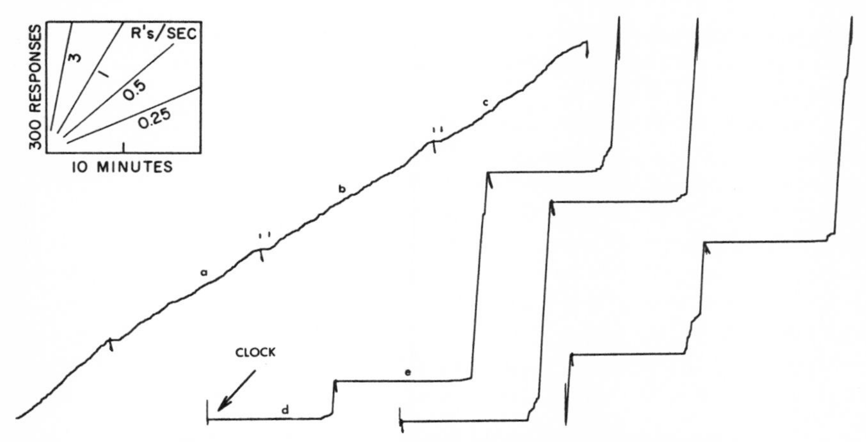

Fig. 329. Transition from FI 10 with slit at large to FI 10 + clock

ment in the intervals *a, b,* and *c* is an artifact; the paper feed was allowed to run during the short TO after each reinforcement, so that the clock could reset.) Under these conditions, little scalloping occurs without the help of the clock, as the segments at *a, b,* and *c* show. Following the reinforcement at the arrow, the slit on the key was reset to its smallest value and began growing through the interval in the manner of a clock. The small slit controls a zero rate of responding immediately; and the bird begins to respond late in the interval and reaches the terminal rate only when the slit becomes almost maximally large. Note that the larger value of the slit now controls a rate of over 10 responses per second, compared with an over-all rate of only 0.35 response per second when the slit was at the same size but stationary. The bird can make a rapid transition between the 2 performances appropriate to the large-size slit under the control of the other stimuli associated with moving or stationary slits. The number of responses in the 1st interval with the clock, at *d,* is small; but beginning with the fol-

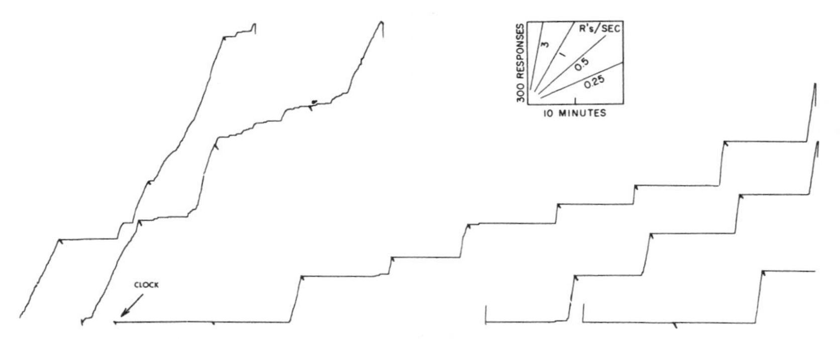

Fig. 330. Transition from FI 10 with slit at large to FI 10 + clock (second bird)

lowing interval, at *e*, the number remains appropriately large. The bird consistently showed a small number of responses during the 1st or the first few intervals following the transition to the clock.

As already noted, the performance of the other bird without a clock was irregular. The beginning of Fig. 330 shows a sample. Scalloping, frequent pauses after reinforcement, intervals in which the terminal rate continues after reinforcement, and second-order effects occur. When the slit resets to small (at the arrow) and begins to grow in the manner of an added clock, the 1st interval is terminated by a single response, even though the slit on the key has grown to large before the response was made. The 2nd segment shows a fair clock performance, however, with a terminal rate of the order of 2.5 responses per second. The remainder of the session shows a low over-all rate, which is the result of both a late start and a lower terminal rate than is usual for this bird on FI 10 with clock. Note that the terminal rates with and

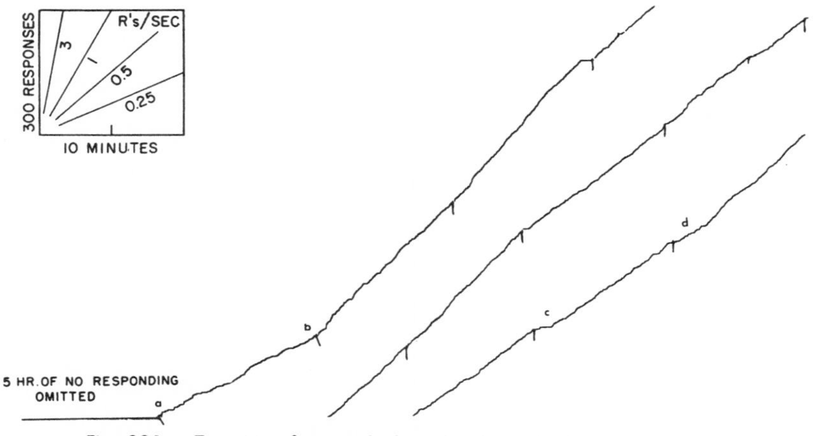

Fig. 331. Transition from a clock to the slit at the least optimal size

without clock are here of the same order of magnitude, whereas in Fig. 329 they differed by a factor of 30. In other transitions from FI 10 with the slit stationary to FI 10 + clock, this bird showed a similar performance, with the rate recovering as the session proceeds.

Transition from FI 10 + clock to FI 10 with the clock stationary at the setting farthest removed from reinforcement. A final clock performance such as that shown in Fig. 312 demonstrates a precise control by the size of the slit. The bird consistently pauses for the major part of the interval and shifts to the terminal rate only when the size of slit is nearly maximum. The extreme control exercised by the small slit in suppressing behavior was shown when the slit remained at its smallest size for the first time (after the series of reversals described in Fig. 326, 327, and 328, in which the slit was sometimes stationary at large from the beginning of an experimental session). The bird remained in the apparatus for 5 hours without making a single response. The 1st re-

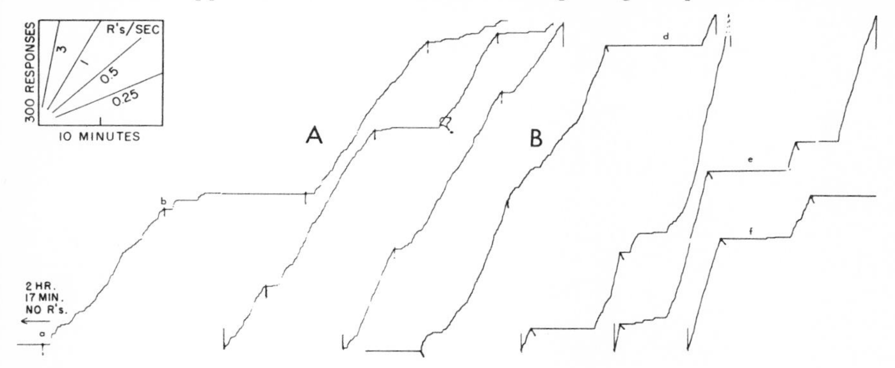

Fig. 332. Transition from a clock to the slit at the least optimal size (second bird)

sponse which occurred was reinforced, as shown in Fig. 331, at *a*. The 1st reinforcement in the presence of the small slit was followed by immediate responding at a low moderate rate, with rather rough grain. The 2nd reinforcement (at *b*) produced a 2nd increase in rate to a value which is roughly maintained for the remainder of the session. Toward the end of the session, at *c* and *d*, the rate tends to be somewhat lower immediately after reinforcement.

The same procedure was carried out with the second bird in the experiment. The bird remained in the apparatus without responding during a complete 12-hour session with the slit small. Placed in the apparatus for a 2nd session, it waited 2 hours and 17 minutes before responding and was then reinforced, at *a* in Fig. 332. The single reinforcement in the presence of the small-size slit produces a substantial rate of responding through the 1st interval. The 2nd reinforcement (at *b*) is followed by a small amount of negatively accelerated responding; and the 3rd reinforcement occurs only after a long pause. Subsequently, further increases occur in the over-all rate of responding, with the rapid development of a lower rate immediately following the reinforcement. The procedure was then changed to FI 10, with the clock growing in the usual manner; and 2 weeks later the experiment was repeated. The 1st response occurs after only 12 minutes of the small slit (at *c* in Record B) and is followed by an immediate increase in rate. The remainder of the session shows a fairly high terminal rate and substantial pausing after reinforcement, which in some instances, as at *d, e,* and *f,* approaches the order of magnitude occurring with the added clock. The irregularity shown by this bird in Fig. 330 is again evident.

We further investigated the effect of changing from FI 10 + clock growing from small to large to FI 10 with the slit at small by making the change in the middle of the session on successive days. Each session began with the same procedure as the end of the preceding; and in the middle of the session the slit was changed to stationary at small if it had been growing, or growing in the manner of a clock if it had been stationary at small. Figure 333 shows a characteristic transition for 1 bird.

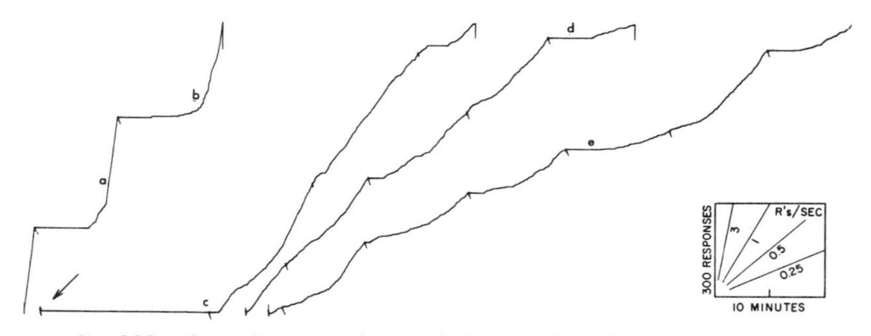

Fig. 333. Second transition from a clock to the slit at the least optimal size

The clock performances at *a* and *b* before the transition show considerable curvature as a result of the many manipulations of the clock. When the slit remains small at the arrow, the rate is zero much longer than in the usual interval. When a response finally occurs and is reinforced at *c,* however, there is an immediate return to the pattern of responding characteristic of this bird on FI 10 without clock. The rest of the figure shows the progressive development of smooth and extended interval curvature, with occasional pauses almost 5 minutes long (*d, e*). The small slit probably favors a lower rate of responding after reinforcement because of its general tendency to control a lower rate in the normal clock performance. Conversely, transitions to FI 10 with the slit at the *optimal* value probably interfere with the development of a lower rate or pause after reinforcement because of the past correlation.

The second bird showed a similar result of the same transition in this experiment. Figure 334A shows the 1st transition. The last performances with clock occur at *a* and *b*; and beginning at the arrow the slit remains small for the remainder of the session. A first response with the slit at small is reinforced at *c,* and the general FI performance being displayed at this stage of the experiment is reinstated. The scallops

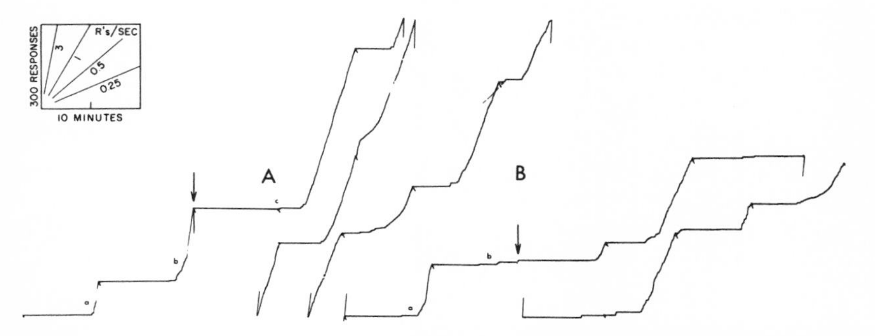

Fig. 334. First and third transitions from a clock to the slit at the least optimal size (second bird)

here are smoother and the pauses more extended and more appropriate to the interval than in the 1st transition for this bird in Fig. 332A. Record B shows the 3rd transition. The segments at *a* and *b* are the prevailing clock performances before the transition, with the interval at *b* showing an unusually small number of responses. When the slit remains stationary at the small size (beginning at the arrow), a well-developed FI scallop is immediately assumed, the pause after reinforcement being of the order of magnitude when an added clock is used. The terminal rate is more irregular, however.

Transition from FI 10 with the slit stationary at small to FI 10 + clock. Interspersed with the transitions described in the preceding paragraphs were changes made during the session from the slit stationary at the smallest size to the clock growing in the normal manner. Figure 335 shows the 1st transition for 1 bird. The final performances with the slit stationary at small occur at *a, b,* and *c*. The various manipulations of the clock prevented any substantial scalloping. When the added clock is rein-

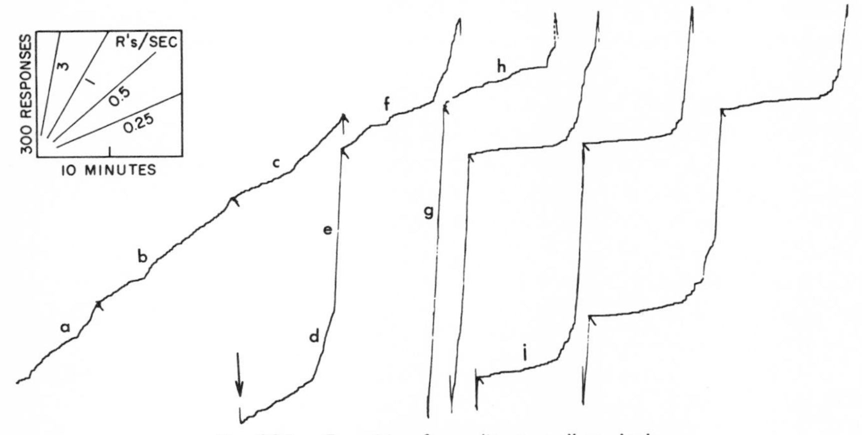

Fig. 335. Transition from slit at small to clock

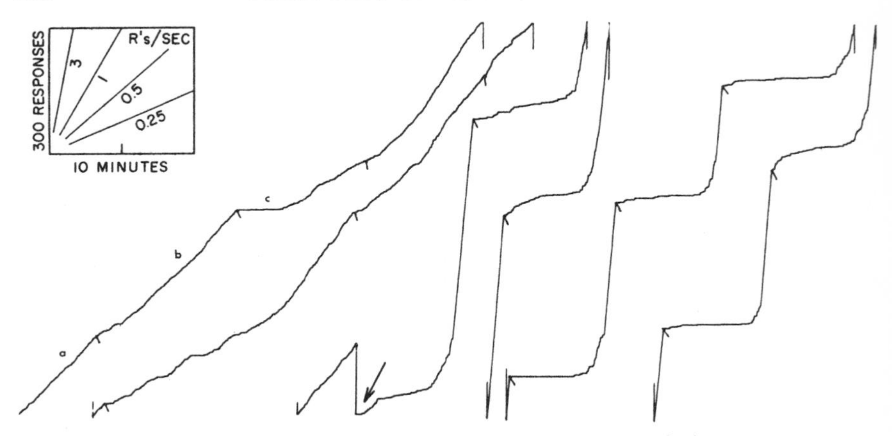

Fig. 336. Second transition from slit at small to clock

stated (at the arrow), the first effect is to continue the rate prevailing under the small-size slit. As the slit becomes larger, however, the rate increases (*d*) and assumes the usual high terminal clock rate, at *e*. The small spot continues to control a substantial rate in the following segment (at *f*); but the terminal rate is reached sooner, and a large number of responses are emitted at the terminal rate (at *g*). The small-size slit again controls substantial responding at *h* and *i*, but the control grows progressively weaker during the remainder of the session.

The following session continued with the usual clock growth, and in the middle of the session the slit was left stationary at small. A 2nd transition from FI 10 with the slit at small to FI 10 + clock occurred in the middle of the session on the following day. In Fig. 336 the performance with the slit stationary at small shows a linear performance (as at *a* and *b*) and an occasional pause and curvature (as at *c*). The return to FI with added clock at the arrow is similar to the 1st transition in Fig. 335. The residual control of the small spot is still to be observed. High terminal rates return immediately, as usual.

Figure 337 shows the beginning of a session with added clock, following a 3rd transition from reinforcement at small to the usual clock. The previous session ended with a good clock performance, in which responding after reinforcement stopped almost completely and the shift to a high terminal rate was abrupt. On the day shown in Fig. 337, however, some control exercised by the small size appears as substantial responding immediately after reinforcement, at *a*, *b*, and *c*. A smooth negative curvature brings the rate to zero as the clock grows to the intermediate values, at which reinforcement has never occurred. The latter part of the interval shows a normal shift to a very high terminal rate.

As we have seen, these variations in the contingencies of reinforcement involving sizes of slit produce an irregular performance in a second bird. Figure 338 shows the 1st transition from FI 10 with the slit stationary to FI 10 with the usual clock growth.

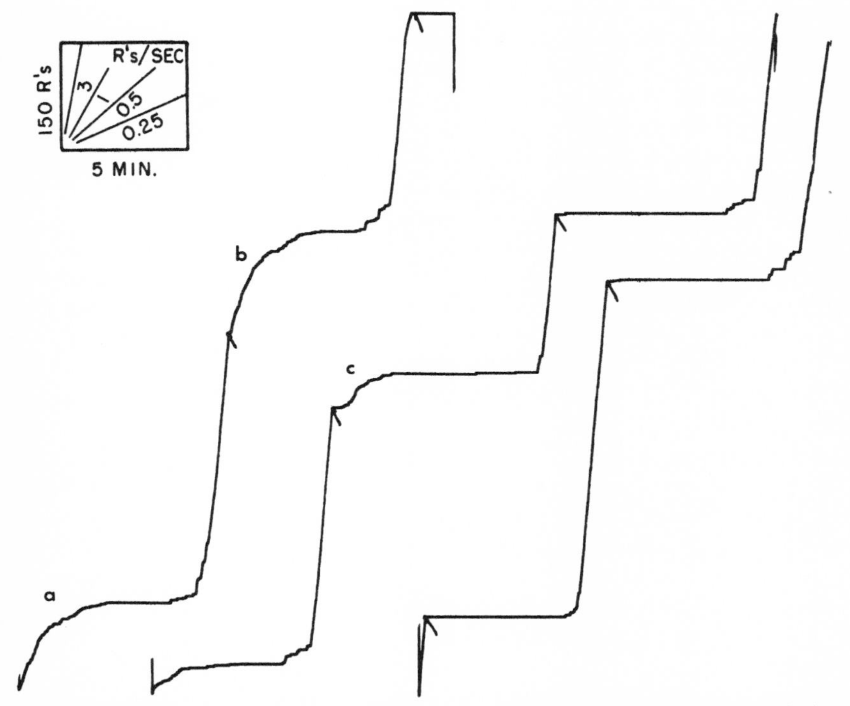

Fig. 337. Overnight loss of scallop on FI 10 + clock after reinforcement on small slit

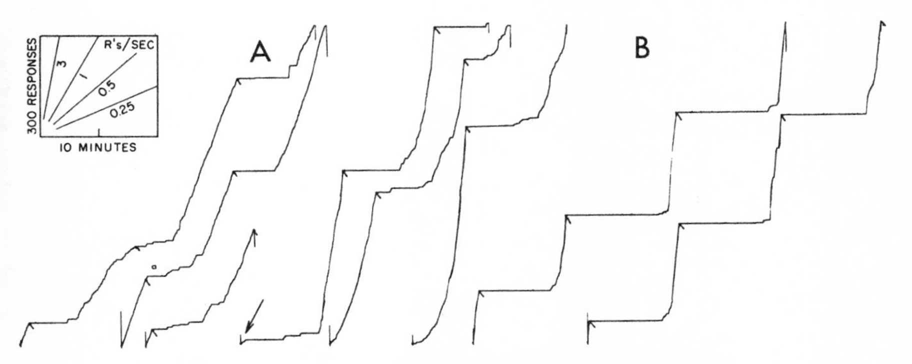

Fig. 338. Transition from slit at small to clock (second bird)

Record A begins with the slit stationary at small. Except at *a*, each reinforcement is followed by a pause, which in turn is followed by a substantial amount of scalloping, although the grain is rough and considerable variation exists from interval to interval. Beginning at the arrow, the clock grew in the normal fashion. The only immediate change in performance is an increase in the terminal rate of responding when the slit reaches the larger sizes. There is considerable irregularity. The middle part of the session has been omitted from the figure. By the end of the session (Record B) a nearly normal clock performance has developed except for small knees and a terminal rate disturbed by occasional pauses.

Figure 339 shows the 2nd to 6th transitions for this bird. The parts of the records before the arrow represent the performances with the slit stationary at small; elsewhere, the clock growth is usual. A more and more effective transition to the clock

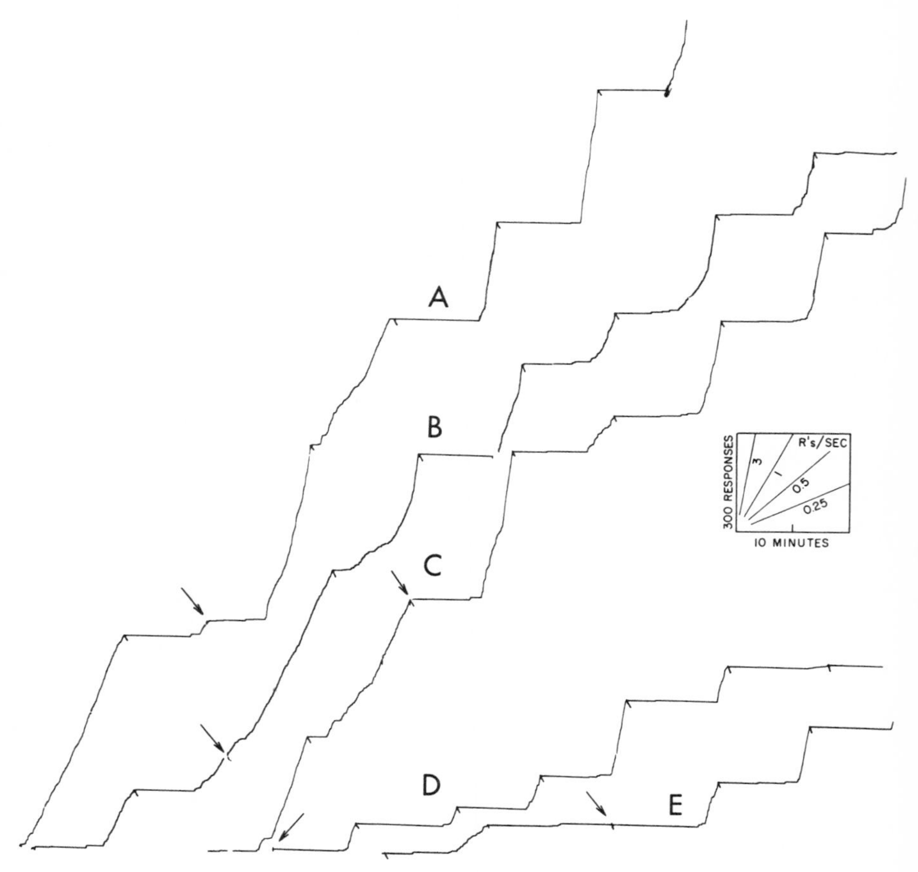

Fig. 339. Second to sixth transitions from slit at small to clock

performance occurs with each successive reversal. The transition in Record A, for example, is similar to that in Fig. 338, the first for this bird. The 3rd transition (Record B) shows a complete absence of clock control in the first segment after the arrow. By Record C the deviations from a final clock performance are less marked, and by Records D and E a fairly good clock performance develops as soon as the slit is made to move. The performance before the clock was added is omitted in Record D.

A discontinuous clock

The physical dimensions of a clock are arbitrary. The clock used in the preceding experiments is only one of many possibilities. A series of discrete color changes, a succession of complicated shapes changing in time, or a pointer moving over a dial in the manner of a timepiece might have been used. In the following experiments the small slit was used as in the previous experiments, but reinforcement was correlated with the middle-size slit. The sequence of events is as follows. Just after reinforcement the slit is at the middle size. It reaches its maximum width halfway through the fixed interval, and at that point changes abruptly to the smaller size. Reinforcement occurs when the middle size has again been reached. The cycle is then repeated during the next interval. In this clock the stimulus on the key just before the reinforcement is almost identical with the stimulus at the start of the next interval. The greatest change in the stimulus on the key occurs abruptly in the middle of the fixed interval.

The subjects were the 2 birds from the preceding experiments. A final performance on FI 10 + clock, consisting of a slit which is small after reinforcement and grows to its maximum size at the end of the fixed interval, had been developed after the transitions described in the preceding section. The clock was maintained without change for 3 full sessions. Figure 340A shows 2 segments from the final performance. Reinforcement was then begun with a "discontinuous" clock. The 1st interval of the session in Fig. 340B begins (at *a*) with the slit at the medium size. A

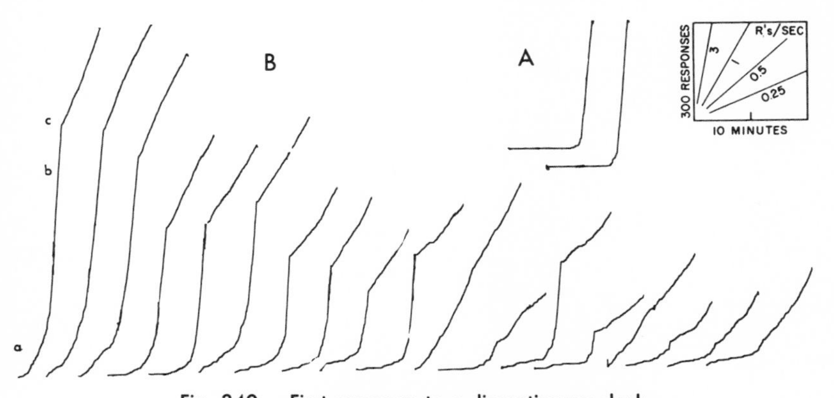

Fig. 340. First exposure to a discontinuous clock

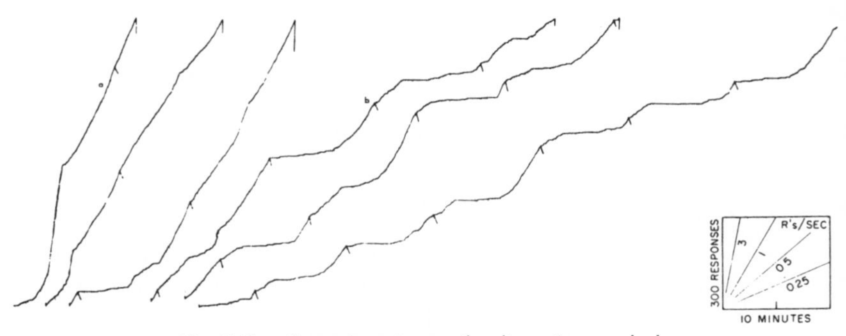

Fig. 341. Second session on the discontinuous clock

moderate rate prevails. A smooth acceleration then leads to the usual high terminal clock-rate at *b* as the slit grows to large. The slit reaches its maximum size at *c*, halfway through the interval, and changes abruptly to the smallest size. The rate drops to a lower but still substantial value which continues unchanged until reinforcement. The bird's performance under the previous continuous clock obviously was not controlled solely by the absolute value of the slit. Later segments in Fig. 341B show a similar control. The intermediate rate during the second half of each interval, where the slit grows from small to middle size, continues practically unchanged for many intervals; but the scallop deepens progressively at the start of the interval, where the slit grows from the middle size toward large. The high terminal rate at the large size holds this high value until the 11th interval. In the last 3 segments of the figure some suggestion of an acceleration continuing throughout the interval appears. However, this factor does not necessarily imply any control by the discontinuous clock, since performances of this order of magnitude can occur on FI 10

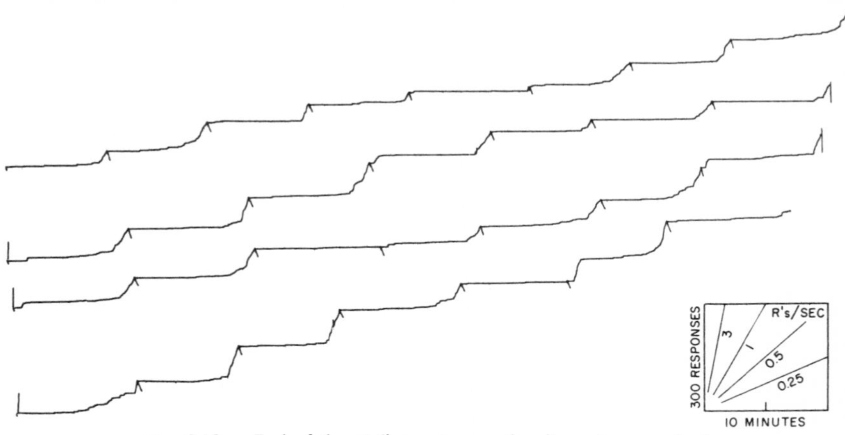

Fig. 342. End of the sixth session on the discontinuous clock

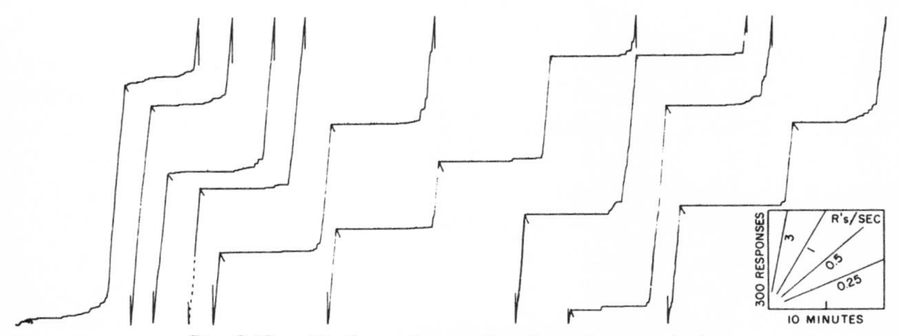

Fig. 343. Ninth session on the discontinuous clock

alone. This exposure to a discontinuous clock has, however, at least eliminated the control over the rate established with the continuous clock. Figure 341 shows the 2nd day of exposure to the discontinuous clock. This performance exemplifies an overnight loss in stimulus control which is commonly observed. The evidence of a developing control by the discontinuous clock is clear. Most of the curves are S-shaped. The stimulus on the key just following the reinforcement is the same as that present at reinforcement. But both large and small slits present during the middle part of the interval are now occasions on which reinforcement never occurs; hence, they control a lower rate of responding. At the end of the session a normal scallop occurs. Figure 342 shows a further development under the discontinuous clock for the last part of the 6th session. Definite control by the clock has now been established. Responding ceases at reinforcement, and the pause is very marked; acceleration to the terminal rate occurs late in the interval, in the manner of the continuous clock except for a low terminal rate. Other stimuli are clearly operating, in addition to the size of slit, to prevent responding to the middle-sized slit just after reinforcement. By the 9th session (Fig. 343) the performance is not very different from that with a continuous clock.

The second bird showed a similar transition to the discontinuous clock, although the rate of development was very different. Figure 344A shows the start of the 1st ses-

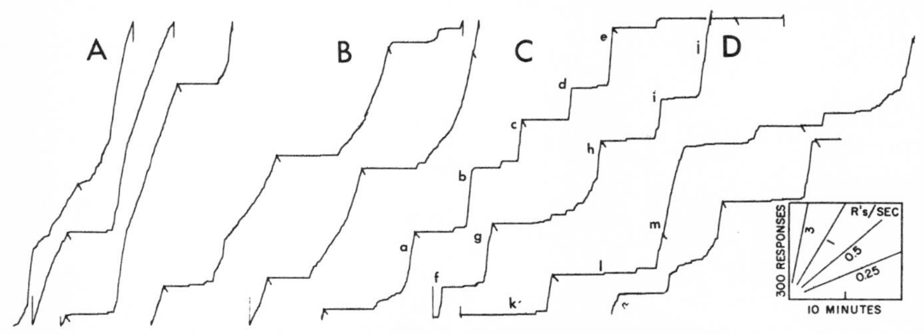

Fig. 344. First exposure to a discontinuous clock (second bird)

sion with the discontinuous clock. The segments are similar to those in Fig. 340. The large slit controls a high rate because of the previous exposure to the continuous clock; and the small slit produces an abrupt shift to a lower but still substantial rate instead of the zero rate previously appropriate to that size. In Record B, in the middle of the same session, the high rate of responding in the presence of the large slit has largely disappeared, and the terminal rate has increased. By the 7th hour of the sessions shown in Record C the rate of responding toward the end of the interval (as the slit approaches its middle value) is very high (at *a, c, e, g, h,* and *j*); but the high rate is recovered in the presence of the large-size slit, appropriate to the continuous clock of the previous session at *b, d, f,* and *i.* By the end of the session, after 12 hours of exposure to the discontinuous clock (Record D), there is less of a tendency to respond when the slit is large in the middle of the interval, as for example, at *k* and *l*; but at *m* the intermediate-size slit controls a high rate. The 3rd session of exposure to the discontinuous clock shown in Fig. 345 shows a further development. At *a* the high rate controlled by the large-size slit returns because of the history on the previous clock; but a fairly normal fixed-interval curve appears at *b.* The appearance of this curve may or may not reflect control by the discontinuous clock. This performance is followed by an interval in which a high rate is maintained substantially throughout. Three intervals then follow which show an extended pause after reinforcement and acceleration to a high terminal rate before the end of the interval. These curves indicate control by the discontinuous clock. Following the reinforcement at *c,* however, a high sustained rate appears throughout the interval, marked by a lower rate in the region of the large and very small slit sizes. Two further instances, at *d* and *e,* also show increased control by the intermediate-size slit. The 5th session of exposure to the discontinuous clock is represented in Fig. 346. Here, the segments are either scalloped, as at *b, e,* and *f,* or show a high rate of response after reinforcement which falls off as the slit reaches the largest size, as at *a, d,* and *g.* Figure

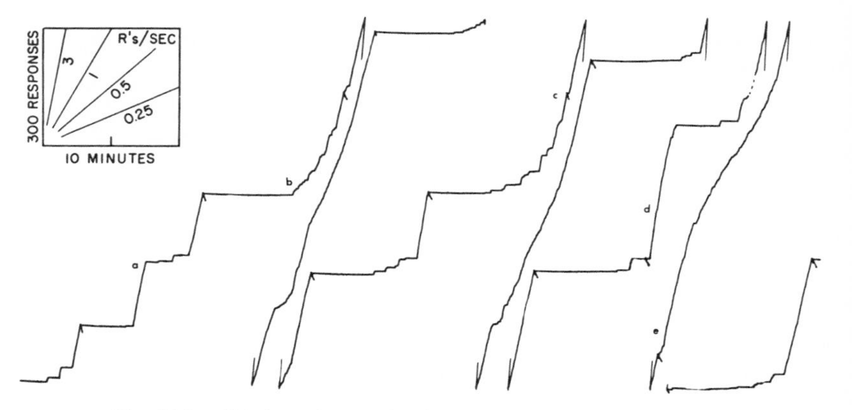

Fig. 345. Third session on the discontinuous clock (second bird)

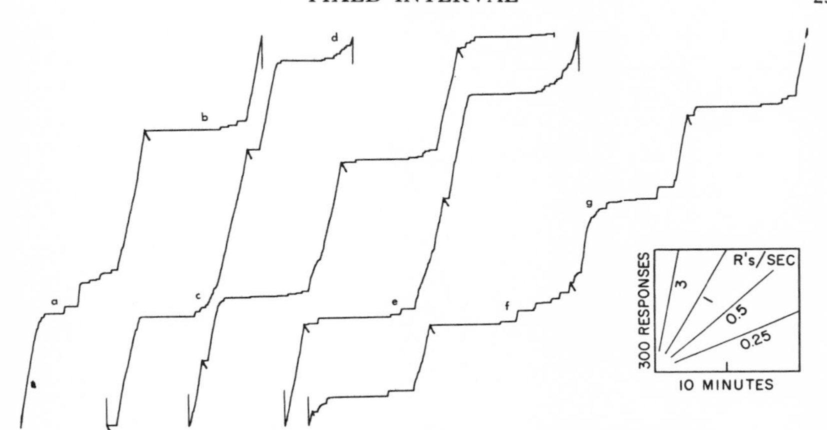

Fig. 346. Fifth session on the discontinuous clock (second bird)

347 shows the final performance for the second bird at the end of 100 hours. The rate is zero for considerably more than half the interval; the terminal rate is fairly high but irregular. Nevertheless, a high rate still occasionally appears near the start of the interval (as at *a*) and the terminal rate at the end of the interval is sometimes not reached (as at *b* and *c*).

Return to the continuous clock on FI 30

These 2 birds were returned to a continuous clock, and the interval was increased to 30 minutes. The transition still gives us some information about the previous clock control. Figure 348 shows the 1st exposure to FI 30 with a continuous clock, where the slit begins at the smallest value following the reinforcement and increases continuously to the large size after 30 minutes. Only 4 intervals occur in the 1st session; they are shown in the order in which they occurred. A response was reinforced at the end of each segment. The general form of all 4 curves is S-shaped. The lowest rates of responding were at the beginning and end of the interval, when the

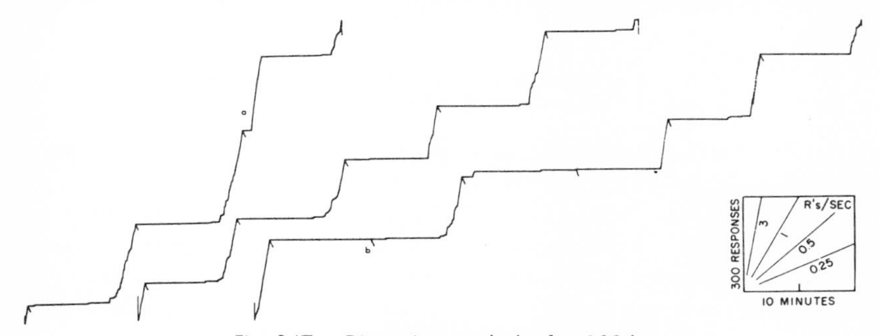

Fig. 347. Discontinuous clock after 100 hr

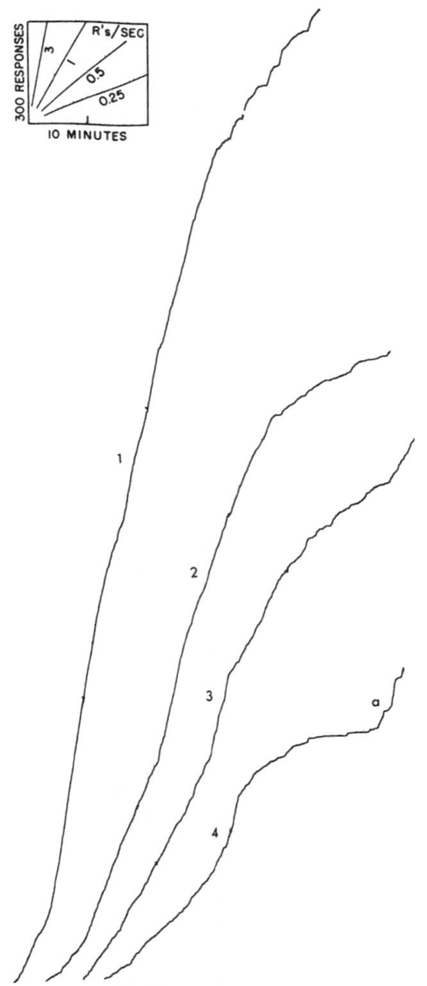

Fig. 348. Return to continuous clock after discontinuous clock

slit was very small and very large, respectively; and the high rates of responding were in the middle range of the interval, where the slit was middle-sized. The form of these curves, of course, corresponds to the likelihood of reinforcement on the previous discontinuous clock. Note, however, that the very large and very small slit values do not control a zero rate of responding, as they would if the absolute size of the slit were the only controlling variable. An increase in rate appropriate to the new schedule appears at *a*.

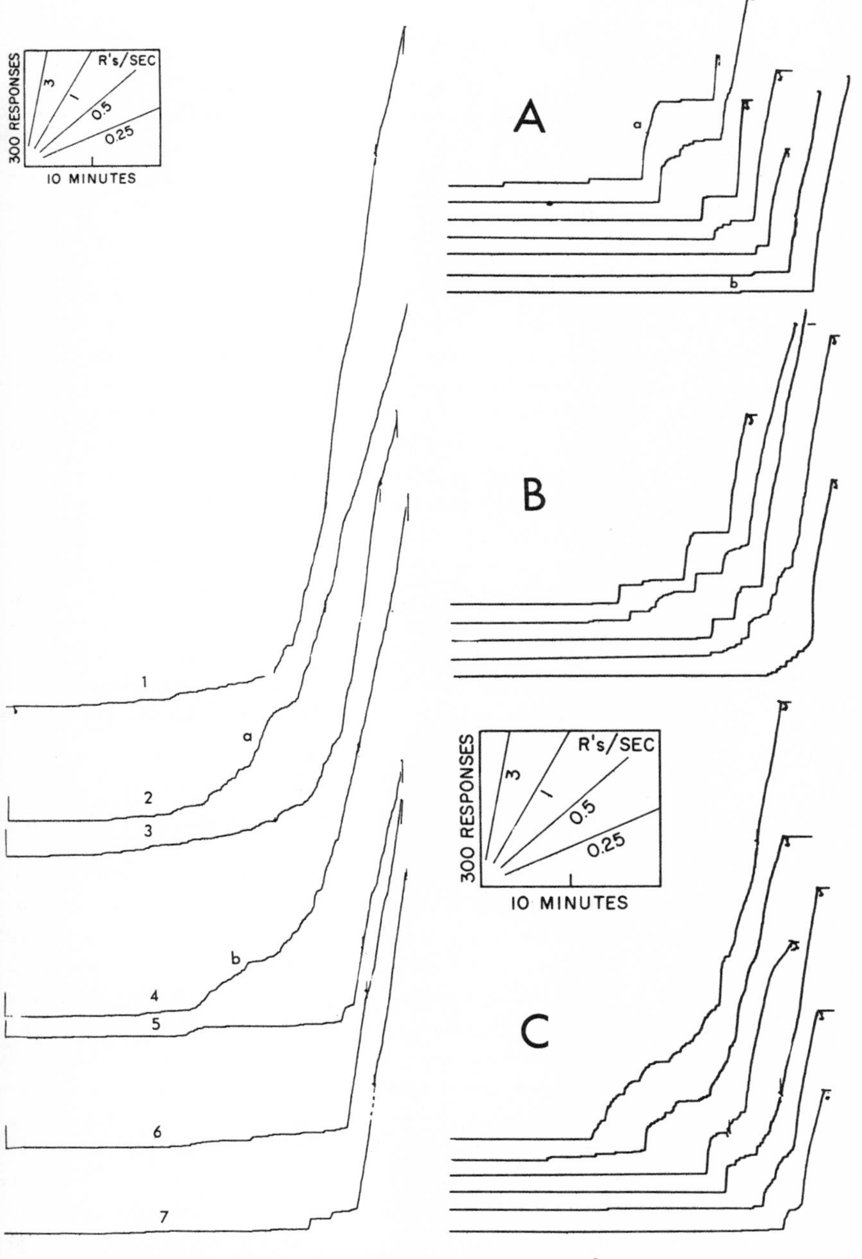

Fig. 349. End of the second session after the return to the continuous clock

Fig. 350. Selected segments during FI 30 + clock after reinforcement on a discontinuous clock

The transition to FI 30 with a continuous clock is almost completed by the end of the 2nd session after the discontinuous clock. The first 4 intervals of the session showed high over-all rates of responding, largely because of the rapid responding in the middle range and the effect of the large slit at the end of the interval. Figure 349 presents the last 7 intervals of the session in the order in which they occurred. Some instances still exist of fairly prominent knees, particularly at *a* and *b*, which are probably a result of the previous reinforcement at the middle slit-size. (A final performance on FI 30 with continuous clock has already been described in Fig. 314.)

The second bird showed a more rapid transition to the continuous clock, but also a more prolonged effect of the previous reinforcement at middle slit-sizes. Figure 350 contains 3 sets of intervals which have been collected in terms of the kinds of rate changes occurring in the middle range of the interval. In order to permit the greater reduction and better comparison of the rate changes in the middle range, horizontal portions of the curves just after reinforcement have been deleted. The group of curves in Record A contains knees consisting primarily of an abrupt shift to a very high rate of responding, followed by a decrease to zero and then a second sudden increase to the terminal rate. The knee at *a* is exceptionally large, and at *b* is probably insignificant. Cases of 2 or more knees have been collected in Record B; included among these is a single curve in which the amount of sustained responding before the terminal rate is reached is so small that the effect is more properly called a rough grain. In Record C the knee consists of a high rate followed by negative acceleration to an intermediate rate, which then accelerates positively and eventually blends into the terminal rate.

THE EFFECT OF TO DURING PART OF A FIXED INTERVAL WITH CLOCK

We attempted to examine some of the properties of a well-developed clock in FI 30 by imposing a 25-minute TO. This TO determined, in particular, whether the progressive change in stimulus was important in the clock control relative to the absolute control of each slit size. The first 25 minutes of the interval was replaced by a TO with the clock running. The schedule actually was FI 5 TO 25, with the clock first seen already advanced to the stage at which it controlled a high rate on FI 30 (e.g., transition from continuous to discontinuous). Figure 351A shows segments for a full session on FI 30 with clock at a given stage of development. These occur in the order in which they were recorded, and horizontal portions of the 30-minute interval have been cut off to facilitate reproduction. Knees are typical for this interval with clock; the terminal rates are well-sustained. Record B represents the 1st day in which the first 25 minutes of the interval were replaced by a TO, all lights in the apparatus being off and the bird not responding. The recorder ran during the TO. The point at which the TO ended is shown by the small vertical dashes. In general, the curvature in the 5-minute period following TO tends to be somewhat more marked than that in

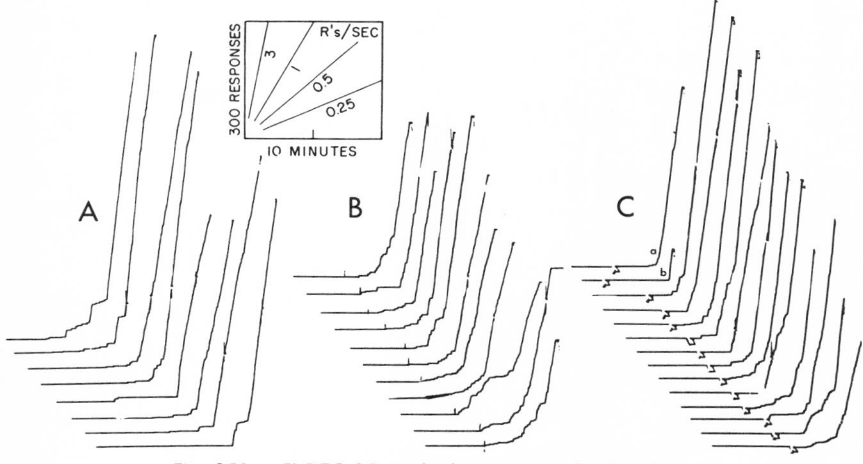

Fig. 351. FI 5 TO 25 + the last portion of FI 30 clock

Record A. Fewer examples of sharp knees exist, and the terminal rates are lower. The total effect is a smaller number of responses per interval. This effect could be in part due to the early disturbances from a prolonged blackout; but it could also be due to the fact that the bird is beginning to develop a 5-minute FI performance with the aid of a clock which is now advancing over only one-sixth of its range during the interval. Record C shows the performance on the 9th and last day of this part of the experiment. The performance is similar to that of FI 5 without the help of the clock, although some intervals, particularly the first 2 (at *a* and *b*), show an excellent clock performance. Higher terminal rates have now been recovered, although the curvature in approaching these rates is still more marked than when the whole interval was run off without a TO.

We further attempted to explore the effect of TO during part of the operation of a clock by alternately replacing 2-minute and 25-minute portions of FI 30 clock by TOs in a single session. The schedule was essentially mult FI 5 FI 28 with residual portions of a clock. Figure 352 shows the results for 1 day. Those cases in which a 25-minute TO occupies the greater part of the interval (up to the dashes) are collected in Record A. The bird can respond only at the end of the interval, and during this interval the clock covers only one-sixth of the total range. Record B shows the segments in which 2-minute TOs (up to the first dashes) leave most of the interval free for responding. Responding begins late because of the operation of the clock, which now covers most of its range. Twenty-five minutes is automatically marked for reference purposes; but no change in procedure was made at these points. No great difference exists between the 2 sets of curves, and the absolute value of the clock appears to control the performance.

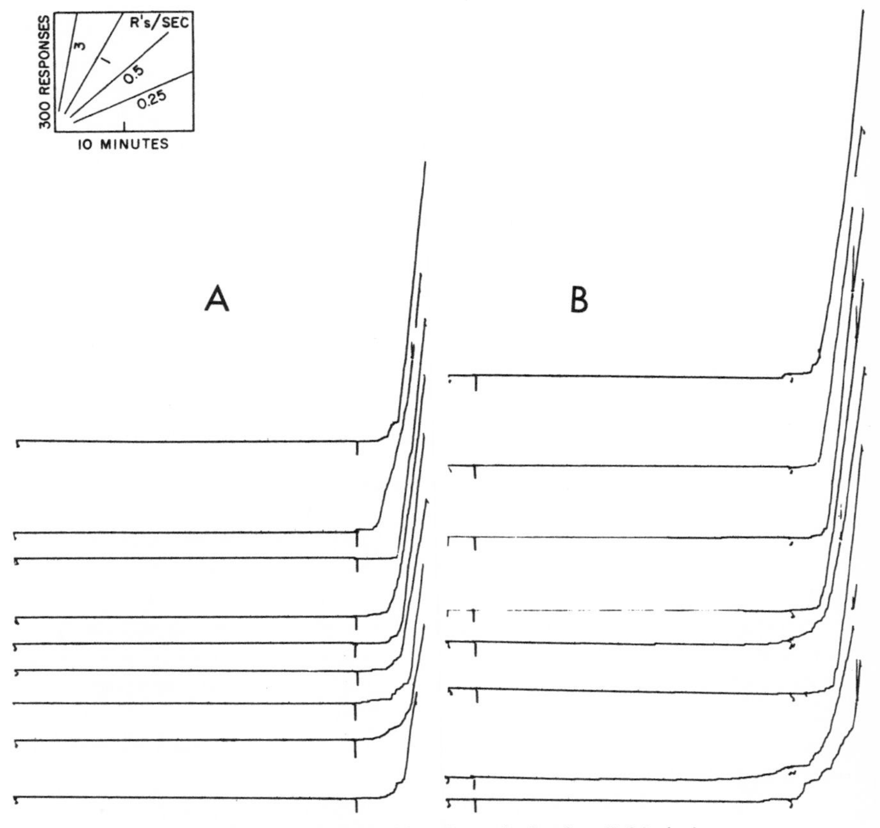

Fig. 352. Mult FI 5 FI 28 with residuals of an FI 30 clock

Magnitude of the clock

We have not made any extensive study of the magnitude of stimuli used for clocks; but 1 bird was exposed to FI 2 with a clock which consisted of only one-fifth the normal growth of the spot of light. This bird had had a long history of FI with added counter and mult FIFI + clock, so that the development of the performance is not significant. After 10 hours of FI 2 with the small-scale clock, however, the performance was no better than that in Fig. 353, which is a complete session. Several instances occur here of no pause following reinforcement, particularly at *a* and *b,* but no case of a pause comparable with those normally observed with the help of the clock. Instead, there tend to be 2 terminal running rates: one is reached shortly after responding is resumed; and the other emerges more or less abruptly, for example, at *c* and *d,* to continue for the balance of the interval.

Another bird with a long and complex history of added counter and clock failed to

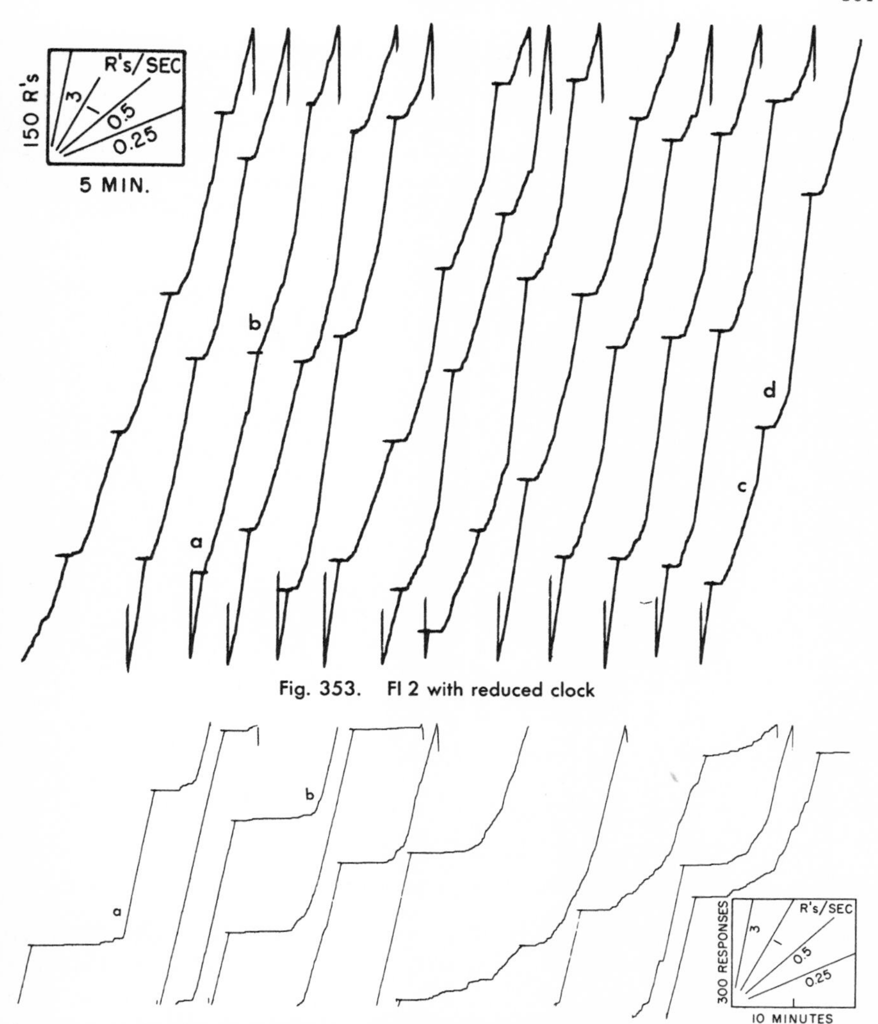

Fig. 353. FI 2 with reduced clock

Fig. 354. Incomplete development of FI 10 + clock after 62 hr

show a good clock performance even with the normal range of stimuli. Figure 354 shows a performance after 62 hours on FI 10 + clock. Although this performance is unquestionably more strongly scalloped than that on FI 10 without the help of a clock, and although many intervals show good clock control (*a* and *b*, for example), the start is generally earlier, and more curvature develops during the interval than is characteristic of the added clock performance. A companion bird with the same complicated history developed a very good clock performance.

FIXED INTERVAL WITH ADDED COUNTER

One explanation of the oscillation in performance in the later stages of an FI performance is that a temporarily stable condition holds the number of responses per interval constant; this fact, as in FR, makes the production of number automatically reinforcing. Therefore, earlier and more rapid responding sets in, and the stable condition is destroyed. Some exploratory experiments on FI with added counter are relevant to this interpretation.

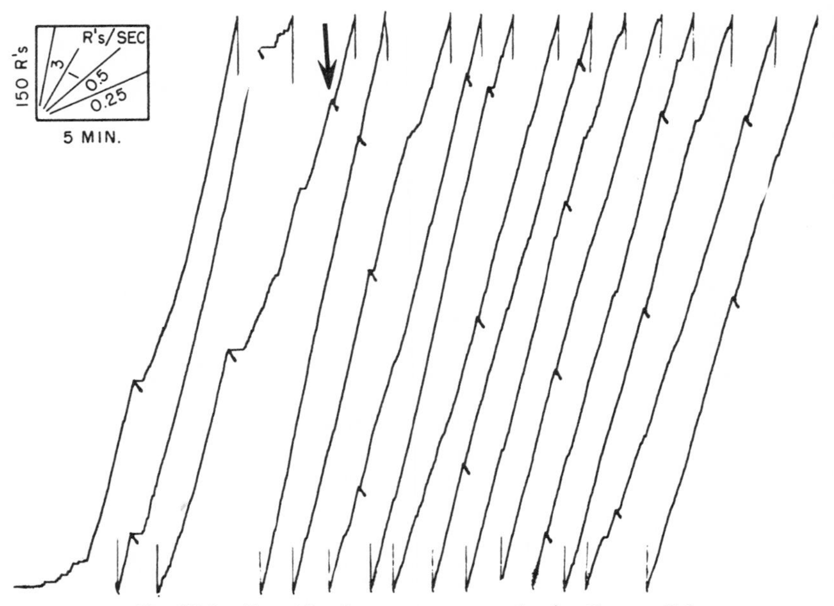

Fig. 355. Transition from counter to optimal setting on FI 5

Two birds used in this experiment had had a history of FI 10. A third had had an extended history of VI and fairly large FIs. The performance of these 3 birds were sampled, one each on FI 5, FI 10, and FI 20, with the spot of light on the key at its smallest size. The spot was arranged to grow with the number of responses emitted as an added counter by the bird. The slit reached its full length at about 600 responses. In an interval containing a small number of responses, reinforcement occurred at a fairly small size of the slit.

FI 5 with added counter

One bird, which had shown a rather rough grain on FI 5, developed some interval curvature when the counter was added. The terminal rate was high, and as many as 700 or 800 responses might occur in each 5-minute interval. The bird was therefore frequently reinforced with the spot at its largest size. The bird did not develop further curvature during 15 hours (extending over 7 sessions) on FI 5 with counter. The size of the slit had nevertheless developed a strong control. Upon changing to the slit that was stationary when at large, a high rate of responding was maintained for the balance of the session. More than 10,000 responses occur without any reduction in this rate. This performance is shown in Fig. 355. The first 3 excursions show a fair sample of the performance with the aid of a counter on FI 5. The main effect of with-

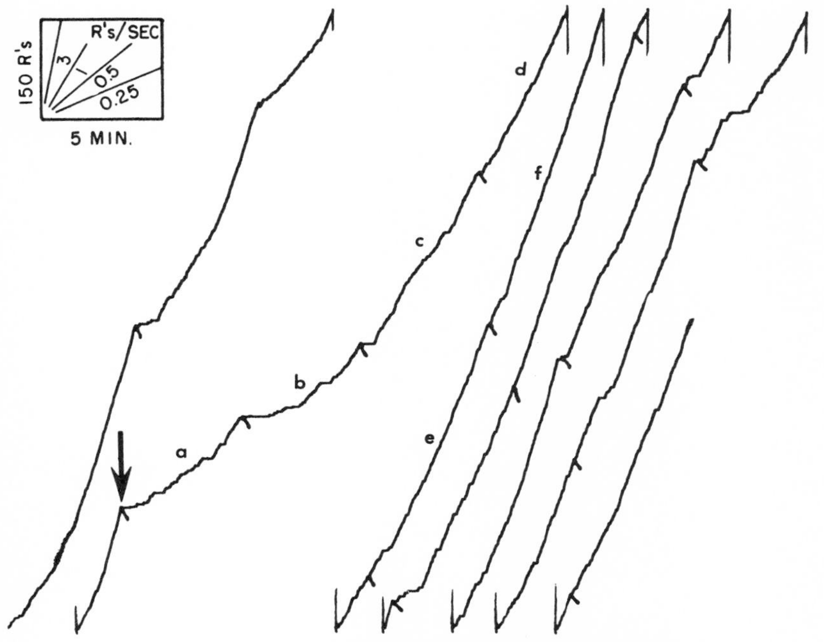

Fig. 356. Transition from counter to unoptimal setting on FI 5

drawing the counter is to eliminate the pause after reinforcement and the intermediate stage of less than the highest terminal rate. However, the other 2 birds at longer FIs showed a very rapid development of the effect of the counter.

As was usual with this bird at this value, marked scalloping occurred but no pronounced pausing after reinforcement. The small size of the counter controls a low rate; but as soon as the spot has begun to grow, the rate steadily increases. At the arrow in Fig. 356 the counter was stopped at small, that is, at the size farthest from the reinforced size. Some scalloping develops in the interval which follows, but the rate remains very much below the terminal rate prevailing with the counter. In other words, some evidence exists here that other factors besides the spot increase the rate during the interval (*a*), but it is slight. The 2nd interval (*b*) also shows curvature, but again the terminal rate is not high. The over-all rate then advances without very much interval curvature for the next 4 intervals (*c, d, e,* and *f*).

Figure 357 shows a later performance on FI 5 + counter. Although the rate following reinforcement is markedly depressed, the prolonged pauses produced by an added clock are lacking. Pauses are also common in short intervals without TO (cf. Fig. 152 for FI 4 after comparable exposure to the schedule). The number of responses emitted per interval is considerably larger with the help of the counter. One interval at *a* shows a low count partly as the result of rough grain; but this performance is not followed by more rapid starting in the subsequent intervals. The production of count acting as a reinforcement appears to get the bird off to an earlier start at a higher rate and results in a larger number of responses per interval.

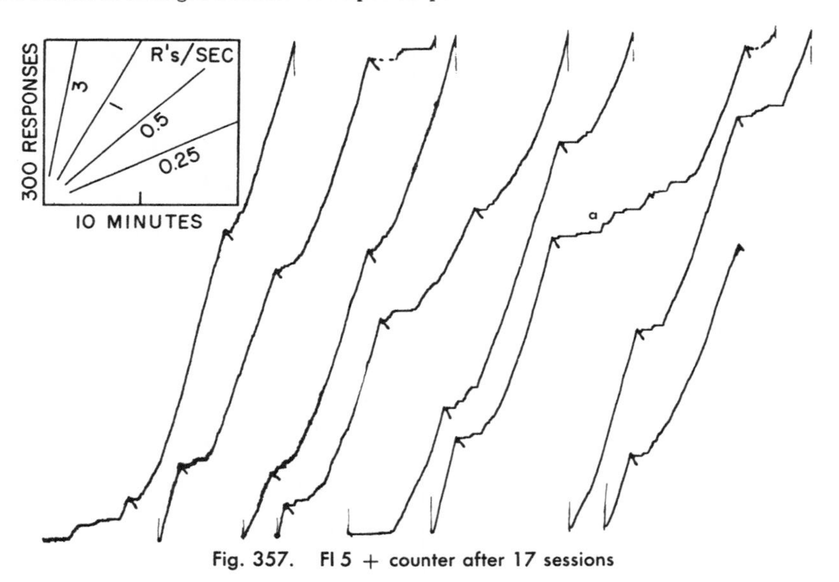

Fig. 357. FI 5 + counter after 17 sessions

FI 10 with added counter

The bird chosen for the study of FI 10 with counter gave the performance on FI 10 alone shown in Fig. 358A. A short pause occurs after reinforcement, but the terminal rate is rapidly reached. The counter was added at the beginning of the session shown at Record B. The first few segments show a progressive increase of slow responding after reinforcement (*a, b,* and *c*). By the 6th interval (at *d*) a well-marked scallop has set in.

The beginning of Fig. 359 shows a later performance on FI 10 + counter. Substantial scalloping may occur. The setting of the counter just after reinforcement is very different from that at most reinforcements; this negative control evidently cancels the reinforcing effect of the production of count. Note that an occasional reinforce-

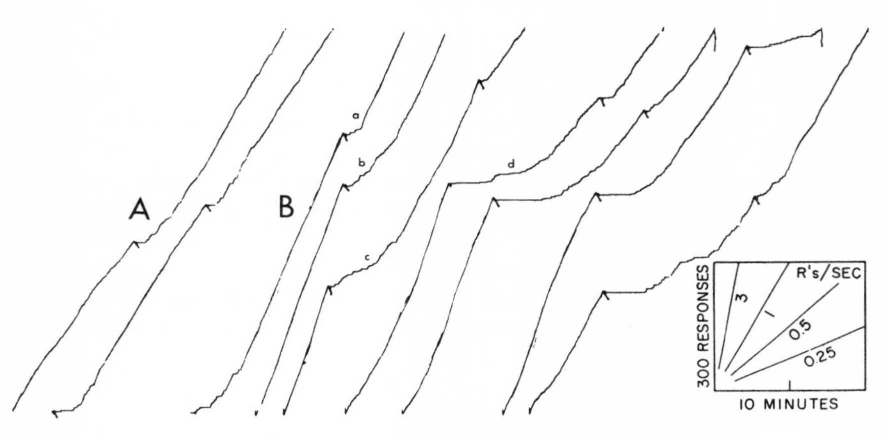

Fig. 358. First effect of the added counter on FI 10

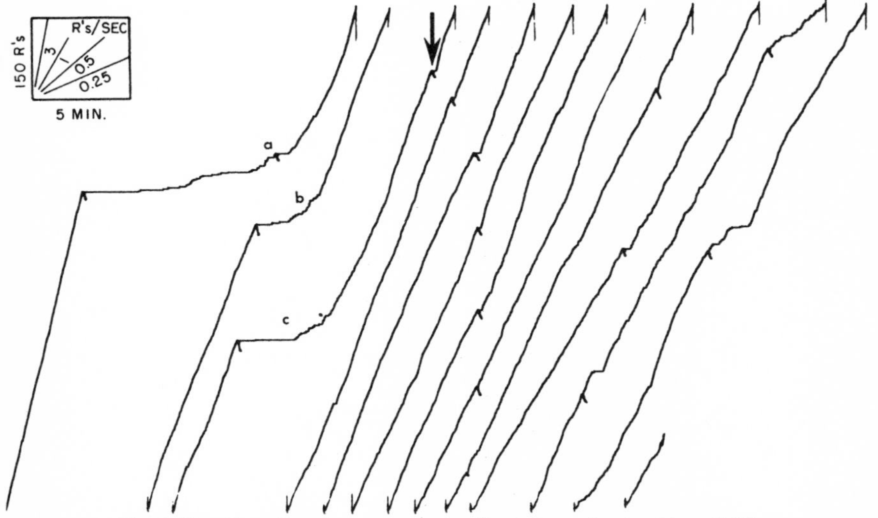

Fig. 359. Transition from counter to slit at the optimal setting: FI 10

ment at a low count (*a*) has the immediate effect of making responding at a low counter reading more probable, and that substantial pausing develops again only during a set of segments showing progressive scalloping (*a, b,* and *c*).

At the arrow the counter did not reset. The high reading controls a high rate for most of the balance of the interval (cf. Fig. 355).

A performance after 95 hours of FI 10 with added counter is shown in Fig. 360, where the segments have been arranged according to number and curvature. A later performance after 115 hours is shown in the order of recording in Fig. 361. In both figures the low reading of the counter after reinforcement delays responding in opposition to any reinforcing effect of the production of count. The curves are, however, by no means as square as with an added clock. Figure 164 shows the same bird 2 months earlier on FI 8. Figure 166 shows another bird on FI 10. Both of these are without TO and without added clock or count. Insofar as these provide a basis for comparison, the effect of the added counter appears to be to deepen the scallop. This

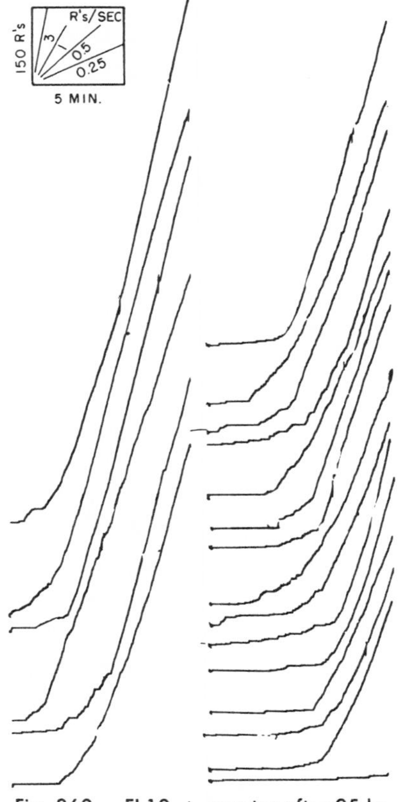

Fig. 360. FI 10 + counter after 95 hr

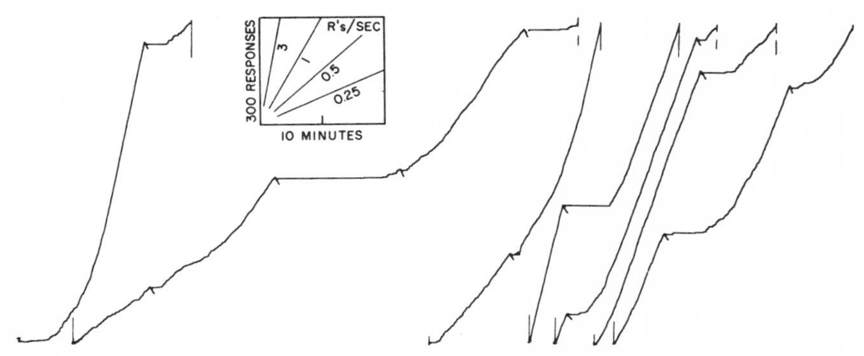

Fig. 361. FI 10 + counter after 115 hr

effect resembles that of a TO. Both provide an anchor point at the beginning of the interval; apart from any change in the size of the slit during the interval, one conspicuous result of a counter is that the interval begins under a stimulus (the smallest size of the spot) which is absent at reinforcement. This should have the effect of eliminating second-order effects in addition to favoring a pause and scalloping after reinforcement. Note, however, that some instances of second-order effects continue to occur under FI with added counter.

The bird in the experiment just described was accidentally reinforced on FI 20 with added counter instead of FI 10 during 1 session (2 sessions before Fig. 359). As the effect of the larger FI, almost 2000 responses were emitted in the 1st interval. The counter, of course, reached its maximal growth-size at about 600 responses. The interval segments in Fig. 362 appear in the order in which they occurred. The segment at *a* shows the beginning curvature previously developed under FI 10 + counter, although the spot is growing only half as fast as before. When the spot reaches maximal size, the terminal rate continues, as in experiments on the transition to a count stationary at maximal size. When the count resets to zero at the 1st reinforcement, a similar performance follows (Segment 2); but the 3rd segment shows some falling off (at *b*) after the spot reaches a substantial size. This performance is somewhat more clearly marked in the 5th interval, at *c*. Similar negative accelerations appear later, particularly at *d* and *e*, and the over-all rate falls off to a value somewhat more appropriate to FI 20.

FI 20 with added counter

The bird assigned to FI 20 + counter had had the longer history mentioned above, including a prolonged exposure to VI. On FI 20 alone, it showed a low sustained rate suggesting a VI performance. Figure 363 shows the 1st day with added counter. The tendency for the spot to control the rate is almost immediate, beginning at the 2nd interval, at *a*; it becomes quite marked before the session has advanced very far.

A later performance (after 28 hours) is presented in Fig. 364, where segments have

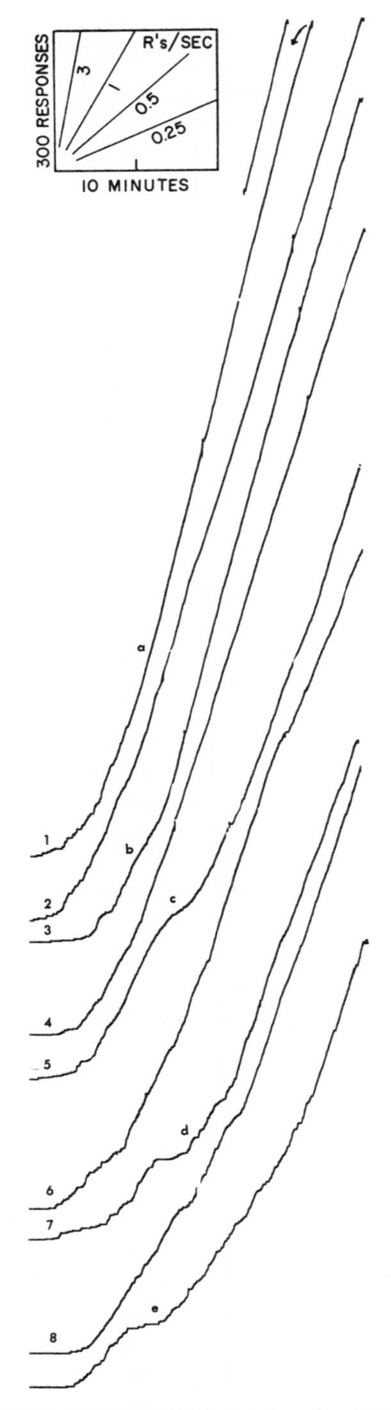

Fig. 362. First session on FI 20 + counter after FI 10 + counter

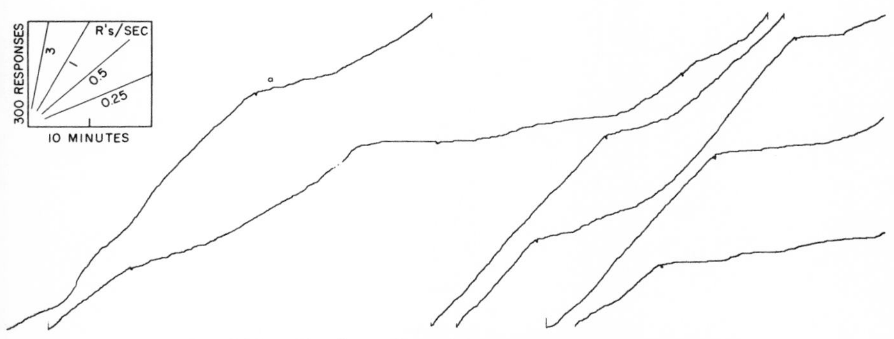

Fig. 363. The first effect of the added counter on FI 20

been arranged in order of the number of responses. Only 1 instance occurs in which the terminal rate is reached during the first 5 minutes, and only 1 instance in which no marked acceleration exists during the interval. In many of these intervals, however, the terminal rate is never reached, and most of them show a fairly sustained slow acceleration. A conspicuous example of a knee exists at *a*; but examples such as this are rare compared with performances with added clock. The prolonged pausing after reinforcement produced by a clock is almost entirely lacking. (Compare Fig. 350 and 352 for a comparable FI with a clock.)

We also checked the control exerted by the maximal setting of the added counter at this value of FI by allowing the counter to remain maximal throughout the latter part of a session. This was done at the arrow in Fig. 365; the result resembled those of Fig. 355 and 359 except for the greater decline in rate produced by the much less frequent reinforcement at FI 20.

A later performance at FI 20 + counter is shown in Fig. 366, which should be compared with Fig. 364. The acceleration is sharpened in Fig. 366. A straight line drawn from *a* to *b* would pass close to the terminations of the periods of low responding after reinforcement.

A second added counter

In one experiment on FI with added counter the counter completed the change from small to large during a fixed number of responses which was usually less than the number generated by the schedule. Under these circumstances the first part of the interval was supplemented by a counter, while the last part occurred with the counter stationary at its largest size.

In the early stages of the experiment, all 3 birds tend to respond at a higher rate while the counter is operating, and to drop to a lower rate shortly before or when it has completed its cycle. Figure 367 shows part of 1 session. The schedule was FI 10, and the counter completed the cycle in 240 responses. Each reinforcement is followed by a substantial pause or period of slow responding, which is in turn followed by a

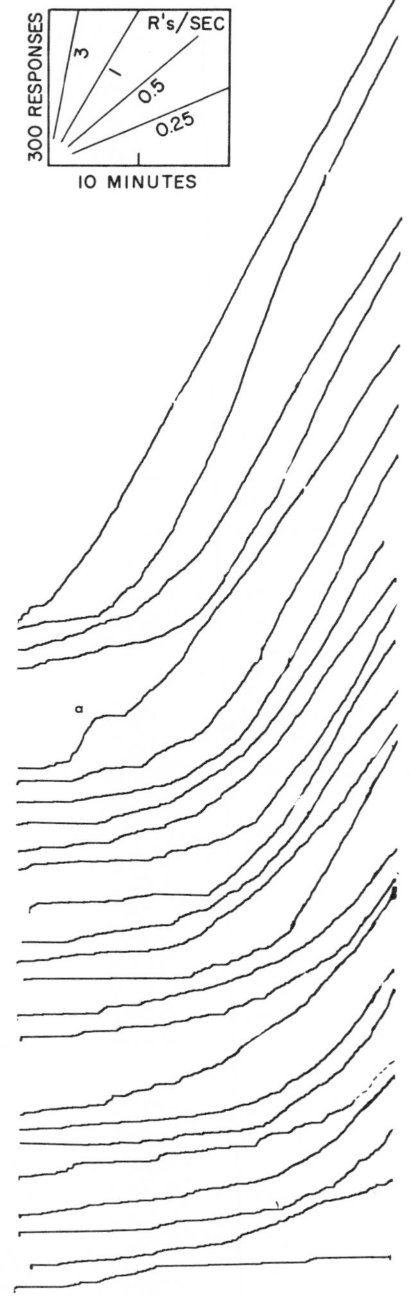

Fig. 364. FI 20 + counter after 28 hr

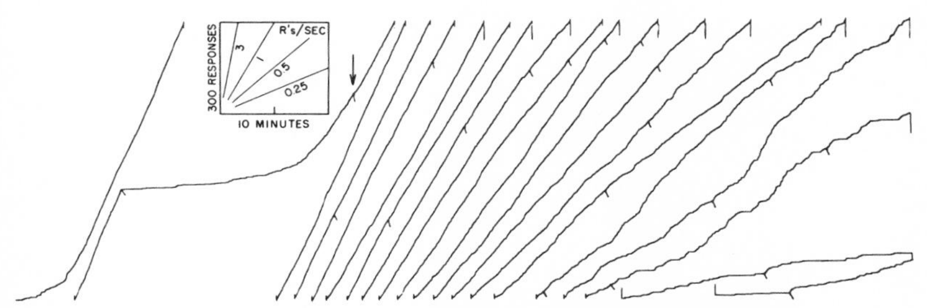

Fig. 365. Transition from counter to slit at optimal setting FI 20

relatively high rate, for from 100 to 250 responses. Thereupon, the rate drops to a fairly stable intermediate value for the rest of the interval. Thus, at this stage the increase in the counter reading is probably reinforcing and produces a high rate; but the counter near or at its largest size probably controls only an intermediate rate, essentially as in an FI 10 alone.

This stage comes and goes and eventually disappears. Figure 368 shows a later performance for another bird after 105 hours on FI 10 with this partial counter. The counter reaches its maximal count in 240 responses. Most of the intervals shown contain more than 240 responses, but no consistent tendency exists for a drop to a lower rate after the counter has ceased to tally. Here, the terminal rate is much higher than that in Fig. 367, and the over-all curvature suggests an FI + clock.

At an earlier stage in the experiment the bird responsible for Fig. 367 was accidentally reinforced on a fixed ratio of 120 for 8 reinforcements. The counter, which completed its count in 120 responses, was present. Figure 369A shows the result. The repeated reinforcement on FR 120 produced a quite uniform acceleration, because of an increase in the running rate which reaches a high value at *a,* the reduction of the pause after reinforcement, and the speeding up of the acceleration to the terminal rate. The apparatus was then corrected, and Record B was made a few minutes later in FI 10 + partial counter. Record B begins with the performance just developed on FR 120 with added counter, and the early high terminal rate drops only slowly from the maximal value at *b* to the substantially lower value at *c,* just before reinforcement. The performance in the 2nd interval starts early and sustains a somewhat lower terminal rate until the 2nd-interval reinforcement at *d.* Thereafter, the performance is essentially similar to earlier performances at FI 10 with a partial counter (cf. Fig. 368).

Fading counter

In speculating about the function of the bird's own behavior as a counter, we assume that the counter resets to zero at reinforcement. This is unlikely unless a substantial TO follows reinforcement. Also, some *continuous* resetting probably goes on, in the sense that a given bit of behavior which may serve as a counter becomes more and

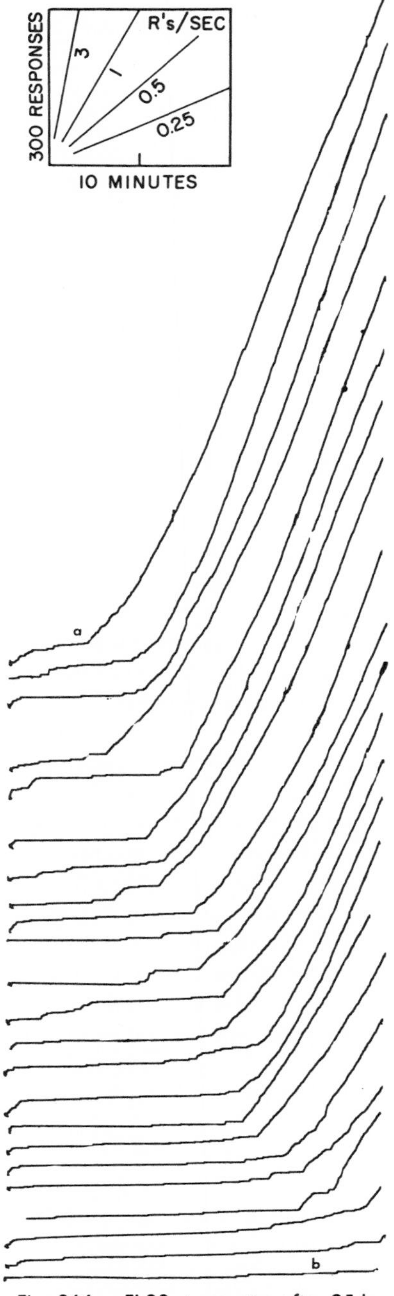

Fig. 366. FI 20 + counter after 95 hr

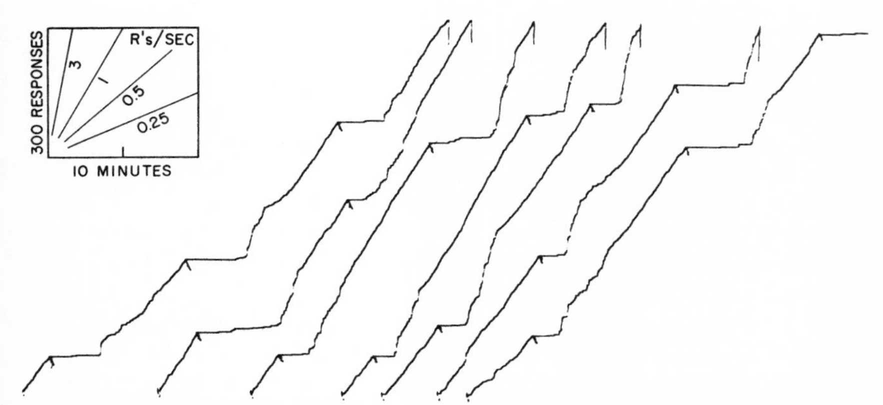

Fig. 367. Early performance on FI 10 + counter reaching maximum size after 240 responses

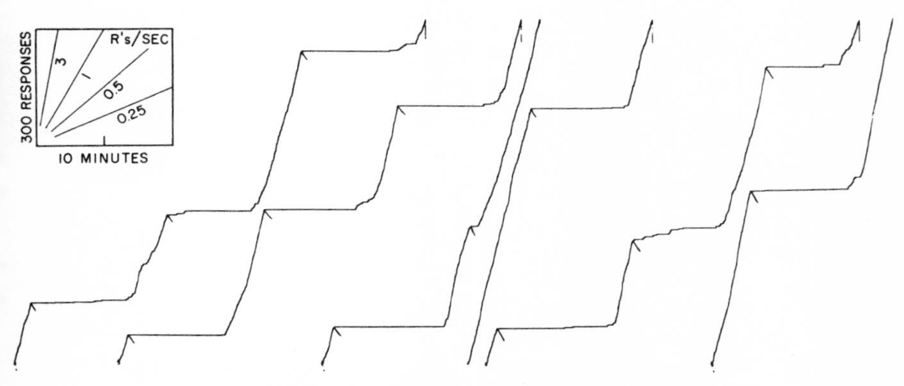

Fig. 368. Final FI 10 + counter (second bird)

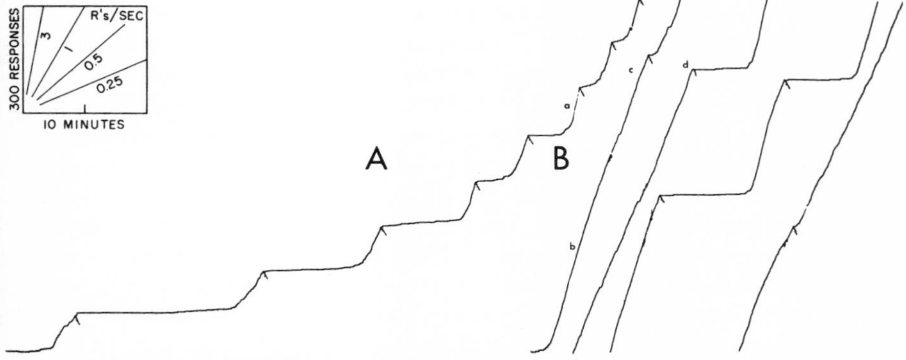

Fig. 369. Eight reinforcements on FR 120 + counter during FI 10 + counter

more remote in time. In a rough attempt to duplicate this characteristic, we have used a "fading" counter. A differential gear controls the slit in such a way that the bird's own responses increase the slit size while a clock slowly reduces it. Values were chosen so that at a rate of 1 response per second the 2 changes were equal (and opposite). Rates above this value would cause a growth of the spot but a slower growth than with a nonfading counter, while rates below this value would permit the spot to shrink slowly toward the smallest size. The counter would reset to zero if the pause were long enough. Although we did not make any exhaustive study of the various properties of such a counter, some evidence exists of its effectiveness.

FI 20. The fading counter was introduced at the beginning of an experimental session after 104 hours on FI 20 with the usual nonfading counter. Figure 370 shows the first result. In the 1st interval the bird quickly reaches a rate (at *a*) at which the counter grows, although more slowly than heretofore. The slit has probably reached its full size by the time of the 1st reinforcement (at *b*). The counter was always re-

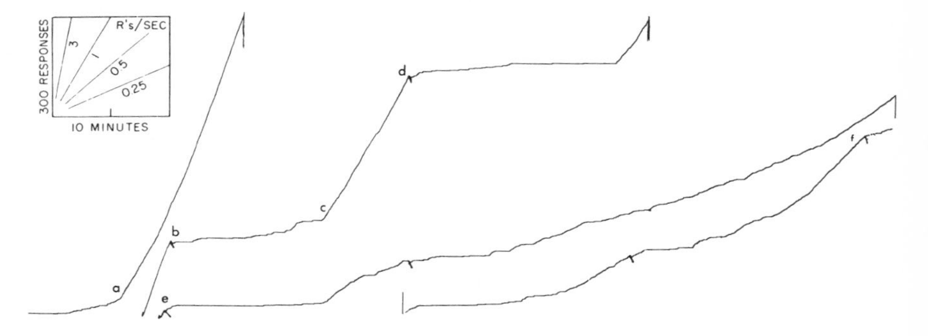

Fig. 370. First exposure to FI 20 + fading counter after FI 20 + counter

set to zero at reinforcement. In the 2nd interval a rate which is just at the critical value is assumed relatively late in the interval (beginning at *c*) and maintained until reinforcement (at *d*). The spot probably did not grow very much during this interval. The following interval shows an even greater delay in reaching a substantial rate, and again the rate is roughly the value at which the spot remained stationary. Whether the counter had any substantial reading at the following reinforcement at *e* is doubtful. Thereafter, the rate was clearly below the critical value, and the performance is similar to that at which the counter is replaced by a setting most unlike that at previous reinforcements. In other words, from point *e* on, the performance would have been the same if instead of using the fading counter, we had simply set the counter stationary at the smallest-size spot. Good interval scallops are shown later in this experimental session, but the critical rate is not reached, except possibly for the end of 1 interval (at *f*). This performance shows the gradual development of a scallop without the help of the counter.

The critical rate was not exceeded in the next 2 sessions of 12 hours each. The rate rose above the critical value in the 2nd interval of the 3rd experimental session, however, and frequently reached that value thereafter. The rate tends to be either substantially above the critical value at which the counter is tallying, or below it, in which case the bird is essentially operating on FI with counter at its minimal setting. Toward the end of this 3rd session the lower value generally prevails, and a great deal of running-through results, followed by marked periods of low rate. In the following session there is again an alternation between rates significantly above the critical rate and very uniform just below it. A later performance for this bird under FI 20 with fading counter is shown in Fig. 371, for the first part of a session beginning 60 hours after the introduction of the counter, and Fig. 372, for the end of the same session. At

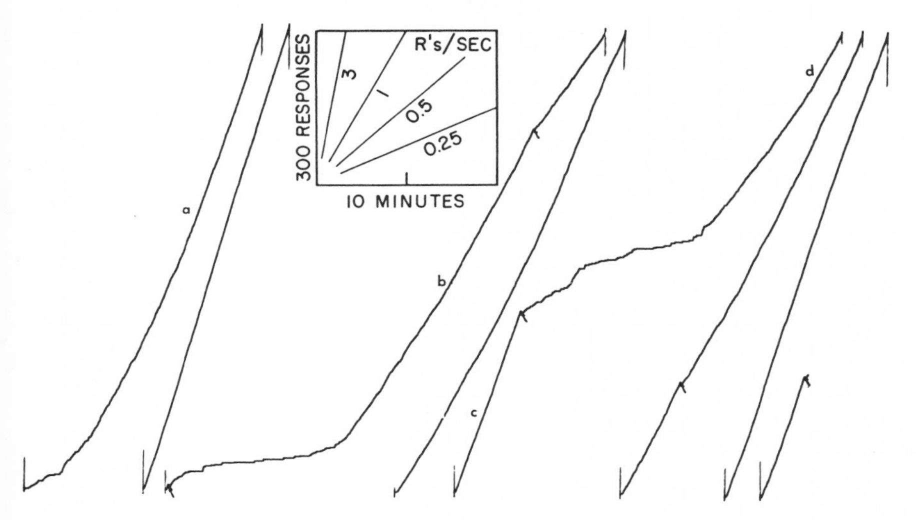

Fig. 371. FI 20 + fading counter after 60 hr

the beginning of the session (Fig. 371), several instances occur of rates which exceed the critical value as during the greater part of the 1st interval, at *a*. The slope at *b* is almost precisely the critical value which maintains the spot, although some growing may have occurred. A clear growth of the spot is indicated at *c*. In the following interval the critical rate is reached at approximately point *d*, and the last interval shown in the figure was run off at a rate considerably above the critical rate throughout the interval. Toward the end of the session, however, as Fig. 372 shows, no instances occur of sustained running above the critical rate, and only occasional instances (as at *a* and *d*) at which the rate that just maintains the size of the spot is approximated. At *b* and *c*, sustained runs occur at lower than the critical value. These intervals must therefore be regarded as FI without benefit of counter. They show well-marked running-through after reinforcement, but long periods of no responding or very slow responding before the assumption of the intermediate terminal rate.

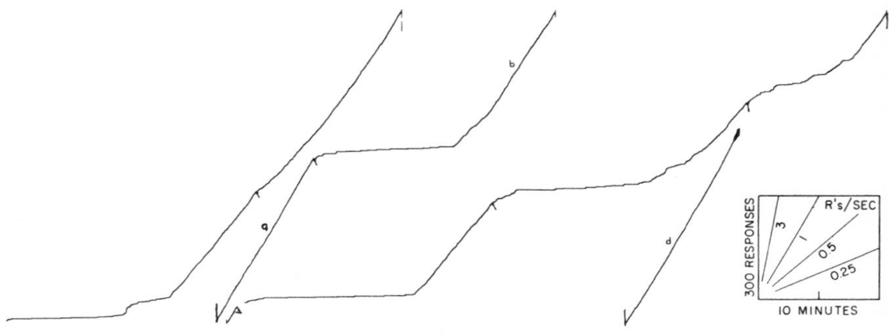

Fig. 372. End of the session begun in Fig. 371

Figure 373 shows the result of removing the fading element 62 hours after it was introduced. The first 2 intervals show the performance at this stage with the fading counter. In the 3rd interval the fading element was removed at the arrow; the immediate effect was to increase the rate to a much higher terminal value (as at *a*). The terminal rate was then maintained throughout the period. The bird's performance with a nonfading counter now resembles Fig. 366.

FI 10. Figure 374 shows another example of the first effect of a fading counter on FI 10. This bird had been giving a well-marked scallop on FI 10 + counter. (See Fig. 361 for a late stage of this performance, and Fig. 360 for an earlier stage. Another sample for this bird occurs at the beginning of Fig. 359.) In Fig. 374 the effect of the fading counter is clearly similar to that when the counter remains small (cf. Fig. 356 for another bird). When the 1st responses fail to increase the size of the spot because they are below the critical rate of 1 per second, the small spot continues to control a low rate, and the earlier acceleration characteristic of this bird (as at *a*) fails to materialize. Only during the 3rd interval, approximately at point *b*, is a rate reached which increases the size of the spot. This rate leads to nearly the characteristic terminal rate at *c*. The next interval shows a very rapid adjustment to this new

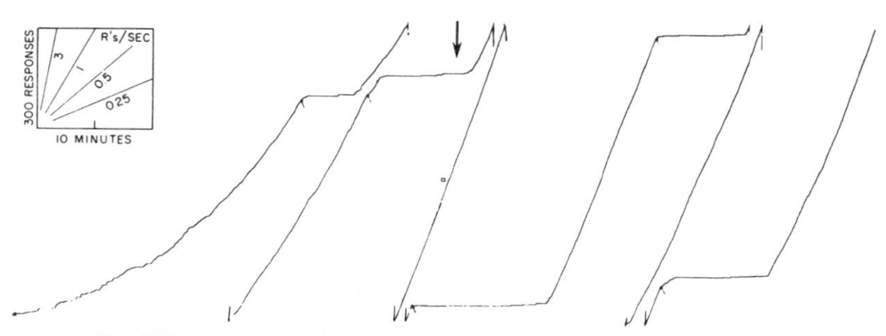

Fig. 373. Transition from FI 20 + fading counter to FI 20 + counter

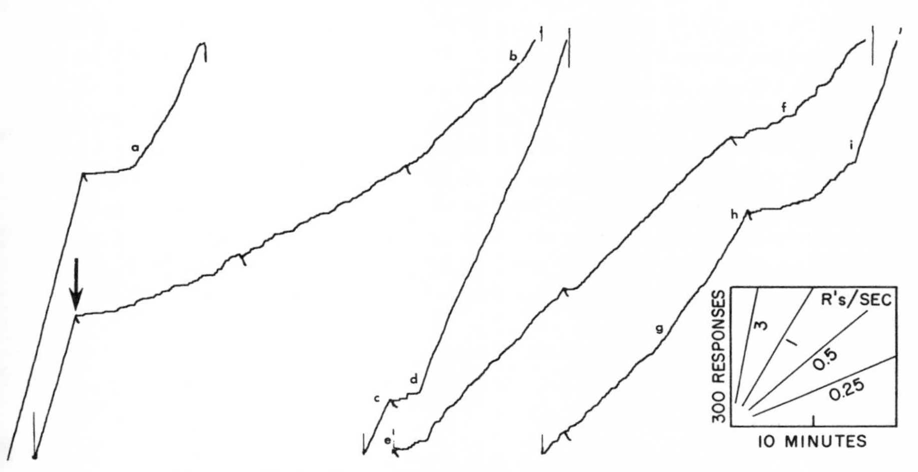

Fig. 374. Transition from FI 10 + counter to FI 10 + fading counter

condition. At *d* a high rate of responding is quickly assumed which rapidly increases the size of the slit. This in turn continues to control a high rate of responding throughout the interval to the reinforcement at *e*. The following 3 intervals, however, show rates which are only very briefly above the critical rate of 1 per second, and the effect is essentially that of FI 10 without counter. Some curvature exists in the 3rd of these intervals (at *f*), and brief runs at this point exceed the critical value. But the value just able to cause the spot to grow is reached again only at *g*. It is barely maintained until reinforcement at *h*. An almost immediate advance occurs at *i* to a rate well above the critical value.

A nonfading counter was then used for 2 sessions, and the bird was then exposed to the fading counter again. Unfortunately, the record for the session was badly re-

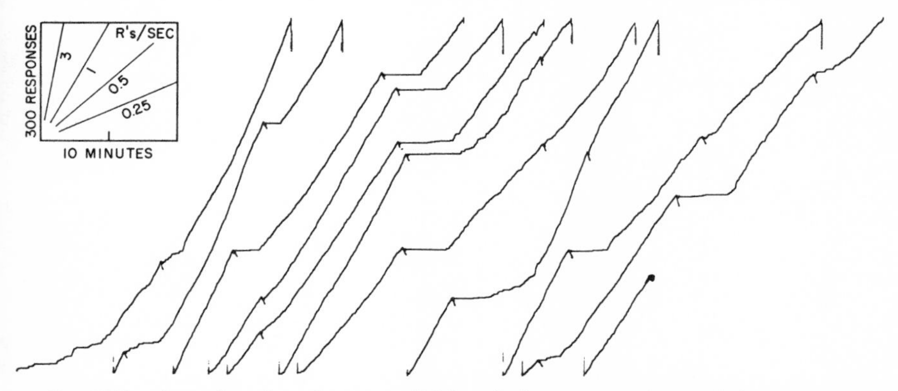

Fig. 375. Second session of return to FI 10 + fading counter after nonfading counter

corded; but it shows a quick return to a performance which was maintained during the complete 2nd session shown in Fig. 375. The terminal rate tends to exceed the critical value. Responding may continue at a rate below the critical value, however, and hence essentially without benefit of counter. Compared with the set of segments in Fig. 360, the segments in Fig. 375 show 2 values of terminal rate, one above and one below the rate sufficient to increase the counter. The terminal rate is assumed more abruptly after the pause following reinforcement. The beginning of Fig. 376 shows a late performance with a fading counter. At the arrow the change to a nonfading counter produces generally higher terminal rates, which are partly reflected in the large numbers of responses per interval. Before the end of the session the terminal rate tends to fall somewhat.

FI 5. The beginning of Fig. 377A shows a performance after 8 sessions of 2 hours each on FI 5 with fading counter. At this size of interval the effect is quite variable,

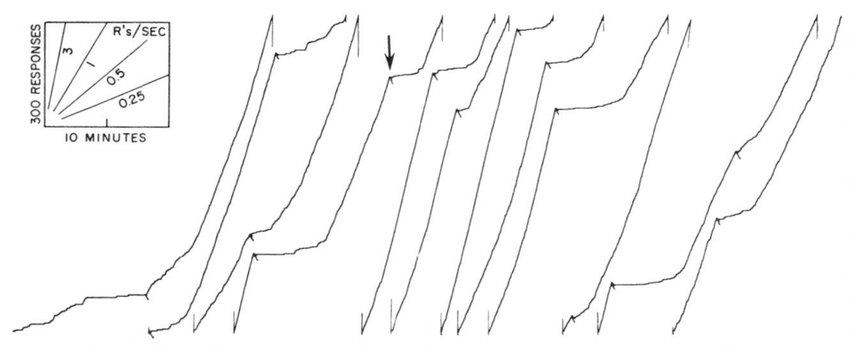

Fig. 376. Final performance on the fading counter and the transition to the nonfading counter: FI 10

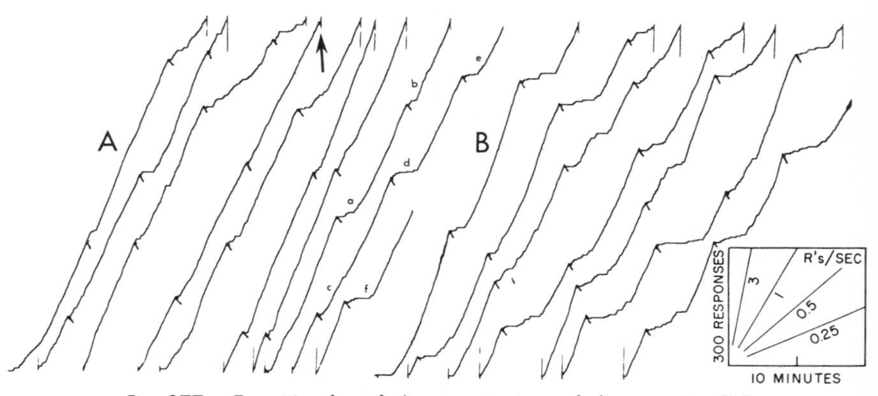

Fig. 377. Transition from fading counter to nonfading counter: FI 5

the grain is rough, pauses after reinforcement may or may not occur, and the terminal rate is of a moderate-to-high value. At the arrow the fading component was removed. Low rates after reinforcement begin to appear; scalloping has become consistent by the end of the session (as at *a, b, c, d, e,* and *f*). An even better interval curvature develops on the following day in the absence of the fading component. Record B represents the end of this 2nd session. The over-all rate is somewhat less here than in Fig. 357 for the same bird, which shows an earlier performance after 58 hours with the usual counter.

THE EFFECT OF A NOVEL STIMULUS ON FI 20

A performance after extended exposure to FI 39 has been reproduced in Fig. 196. The bird was showing a fairly stable performance, with well-marked scallops and a

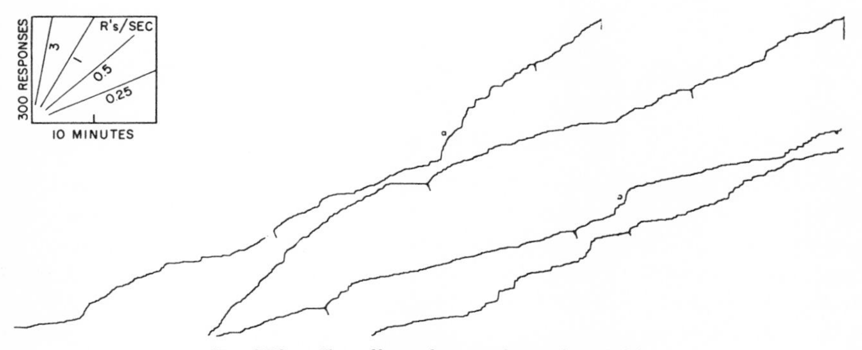

Fig. 378. The effect of a novel stimulus: FI 20

fairly regular acceleration to a terminal rate. In preparation for an experiment with added counter, an opaque mask was placed over the key, so that only a small slit of light was exposed. At the same time, the schedule was changed to FI 20. Figure 378 shows the effect of the novel stimulus. The fact that almost every interval has marked negative curvature suggests the early stages of crf to FI. Occasional breakthroughs occur, as at *a* and *b,* at higher rates. The grain is unusually rough. During the remainder of this session, the performance was essentially linear, with some slight continuing negative curvature leading to a low rate just before reinforcement. This rate suggests the 2nd phase in the transition from the crf. The 2nd session (Fig. 379) maintains the parallel with the transition to FI after crf, with an over-all linear performance and a slight tendency to respond faster immediately after reinforcement. This constant rate of responding occurs despite a long history in which the bird showed a marked scallop.

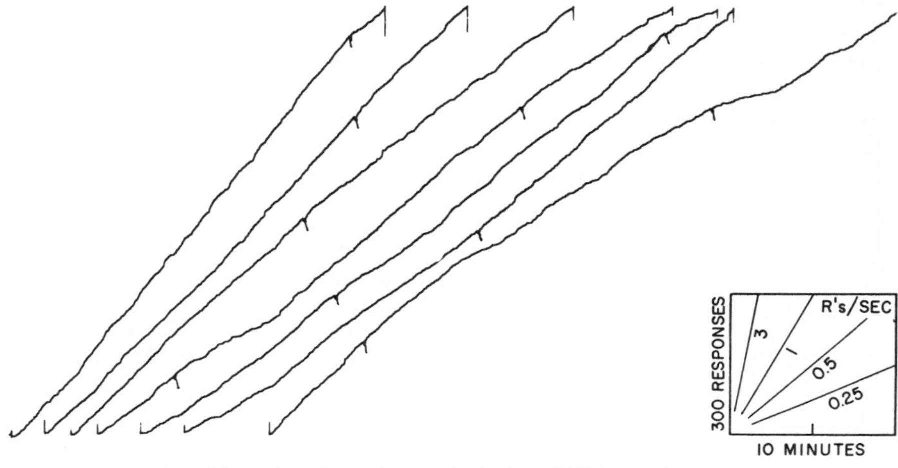

Fig. 379. The effect of a novel stimulus: FI 20 (second session)

SATIATION UNDER FI 2

Figure 380 is a plot of responses emitted during an 18-hour session, when a bird received over 500 reinforcements on FI 2. The extent of this degree of satiation is suggested by the fact that 60 reinforcements maintained the bird at approximately its experimental weight. The over-all curve is negatively accelerated over the first 9 hours, with very few deviations from an orderly decline in rate. For the last 9 hours of the session, it is roughly constant at about 0.35 response per second. Figures 220 and 221 showed this bird's transition to FI 2 TO 5, following a history of 1300 reinforcements on FI. The satiation curve was taken without TO after reinforcement during the session following Fig. 221, after a history of 450 reinforcements with TO. Details of Fig. 380 are presented in Fig. 381. A segment has been taken from the larger curve every 63 minutes. The top segment, after 40 reinforcements, shows a performance similar to the previous session, in Fig. 221. As satiation proceeds, however, the effect is an increase in the pause following the reinforcement and the extent of the curvature. By the 4th segment, strong negative acceleration occurs. From Records E to H the over-all rate of responding falls sharply, largely because of a decline in the terminal rates. By Record H, intervals occur in which the grain is very rough and a terminal rate is not reached. Toward the end of the session, however, the few responses per interval occur toward the end of the interval, and most reinforcements are received almost as soon as scheduled.

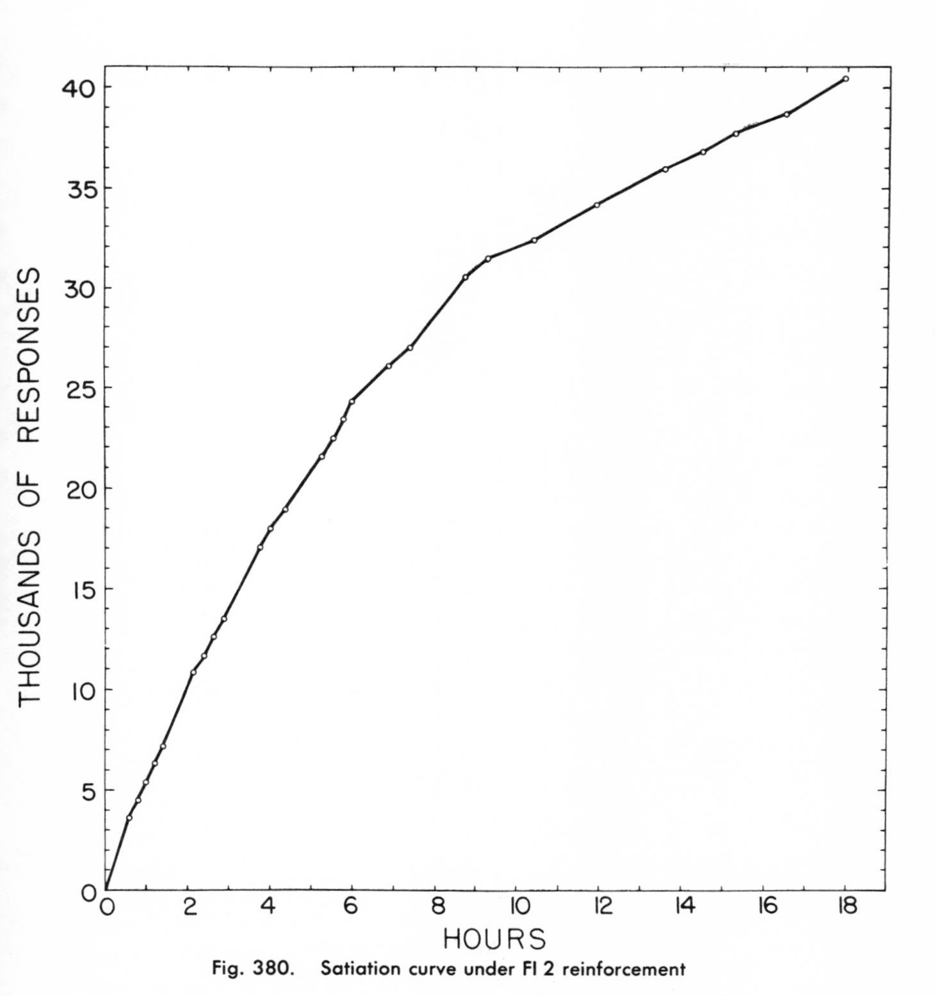

Fig. 380. Satiation curve under FI 2 reinforcement

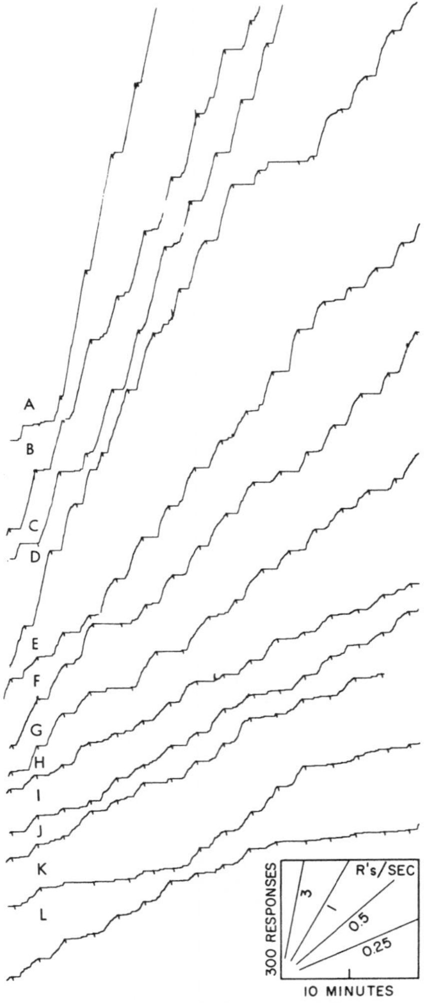

Fig. 381. Segments excerpted from a satiation curve under FI 2 reinforcement

THE EFFECT OF A LOSS OF BRAIN TISSUE ON FI

We studied FI performances in 2 birds after various parts of the brain were removed. (See p. 85.) The records of these birds are presented merely as examples of the kinds of disruptions that can occur in FI performances because of brain lesions but not particular lesions. The final performance after a long history of multiple schedules, the last of which was mult FI 15 tand FI 15 FR 10, was similar to that of another bird, shown in Fig. 667 in Chapter Ten. The bird was then submitted to an operation, the postmortem result of which was described as a "cyst in forward part of brain 9 mm in depth by 3 mm diameter extending to the right cerebral hemisphere . . . marked loss of tissue with evidences of trauma to the other half."

One month later it was run on FI 15. The key was green—the color present during FI 15 on the multiple schedule. Figure 382 shows the 1st session. The standard FI performance is severely disrupted; the over-all rate is low; the grain is irregular; and many intervals show negative curvature. Figure 383 shows a session beginning 20 hours later. Segments are arranged in order, beginning at the top. A high rate of responding occurs for about 1 hour in the middle of the session, but the over-all rate is otherwise low, with rough grain and the occasional brief emergence of the terminal rate at any time during the interval (as at *a* and *b*). The next sessions showed only sporadic responding, with many intervals containing only reinforced responses. During the next 4 sessions, the bird was tested on mult FI 15 FI 60; but the over-all level of responding was so near zero that the multiple schedule was discontinued by a change to FI 1 at the end of the 4th session. Figure 384 shows the result. The record begins at the low rate from the previous mult FI 15 FI 60 schedule, and the 1st response on the new schedule is reinforced at *a*. After 7 reinforcements on FI 1, a substantial rate develops (at *b*). Some oscillation occurs in the over-all rate (as at *c*), and a long segment (at *d*) shows an intermediate rate. Before the end of the session, pauses and lower rates follow reinforcement, and most intervals show some positive acceleration.

In the 3rd and 6th sessions on FI 1, Fig. 385A and B, respectively, each reinforcement is followed by a substantial pause and slow acceleration. The over-all rate is considerably lower than normal. In spite of this very low over-all rate, reinforcements are not postponed; and a substantial terminal rate usually gives the effect of an extended scallop.

In the following session (Fig. 385C) the interval of reinforcement was increased to 5 minutes. The beginning of each interval resembles the performance in Records A

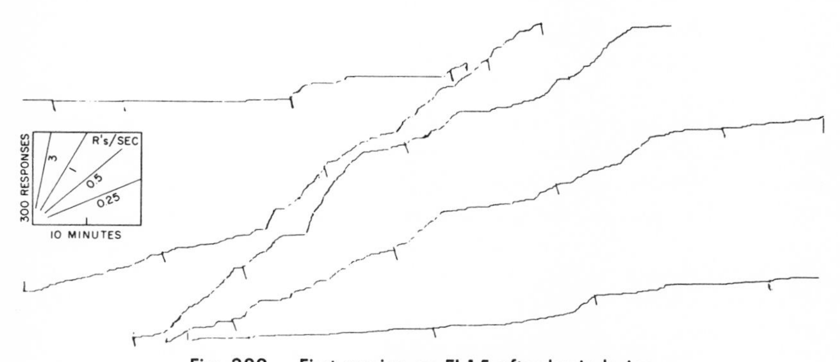

Fig. 382. First session on FI 15 after brain lesion

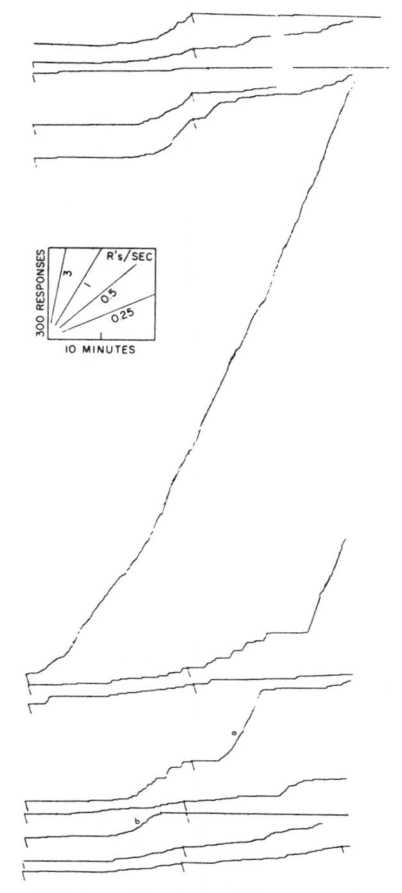

Fig. 383. Performance after 20 hr of FI 15 after brain lesion

and B; and the low terminal rate, once assumed, is maintained for most of the interval. Negative curvature sets in at *a* and *b*, however, and the over-all curve shows a decline in rate which nearly reaches zero before the end of the session (not shown in the figure). The bird could sustain a performance on FI 1, but it was unable to sustain a performance on FI 5. This performance is similar to that when smaller amounts of reinforcements are given, and is probably related to the "lack of motivation" that occurs as a result of brain damage by tumors. A second bird in the experiment confirmed this picture very closely.

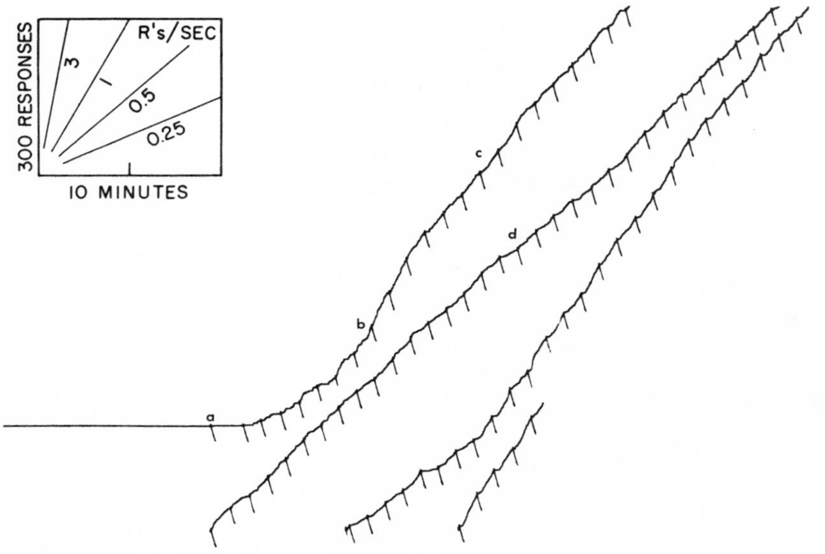

Fig. 384. Increase in the over-all rate on FI by reducing the interval to 1 min

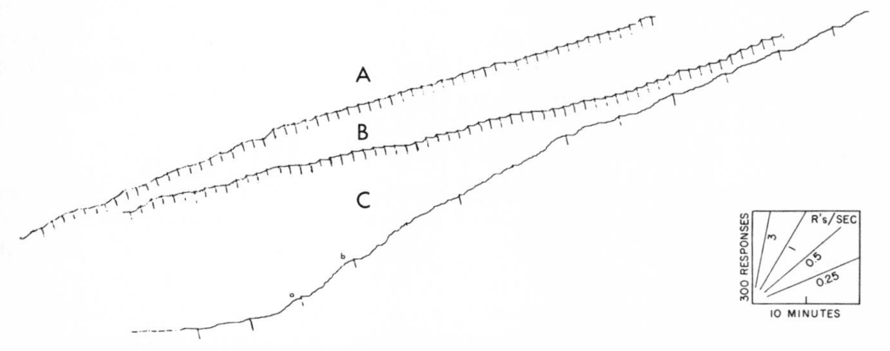

Fig. 385. Later performance on FI 1 and transition to FI 5 after brain lesion

Chapter Six

• • •

VARIABLE INTERVAL

INTRODUCTION

A VARIABLE-INTERVAL (VI) schedule is one in which the intervals between reinforcements vary in a random or nearly random order. The VI schedule is designed to produce a constant rate by not permitting any feature of the bird's behavior to acquire discriminative properties. In contrast, in fixed-interval and fixed-ratio schedules the fixed pattern establishes a correlation between behavior and reinforcement.

In some VI schedules, however, the intervals in the series are not arranged to produce a constant rate. As a limiting case, only 2 intervals of reinforcement occur in random alternation. (This schedule produces the effects discussed more generally under mixed FIFI.)

Since the VI reinforcements are determined by the passage of time, such a schedule involves the same differential reinforcement of low rates as FI. (See Introduction to Chapter Five.) Once a VI has generated a constant rate, the fact that reinforcements occur at one rate could establish this rate as a discriminative stimulus. There is no explicit differential reinforcement in respect to this rate, however.

Specification of the variable-interval schedule of reinforcement

The most important specification is the mean interval of reinforcement. Different distribution of intervals may produce the same mean interval, however, but still have diverse effects. The most frequently used distributions are arithmetic series, adjusted to give a slightly higher frequency of reinforcement at shorter intervals.

An arithmetic schedule is not specified completely until we state the largest and smallest intervals. Most VI schedules will have zero or near zero as the shortest interval of reinforcement, in order to prevent the development of a pause after reinforcement. Once the mean interval of reinforcement and the largest and smallest intervals have been specified, the size of the steps still has latitude. For example, one arithmetic VI schedule begins 0, 20, 40, 60, 80. . . . A schedule with similar mean and extremes could begin 0, 10, 20, 30. . . . The number of steps in a series and the number of series "scrambled" in a program are usually practical matters, limited by the

VI programmer and the length of the daily experimental session. In a geometric variable-interval schedule the schedule is specified by the mean interval of reinforcement, the shortest interval, and the progression for generating the intervals in the series.

The final step in constructing an actual variable-interval schedule is the randomization of several sets of the intervals. Strict randomization is not attempted because it always entails the probability of many short or long intervals in a row. Only with infinite experimental time could true randomization be used.

Development of a stable performance under variable-interval reinforcement

VI 1

In one experiment the schedule of reinforcement consisted of an arithmetic series of intervals ranging from 0 to 120 seconds in steps of 10 seconds. These were rearranged

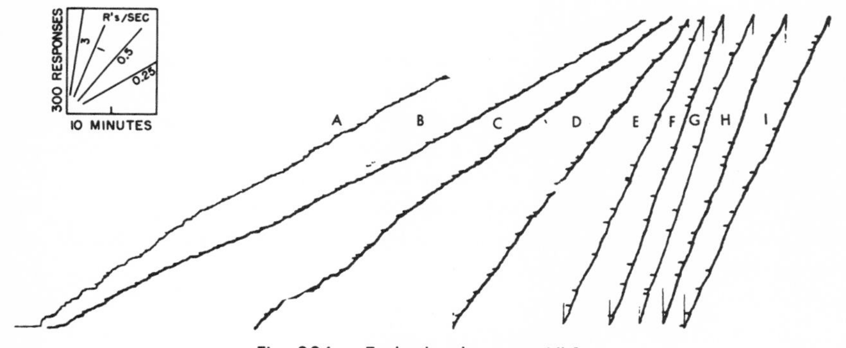

Fig. 386. Early development: VI 1

in "random" order. The effect of this schedule after crf is shown in Fig. 386, which contains the first 1000 responses of the 1st, 2nd, 4th, 6th, 9th, 12th, 14th, and 16th sessions after crf. Record A (the entire session following crf) shows slight negative curvature and an over-all rate of responding of approximately 0.4 response per second. The over-all rate falls to 0.3 response per second at the start of the 2nd session in Record B. Through Records C, D, E, and F, the over-all rate increases to 1.5 responses per second, which is roughly maintained during the remainder of the sessions shown in the figure. Except for slight oscillations, the final performance after 16 sessions is an approximately constant rate of responding with no rate change correlated with any feature of the schedule.

Figure 387 is a plot of the average rates of responding during the first 10 sessions on a 1-minute, variable-interval schedule after continuous reinforcement for 2 other birds on the same schedule as in Fig. 386.

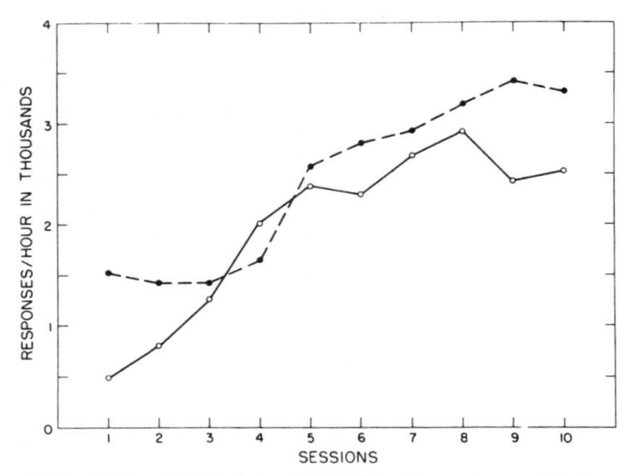

Fig. 387. Mean rate during first 10 sessions of VI 1

A bird which had received crf in a separate apparatus gave the curve shown in Fig. 388 when subjected for the first time to an arithmetic VI 1. The transition from crf to VI is disturbed by novel stimuli.

No responding occurs for the first 9 minutes. The first response, at *a,* is reinforced. Successive reinforcements occur at *b, c,* and elsewhere as marked in the usual way. A continuous acceleration in rate leads to an over-all rate of approximately 0.8 response per second, at *d,* which is maintained for the remainder of the session. Most of the positive acceleration represents a reduction in the disturbance caused by novel stimuli in the new apparatus.

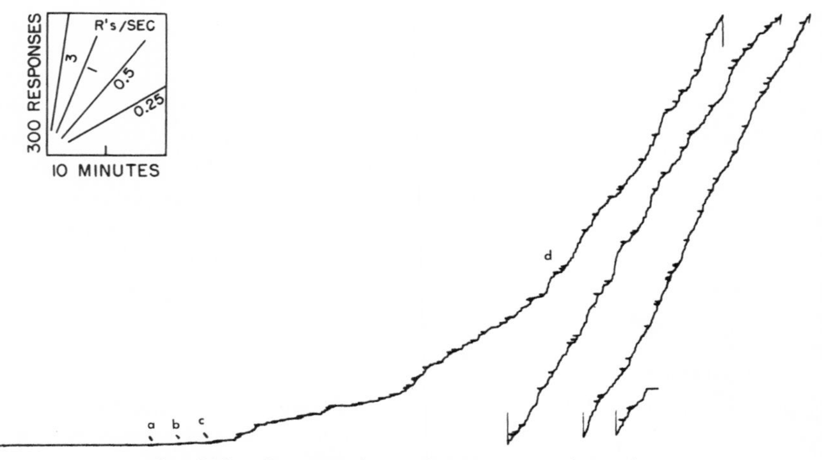

Fig. 388. Transition from crf to VI 1 in novel stimulus

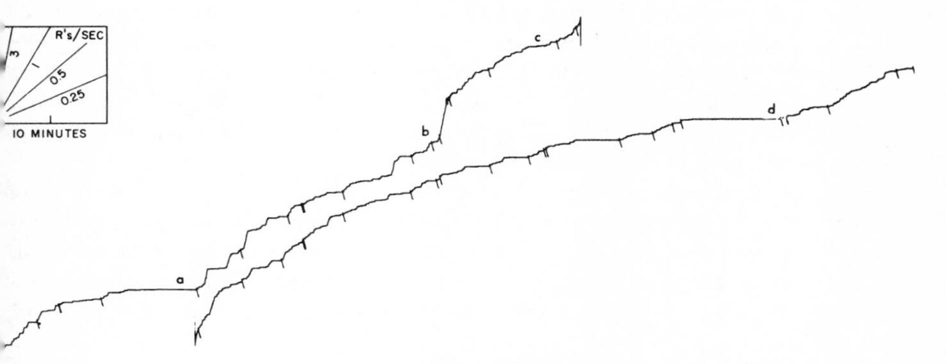

Fig. 389. Transition from crf to VI 2

VI 2

Figure 389 shows a transition from crf to VI 2. The schedule of reinforcement was composed of an arbitrary selection of intervals in the following order: 0, 30, 120, 60, 240, 230, 120, 10, 240, 180, 0, and 210 seconds. The over-all curve for the 1st session consists of 4 negatively accelerated portions ending at *a, b, c,* and *d.* The grain of the record is rough. Most of the responding occurs in bursts at rates much higher than the prevailing rate. The highest local rates of responding tend to occur after reinforcement. The over-all curve is also negatively accelerated, reaching a very low over-all rate by the end of the session. By this time, there are no consistent rate changes correlated with reinforcement.

Figure 390 shows the complete 2nd and 3rd sessions for this bird. The 2nd session (Record A) begins with a negatively accelerated portion (ending at *a*) similar to those in the 1st session. The rate then increases, and a 2nd negatively accelerated portion appears between *a* and *b*. The over-all rate of responding remains approximately 0.45

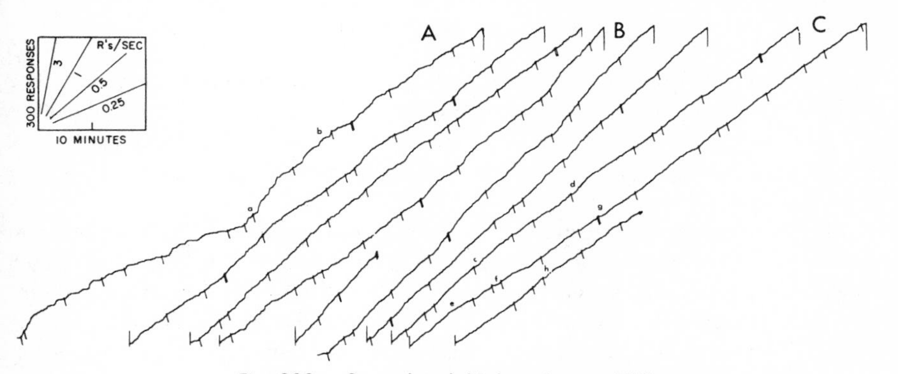

Fig. 390. Second and third sessions on VI 2

response per second for the remainder of the session, with rough grain, marked oscillation in local rates, and no consistent rate change correlated with reinforcement. The 3rd session (Record B) begins at approximately 0.6 response per second and declines gradually to slightly more than 0.4 response per second toward the end of the session. Beginning at *e* the over-all rate is fairly constant. A pause begins to appear after some reinforcements, as at *c, d, f, g,* and *h.*

The entire 4th session, shown in Fig. 391, begins with a very low rate of responding for the first 6 minutes which we cannot now account for. Following the 2nd reinforcement, however, a normal performance similar to Record B of Fig. 390 is reinstated. The bird responds at a roughly stable over-all rate thoughout the session, beginning at approximately 0.6 response per second and falling to approximately 0.4 by the end

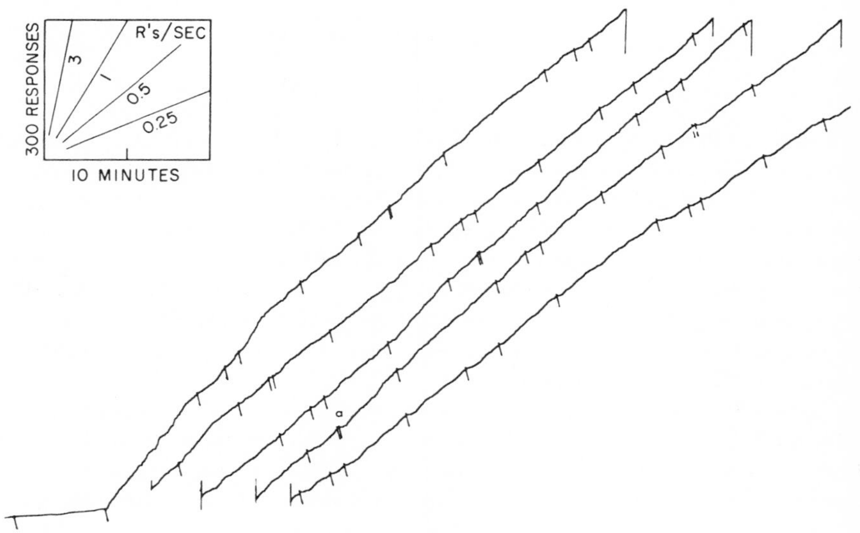

Fig. 391. Fourth session on VI 2

of the session. The tendency to respond slowly just after reinforcement becomes more marked, and beginning at *a,* every reinforcement is followed by a pause of the order of 10 seconds. This performance is almost certainly due to the abbreviated form of the variable-interval series, which does not provide for enough reinforcements after short intervals.

The development of a second VI 2 performance with the same series is shown in Fig. 392, where segments from various stages of the first 3 sessions have been chosen to show the character of the local rate changes. Record A, which shows the performance immediately following continuous reinforcement, should be compared with Fig. 389. The over-all curve is negatively accelerated, and reinforcements tend to be followed by higher rates of responding which fall off sharply to lower values. Reinforced

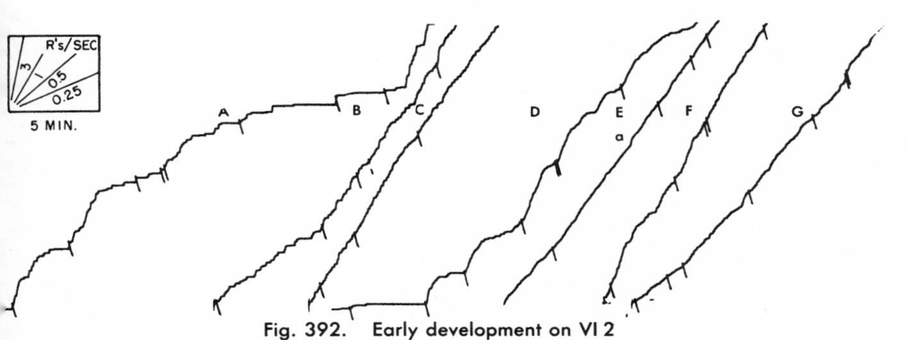

Fig. 392. Early development on VI 2

responses tend to occur after pauses. Record B shows a segment from the 2nd hour of the 1st session following crf. The over-all rate has increased, the grain of the record is rough, and the rate has a slight tendency to be higher immediately after the reinforcement. Record C contains the end of the 1st session, where the over-all rate has increased and the local rate is fairly constant except for the fine grain of the record. Record D is the start of the 2nd session after crf. The earlier negative acceleration after each reinforcement returns; but in the 2nd hour of the session (Record E) the over-all and local rates are considerably smoother. Bursts of a few responses at a high rate occasionally occur, as at *a*. Record F, a segment from the start of the 3rd session, shows only a very slight return to negative curvature, and there is no consistent rate change correlated with reinforcement. By the end of the 3rd session, in Record G, after a total of 8 hours of exposure to VI 2, the rate is lower and shows a slight tendency to fall after reinforcement. Figure 393 shows the entire 4th session on VI 2. The over-all rate is roughly constant. (Both birds show a slow oscillation in rate.) Short bursts of responding at a high rate occur at *a* and *b*. Slow responding follows many reinforcements toward the end of the session, as at *c, d, e,* and *f.*

An increase in the frequency of shorter intervals eliminates the pause after reinforcement developed by those 2 birds. Figure 394 represents the session following Fig.

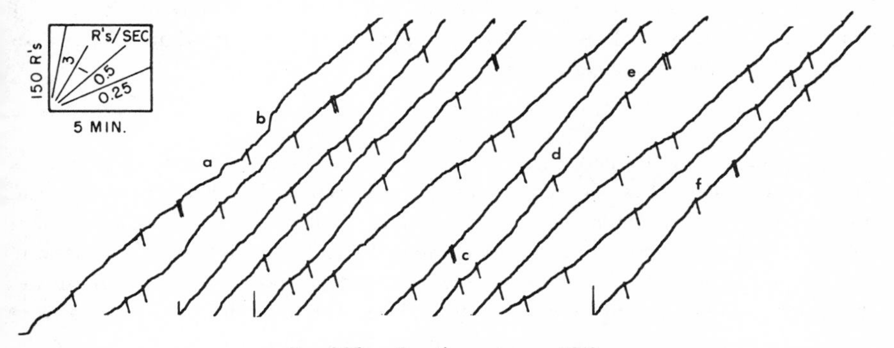

Fig. 393. Fourth session on VI 2

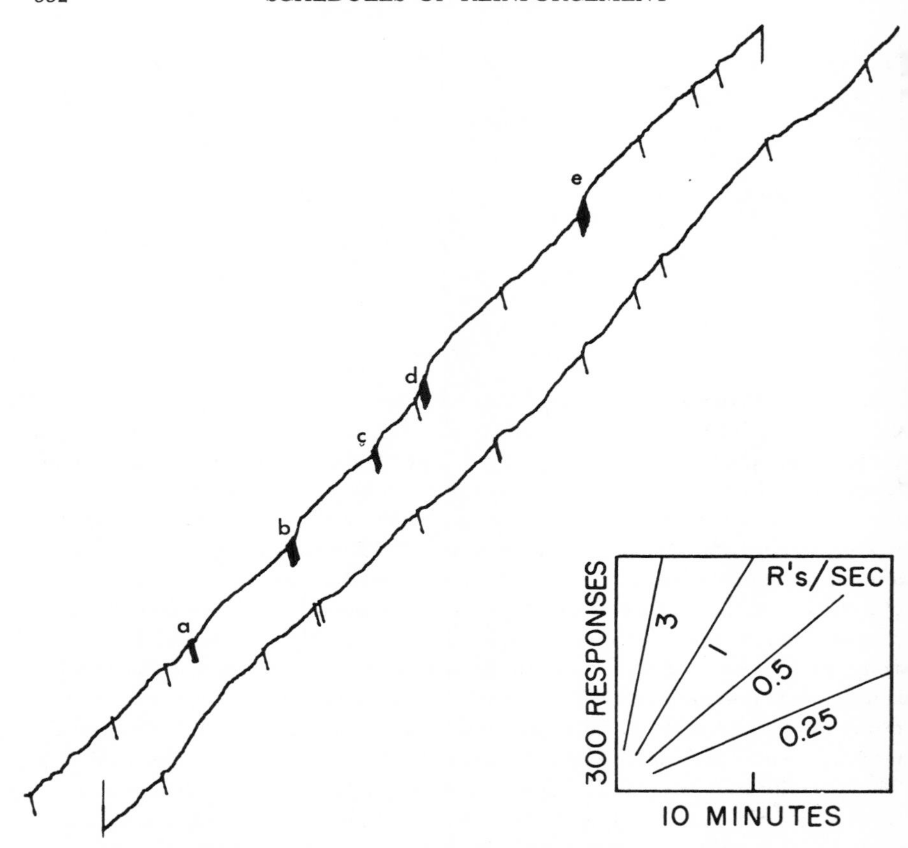

Fig. 394. Disappearance of the pause after reinforcement with higher frequency of short intervals

391. At *a, b, c, d,* and *e* an apparatus failure produced successive reinforcements of from 3 to 10 responses. These are all followed by brief rapid responding. The pause after reinforcement no longer occurs.

VI 3

The development of a performance on VI 3 during the first 17 sessions after crf is shown in Fig. 395, where the final segment from each session is reproduced. The schedule was composed of intervals as follows: 300, 30, 280, 120, 360, 300, 0, 240, 220, 180, 10, 280, 100, and 60 seconds. By the end of the 1st session in Record A a constant rate of approximately 0.35 response per second had developed. The over-all rate increased progressively during the next 5 sessions (Records B through F), reaching an over-all rate of approximately 1 response per second by Record F. Reinforcements are not marked in Records B, C, and D. At this stage approximately 20 responses are emitted rapidly just after reinforcement; they are followed by an abrupt shift to a lower rate before the prevailing variable-interval rate is reached. This is more pronounced by the end of the next session (Record G). The rate just after reinforcement is now about 3.8 responses per second, compared with a prevailing rate of approximately 1.4 responses per second. For the remaining 10 sessions (Records H through P) the over-all rate remains at approximately 1.25 responses per second. But the high rate immediately after the reinforcement and the abrupt shift to a lower rate become more pronounced.

Figure 396 shows the performance for a second bird after 45 hours on the same schedule of reinforcement. The bird maintains a constant over-all rate throughout the session, with marked local changes, and shows the same tendency to respond at a higher rate immediately after reinforcement.

Another bird on a variable-interval schedule similar to the VI 3 just described showed the performance recorded in Fig. 397. This figure contains segments from alternate sessions during the first 26 sessions on the schedule. The actual series of intervals was: 5, 50, 190, 190, 170, 10, 300, 80, 50, 280, 20, 290, 120, 355, and 310 seconds. The average interval is 164 seconds, although for brevity the schedule is here called VI 3. The first segment of the figure occurs after 12 hours on VI 3 after crf, and already shows a marked tendency for the rate of responding to be highest immediately following the reinforcement. Only the reinforcement at *a* is not followed by an increase in the rate. The over-all rate is approximately 0.75 response per second, very nearly as high as the final rate generated under this schedule in the last segment of the figure. At *b* the over-all rate has reached a value comparable with the high rate typically following reinforcement. This continues after reinforcement and shows no break to a lower rate. Elsewhere, however, very "square" changes in rate occur after reinforcement. Toward the end of the session, instances of abrupt shifts to a higher rate occur at other points in the schedule. Figure 398 shows a larger sample of the performance on this variable-interval schedule from a session following the last in Fig. 397.

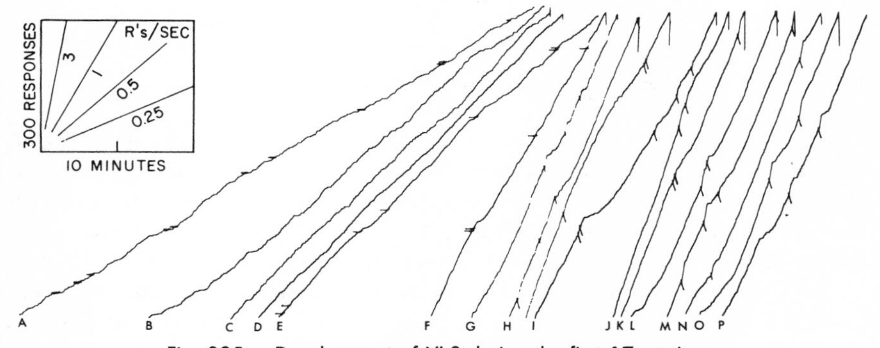

Fig. 395. Development of VI 3 during the first 17 sessions

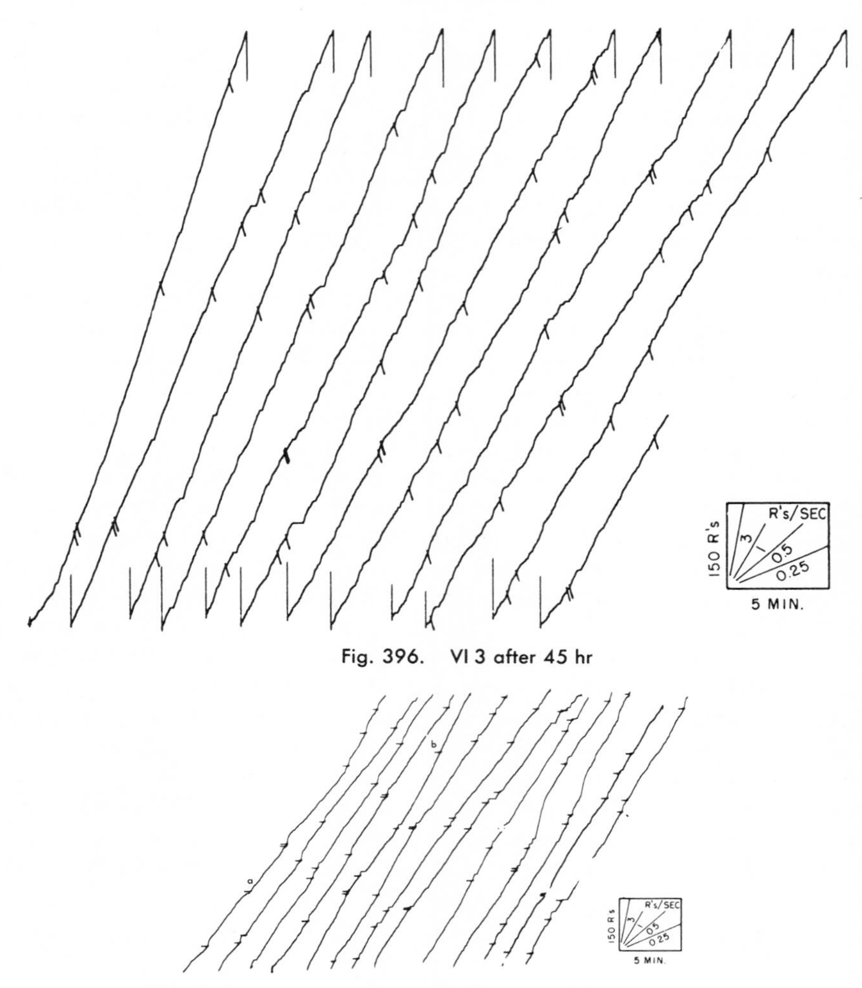

Fig. 396. VI 3 after 45 hr

Fig. 397. Development of VI 3 during the first 26 sessions

The bird has had 72 hours of exposure to VI 3. Although the local rate frequently changes considerably, particularly after reinforcement, the over-all performance is stable, and the effects are uniform from segment to segment. A high rate immediately after reinforcement may continue for from 20 to 40 responses at a rate of about 2.7 responses per second. This rate is followed either by an abrupt shift to a rate of 1.1 responses per second, which prevails throughout most of the bird's performance, or to a pause of a few seconds followed again by the higher rate. Where a second reinforcement occurs soon after the first, while the rate of responding is still high, a pause may

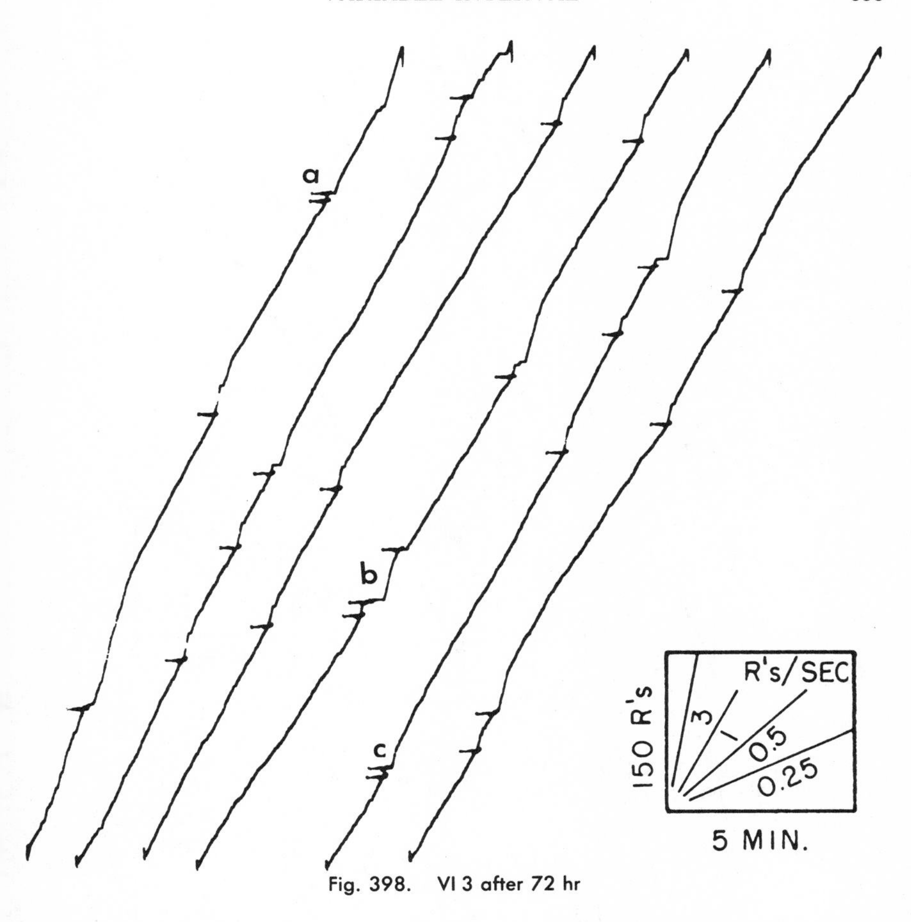

Fig. 398. VI 3 after 72 hr

follow the second reinforcement instead of the usual high rate (for example, at *a, b,* and *c*). When the bird begins responding at the end of the pause, however, a brief period of high rate comparable with that after reinforcement follows.

Figure 399 shows the performance of a second bird after 80 hours on the same schedule. The bird shows several kinds of rate changes after reinforcement, but the small square curve is readily observed.

After the performance on VI 1 shown in Fig. 386, the bird was reinforced for 29 hours on VI 2 and VI 3 schedules similar to those already described. The variable-interval series was then changed to one of the form: 0, 1, 1, 2, 3, 5, 8 . . . , in which successive terms are determined by adding the preceding two (the so-called Fibonnacci numbers). The largest interval was 987 seconds, and the mean, 152 seconds. Figure 400 shows the performance after 16 hours on this schedule. The sustained over-all

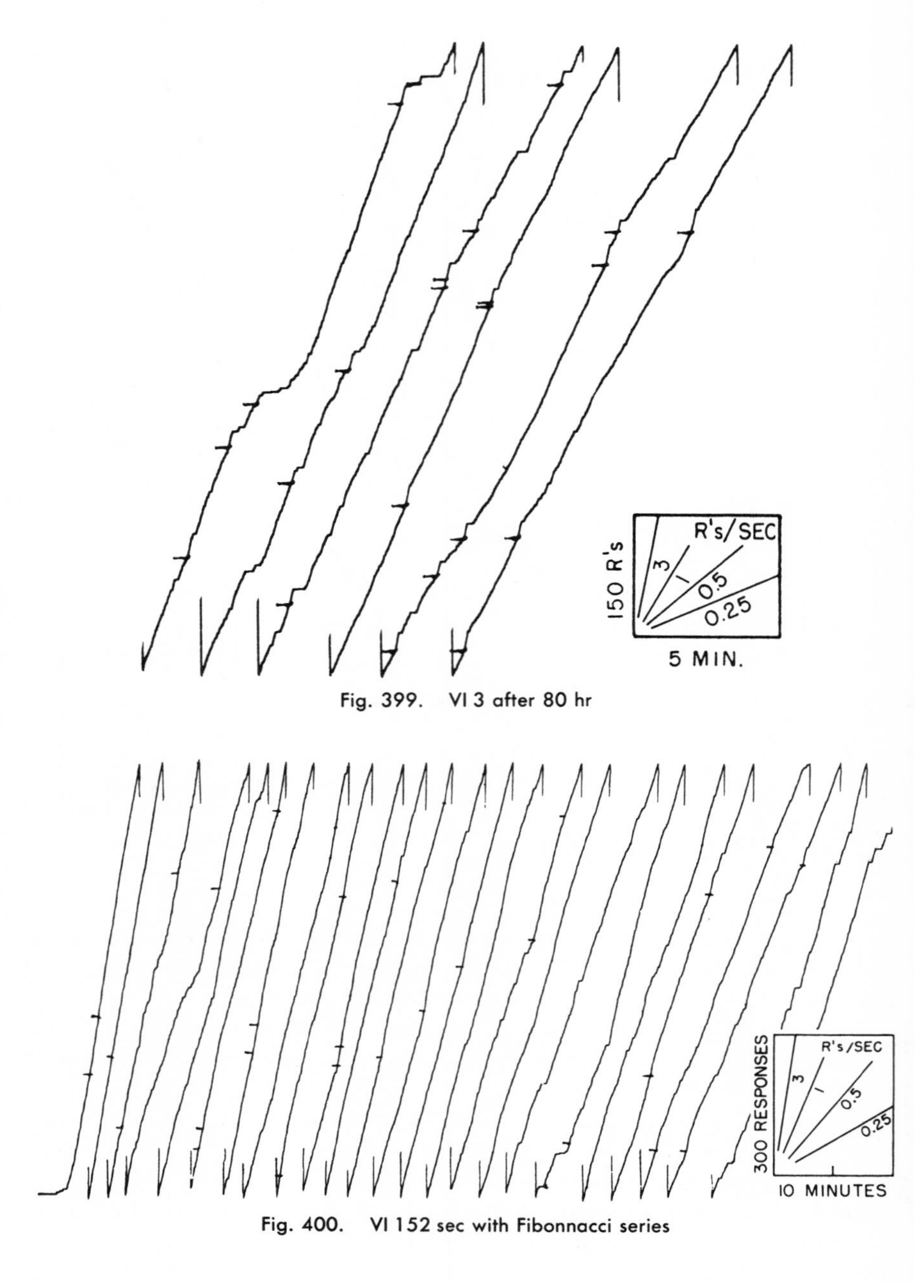

Fig. 399. VI 3 after 80 hr

Fig. 400. VI 152 sec with Fibonnacci series

rate of responding is more than 2.5 responses per second. Each reinforcement is followed by an unusually high rate; but instead of the abrupt decline to the lower rate, the rate falls slowly to an only somewhat lower value.

The lack of local effects just after reinforcement is presumably due to the preponderance of short intervals and the gradualness with which longer intervals appear in the series.

In a study of the effects of deprivation level on performance when a VI 3 schedule remained unchanged over a long period, the birds showed several types or stages of performance. Figure 401A reproduces a segment for each of the 3 birds, showing an almost linear performance. Although slight local changes in rate occur, they are not associated with the reinforcement or any other feature of the schedule. These curves represent temporary states interspersed between the other performances shown in the figure. Record B contains one segment for each bird showing a common deviation in

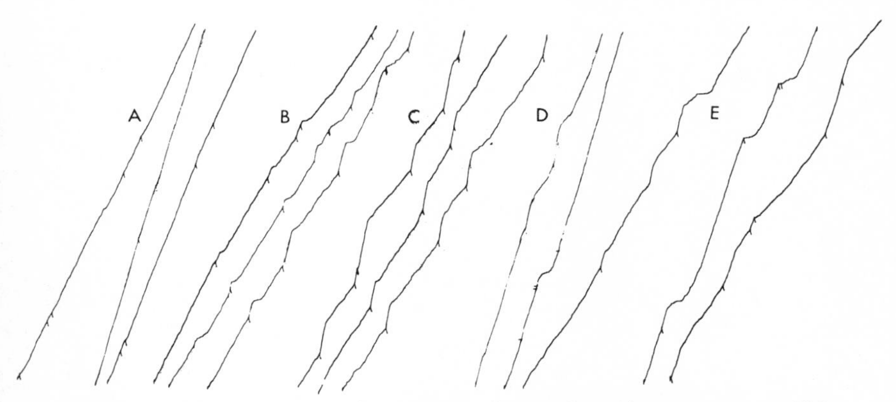

Fig. 401. Varieties of local rate changes occurring during long exposure to VI 3

which the rate increases briefly after reinforcement and then declines rapidly to zero with a later return to an intermediate value. Occasionally, the higher rate after reinforcement is not followed by a pause, so that the whole curve is displaced upward. Thus, in the segments for the 3 birds in Record C the high rate immediately after reinforcement is followed by an abrupt shift to an intermediate rate, which is maintained until reinforcement. In Record D a segment for each bird illustrates similar changes executed more slowly and with fairly smooth curvature. In Record E segments for 2 of these birds show an unusual decrease in rate after reinforcement. Other types of change follow adjacent reinforcements.

In spite of the local changes in rate produced by a particular set of intervals, VI schedules almost invariably produce sustained and relatively constant over-all rates. Figure 402 shows a very rare exception which we cannot explain. The bird was showing a fairly linear performance on VI 3, as the 1st segment shows (followed immedi-

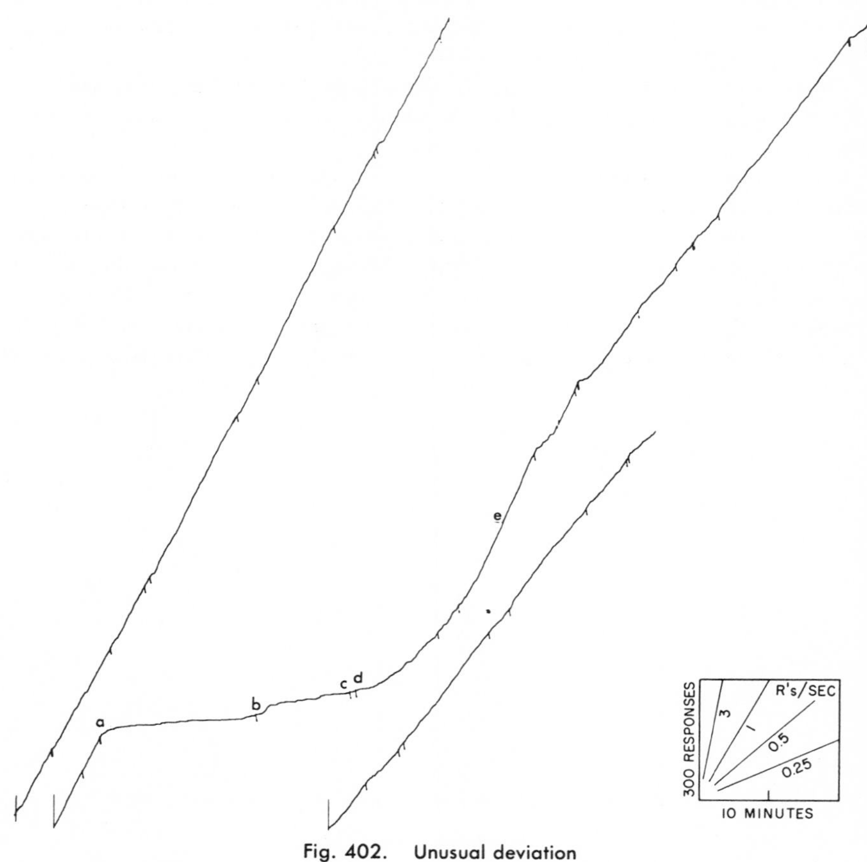

Fig. 402. Unusual deviation

ately by the second). At *a,* however, the rate falls over a period of a minute or two to a very low value, which is maintained in spite of reinforcements at *b, c,* and *d.* A smooth acceleration then leads to a rate which is slightly higher than normal at *e,* but drops to a normal value thereafter.

Variable-interval reinforcement sustains a fairly uniform rate for long periods. Figure 403 shows a complete 14-hour session on VI. The over-all rate varies slightly, and in three instances a low rate is in force for from 5 to 10 minutes. Most of the time, however, the bird is fully under control of the schedule. It emits approximately 87,000 responses during the session. A performance of this magnitude may be repeated on a daily schedule. Another bird in a 9-hour session on the same schedule gave a performance represented by Fig. 404.

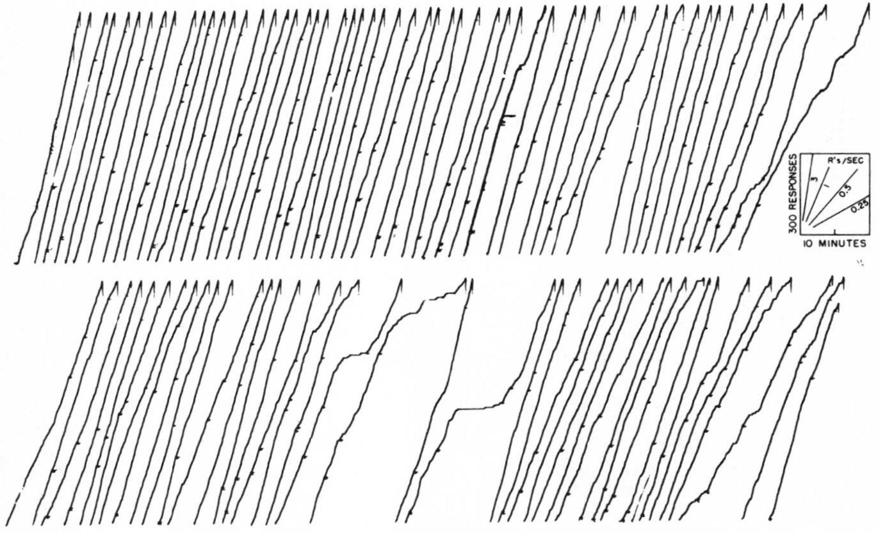

Fig. 403. Sustained VI performance during a 14-hr session

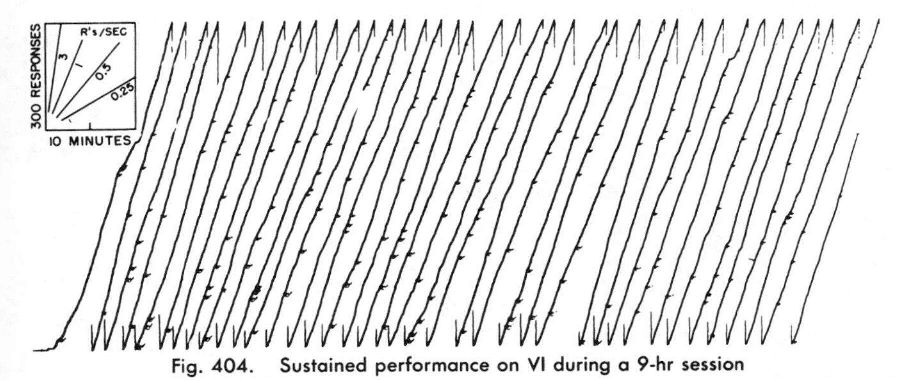

Fig. 404. Sustained performance on VI during a 9-hr session

The behavior of the rat on VI 3.5

An arithmetic VI 3.5 schedule of reinforcement was used in a study on pre-aversive stimuli in the rat. Figure 405 shows 2 complete daily sessions. The deviations from a roughly constant rate show a sort of long-term oscillation, but show no relation to the schedule.

Geometric VI 7

We modified the series of intervals in the experiment shown in Fig. 400 by omitting the zero interval and one 1-second interval, and by extending the series until the

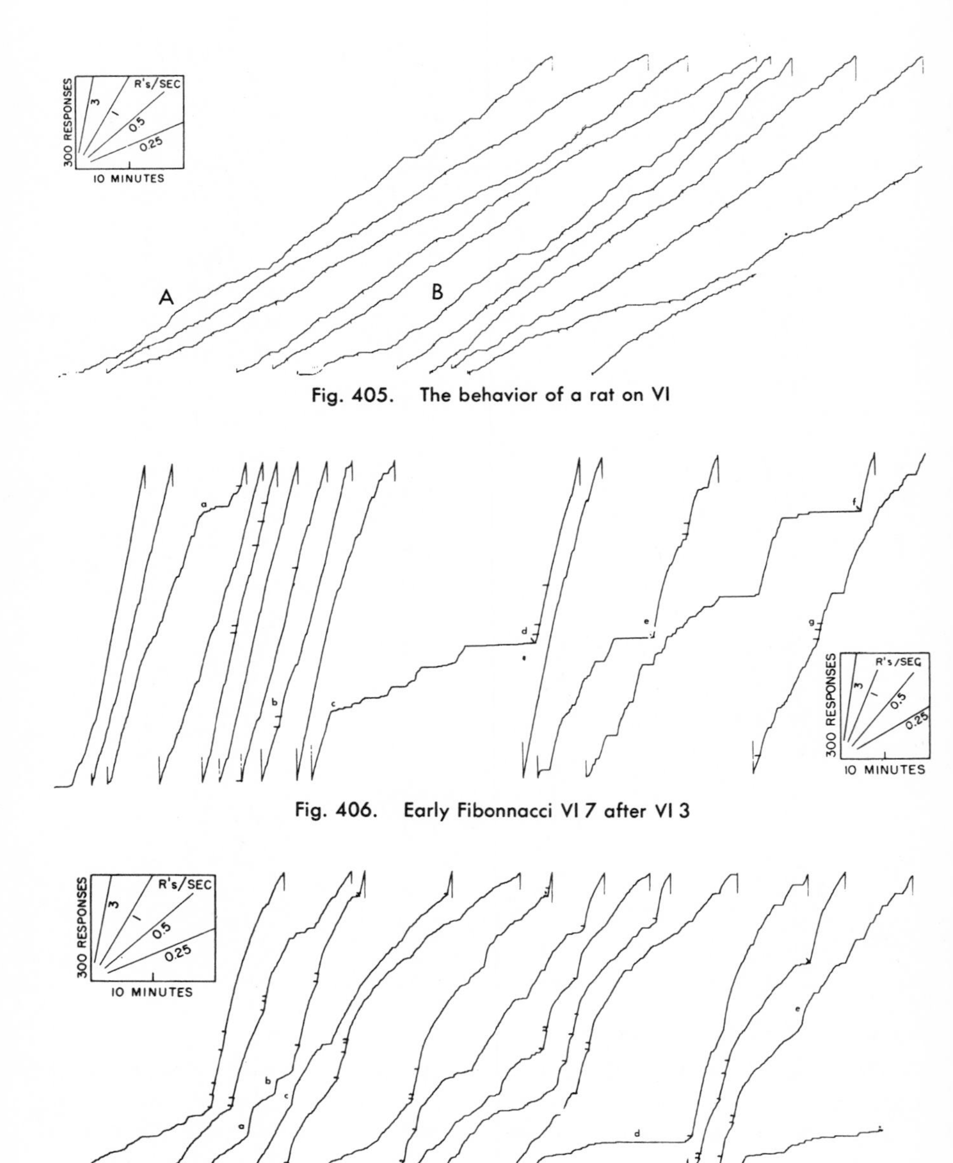

Fig. 405. The behavior of a rat on VI

Fig. 406. Early Fibonnacci VI 7 after VI 3

Fig. 407. Fibonnacci VI 7 after 162 hr

longest interval was approximately 45 minutes. The mean interval was 7 minutes. Figure 406 presents the transition from the series in Fig. 400. The rate is at first appropriate to the previous VI 3; but the 1st interval in the schedule is unusually long and the rate breaks before reinforcement at *a*. The next few reinforcements occur after intervals of the same order as under the preceding schedule. The longest interval in the series follows the reinforcement at *b*. A high over-all rate is sustained for approximately 10 minutes, breaking sharply at *c* and showing a long pause before reinforcement 45 minutes later at *d*. This reinforcement reinstates the prevailing variable-interval rate; and for the remainder of the session the rate begins high following each reinforcement and declines smoothly until the next reinforcement. During the longer intervals the reinforcement occurs after pauses, as at *e* and *f*. During the shorter intervals, the reinforcement occurs at a high rate, as at *g*.

Figure 407 shows a later performance on this VI 7 after 162 hours of exposure to the schedule. The figure contains the entire session except for the first 2½ hours. Each reinforcement is followed by a maximal rate, which falls off smoothly to a lower value.

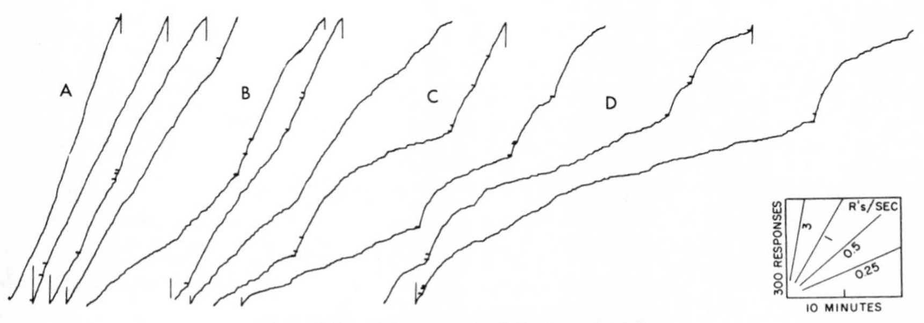

Fig. 408. Fibonnacci VI 7 after 240 hr

Occasionally, the higher rate appears spontaneously during a longer interval, as at *a, b, c,* and *e*. The intermediate rate, showing some negative acceleration, is usually sustained throughout the interval. At *d,* however, a pause of more than 5 minutes occurs. The next reinforcement is then followed by over 200 responses at 3 responses per second, a considerably more sustained run than typically occurs.

Figure 408 presents a later performance for a second bird on the same schedule and after a similar history, after 240 hours of reinforcement on VI 7. Record A shows the start of the session; Record B shows the 3rd hour; Record C, the 7th hour; and Record D, the end of the session at 10 hours. The over-all rate declines from approximately 1.5 responses per second in Record A to 0.35 response per second at the end of Record D. The decline is executed mainly as an acceleration in the change to a lower rate after reinforcement.

A third bird with a similar history was unable to sustain comparable rates on this variable-interval schedule. Figure 409 is a sample of its performance after 83 hours of VI 7. The session starts (Record A) with a roughly linear performance; and the

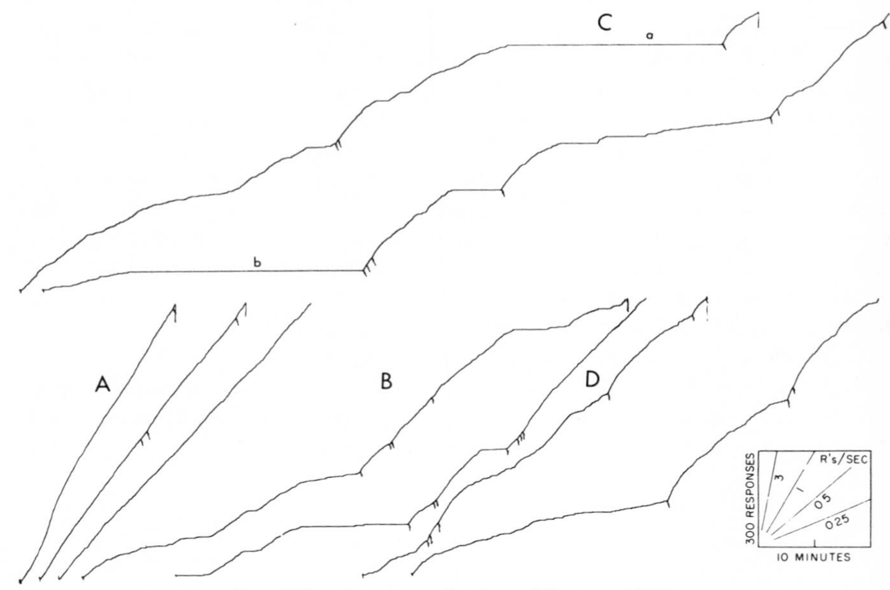

Fig. 409. Low over-all rate on Fibonnacci VI 7

over-all rate of responding declines slowly, reaching 0.6 by the 3rd segment, a decline of approximately 30%. By the 5th and 6th hours of the session (Record B), the over-all rate has fallen to 0.3 with the highest rates immediately following the reinforcement. The 7th and 8th hours of the session (Record C) show a further decline in the over-all rate to 0.2 response per second, with pauses occurring during the longer intervals, 19 minutes at *a* and 21 minutes at *b*. In spite of the very low over-all rate and the long pauses, each reinforcement is followed by rates of the same order as those in Record A. Record D shows the 14th and last hour of the session. The over-all rate has increased but has not reached the value at Record A. Responding is sustained throughout the longer intervals, although at very low rates.

After 138 hours of reinforcement on this schedule, an almost linear performance developed; the over-all rate was 0.6 response per second, with only a slight drop during the session. Figure 410 shows an entire session. When the over-all rate is sufficiently high, as at *a* and *b*, the reinforcement is not followed by an increase in rate. At lower prevailing rates most reinforcements are followed by a slight increase (as at *c*). Unlike the earlier performance for this bird, the intermediate rate of responding is maintained, with only a slight and orderly decline through the longest interval in the series, ending at *c*.

After 252 hours of reinforcement on this VI 7 schedule the over-all rate shows a further increase of the order of magnitude shown in Fig. 411, which contains the first 7 hours of a daily session. The over-all rate declines only slightly from 1.5 responses per

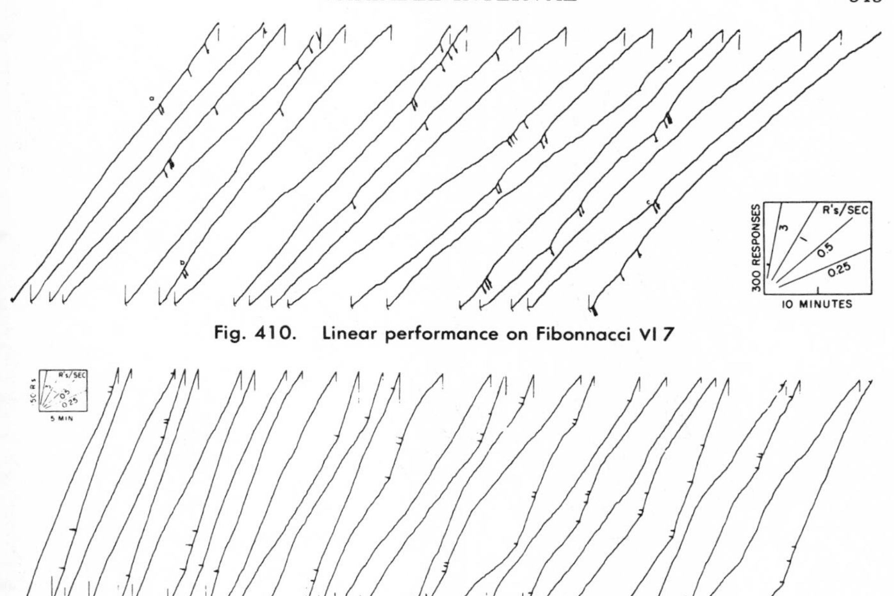

Fig. 410. Linear performance on Fibonnacci VI 7

Fig. 411. Fibonnacci VI 7 after 252 hr

second at the start of the session to 1.25 responses per second at the end. Marked rate changes after reinforcement are still prominent, however, and the prevailing performance is a more or less rapid decline to an intermediate rate which is maintained through the interval. Occasionally, the highest rate of responding reappears after the rate has fallen to an intermediate value.

The experiments just described were followed by others on the effect of a TO on VI to be described later. In the experiments on TO a new VI 7 schedule was in force, in which, as in the preceding schedules, the intervals comprised a roughly geometric series. The longest interval was 1200 seconds, and the mean, 7 minutes. (The actual intervals and the order were as follows: 360, 960, 10, 180, 20, 720, 1, 40, 960, 720, 240, 360, 360, 20, 480, 240, 960, 720, 40, 1200, 1, 80, 120, 480, 40, 360, 1200, 5, 480, 40, 120, 10, 1200, 200, 80, 1, 120, 5, 80, 240, 5, 80, 10, 20, 240, 960, 120, 20, and 720 seconds.) This schedule produced a much more linear performance. Figure 412 shows the first performance after the removal of the TO. Except for very brief local rate changes, the over-all rate is quite constant through the entire 8-hour session. Figure 413 shows the performance of a second bird in the first session without TO after a similar history. The longest pauses occurring during the 40,000 responses in the figure are of the order of 30 seconds. The over-all rate shows some decline during the session, and this bird continues to show a slightly higher rate immediately after the reinforcement.

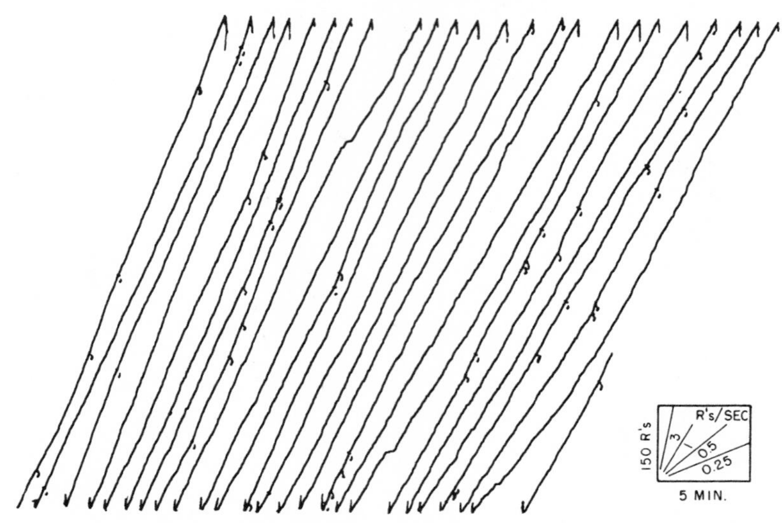

Fig. 412. Linear performance on geometric VI 7 after exposure to time out

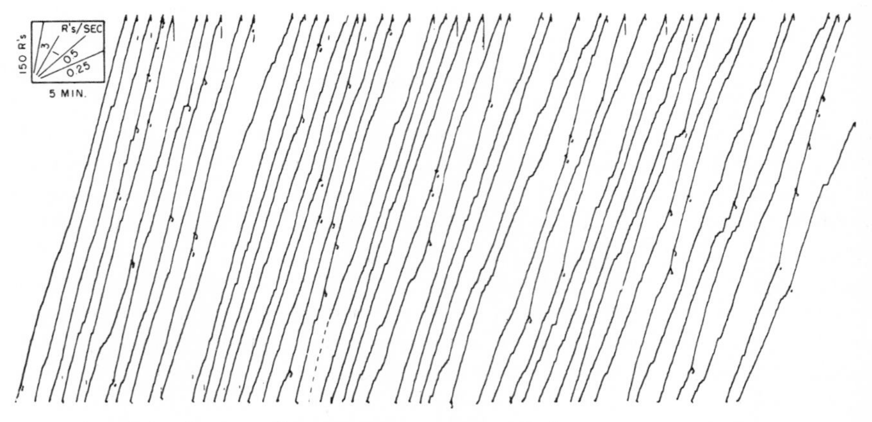

Fig. 413. Linear performance on geometric VI 7 after exposure to time out

Two more pigeons developed stable performances on the same roughly geometric VI schedule. Figure 414 shows an intermediate stage of development after 105 hours of exposure to the schedule. The figure contains the 5th and 6th hours of the session. The bird responds at either one of two rates. Immediately after reinforcement, a high rate is maintained for 60 to 100 responses, after which there is a rapid shift to an intermediate rate of approximately 0.3 response per second maintained until reinforcement. Occasionally, the reinforcement is not followed by an increase in rate of responding, as at *a*; a high rate sometimes appears elsewhere, as at *b*.

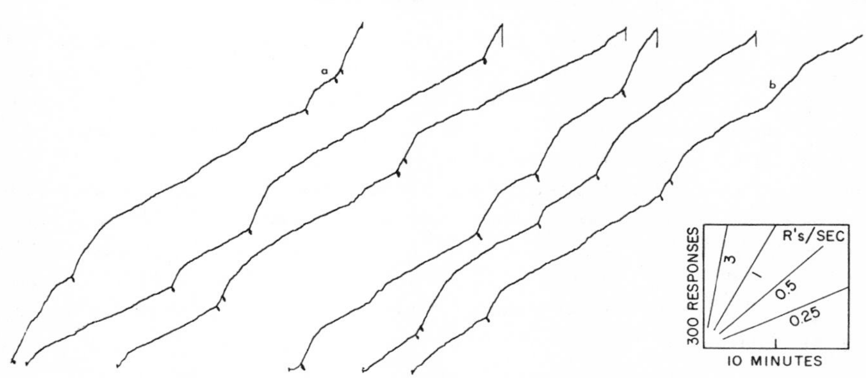

Fig. 414. Intermediate development of geometric VI 7

The second bird showed a much higher over-all rate of responding after a comparable exposure to VI 7. Figure 415 shows a complete session after 132 hours on the schedule. Rate changes are lacking in the vicinity of reinforcement so long as the over-all rate is high. A marked decline toward the end of the session results from the appearance of a second and lower rate, and reinforcements are now followed by increases.

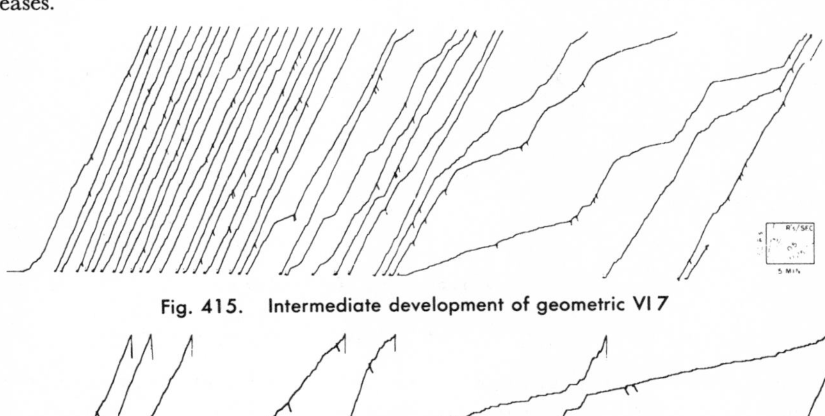

Fig. 415. Intermediate development of geometric VI 7

Fig. 416. Geometric VI 7 215 hr after crf

In another experiment 4 birds passed directly from crf to a roughly geometric VI schedule. Two birds placed on the geometric schedule discussed above gave performances similar to that in Fig. 416 for a session beginning 215 hours after crf. Two or more stable rates shift fairly abruptly from one to the other. The shifts are only occasionally correlated with reinforcement. The performance suggests the special case of VI called a two-valued interval discussed in the sections on mixed schedules. Two of these birds were on another schedule, composed of intervals in the following order: 1343, 4, 10, 69, 10, 712, 38, 32, 224, 47, 56, 136, 47, 1800, 0, 537, 69, 104, 224, 1800, 20, 978, 172, 4, 392, 20, 32, 978, 82, 136, 0, 712, 394, 1, 82, 2700, 15, 15, 104, 56, 38, 72, 25, 2700, 25, 537, and 299 seconds. Figure 417 is a sample of the performance after 105 hours. Except for a higher over-all rate, it resembles Fig. 414.

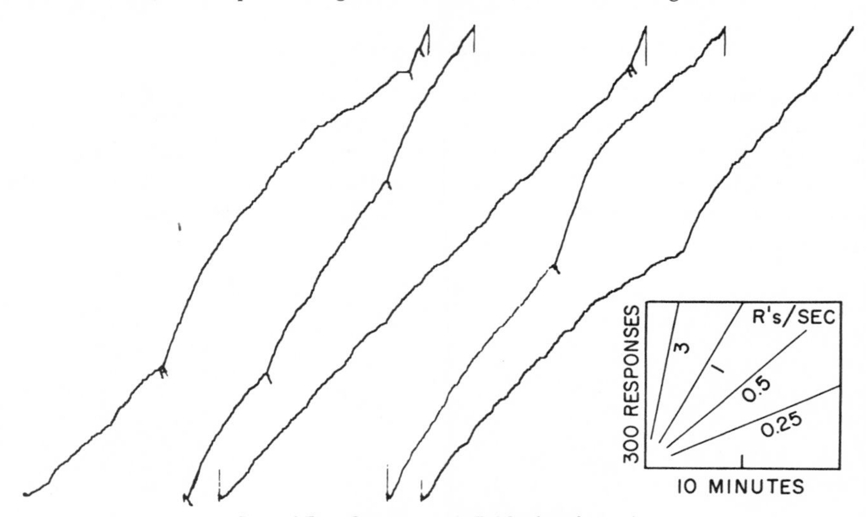

Fig. 417. Geometric VI 7 105 hr after crf

EXTINCTION AFTER VARIABLE-INTERVAL REINFORCEMENT

The extinction curves to be reported were mainly the results of apparatus failures and do not represent any systematic study of the extinction curve after variable-interval reinforcement. The general type of curve, however, can be compared with those under other schedules. Figure 418 shows a complete extinction curve after 30 hours of reinforcement on the Fibonnacci schedule for the bird whose later performance on this schedule was shown in Fig. 408. The extinction curve contains slightly more than 1000 responses; and except for the abrupt shift to the long pause at *a* and a compensating rate at *b,* the curve shows a rough negative acceleration. Figure 419 shows an extinction curve for the same bird 50 hours later on the same schedule. A portion of the performance on the VI 7 schedule is shown before the arrow. The prevailing rate is now higher than that before the extinction curve in Fig. 418. The curve in

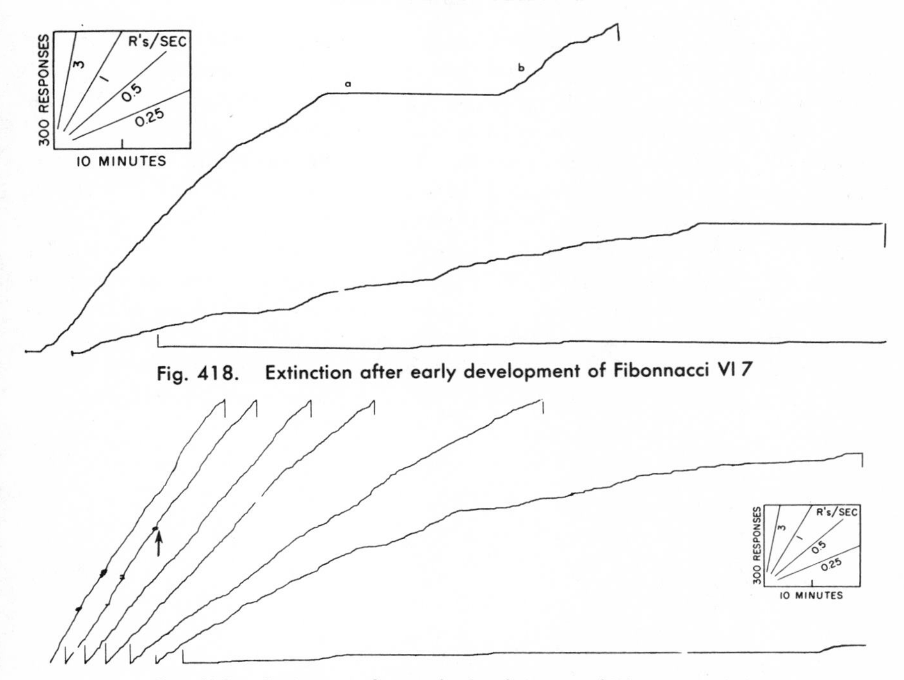

Fig. 418. Extinction after early development of Fibonnacci VI 7

Fig. 419. Extinction after early development of Fibonnacci VI 7

Fig. 419 contains approximately 4000 responses distributed through 2¼ hours, when the rate of responding fell continuously and smoothly to a very low value.

Figure 420 contains the extinction curve of a second bird after 225 hours on Fibonnacci VI 7. Figure 411 showed the irregular performance for this bird at this stage of the schedule. A high rate follows reinforcement, and the rate oscillates considerably during the longer intervals. Figure 420 follows immediately upon Fig. 411, both figures comprising one session. During the first part of extinction the rate oscillates in a manner similar to that under the previous variable-interval reinforcement.

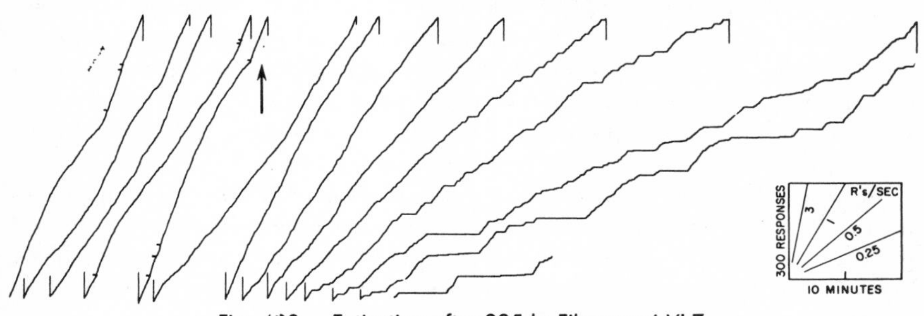

Fig. 420. Extinction after 225 hr Fibonnacci VI 7

As the rate begins to fall during extinction, however, the decline in the rate becomes more continuous. Pauses become longer and are separated by responding at approximately 0.8 response per second, though with rough grain. Even at the end of the curve, if the bird responds at all, it is at approximately this rate.

Figure 421 illustrates a later extinction curve for the same bird after approximately 25 hours on the geometric schedule which produced the performance shown in Fig. 412. This bird had had 325 hours of reinforcement on the VI 7 schedules. The extinction curve has been broken into two parts. Between Record A and Record B a low over-all rate of approximately 200 to 250 responses per hour was sustained for 8 hours. The extinction curve begins at approximately 1.5 responses per second and shows a decline to approximately 1 response per second by the segment of the record at *a*. Thereafter, the rate of responding falls sharply, reaching approximately 0.2 response per second at *b*. The rate returns briefly to 1 response per second for approximately 250 responses at *c*, after which it again falls off sharply. A reduced graph of

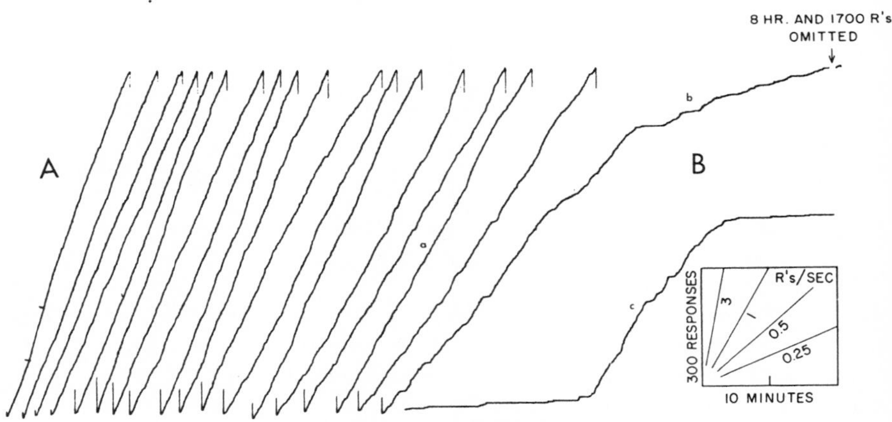

Fig. 421. Extinction after 255 hr Fibonnacci VI 7

this extinction curve, showing the whole curve in a continuous graph, would show an initial leg consisting of about 16,000 responses, declining from 1.5 responses per second to 1 response per second, followed by an abrupt shift to a very low but sustained rate for over 8 hours. The return to the initial high rate of responding similar to that in the start of the session, shown in Record B, would appear as a slight deviation from the long period of sustained slow responding. The initial leg is evidently the effect of the VI schedule in building up a sustained high rate during a long interval.

Figure 422 shows a later extinction curve after 168 hours of reinforcement on the geometric VI. This bird had been showing a fairly linear performance except toward the end of the session, when the over-all rate was reduced by low rates during the longer intervals. Extinction begins at 1.25 to 1.5 responses per second, which is maintained for the first 8000 responses to *a*. Thereafter, the rate of responding falls fairly continuously, reaching a very low value at the end of the figure. Figure 423 shows the

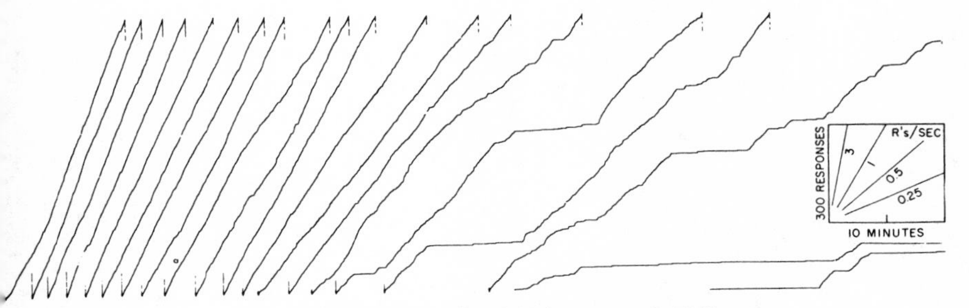

Fig. 422. Extinction after 168 hr geometric VI 7

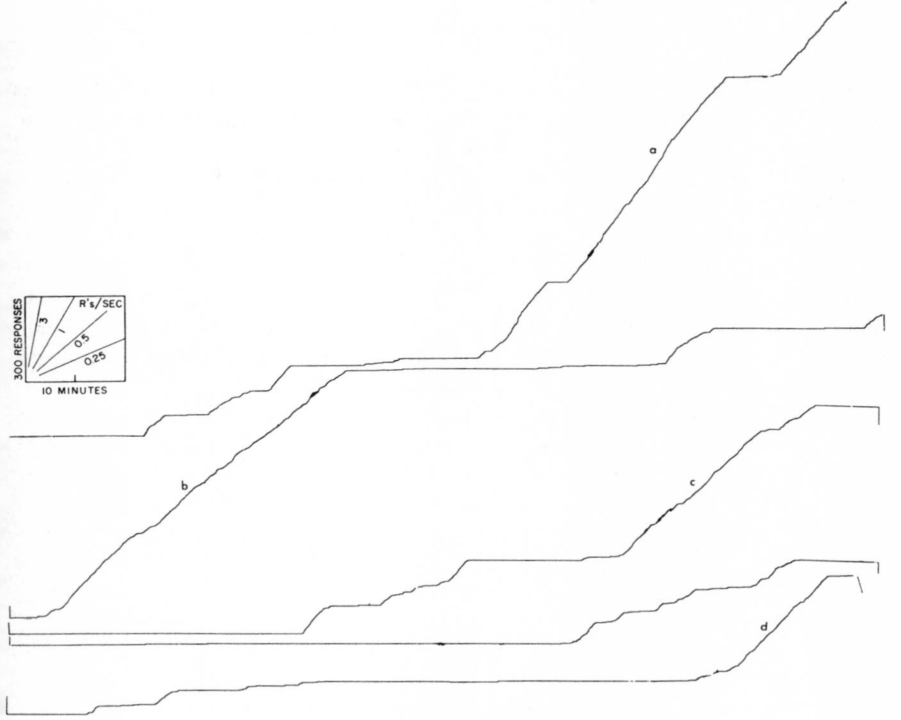

Fig. 423. Continues Fig. 422

Fig. 424. Extinction after 127 hr geometric VI 7

remaining 7 hours of the extinction period. The low rate at the end of Fig. 422 gives way to periods of sustained responding (as at *a, b, c,* and *d*) separated by periods of long pauses or very low rates. The sustained rate, however, does not reach the original variable-interval value or the earlier rate in extinction in Fig. 422. As extinction progresses, the over-all rate again declines, because of the increasing length of the pauses separating the periods of sustained responding as well as a lower actual rate.

Figures 424 and 425 contain extinction curves for 2 birds reinforced on the geometric VI. The final performance of the bird in Fig. 424 has already been shown in Fig. 417. Both birds were exposed to VI 7 immediately after crf. The extinction curve in Fig. 424 was taken after 127 hours after crf. A portion of the performance on VI 7 is shown before the arrow. This was recorded toward the end of the session, when the over-all rate was characteristically low for this bird. In extinction the rate continued to fall fairly continuously; and rates comparable with those which occurred during the first segment of the figure do not appear at any point for the remainder

Fig. 425. Extinction after 102 hr geometric VI 7

of the curve. Thus, changes in extinction may be related to changes occurring during the variable-interval session.

Figure 425 shows only the first 4000 responses of the second extinction curve. The rate falls off sharply after the first excursion, as it did during the longer intervals in the geometric series. Thereafter, high rates of the order of those which formerly followed reinforcement alternate with near-zero rates. Extinction was continued for a total of 14 hours, when the over-all rate of responding fell continuously. The curve continued to show periods of sustained responding at a high rate, separated by long periods of very slow responding.

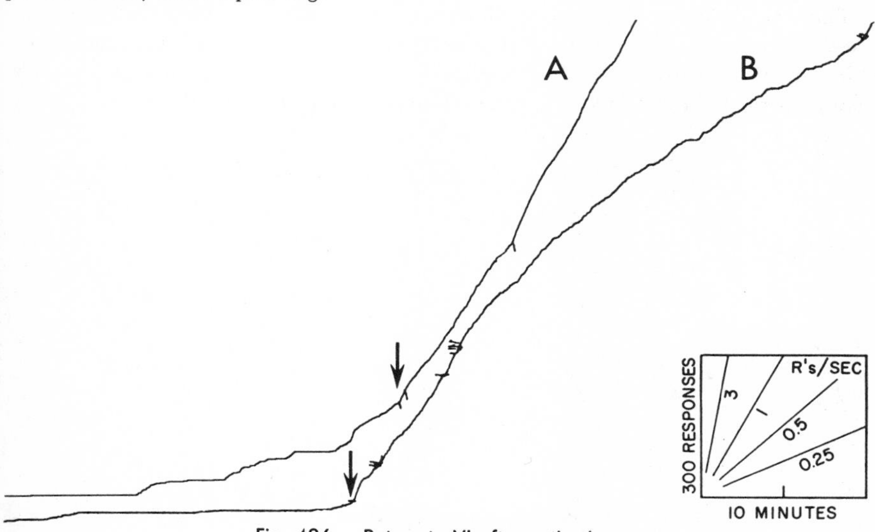

Fig. 426. Return to VI after extinction

Return to variable-interval reinforcement after extinction

Figure 426 shows the return to variable-interval reinforcement immediately after the extinction curves described in Fig. 422 and 419. The figure begins with the final portions of the extinction curves. The magazine was re-connected, so that reinforcement occurred at the arrows. A single reinforcement reinstates the highest rate that was normally observed under this schedule, and in both records the ensuing performance is not distinguishable from the previous variable-interval performance.

THE EFFECT OF TIME OUT ON A VARIABLE-INTERVAL PERFORMANCE

If a variable-interval schedule is successful in generating a nearly constant rate of responding, a time out after reinforcement should have no effect on the performance. The time out makes a difference only to the extent that the bird's own behavior controls local rates because of differential reinforcement. This possibility permits us to explore the failure to generate a linear performance in every case. In one experiment to be described in this section, we studied the effect of TO on VI in 2 birds by attaching a TO to each reinforcement. In another experiment on 3 birds, TO occurred independently of the reinforcement.

Time out after reinforcement

A 7-minute TO was introduced after every reinforcement in a performance under a geometric schedule similar to those already described in Fig. 415 and 416. The first part of Fig. 427 shows a performance after 220 hours of exposure to a geometric VI 7 TO (following a previous history of 227 hours of a geometric VI without TO). The rate oscillates radically between a high value immediately following TO to intermediate rates, with rough grain sustained during most of the longer intervals. Frequently, reinforcement occurs when the bird is responding at a high rate, as at *a* and *b,* even though these reinforcements occur after relatively long intervals. When the TOs no

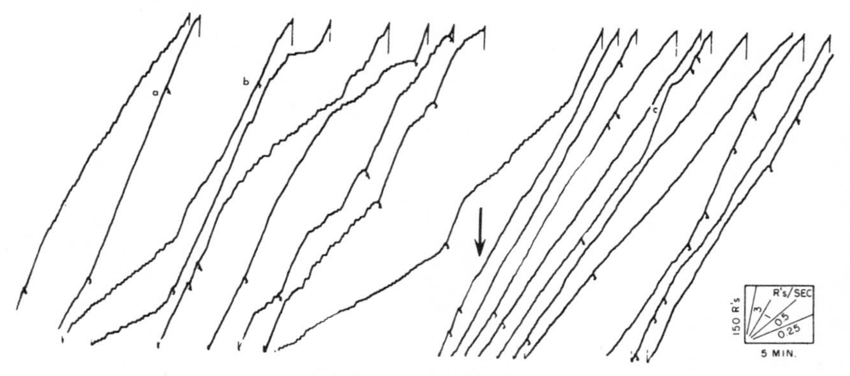

Fig. 427. Removal of TO after reinforcement on geometric VI 7

longer follow reinforcement (beginning at the arrow), the rate remains relatively constant, at approximately 1 response per second, or slightly below the maximal rate under VITO. A single break-through at a high rate occurs at *c,* where approximately 150 responses are emitted at 1.7 responses per second. The effect of the removal of the TO in producing a more orderly and linear performance is temporary, however; and in subsequent sessions the performance became as irregular as that in the first part of Fig. 427.

The 2nd bird in the experiment showed a more linear performance under VI 7 TO 7, as Fig. 428 shows, after 200 hours of reinforcement on VITO. The TO after reinforcement was discontinued at the start of the following session, and the performance is shown in Record B. The over-all rate and the character of the changes in rate remain about the same, but the records give the impression of a narrower range of rates and a generally more orderly performance.

Although adding or removing a TO after every reinforcement usually has an immediate effect, the performance generally settles down to a common pattern. The sched-

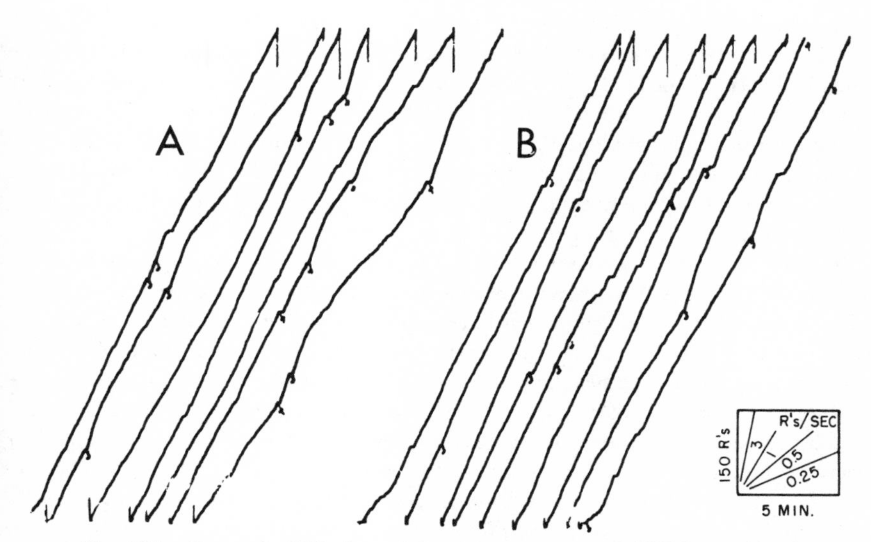

Fig. 428. Removal of TO after reinforcement on geometric VI 7 (second bird)

ule producing the wide rate changes of Fig. 407 and 408 continued to give the same patterns when a 2-minute TO was added to every reinforcement. Figure 429B should be compared with Fig. 408, and Fig. 429C with Fig. 407. Record A is for a 3rd bird with the same history.

Time out not correlated with reinforcement

The effect of a TO not correlated with reinforcement was tested on the 3 birds in Fig. 429. Instead of the Fibonnacci series in the earlier studies, the schedule was now a geometric series with a mean of nearly 10 minutes and a largest interval of 45 min-

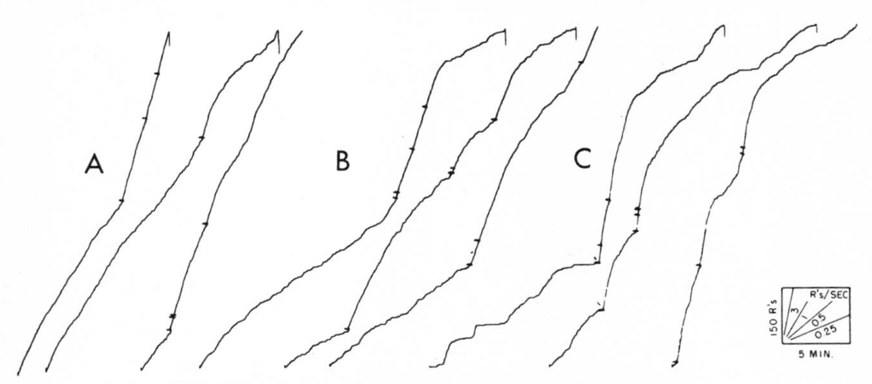

Fig. 429. Late performance with TO after reinforcement

utes. The schedule generated a much more linear performance. Figure 430 shows the first effect of the 2-minute TO occurring independently of reinforcement. The birds had been exposed to the new geometric variable-interval series for approximately 100 hours. The TOs are indicated by the dashes above the curves. The recorder and programming equipment did not, of course, operate during the TO. Reinforcements are recorded in the usual manner. The TOs have little effect. Records B and C may be slightly more irregular than the prevailing performance without TO, although the effect, if any, is slight.

The TO procedure was continued for a total of 44 hours, during which no effect appeared except possibly when the over-all rate had fallen toward the end of the session, so that each reinforcement was followed by a negatively accelerated curve. At this phase the occurrence of a TO where the rate is changing from one value to another provides a check on whether the primary variable responsible for the change is time or some other condition of the experiment. Figure 431 represents a continuous record which has been broken up into segments beginning with reinforcement. The TOs occurring every 7 minutes are indicated by the dashes above the curves. The rate changes are not sufficiently predictable to enable us to judge whether any particular change is due to TOs. However, TOs (at *a, b, c, d,* and *j*) occur at substantial rates and always show a slight decline in rate. At *g, h,* and *i,* where the rate of responding is very low, the TO has no observable effect. The most extreme breaks in the curve occur at *c* and *e,* where TO is followed by a zero rate for many minutes. The TOs during these pauses do not reinstate responding. At *f* there is little, if any, effect.

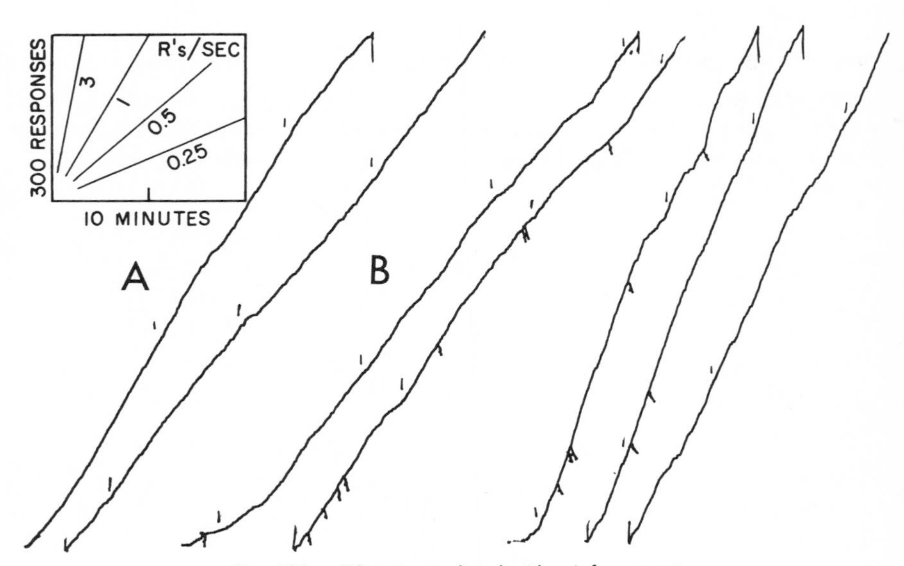

Fig. 430. TO not correlated with reinforcement

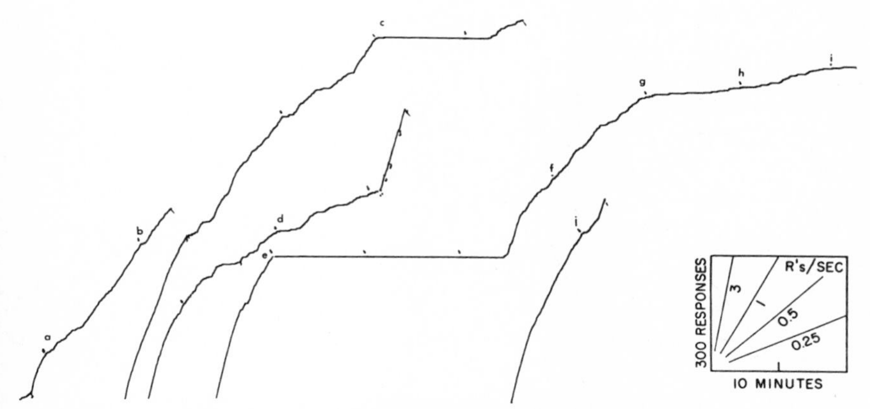

Fig. 431. The effect of TO on over-all rate changes

VARIABLE-INTERVAL REINFORCEMENT WITH LIMITED HOLD

In the interval schedules already discussed, reinforcement is determined by a programming clock which closes the circuit from the key to the magazine whenever a reinforcement is to be made available. The circuit remains closed until a response occurs and is reinforced. The reinforcement need not remain available indefinitely, however. When the programming clock "sets up" a reinforcement, the circuit from the key to the magazine may close only for a designated interval of time. If the bird responds within this interval, it is reinforced; if it does not, that reinforcement is lost and responses will continue to go unreinforced until the programming clock once again sets up a reinforcement. Such a schedule may be qualified as "limited hold."

In the following experiments on VI limited hold the intervals during which reinforcement is available are of the order of fractions of a second. Later experiments by W. H. Morse and R. J. Herrnstein have shown that limited holds of the order of 10 seconds produce major changes in fixed-interval performances.

Development of VI 1 limited hold

The transition from VI to VI limited hold was made from an early performance on an arithmetic VI 2. A session under this schedule has already been described (Fig. 391). The over-all rate is roughly linear between 0.4 and 0.5 response per second, with a slight pause following reinforcement. In the next session the schedule was changed to VI limited hold; each reinforcement was set up for only 0.75 second. A decline in the over-all frequency of reinforcement was anticipated because the bird might not be responding during the hold. The mean interval of reinforcement was therefore reduced to 1 minute. We determined this figure by estimating the number of reinforcements that would be received on limited hold if the bird continued responding as in Fig. 391. We estimated that the bird would receive about half the reinforce-

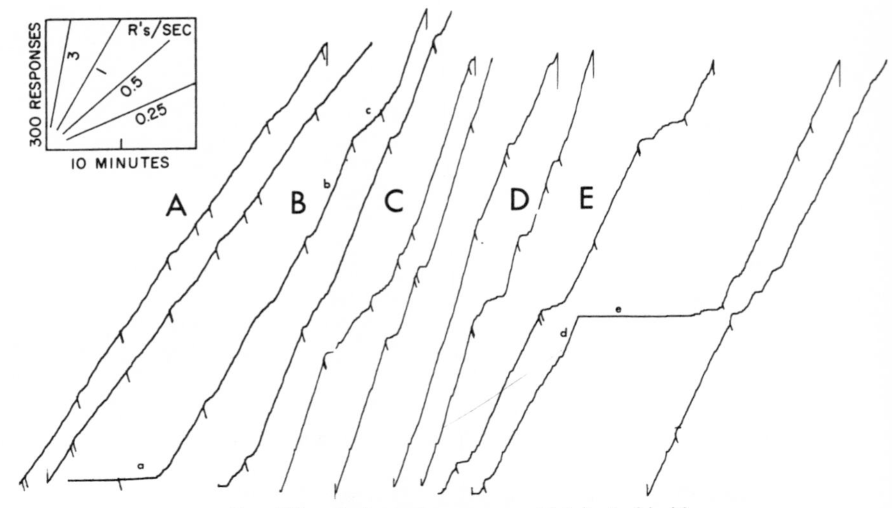

Fig. 432. Early performance on VI 1 limited hold

ments designated by the schedule; this proved to be the case. During the first session under VI limited hold, 34 out of the 64 possible reinforcements were received.

An early performance is shown in Fig. 432, where representative segments from the first 5 sessions with limited hold are given in Records A through E. During the 1st session (Record A) the over-all rate is approximately the same as that under variable-interval reinforcement. The pause following reinforcement, however, has now given way to a brief burst of responding at a higher rate. Local changes in rate are considerably smoother than those in the previous session, and the long-term oscillations in the over-all rate are missing. By the 2nd session (Record B), instances appear where the local rate increases very markedly. For example, the rate at *b* reaches 1.5 responses per second. At the same time, however, long pauses and periods of slower responding also appear, as at *a*, as well as intermediate rates, as at *c*. The bird continues to respond faster immediately after the reinforcement, and bursts are in turn followed by a pause and scallop or an abrupt shift to the variable-interval rate. The 3rd and 4th sessions (Records C and D) show similar performances. The 5th session (Record E) shows a severe decline in the over-all rate with a long pause at *e*, after an abrupt shift from the terminal variable-interval rate at *d*.

While the terminal variable-interval rates of responding are of the same order seen after comparable exposure to a variable-interval schedule alone, it is clear that the limited-hold contingency has an effect beyond that of a change in the number of reinforcements. Pauses as long as those occurring at *a* and *e* would not ordinarily appear.

The following session (Fig. 433) begins with a pause of about 15 minutes after which the rate accelerates briefly to approximately 1.25, about the same value as the highest rates in the previous session. Responding continues at about this level until just after the reinforcement at *a*; here, following the short period of higher rate after the reinforcement and a short pause, the bird begins to respond at a rate of approximately 4 per second. The highest local rate was previously about 3 per second. The exceptionally high rate continues for almost 1000 responses, until, at *c*, it abruptly shifts to a lower value; for the remainder of the session the rates oscillate between 4 per second and from 1.25 to 1.5 per second. The shift from one rate to another sometimes occurs at reinforcement.

The arrows in the figure show the 2 shortest intervals in the series. Reinforcements were received here because the rate of responding immediately after one reinforcement is high enough to guarantee a response within the 0.75-second hold. The high frequency of these successive reinforcements relative to the other intervals of reinforcement, where the limited hold is more costly, presumably explains the continuation of a high rate immediately after reinforcement.

Figure 434 gives the 8th session on VI limited hold, showing an intermediate development of the performance. The mean rate of the first $3\frac{1}{2}$ segments is 1.52 responses per second. At *a* the rate shifts abruptly to 3 responses per second. For the remainder of the session the rate is high after the reinforcement, but drops fairly rapidly though smoothly to a lower value before the next reinforcement. Following the reinforcement at *b*, the next reinforcement is not received for almost 20 minutes. The lower rate of responding is sustained for a full excursion of the pen. The higher rate then reappears.

Figure 435 shows the performance after 34 sessions (102 hours) on VI 1 with limited hold. Responding is maintained at a fairly constant rate of from 2 to 2.5 responses per second. The session begins with a slight warm-up at *a*, but after a brief acceleration a maximum rate is maintained except just after reinforcement. The arrows indicate the 0-second intervals of reinforcement, where 2 successive responses are rein-

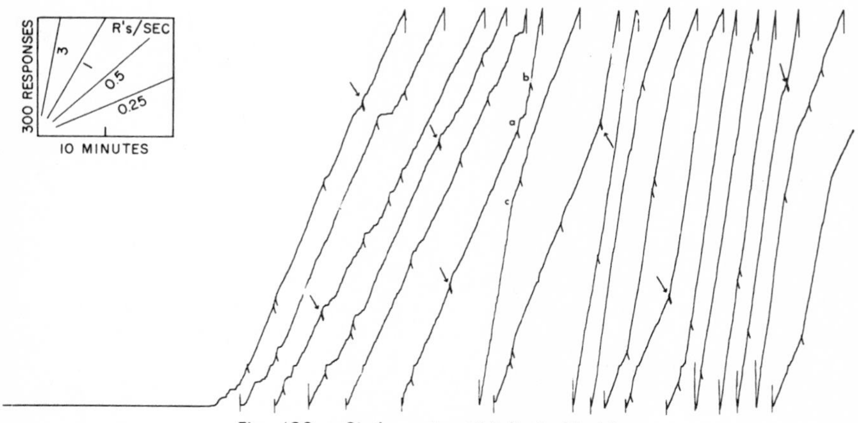

Fig. 433. Sixth session VI 1 limited hold

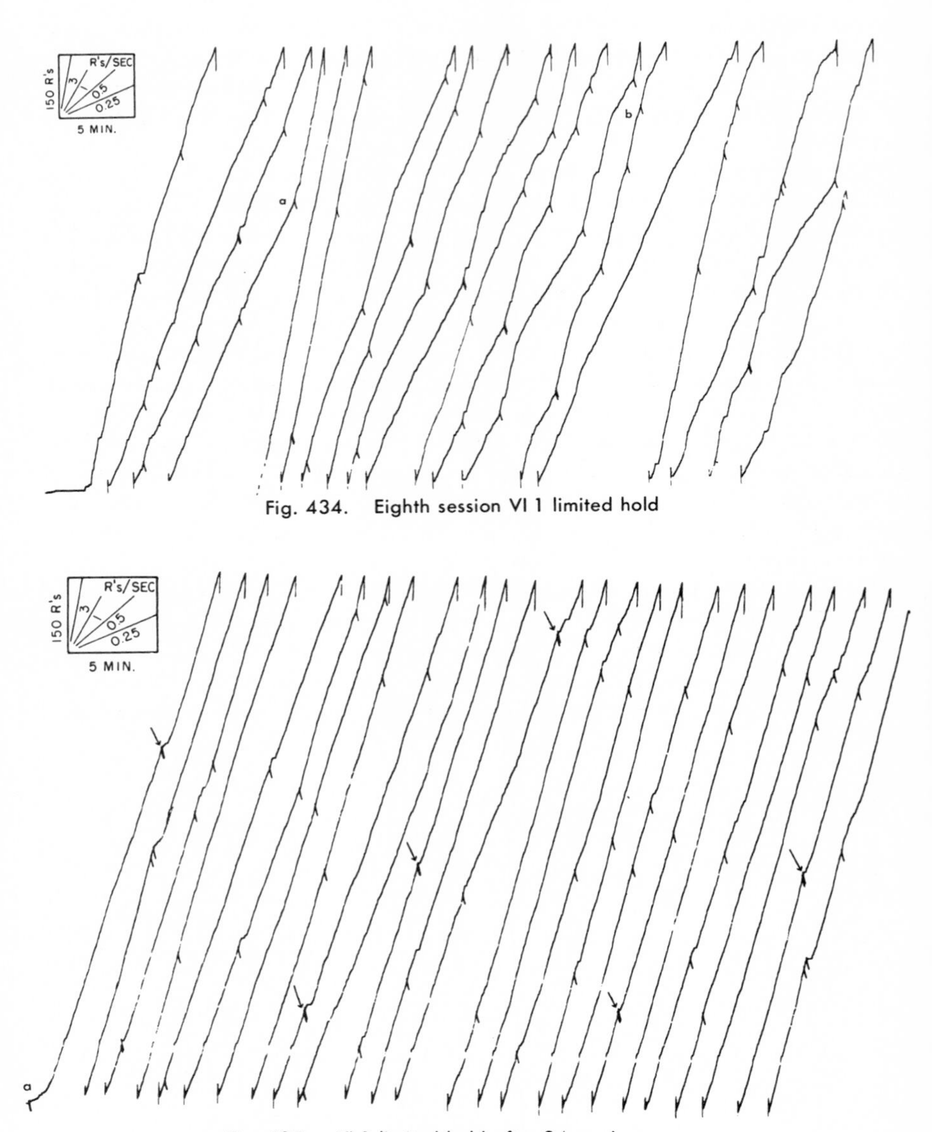

Fig. 434. Eighth session VI 1 limited hold

Fig. 435. VI 1 limited hold after 34 sessions

forced. The double reinforcement now controls a lower rate of responding, since no instances exist where 3 successive responses are reinforced. By this time the value of the limited hold had been reduced to 0.24 second. We were unable to detect any effect of the change from 0.75 second. Although the schedule set up 60 reinforcements, only about 15 were received, because of the limited hold. The initial "warm-up" apparent in Figs. 433, 434, and 435 is characteristic of the performance under VI

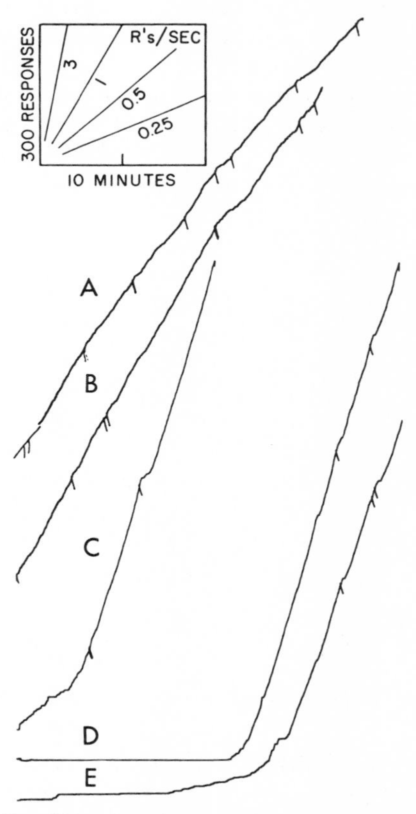

Fig. 436. Development of warm-up on VI 1 limited hold

limited hold. On VI without limited hold the rate is usually highest at the start of the session. Figure 436 shows the development of the warm-up. The first 1000 responses of each of the first 5 sessions are shown in Records A through E. These are the same sessions from which the segments in Fig. 432 were taken. In Record C the rate has some tendency to be lower at the beginning of the session; this lower rate becomes very pronounced in the 4th and 5th sessions, where the prevailing rate of responding is not reached until almost 15 minutes after the start of the session. The scattered responding early in Record E indicates that the bird is facing the key and occasionally pecking it; nevertheless the rate remains low. The development of a slow start coincided with the development of pauses and abrupt shifts in the local rate during the session. During the rest of the experiment a warm-up has not always been observed. Figures 434 and 435 show slight examples. The limited-hold contingency would tend to perpetuate any tendency to begin the session with a low rate of responding, since once a low over-all rate is established, the frequency of reinforcement decreases correspondingly.

When the schedule of reinforcement was changed to VI 3 without the limited-hold contingency, the effects of the previous limited-hold reinforcement were found to be

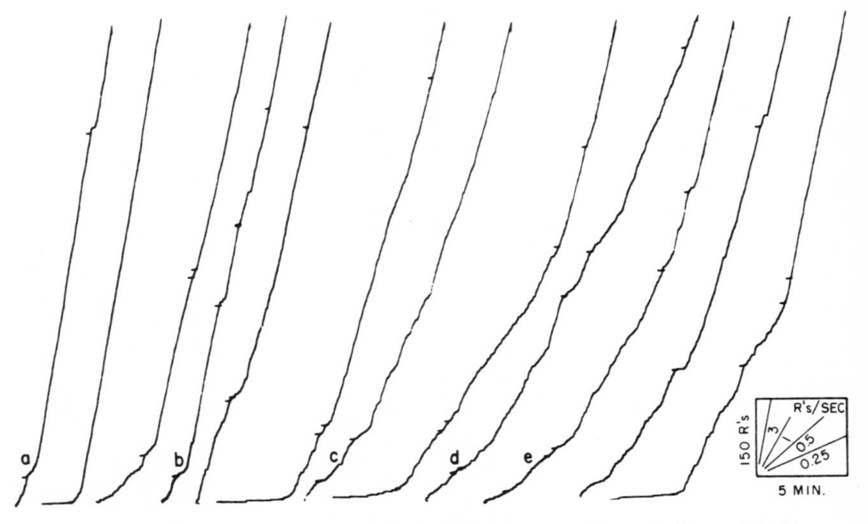

Fig. 437. Persistence of start of session warm-up after removal of limited hold

only partially reversible during 28 sessions. Figure 437 shows the first 920 responses of each of the last 12 of these sessions on VI 3. A warm-up at the start of the session persists, although reinforcements now occur with equal likelihood during the lower rates of responding.

The performance which has just been described was confirmed by a second bird in the experiment, except that the rate increased a little less abruptly when the limited-hold contingency was added. The adjustment was quicker when the limited-hold contingency was removed; also, a decline in rate of the order of 50% resulted, as well as a loss of warm-up. Effects of the previous limited-hold reinforcement continued to show, however, in the continuing oscillation between high and intermediate rates.

Development of VI 1.5 limited hold

Figure 438 shows a late performance on VI 1.5 limited hold for the birds whose final performance on VI 3 was shown in Fig. 398 and 399, respectively. The limited hold was 0.5 second. We carried out the transition to the limited hold by doubling the frequency of reinforcement from VI 3 to VI 1.5 so that the actual over-all frequency of reinforcement would not change radically under limited hold. The complete daily sessions shown in Records A and B of Fig. 438 were recorded after approximately 100 hours of reinforcement on VI 1.5 limited hold. The over-all rate of responding has increased from about 1.25 responses per second under the previous VI 3 to about 2.5 responses per second. This order of increase might have occurred normally as the result of prolonged exposure to the original VI 3 reinforcement. The effect of the limited-hold contingency is evident, however, in the general lack of intermediate rates characteristic of later VI limited hold. Pauses generally occur a few responses after the reinforcement, and the rate changes before and after are generally abrupt. If for

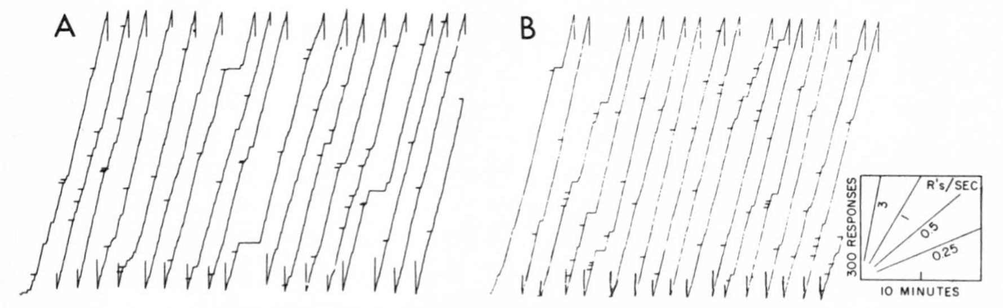

Fig. 438. Late performance on VI 1.5 for two pigeons

any reason the over-all rate of responding drops to a value so low that the likelihood of a response occurring during the limited-hold interval is very small, the bird may stop responding entirely.

At one stage during the development of VI limited hold, the birds were extinguished. (See Fig. 441 and 442 for extinction after VI limited hold.) Late in the session the magazine was re-connected, but the rate had reached so low a value that no responses coincided with a limited-hold interval. In one case the following session began with a 3-hour period during which no response occurred. The beginning of Fig. 439 shows the end of this 3-hour period. Before the arrow a single reinforcement had been set up in the normal fashion so that the first response occurring at any time thereafter would be reinforced. The first reinforcement (at the arrow) is followed by a high rate, which falls off sharply to an intermediate value with rough grain. A second response coincided with a limited hold at *a*. This reinforcement reinstated a substantial rate, under which the frequency of reinforcement was sufficient to maintain the behavior. This figure illustrates one of the major differences between a limited-hold con-

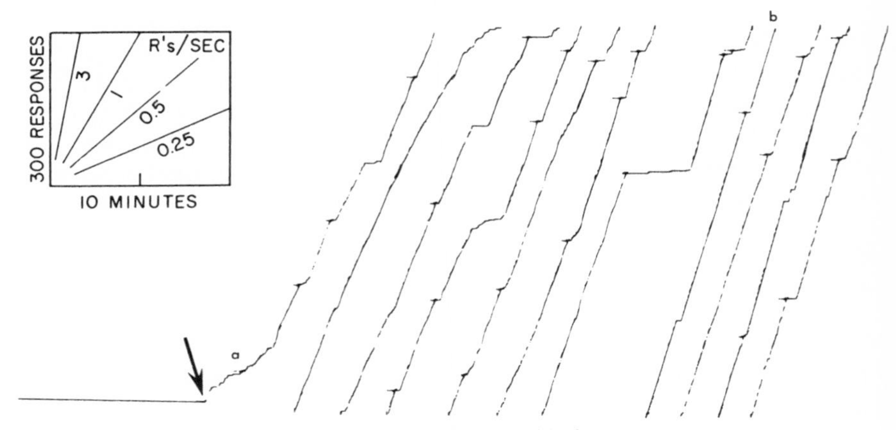

Fig. 439. Return to VI limited hold after extinction

tingency and a normal variable interval. The normal variable-interval schedule is "self-correcting"; that is, any tendency to slow down causes relatively frequent reinforcement. This condition has the effect of increasing the rate. The figure also shows an example of a VI limited-hold performance at an earlier stage than that in Fig. 438. Because of a consistent pause after reinforcement, the shorter intervals of reinforcement in the variable-interval series are missed. This fact in turn maintains the pause following the reinforcement. Whether the performance shows runs after reinforcement, as in Fig. 433, or pauses, as in Fig. 439, depends upon the early history which brings about one or the other condition. Both are self-perpetuating.

Figure 440 shows the return to VI limited hold after extinction for the second

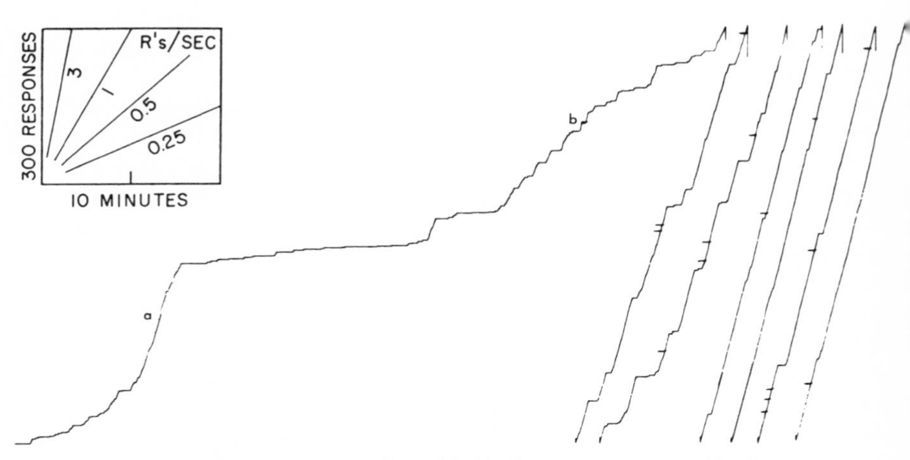

Fig. 440. Return to VI limited hold after extinction (second bird)

bird in the experiment. During the session preceding the figure, extinction had produced a low rate at which few reinforcements would occur. Only 350 responses were emitted during 5 hours and only 1 reinforcement occurred toward the end of the session. The following session (Fig. 440) begins with a positively accelerated segment to a moderately high rate at *a*. The schedule specified a reinforcement during the 10-minute period represented by this first segment, but the bird was pausing during the hold. The rate then falls to a low value followed by a rough acceleration to an intermediate rate at *b*, where a reinforcement is received despite the low over-all rate and rough grain. This first reinforcement at *b* leads to the characteristic performance under VI limited hold after a short period at an intermediate rate.

Extinction after variable-interval reinforcement with limited hold

The response was extinguished after 25 sessions (60 hours) on VI 1 limited hold, with the hold varying from 0.75 to 0.24 second. Figure 441 shows the resulting curve

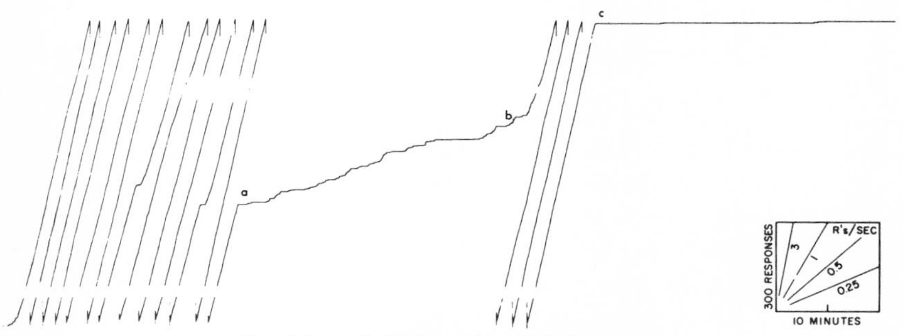

Fig. 441. Extinction after VI limited hold

for one bird. The tendency to respond at a single rate is strong. During 3 hours 16,000 responses occur; several hundred of these, in the range from *a* to *b*, occur at other than the prevailing rate. The rate changes at *a* and *c* are very sharp. (The breaks near the bottoms of the records are due to a defective recorder.)

A later extinction curve, after 7 further sessions (21 hours) on VI 2 and VI 3 without limited hold still shows the effect of the limited-hold reinforcement. Figure 442 begins on a VI 3 schedule of reinforcement. Extinction begins at the arrow. The resulting curve is similar in over-all form to Fig. 441. The bird responds either at the prevailing variable-interval rate or not at all, except for brief accelerations at *a* and *b* and a brief period of slow responding at *c*. These curves resemble extinction after fixed-ratio reinforcement in the general absence of intermediate rates of responding. Both FR and VI limited hold reflect the fact that the probability of reinforcement does not increase with the passage of time during a pause.

Figure 443 describes the return to a variable-interval performance following the extinction shown in Fig. 442. This figure simply continues the curve in Fig. 442.

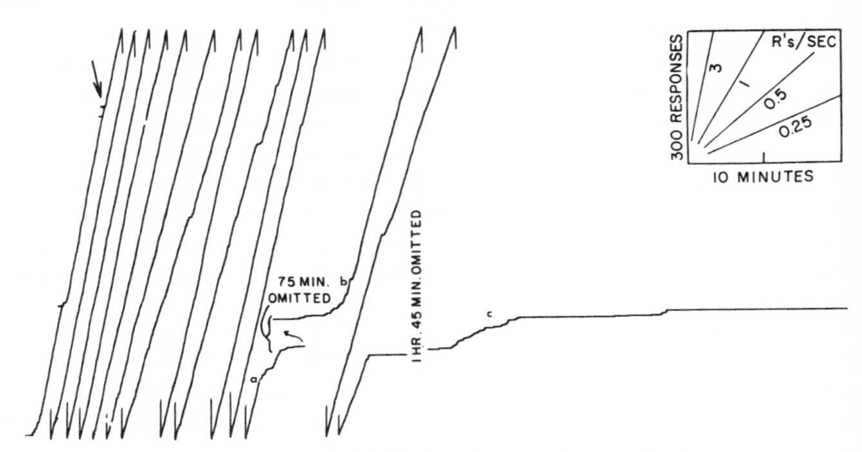

Fig. 442. Extinction after VI limited hold (second bird)

Fig. 443. Return to VI 3 after extinction

The magazine had been re-connected, and the first response at *a* was reinforced. By the third reinforcement thereafter a variable-interval performance is reinstated essentially unaffected by the prolonged extinction.

THE EFFECT OF DEPRIVATION ON PERFORMANCE UNDER VARIABLE-INTERVAL REINFORCEMENT

We conducted two experiments in which the rate of responding under variable-interval schedules of reinforcement was measured as a function of the body-weight of the bird. A single schedule of reinforcement was maintained throughout the experiment, and the body-weight of the bird was changed from day to day during the experiment. As already noted, body-weight is affected by recent drinking, eating of grit, and defecation. When the bird is fed at the same time and the same amount each day, these factors become fairly stable; but with the radical shift in metabolism that occurs when gross changes are made in the body-weight of the bird, they are likely to have a disturbing effect. In interpreting the following results it is also important to note that a given body-weight represents a different condition of deprivation if it follows higher weights than if it follows lower ones.

We altered body-weights by daily feeding or deprivation until a given level was obtained. The successive sessions of the experiment therefore do not always imply successive days.

Rate of responding as a function of body-weight

Experiment I. Final performances for two birds on a geometric variable-interval schedule with a 7-minute mean have already been described in Fig. 408 and 411. These birds were then run on this schedule while their weights were varied over a fairly wide range. Figure 444 shows the results. The numbers beside the points show the order of the sessions. The bird represented by the circles showed the higher rate of responding. Its body-weight ranged from 420 to 569 grams. The *ad lib* weight

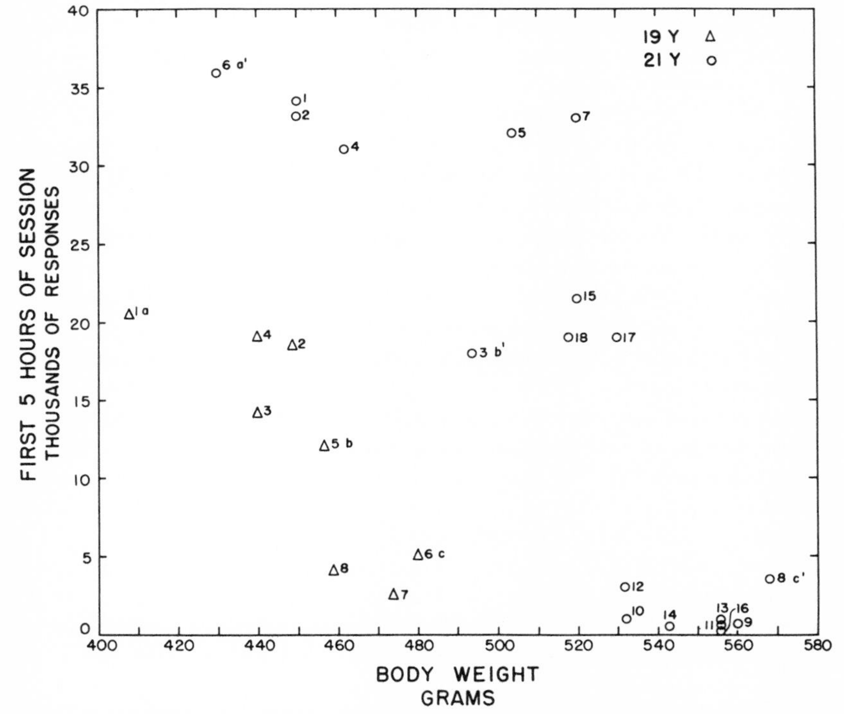

Fig. 444. Plot of mean daily rate of responding as a function of body-weight

of the bird determined 2 years earlier was 522 grams. However, the highest weight recorded, 568 grams, is probably very close to the weight which this bird would have reached at the time of the experiment on a free-feeding schedule. The graph shows a rough relation between the weight of this bird and the rate during the first 5 hours of the session. Between 490 and 530 grams the rates scatter widely. The rate is low between 530 and 568 grams, but increases sharply as the weight falls from 530 to 490 grams. The rate increases more slowly as the weight is further reduced. Samples of variable-interval performances at the lowest, middle, and highest body-weights are shown in Fig. 445 from the sessions marked a′, b′, and c′ on the graph. In Record A deviations from a constant rate are minor. At an intermediate body-weight, Record B, the maximum rate has fallen slightly, largely because of the appearance of grain and slight pauses; and the over-all rate has fallen considerably, largely because of the marked negative curvature which appears during the longer intervals. Each reinforcement characteristically reinstates the maximum rate. In the performance at the highest body-weight (Record C) high rates still appear after reinforcement, but these are sustained for only a few responses.

The second bird, represented by the triangles in Fig. 444, shows roughly the same relative curvature, although there are fewer points and the range of body-weight is smaller. The bird's *ad lib* weight, determined 2 years earlier, was 500 grams; but the range in the present experiment is from 408 to 480 grams. Figure 446 shows segments from the lowest, middle, and highest body-weights, at *a, b,* and *c,* respectively, in Fig. 444. The performance at the lowest body-weight, Record A, is similar to that of the other bird except for the lower over-all rate. However, at the middle body-weight, Record B, both the maximum rate and the average rate fall. At the highest body-weight recorded, Record C, the performance resembles that in Fig. 445.

Experiment II. We conducted a second experiment on the effect of deprivation on the performance under variable-interval reinforcement with the birds which had reached a final performance under VI 3 after the history of VI limited hold already described. The experiment comprised approximately 120 sessions each, during which the body-weights were varied from 370 to 505 grams in one bird and from 382 to 540 grams in another. Each dot in Fig. 447 and 448 represents a daily session at the beginning of which the body-weight was as indicated on the abscissa and during which the average rate of responding was as indicated on the ordinate. Weight and corresponding rates have been averaged for successive blocks of 10 sessions and presented in the solid line in the figures. The numbers to the right of the points indicate the order of occurrence of the blocks of 10 sessions.

During the first 50 sessions in Fig. 447 the relationship between the average rate of responding and the weight of the bird is roughly linear. Near-zero rates of responding were recorded when the weight approached 500 grams. Maximal rates occurred at 420 to 425 grams. Between the 50th and 90th sessions, however, when the weight of the bird was reduced further to 380 grams, the rates of responding generally fell to lower values; and between the 90th and 120th sessions when the weight of

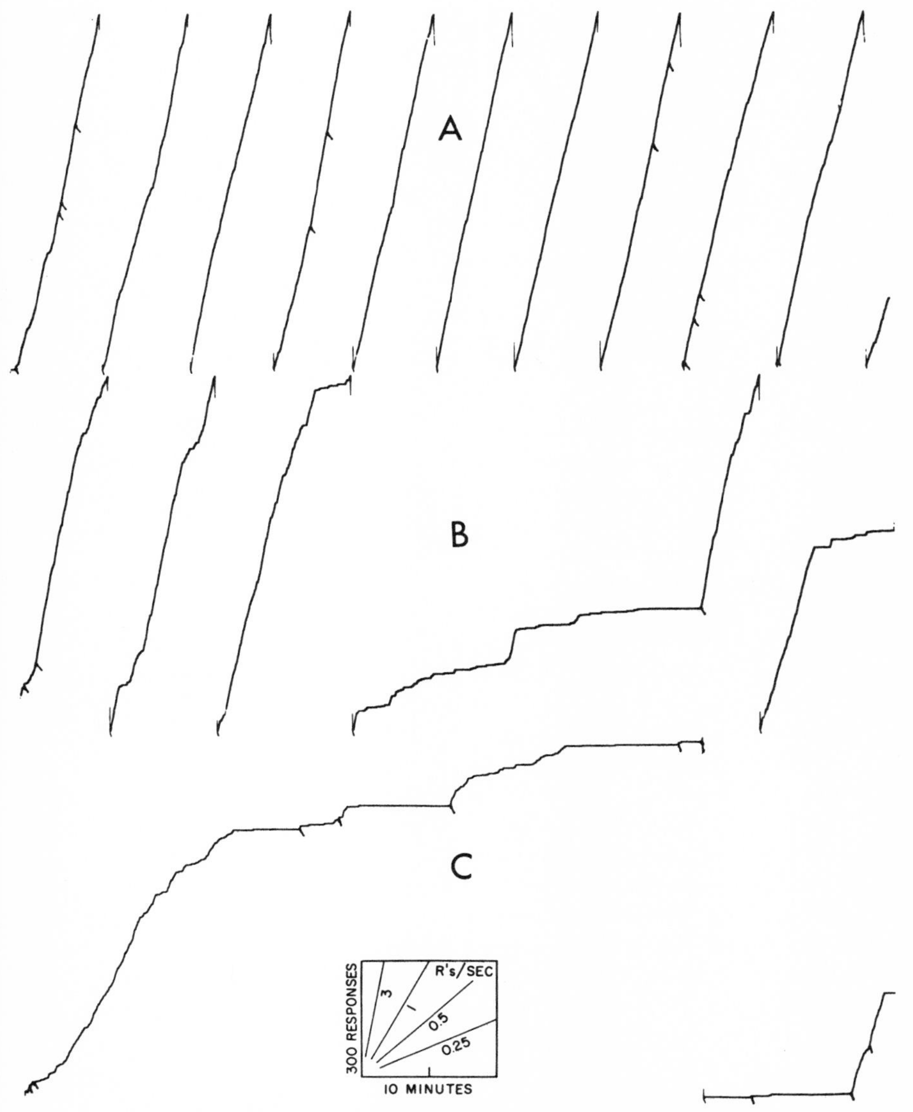

Fig. 445. Sample VI performances at three body-weights

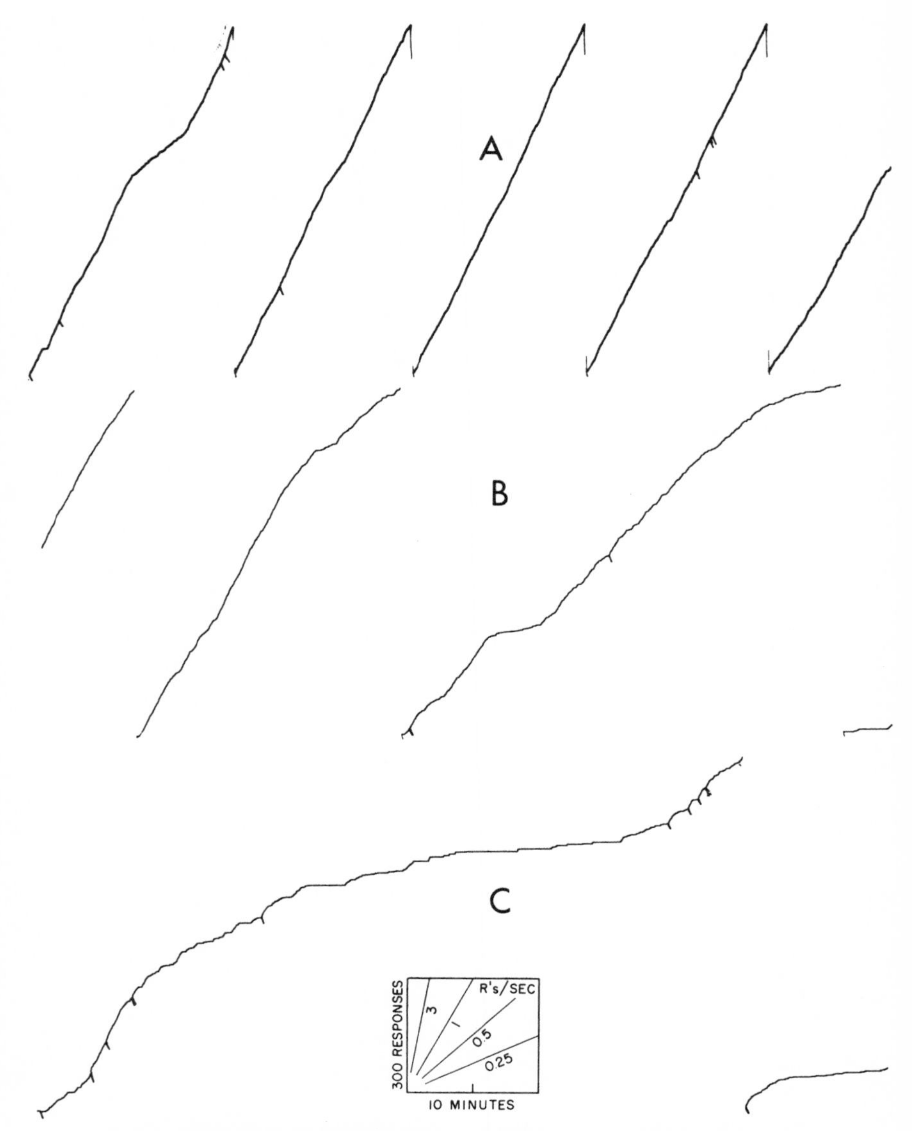

Fig. 446. Sample VI performances at three body-weights (second bird)

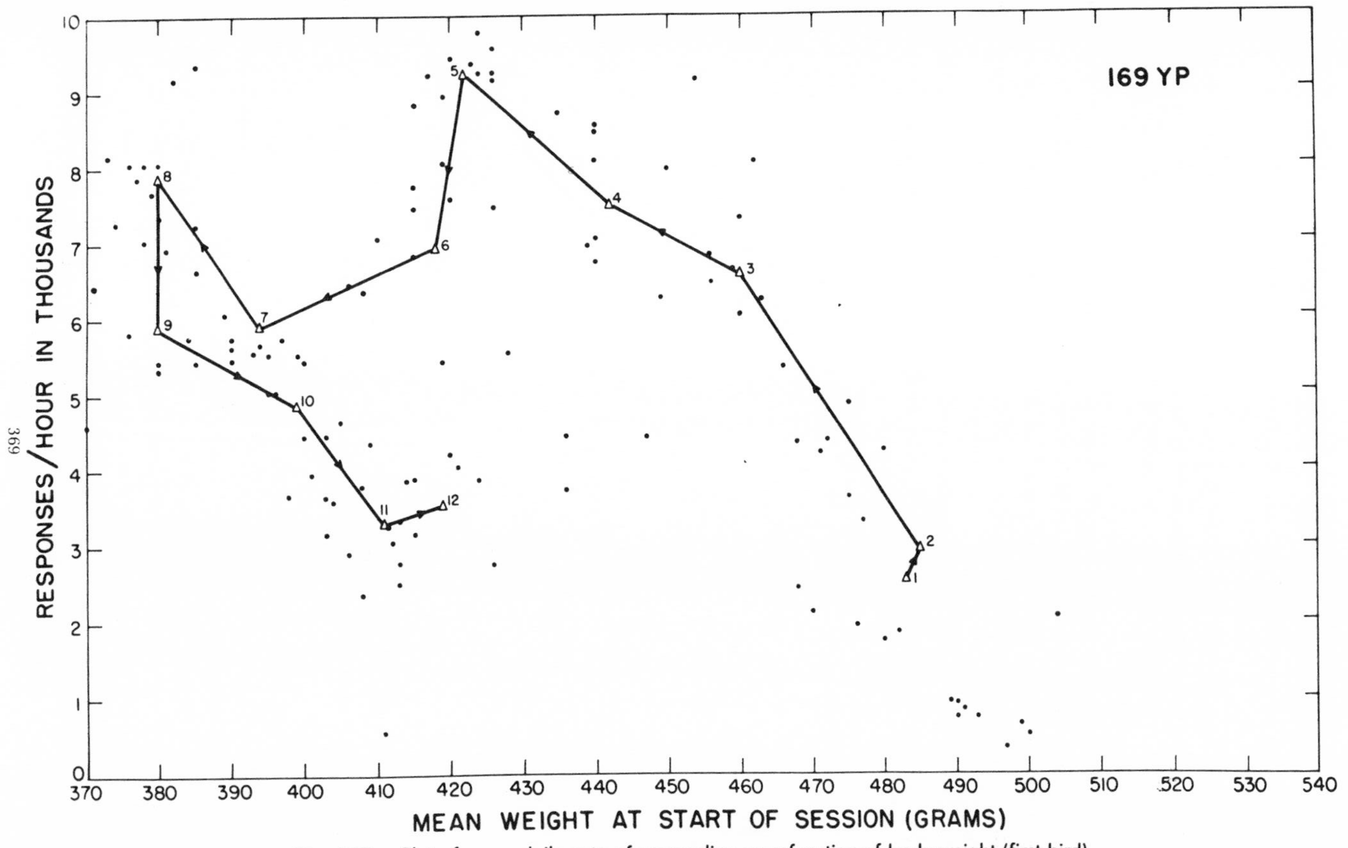

Fig. 447. Plot of mean daily rate of responding as a function of body-weight (first bird)

the bird was now increased again to 420 grams, the rate of responding continued to fall.

The sequence in both birds shows that abrupt and gross changes in body-weight are likely to produce greater changes in rate than a slow gradual change. Some of the variability in the two figures is undoubtedly due to the relatively rough manipulations of weight.

The decline in rate at the very low body-weight is possibly due to inanition. The continued decline in rate as the weights were increased toward the end of the experiment suggests either a progressive effect of deprivation or a change in the variable-interval baseline, because of prolonged exposure under conditions of extreme deprivation.

The maximum change in rate produced by the change in deprivation is by a factor of 4 for the bird in Fig. 447 and of $2\frac{1}{2}$ for the bird in Fig. 448. The range of rates oc-

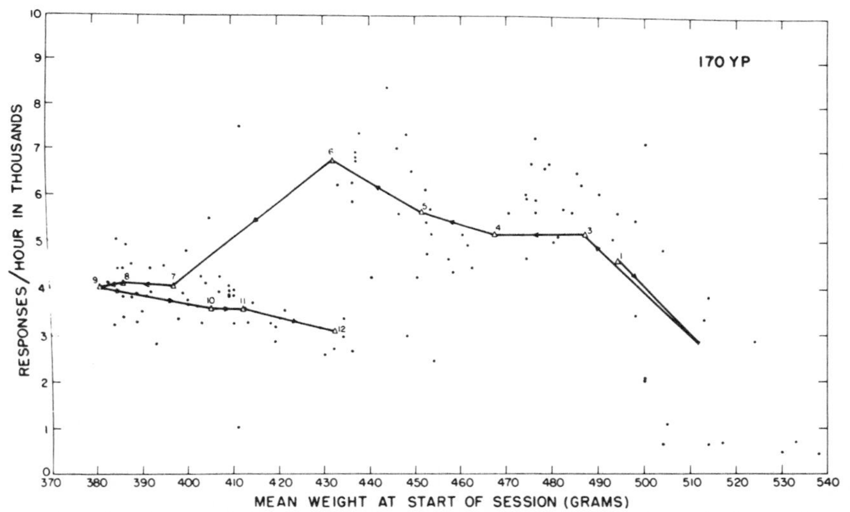

Fig. 448. Plot of mean daily rate of responding as a function of body-weight (second bird)

curring in individual sessions, however, is considerably greater. Figure 449 shows segments from 4 sessions at 4 extreme rate ranges. Record A at 423 grams shows one of the highest rates recorded, in the vicinity of Point 6 in Fig. 448. Record B at 463 grams in the vicinity of Point 4 shows an in-between rate. Record D at 499 grams shows one of the lowest rates recorded in the experiment, to the right of Point 1 in the figure. Record C shows an intermediate rate which occurred at a weight of 408 grams (lower than the body-weight which produced the high rate of responding in Record A). Record C occurred in the vicinity of Point 11, where the curve shows a decline in rate as the weight is increased.

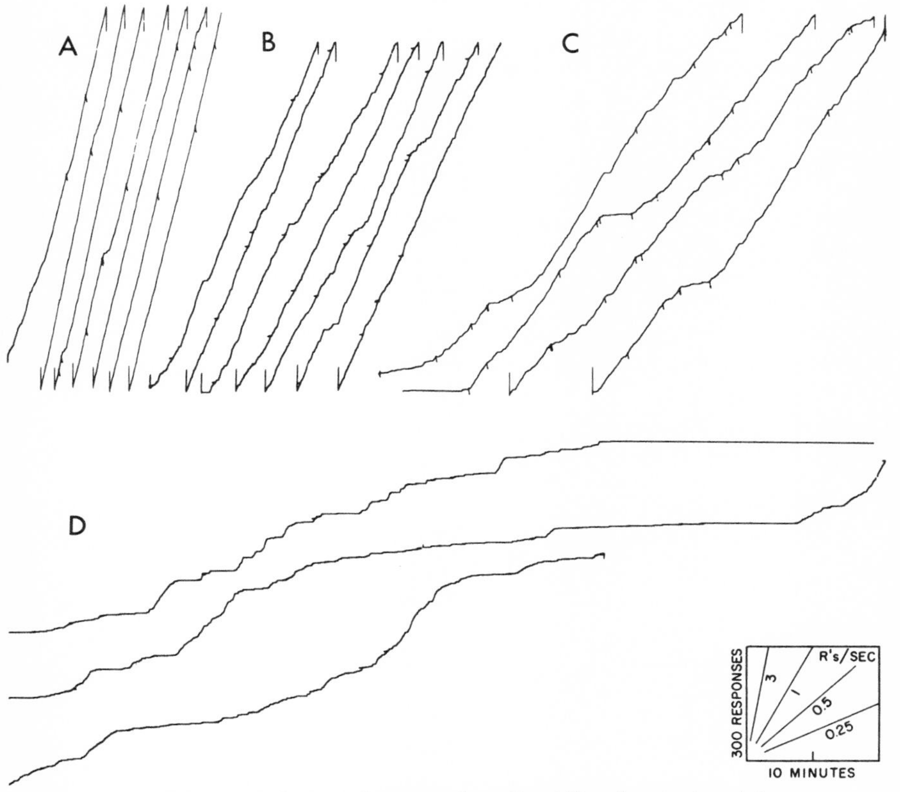

Fig. 449. Sample VI performances from Fig. 448 at four body-weights

Rate of responding as a function of pre-feeding

Various amounts of food, ranging from 10 to 80 grams, were fed to a normally deprived bird just before the start of an experimental session in an attempt to vary the deprivation-level in another way. Two birds showing stable performances on a geometric VI 7 schedule of reinforcement were used. (These are the same birds whose variable-interval performances were recorded as a function of body-weight in Fig. 444.) Figure 450 shows the rates during a 5-hour session. Except at the largest amount of pre-feeding used, the procedure had no effect on the variable-interval performance. The only substantial decline in rate was produced by feeding 80 grams to one bird (*b*). The other bird showed an over-all rate of responding well within the normal range, although it was fed 80 grams or 20% of its body-weight. This result differs from an earlier report of the effect of pre-feeding on FI performances in rats (Skinner, 1938). We have not repeated this experiment with FI reinforcement, but the difference may

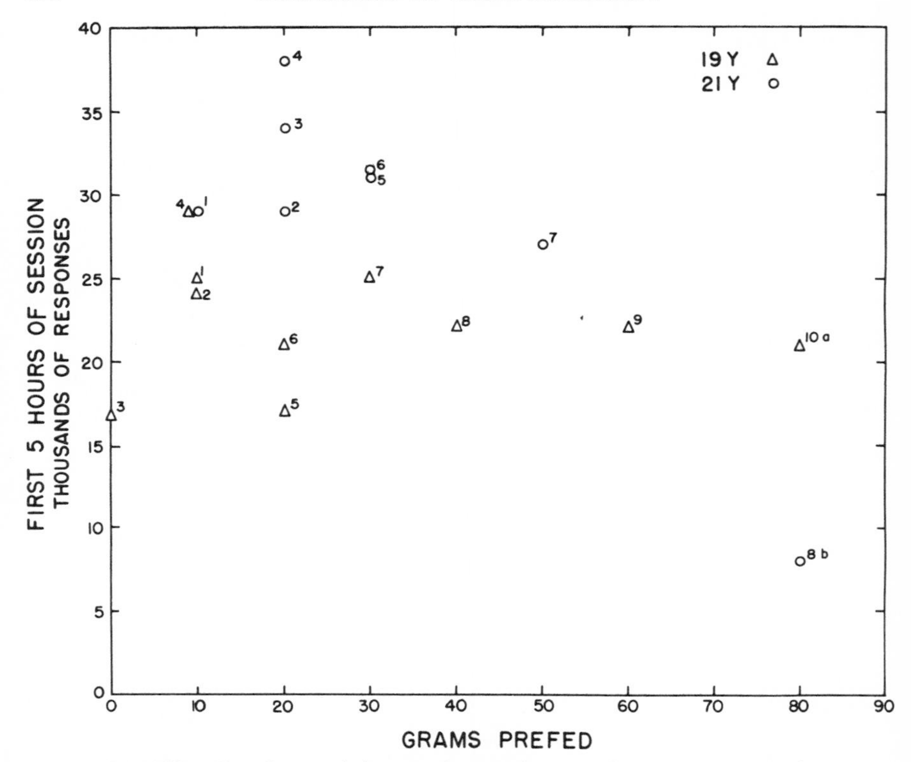

Fig. 450. Plot of mean daily rate of responding as a function of pre-feeding

be due to the use of a VI schedule, which appears, in general, to be less sensitive to variables such as level of deprivation and drugs.

Feeding a pigeon immediately before an experiment and changing its weight over a matter of days with a given feeding regimen are clearly very different operations. But the negative result of pre-feeding is surprising, even so. Most long experimental sessions show a decline in rate, which appears to be due in part to the food ingested during reinforcement. The temporal relations between the ingestion and the effect on rate are similar to those in the pre-feeding experiments. The gradual intake of food might be more effective than the ingestion of a single large amount.

One factor to be considered in all discussions of the effect of deprivation on VI is the rate contingency generated by the schedule. The preceding experiments involved long-standing performances on VI. Reinforcements usually occur at a single rate, or at a rate varying within narrow limits. Hence, once this rate is assumed, it tends to perpetuate itself. Extinction curves after brief VI show more gradual acceleration. Extinction curves after prolonged VI show a stepwise structure in which a rough high rate predominates.

The severe decline in over-all rate during a long session on a well-developed VI appears to show a process of satiation in conflict with a tendency to maintain a single rate.

An extreme example is given in Fig. 451 and 452, which were recorded continuously in this order. The fairly stable VI performance at the beginning of Fig. 451 gives way to performances suggesting the records taken at different body-weights (Fig. 445 and 446). Here, immediate feeding (in a series of reinforcements) produces the same result as a long-term modification of body-weight.

Variable-interval performance under water deprivation

We studied the performance on an arithmetic VI 3 in 3 birds in an apparatus which contained magazines for both food and water reinforcement. In the early stages of the experiment the effective conditions of deprivation were not known, and factors such as the amount of water per reinforcement, the number of hours of water deprivation, and the amount of water fed in the home cage were varied freely until values were discovered which produced stable responding under the variable-interval reinforcement. The birds were deprived of both water and food, but these operations are not independent. Extreme water deprivation reduces food consumption, and both of these reduce weight. However, stable conditions for water reinforcement can be set up.

Figure 453A shows a complete daily session after 12 sessions of water reinforcement on VI 3. The session begins at a high rate, which falls off during the first 2000 re-

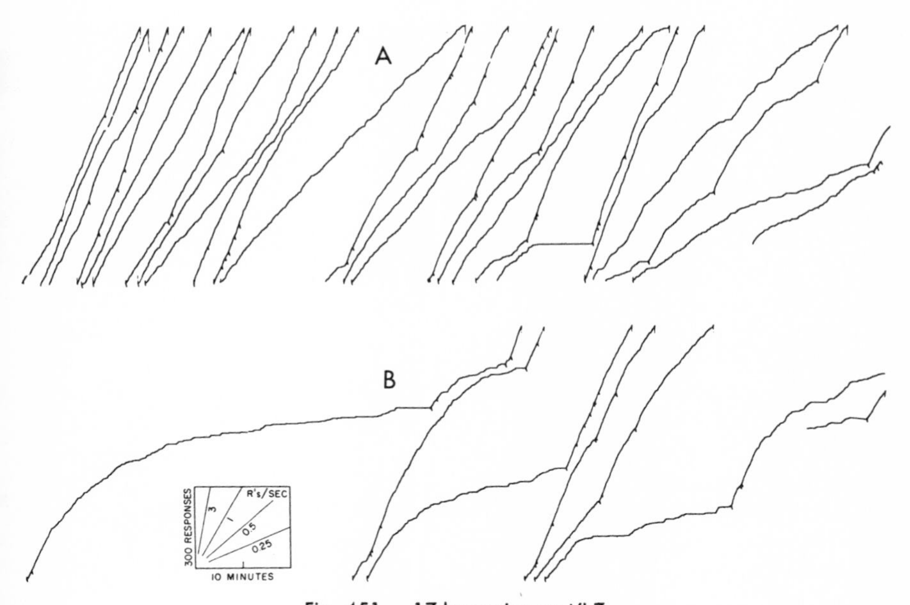

Fig. 451. 17-hr session on VI 7

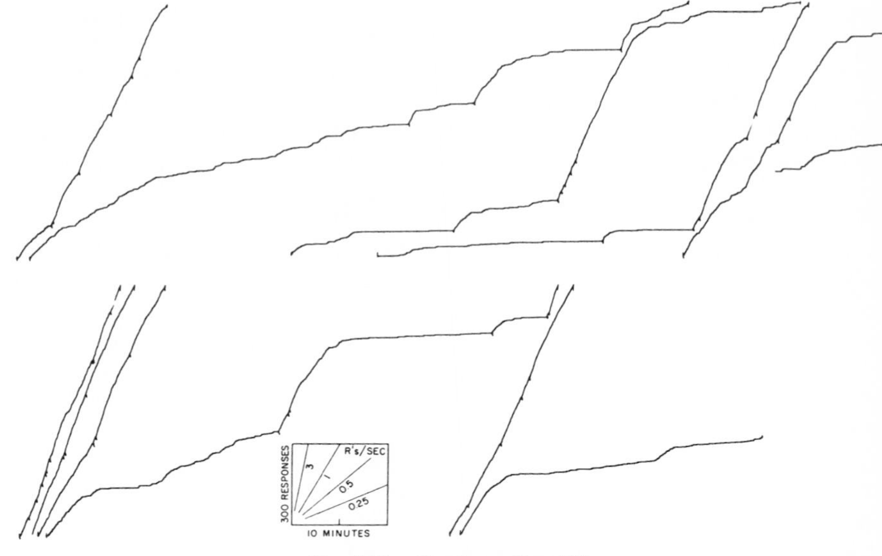

Fig. 452. Continues Fig. 451

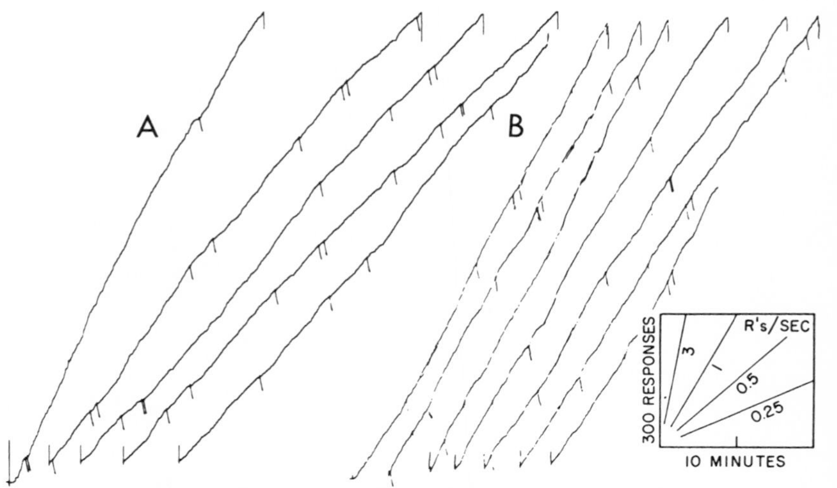

Fig. 453. Performances on food and water deprivation (first bird)

sponses to a value which is maintained for the rest of the session. The rate has a slight tendency to be low after reinforcement with the schedule used. After 6 further sessions of water reinforcement, the water magazine was disconnected and the food magazine substituted. The performance in Record B was recorded after 3 sessions of food reinforcement. The rate of responding is higher than that under the preceding water reinforcement, and the rate immediately after the reinforcement is now relatively high. The performance in Record A represents the highest rate which we could generate under water deprivation. Food deprivation and reinforcement appear to be somewhat more effective in controlling a high level of activity.

Figure 454 shows the transition from food to water reinforcement under conditions of both food and water deprivation. At the arrow, food reinforcements are discontinued

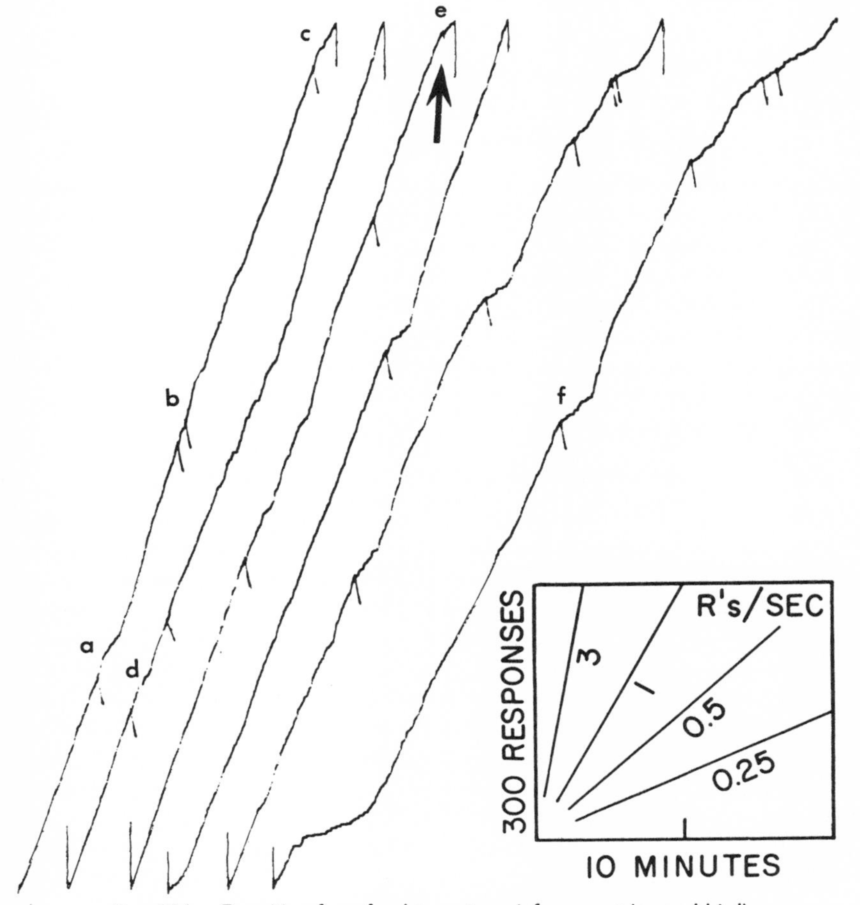

Fig. 454. Transition from food to water reinforcement (second bird)

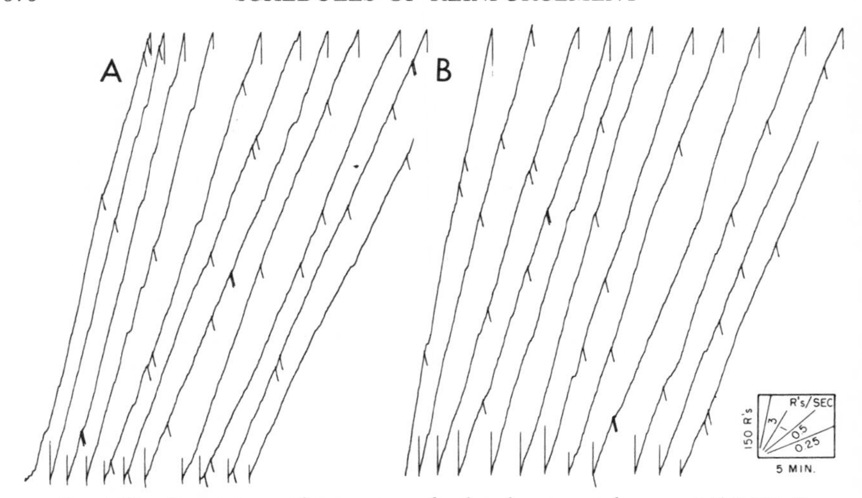

Fig. 455. Successive performances on food and water reinforcement (third bird)

and water reinforcements substituted. The first water reinforcement at *e* is followed by a decline in rate, which becomes most marked at *f*. This performance should be contrasted with that at the start of the record where the rates following the reinforcement with food at *a*, *b*, *c*, and *d* increase slightly over the prevailing rate.

Figure 455 compares food and water performances on VI reinforcement for another bird. This bird had a higher over-all rate on both food and water. This figure also shows a higher rate of responding under food reinforcement (Record B) than under water reinforcement (Record A). However, the rate immediately after reinforcement is roughly the same in both cases.

THE EFFECT OF A PRE-AVERSIVE STIMULUS ON A VARIABLE-INTERVAL PERFORMANCE

A pre-aversive stimulus is a stimulus which characteristically precedes an aversive stimulus. An example is a buzzer which is characteristically followed by a shock, as in the procedure used by Estes and Skinner (1941). A hungry rat is reinforced on a fixed-interval schedule. During the session a buzzer is presented for 3 minutes; the rat is then shocked through a grid floor. As a result, the rate of responding declines—possibly to zero—when the buzzer, the pre-aversive stimulus, is present. The effect is similar to the kind of disruption in human behavior called "anxiety." The effect of a pre-aversive stimulus of food-reinforced behavior has been used extensively by Hunt, Brady, Lindsley, and others to study the effects of electroconvulsive shock, drugs, and radiation.

The effect of the pre-aversive stimulus on the performance of the pigeon on VI was studied in the following way. A stable performance on VI 3 was established. Then, the general illumination in the apparatus was dimmed for 30 seconds. This procedure was repeated every 5 minutes, until the novel stimulus no longer had an effect on the VI rate. Each 30-second period during which the light was dim was then followed by an 8-second electric shock of approximately 400 volts, 60 cycles a c, delivered through a floor grid (independent of a response). The pigeon's feet were coated with a light film of graphite paste, and the polarities of the grid were constantly shifted at a high rate in order to prevent serious shorting of the circuit. We determined the level of the shock by adjusting the voltage to elicit vigorous leg-lifting movements. The intensity was kept below values which produced violent reactions.

Development of a final performance with the pre-aversive stimulus

Figure 456 (the 2nd session in which the pre-aversive stimulus was followed by shock) shows the development of a lower rate of responding during the pre-aversive stimulus. (The vertical lines below the record refer to the start of the pre-aversive stimulus and the occurrence of the shock at the end of the pre-aversive stimulus.) Complete suppression of the variable-interval performance is evident at *c* and *d,* and partial suppression at *a* and *g,* where some responding occurs during the 30-second interval. At *b* the rate remains low beyond the termination of the pre-aversive stimulus and shock; but, in general, the normal VI rate is assumed immediately after the shock, at the end of the pre-aversive stimulus. Note that the schedule of reinforcement is maintained through the pre-aversive stimulus, and that reinforcements

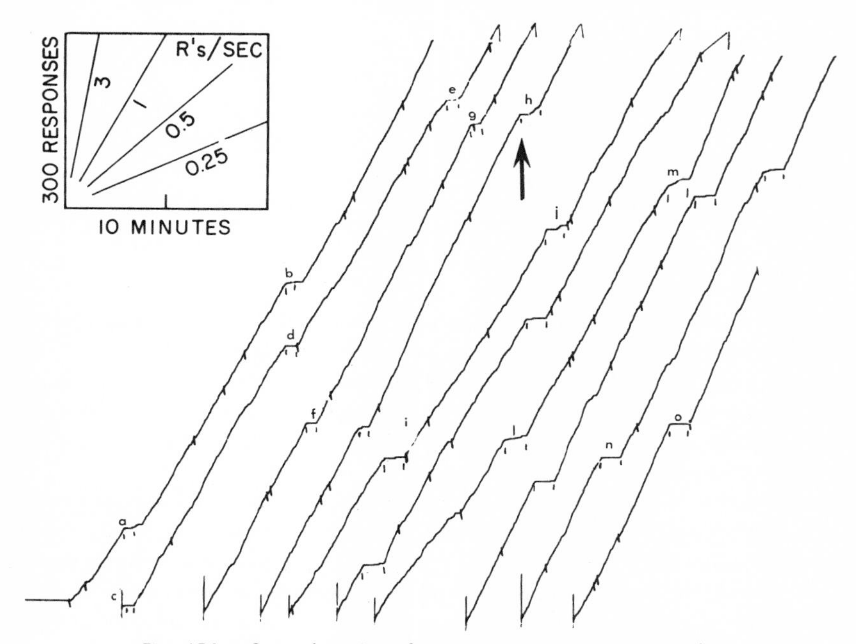

Fig. 456. Second session of exposure to pre-aversive stimulus

occasionally occur at that time, as at *g*. At the arrow the duration of the pre-aversive stimulus was increased to 60 seconds. A zero rate is first maintained for about the previous duration of 30 seconds, and is followed by some responding during the second half of the pre-aversive stimulus, as at *h*, *i*, and *j*. For the remainder of the session, however, either the rate during the pre-aversive stimulus is lower but sustained, as at *l* and *m*; or (toward the end of the session) responding completely stops, as at *n* and *o*. When the pre-aversive stimulus was not present, the performance remained appropriate to the schedule of reinforcement.

Figure 457 shows the performance on VI with a 60-second pre-aversive stimulus every 5 minutes after a total of 62 sessions (240 hours). The rate during the pre-aversive stimulus is now very close to zero, although some responding may occur, as at *e*, *f*, and *i*, during the first few seconds. The rate under VI is often suppressed after

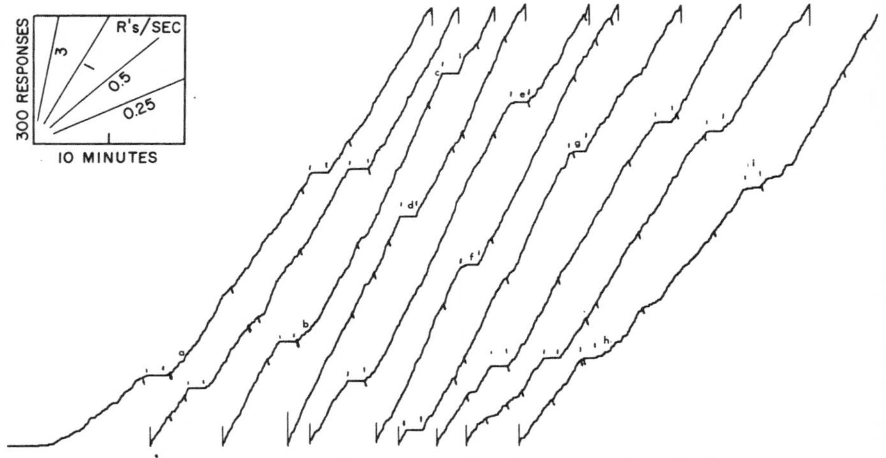

Fig. 457. Late performance on VI with a pre-aversive stimulus

the shock, as at *b*, *e*, *h*, and *i*, although an immediate shift to the highest observed rate may take place, as, for example, at *c*, *d*, and *g*. The session also begins with some warm-up, and the prevailing VI performance is not reached until possibly 15 minutes after the start of the experiment. The grain is much rougher than that in the earlier VI performance in Fig. 450, and the over-all rate has fallen, largely because of the lower rate immediately following the shock.

Occasionally, as at *d* in Fig. 457, the higher rate after the shock suggests compensation for the pause during the pre-aversive stimulus. Figure 458 contains several similar instances from the immediately preceding and following sessions for the same bird. The dotted lines in Records A and B are intended to suggest the extrapolation of earlier parts of the curves. The rate after the pre-aversive stimulus and shock exceeds the prevailing VI rate until the extrapolation is roughly reached.

The presentation of the pre-aversive and aversive stimuli on an FI schedule (e.g., every 5 minutes) eventually sets up a temporal pattern. The buzzer is no longer the only pre-aversive stimulus. The bird's own behavior is a stimulus which can enter into the contingencies of the schedule of shocks and acquire the power to suppress the rate. Such stimuli show temporal gradients.

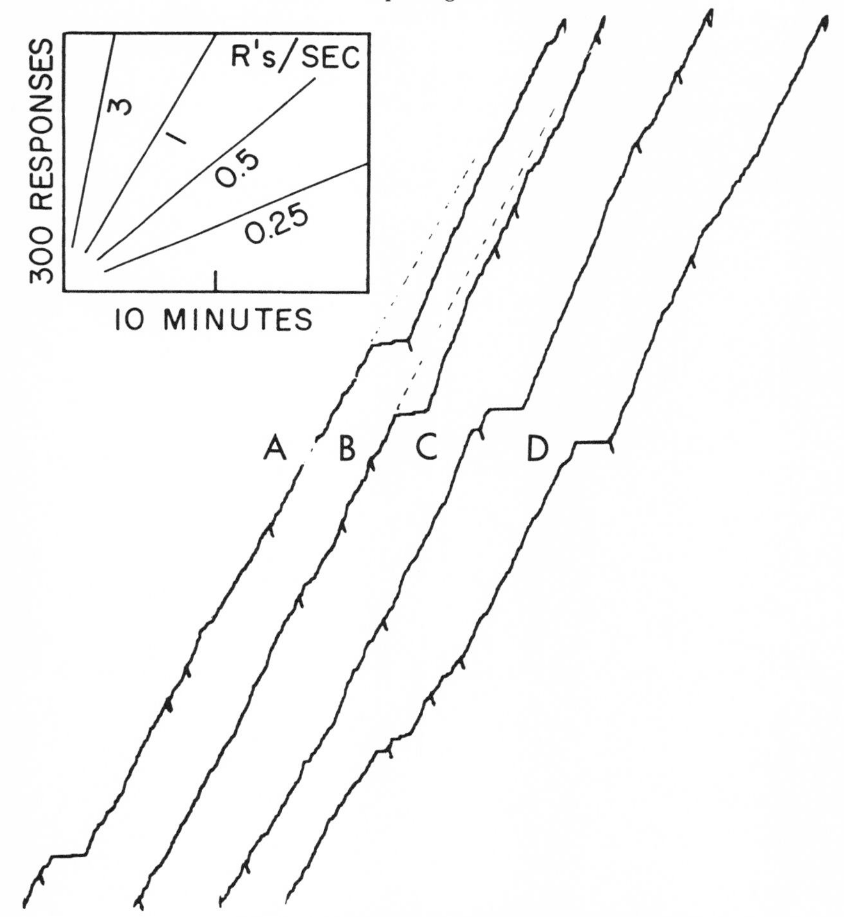

Fig. 458. Compensation following the pre-aversive stimulus

Figure 459 shows a later performance for another bird on VI 3, in which a 60-second pre-aversive stimulus occurred every 10 minutes. This is the 67th session of the pre-aversive procedure. The intervals between aversive stimuli had been 5 minutes until the 11th session preceding the figure. The particular VI schedule produces a brief run after reinforcement. The rate during the pre-aversive stimulus is sometimes zero, although responding is usually slow but substantial. The session characteristically begins with a very low rate which increases gradually as the session progresses. The prevailing VI performance is not reached until at least 45 minutes after the start of the session. The rate during the 10 minutes between the presentations of the pre-aversive stimulus frequently shows a marked decline, as at *a, b, c, d,* and *e.* At the end of the pre-aversive stimulus the rate quickly accelerates to a maximal value, and then declines continuously until the next presentation of the pre-aversive stimulus. This effect is most marked at the beginning of the session. A similar performance 5 sessions later, shown in Fig. 460, illustrates the reproducibility of the main features.

The effect of the pre-aversive stimulus during extinction

Figure 461 shows the effect of pre-aversive and aversive stimuli on extinction after VI 3. After 74 sessions of VI reinforcement with a periodic pre-aversive stimulus (Fig.

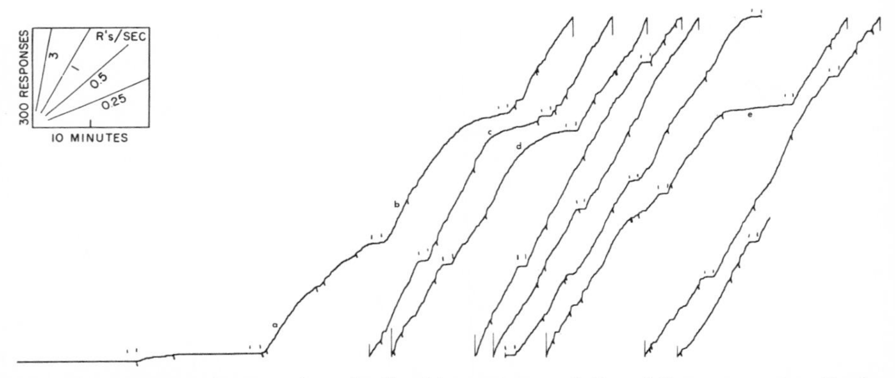

Fig. 459. Temporal pattern from the fixed-interval presentation of the pre-aversive stimulus

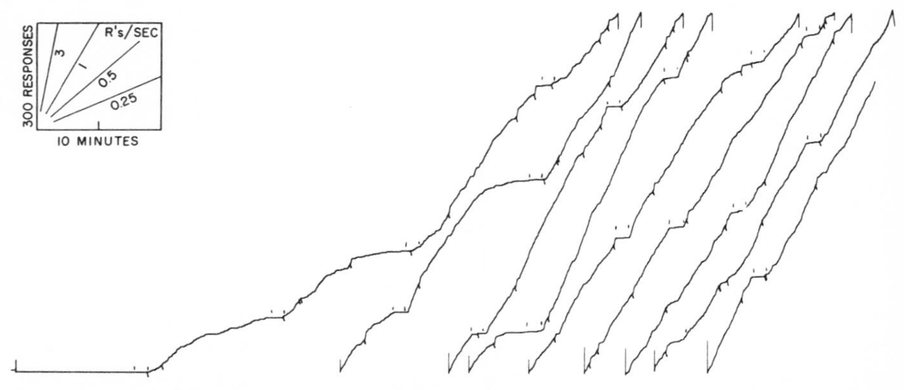

Fig. 460. Temporal pattern from the fixed-interval presentation of the pre-aversive stimulus

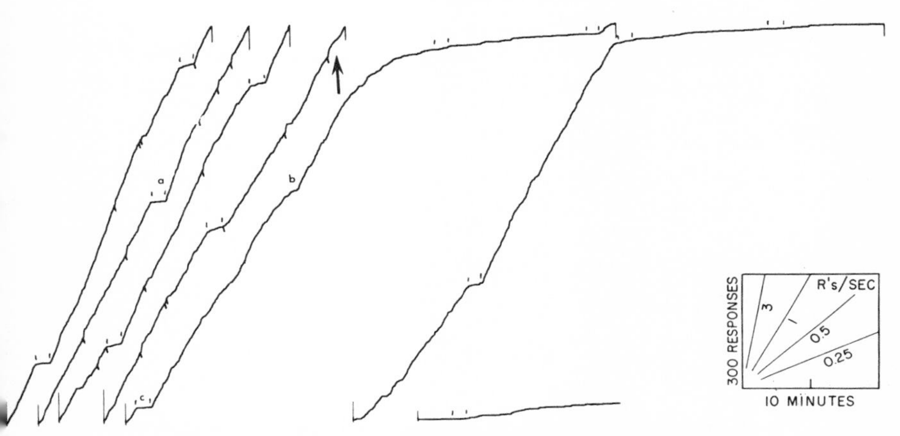

Fig. 461. The effect of the pre-aversive stimulus during extinction

459 and 460 are samples), reinforcement was discontinued. The warm-up part of the session has been omitted in Fig. 461. This figure shows compensatory fast responding after the termination of the pre-aversive and aversive stimuli, as at *a*. (Cf. Fig. 458 for the other bird.) The magazine was disconnected at the arrow, but the pre-aversive and aversive stimuli continued to appear every 10 minutes as before. The resulting extinction curve contains 2 segments of approximately 1000 responses each at the prevailing VI rate, separated and followed by segments where the rate is constant but near zero. The second pre-aversive stimulus after extinction has begun (at *b*) shows only a slight effect. However, the first and fifth presentations, occurring against a background of a high rate, show the suppression effect. The fourth, occurring during slow responding, may be followed by a "rebound." In general, this is very sharp curvature after so long a history of VI 3. It suggests that the aversive stimuli are hastening the extinction process.

Removal of pre-aversive stimulus

Following the final performances shown in Fig. 457 and 459, the pre-aversive stimulus was omitted. The same VI 3 was in force throughout the session, and a shock occurred every 5 minutes, although it was no longer preceded by a pre-aversive stimulus. This procedure was designed to show the disruptive effect of the preceding shock on the VI performance apart from any effect of the pre-aversive stimulus. Figure 462 shows a performance after 9 sessions during which the shock appeared "unannounced" every 5 minutes. The major effect has been severe suppression of the over-all rate during the early part of the session as an intensification of the earlier warm-up. The session represented shows a continuous acceleration of the over-all rate during the first 4 or 5 excursions of the recording pen which cover about 2 hours of the session. The periodic shocks were not marked, but no rate changes having a

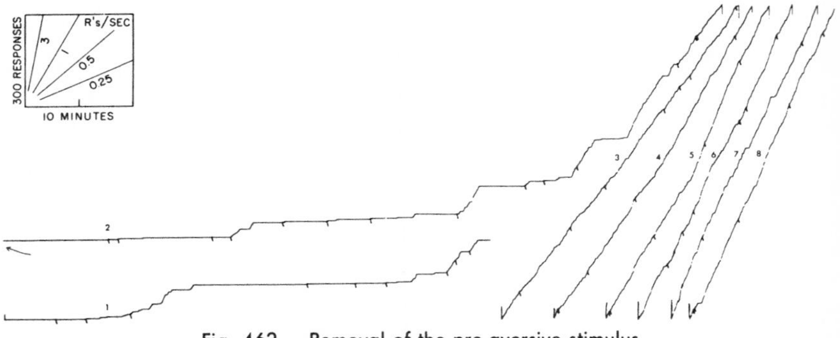

Fig. 462. Removal of the pre-aversive stimulus

period of 5 minutes are apparent. There is no evidence that the periodic presentation of the shock has any consistent effect on the rate comparable with the curvature of Fig. 459 and 460. This procedure shows that the pre-aversive stimulus tends to restrict the effect of the shock to the pre-aversive period.

Removal of the aversive stimulus

The shock was removed from the program in the following session. The schedule of reinforcement was now simply VI 3. Figure 463 shows the same low rate at the start of the session and the acceleration to the normal VI performance, although these are considerably less marked than those in Fig. 462. A normal VI performance is reached within an hour; and the over-all rate is higher (1.5 responses per second compared with maximum over-all rates of 1.25 in the previous session).

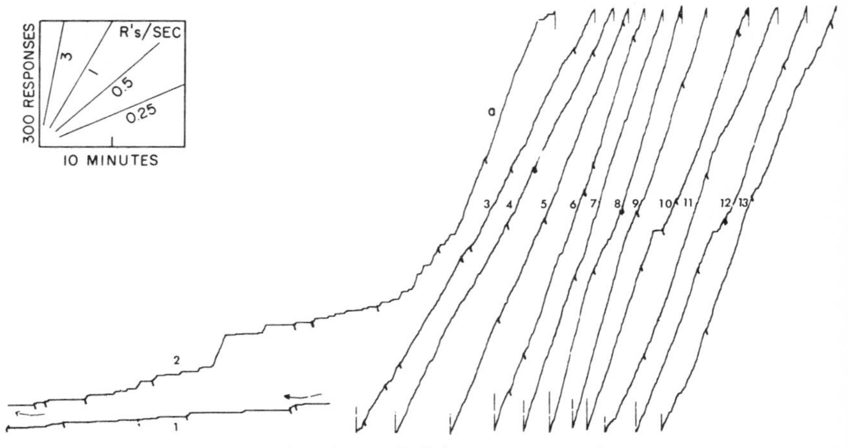

Fig. 463. Removal of the aversive stimulus

Extinction without aversive or pre-aversive stimuli

Figure 464 shows extinction in the absence of both pre-aversive and aversive stimuli. This occurs in the session immediately following Fig. 457. The session begins with the slight warm-up which this bird has been characteristically showing. In the absence of aversive stimuli the rate reaches a higher than usual value, however. The highest rate (at *a*) is approximately 1.1 responses per second. During the first 6000 responses the rate declines fairly smoothly; it falls nearly to zero for 10 hours (omitted from the figure), during which only 350 responses occurred. Extinction continues with the low over-all rate in the last 3 segments of the figure. Oscillations in rate at *b* through *h* are more marked than usual in extinction after VI reinforcement, and may be the result of the previous aversive and pre-aversive stimulation. The periodicity could reflect the period of aversive stimulation; or it could mean that the

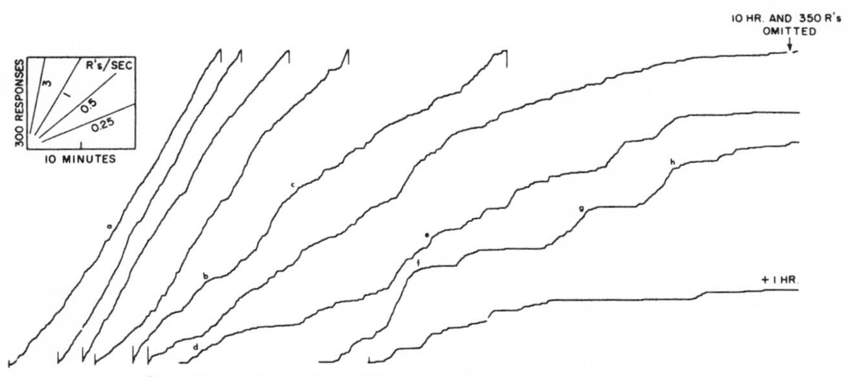

Fig. 464. Extinction without aversive or pre-aversive stimuli

presence of a rate at which shocks have frequently occurred suppresses the rate, but that the new rate automatically eliminates the suppressing stimulus.

Pre-aversive stimulus in the rat

For comparison, Fig. 465 shows the effect of a single pre-aversive stimulus on performance on VI in the rat. The pre-aversive stimulus was 5 minutes long and occurred in the middle of each daily session. The aversive stimulus was a shock. The figure shows 15 successive daily sessions in which the pre-aversive and aversive stimuli occurred. The lever pressing stopped almost completely at the end of the 5-minute period, but nearly all segments show responding for the first few moments of the pre-aversive stimulus. This performance represents a temporal discrimination. The VI rate is considerably depressed immediately following the shock, and the over-all rate remains lower during the latter part of the session. The normal decline in the VI schedule of reinforcement makes it impossible to attribute this effect entirely to the pre-aversive program.

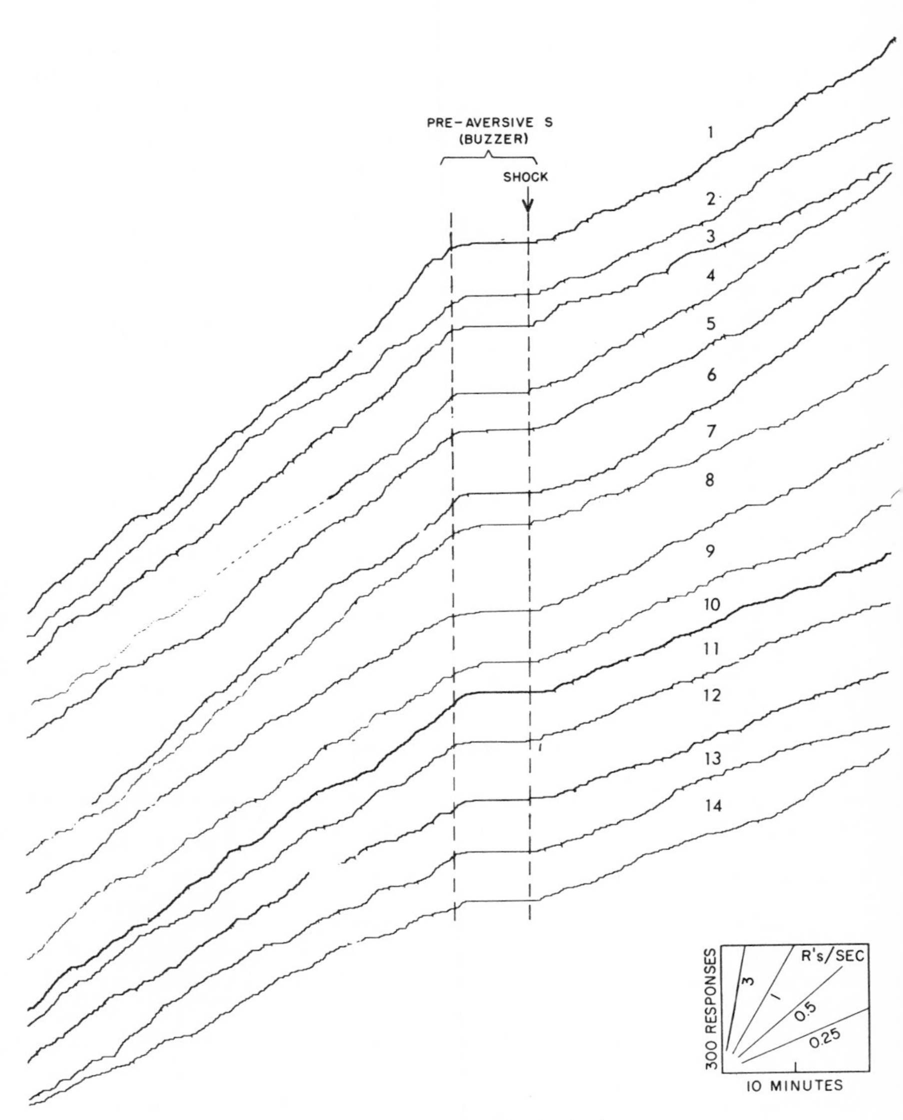

Fig. 465. Pre-aversive stimulus in the rat

CONTINUOUS ADJUSTMENT OF DRUG LEVEL IN TERMS OF BEHAVIOR

We used variable-interval performances in studying the effects of certain drugs. In one experiment we explored the possibility of controlling the level of sodium pentobarbital during a session by automatically administering the drug in relation to the rate of responding on a variable-interval schedule. This possibility was suggested as a convenient alternative to other criteria for the repeated administrations of a drug, such as the amount of the drug in blood samples, or collateral effects such as rate of breathing. The plan was to establish a stable performance on VI and to administer sodium pentobarbital slowly and continuously until the rate fell below a certain value. The drug would be discontinued whenever the rate of responding was below the given value and re-administered whenever the rate of responding was above it. (With a drug which had an excitatory effect the procedure would be the opposite; the drug would be administered whenever the rate fell below a certain value and withheld when it increased above it.)

In the present experiment we administered the drug orally by infusing the grain with an aqueous solution of sodium pentobarbital. A given amount of grain was infused with a specified solution and allowed to dry quickly. Some of the drug was undoubtedly lost through oxidation under these conditions. We were not interested at this stage in specifying the amount of sodium pentobarbital actually given from day to day but rather in amounts relative to the concentrations of the infusing solution. In order to evaluate the effect of a specific drug, dose levels would, of course, have to be controlled more accurately. This control could be achieved with better methods of preparing the drug for oral administration, by adjustment of the concentration of a gas in an ambient atmosphere, or by a continuous method of intravenous injection.

The experimental chamber contained 2 food magazines, one of which contained clean grain and the other grain infused with sodium pentobarbital. An arbitrary rate of responding was specified to determine whether reinforcement would be drugged or clean. We tried several values of criterion rates of responding. In the experiment to be reported here, reinforcements were drugged whenever the rate was above 0.3 response per second for 3 minutes. Conversely, reinforcements were made with undrugged grain whenever the rate was below 0.3 response per second for 3 minutes. Figure 466 contains graphs of complete daily sessions under the various procedures carried out in the experiment. The curves are reduced in order to emphasize the over-all rate changes occurring during the sessions. Record A shows a control VI performance without sodium pentobarbital. This was taken after approximately 34 sessions on VI during most of which the drug had been automatically adjusted. The over-all rate is roughly constant, with the usual slight negative curvature. A second bird in the experiment showed a similar curve under these conditions. Record B shows the effect of a small amount of sodium pentobarbital delivered in all reinforcements. The curve in Record B is negatively accelerated throughout and shifts abruptly to a low rate before the end of the session. Record H shows a perform-

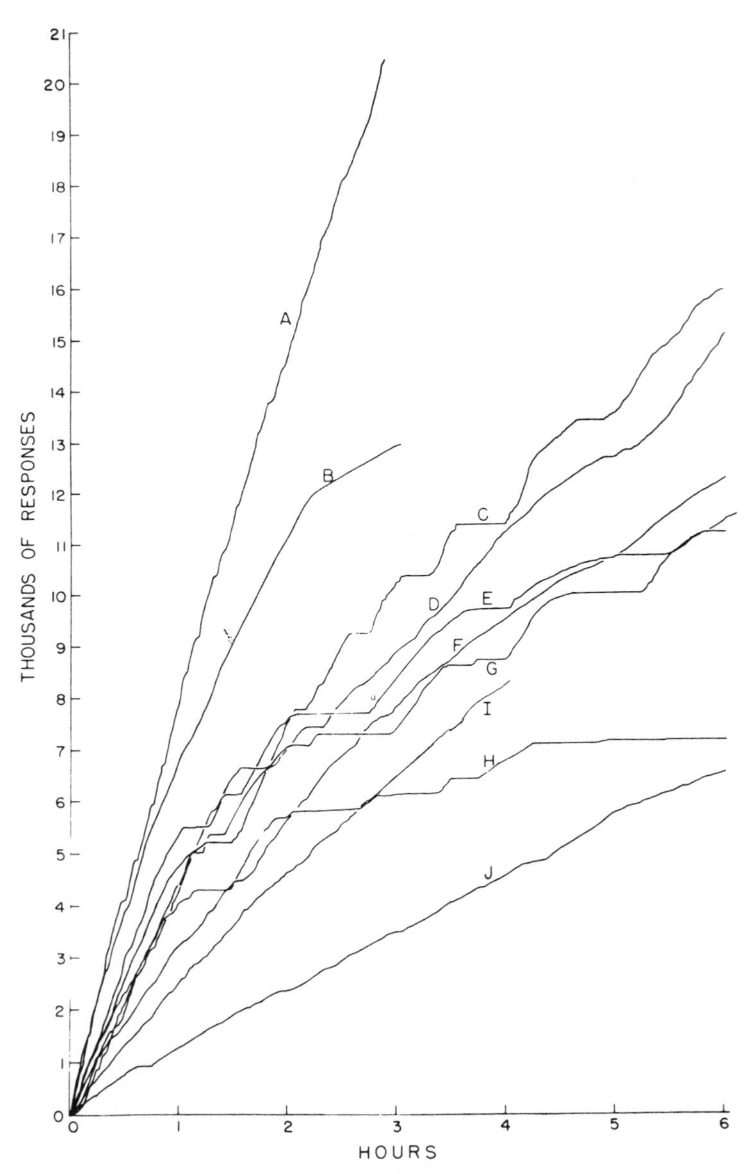

Fig. 466. Daily sessions under various conditions of drug administration

ance in which all reinforcements were drugged with a larger amount of sodium pentobarbital. This dose level produced the most markedly negative acceleration and the lowest number of responses for the session. (In Curve J a slow-down contingency has been added to the variable-interval schedule. See Chapter Nine.) In Curve H the decline in rate is not continuous. The over-all negative curvature is produced by increasingly longer pauses and periods of low rate. In Record C the adjusting pentobarbital doses produce a rapid oscillation in rate. Here, the bird is either not responding and receiving no drug, or it is responding at a rate higher than the critical value and receiving all drugged reinforcements. Figure 467 shows the same curve enlarged. The first 5000 responses of the session are run off at essentially a normal VI rate except for brief runs at an intermediate rate after some reinforcements. All reinforcements are drugged. The first response following the pause at *a* is reinforced with clean grain, and a substantial rate quickly develops. Drugged grain is then received until *b,* where the rate again falls off to zero for approximately 5 minutes. As the session progresses, the oscillation becomes more marked and the pauses longer. The pause at *c,* for example, is 25 minutes. A more prolonged acceleration brings the rate back to its highest value at *d,* and extended negative curvature over the next 1300 responses leads to a zero rate again at *e.*

Record D in Fig. 466 shows a daily session in which the adjusting drug level produces an intermediate rate of responding with less rapid oscillations than those in

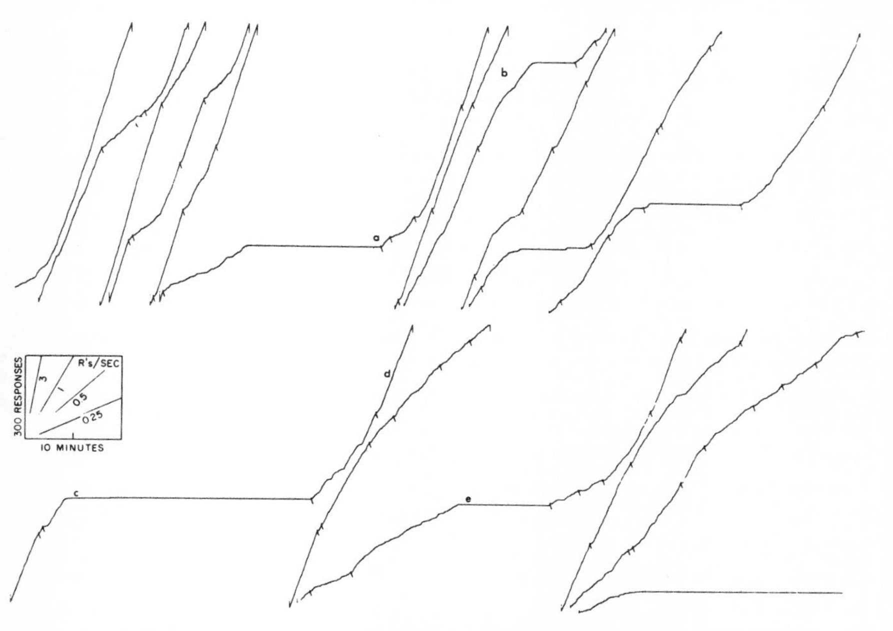

Fig. 467. Daily session showing rapid oscillation under adjusting drug level

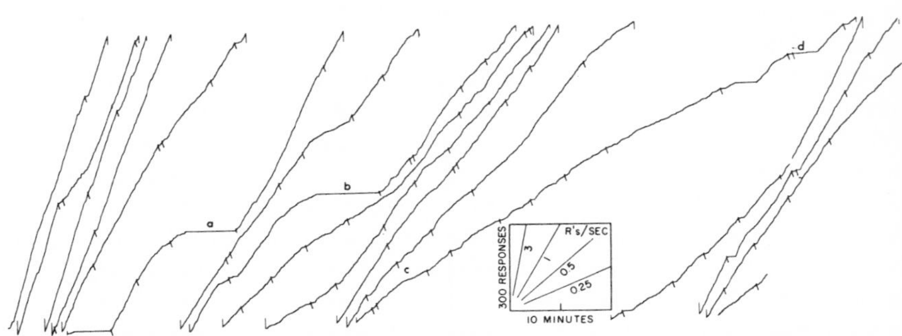

Fig. 468. Daily session showing intermediate rate under the adjusting drug level

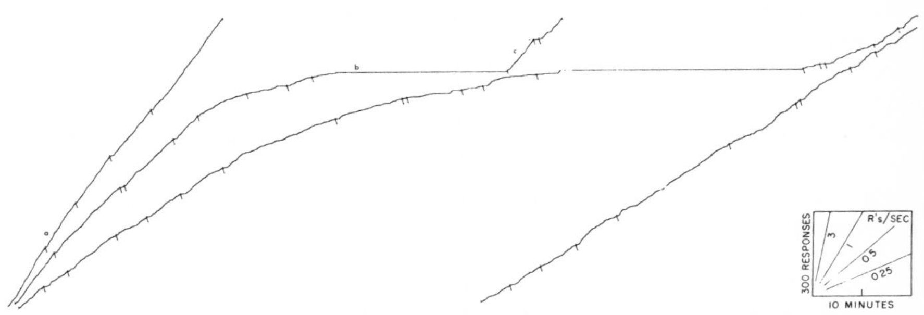

Fig. 469. Last part of a daily session showing oscillation and over-all negative curvature under larger amounts of adjusting drugs

Record C. The over-all curve is negatively accelerated. Figure 468 shows the entire record in detail. At *a* and *b* the rate falls to zero under the influence of the sodium pentobarbital in most of the preceding reinforcements. The section of the record from *c* to *d* shows an intermediate rate of responding of the order of 0.3 response per second. This is the rate above which reinforcements are drugged and below which they occur without the drug. The procedure does not, however, maintain a steady state. Later, the rate accelerates continuously as the effect of the now only occasionally administered drug wears off.

Record E in Fig. 466 shows a daily session with a larger dose per reinforcement, where the criterion rate was 0.5 response per second for 3 minutes rather than 0.3 response per second as in the rest of the sessions described. The session shows a more marked over-all negative acceleration and more responding at intermediate rates below the critical value. Figure 469 shows part of Record E, beginning at *a* in Fig. 466, in greater detail. The change from a high rate at *a* to a zero rate at *b* is fairly smooth. A quick return to an intermediate rate at *c* is followed by a continuous decline in the over-all rate for more than an hour. Here, a more graded effect has been achieved, but by no means a stable state.

A closer approach is made in Record F, which shows a daily session where scattered drugged reinforcements produce an intermediate rate of responding that declines continuously during the session. Record G shows a performance similar to Record C, with more extended pauses and a more gradual acceleration from one rate to another.

Ordinarily, the rate of responding is decreased when reinforcements are discontinued. When reinforcements have contained a drug, however, a cessation of reinforcement may produce the unusual effect of a temporary increase in rate. Figure 470 shows the end of a daily session, in which oral administration of sodium pentobarbital at reinforcement had produced a very low rate of responding. After the reinforcement at *a*, the magazine was disconnected. The effect is a continuous increase in the rate during the next 2500 responses as the effect of the drug wears off. At *b* the rate begins to fall in extinction, but a reinforcement at *c* reinstates an intermediate variable-interval rate.

We reduced the amount of oscillation in this experiment by adding a drl contingency to the variable-interval schedule. (See Chapter Ten.) Record I in Fig. 466 shows a performance under a schedule in which reinforcements were available on VI but a response was not reinforced unless 2 seconds had elapsed since the preceding response. The over-all rate is approximately 0.6 response per second and the curve is roughly linear, although the rate of responding has some tendency to be higher at the very start of the session. The rate just before the drl contingency was added was 1.3 responses per second. (Details of the performances before and after drl was added appear in Fig. 569 in Chapter Nine.) When the adjusting drug procedure was added to this schedule, it produced the stable intermediate rate of responding seen in Record J, Fig. 466. Here, the over-all rate is 0.3 response per second—the criterion value above which reinforcements are drugged and below which they are not. A slight amount of "hunting" is evident in the small grain of the curve. This curve shows the successful maintenance of a constant rate of responding by an adjustment of the drug dosage in terms of the rate of responding.

The infusion of the grain with sodium pentobarbital solution could possibly affect the taste of the grain for the bird. Human subjects find sodium pentobarbital very bitter. This factor may have been one cause of lower rates under drugged reinforcements. But a result such as that of Fig. 470, in which the effect of the drug wore off during extinction, indicates that the major depressive effect of the drug is due largely to action on the central nervous system.

Other schedules might be more effective in providing a baseline for an adjusting drug level, particularly because of the rate contingency which develops under VI and which may have disturbed the experiments on the effect of the level of deprivation. A large fixed ratio showing some pause after the reinforcement, or a fixed-interval schedule with an extended scallop, might be preferred. The fixed-interval schedule has the advantage that reinforcements will usually continue to be delivered on a fixed schedule. This factor might eliminate some of the oscillations in rate that are due

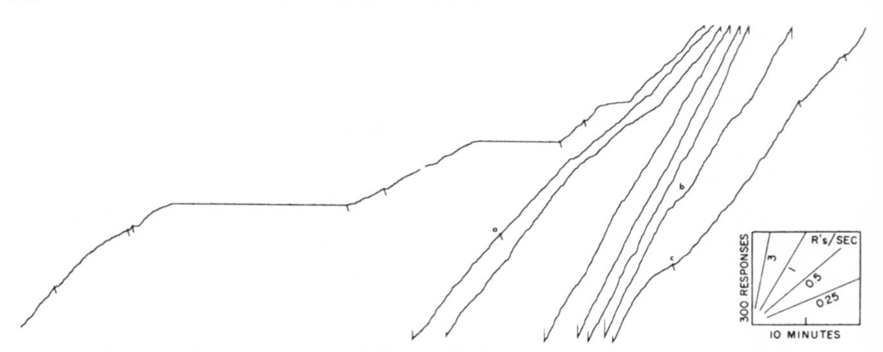

Fig. 470. Increased rate of responding during extinction when drug is discontinued

to the groupings of reinforcements on a variable-interval schedule. Another possible baseline for adjusting a drug level might be the rate of responding during a pre-aversive stimulus, provided a given dose of a given drug changes the amount of responding that occurs during such a stimulus, as is often the case. A program of adjusting drug administration could be used ultimately in a multiple schedule in which the schedule of reinforcement in the presence of one stimulus would be used to adjust the dose level, while the effect of the given dose level would be observed on the second schedule in the multiple program.

Chapter Seven

• • •

VARIABLE RATIO

INTRODUCTION

IN A VARIABLE-RATIO (VR) schedule of reinforcement the reinforcement occurs after a given number of responses, the number varying unpredictably from reinforcement to reinforcement. The schedule stands in the same relation to the FR as VI does to FI.

Like FR it arranges for differential reinforcement of high rates. (See Chapter Five, p. 133.) A cursory examination of the figures in this chapter will reveal very few instances where a reinforcement occurs after a pause.

Variable-ratio schedules produce a variety of performances, depending upon the distribution of the numbers of responses required for reinforcement. As in all schedules requiring a number of responses, the bird will stop responding altogether if the average number goes beyond a certain value. We have attempted to study variable-ratio schedules which produce stable performances with a constant over-all rate of responding. As in the design of VI schedules, various systems may be used to generate the numbers of responses in the series. A mean number of responses per reinforcement is usually fixed. The smallest number is usually 1; i.e., in some cases the first response after reinforcement is reinforced. The largest number and the progression specifying the successive values in the series are other arbitrary points to be selected.

As with fixed-ratio reinforcement, we cannot observe the transition from crf to VR with a large mean, because extinction takes place. The larger mean ratios are reached after stable performances are established on lower means. The mean must be carefully increased until the desired value is obtained.

DEVELOPMENT OF STABLE PERFORMANCES ON VARIABLE-RATIO SCHEDULES

A stable performance was generated on several variable-ratio schedules in 2 separate experiments. In the 1st the birds went directly to VR after crf; in the 2nd they were exposed to 3 sessions of VI before VR.

Figure 471 shows the transition from crf to an arithmetic VR. In Record A the mean was about 40; in Record B it was 50. Sometimes, successive responses were reinforced, and the largest ratio was approximately 100. Both birds show a rapid

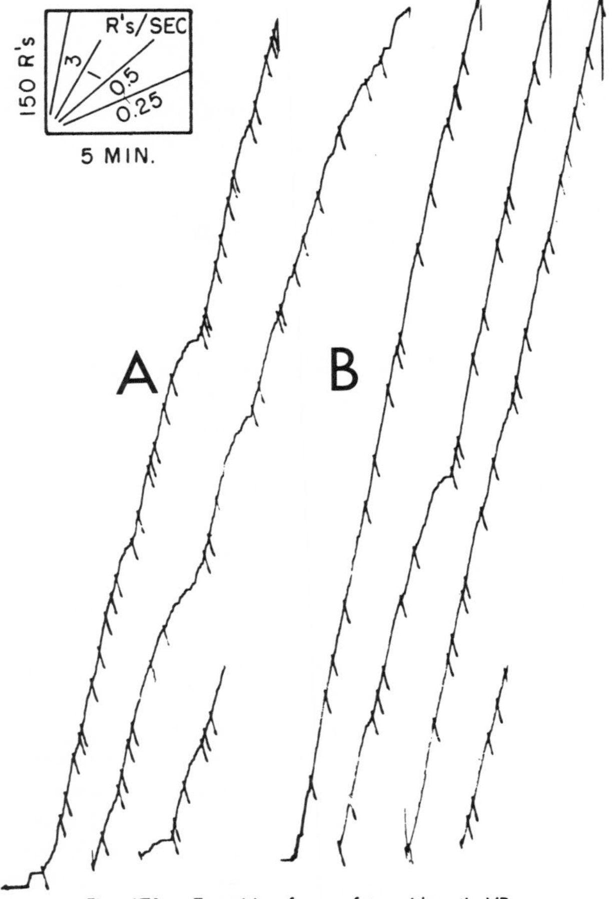

Fig. 471. Transition from crf to arithmetic VR

development of a high over-all rate. The high rate from the extinction of crf produces a high frequency of reinforcement, which in turn sustains a very high rate of responding. The 2nd segment of Record A shows about 1.75 responses per second, the last segment of Record B almost 3 responses per second. This is an extremely rapid development; but the frequency of reinforcement is, of course, high compared with the transitions from crf to VI.

During the next 16 sessions we increased the mean ratio from the values shown in Fig. 471 to 120, 240, and 360 responses by adding successively larger numbers of responses in the series. Except for a relatively higher density of shorter ratios, the numbers formed an approximately arithmetic progression: 1, 10, 20, 30, 60, 100, 180, 240, 300, 360, 420, 480, 540, 600, 660, 690, 690, 720, and 720 responses. We de-

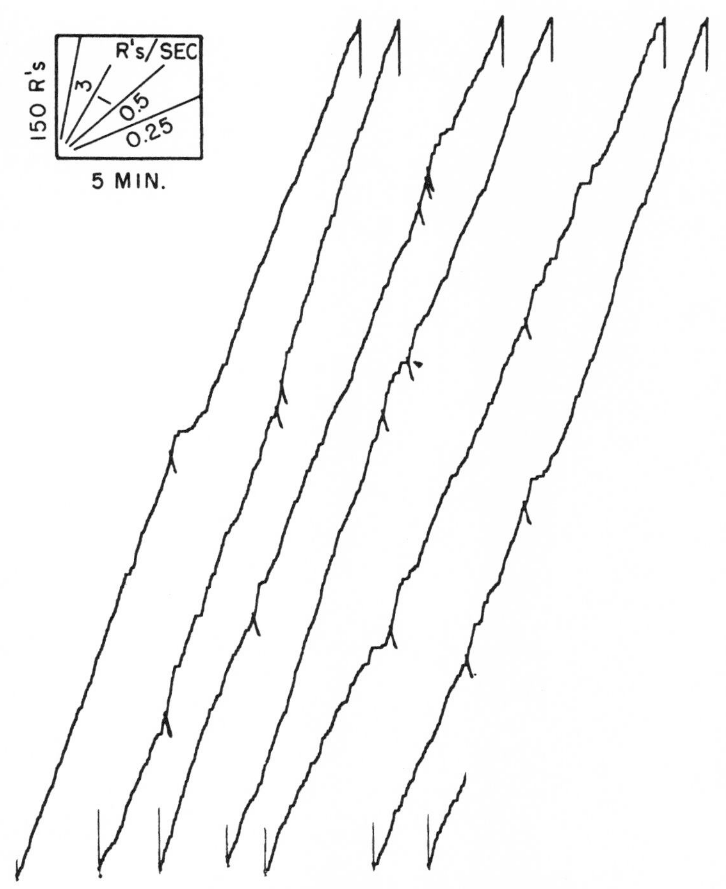

Fig. 472. VR 360 after 6 sessions (first bird)

signed the actual schedule by scrambling several sets of these numbers. Performances after 6 sessions of exposure to VR 360 are shown in Fig. 472 and Fig. 473. Both birds show a higher rate immediately after reinforcement. From 25 to 75 responses are emitted at rates varying from 2.8 to 3.5 responses per second in Fig. 472, and from 3 to 3.5 responses per second in Fig. 473. The over-all rate, however, is much lower for both cases—approximately 1.5 responses per second in Fig. 472 and 2 to 2.5 in Fig. 473. The short period of a high rate of responding immediately following the reinforcement is followed either by an abrupt shift to a lower rate which is maintained until the next reinforcement, by a pause and acceleration to that rate, or by a pause and an abrupt shift to the lower rate of responding. Figure 472 shows a lower over-all rate and more curvature at intermediate rates.

Figure 474 shows examples of the lowest sustained rates occurring at this time; one example is given for each bird. These occurred 4 sessions after the performance already given in Fig. 473. Record A now shows a marked scallop following the period of high rate after reinforcement.

The run after reinforcement in these VR schedules is similar to that in certain variable-interval schedules where the density of short intervals is relatively large. This is effectively a two-valued schedule of reinforcement. The performance immediately after the reinforcement is appropriate to a short ratio. If no reinforcement

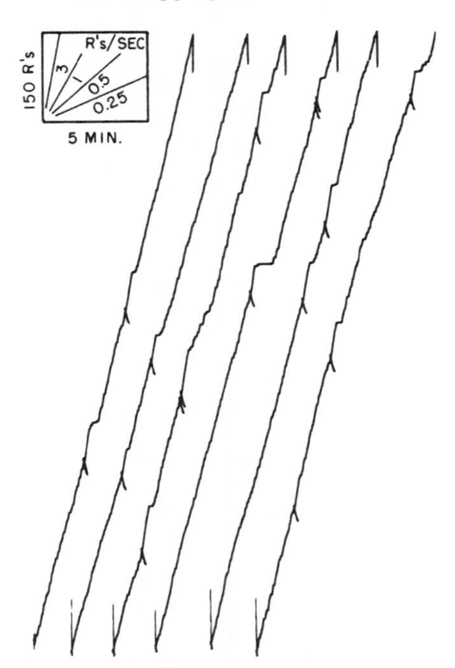

Fig. 473. VR 360 after 6 sessions (second bird)

occurs within, say, 75 responses, a shift to a lower rate appropriate to the larger ratios occurs. The VR schedule remained in force for 60 sessions (except for a few sessions in which the schedule was changed to FR 360). Some variation existed from session to session in the ability to sustain a high rate under VR. Figures 475 and 476 show the extremes of performances on VR 360 for one bird. Figure 475 shows a segment from a daily session in which the over-all rate is high. Most rapid responding still occurs immediately after reinforcement (of the order of 10 responses per second). Sustained rates of responding, as in the first excursions of the figure, are of the order of 7 responses per second, with local rates reaching 10 responses per second during 20 to 30 responses.

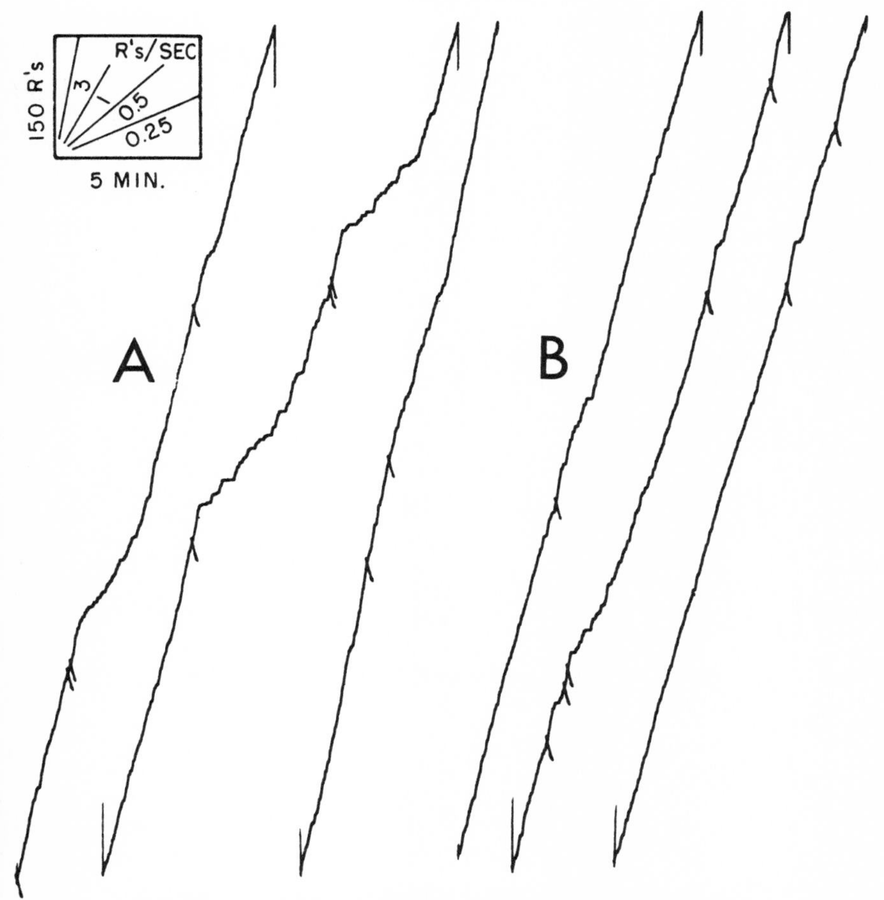

Fig. 474. Lowest sustained rates during early development of VR

The performance shown in Fig. 476 is for the session following that in Fig. 475. Here, the rate in the few responses immediately following the reinforcement is of the order of 8 responses per second, and the relatively abrupt shift to a high rate gives way to long pauses. One of these (omitted from the figure) was more than 1 hour. After the pause, several hundred responses may be emitted before the terminal rate is reached. The performances shown in Fig. 475 and 476 represent the limits of a "stable" state, in the sense that the bird will maintain a performance lying between these two extremes indefinitely.

The second bird never developed a sustained high rate on this mean value of VR. Figure 477 shows the performance after 60 sessions on VR 360, including several sessions on FR 360. The entire performance is executed in bursts of from 10 to 75

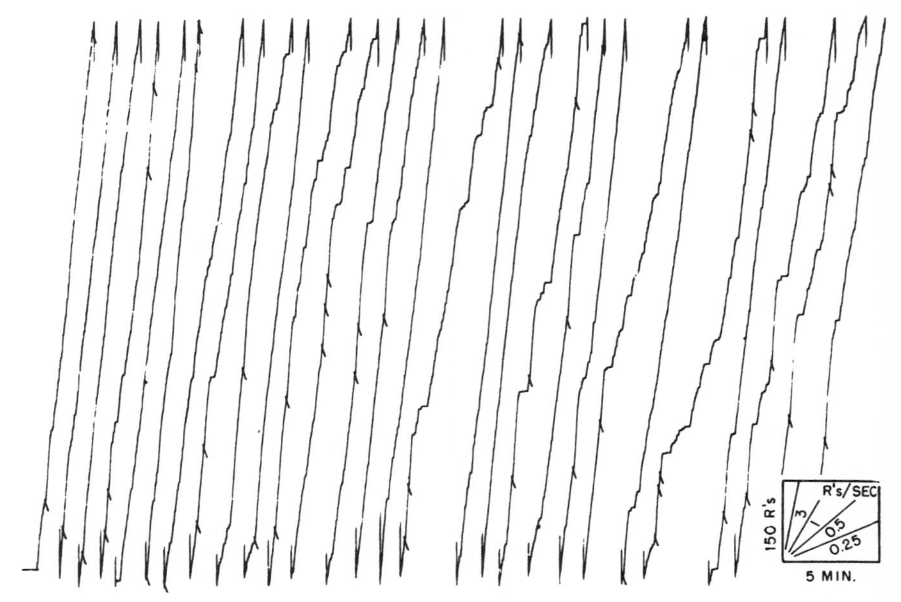

Fig. 475. Late VR performance showing high over-all rate

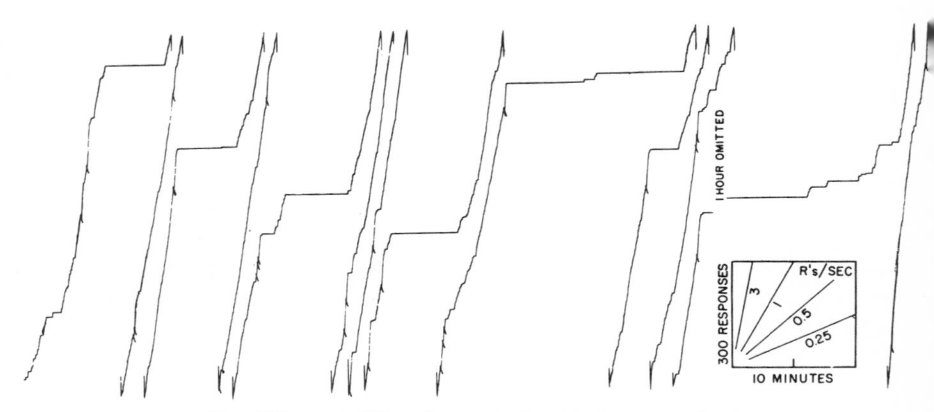

Fig. 476. Late VR performance showing low over-all rate

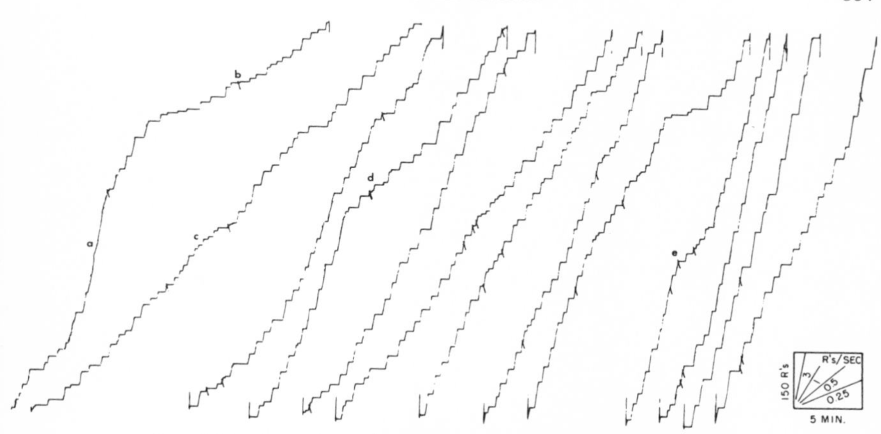

Fig. 477. Late VR performance without sustained responding

responses, at a very constant rate of 4 to 5 per second, separated by pauses varying from 10 to 70 seconds. Reinforcements no longer generate a high rate as in the earlier sessions of the experiment. For example, note the reinforcements at *b, d,* and *e.* The only sustained responding occurs at *a,* where slightly over 200 responses are run off with only a slight suggestion of pauses. In the region at *c* the groups of responses are consistently of the order of 10 to 20, while in the region of *d* they are of the order of 75 to 125 responses.

This performance is similar to that on a variable-interval schedule with differential reinforcement of high rates (see Chapter Nine), and may be considered as the effect of the differential reinforcement of high rates resulting from ratio reinforcement. This part of the contingency is here acting to offset any other factors in the variable-ratio schedule that sustain responding for longer periods of time without pauses.

Two birds were reinforced on VI 1 following crf for 3 sessions of 60 reinforcements each. They were then exposed to a variable-ratio schedule. Figure 478 shows the transition for one bird. Record A shows the last session on VI 1, where the over-all rate is approximately 0.35 response per second. The mean number of responses emitted per reinforcement was 20. At the beginning of the following session (Record B) the schedule was changed to VR 15. The result is both an increase in over-all rate to approximately 0.8 response per second and, necessarily, a marked increase in frequency of reinforcement, resulting simply from the mechanics of the schedule. The second bird showed a similar performance on the transition to VR, except that the higher rate developed much more rapidly.

After 3 sessions on VR 15 the schedule was changed to VR 60, and pauses after reinforcement and occasional low rates of responding between reinforcements began to appear. For 35 sessions the mean variable ratio was then kept at 82. The over-all and local rates of responding increased gradually, and the pauses after reinforce-

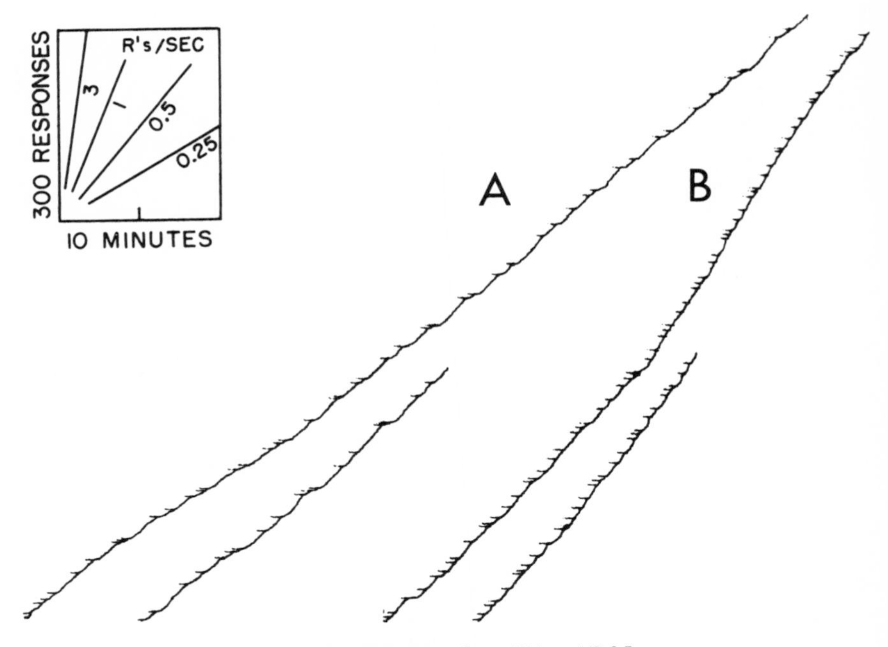

Fig. 478. Transition from VI 1 to VR 15

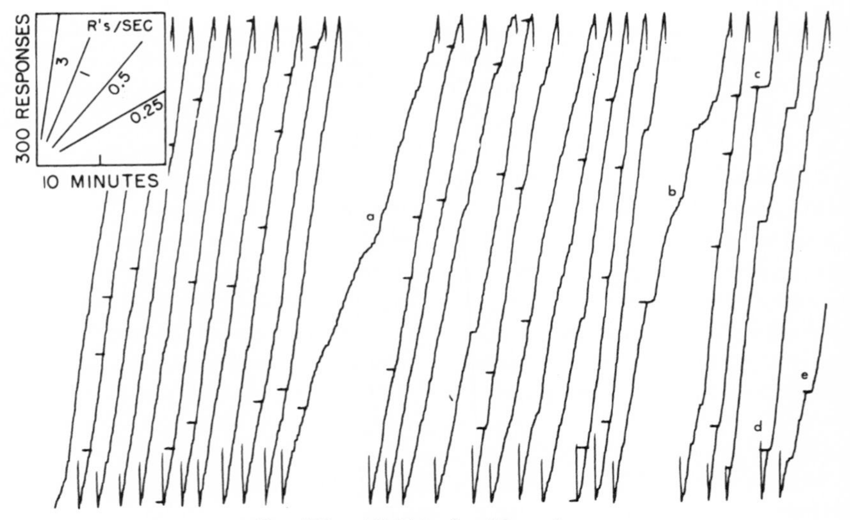

Fig. 479. VR 110 after 12 sessions

ment and deviations from a constant rate eventually disappeared. Because of this slow "push-up," both birds show a stable high rate. The mean ratio was then increased to 110. Figure 479 shows the 12th session following. A stable over-all rate of approximately 4 responses per second is maintained except for occasional periods of intermediate rates, as at *a* and *b*, and the appearance of a pause after reinforcement toward the end of the session, as at *c*, *d*, and *e*. After 11 further sessions on VR 110, the performance shows less responding at intermediate rates and fewer pauses after reinforcement.

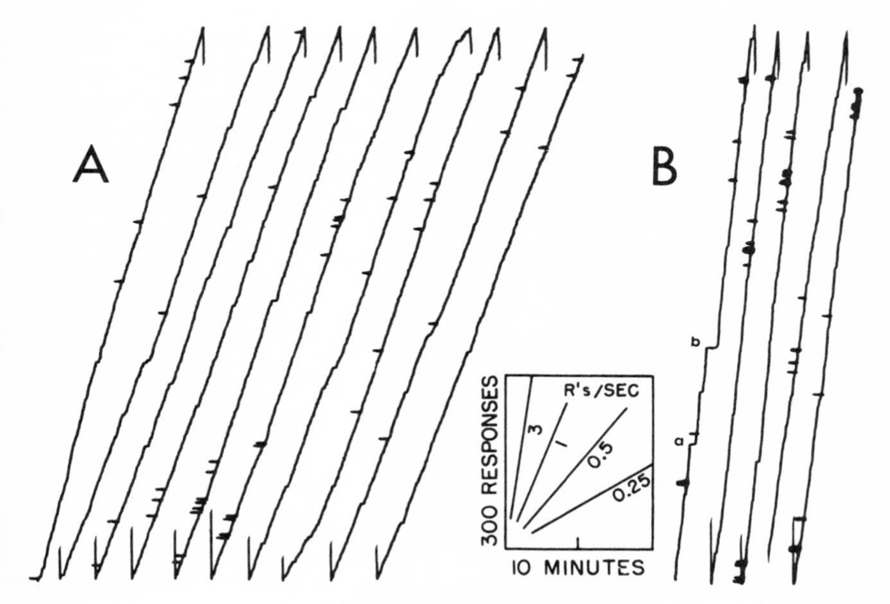

Fig. 480. Development of VR 173

Figure 480A shows the 1st session of the transition from VR 110 to VR 173. The rate of responding falls considerably below the final value on VR 110, and even below the early development on VR 110 in Fig. 479. After 12 sessions on VR 173, however, Record B shows a three-fold increase in the rate of responding. By this time, intermediate rates of responding have again disappeared, and any deviations from the sustained high rate of responding are abrupt shifts to a zero rate, as at *a* and *b*.

THE CONTRIBUTION OF FREQUENCY OF REINFORCEMENT TO RATE OF RESPONDING UNDER VARIABLE-RATIO REINFORCEMENT

When reinforcement is determined by the number of responses, as it is on any ratio schedule, the frequency of reinforcement increases with the rate of responding. We cannot be sure that the high rates generated by variable-ratio schedules are not due to increased frequency of reinforcement rather than to the differential reinforcement of rates or groups of responses. The following experiment was designed to separate frequency of reinforcement from other factors in the variable-ratio schedule.

The problem was to design a control experiment in which frequency of reinforcement would be identical with a variable ratio, but in which none of the other factors of a variable-ratio schedule would be present. The procedure was to "yoke" two experimental boxes, as described in Chapter Three. When a reinforcement occurs in one box, it automatically sets up a reinforcement in the second box. The apparatuses were in separate rooms and hence well-insulated from each other, except for the electrical connection which set up reinforcements in the second box. The bird in the first box was reinforced on a variable-interval schedule. The same schedule was set up in the second box, since every time the first bird was reinforced, a reinforcement was set up for the second bird. We matched rates of responding in the two situations by varying levels of deprivation. When rates of responding were approximately the same, the schedule of reinforcement in the first box was changed to *variable-ratio.*

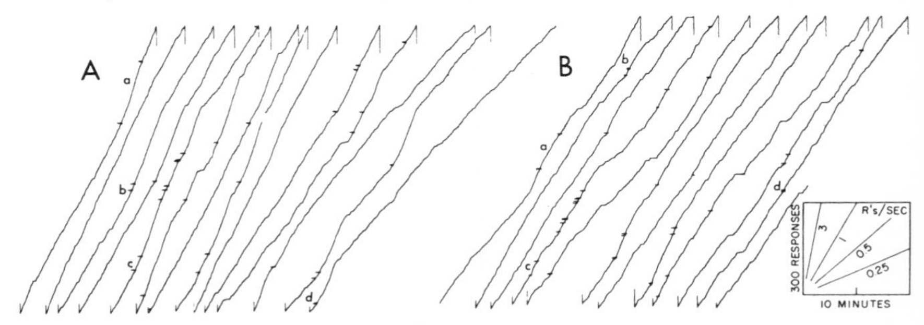

Fig. 481. Final VI performances for matched pair

The values of responses per reinforcement were chosen to match the actual numbers appearing in the performance on the variable-interval schedule. The frequency of reinforcement in the lead bird was thus unchanged at the start of the transition from variable-interval to variable-ratio. Any initial increase in rate must result from some other factor, such as the differential reinforcement of high rates. As soon as the rate has increased, however, the frequency of reinforcement is increased, and the process may continue in an "autocatalytic" fashion. The bird in the second yoked apparatus, however, still remains on a variable-interval schedule. The actual intervals are determined by the performance of the first bird on the variable-ratio schedule, but have no relation to the number of responses emitted by the second bird. The schedules of reinforcement of the 2 birds are variable-interval and variable-ratio, respectively, while the frequency of reinforcement is identical. The extent to which the increased frequency of reinforcement in the variable-interval schedule is responsible for the increased rate can be determined from the increase in rate under the variable-interval schedule in the yoked apparatus.

We conducted the experiment with 2 pairs of birds. They were placed on a geometric VI 5 immediately after crf. The intervals in the schedule ranged from a few seconds (when successive responses were reinforced) to 33 minutes. In matching the rates of the 2 birds of each pair by adjusting the level of deprivation, we specified an arbitrary rate between the 2 rates first observed; and we adjusted the birds' weights in terms of whether the over-all rate for a session was above or below this value. One bird in one pair showed signs of illness during the experiment, and to save time, another was substituted. The new yoked pair had a relatively short history on VI before the transition to VR.

Matched pair I

Figure 481 shows final performances on VI for 1 pair of birds. The figure shows the middle parts of the session just preceding the transition to a variable-ratio schedule.

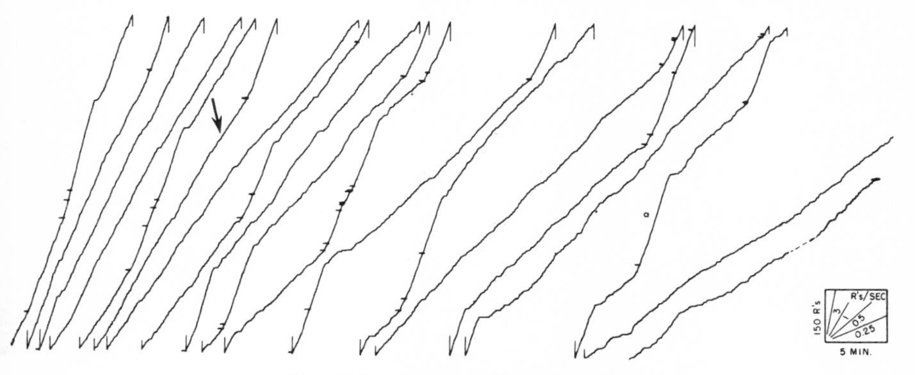

Fig. 482. Transition from VI to VR

Record A for the "first" bird shows a typical geometric VI performance with an overall rate slightly lower than that in Record B. Corresponding points in the schedules are indicated by the small letters. The over-all rate during the session is roughly constant. Marked short-term oscillations occur, and the rate tends to be slightly higher after reinforcement. Figure 482 shows the transition for the first bird from VI to VR. The record shows the start of the session; and the part of the record before the arrow shows the final VI performance. The numbers of responses required for reinforcement ranged from 1 (successive responses reinforced) to 1980. Because the rate of responding following the reinforcement tends to be somewhat higher than elsewhere, the first effect of the variable-ratio reinforcement is a marked increase in the number of reinforcements occurring in groups. This effect, in turn, produces wide oscillations in the over-all rate and sustained responding at rates higher than those under VI. At *a,* for example, the rate is above 2 responses per second for more than 200 responses; the highest sustained rate on VI was approximately 1.5 responses per second. We cannot be sure that this increase in the local rate of responding was produced by the increased frequency of reinforcement, since differential reinforcement of high rates may have produced some effect.

Figure 483 shows the performance of the yoked bird. The record shows the usual high starting rate, which quickly falls to the usual value. The point at which the first bird's schedule was changed to variable ratio is indicated by the arrow. Subsequently, the over-all rate oscillates somewhat, although not so markedly as in Fig. 482.

Figures 484 and 485 describe the 2nd session for the yoked pair. The over-all rate on VR in Fig. 484 shows a further increase, while the performance of the yoked bird shows roughly the same rate as when under the previous variable-interval reinforcement. However, the higher density of reinforcements after short intervals, resulting from the increase in rate of reinforcement in the first bird, produces a higher rate of responding immediately following reinforcement.

The difference in over-all rate between the yoked birds becomes even more marked by the 4th session. Figure 486 shows the first parts of the session. The variable-ratio performance in Record A now shows an over-all rate of approximately 2 to 3 responses per second with sustained rates reaching 3.5 responses per second. Intermediate rates of responding are infrequent, and most of the changes in rate involve

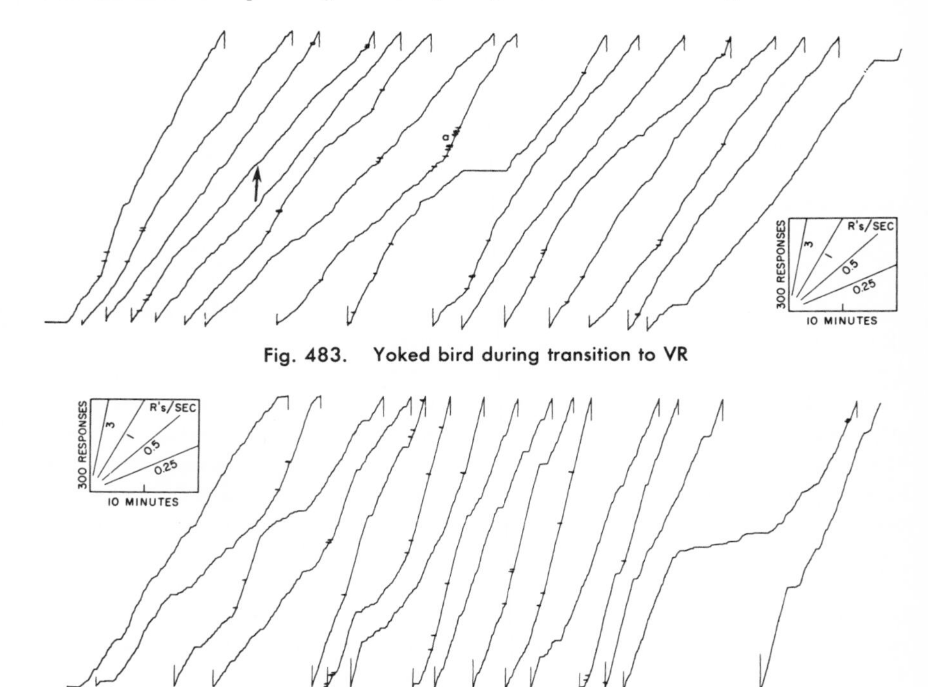

Fig. 483. Yoked bird during transition to VR

Fig. 484. Second session on VR

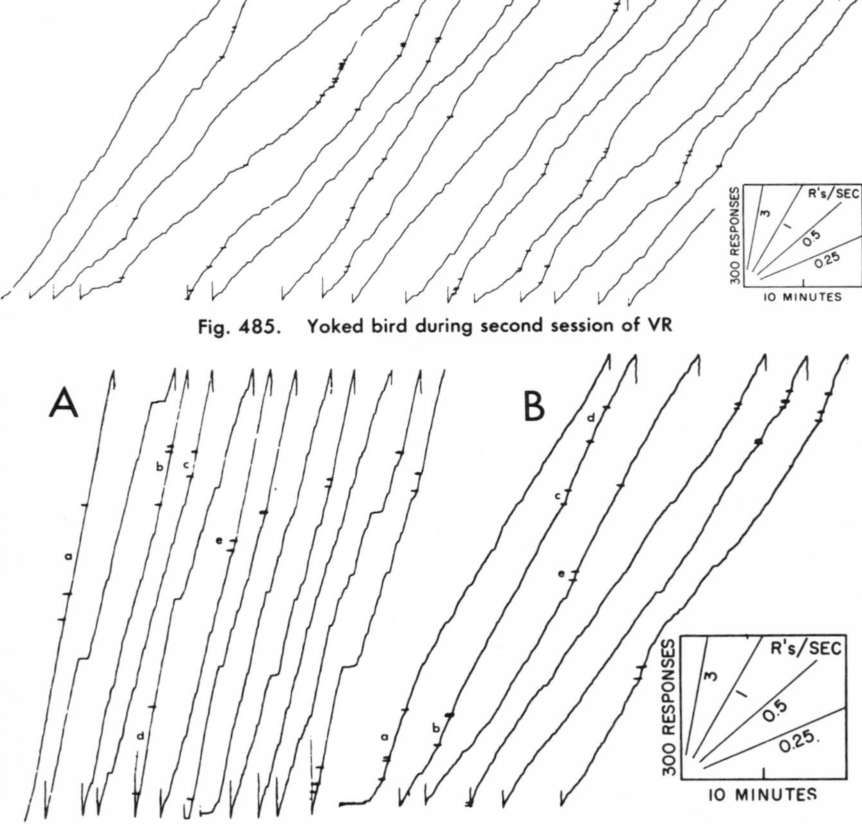

Fig. 485. Yoked bird during second session of VR

Fig. 486. Yoked pair during the fourth session of VR

abrupt shifts between a high prevailing rate and pauses. The yoked bird, whose concurrent performance is shown in Record B, displays an over-all rate of slightly less than 1 response per second, even though the frequency of reinforcement is identical with that of Record A. (The small letters on the record indicate corresponding points in the schedules.) The over-all rate is roughly constant; the rate has a slight tendency to be higher after the reinforcements. Pauses of more than a few seconds' duration are rare.

The VR performance after 11 sessions is shown in Fig. 487A taken from the middle part of the session. While the local rate of responding is only 3 responses per second, rather than 3.5 responses per second as in Fig. 486, responding is sustained at this rate for longer periods. There are clearly only two rates: the prevailing rate of 3 responses per second and zero. The over-all rate is about 2 responses per second. The cor-

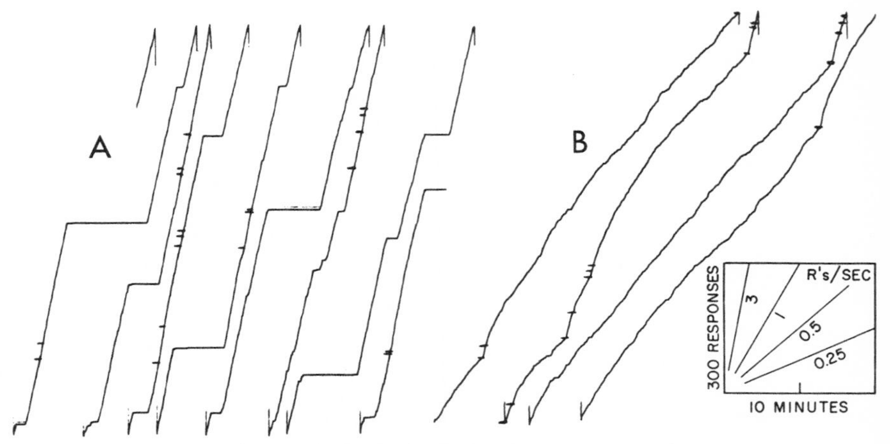

Fig. 487. Yoked pair during the eleventh session of VR

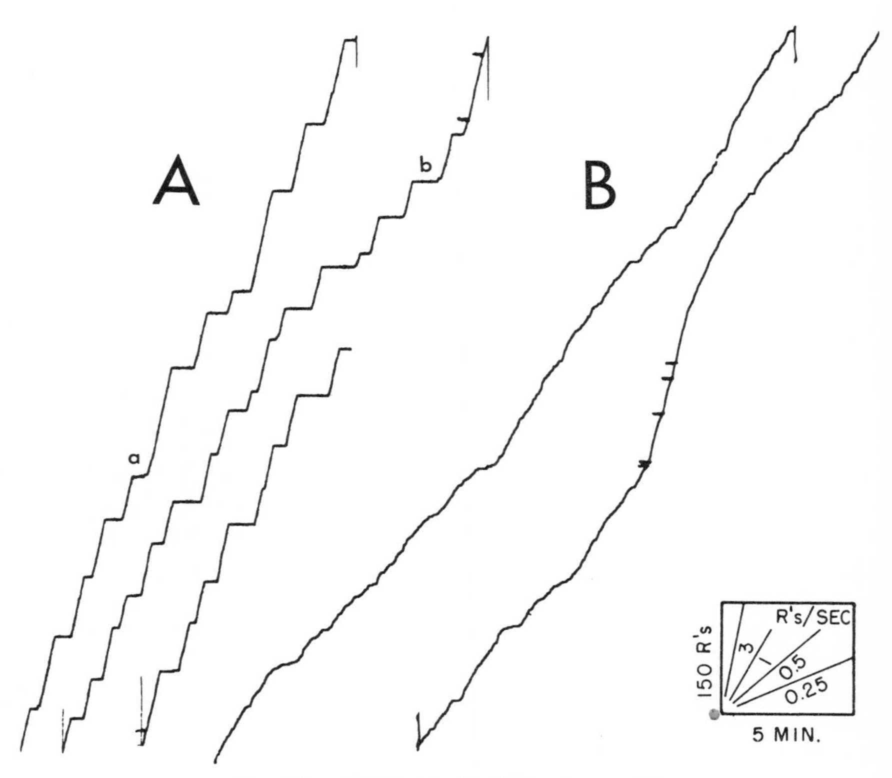

Fig. 488. Yoked pair after 29 sessions on VR

responding record for the yoked bird (Record B) continues to show over-all rates of less than 1 response per second. Also, the rate changes continue to reflect the VI, with smooth transitions from the higher rate following the reinforcement to lower rates elsewhere.

Figure 488 records the final performance in this phase of the experiment, after 29 sessions of VR for the first bird and 29 sessions of VI for the second. The final VR performance consists of sustained runs averaging about 75 responses each at 3 responses per second, separated by pauses of from 1 to 2 minutes. The shift from pausing to the prevailing rate is usually instantaneous. Occasionally, slight curvature appears, as at *a* and *b*. The yoked bird on VI shows rates as high as approximately 0.75 response per second. The rate of responding immediately following the reinforcement remains approximately 2.5 responses per second and is sustained slightly longer than that in the earlier performance (for example, Record B of Fig. 487).

Matched pair II

The second pair of birds did not constitute a repetition of the preceding experiment, because the first bird did not sustain substantial rates of responding under the variable-ratio schedule of reinforcement. Figure 489A shows the transition from VI to VR. Record B gives the corresponding performance for the yoked bird. When the sched-

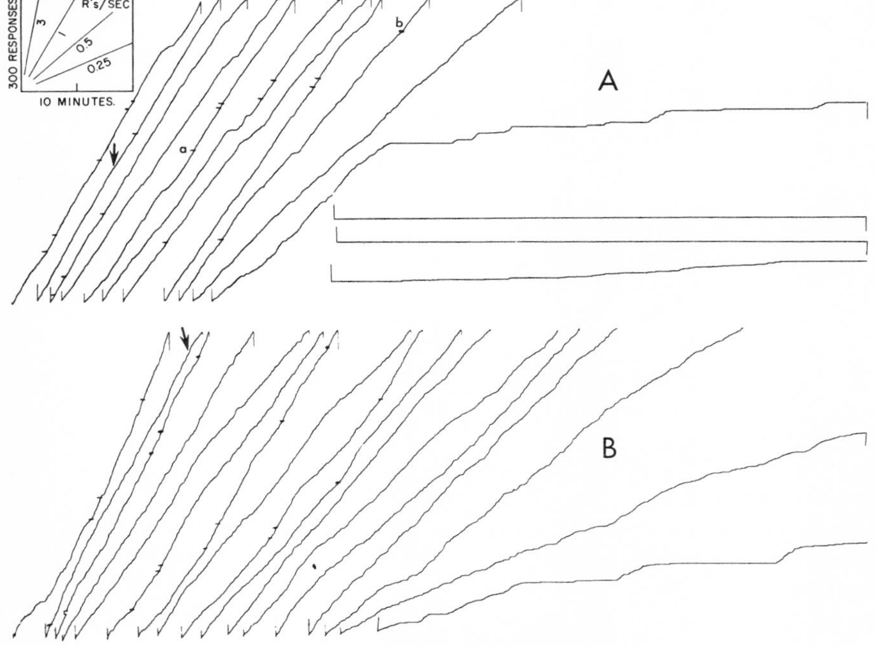

Fig. 489. Transition from VI to VR

ule was changed to VR at the arrow, the special (and at this stage largely accidental) contingencies failed to take effect. The over-all rate continuously declined. Responding reached so low a value following the last reinforcement at *b* that the 1980 responses required for the next reinforcement in the VR schedule were not emitted during the remaining 4 hours of the session. Note that this same ratio was completed earlier in the series at *a*. The yoked bird (Record B) shows a much larger extinction curve than Bird 50G, with a smooth decline in over-all rate. This bird had 47 sessions of reinforcement on VI, compared with only 8 sessions for the bird in Record A, which, as noted above, was added to the experiment at a late stage. The bird with

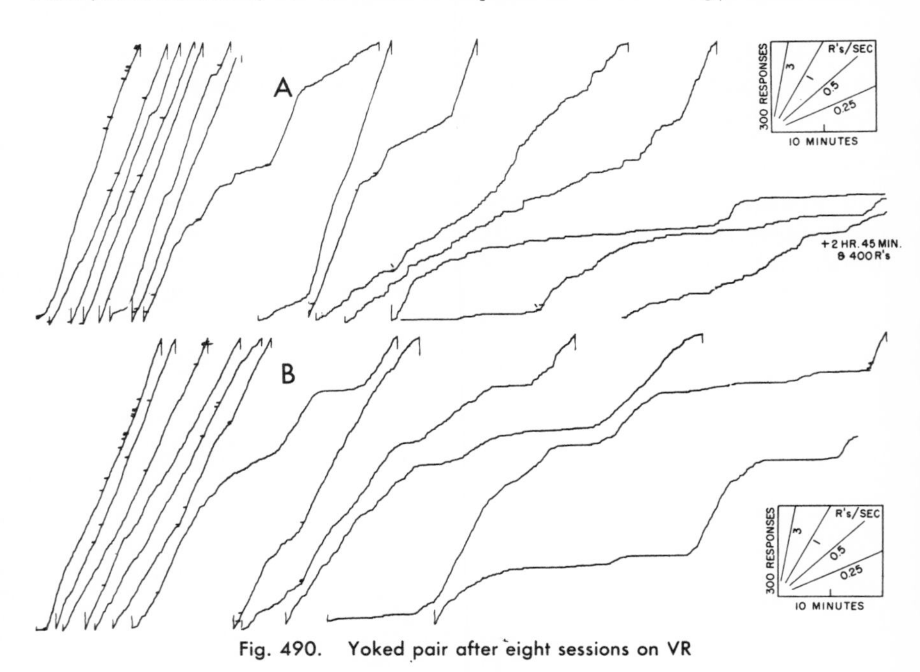

Fig. 490. Yoked pair after eight sessions on VR

the shorter history was assigned to the VR schedule in anticipation of an increase in rate. This increase could then not be attributed to a relatively long history; but it failed to occur.

In the following session the schedule was changed to VI in order to reinstate a substantial over-all level of responding. This change was successful; and a second return to the variable-ratio schedule produced a more successful transition, at least to the extent that the bird maintained a substantial over-all level. Figure 490A shows the performance after 8 sessions on the variable-ratio schedule. Record B shows the performance on VI in the yoked apparatus. While the variable-ratio reinforcement has not yet had its final effect, the performance in Record A begins to show a change in that direction. The over-all rate is roughly of the same order of magnitude in both curves: the only difference is in the character of the changes. The rate changes in Record A are relatively abrupt compared with those in Record B, and the curve shows considerable grain. Groups of from 5 to 10 responses emitted at high rates and separated by pauses appear. Similar pauses are relatively few in Record B under the yoked VI. The bird in Record A is still unable to sustain substantial rates of responding during the larger values of ratios or toward the end of the session, and the yoked bird also shows a marked decline in rate during the session corresponding to the growing infrequency of reinforcements.

TRANSITION FROM VARIABLE RATIO TO FIXED RATIO

Figures 475, 476, and 477 recorded final performances on VR 360, reached after crf and a series of VRs of lower means. The lowest over-all rates of responding were

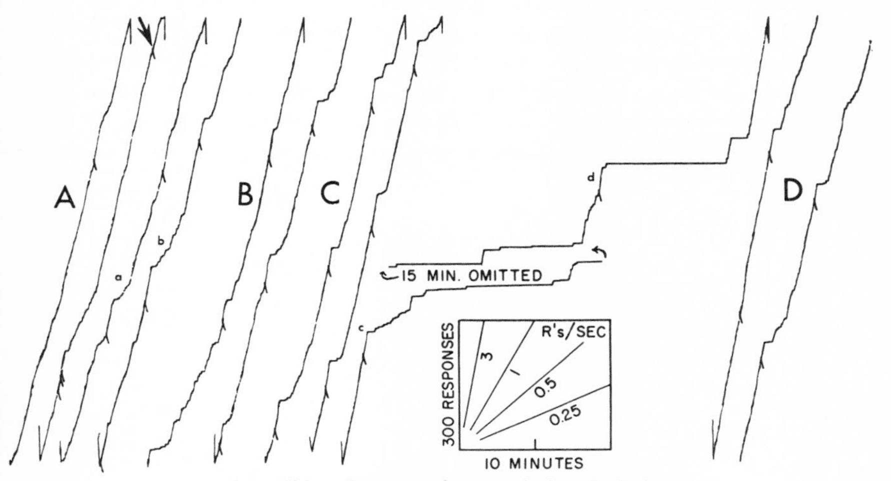

Fig. 491. Transition from VR 360 to FR 360

still substantial compared with rates on FR, in which long pauses frequently occur after reinforcement on ratios of the order of 300 responses and the over-all rate of responding may approach the point where responding ceases altogether. We studied this difference in performance on VR and FR by changing the VR 360 schedule to FR 360 and observing transitional effects. Theoretically, the schedules are similar except for the possibility that the bird's own behavior may serve as a counter with FR. Differential reinforcement of high rates should be the same in both schedules.

Figure 491 shows a transition from VR to FR. The first excursion of Record A is the fourth in the session. The first 2 excursions show the prevailing performance on VR. At the arrow the schedule became FR 360. The lower rate and acceleration at *a* and *b* are characteristic of the variable-ratio performance during longer ratios. Record B shows the 11th and 12th excursions, with a further development of the fixed-ratio performance. The pause and acceleration near the start of the ratio segment is becoming more marked and occurs uniformly. Reinforcement is first followed by a "priming" run, however. Record C shows the 18th through the 21st excursions of the recording pen. A brief period of responding at a high rate follows the reinforcement at *c*, but 45 minutes is required for the next 360 responses. A long pause also occurs near the start of the fixed-ratio segment at *d*. Thereafter, 3 ratios are run off at maximum rate, and a remaining ratio shows the character of the first excursion in Record C. The end of the session (Record D) shows consistent scalloping following the priming run, but no more pausing than ordinarily occurs under VR 360.

Figure 492 shows the entire 3rd session on FR 360. It begins with instances of responding at a high rate immediately after reinforcement before pausing, as at *a, b,* and *c*. But these priming runs become less frequent as the session continues. Most of the fixed-ratio segments now show a pause immediately after reinforcement. The over-all rate of responding falls progressively during the session as the pause and lower rate following reinforcement become extended. During the final 70 minutes of the session, only 250 responses occur, and the last ratio is not completed. Many of the ratio segments are identical with a standard fixed-ratio performance, as at *d, e, f,* and *g*. Toward the end of the session (lower curves), segments begin to show a prolonged

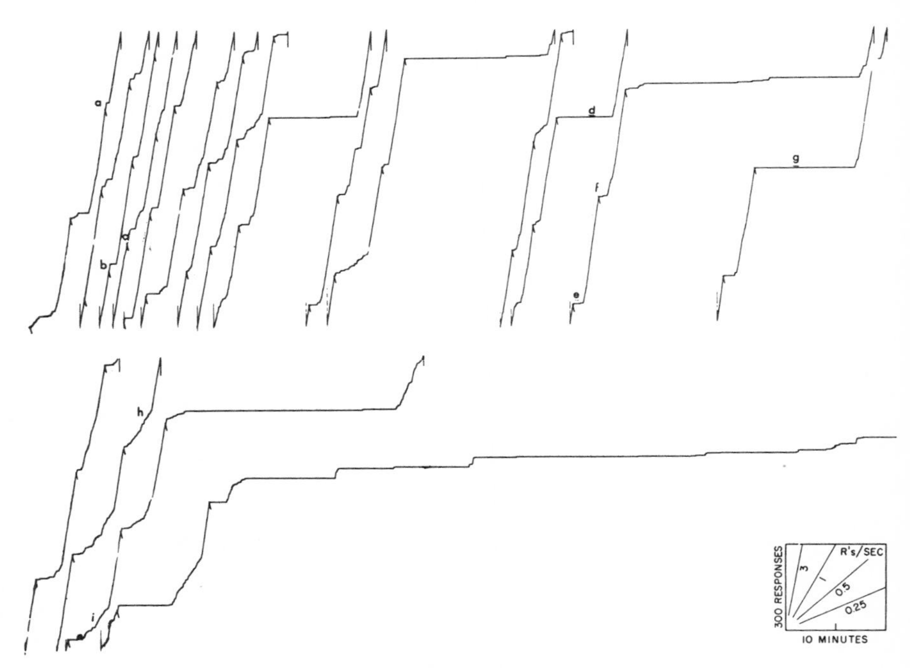

Fig. 492. FR 360 three sessions after VR 360

acceleration with rough grain, as at *h* and *i*. When FR 360 was maintained for 6 sessions, the trend in Fig. 492 continued. The over-all rate remained low, with substantial pauses following reinforcements.

Changing the schedule of reinforcement back to variable-ratio eliminated the longer pauses and reinstated the high rates immediately after reinforcement. The transition was not immediate, however. In Fig. 493 the schedule of reinforcement was changed to VR at the arrow. A long pause occurs because of the recent FR, and the first reinforcement on the variable-ratio schedule occurs at *a*. A second reinforcement occurs immediately after a few responses. The reinforcement at *b* occurs after 750 responses and is followed by a pause of about 2 minutes. A similar pause follows other reinforcements at *c* and *d*. But by the reinforcement at *e*, the transition to the variable-ratio performance is practically complete, with consistent responding immediately after reinforcement. The rate is not yet as high as will be characteristic of this bird under the final VR. The remainder of the session (Fig. 494) shows the reappearance of pauses and accelerations following priming runs after reinforcement. The marked curvature and extended periods of intermediate rates appear to be due to the preceding FR, since further exposure to VR reduced them greatly. The second bird had shown a considerably lower over-all rate under VR and much more pausing under FR. In the transition from FR to VR the change in performance was not completed during the first session.

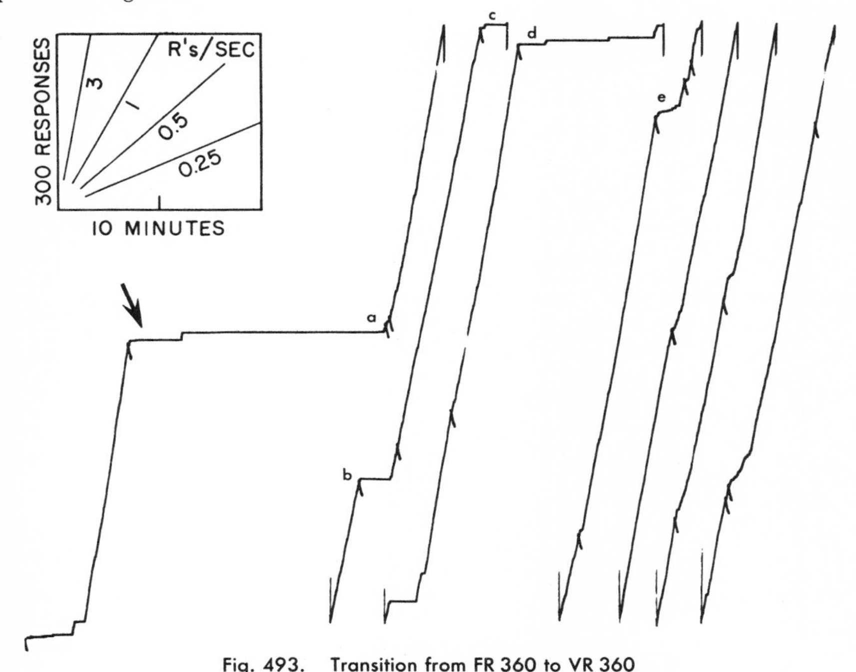

Fig. 493. Transition from FR 360 to VR 360

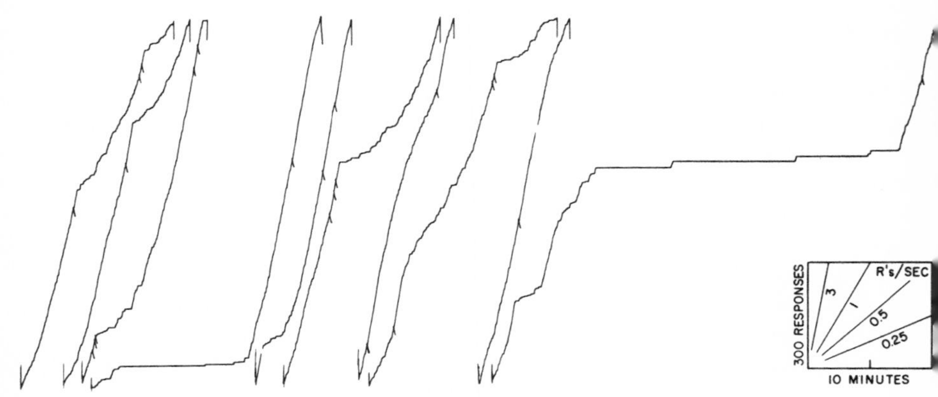

Fig. 494. Continuation of Fig. 493

A second transition from VR to FR was made after 18 sessions of reinforcement on VR 360, after the transition from FR shown in Fig. 493. At the arrow in Fig. 495 the schedule of reinforcement is changed to FR 360. Through the remainder of the session, the pause after reinforcement and acceleration to a terminal rate develop progressively. The bird continues to respond at the terminal rate for from 25 to 75 responses after reinforcement, although short ratios are no longer being reinforced. By the end of the second session, shown in Record B, the pause and acceleration during the ratio segment have become marked, but responding at the terminal rate immediately following the reinforcement is disappearing.

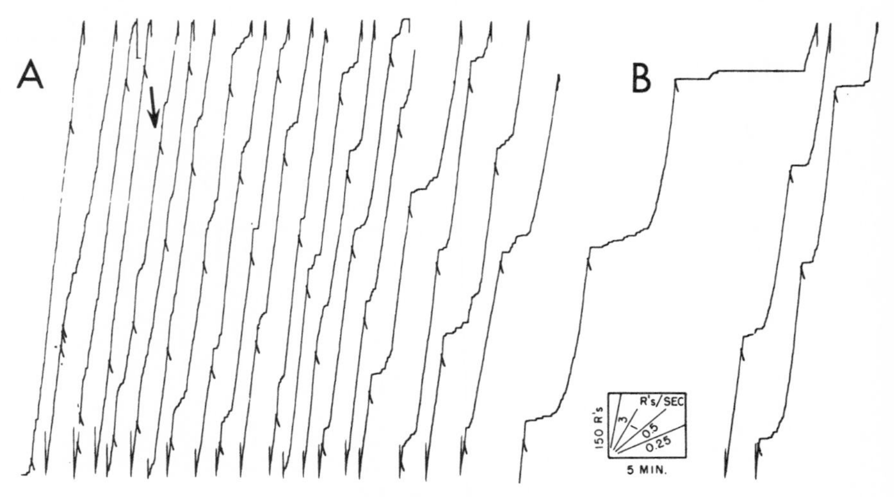

Fig. 495. Second transition from VR 360 to FR 360

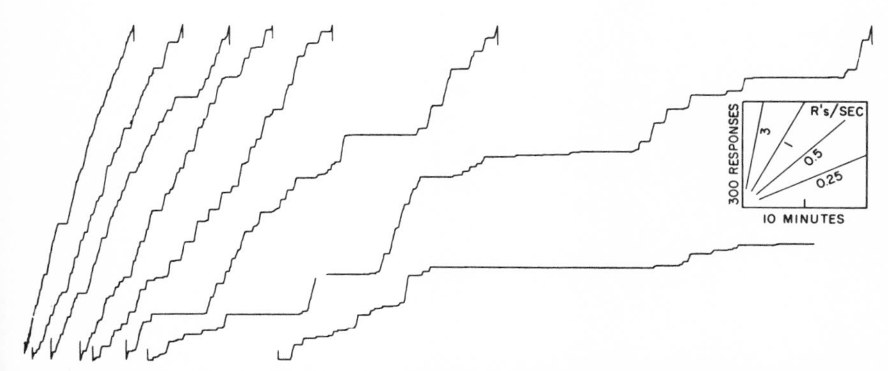

Fig. 496. Extinction following a VR performance showing lack of sustained responding

Extinction after variable ratio

The variable-ratio performances described in this chapter differ considerably from one another, depending upon the particular schedule of reinforcement, the length of exposure to the schedule, and the previous history of the bird. Extinction curves following variable-ratio performances reflect the prior performance. Figure 496 is an extinction curve taken a few sessions after Fig. 487A and a few sessions before Fig. 488A, where the bird responded typically in rapid bursts of from 50 to 75 responses at a rate separated by pauses of from 60 to 90 seconds. The extinction curve in Fig. 496 is roughly negatively accelerated and reaches a low over-all rate of responding after 7000 responses. Throughout the extinction the character of the performance from the preceding variable-ratio reinforcement is preserved, with most of the responding occurring in sustained runs separated by pauses of from 1 to 2 minutes. Toward the end of the session, longer and longer pauses appear between groups of these repeated short segments. Another bird was extinguished 4 sessions after the performance shown in Fig. 474B. The curve (Fig. 497) shows a rough negative acceleration reaching a very low rate after about 8000 responses. The early part of the extinction curve also consists of sustained responding in short bursts, separated by short pauses. The decline in over-all rate as extinction proceeds follows from the increasing length of pause separating these bursts of responding.

Figure 498 contains an extinction curve after 47 sessions of variable ratios from 15 to 110. The curve was recorded several sessions before that of Fig. 480B. A segment of the performance on VR 110 appears before the arrow. The high rate of responding that this bird eventually developed in Fig. 480 under a larger mean has not yet occurred. Extinction continues at essentially the variable-ratio rate for the first 3000 responses. In the segment at *a* the rate shifts to a somewhat lower value for approximately 600 responses; after this the remaining $2\frac{1}{4}$ hours of the session show mostly

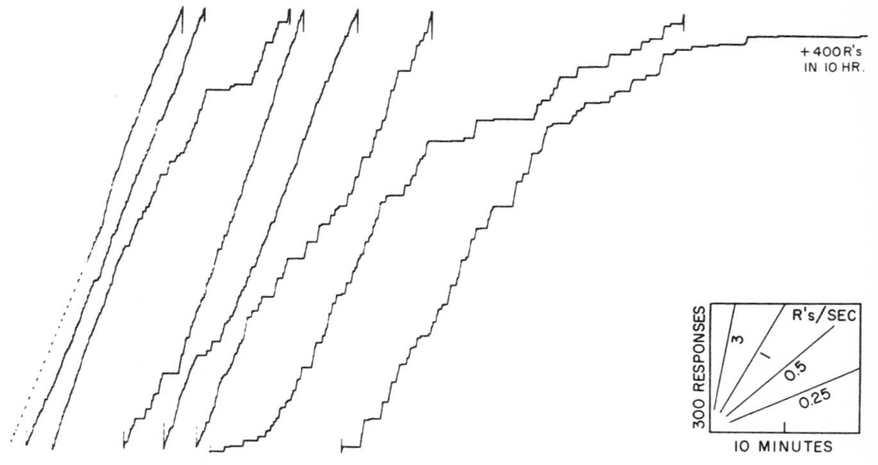

Fig. 497. Extinction after early VR 360 performance

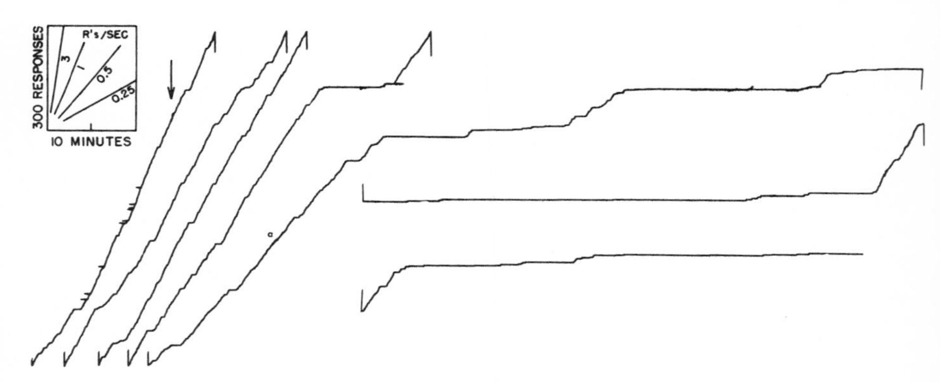

Fig. 498. Extinction after VR 110

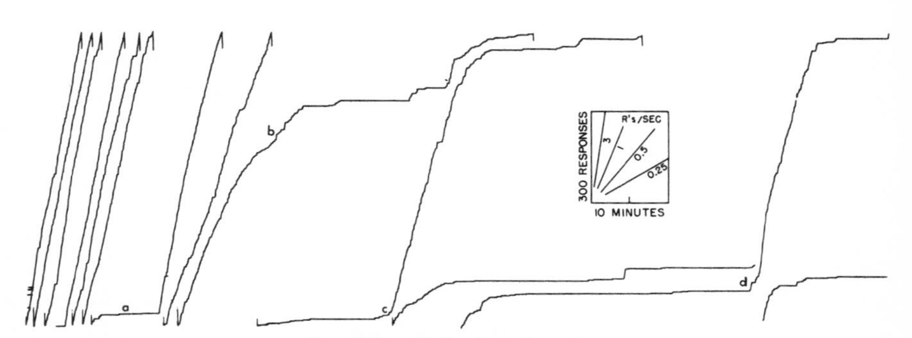

Fig. 499. Extinction after VR 173

long pauses separated by brief periods of responding at approximately the original variable-ratio rate.

Figure 499 shows an extinction curve after a stable performance had developed on VR 173, after a history of 56 sessions with various mean values after crf. The record begins with a small portion of the VR 173 performance. This performance is followed by about 5000 responses at the original variable-ratio rate. An abrupt pause occurs at *a*. The remainder of the extinction curve consists of longer and longer pauses, separated by segments at the beginning of which the bird responds at near the original variable-ratio rate. The transition to lower rates in these segments shows fairly extended curvature (*b*). Changes to higher rates are usually fairly abrupt, as at *c* and *d*.

While these curves differ considerably in detail, they all represent fairly rapid extinction, most of the responses being emitted early with the rate falling sharply to zero. This performance contrasts with extinction after VI, in which the rate declines fairly continuously from the original variable-interval rate through intermediate rates to a very low value. In Fig. 422, for example, 18,000 responses are emitted at a rate which is falling continuously. Short-term rate changes also tend to be more abrupt in variable-ratio extinction.

THE EFFECT OF SODIUM PENTOBARBITAL ON A VARIABLE-RATIO PERFORMANCE

Figure 500 shows the effect of an intramuscular injection of 5 milligrams of sodium pentobarbital on the performance of a bird on VR 360. The record contains the complete session. The prevailing performance under VR is represented by the first 7 excursions of the record. Following the reinforcement at the arrow the apparatus was shut off, the bird was removed from the apparatus, injected, and replaced, and the apparatus was started again. The bird was out of the apparatus for approximately 60

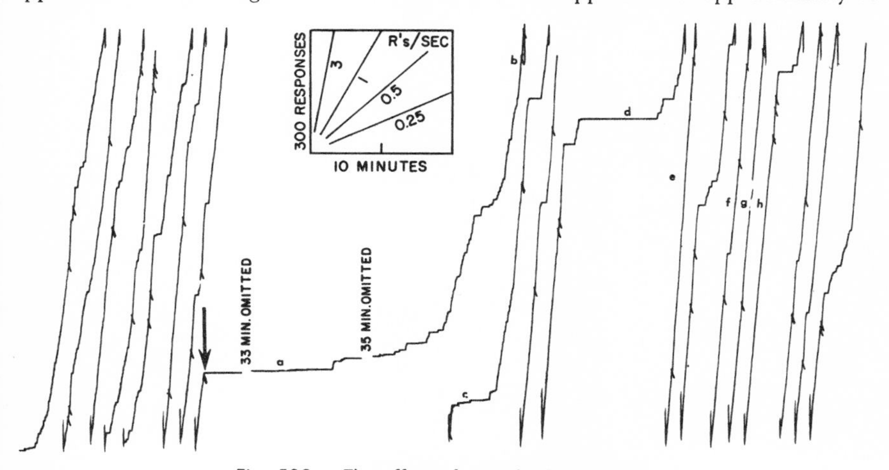

Fig. 500. The effect of pentobarbital on VR

seconds, during which the recorder did not run. The injection produces an immediate cessation of responding, probably because of both the injection and drug effect. (The effect of the injection *per se* is very slight, judging from the effect of a saline injection in Fig. 88. The bird was on FR 400 with added counter and the 0.5-cubic centimeter injection immediately after reinforcement is followed by a pause of 2½ minutes before responding begins normally. Since pauses usually occurred following the reinforcement, this performance can be taken as the maximum disruptive effect of the injection *per se*.) The first response occurs at *a*, 38 minutes after the injection. Another pause of 35 minutes soon occurs; a period of acceleration follows, during which a large ratio is completed. Reinforcement occurs at *b*, almost 800 responses later. A fairly normal variable-ratio performance then appears. The break at *c* and the long pause at *d* possibly indicate some surviving effect of the drug. The segments at *e*, *f*, *g*, and *h* show average rates of responding of almost 8 responses per second, compared with from 4 to 5 responses per second at the start of the session before the injection. This performance represents an excitatory phase of the drug, which occurs after subdepressive doses and some time after the major depressive action of larger doses of the drug have disappeared.

Chapter Eight

• • •

TANDEM SCHEDULES

INTRODUCTION

A TANDEM SCHEDULE is one in which a single reinforcement is programmed by 2 schedules acting in succession without correlated stimuli. For example, in tand FI 45 FR 10 the 10th response counted after 45 minutes elapses is reinforced.

Three types of tandem schedules may be distinguished, depending upon the values of the component schedules. The first component may be substantial and the second very brief (tand *FR* 150 FI 5 sec). The second component may be substantial and the first very brief (tand FI 10 sec *FR* 150). Or, both components may be substantial (tand *FI* 10 *FR* 150).

The component schedules may be any of those already examined. Most tandem schedules where both component schedules are of the same sort are trivial. A tandem FRFR schedule is simply a larger fixed-ratio schedule. A tandem VIVI is usually a variable-interval schedule with at least a slightly larger shortest interval. A tandem VRVR is simply a variable-ratio schedule with at least a slightly larger smallest number. Tandem FIFI, however, has interesting possibilities. (The 2nd interval is assumed to be measured from the 1st response after the expiration of the 1st interval.) If the first effect is similar to a single-interval schedule, the pause developing in the appropriate performance may at some point exceed the first of the tandem intervals. The result is then an increase in the next interval, a still longer pause, and so on. In this respect it has the effect of tand FR 1 *FI*.

When a ratio schedule is followed by an interval schedule, we have three important cases:

1. The initial ratio schedule is substantial and the following interval schedule very brief (e.g., tand *FR* 350 FI 5 sec). This differs from FR 350 alone in that it eliminates the differential reinforcement of high rate associated with FR. The FI component is so short that the number of responses from reinforcement to reinforcement does not vary enough to greatly modify the fixed number of responses at reinforcement.

2. The initial ratio schedule has a small value and the 2nd-interval component is substantial (e.g., tand FR 10 *FI* 10). Here, the clock which times the fixed-interval schedule starts only after the responses in the small ratio have been emitted.

3. Both the initial fixed-ratio schedule and the following interval schedule have sub-

stantial values (e.g., *FR* 100 *FI* 5). This is roughly equivalent to a variable-interval schedule with a larger smallest interval of reinforcement. The substantial initial fixed-ratio component, however, eliminates the regenerative feature present in interval schedules, in which single responses begin to be reinforced whenever the rate becomes very low.

When the 1st component is interval and the 2nd component is ratio, we have 3 major cases:

1. The beginning interval schedule is substantial, and the following ratio schedule has a small value (e.g., *FI* 10 FR 5). This schedule provides a technique for adding a differential reinforcement of high rates to an FI schedule without otherwise disturbing the major properties of the schedule.
2. The initial interval schedule is brief, and the tandem fixed-ratio schedule is substantial (e.g., FI 30 sec *FR* 200). This is a schedule in which the actual contingencies at reinforcement are very sensitive to the behavior generated. Responses made during the short interval increase the effective ratio until with a very large ratio, a pause develops after reinforcement. Fewer responses will then be required and the schedule should stabilize at some value where the pause after reinforcement is roughly the same as the interval in the initial component. But pausing in a "strained" ratio is a very subtle feature of the performance.
3. Both the initial interval schedule and the following ratio schedule are substantial (e.g., tand *FI* 5 *FR* 200). This is substantially a VR with a large smallest ratio of responses. The total number of responses emitted per reinforcement is again sensitively related to the pause after reinforcement.

The differential reinforcement of rates (Chapter Nine) supplies special cases which might be considered tandem schedules. For example, a schedule in which a response will be reinforced at the end of an FI whenever the rate satisfies a drh contingency is essentially a case of tand *FI*crfdrh.

TANDEM FIXED-INTERVAL FIXED-RATIO SCHEDULE

Tand FI 45 FR 10

Development to a final performance. Two birds were placed on tand FI 45 FR 10 whose transition from crf to FI 45 has already been described in Chapter Five. The performance of one bird will be described in detail, and that of the second used to confirm the main points of the experiment. Figure 501 shows the transition from FI 45 to tand FI 45 FR 10. The performance of FI 45 was not well-developed. The bird was showing a low over-all rate of approximately 0.2 response per second, with rough grain, and little or no consistent scalloping. In the 2nd session following (shown complete in Fig. 502 except for the first hour) the tandem ratio has produced both local and over-all changes. The grain now has a stepwise character; the bird responds at about 3.5 responses per second in short bursts of about 10 responses. The over-all rate increases from 0.2 to 0.6 response per second. Even at the highest rates the performance consists

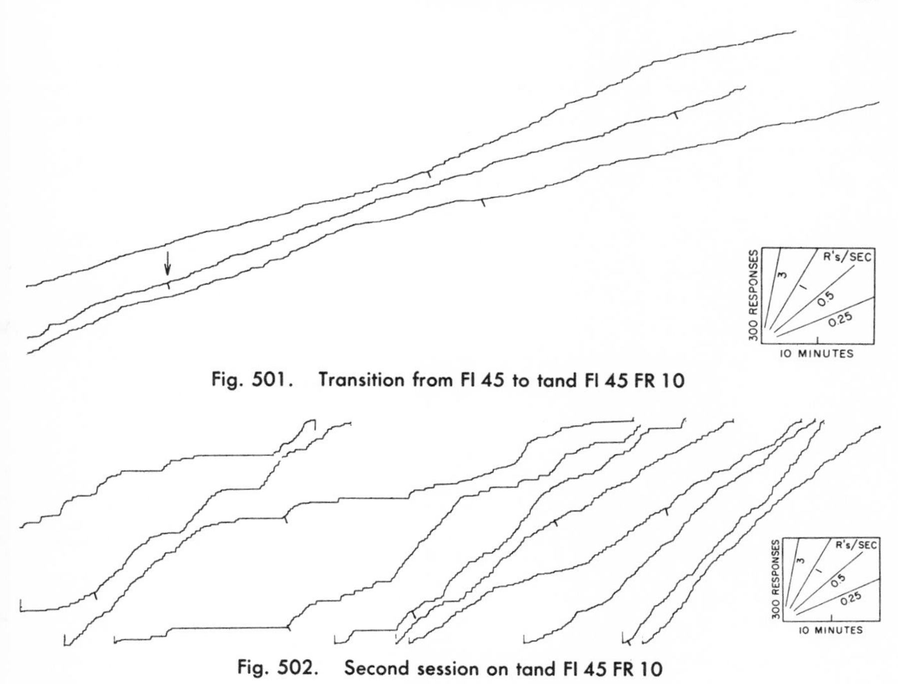

Fig. 501. Transition from FI 45 to tand FI 45 FR 10

Fig. 502. Second session on tand FI 45 FR 10

of short bursts of responses. No evidence exists yet of the development of an interval scallop. Figure 503 shows 45-minute segments, one each from the 3rd, 4th, 5th, and 6th days. Record A is from the first hour omitted in Fig. 502. Records B and D exhibit a slight tendency toward interval scalloping. Records A and C, however, show a roughly linear performance. Fifteen sessions later, after a total of 170 hours of reinforcement on the tandem FI 45 FR 10 schedule, the over-all rate of responding has been increased and the scallop has become prominent. This stage is shown in Fig. 504, which contains one 45-minute segment from each of 4 successive sessions. In Records A and B the over-all rate accelerates rapidly until the middle of the interval, where it begins to decline to a lower terminal value. In Record C the acceleration is slower but continuous, reaching a final over-all rate of approximately 0.6 response per second. In Record D the acceleration is rapid to an over-all rate of responding of the order of the high rate reached in the middle of the interval in Records A and B, which is maintained throughout the interval, except for slight local changes at *a* and *b*. All of these over-all interval performances are executed even though the local rate is discontinuous. A later performance on tand FI 45 FR 10 (after 290 hours of reinforcement) is illustrated in Fig. 505 and 506, which comprise a complete session. In Fig. 505 a period of from 5 to 10 minutes follows each reinforcement (except at *c*) during which the rate increases to a terminal value of approximately 2.5 responses per second.

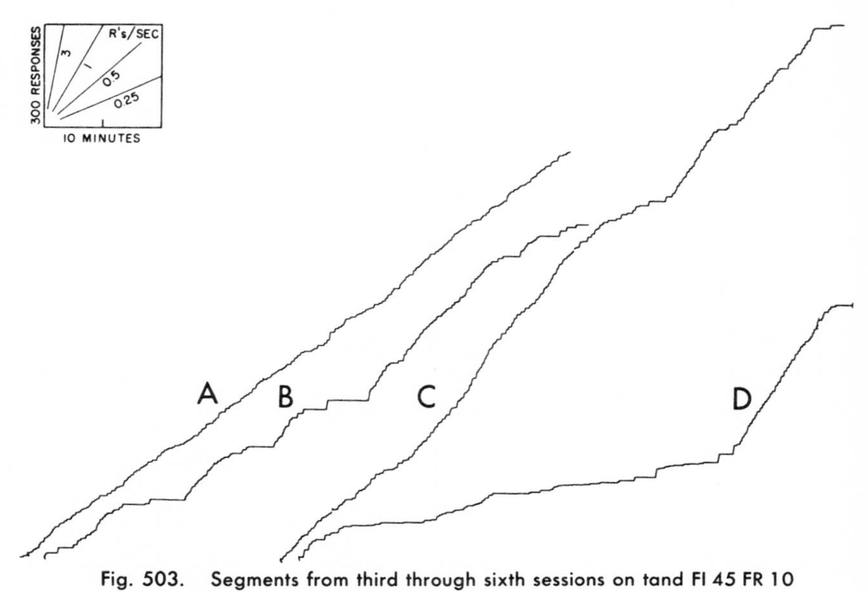

Fig. 503. Segments from third through sixth sessions on tand FI 45 FR 10

Fig. 504. Segments from the 21st to 24th sessions on tand FI 45 FR 10

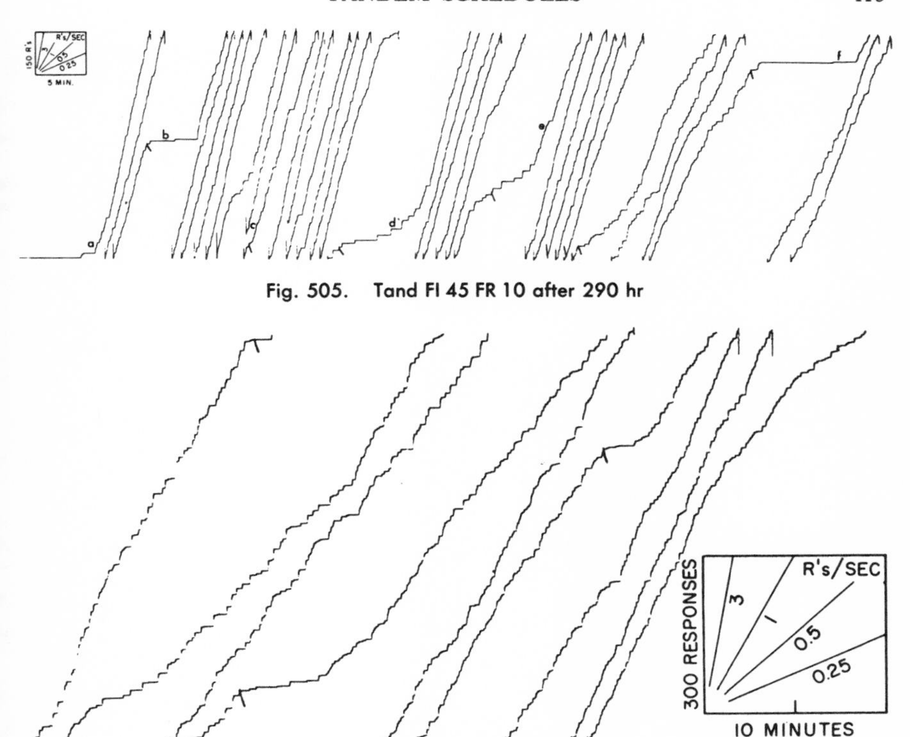

Fig. 505. Tand FI 45 FR 10 after 290 hr

Fig. 506. Continuation of Fig. 505

The terminal rate is maintained for the remainder of the 45-minute interval. At *a*, *b*, and *f* there is a fairly abrupt shift to the terminal rate, while at *d*, *e*, and elsewhere, the over-all curvature is smooth. The curvature results from a series of increasingly shorter pauses between bursts of responses. Even at the terminal rate, the fine grain of the record consists of small bursts separated by pauses, although by this time the pauses are of the order of from 2 to 5 seconds. The running rate, therefore, is considerably higher than the over-all rate of about 2.5 responses per second. A rate of 8 responses per second may be reached for runs of 25 to 50 responses.

Toward the end of the session (Fig. 506) the over-all and terminal rates are lower. The running rate in each burst, however, is little changed.

The last 4 recorded sessions on tand FI 45 FR 10 are represented in Fig. 507 by the 3rd interval segment from each session. The curves at Records A and B contain 3300 and 4200 responses, respectively, and have been broken for better reproduction. The terminal rate is high until the end of the interval in Record A, but falls somewhat in Record B. Prolonged interval scallops are evident in Records C and D. The smooth over-all increase in rate during the first part of the fixed-interval segment is accom-

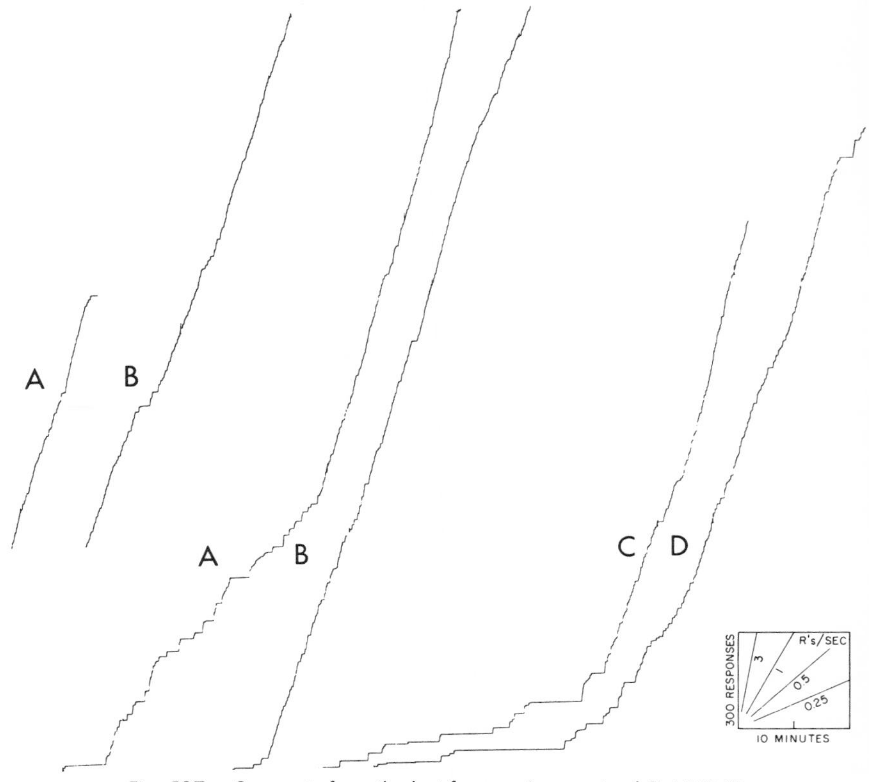

Fig. 507. Segments from the last four sessions on tand FI 45 FR 10

plished by the shortening of the pauses between bursts of responses. The sustained rapid responding during the latter part of each interval is composed of short bursts of very fast responding, separated by pauses of from 1 to 2 seconds.

The second bird showed a similar first effect of the tandem FR 10. The transition occurred after 30 hours of reinforcement on FI 45, in which the final performance was a low over-all but steady rate of approximately 0.03 response per second. Figure 508 shows the entire 1st session on tand FI 45 FR 10. Because of the very low over-all rate, the tand FR 10 postpones many reinforcements far beyond the designated interval. The reinforcement at *a,* for example, occurs 87 minutes after the previous reinforcement, and that at *b* occurs 59 minutes after. The performance developed much more slowly than with the first bird. After 162 hours (Fig. 509), the curves are composed almost entirely of runs of from 10 to 30 responses at from 3 to 4 responses per second, separated by pauses of approximately 1 minute. High over-all rates occasionally emerge in which the long pauses separating bursts of responses are absent. The

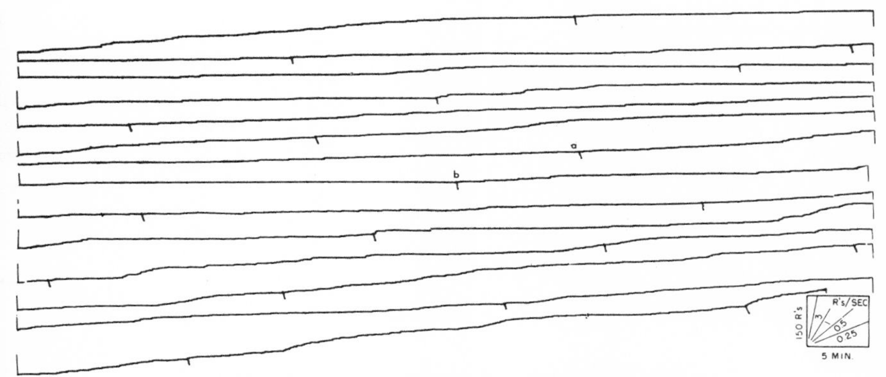

Fig. 508. First session on tand FI 45 FR 10 (second bird)

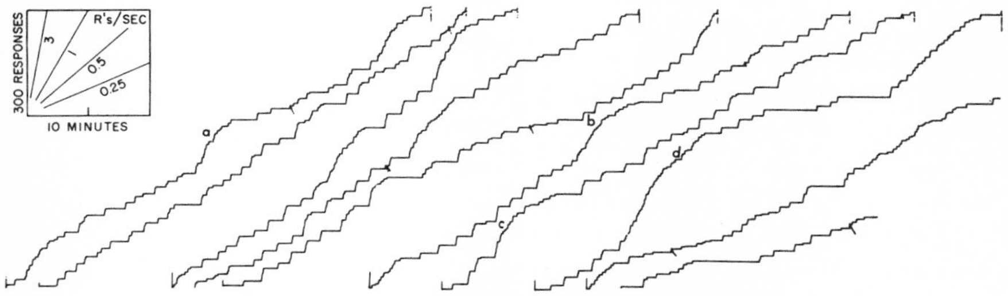

Fig. 509. Tand FI 45 FR 10 after 162 hr (second bird)

change is gradual, however, and the over-all curvature is fairly smooth, as at *a, b, c,* and *d.* A slight tendency toward the development of interval curvature is evident.

Figure 510 shows the final recorded performance after a total of 754 hours on this schedule. The bursts of responses at from 4 to 5 responses per second are now shorter and usually separated by shorter pauses. The over-all rate has declined from the earlier performances. Little, if any, consistent interval curvature exists.

Transition from tand FI 45 FR 10 to FI 45. In the session following Record D in Fig. 507 the tand FR 10 was removed, the bird being reinforced simply on FI 45 for a total

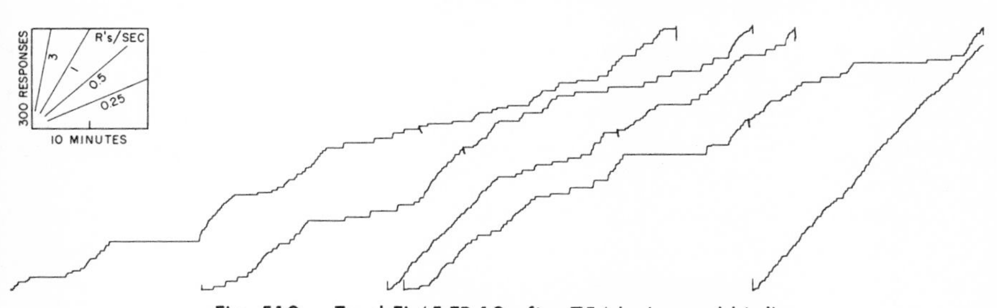

Fig. 510. Tand FI 45 FR 10 after 754 hr (second bird)

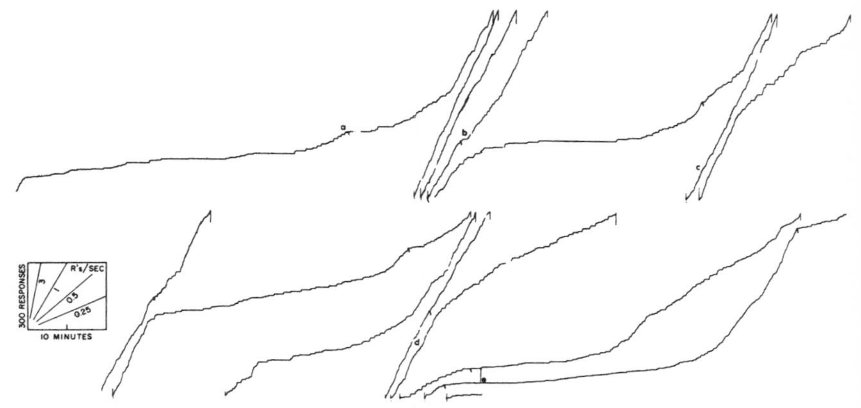

Fig. 511. FI 45 ten sessions after tand FI 45 FR 10

of 10 sessions. Figure 511 shows the entire last session after 61 hours of reinforcement on FI 45. The removal of the tand FR 10 has reduced the over-all rate. The highest terminal rate is now of the order of 1.25 responses per second, compared with 2.5 responses per second in Fig. 505. The acceleration during the early part of the interval is occasionally very prolonged, as at *a* and *e*. The whole record gives the appearance of more curvature extending over longer periods of time. Substantial responding may occur immediately after reinforcement, as at *b*. In that case the terminal rate is reached very quickly, but may give way to lower rates with fairly smooth negative curvature. A second acceleration before the reinforcement frequently appears. While the grain of the record during the period of low or intermediate rates of responding still consists of pauses separated by runs, this performance has become less pronounced; and instances now occur, at *c* and *d*, where responding is sustained without pauses.

When the tandem FR 10 is again added to the schedule, the previous performance on this schedule is quickly recovered. Figure 512 shows the 11th session, 72 hours after the schedule was changed a second time to tand FI 45 FR 10. The over-all rate of responding and especially the terminal rate of responding in the interval have increased to the same order of magnitude as before. The tendency to respond immediately after the reinforcement is marked, even though this responding is followed by a long pause and acceleration as in a normal fixed interval. Occasionally, the terminal rate of responding is sustained throughout the whole 45-minute period, as at *a*, although most segments show a long pause followed by an eventual acceleration to a terminal rate.

Tand FI 45 FR 10 with TO 15 after reinforcement. A 15-minute TO after reinforcement was added to the schedule in the session immediately following Fig. 512. The result (Fig. 513) is a marked decline in the over-all rate. From 10 to 15 responses usually occur just after the termination of the TO 15, as at *a*, *b*, *c*, and *e*; and following

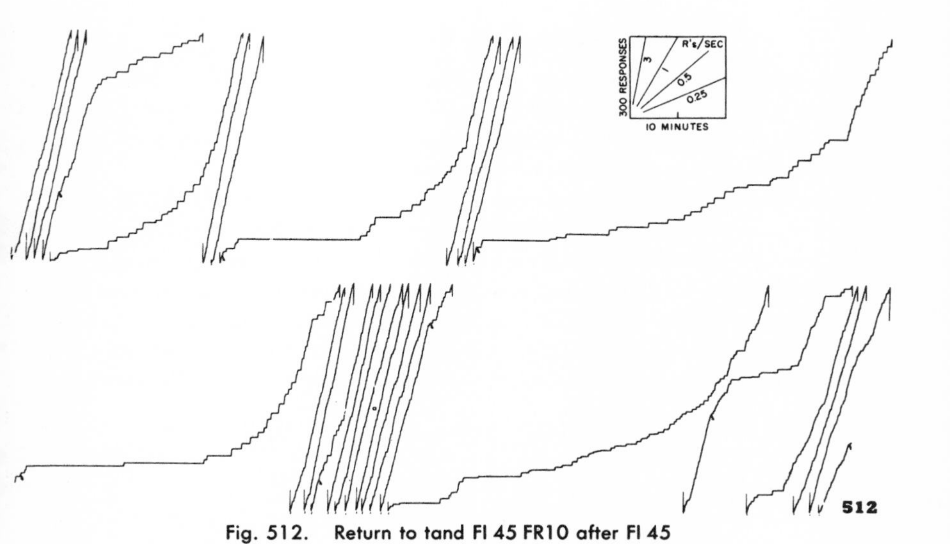

Fig. 512. Return to tand FI 45 FR10 after FI 45

Fig. 513. First effect of TO 15 after reinforcement on tand FI 45 FR 10

the last reinforcement in the figure, at *f*, responding is fairly sustained. Only at *d* does the time out have the effect of eliminating responding at the beginning of the interval, as in the experiments reported in Chapter Five. The time out does, however, postpone the development of the terminal rate.

These effects are temporary, however. By the 2nd session on tand FI 45 FR 10 TO 15 (Fig. 514), the over-all rate of responding has increased to the earlier order of magnitude, and fairly consistent scallops occur in most of the fixed-interval segments. Some responding begins soon after reinforcement; and most interval segments show a fairly smooth acceleration to a terminal rate, which is maintained until the reinforcement. One marked deviation occurs at *a*, where the terminal rate is reached early. The rate shifts abruptly to a very low value, and a new scallop begins. The terminal rate

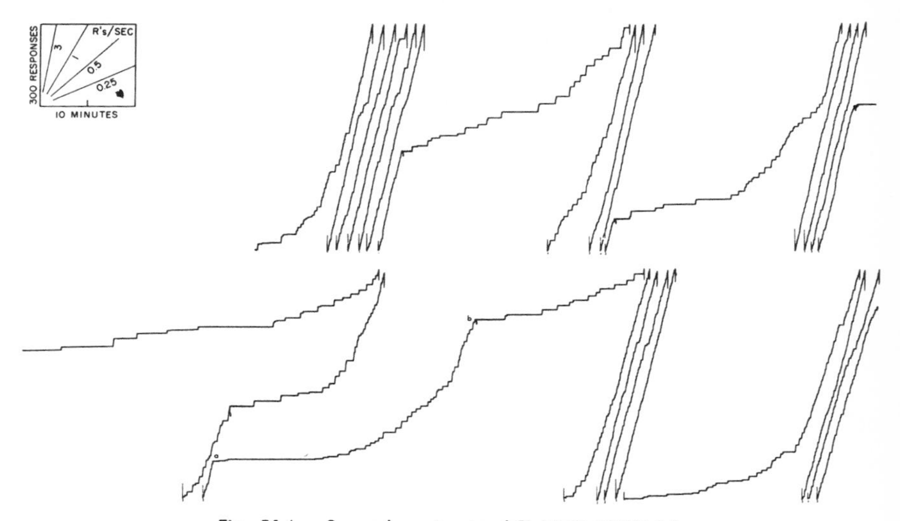

Fig. 514. Second session tand FI 45 FR 10 TO 15

is reached a second time, before reinforcement at *b*. Seven sessions later, after a total of 69 hours of reinforcement on the tandem schedule with time out after reinforcement, the effect has become more consistent (Fig. 515). Here, a reasonably long pause occurs after reinforcement and a smooth over-all acceleration to a terminal rate, which is maintained steadily until the next reinforcement. Especially good examples of extended and smooth curvature occur toward the end of the session, at *a*, *b*, and *c*. The acceleration is caused by a change in the duration of the pause rather than in the actual local rate.

The pause after reinforcement and the period of acceleration to the terminal rate of responding become longer as the experiment continues. Figure 516 shows the performance 4 sessions later. The over-all rate of responding has fallen slightly, because of the longer pauses and more extended intermediate rates.

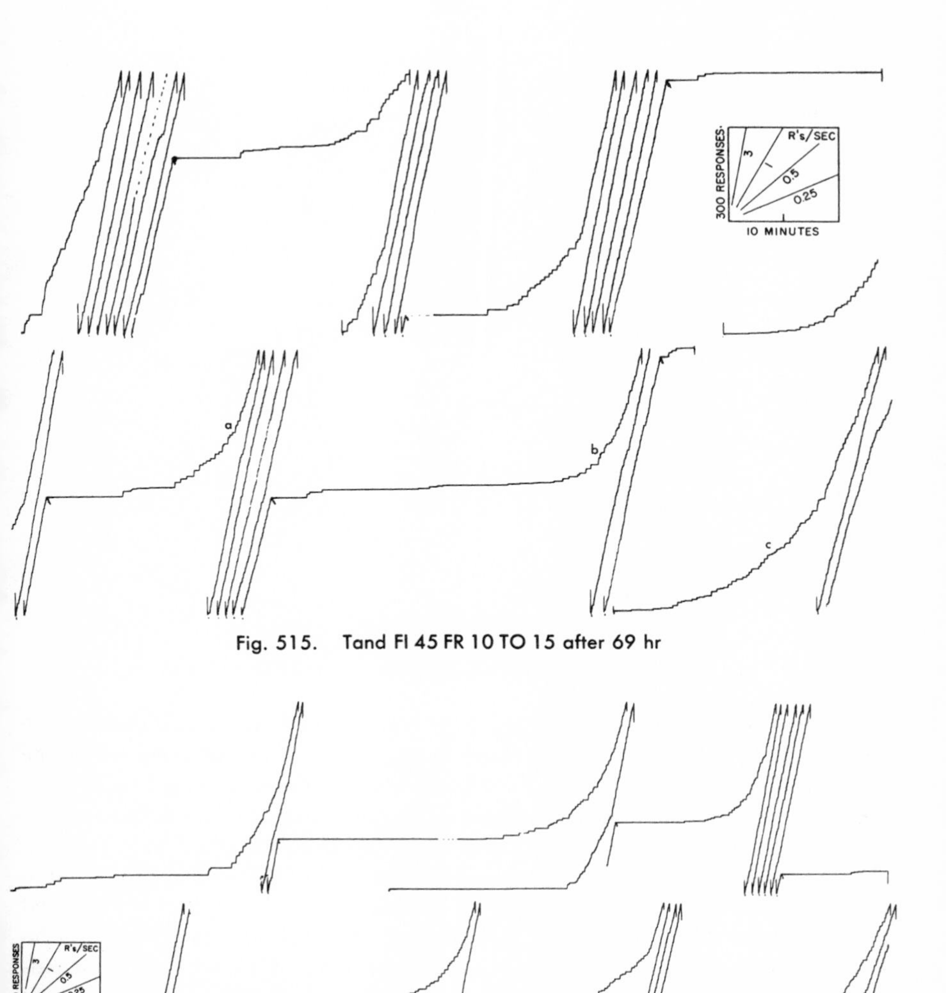

Fig. 515. Tand FI 45 FR 10 TO 15 after 69 hr

Fig. 516. Tand FI 45 FR 10 TO 15 showing extended pauses and curvature

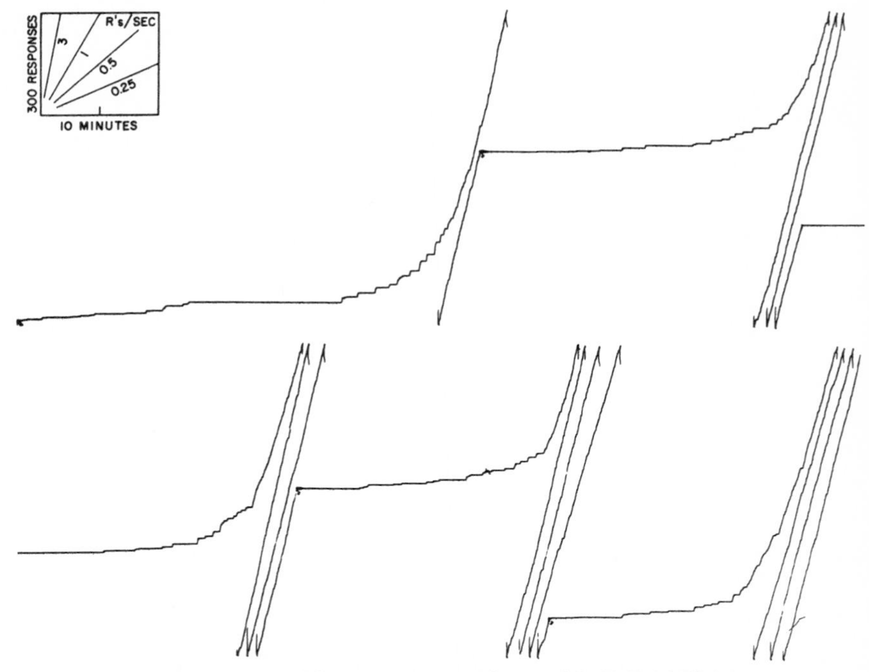

Fig. 517. FI 45 TO 15 ten sessions after tand FI 45 FR 10 TO 15

Transition from tand FI 45 FR 10 TO 15 to FI 45 TO 15. The final recorded performance under tand FI 45 FR 10 TO 15 was shown in Fig. 516. In the following session the tandem FR 10 had been removed. The schedule had now become FI 45 TO 15. This schedule was maintained for 10 sessions, during which no noticeable change occurred in the performance. The entire 10th session is shown in Fig. 517, which closely resembles Fig. 516. In the transition from tand FI 45 FR 10 to FI 45 without time out, however, both the over-all rate and the character of the curves changed in the direction of the usual FI 45 performance. Possibly, further exposure to the FI 45 reinforcement in this experiment would have produced a lower rate of responding and rate changes more characteristic of an FI 45 performance. A difference is to be expected from the higher terminal rates in the present experiment. These rates make it less likely that reinforcement will occur after a pause. The marked run-and-pause character of the tandem performance without time out makes it more likely that reinforcement will be set up during a pause.

The final performance of the second bird on tand FI 45 FR 10, already described in Fig. 510, showed little evidence of interval curvature after 754 hours of exposure to the tandem schedule. A marked pause and scallop appeared when a 15-minute TO was added after reinforcement. Figure 518 shows the entire 2nd session with TO 15.

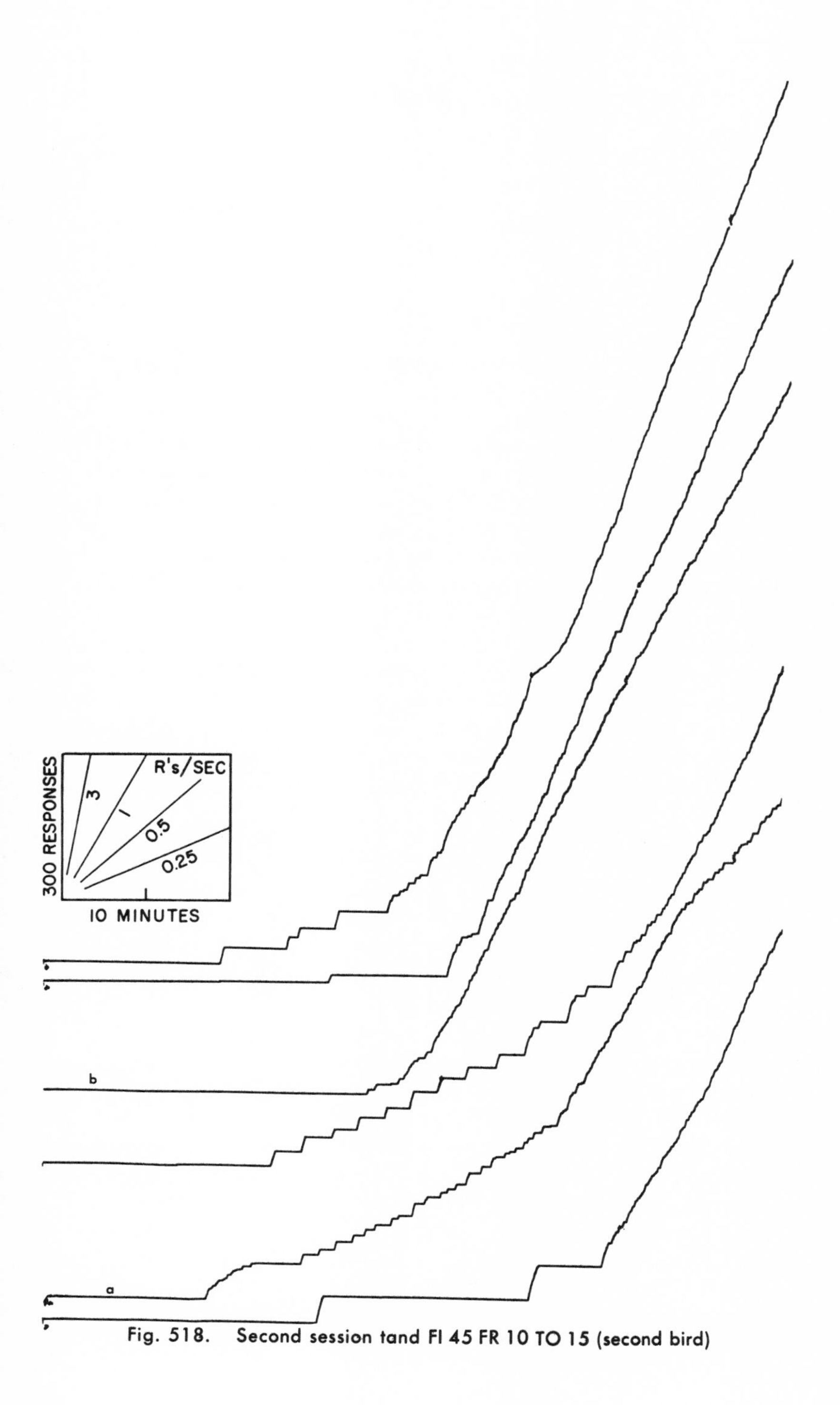

Fig. 518. Second session tand FI 45 FR 10 TO 15 (second bird)

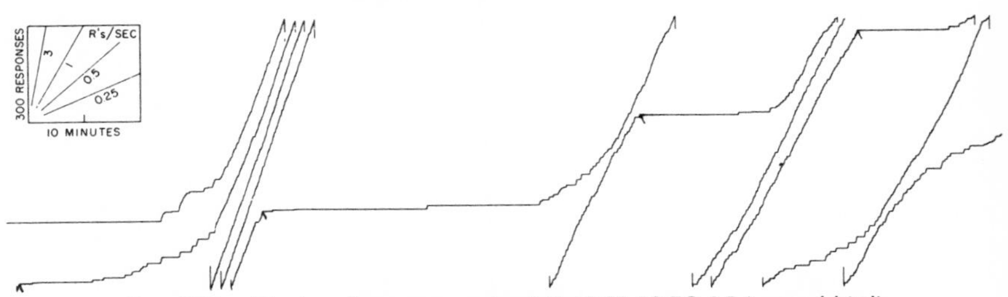

Fig. 519. Final performance on tand FI 45 FR 10 TO 15 (second bird)

The fixed-interval segments, which have been arranged in order of the number of responses in the interval, show that the over-all rate is roughly the same as before time out; but the terminal or highest sustained rates of responding are considerably higher. Pauses after reinforcement vary from 9 minutes at *a* to 19 minutes at *b*. Figure 519 shows the final recorded effect of the TO, 10 sessions later. The highest terminal rate has increased to almost 2 responses per second, compared with from 1.2 to 1.5 responses per second in Fig. 518. Reinforcements are still uniformly followed by pauses, and the fine grain of the record consists of short bursts of responding at from 3 to 4 responses per second.

Extinction after tand FI 45 FR 10. The response was extinguished 30 hours before the final performance of tand FI 45 FR 10 shown in Fig. 512. Figure 520 contains the entire extinction curve. It begins with the slow acceleration to the terminal rate of responding at *a* which normally characterizes the start of the session. The terminal rate is sustained for almost 8000 responses before it falls off, at *b*. For the remainder of

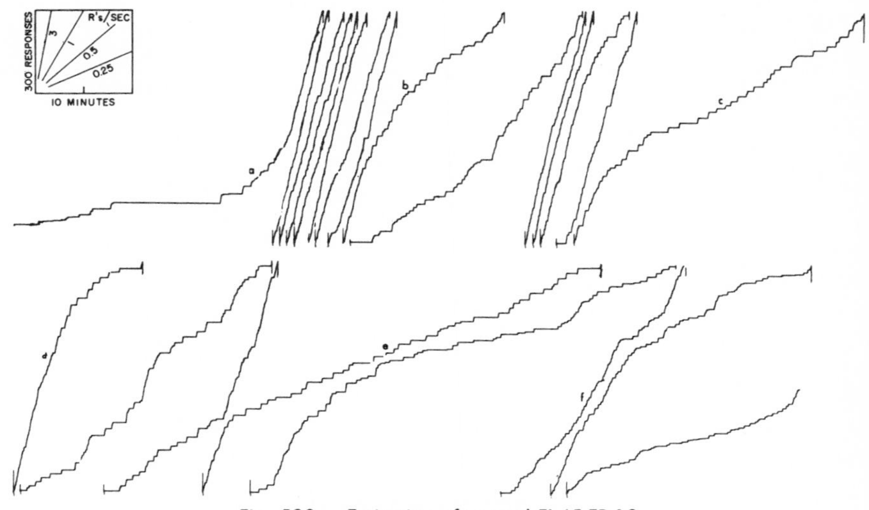

Fig. 520. Extinction after tand FI 45 FR 10

the curve the bird oscillates between intermediate rates of responding, consisting of bursts of from 10 to 30 responses separated by pauses (as in the segments at *c* and *e*), and periods of more sustained responding at rates approaching the former terminal rate at *d* and *f*. The higher rate is achieved by the shortening of the pauses between bursts.

A second extinction curve taken in the session following Fig. 517 confirms the fact that the performance still retains the effect of the earlier tandem ratio. Figure 521 shows the curve. It begins with the normal acceleration to a terminal rate at *a*, which is sustained for 20,000 responses until it falls off at *b*. The remainder of the curve is characterized by long periods of zero or near-zero responding (many of which are omitted from the figure), separated by periods of responding at near-terminal rates for a few thousand responses. The fine grain of the record shows no intermediate rate of responding. The responses which occur in bursts show a rate of the order of 5 responses per second. Except for the longer pauses, this curve resembles Fig. 520, taken after tand FI 45 FR 10 without TO.

The extinction curve after tand FI 45 FR 10 for the second bird appears greatly reduced in Fig. 522. The rate declines fairly steadily for the first 3 or 4 hours, and thereafter the level of performance becomes relatively stable. The grain throughout shows the effect of the tandem ratio. A sample toward the end of the curve has been reproduced on the usual scale.

Large FR following tand FI 45 FR

Following the experiments on tand FI 45 FR 10 described above, the tandem ratio was progressively extended up to FR 400. Although this extension postponed reinforcement, the terminal rate was so high and so consistent that little relative change in the 45-minute interval was produced. The performance remained unchanged. Many thousands of responses per reinforcement were being emitted. A ratio of this magnitude could not be held on a straight FR basis. In order to investigate the special powers of this tandem schedule to sustain this level, we changed the schedule to FR 875, the TO remaining in force. Now the bird could complete the ratio required for reinforcement in a short time. Figure 523 shows the performances for both birds. Record A follows 20 reinforcements on FR 875 TO 15, and Record B, 17 reinforcements. The periods of acceleration before the terminal rate is reached are effects of the earlier interval component. The terminal rate is of the same order as before, and relatively low compared with FR performances reached directly from crf.

Similar performances were maintained for a total of 22 hours. The TO 15 was then removed. The ratio continued to be "held" for the remaining 77 hours of the experiment, though pauses increased and the general picture closely resembled a larger-ratio performance. The second bird sustained the ratio less successfully. (Cf. Fig. 113 and 114 for the later performances of the bird in Fig. 522, and Fig. 115 for that of the other bird.)

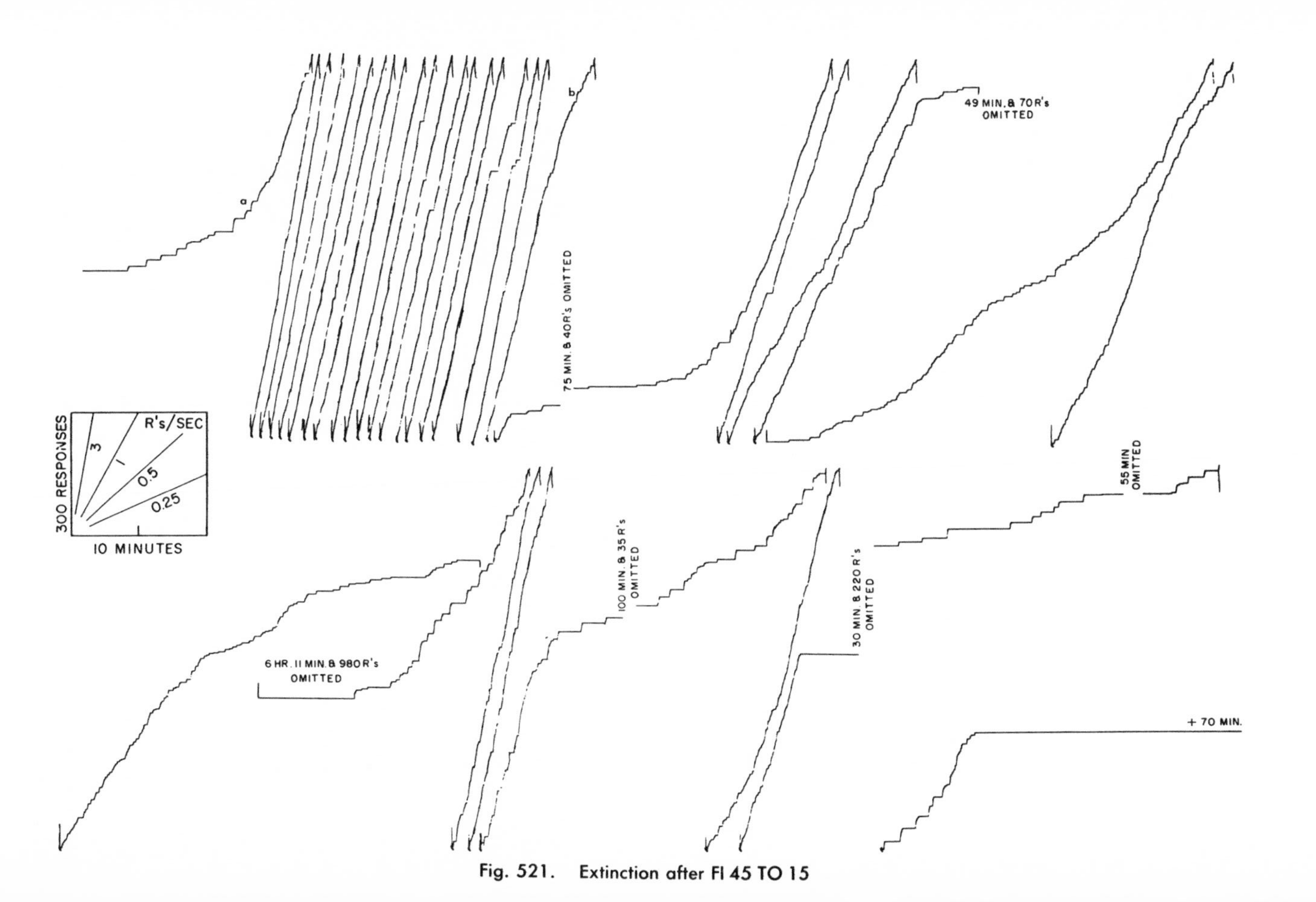

Fig. 521. Extinction after FI 45 TO 15

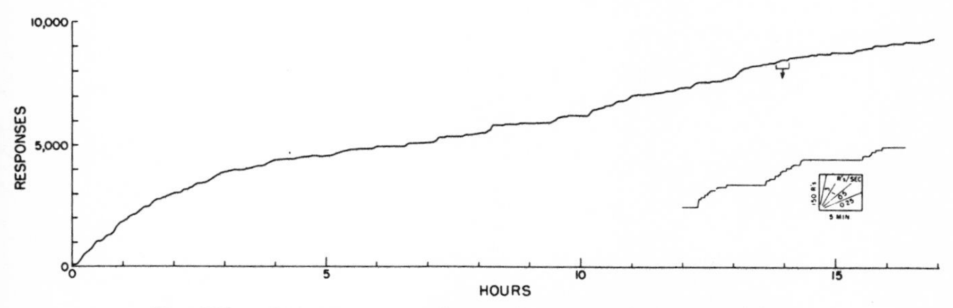

Fig. 522. Extinction curve after tand FI 45 FR 10 in reduced form

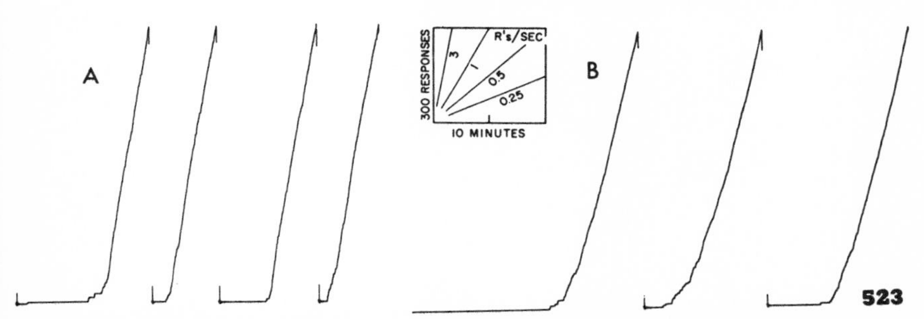

Fig. 523. Early performance on FR 875 TO 15 after tand FI 45 FR 10 TO 15

Tand FI 8 FR 3 and tand FI 10 FR 3

We studied the effect of a small, tandem fixed ratio added to a shorter fixed interval with 2 birds with the following history: (1) crf to FI. This has been described in Fig. 166 and 167. (2) FR with ratio determined by the mean number shown in the interval. The birds failed to develop ratio performance. Pauses were extended indefinitely. (3) FI. Recovery of the earlier performance, one bird on FI 8, the other on FI 10.

A tandem ratio of only 2 responses was then added. After 6 sessions, this was changed to 3 responses. For convenience, the schedule will be called FI FR 3 throughout.

Figure 524 shows representative parts of 4 sessions at various stages in the development of the performance on tand FI 8 FR 3 and of a 5th session after the return to FI 8 alone. The interval segments are in the recorded order. Records A, B, C, and D contain 8 successive interval segments at 25, 45, 80, and 150 hours of exposure to the tandem schedule. The terminal rate, the rapidity of the acceleration to the terminal rate, and the pauses after reinforcement all increase progressively during this period. When the tandem ratio was removed, however, the trend continued. Record E shows segments from the 16th session, or 115 hours after the tandem ratio was dropped. The terminal rate is approximately 2.5 responses per second, and pauses of 3 to 5

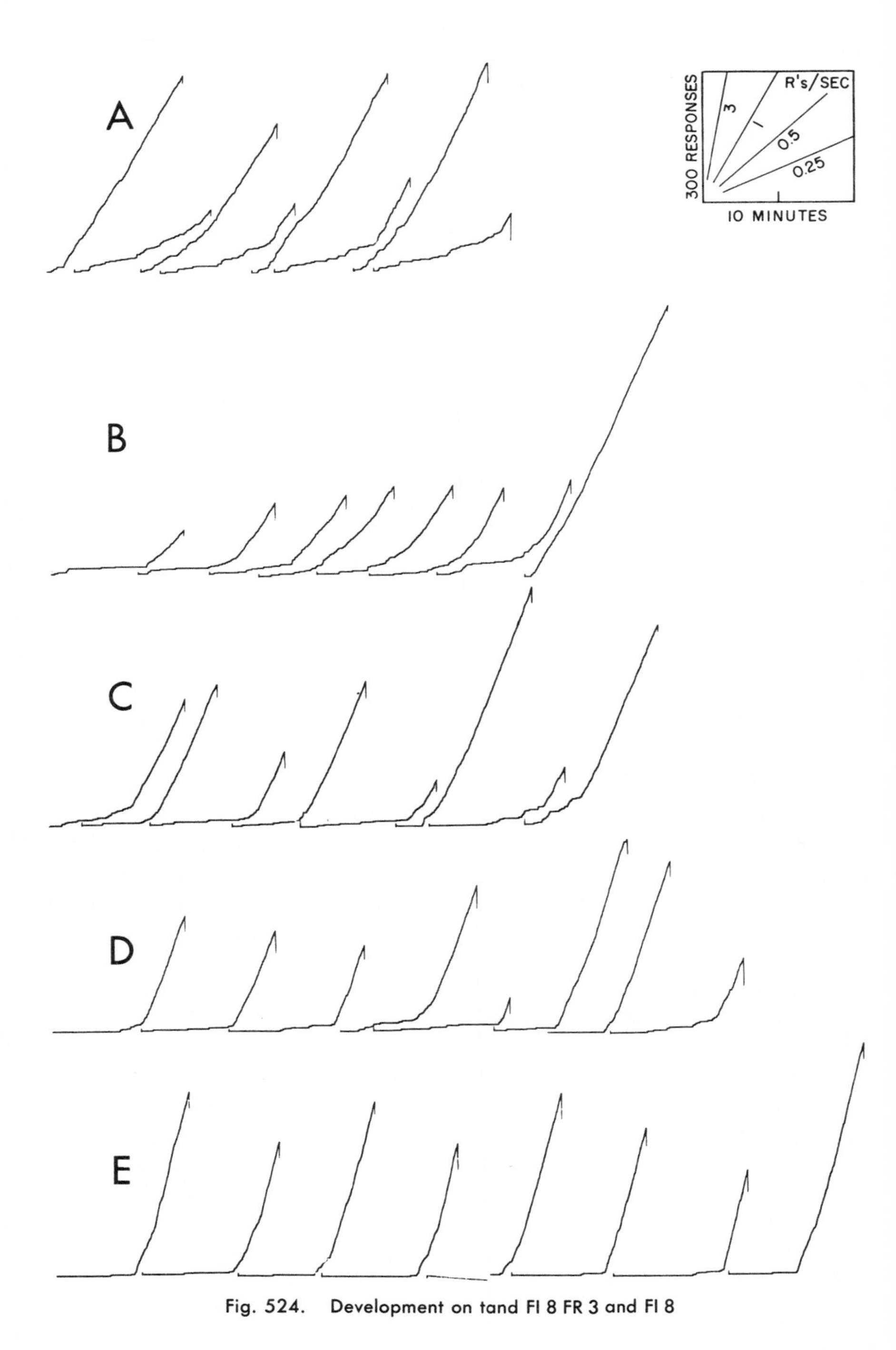

Fig. 524. Development on tand FI 8 FR 3 and FI 8

minutes follow the reinforcements. Although we cannot be sure that the progressive change in the character of the performance in Records A, B, C, and D was due to the tandem ratio rather than simply a result of the continuous exposure to the fixed interval, the performance in Record D with its high terminal rate is unusual.

The performances in Fig. 524 represent various stages of the experiment fairly accurately. Deviations in the form of "knees" sometimes occurred, however. Some of these are shown in Fig. 525, which gives successive interval segments from 4 sessions after the return to FI 8.

Figure 526 presents an extinction curve after 95 hours of reinforcement on tand FI 8 FR 3, 3 sessions after Record C in Fig. 524. A continuous decline in over-all rate is marked by a strong oscillation. No prolonged pauses occur in the curve, even when the over-all rate of responding reaches a very low value. The curve suggests that the tandem component has had a more marked effect than appears in the tand FI 8 FR 3 performance. The grain is ratio-like, and the performance is sustained much beyond extinction after FI.

Figure 527 shows a second extinction curve, taken shortly after Record D in Fig. 524, when the terminal rate and pause after reinforcement on FI 8 had increased considerably. The smooth oscillation between low and high rates is even more marked. The oscillation is between the near-zero and high terminal rates of the tand FIFR performance. Note that 4 hours in which 840 responses occurred are omitted from the record.

The second bird, on tand FI 10 FR 3, also showed a progressive increase in the pause after reinforcement and in the terminal rate. The final tandem performance shows a large pause, a higher terminal rate, and more responses per interval than a normal FI performance. The effect is again not reversible. Figure 528 presents the final performance on FI 10 after 370 hours on FI 10, tand FI 10 FR 3, and FI 10. Except for occasional knees, at *a*, *c*, *d*, and *e*, the bird pauses or responds at a very low rate for the greater part of the interval and then changes abruptly to the terminal rate. The terminal rate is between 3 and 4 responses per second, and the pauses following the reinforcement may be as long as 8 minutes (at *b*, for example). Many of the segments are similar to those under a fixed interval with clock. The clock performance is less variable. (Note that the performance in Fig. 528 is without benefit of TO.) The irreversibility of the tandem effect is not surprising. When a terminal performance of this sort is maintained, whether the last 3 responses are timed or counted is of little significance. We could have emphasized the missing tandem ratio component by differentially reinforcing low rates and thus slowing down the whole performance.

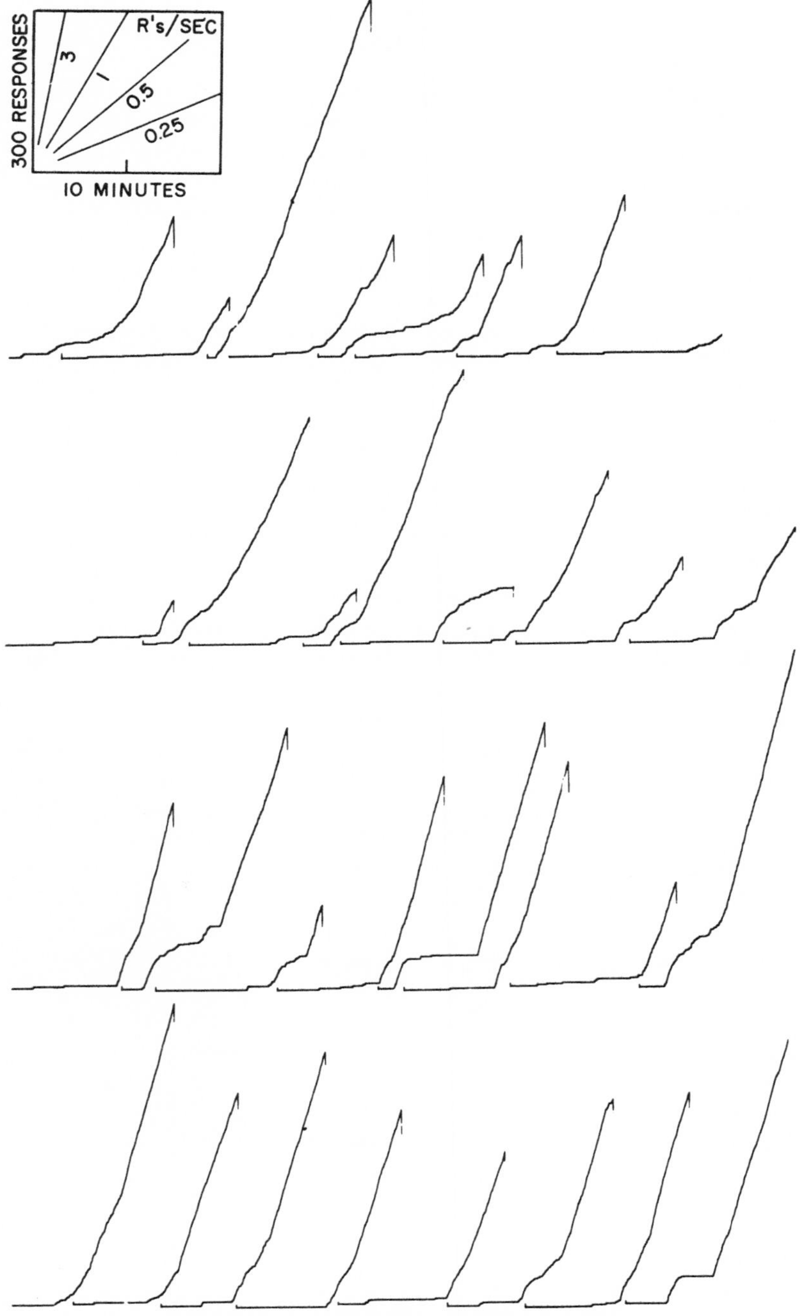

Fig. 525. Deviations from a smooth interval performance

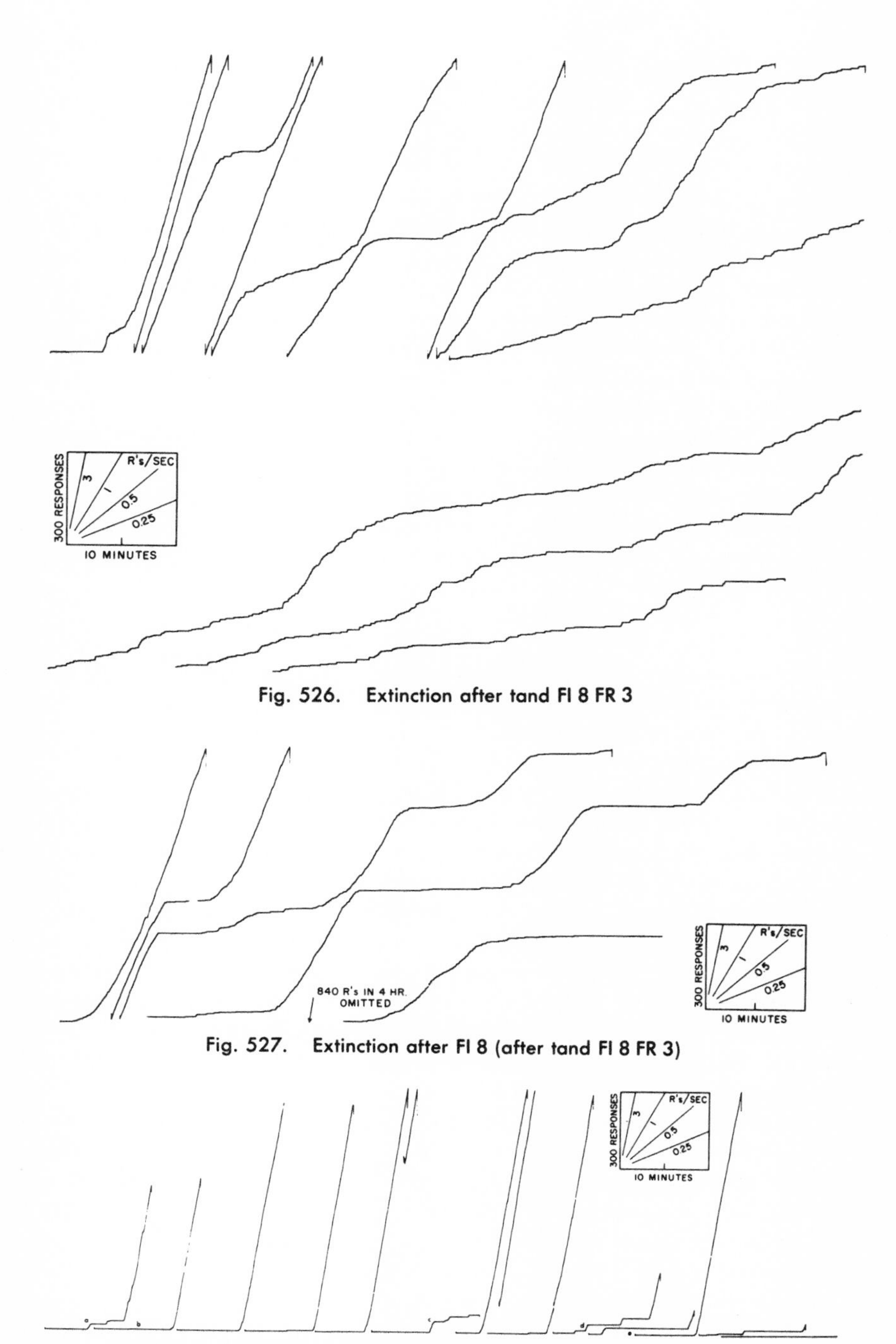

Fig. 526. Extinction after tand FI 8 FR 3

Fig. 527. Extinction after FI 8 (after tand FI 8 FR 3)

Fig. 528. Final performance on FI 10 after tand FI 10 FR 3 (second bird)

TANDEM FIXED-RATIO FIXED-INTERVAL SCHEDULE

In the tand FRFI schedule the reinforcement is based on a fixed number of responses, but the completion of this number does not produce reinforcement; instead, it starts a clock which, after an interval, sets up a reinforcement. The next response is reinforced. It provides a technique for eliminating the differential reinforcement of high rates associated with fixed-ratio reinforcement while maintaining an approximately constant fixed number of responses per reinforcement. If the tandem fixed interval is only a few seconds long, the variation in the number of responses from fixed-ratio segment to fixed-ratio segment will be slight. But the relation between reinforcement and immediately preceding responses will be that of an interval schedule, where the probability of reinforcement increases with time since the last response. The likelihood of reinforcement after any given pause is slight, however, since only responses are reinforced which occur after pauses beginning after the fixed ratio has been counted out. A schedule of this type is important in an analysis of the high rate in a fixed-ratio schedule.

Experiment I

1. Early development after crf. In a first experiment a brief interval of 10 seconds was added to a fairly large ratio of 240. The schedule was put in force immediately after crf. The bird probably could not have made the transition directly to FR 240. The tandem schedule, however, takes immediate effect.

Figures 529 and 530 comprise the entire (13 hour) 1st session of the transition from crf to tand FR 240 FI 7 sec. Figure 529 begins with a succession of negatively accelerated curves, common in transitions to FI and FR after crf. The over-all rate increases gradually during the first 8 reinforcements, to *a,* after which any further change is only moderate. The over-all rate reached approximately 0.75 response per second by the last segment of the figure. During the remaining 9 hours of the session (Fig. 530), both the over-all and local rates of responding increased progressively, reaching 1.9 and 2.5 responses per second, respectively, by the last excursion of the record. Pauses began to develop after reinforcement, as at *a, b,* and *c,* though they disappear temporarily in the region of *d.* In the middle part of the figure the rate oscillates and some curvature occurs during many of the ratio segments. Some responding occurs immediately after reinforcement at *e,* before the pause, and acceleration frequently leads to the terminal rate. Curvature in a normal fixed-ratio performance usually foreshadows long pauses after reinforcement, and two of these occur at *f* and *g,* 167 minutes and 34 minutes long, respectively. An almost standard fixed-ratio performance emerges by the end of the session. The last reinforcement of the session was followed by a 40-minute pause, after which the session was terminated.

Why the tandem interval helps the bird to maintain a substantial level of responding under this large fixed ratio is not clear. The very late development of a high terminal rate may be the explanation. Normally, a high terminal rate is quickly

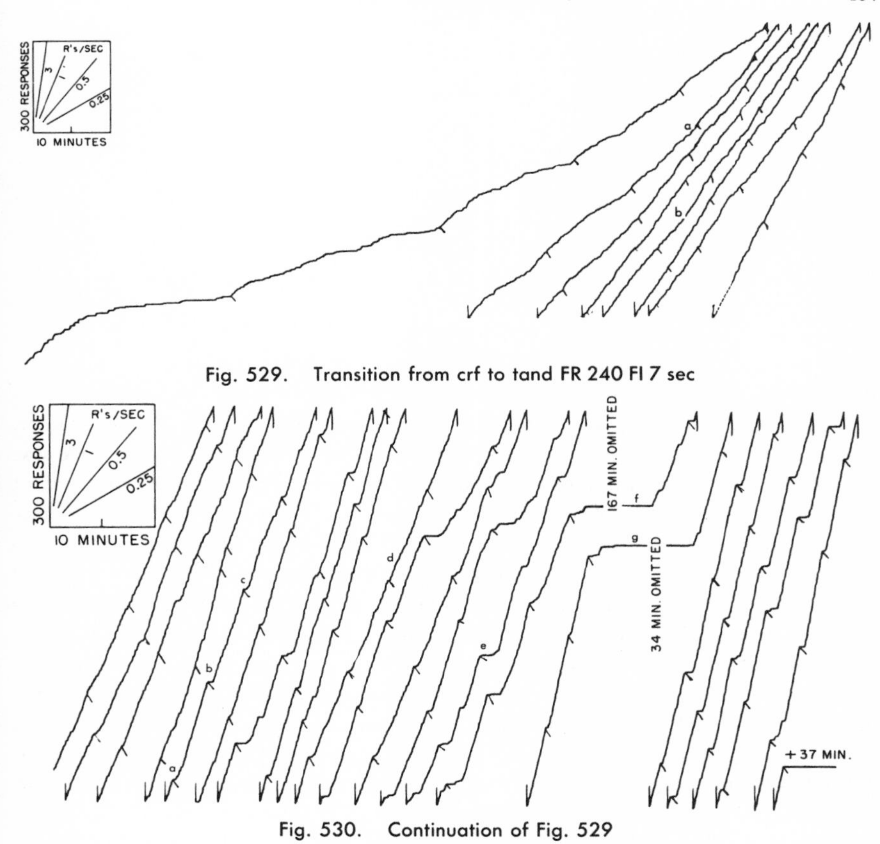

Fig. 529. Transition from crf to tand FR 240 FI 7 sec

Fig. 530. Continuation of Fig. 529

reached. This produces pausing, possibly because of the contrasting count just after or just before reinforcement. If the small tandem interval delays the development of the high terminal rate until the ratio schedule is more effective in establishing count as reinforcement, the present result would follow.

The second bird in the transition from crf to FR 240 FI 7 showed too low a rate to produce even 1 reinforcement on the tandem schedule during the first 2 hours of the session. The size of the fixed-ratio component of the tandem schedule was therefore reduced to 80, and then later increased to 140. Both the local and over-all rates of responding increased during the first 6 sessions, resulting in the performance shown in Fig. 531 for the 6th session after continuous reinforcement on tand FR 140 FI 7 sec. The terminal rates of responding are characteristically between 2.5 and 3 responses per second, although they vary considerably, with instances as low as 1.4 responses per second, as at *a*. A pause after the reinforcement is sometimes absent, as at *c*. The change from pause to terminal rate varies from the smooth curvature at *b* and *d* to an

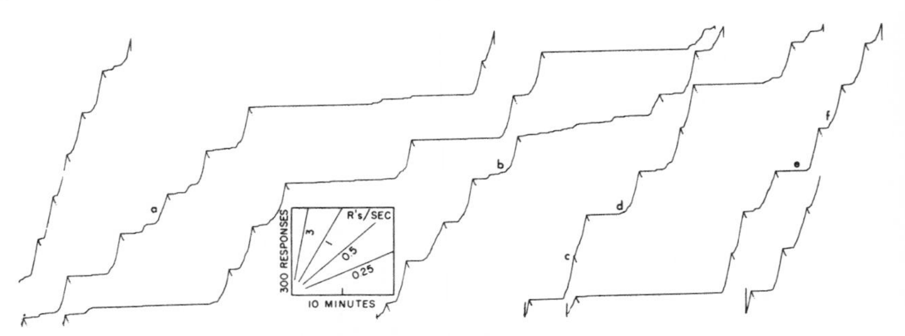

Fig. 531. Early performance on tand FR 140 FI 7 sec

abrupt shift, as at *e*, or an abrupt shift to an intermediate rate, which then accelerates to the terminal rate, as at *f*. Figure 532 shows the performance 13 sessions later, after the size of the fixed ratio has been increased to 200. This performance was fairly stable and was maintained for the next 30 sessions. Pauses after reinforcement vary from a few seconds, as at *d*, to 12 minutes, as at *a*. The terminal rate of responding is stable at about 2.8 responses per second. Most of the changes from pause to terminal rate are now relatively abrupt. The performance differs from a simple fixed-ratio schedule in the frequent appearance of intermediate rates of responding, as at *b* and *e*, in the failure to maintain the terminal rate once it is reached, as at *c*, and in the grain of the terminal rate, as at *e*.

The pause after reinforcement gradually disappeared during 3 sessions 30 hours later, when the fixed-ratio component of the tandem schedule was reduced to 80. Figure 533 shows the 3rd session on tand FR 80 FI 7 sec. The terminal rate of responding remains unchanged, and sustained intermediate rates, as at *a* and *b*, still occur.

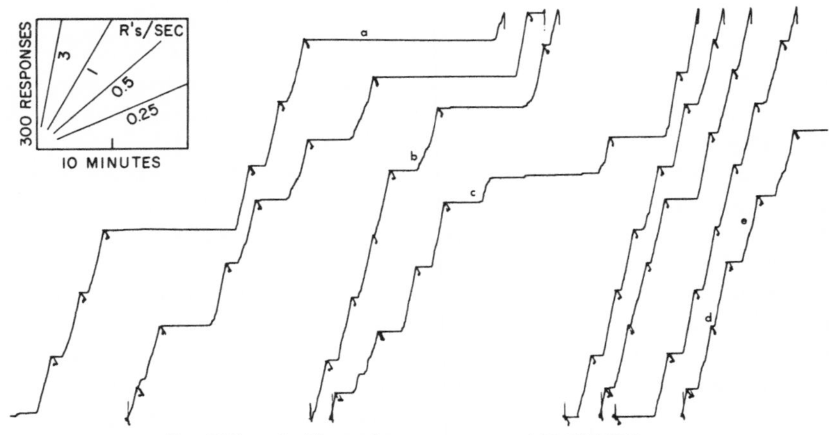

Fig. 532. Stable performance on tand FR 200 FI 7 sec

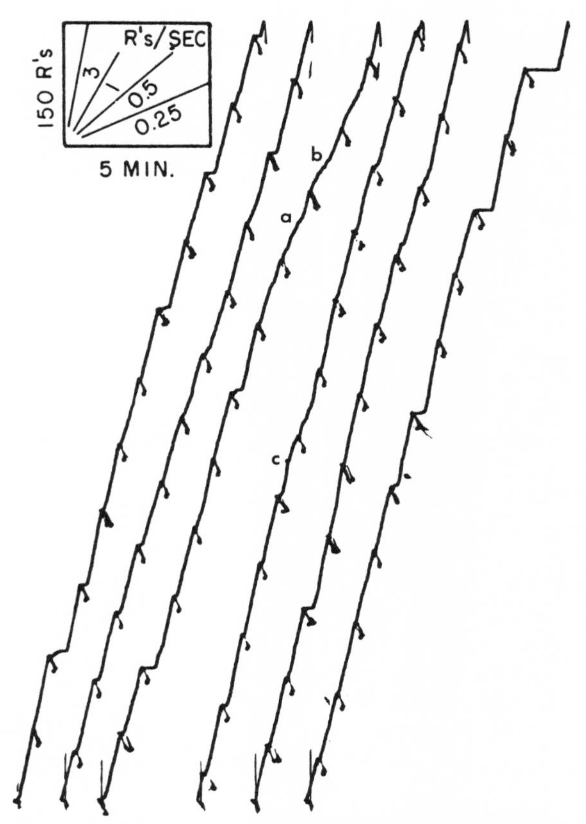

Fig. 533. Third session on tand FR 80 FI 7 sec after tand FR 200 FI 7 sec

Note the S-shaped curve at *c*. The bird later failed to maintain a sustained performance on the tandem schedule, although the fixed-ratio component was reduced to as low as 60. During the last sessions of the experiment, not enough responding occurred to produce reinforcements. The schedule was then changed to tand FR 10 FI 7 sec in order to restore the response. Figure 534A gives the result. The first fixed ratio was completed at *a*, within the first half-hour of the session. A second reinforcement occurs at *b*, 85 minutes later. The size of the fixed ratio was then increased to 20 for the remainder of the session. The rate of responding increased during the next 8 segments. During the next 9 sessions of the experiment, the size of the ratio component of the tandem schedule was gradually increased to 100. Records B through J show the first 10 reinforcements of each session. A pause after reinforcement develops as the size of the ratio is increased, and the terminal rate of responding increases to from 2 to 3 responses per second. Even in Record J, however, the local

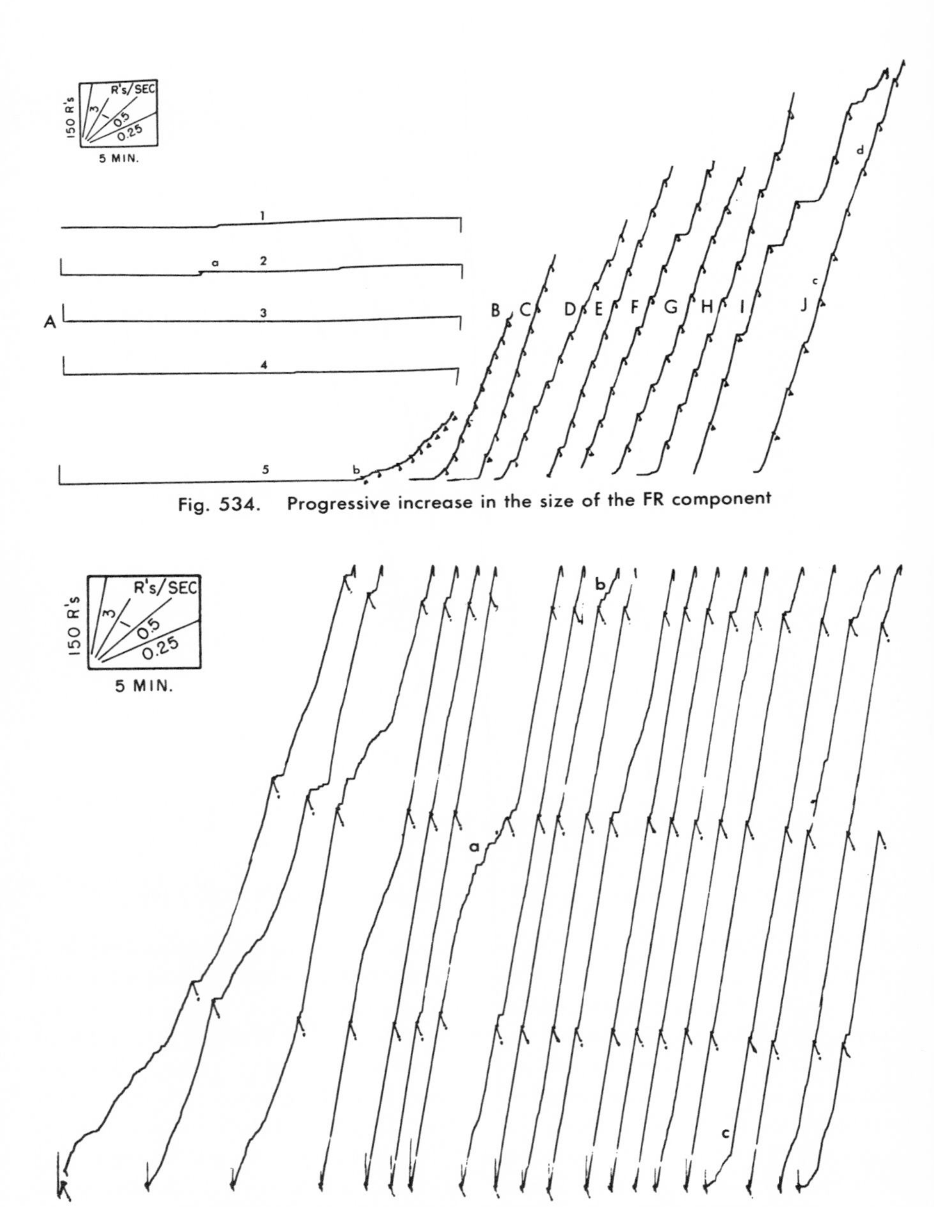

Fig. 534. Progressive increase in the size of the FR component

Fig. 535. Sudden emergence of high rates on tand FR 335 FI 7 sec

rate varies considerably, from approximately 2.2 responses per second at *c* to 1.7 responses at *d*.

In this second bird the tandem 7-second interval seems to have given little if any help in maintaining a ratio performance.

Later development. The first bird in the experiment was maintained on tand FRFI for more than 60 sessions, during which the fixed ratio varied between 170 and 335. The performance shown at the end of the session represented in Fig. 530 continued essentially unchanged for 11 sessions except for a slight increase in the rate of responding. The size of the fixed ratio was then increased to 335. During the 3rd session of this tand FR 335 FI 7 sec, sustained responding at 4 responses per second suddenly occurred, with no pausing after reinforcement and few deviations from a constant local rate. Figure 535 shows the session. In comparing this figure with Fig. 529 and 530, note that the recorder scale has been changed. A given slope in these first 2 figures represents a lower rate than that in the figures for the rest of the experiment.

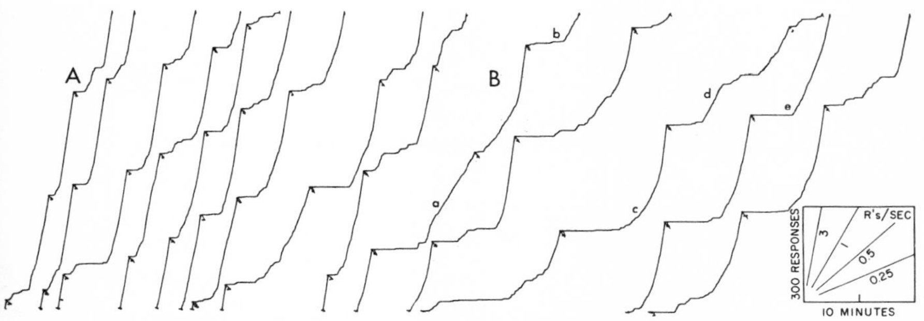

Fig. 536. First appearance of pause and curvature on tand FR 335 FI 7 sec

The bird had been showing a performance similar to the first few excursions of the record, with only an occasional emergence of sustained responding through several ratio segments. Brief responding at an intermediate rate with rough grain continues to occur occasionally at *a, b,* and *c*. Rate changes of this order are rare on a simple fixed-ratio schedule except immediately after reinforcement.

During subsequent sessions on tand FR 335 FI 7 sec, pausing after reinforcement increased; and in the 9th session a performance such as that in Fig. 536 developed. Record A shows the first 2 hours of the session, Record B, the 5th and 6th hours. The pause after reinforcement increases progressively throughout the session, leading occasionally into smooth and extended curvature, as at *c,* and in any case to extended scallops. A terminal rate of responding of the order of magnitude of 6 responses per second is usually reached before reinforcement, but not very long before. At *d* an intermediate rate is sustained in a "knee," and at *a* an intermediate rate of responding is maintained until reinforcement. Occasionally, an abrupt shift occurs from the pause after reinforcement to an intermediate rate, as at *b* and *e*.

Pausing and heavy scalloping continue to develop in the following session, all 15 hours of which are shown in Fig. 537 to 540. Toward the end of Fig. 537, the segments at *a, b,* and *c* show longer pauses and more extended curvature than those in the previous session (Fig. 536). Figure 538 shows the 29th through 52nd reinforcements, where the over-all rate remains about the same, although the individual segments show a variety of rate changes, including 2 marked knees at *a* and *b*. By the 53rd reinforcement in Fig. 539, the pause following the reinforcement becomes much more extended than heretofore, producing the lowest over-all rate yet seen. Except

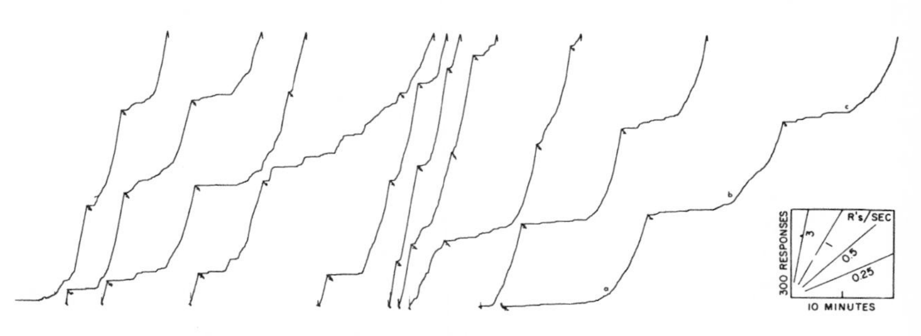

Fig. 537. Progressive development of pause and curvature

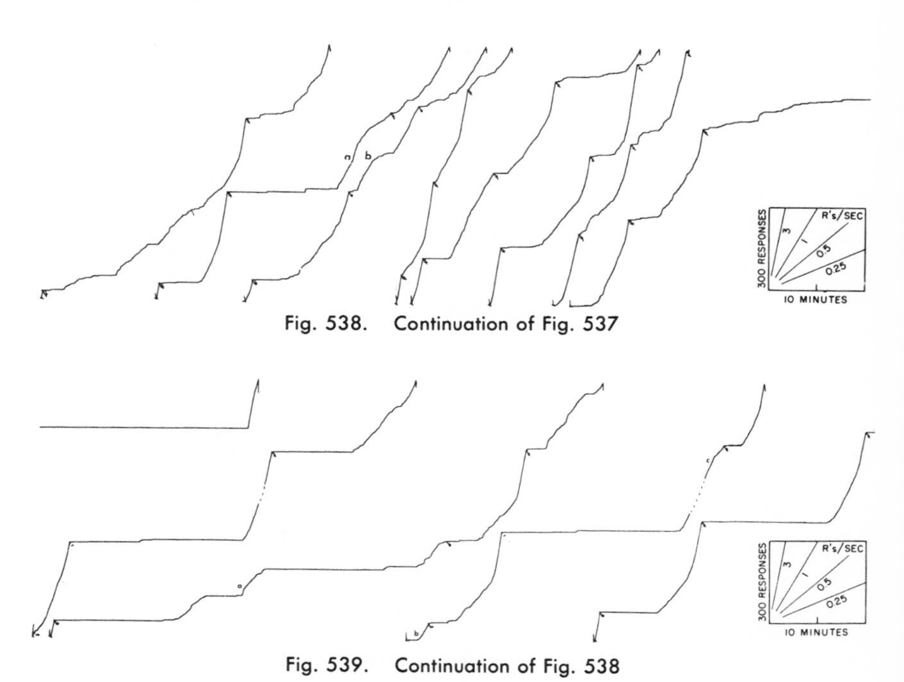

Fig. 538. Continuation of Fig. 537

Fig. 539. Continuation of Fig. 538

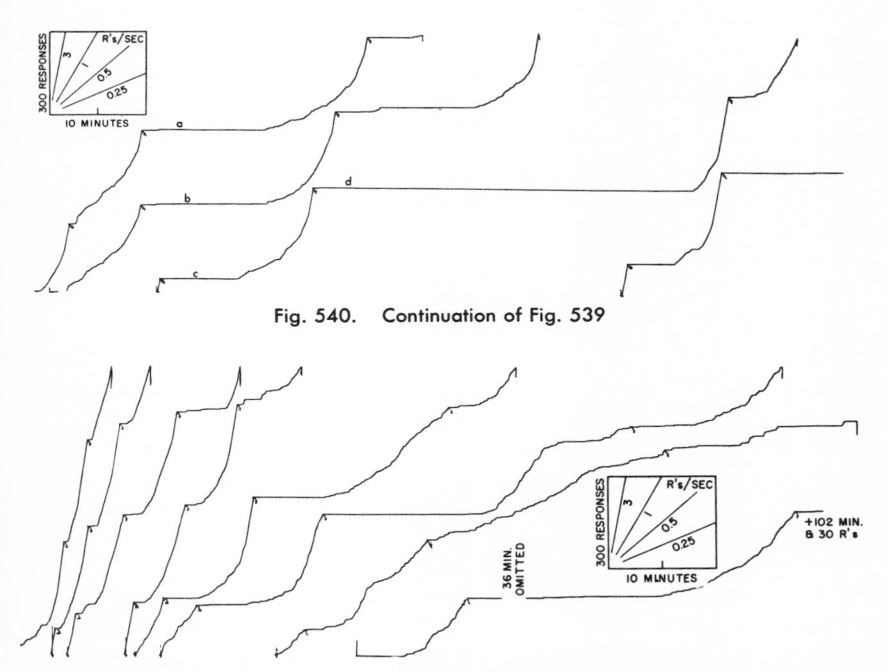

Fig. 540. Continuation of Fig. 539

Fig. 541. Later performance on tand FR 335 FI 7 sec showing development of pause and curvature

for the segments at *a* and *b,* where a terminal rate of responding is never reached, and *c,* where the rate falls off just before the reinforcement, all segments accelerate to a rate of more than 4 responses per second before the end of the ratio. The low over-all rate continues in the final part of the session, which is shown in Fig. 540. The segments beginning at *a, b,* and *c* show a very prolonged period of acceleration to the terminal rate, which has fallen to the order of 3 responses per second. After the reinforcement at *d,* 40 minutes elapse before the bird begins to respond. The 75 reinforcements received during the entire 15-hour session probably did not produce any substantial satiation to explain the decline in the over-all rate.

The over-all rate increased in the following session, with a performance similar to that in Fig. 535, except for slightly longer pauses after reinforcement. Five sessions later, however, pausing and marked scallops reappear toward the end of the session. In the next session, shown complete in Fig. 541, the performance begins normally but goes through a decline in over-all rate similar to the changes represented in Fig. 537 through 540. By the end of the session the bird frequently fails to reach the terminal rate, and long periods occur in which the rate is zero or near zero.

The character of the performance on the tand *FR*FI was explored at lower values

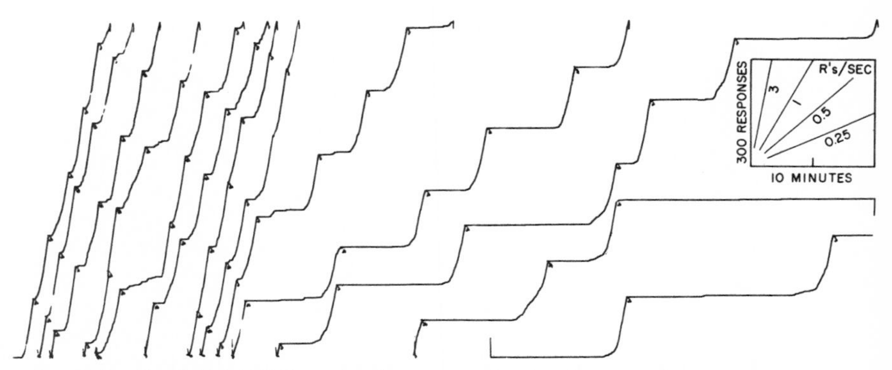

Fig. 542. Tand FR 170 FI 7 sec showing progressive development of pause and curvature

of the ratio component. Occasionally, the bird sustained a daily session of 60 reinforcements with very little decline in over-all rate, but a more characteristic performance was a progressive decline resulting from the elongation of the pause after reinforcement, and to some extent by a more extended period of curvature. Figure 542 contains a representative sample of this performance, when the schedule of reinforcement was tand FR 170 FI 7 sec, 30 sessions after the fixed-ratio component had been reduced from the FR 335 of Fig. 541. The schedule of reinforcement was then changed to FR 170 without a tandem interval. Figure 543 shows an entire daily session, 250 reinforcements later. The over-all rate still falls off during the session. The elimination of the tandem interval has increased the local rate of responding and the number of ratio segments sustained without pausing. Curvature and intermediate rates of responding are reduced. Toward the end of the session, for example, where nearly all segments in the tand *FR*FI showed marked curvature, instances of abrupt shifts to the terminal rate occur (as at *f* and *g*). The performance still shows effects of the tandem schedule in the knees at *a*, *b*, and *c*, the intermediate rate at *d*, and occasional marked curvature, as at *e*.

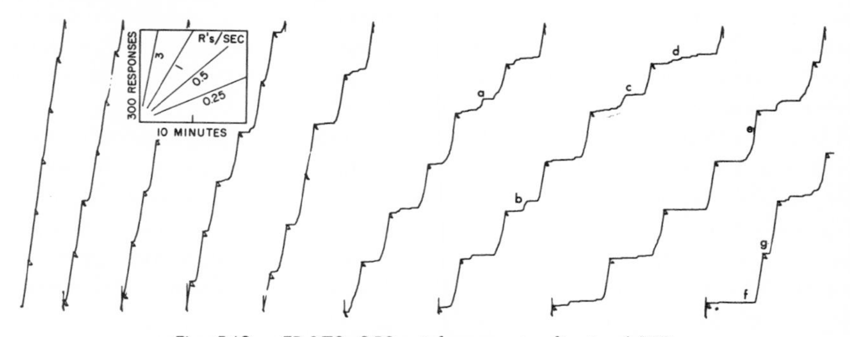

Fig. 543. FR 170, 250 reinforcements after tand FRFI

Experiment II

In a second experiment on tand *FR*FI the size of the fixed-ratio component was maintained at values which did not produce marked pausing or prolonged curvature. Figure 544 illustrates a transition from crf to tand FR 150 FI 6 sec. This is similar to the 1st transition observed in the previous experiment, except that the grain is somewhat rougher. The transition begins with negatively accelerated curves after reinforcements (at *a*, *b*, and *c*) leading into a linear phase (at *d*). High rates after reinforcement return at *e*, *f*, *g*, and *h*. The end of the session is roughly linear, except for one break-through at *i*. The over-all rate reaches approximately 0.75 response per second by the last excursion, and brief pauses appear after the reinforcements at *j*, *k*, and *l*.

The 2nd session on the tandem schedule shown in Fig. 545 begins at roughly the same rate as that at the end of the preceding session. A pause of from 5 to 10 sec-

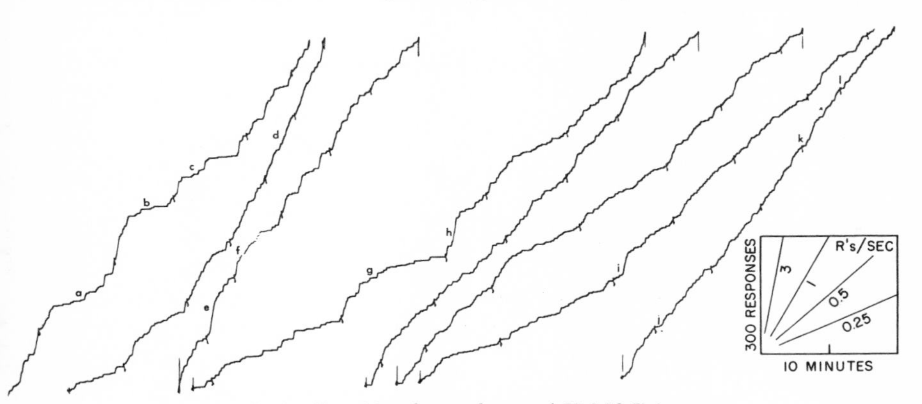

Fig. 544. Transition from crf to tand FR 150 FI 6 sec

onds frequently follows reinforcement. By the 3rd excursion, pauses appear consistently, and the acceleration to the terminal rate becomes more extended. Marked scallops occur at *a* and *b*; and in a final segment, at *c*, a very low rate for approximately 15 minutes is followed by a 55-minute pause, after which the session was terminated. The over-all rate of responding passes through a maximum in the middle of the session, with the highest terminal rate of responding slightly over 1 response per second. By the end of the session, both over-all and terminal rates fall to about the same values as those at the start of the session.

This bird was not able to maintain a sustained performance on tand FR 150 FI 6 sec. The performance shown in Fig. 546 for the end of the 3rd session shows a continuing decline. Note that in spite of the low over-all rate and the marked curvature, the terminal rates are still 0.75 response per second, as in the previous 2 sessions. The next session shows a progressive increase in curvature and pausing and a lower over-all

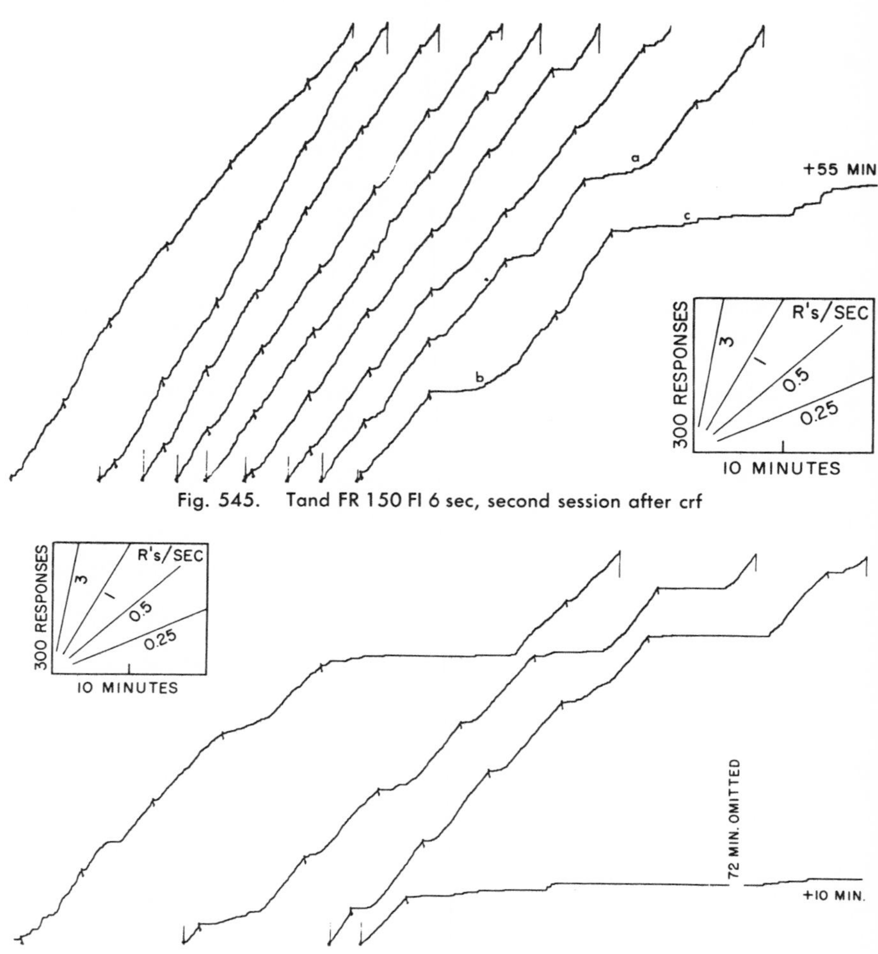

Fig. 545. Tand FR 150 FI 6 sec, second session after crf

Fig. 546. Tand FR 150 FI 6 sec, third session after crf

rate. Two sessions later, responding essentially ceased. The fixed-ratio component of the tandem schedule was then reduced to 50. Figure 547A shows the entire 1st session, which begins with only a slight pause after reinforcement and a constant rate during each ratio segment. The terminal rate increases from 0.9 response per second at *a* to 0.95 response per second at *b,* 1.5 responses per second at *c,* 1.7 responses per second at *d,* and the highest value in the session—2.2 responses per second—at *f.* Thereafter, the terminal rate falls to the order of 1.25 responses per second. The highest *over-all* rate occurs in the excursion at *e.* The decline in over-all rate beginning at *g* is in part probably a result of satiation from the 180 reinforcements received during the session. (Sixty reinforcements ordinarily keep the bird at a constant daily weight.) The start of the next session (Record B) shows occasional instances of curvature to the terminal rate, as at *h* and *i*, and S-shaped segments, as at *j*.

Three sessions later, the fixed-ratio component of the tandem schedule was increased to 100, and the pause following the reinforcement and the period of acceleration to the terminal rate of responding became longer. The various types of performances present at this stage appear in groups (Fig. 548). Segments containing very uniform curvature are evident at *a*, *b*, *c*, and *d*. The segments from *e* through *f* show similar though squarer types of performance, while the segments at *g*, *h*, and *i* show only slight pausing. Beginning at *i*, many segments show marked negative

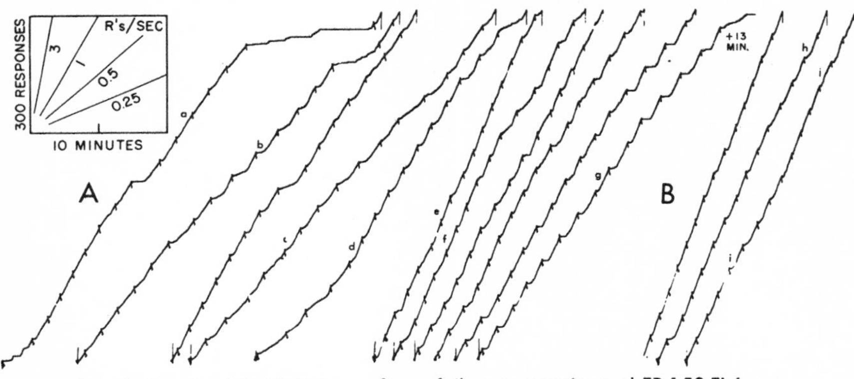

Fig. 547. Tand FR 50 FI 6 sec after a failure to sustain tand FR 150 FI 6 sec

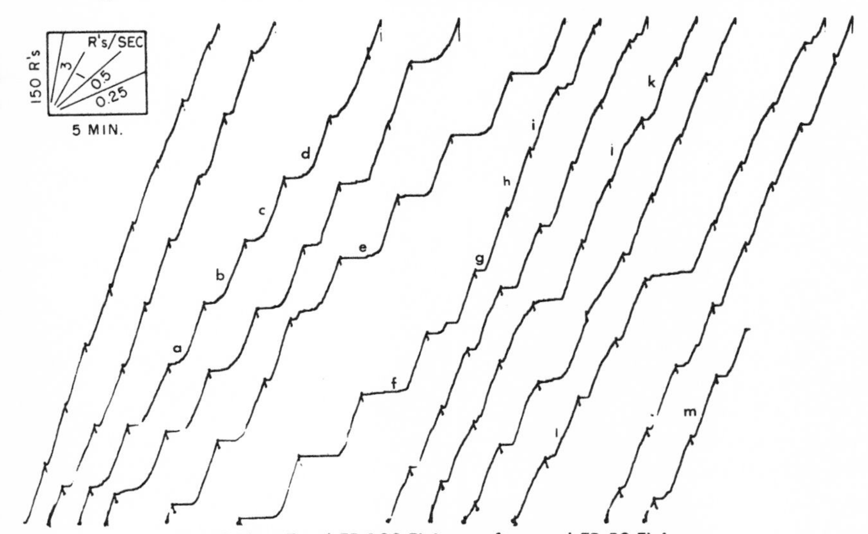

Fig. 548. Tand FR 100 FI 6 sec after tand FR 50 FI 6 sec

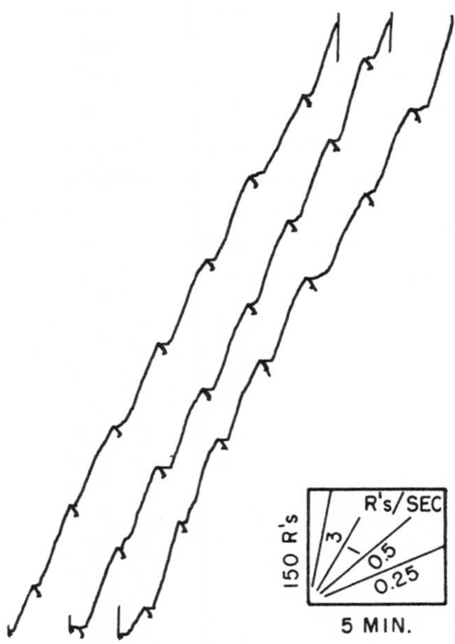

Fig. 549. Tand FR 100 FI 6 sec showing S-shaped segments

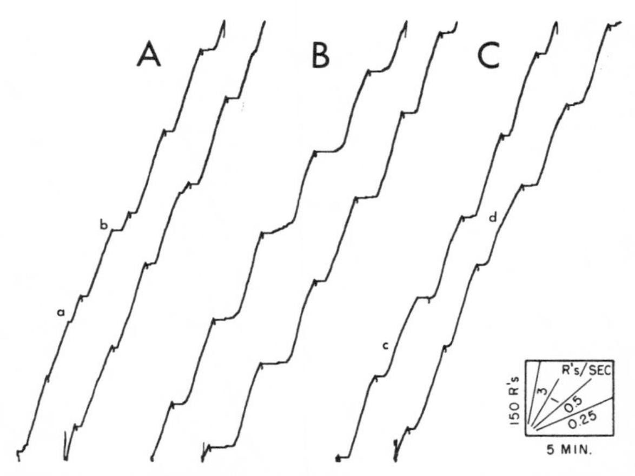

Fig. 550. Tand FR 150 FI 6 sec showing S-shaped segments

curvature. These are particularly marked at *j* and *k*, although instances still occur where the terminal rate is maintained until the reinforcement, with no decline, as at *l* and *m*. The tendency for the rate to fall during the latter part of the session on FRFI increases during the next 2 sessions. Figure 549 shows the 1st hour of the session, in which S-shaped curves are particularly marked. The higher rate of responding in the middle of the fixed-ratio segment is now being maintained for a larger number of responses, and the drop to a lower rate just before the reinforcement has become more abrupt.

The size of the ratio component of the tandem schedule was then increased to 150 for 3 sessions. The last of these is represented in Fig. 550. The 1st and 2nd, 8th and 9th, and 16th and 17th excursions are shown in Records A, B, and C, respectively. The pause and acceleration following reinforcement become more extended. The S-shaped curves occur throughout the session, although they become more frequent and prominent toward the end in Record C. In Record A, where S-shaped curves are not present, two instances of pauses occur at *a* and *b*, at about the same position in the ratio segment as the point of maximum curvature in the later negatively accelerated segments. At *c* and *d* the shift to the lower rate occurs more abruptly and earlier in the fixed-ratio segment than heretofore.

In the session following Fig. 550 the schedule of reinforcement was changed to tand FR 125 FI 6 sec and maintained for 34 hours (except for a period of 6 hours during which the tandem interval was removed). The segments in Fig. 551 show the final performance after this exposure to the tandem schedule. The pause after reinforcement varies from zero to approximately 1 minute and is followed by an instantaneous shift to the highest rate of responding, about 4.5 responses per second. This rate is maintained for from 40 to 75 responses, after which a relatively abrupt shift occurs to a rate of approximately 2.5 responses per second. The higher rate following the pause brings the end of each segment close to an extrapolation of the end of the preceding segment. Where the reinforcement is not followed by a pause, as at *a*, responding is resumed at the lower rate, maintaining the over-all linear envelope. The short period of higher rate has the effect of compensating for the pause, which is seldom seen in simple FR performances.

The other bird maintained less substantial responding, and stopped responding altogether soon after the first transition from crf to tand FR 150 FI 6 sec. When the ratio component of the tandem schedule was reduced to 65 responses, a stable over-all rate developed, with a performance very similar to that already described for the other bird in Fig. 547. When the size of the fixed ratio was later increased to 150, the bird was able to maintain a sustained performance, as shown in Fig. 552, the 8th session following the transition from crf. The pause following the reinforcement varies from zero to 11 minutes, with only moderate acceleration to a low terminal rate of the order of 1 response per second. The performance gives the over-all impression of a fixed-ratio schedule, except for the low terminal rate.

During the 2nd and 3rd sessions following Fig. 552, the terminal rate of respond-

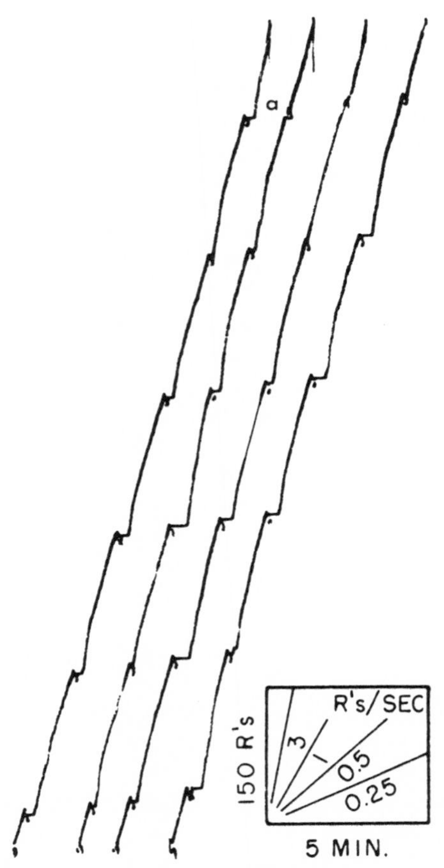

Fig. 551. Final tand FR 125 FI 6 sec showing S-shaped segments

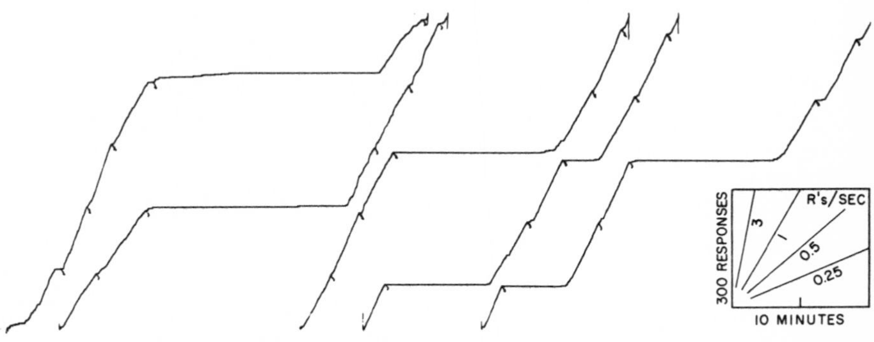

Fig. 552. Sustained performance on tand FR 150 FI 6 sec (second bird)

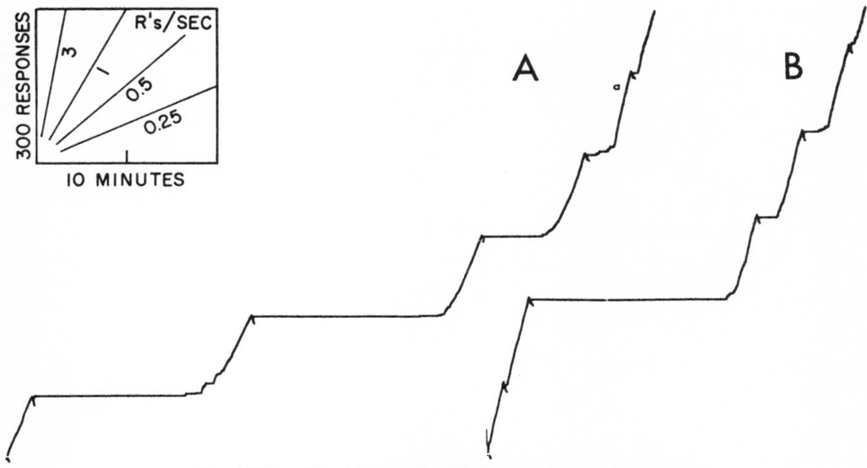

Fig. 553. Final tand FR 150 FI 6 sec (second bird)

ing increased. A few ratio segments from the 2nd session are shown in Fig. 553A, where the terminal rate at *a* reaches 2.5 responses per second. The 3rd session (Record B) shows a somewhat lower terminal rate of 2 responses per second; but by this time it is maintained more uniformly with less extended curvature. The experiment was terminated after the performance in Fig. 553.

TANDEM VARIABLE-RATIO FIXED-INTERVAL SCHEDULE

In tand *FR*FI we modify a fixed-ratio schedule by minimizing the differential reinforcement of high rates. The relation between reinforcement and the few preceding responses is that of an interval schedule, while all other properties are those of a fixed-ratio schedule. In a tandem *variable-ratio* fixed-interval schedule a variable-ratio schedule is specified; but a reinforcement becomes available only after a few seconds have elapsed since the last response in a ratio is emitted. The major features of the VR are preserved; that is, the reinforcement is primarily determined by a number of responses which varies from reinforcement to reinforcement. The differential reinforcement of high rates resulting from the relation between reinforcement and the few preceding responses is, however, reduced or eliminated. Tand *VR*FI differs from VI in that long pauses are usually not reinforced (unless they begin during the relatively short FI 6 sec component of the tandem schedule). In VI whenever the over-all rate of responding is low for any reason, reinforcements will occur on schedule or close to schedule if any responding occurs at all. In tand *VR*FI, however, whenever the over-all rate becomes low, no further reinforcements will occur for possibly a long time.

Experiment I

In a first experiment tand *VR*FI was studied directly after crf. The variable-ratio program was as follows: 50, 5, 130, 100, 20, 120, 140, 100, 5, 120, 80, 160, 10, 60, 20, 155, 30, 70, 40, 150, 40, 120, 5, 90, 80, 155, 1, 80, 100, 10, 60, 70, 110, 1, 90, 50, 60, 130, 40, 150, 90, 140, 110, 1, 150, 130, 10, 50, 140, 30, 160, 70, 110, 155, 20, 160, and 30 responses. The mean is 80. The tandem fixed interval was 6 seconds. The performance during nearly 100 hours on tand VR 80 FI 6 sec is shown in Fig. 554, where 1000-response segments have been selected from various stages of the development. Each excursion is the 2nd segment of its session. Records A and B, at slightly over 1 response per second, show the performance 9 hours and 16 hours, respectively,

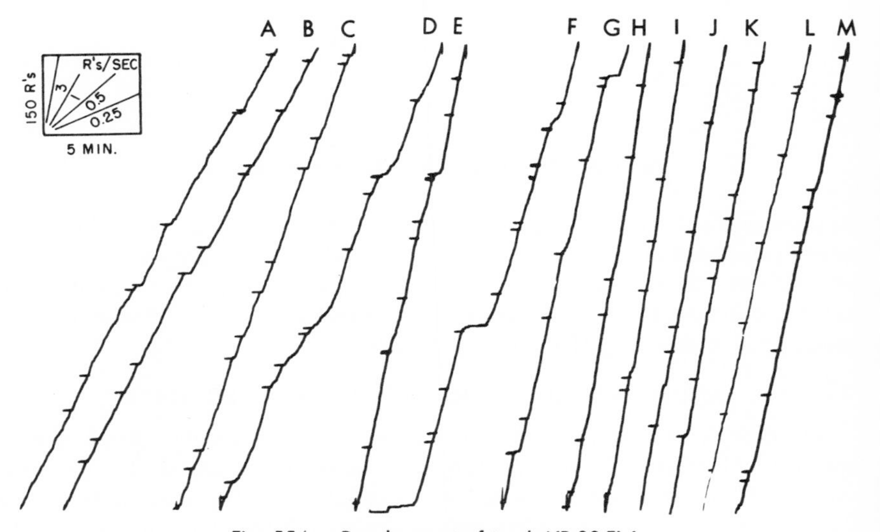

Fig. 554. Development of tand VR 80 FI 6

after crf. The local rate is considerably irregular; and reinforcements are variously followed by pauses, continuations of the preceding rate, or increases in the rate. In Record C, 44 hours after continuous reinforcement, the rate has increased to 1.7 responses per second, with frequent pauses after the reinforcement. At this stage of tand *VR*FI the rate fell to very low values for periods up to 2 hours, and in one case more than 10 hours. (This exceptional performance is shown in Fig. 555 for the entire session, after 35 hours of reinforcement on the tandem schedule—between Records B and C of Fig. 554. The session begins with a performance similar to that in Record B. After reinforcement at *a,* however, the rate falls off sharply but smoothly to an intermediate value maintained for more than 2 hours, during which reinforcements occur relatively infrequently. At the end of the figure the rate of responding

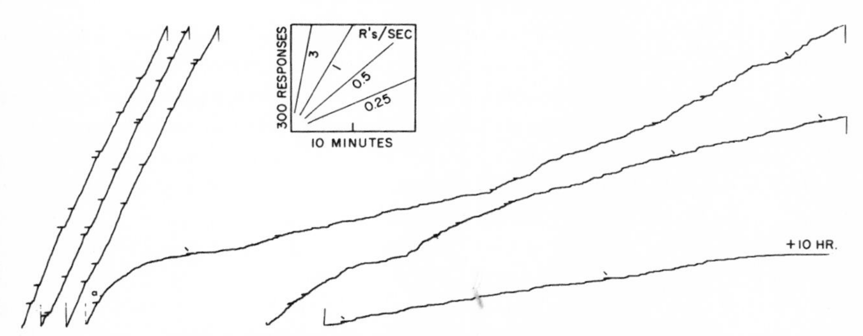

Fig. 555. Temporary low over-all rate preceding increase

falls to zero. The bird remained in the box for 10 hours without responding to the key.) In Record D of Fig. 554, after 54 hours on tand VRFI, the over-all rate falls below that of Record C, but sustained local rates of responding now occur of the order of 2.5 responses per second. In Record E, at 55 hours, the over-all rate reaches 2.8 responses per second; but this falls to a lower rate in the next session (Record F at 57 hours), as a result of a lower local rate. In Record G, 64 hours after crf, the over-all rate again reaches 2.8 responses per second, with pauses or periods of lower rate occurring only after a few reinforcements.

An extinction curve shown in Fig. 556 was taken between Records F and G, after 62 hours of reinforcement on the tandem schedule. The curve consists of sustained responding for almost 2000 responses in the manner of the current tand VRFI performance followed by an instantaneous shift to a zero rate of responding at *a*. The

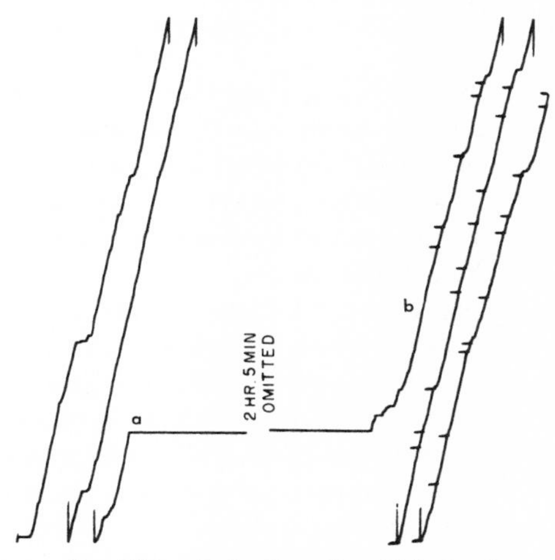

Fig. 556. Extinction after 62 hr tand VR 80 FI 6 sec

bird does not respond for over 2 hours. When responding resumes, the magazine is reconnected at *b,* and a single reinforcement reinstates the typical variable-ratio performance. By Record H in Fig. 554 the over-all rate reaches the maximum value seen in the experiment, 4 responses per second, 74 hours after crf; and it remains in this region for the remainder of the tand VRFI. The FI component was removed starting at Record K, and simple VR prevailed through Records L and M to a total of 94 hours after crf. The performance shows no appreciable change, except possibly a decrease in over-all rate. A second extinction curve was taken after 87 hours of reinforcement on tand VRFI and a short exposure to VR alone, just after Record K. Figure 557 shows the entire curve, which is much larger than that in Fig. 556. An initial period of sustained responding at the prevailing rate is maintained for more than 9000 responses before the rate falls at *a.* The remainder of the curve consists of 3 segments, at *b, c,* and *d,* of 800, 600, and 150 responses, respectively, executed at the prevailing variable-ratio rate, separated by long periods of zero or near-zero rates. While the

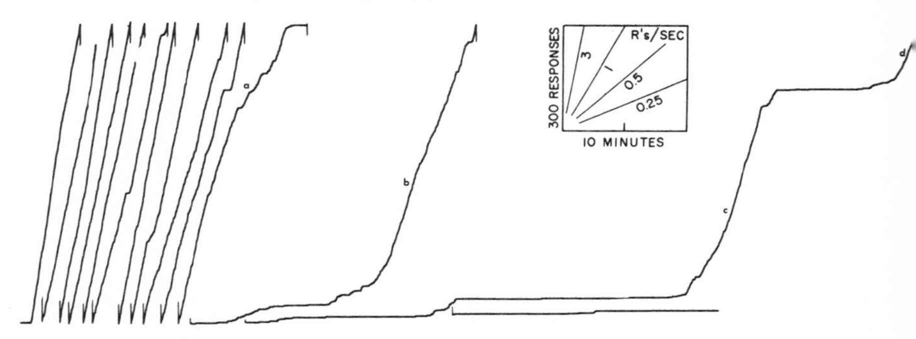

Fig. 557. Extinction after 87 hr tand VR 80 FI 6 sec and VR 80

transitions from one rate of responding to another are not so abrupt as in the previous extinction curve, the transitions are considerably more abrupt than in the usual variable-interval extinction curve. If plotted in a single continuous segment, the over-all curve would consist mainly of a long leg at the prevailing variable-ratio rate, falling sharply to a zero rate, punctuated by relatively small deviations at the higher rate. Compare Fig. 496 through 499.

The second bird in the experiment developed a similar final performance on tand VRFI, although the development was slower and followed a somewhat different course. Figure 558 shows the change in the over-all rate of responding during the first 7 sessions (34 hours) of the experiment. The over-all rate remains very low for the first 14 hours, after which it increases more rapidly to almost 0.5 response per second by the end of the figure. By this time, however, sustained local rates of responding appear which are considerably higher than the prevailing over-all rate. Figure 559 gives the entire session represented by point *a* in Fig. 558. This is the 6th session, after 33 hours of reinforcement on the tandem schedule. It begins with a performance similar

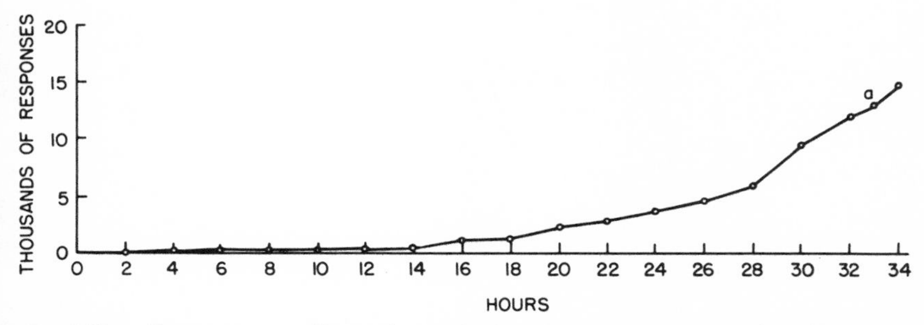

Fig. 558. Change in over-all rate during the first seven sessions on tand VR 80 FI 6 sec (second bird)

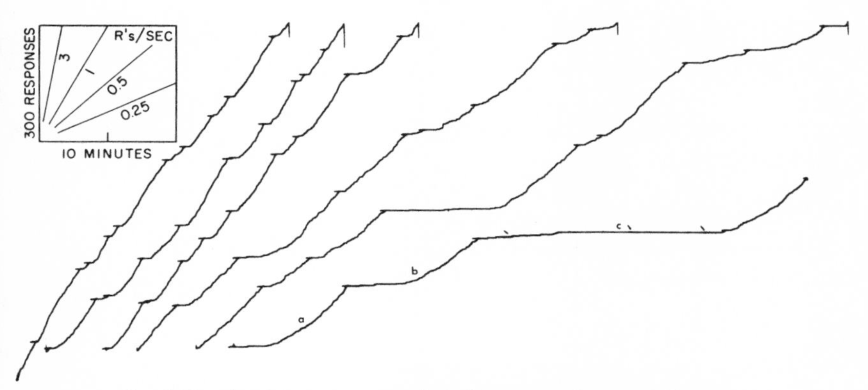

Fig. 559. Sixth session on tand VR 80 FI 6 sec showing fall in rate

to the early stages of the other bird, with an over-all rate of approximately 0.8 response per second. Pauses after reinforcement and subsequent acceleration to the terminal rate are somewhat more extended, however. As the session progresses, both the pause and the period of acceleration become much more extended, leading to marked scallops, as at *a* and *b*, a near-zero rate of responding, as at *c*, and a final over-all rate for the last segment of the figure of approximately 0.2 response per second. This performance may represent a stage of development of tand VRFI similar to that of Fig. 555, where a later increase in over-all rate was preceded by a period where the rate, and consequently the frequency of reinforcement, fell to a low value.

Figure 560 represents the remainder of the experiment on tand VRFI, from the 8th to the 67th sessions. The second pen excursion has been selected from the record for every 4th session. Records A through N show a progressive increase to a value in the vicinity of 3 responses per second. The over-all rate varies from segment to segment, and pauses are variously present or absent after reinforcement. After the session represented by Record N, the fixed-interval component of the tandem schedule was

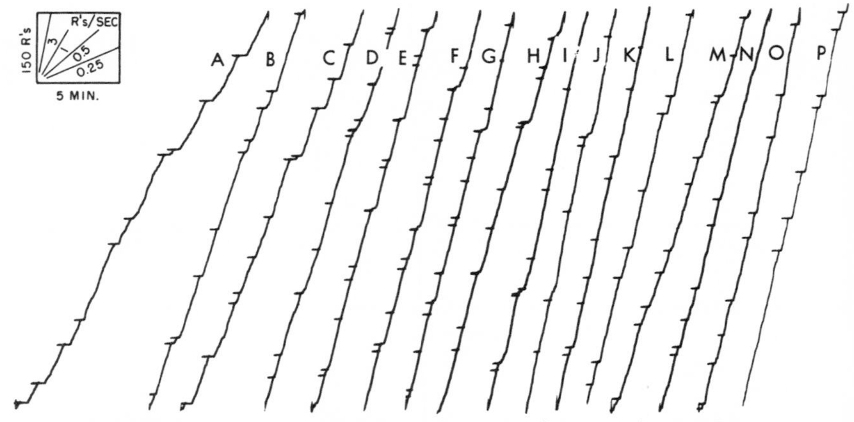

Fig. 560. Development of tand VR 80 FI 6 sec performance during 67 sessions

removed. Records O and P for the 4th and 8th sessions on the simple VR 80 show no significant change.

Both birds in this experiment developed rates and performances similar to those on VR. If the tandem FI 6 sec has any effect, it is probably in the greater variation in the rate from segment to segment and in the phase where over-all rate falls far below the rates prevailing at the time.

Experiment II

The development of a final performance on VR 360 has already been described in Chapter Seven. Figure 475 shows a well-sustained level; Fig. 476, on the following day, shows considerable pauses; and Fig. 477, for the second bird, shows a marked stepwise performance. Both birds were placed on a tand VR 360 FI 6 sec. The conditions at the moment of reinforcement have thus changed, at least potentially, to those of an interval schedule, while the other contingencies of the variable-ratio schedule of reinforcement were maintained. The previous performance in Fig. 477 was optimal for transition to the tandem schedule, since the brief sustained responding and the frequent pauses must increase the likelihood that the variable-ratio requirement of the tandem schedule will be met just before a pause, so that the first response after a pause will be frequently reinforced. Figures 561 and 562 show the 7th session after transition. The early effect of the FI 6 sec component of the tandem schedule is a decline in the over-all rate and the appearance of smooth curvature from the prevailing rate to the pause. The transition from pausing to responding still remains instantaneous. Toward the end of the session, in Fig. 562, intermediate rates of responding begin to appear, as at *a,* where sustained runs of approximately 50 to 75 responses are separated by short periods of intermediate rates rather than pauses. The rate immediately after reinforcement frequently takes an intermediate value, as at *a* and *b* in Fig. 561 and *b* in Fig. 562, possibly because of the reinforcements after pauses. The reinforcement at *b* in Fig. 562 is an instance in which the variable ratio was counted out just before the bird was about to pause, and the reinforcement occurred just after the pause. Figure 563 contains the latter part of a later experimental session beginning 137 hours after the change to the tandem schedule. Both the over-all and local rates decline markedly. Reinforcements occur after pauses or at lower rates of responding, as at *b*, for example. Pauses now occur after reinforcements, as at *b*, *c*, *e*, and *f*, although these were rare in the previous VR performance. The pauses presumably occur because successive responses are no longer likely to be reinforced, as under the previous variable-ratio schedule, since the minimum time elapsing between reinforcements is 6 seconds. The variable-ratio character of the performance is preserved in the many fairly abrupt shifts to zero rates at times other than at reinforcement, as at *a*, *d*, and *g*. In general, the addition of the FI 6 sec schedule has produced a more irregular performance.

The second bird, which had been showing the performance in Fig. 475 and 476 on VR, took longer to show the effect of the FI 6 sec component of the tandem schedule. Under sustained responding at high rates with but few shifts to lower rates, the probability that the variable ratio would be counted out just before a pause was very small.

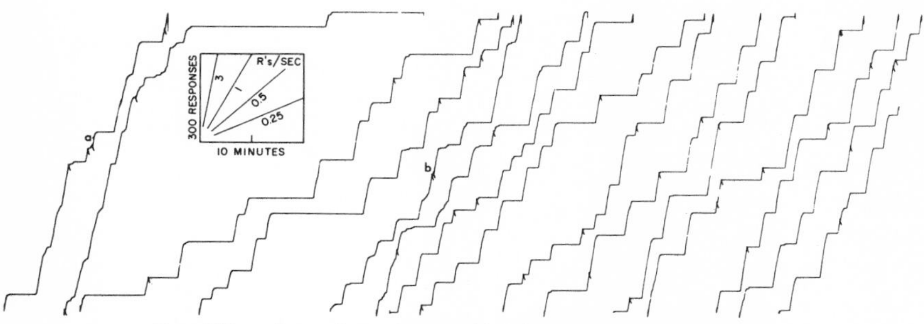

Fig. 561. Seventh session tand VR 360 FI 6 sec after VR 360

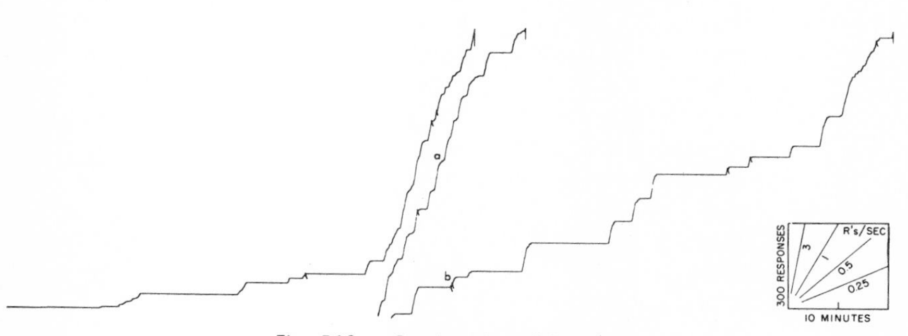

Fig. 562. Continuation of Fig. 561

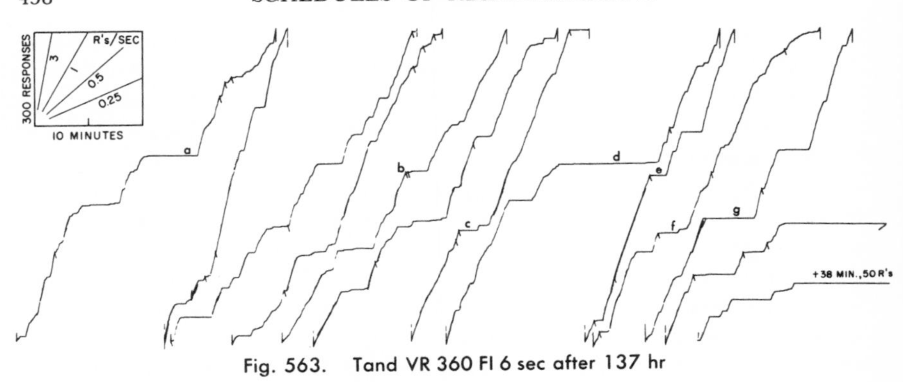

Fig. 563. Tand VR 360 FI 6 sec after 137 hr

Figure 564 shows a performance after 8 sessions on tand VR 360 FI 6 sec. The principal change from Fig. 475 is the appearance of sustained intermediate rates. Even in the poorly sustained performance of Fig. 476, comparable portions of an intermediate rate are lacking.

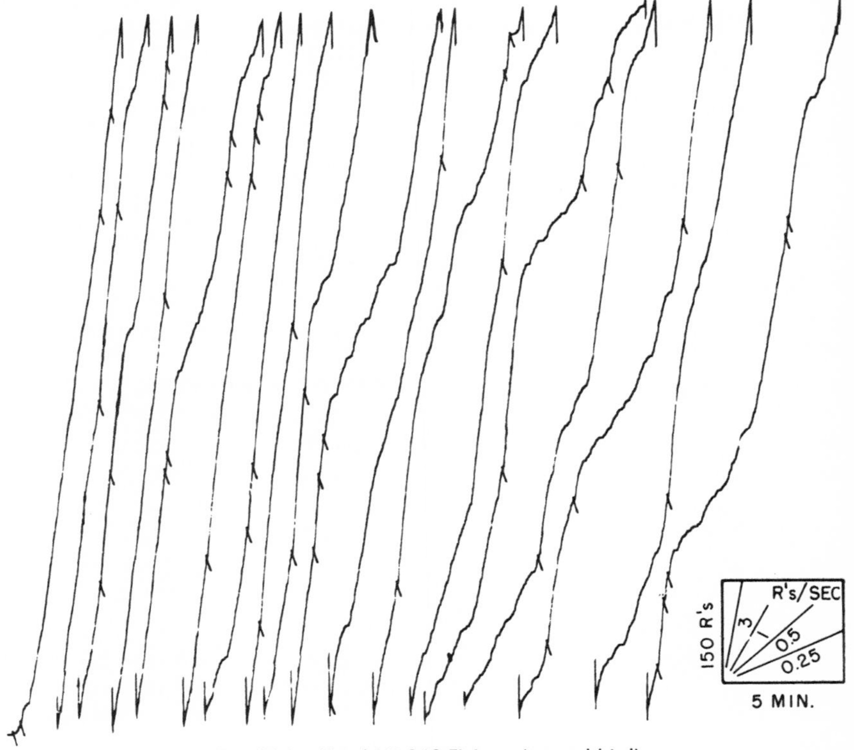

Fig. 564. Tand VR 360 FI 6 sec (second bird)

Chapter Nine

• • •

DIFFERENTIAL REINFORCEMENT OF RATE

INTRODUCTION

In our analysis of the effects of schedules the rate at the moment of reinforcement is an important variable. For example, in FI the probability is much greater that a reinforced response will follow a pause than in the otherwise not-so-different tand *FI*FR. But a given schedule does not guarantee such a relationship at every reinforcement. Indeed, under exceptional performances the relationship may vanish or may even be reversed.

The importance of this variable can be checked by direct control of rate conditions at reinforcement. The resulting performance will enable us to estimate the extent to which momentary rates function as a stimulus in any set of contingencies of reinforcement.

By "rate at reinforcement" we may mean the reciprocal of the inter-response time preceding a single reinforced response, or of the elapsed time required by the last *n* responses. At this stage, practical considerations have been most important in dictating procedures. The differential reinforcement of low rates (drl) is easiest to arrange in terms of a single inter-response time (IRT); the differential reinforcement of high rates (drh) is easiest when a given number of responses must be executed within a set period.

A device which sets up a reinforcement only when some such specification has been met can be added to any schedule. Thus, in FI 5 drl 6 a response is reinforced approximately every 5 minutes but only when it follows the preceding response by at least 6 seconds. In VI 5 drl a response is reinforced on a VI schedule with a mean of approximately 5 minutes but only when a few responses have met a given speed requirement. In crfdrl 3 every response is reinforced which follows the preceding one by at least 3 seconds. (The completion of the preceding reinforcement may or may not be used in establishing the criterion for the following response. If it is not, the schedule is technically FR 2 drl 3.)

VARIABLE INTERVAL WITH DIFFERENTIAL REINFORCEMENT OF LOW RATES

VI 1 drl

In one experiment, drl 6 (no response reinforced unless preceded by a 6-second IRT) was added to an early VI 1 performance, 3 sessions after crf. Figure 565 shows the last of this performance before the arrow in Record A. A roughly linear over-all rate of 1.1 responses per second prevailed. The drl 6 contingency postponed the next reinforcement until *a,* about 10 minutes later. (The largest interval in the VI series was 200 seconds.) For the remainder of the session the over-all rate fell continuously, reaching approximately 0.35 response per second by the last segment of the record at *b.* A final performance after 9 sessions on this VI 1 drl 6 is shown in Record B, where

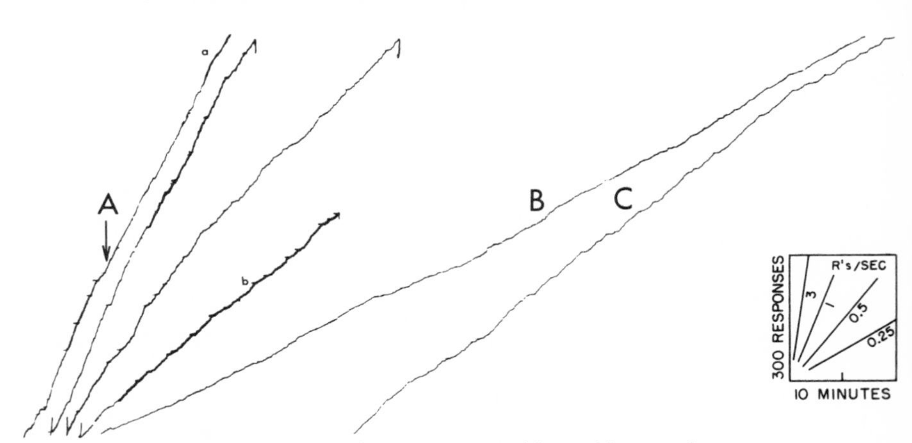

Fig. 565. Transition from VI 1 to VI 1 drl 6 and later performance

the rate of responding is constant at 13 responses per *minute.* If the bird were responding exactly once every 6 seconds, the rate would, of course, be 10 responses per minute. Local features of the curve in Record B show that the extra 3 responses per minute are the result of occasional small groups at higher rates.

A second bird showed a similar effect of VI 1 drl 6. Record C shows a final performance after a similar history. The over-all rate of responding is higher, and the tendency to pause after the reinforcement is more pronounced.

Figure 566A begins with the performance under VI 1 drl 6 seen in Fig. 565B. At the arrow the drl was decreased to 1 second. The rate of responding increases slowly during the remainder of the session, the final rate in the session being about doubled. With further exposure to drl 1, the rate continues to increase, reaching the final performance shown in Record B after 11 sessions. The last segment in Record B shows a linear performance at slightly less than 0.8 response per second, compared with the rate

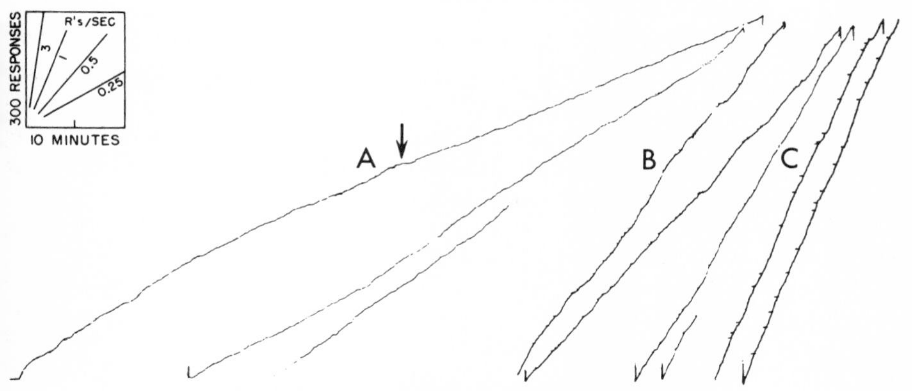

Fig. 566. Transition from VI 1 drl 6 to VI 1 drl 1

of 1.1 reponses per second for the bird on VI 1 before drl. The second bird again showed a slightly higher rate. Record C shows its performance on VI 1 drl 1 after a similar history. The over-all rate is also lower than that under VI 1 without drl, although it should be noted that the VI 1 would probably have continued to rise.

The drl was again increased to 6 seconds and final performances for both birds are shown in Fig 567A and B. Record A shows an increase above the earlier performance in Fig. 565B, while Record B may show a slight decrease below the earlier performance of Fig. 565C. In Record A a burst of about 75 responses at a rate appropriate to the variable-interval schedule without drl occurs at *a*, displacing the curve vertically. These break-throughs occur frequently at this stage, possibly because of frequent transitions between various rate contingencies.

The size of the drl was then increased in the following session to 12 seconds, and Records C and D show performances 5 sessions later. In Record C the over-all rate of responding falls to 7.5 responses per minute, and in Record D, to 5.8 responses per minute. If the bird were responding at a rate exactly satisfying the drl contingency, the rate would be 5 responses per minute. The brief bursts of responding at *b, c,* and *d* resemble the earlier performance of the same bird in Record A.

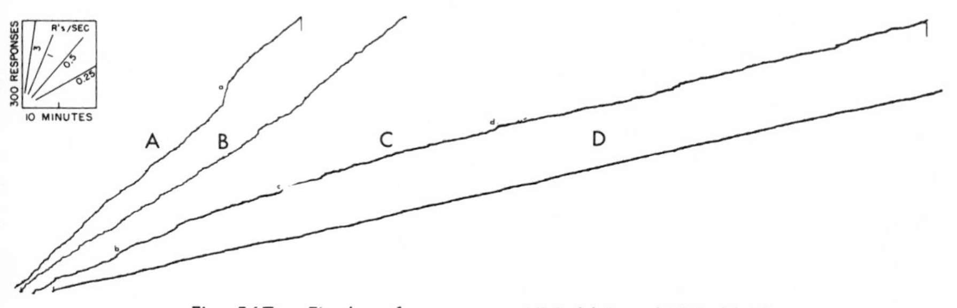

Fig. 567. Final performance on VI 1 drl 6 and VI 1 drl 12

Extinction after VI 1 drl 12

The magazine was disconnected at the start of the session following the final performance on VI 1 drl 12 shown in Fig. 567C. Extinction was carried out for 16 hours. The entire curve (shown in Fig. 568) contains 800 responses, and shows a smooth and orderly decline to a near-zero value. The only major deviation is at *a*, where the original VI 1 drl 12 rate reappears for about 15 minutes.

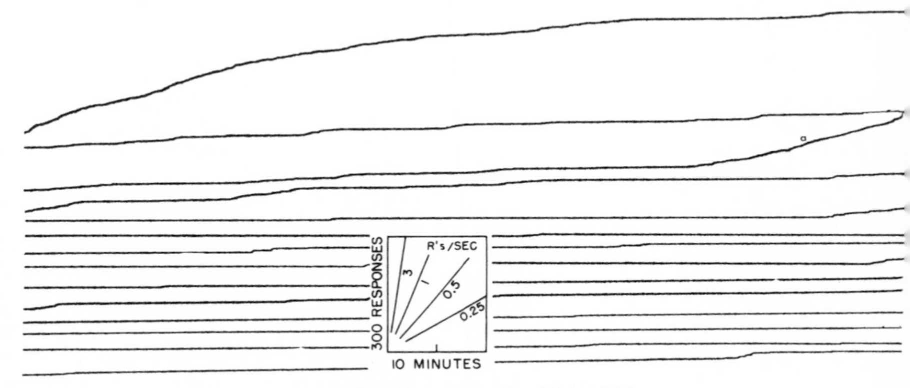

Fig. 568. Extinction after VI 1 drl 12

VI 4.5 drl 2

An experiment in which a drl 2 was added to a well-developed VI 4.5 has been mentioned in Chapter Six, where dosages of sodium pentobarbital were regulated in terms of rate under VI. Figure 569 shows one transition from VI 4.5 to VI 4.5 drl 2 in that experiment. The adjusting drug procedure was not in force at this stage. Record A

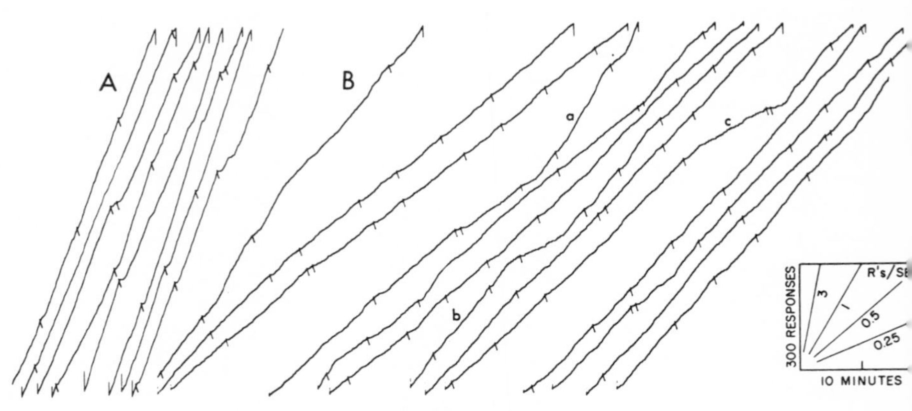

Fig. 569. Transition from VI 4.5 to VI 4.5 drl 2

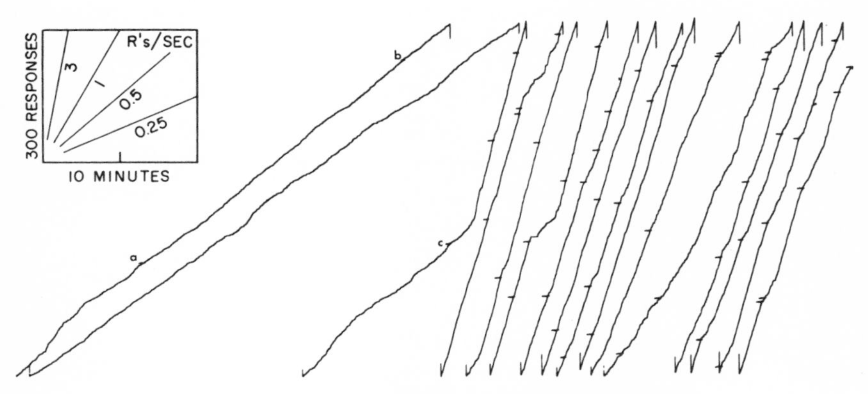

Fig. 570. Limited-hold reinforcement after VI drl

is a sample of the final performance under VI 4.5 after a long history of VI with and without continuous administration of sodium pentobarbital. Record B shows the 10th session under VI 4.5 drl 2. The drl produces a fairly stable over-all rate of responding of about 0.5 response per second. This is twice the rate which just satisfies the drl contingency at every response, but reinforcements are not significantly postponed. Each session begins characteristically at a relatively high rate, and occasional periods of sustained responding occur at rates considerably higher than the prevailing rate, as at *a* and *b*. An occasional shift to a lower-than-usual value is exemplified at *c*. Slight pausing or slower responding just after the reinforcement reflects the low density of short intervals in the variable-interval schedule. A second bird showed a similar transition.

Limited hold after drl

The limited-hold contingency discussed in Chapter Six was accidentally put in force against a background performance which involved the differential reinforcement of slow responding. A bird had been on a chained VI 3 VI 3 drl schedule. Figure 570 begins with the low rate established on the VI 3 drl part of this schedule. The limited-hold contingency was, however, in effect. The first 3 reinforcements, at *a*, *b*, and *c*, are widely spaced, because the low rate from the drl makes the occurrence of a response in the 0.5-second period of limited hold unlikely. Following the 3rd reinforcement, at *c*, a rapid acceleration occurs to the VI rate formerly shown by this bird in the absence of the drl. This is a much more rapid transition than that which results from restoring VI after VIdrl. The low rate and pattern of responding under VIdrl may make the limited-hold contingency especially effective by providing maximum contrast between occasions for reinforcement and nonreinforcement while sustaining the behavior at the over-all level required if reinforcements are to occur to maintain the behavior. A slight differential reinforcement of rapid responding is assumed to have occurred at *a, b,* and *c,* on a scale much too small to show in the record.

FIXED INTERVAL WITH DIFFERENTIAL REINFORCEMENT OF LOW RATES

Well-developed performances on several values of FI have been shown in Fig. 252, 260, 261, 264, and 266. We studied the effect of the differential reinforcement of low rates on these performances. Reinforcement occurred whenever the bird paused for at least 6 seconds after the designated fixed interval had elapsed. A 5-minute TO followed all reinforcements.

FI 1 drl 6 TO 5

The record of the transition from FI 1 to FI 1 drl 6 was lost because of an apparatus failure. A well-developed performance is evident in a later figure (Fig. 576).

Fig. 571. Transition to FI 2 drl 6 TO 5 and later performance

FI 2 drl 6 TO 5

Figure 571A contains the transition to FI 2 drl 6 TO 5. The session begins with a performance appropriate to FI 2 TO 5, and the rate is above 1 response every 6 seconds at the end of the first 2-minute interval. Responding continues at the terminal value under the previous FI 2 schedule until a break in the performance satisfies the drl contingency for a reinforcement at *a,* after 900 responses and 20 minutes. The second interval segment following the reinforcement at *a* begins with a pause of almost 2 minutes, but the rate exceeds the critical value and reaches the terminal value appropriate to the FI performance. The rate of responding falls momentarily at *b,* and the second reinforcement is received after approximately 5 minutes and 100 responses. By the end of the session, responding occurs in bursts separated by pauses which occasionally satisfy the drl requirement, and reinforcements begin to occur more frequently. By the 3rd session the terminal rate during the interval has fallen to 0.4 response per second. Record B shows a final performance from the middle of the 12th session after a total of 230 reinforcements under FI 2 drl 6 TO 5. The over-all rate has fallen to 0.15 response per second, with a pause after the reinforcement and acceleration to a terminal rate of 0.2 response per second. Although the terminal rate of responding is slightly higher than the 10 responses per minute which just satisfies the drl requirement, slight variations in the local rate provide instances where 6-second pauses occur. At *c, d,* and *e* responding continues for from 10 to 15 minutes before a pause long enough to meet the drl requirement occurs. The differential reinforcement has obviously been effective in eliminating the high terminal rates under FI 2, although interval curvature remains.

FI 4 drl TO 5

Figure 572A shows the transition from FI 4 TO 5 to FI 4 drl 6 TO 5. The drl contingency greatly lengthens the intervals between 3 of the first 4 reinforcements. Fol-

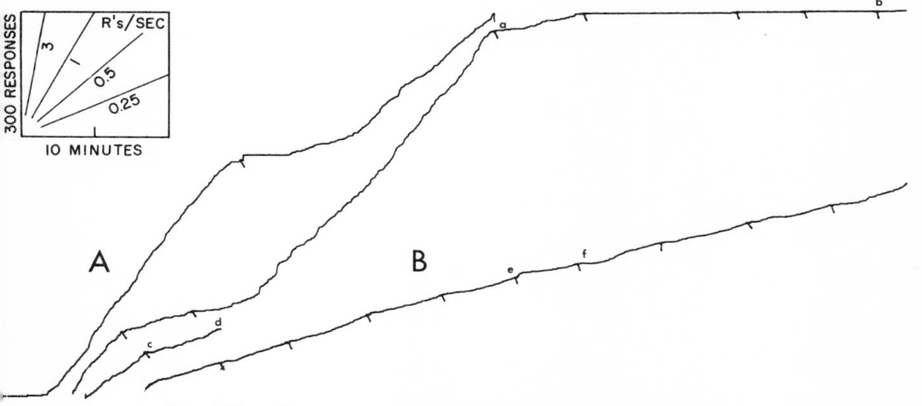

Fig. 572. Transition to FI 4 drl 6 TO 5 and later performance

lowing the reinforcement at *a,* however, the over-all rate of responding remains low. The next 4 reinforcements are delayed by pauses rather than by an excessive rate. Following the pause after reinforcement at *b,* the rate increases to 0.4 response per second; the reinforcements at *c* and *d* are not postponed, because responding now occurs in small bursts.

The rate during the next few sessions remained generally low, with frequent instances of reinforcements postponed because no responses occurred. Eventually, the over-all rate of responding increased and became stable from segment to segment. Record B shows a segment from a daily performance after a total of 460 reinforcements on FI 4 drl 6 TO 5. The over-all rate of responding has fallen to approximately 0.15 response per second, and most reinforcements now occur on schedule. The overall pattern of responding is linear, with some slight interval scalloping. A small burst follows the reinforcement at *e,* and a pause with compensation follows reinforcement at *f.*

FI 8 drl 6 TO 5

Figure 573A shows the transition from FI 8 TO 5 to FI 8 drl 6 TO 5. This is the first part of the session. The 1st interval begins as usual and accelerates to the terminal rate of responding, which is maintained for approximately 10 minutes. No pauses as long as 6 seconds occur, and the 1st reinforcement does not occur until the rate falls off, at *a*. The 2nd and 3rd reinforcements, at *b* and *c*, occur after similar segments. The reinforcement at *c* is followed by a 6-minute pause and a slow acceleration to a lower terminal rate. The rate is still too high to satisfy the drl 6 requirement, however, and the reinforcement at *d* occurs only after 23 minutes. The rate is near zero during the next segment, but recovers during the last 2 reinforcements of the record, at *e* and *f*. The reinforcement at *f* occurs within 1 minute of the designated schedule, largely because of the grain of the record.

By the 5th session of FI 8 drl 6 both the over-all and terminal rates have declined; and,

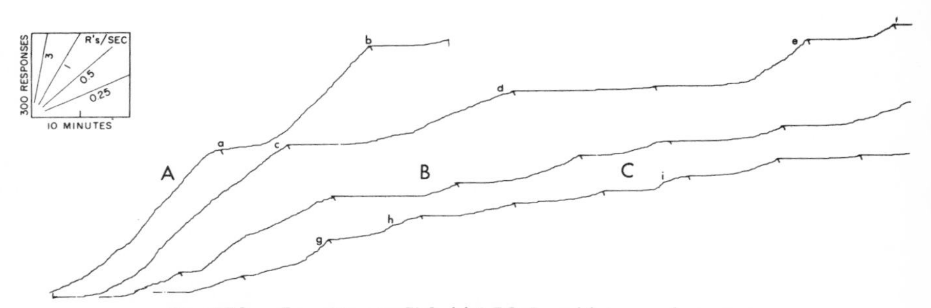

Fig. 573. Transition to FI 8 drl 6 TO 5 and later performance

in general, the performance consists of a long pause after the reinforcement and a slow, shallow acceleration to a low terminal rate. The start of the session (Record B) shows a generally higher rate than the remainder of the session. Record C illustrates the final performance on the schedule after 14 sessions of exposure. The higher rate at the start of the session has become much less pronounced, and most of the intervals are scalloped, with an even lower terminal rate of responding than that in Record B. Most reinforcements are received within 10 minutes. Three break-throughs at higher rates of responding occur at *g*, *h*, and *i*. (Note that the reinforcement at the beginning of the segment at *e* was delayed by an apparatus failure rather than by the bird's failure to satisfy the drl.)

FI 16 drl 6 TO 5

Figure 574 shows the transition from FI 16 TO 5 to FI 16 drl 6 TO 5. Here, the first performance with drl is not very different from the preceding FI schedule, because the terminal rate had contained a sufficient number of 6-second pauses to avoid severe postponement of reinforcement. The drl still reduces the terminal rate, however, as Record B, a segment taken 2 sessions following Record A, shows. As the rate of responding falls, the grain of the record becomes smoother. This smoother grain works against satisfying the contingency. The intervals between reinforcements become longer, reaching 30 minutes after the reinforcement at *a* and 32 minutes after the reinforcement at *b*. The period of curvature becomes more extended, as in the segment ending with the reinforcement at *c*. Record D shows the 8th session. (Record C is for a later session described below.) The pause after the reinforcement is now longer. The terminal rate of responding has fallen from 0.25 to 0.3 response per second, and the scallop extends through most of the interval segment. The terminal rate of responding is now low enough so that most reinforcements occur practically on schedule. The slow-down contingency shows a further effect 4 sessions later (Record E), where the terminal rate has fallen to the order of from 1.5 to 1.8 responses per second. The changes within the interval segments, however, are no longer so smooth as those in Record D. Further exposure to the slow-down contingency results in an even more irregular performance by the 13th session, as Record C shows. (Note again that the segments in the figure have been arranged, for best reduction, in a different order.) While the over-all rate has increased considerably because of sustained periods of higher rates of responding, as at *d, e, f,* and *g,* reinforcements are not greatly postponed, because of the marked oscillations in over-all rate and the frequent instances of 6-second pauses in the fine grain of the record.

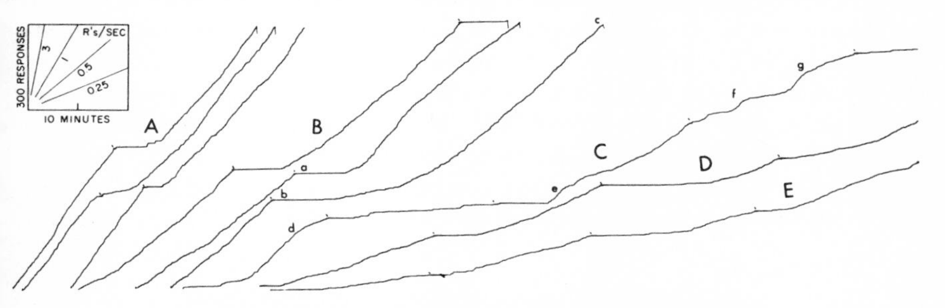

Fig. 574. Transition to FI 16 drl 6 and later performance

FI 32 drl 6 TO 5

A final performance on FI 32 drl 6 TO 5 was recorded for one bird after a well-developed performance on FI 1 drl 6. Figure 575 shows the 11th session on the FI 32 drl 6 TO 5 schedule.

Transition from FI drl 6 TO 5 to FI TO 5

When a final performance on each of the five values of FI drl 6 had stabilized, the drl was removed, and the transition to a simple FI was observed. (The bird chosen for

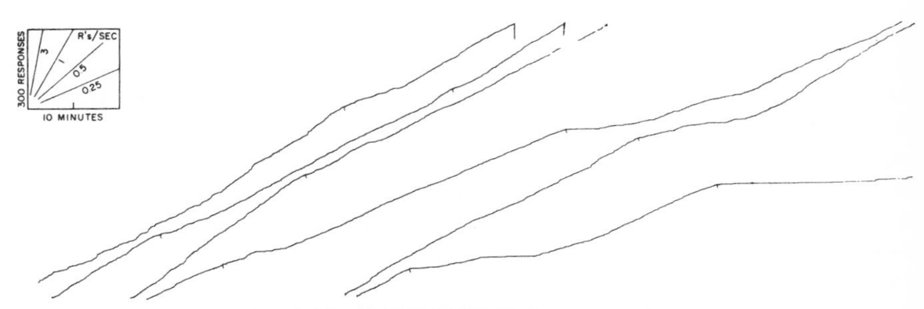

Fig. 575. FI 32 drl 6 TO 5 after ten sessions

the final performance on each value of FI drl 6 was the same as that in the first transition from crf. We have no case to report for the transition to FI 2, since the bird originally at this value had been used in studying FI 32 drl.)

FI 1. The final performance after 13 sessions on FI 1 drl 6 TO 5 shown in Fig. 576 has already been mentioned. The over-all curve is roughly linear, with a slight tendency for a higher rate at the start of the interval. The drl contingency is satisfied for most reinforcements by the rough grain of the record, which includes many 6-second pauses, rather than by a low over-all rate. When the slow-down contingency is removed in the next session (Record B), the over-all rate increases rapidly to 1.5 responses per second by the 5th reinforcement in the session. This rate may be compared with an over-all rate of 0.4 response per second in the previous session with drl. Toward

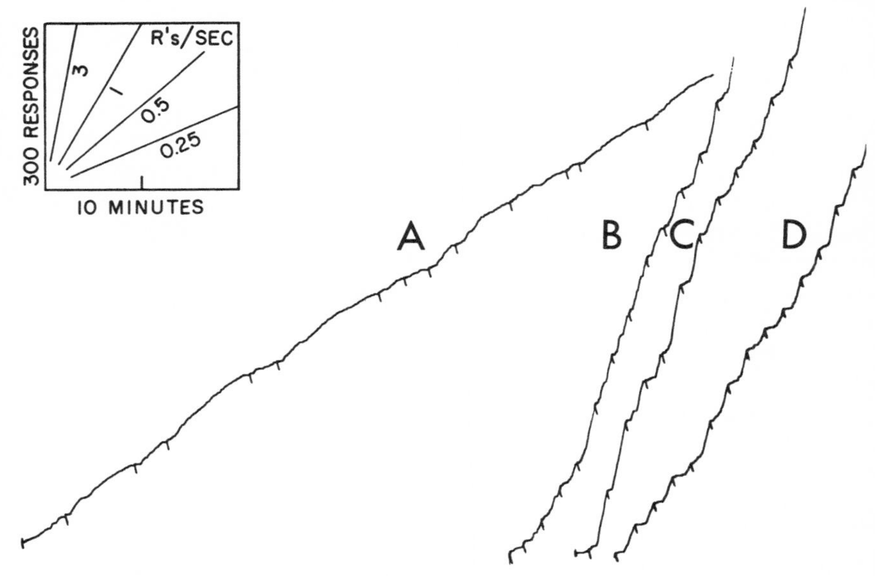

Fig. 576. Transition from FI 1 drl 6 TO 5 to FI 1 TO 5 and later performance

the end of the segment shown in Record B, a period of lower rate of responding develops at the start of the interval segment, and a terminal rate of almost 4 responses per second is reached with few pauses or deviations in rate. In the next session (Record C) the fixed-interval segment becomes more sharply scalloped, and reinforcement uniformly occurs after a sustained terminal rate has been reached. By the 9th session (Record D) the over-all rate has declined as the pause and acceleration following the reinforcement have become more marked. The terminal rates range from 2 to 3.5 responses per second.

FI 4. Figure 577A shows the final performance under FI 4 drl 6 TO 5 when the bird was placed on that schedule for the second time after being tested at other values. The over-all rate is higher than it was during the first exposure to this schedule, described in Fig. 572. The contingency was being satisfied in part by a rough grain. Records B and C show the development of the performance on FI 4 after drl was discontinued. In Record B the terminal rate increases almost immediately. The over-all performance remains roughly linear, and strong scalloping develops. By the 9th session in Record C, the over-all rate of responding has fallen even farther. None of the interval segments begin with substantial pauses, and two (at *a* and *b*) begin with the full terminal rate from the preceding interval in spite of the TO 5.

FI 8. When the drl contingency was omitted from the FI 8 performance, the first session, shown complete in Fig. 578, shows little change. Both the pause at the start of the interval segment and the magnitude of the terminal rate of responding increase somewhat. Compare, for example, the segments at *a, b,* and *c* with the earlier performance in Fig. 573. By the 4th session on FI 8 TO 5, shown in Fig. 579A, the termi-

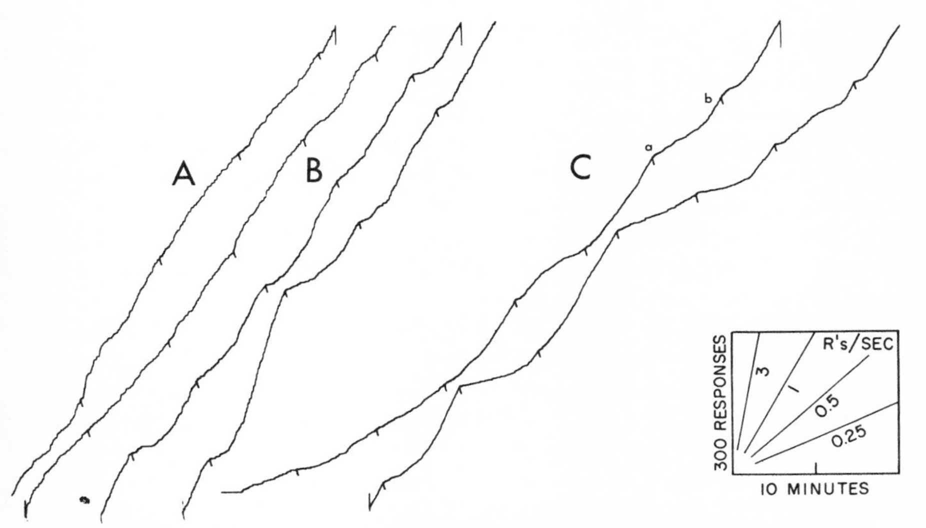

Fig. 577. Transition from FI 4 drl 6 TO 5 to FI 4 TO 5 and later performance

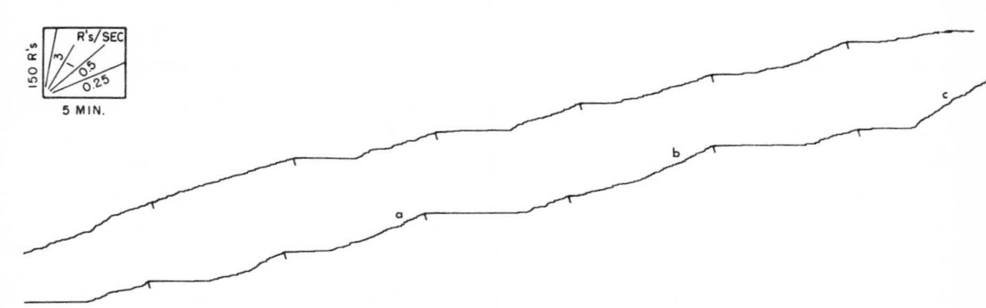

Fig. 578. Transition from FI 8 drl 6 TO 5 to FI 8 TO 5

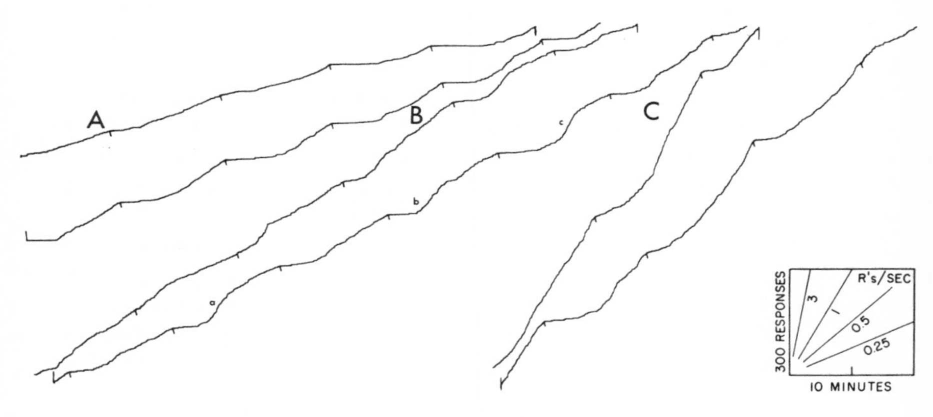

Fig. 579. Later performance on FI 8 TO 5 after drl

nal rate of responding has increased further, with a corresponding decrease in the length of the pause at the start of the interval. By the 6th session (Record B) the over-all rate has increased markedly, with the highest rates of responding occurring at the middle of the interval segments, as at *a, b,* and *c,* so that the curves take on an S-shaped character. The performance 10 sessions after removal of drl is shown in Record C. The over-all rate is now fairly high. The pause following the reinforcements is either absent or brief, but the successive segments show consistent curvature to terminal rates varying from 0.8 to 1.5 responses per second.

FI 16. The effect of removing the drl contingency on FI 16 is shown in Fig. 580, where the course of development is represented by 2 consecutive interval segments each from the 1st, 2nd, 4th, and 9th sessions. (The order of the records is inverted to permit a more compact arrangement.) Record A, taken from the end of the 1st session, shows about the same rate of responding as with drl; but local oscillations in the rate, as at *a* and *b,* are becoming more marked. A smooth and continuously accelerated scallop emerges by the end of the 2nd session (Record B); but by the 4th session (Record C) the curve shows a smooth but rapid acceleration to a much higher terminal rate than that which occurred heretofore. Finally, by the 9th session (Record D) the period of lower rate of responding during the first part of the fixed-interval segment becomes more extended, and the shift to the terminal rate of responding more abrupt.

FI 32. The effect of removing the drl contingency on FI 32 is shown in Fig. 581, where the order of the segments has also been inverted. Each record contains 2 successive segments from the end of the respective session. No effect of the removal of the drl contingency is evident at the end of the 1st session (Record A). The over-all curve reveals a sustained intermediate constant rate with a very slight increase during each interval segment. By the 2nd session (Record B) the terminal rate has increased above the over-all rate under drl and the scallop has become more marked. This trend continues in the subsequent sessions in Records C through G for the 3rd, 4th, 6th, 7th, and 10th sessions, respectively. Interval segments frequently begin with a pause of the order of 5 minutes; then the rate of responding increases to approximately

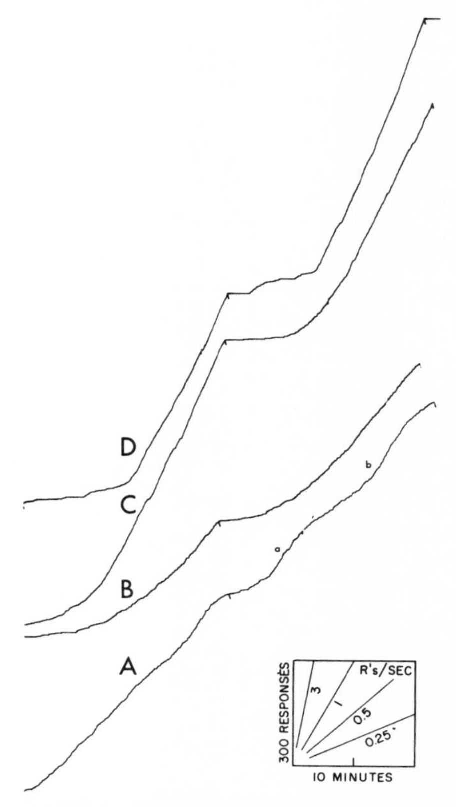

Fig. 580. Transition from FI 16 drl 6 TO 5 to FI 16 TO 5 and later performance

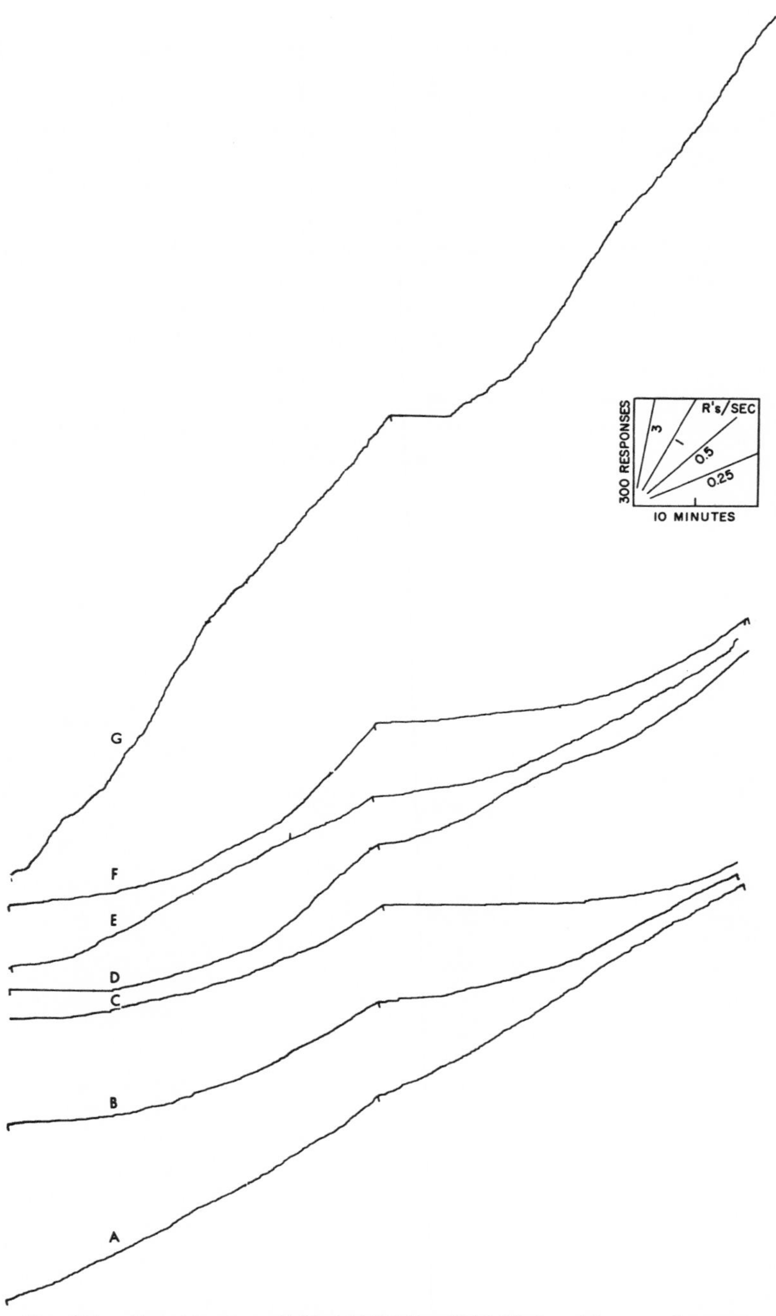

Fig. 581. Transition from FI 32 drl 6 TO 5 to FI 32 TO 5 and later performance

0.75 response per second, which is maintained with rough grain and some oscillation until the next reinforcement.

A COMPARISON OF EXTINCTION AFTER FIXED INTERVAL AND FIXED INTERVAL WITH DIFFERENTIAL REINFORCEMENT OF LOW RATES

We examined the effect of differential reinforcement of low rates on the fixed-interval performance by taking extinction curves after final performances had been recorded on FI 1, 4, 8, 16, and 32 with drl and by taking a second set of extinction curves for the same birds following the development of a normal FI pattern after the drl contingency had been removed.

FI 1. Figure 582 contains an extinction curve after FI 1 TO 5 (Record A) and an earlier extinction curve after FIdrlTO 5 (Record B). The curves reflect the immedi-

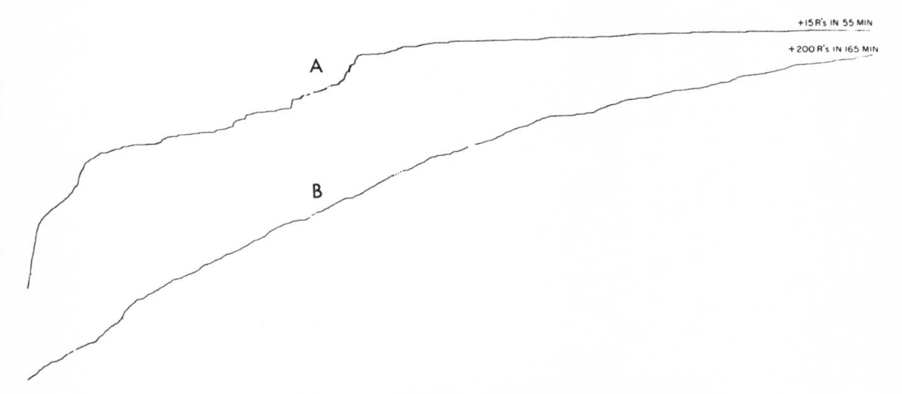

Fig. 582. Extinction after FI 1 TO 5 and FI 1 drl 6 TO 5

ately preceding performances. Record A begins at the high terminal rate of the FI 1 performance and falls off sharply with frequent re-occurrence of brief periods at a high rate. Record B continues the low, sustained intermediate rate of the final performance under FI 1 drl 6 in Fig. 576A. The over-all rate falls very slowly during the session, averaging 2.5 responses per second for the last half of the curve compared with 3.5 responses per second for the first part.

FI 4. Figure 583 contains extinction curves after FI 4 TO 5 and FI 4 drl 6 TO 5. In Record A the extinction after FITO 5 shows marked S-shaped features, with the rate oscillating between almost zero and the terminal fixed-interval rate. Extinction after FIdrl (Record B) begins at a moderate rate which declines continuously and slowly, showing only minor oscillations.

FI 8. Extinction curves in Fig. 584 after FI 8 with and without drl tell the same story. Extinction after FI 8 without drl produces a curve consisting of 3 major segments at rates near the terminal fixed-interval rate, separated by long pauses and pe-

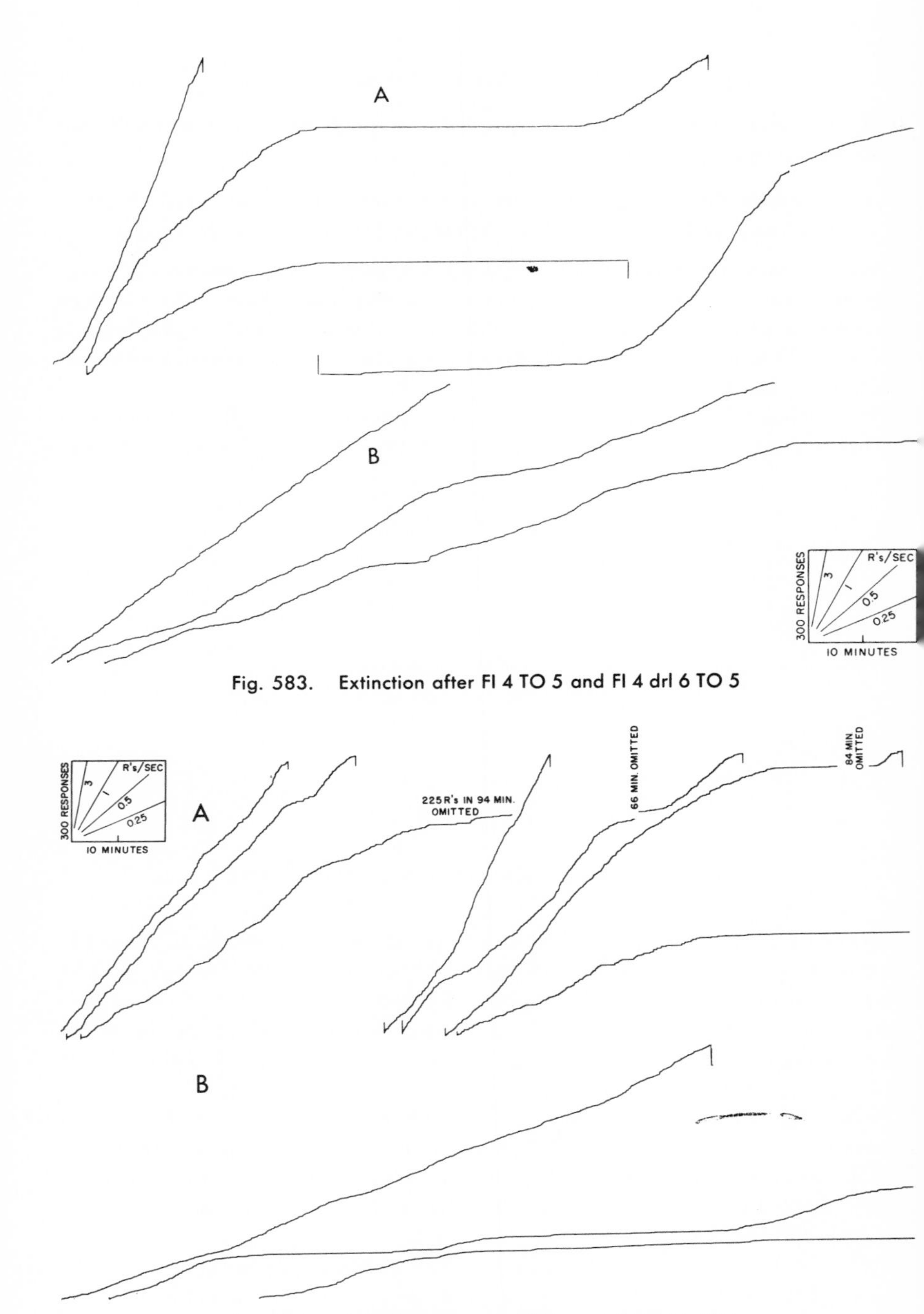

Fig. 583. Extinction after FI 4 TO 5 and FI 4 drl 6 TO 5

Fig. 584. Extinction after FI 8 TO 5 and FI 8 drl 6 TO 5

riods of slow responding. In contrast, extinction after FIdrl begins at a moderate rate sustained for more than 1 hour, after which the rate oscillates between very low and moderate values.

FI 16. Figures 585 and 586 are a comparison of the extinction curves after FI 16 with and without drl, respectively. The beginning of extinction after FI is shown in Fig. 585A and continued in Fig. 586. The curve begins with approximately 4000 responses at the high terminal rate from the previous FI 16 reinforcement. The rate then falls to an intermediate value, after which the original high rate of responding re-

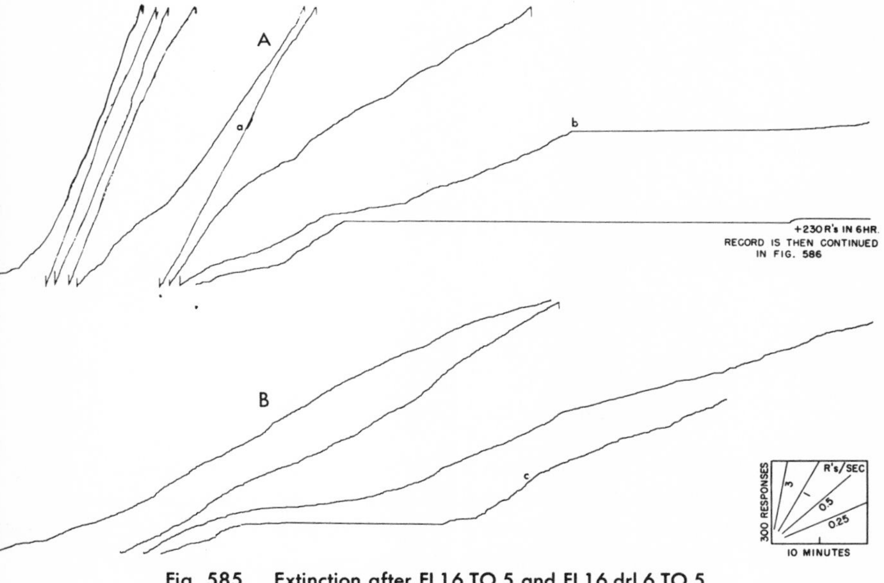

Fig. 585. Extinction after FI 16 TO 5 and FI 16 drl 6 TO 5

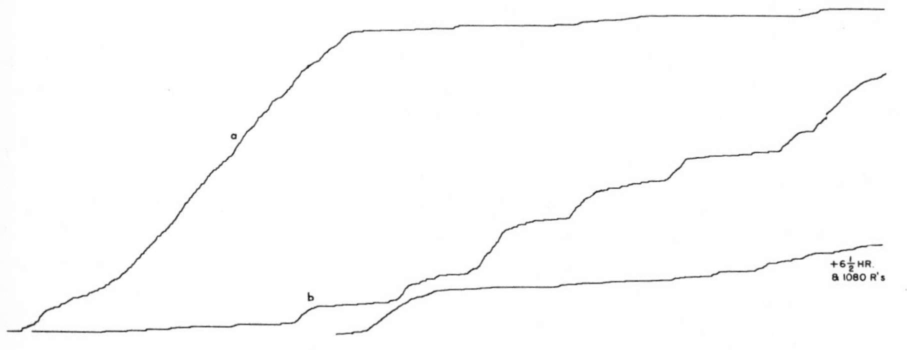

Fig. 586. Continuation of Fig. 585A

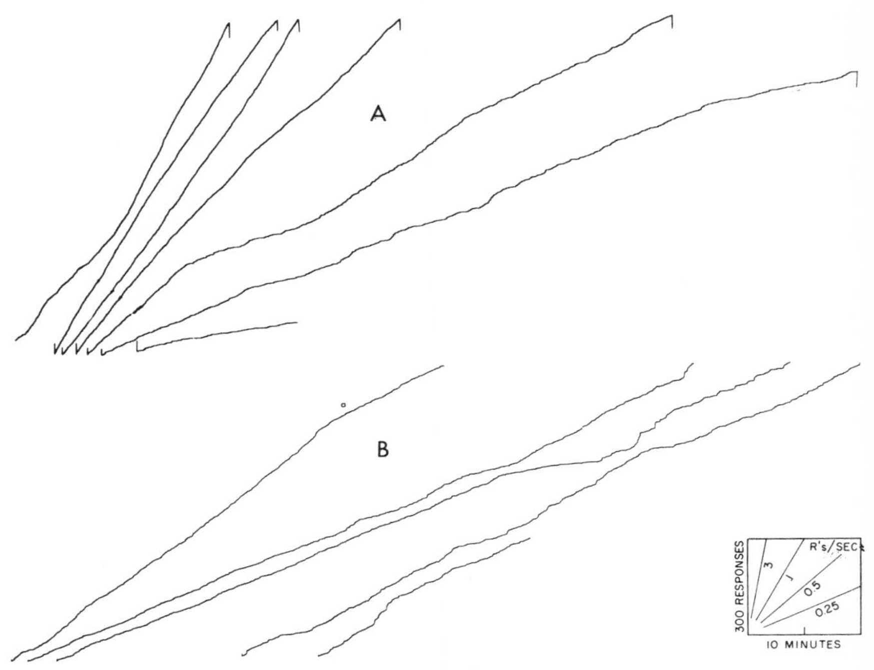

Fig. 587. Extinction after FI 32 TO 5 and FI 32 drl 6 TO 5

turns once more, at *a*. The rate then falls for the remainder of the session, reaching zero for 20 minutes after *b*. Much later in the session, after 7 hours in which the rate was very near zero, the rate of responding increased temporarily, at *a* in Fig. 586. The rate did not reach the high FI terminal value, however, and the grain is rough. Beginning at *b* the curve oscillates with a fairly stable frequency. The highest rate is of the same order of magnitude as at *a*. The persistence of intermediate rates of responding in prolonged extinction may show some survival of the effect of the earlier drl performance. Figure 585B shows the parallel extinction curve after FIdrl. The session was shorter and the record is comparable only with Record A without the continuation in Fig. 586. The over-all rate begins at approximately 0.25 response per second, approximately one-sixth the beginning rate in Record A. The over-all rate falls slowly with 2 marked oscillations. Near the end of the session, sustained responding occurs at the same rate as that at the start of the session. The highest rate during extinction occurs at *c*, at 0.4 response per second.

FI 32. Figure 587 shows extinction curves after FI 32 with and without drl. Both curves begin with the rate appropriate to preceding performances. The rate in Record A falls fairly continuously, while in Record B, after drl, the decline is less orderly. The starting rate falls rather abruptly from 0.4 response per second to about 0.2 response per second at *a*. This rate is then maintained approximately for the remainder of the session, with local deviations.

FIXED RATIO WITH DIFFERENTIAL REINFORCEMENT OF LOW RATES

Reinforcement occurs in FRdrl when the bird pauses for a given period after having completed the fixed ratio. Differential reinforcement of low rates on a ratio schedule is difficult because of opposing tendencies generated by the schedule: the schedule produces high rates just before reinforcement, largely because completing the ratio is reinforcing. The schedule is similar to tand FRFI in that the main features of FR are preserved while the differential reinforcement of high rates normally present in the schedule is eliminated. The drl has the added property, however, of postponing the reinforcement until the rate falls below a critical value. The schedule theoretically should generate an oscillating state, at least with small ratios. A high rate postpones reinforcements and increases the number of responses per reinforcement. As the rate falls, the number of responses per reinforcement approaches the value of the fixed ratio, and reinforcement at this value again tends to increase the rate of responding; and so on.

The effect of differential reinforcement of low rates on an established fixed-ratio performance

Early exposure. In one experiment a performance on a moderate size of fixed ratio was first established. The differential reinforcement of low rates was then added. Figure 588 contains a transition from FR 150 to FR 150 drl 3. Record A shows the last 2 excursions on FR 150, after 11 sessions during which the size of the fixed ratio was advanced from 20 to 150 following crf. The final performance consisted of a pause

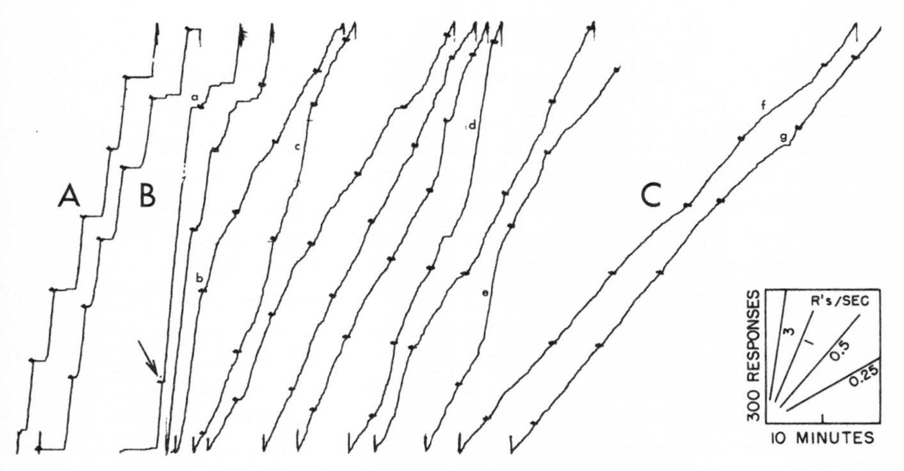

Fig. 588. Transition from FR 150 to FR 150 drl 3

after reinforcement up to 2 or 3 minutes and a shift to a terminal rate, either abruptly or with a small amount of curvature. The following session in Record B begins with 1 reinforcement on FR 150 at the arrow. The schedule then becomes FR 150 drl 3. The terminal rate continues for more than 600 responses before a 3-second pause sets up the reinforcement at *a*. A lower terminal rate is assumed at *b*. The rest of the session is characterized by an over-all rate of slightly more than 1 response per second with rough grain. At *c, d,* and *e* the earlier rate and smoother grain return. These rate changes are independent of the reinforcement. Record C contains a segment from the 3rd session after the introduction of the drl 3. The performance is roughly linear at about 0.6 response per second. The rate frequently shifts widely, as at *f* and *g,* and the drl 3 requirement is being satisfied mainly by the rough grain.

Figure 589 shows a transition from FR 60 to FR 60 drl 6. The bird had had only

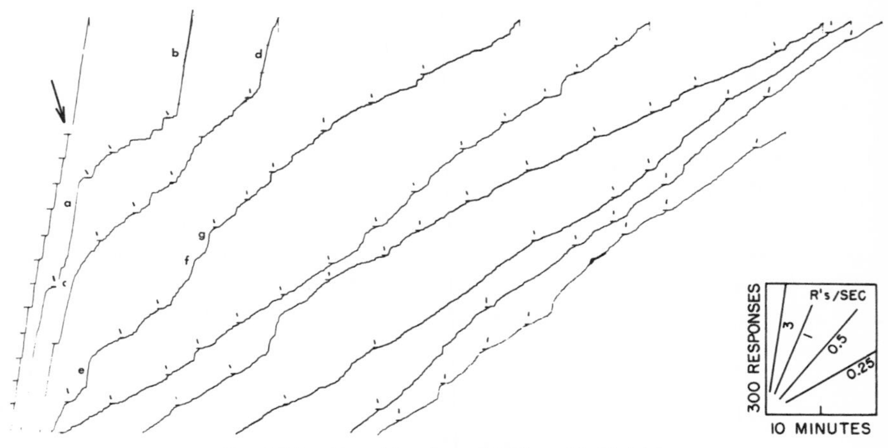

Fig. 589. Transition from FR 60 to FR 60 drl 6

3 sessions on FR 60. A sample of the performance appears before the arrow. Approximately 600 responses were then emitted at the ratio terminal rate before a slight decline satisfied the added drl contingency. The same high rate appears again at *a, b,* and *c,* separated by segments at much lower rates. Later in the session an over-all rate of approximately 0.4 response per second is reached, with most reinforcements being received within 20 or 30 responses after the ratio has been completed. Occasional short spurts of rapid responding occur midway through the session, as at *d, e, f,* and *g.* Other examples appear after the over-all performance has become linear. The 2nd session on FR 60 drl 6 (Fig. 590) shows an overnight loss of the control by the drl contingency. The record begins with a sustained run at the terminal fixed-ratio rate at *a*; the first 6-second pause occurs some 850 responses later, at *b*. Thereafter, the drl is effective in preventing the emergence of higher rates of responding, except at *c* and *d.*

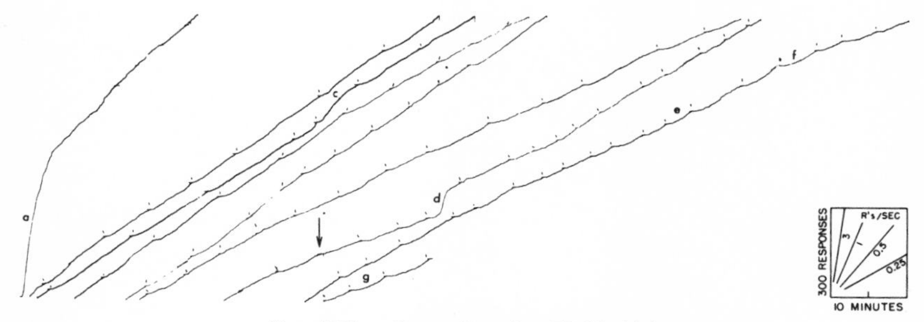

Fig. 590. Second session FR 60 drl 6

At the arrow the size of the ratio was reduced to 30. The rate falls to 0.3 response per second by the end of the session, and a pause and period of smooth curvature develop after reinforcement. Some segments, at *e, f,* and *g,* show quite smooth and continuous curvature. The performance has adjusted to the decrease in size of ratio, and reinforcements now occur more frequently and after fewer numbers of responses than under the previous FR 60 schedule.

Further exposure to fixed ratio with differential reinforcement of low rate. Two birds were reinforced on FRdrl 6 for 60 sessions with a total of nearly 450 hours on the schedule. During this period the drl contingency was occasionally removed, and the value of the fixed ratio occasionally changed. No single type of performance emerged as a general pattern.

The extent of the variation in performance throughout the experiment is illustrated in Fig. 591, which contains excerpts from various stages. The segments in the figure do not occur in any order and simply represent the kinds of rate changes observed. Records A and B show a linear performance in which only a very slightly lower rate follows reinforcement. The rate exceeds the critical value and segments occur containing large numbers of responses because the drl requirement is not met. Records C and D illustrate another type of performance which appeared frequently. The fixed-

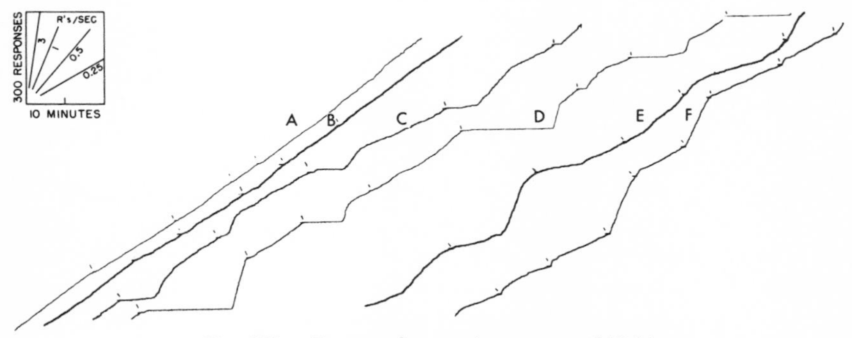

Fig. 591. Excerpts from various stages of FRdrl

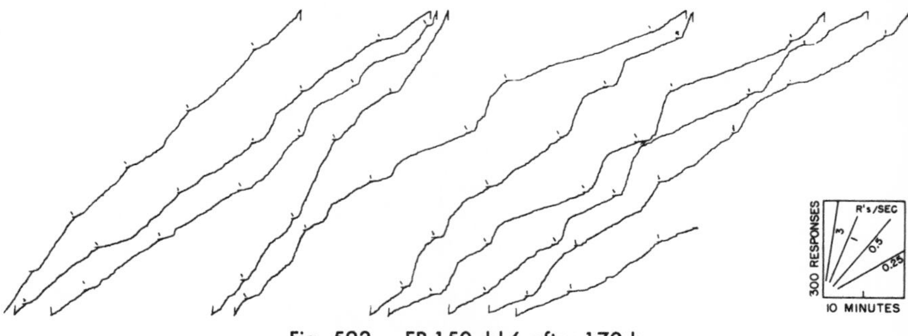

Fig. 592. FR 150 drl 6 after 170 hr

ratio segments begin with a low or zero rate. Toward the middle of the segment the rate becomes high; then an abrupt shift usually occurs to intermediate rates, which are maintained until a pause occurs that satisfies the drl requirement. The extent and rate of the fast responding vary considerably. These kinds of rate changes are often intermixed with linear performances. The performance appears to show a rate related to each of the components of the schedule—a scallop for the FR, a low terminal rate for the drl. Records E and F illustrate a performance where somewhat similar changes in the rate occur independently of reinforcement. These sometimes include continuous rate changes.

All these kinds of changes appear at various stages in the performances of all the birds studied on FRdrl. The sequence does not appear to be orderly, however. Figure 592, for example, shows a single session on FR 150 drl 6 after 170 hours of reinforcement. All varieties of rate changes during the ratio segment described in connection with Fig. 591 may be found.

In Chapter Eight it was noted that on tand FRFI the bird may not be able to maintain a sustained performance on a large fixed-ratio component, but that a pro-

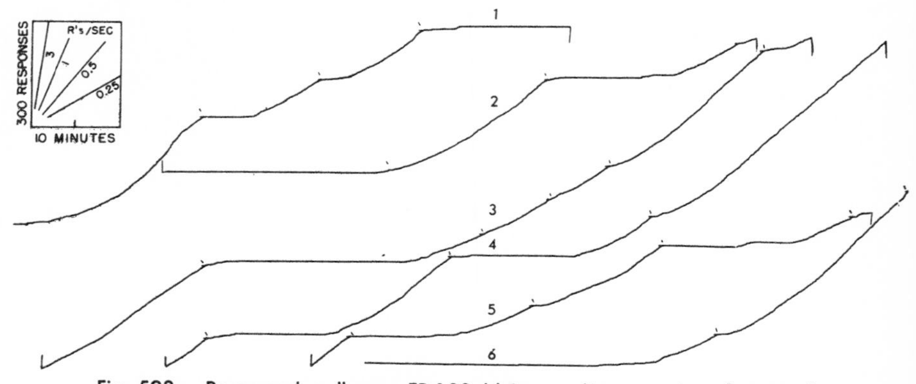

Fig. 593. Pause and scallop on FR 100 drl 6 preceding cessation of responding

longed pause and period of acceleration followed reinforcement, rather than the complete and abrupt cessation of responding in a simple FR. Figure 593 shows a similar effect in the performance after 207 hours of reinforcement on FR at ratios of from 30 to 100 with a 6-second drl. This performance preceded a complete cessation of responding. We subsequently reinstated the performance on this schedule by reducing the size of the ratio component. Throughout nearly all segments in Fig. 593 the rate increases fairly continuously. Therefore, it usually becomes too high to satisfy the drl requirement when the number of responses in the ratio has been emitted. The large number of responses per reinforcement and the absence of any reinforcements after small numbers of responses produce prolonged pauses after reinforcement, as in a "strained" fixed ratio.

The effect of removing the drl contingency. Figure 594 shows the effect of removing the

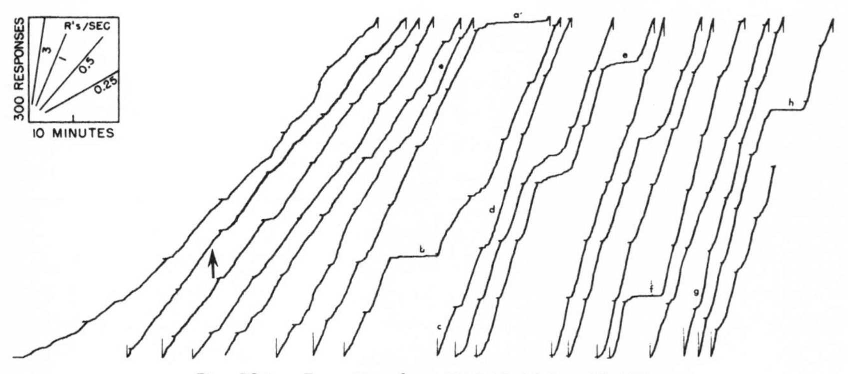

Fig. 594. Transition from FR 150 drl 6 to FR 150

drl contingency for the bird whose earlier performance on FR 150 drl 3 has been shown in Fig. 588. The 1st excursion shows the performance after 17 sessions of reinforcement on the FRdrl schedule, with an over-all rate of approximately 0.6 response per second, an occasional brief pause following the reinforcement, and frequent shifts in the local rate. When the 3-second drl contingency was removed, at the arrow, the local and over-all rates of responding gradually increased over the next 7 excursions of the recording pen. The over-all rate of responding became 1.5 responses per second during the excursion (at *c*), and the terminal rate of responding, almost 2 responses per second (at *d*). The terminal FR rate of responding varied considerably during the remainder of the session, but it reached a value as high as 4.5 responses per second, as at *g*. Along with the increase in the terminal rate, abrupt shifts to zero or near-zero rates of responding began to occur following the reinforcement, as at *a*, *b*, and *e*, or elsewhere, as at *f* and *h*.

The following session, shown in Fig. 595, begins with a recovery of some of the properties of the previous FRdrl performance. Following the reinforcement at *a*,

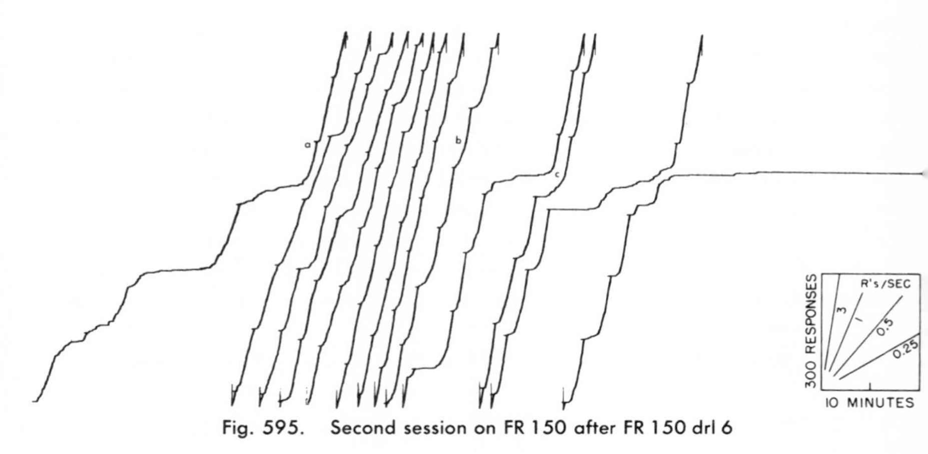

Fig. 595. Second session on FR 150 after FR 150 drl 6

however, a fixed-ratio performance is reached comparable with the performance at the end of the previous session. As the session proceeds, however, the terminal rate of responding becomes delayed, as the pause and period of acceleration following the reinforcement become a standard feature of the segment. The smooth curvature at *b* and *c* and extended intermediate rates preceding the terminal rate foreshadow substantial pauses at the final segments in the figure. The removal of the drl contingency has generated prolonged pauses after reinforcement and elsewhere, instead of the previous performance, where responding was sustained continuously at low rates.

Figure 596 shows a similar transition for a second bird in the experiment. The drl contingency was removed at the start of the session. The bird had had a history of 170 hours of reinforcement on various fixed ratios up to 150 with differential reinforcement of responses following 6-second pauses. The over-all rate of responding accelerates steadily, reaching a near-terminal performance by the 3rd excursion of the figure, with an over-all rate of about 2.2 responses per second and a terminal fixed-ratio rate of 3 responses per second. As in the previous transition, a continuous development of a pause following the reinforcement occurs as the fixed-ratio performance

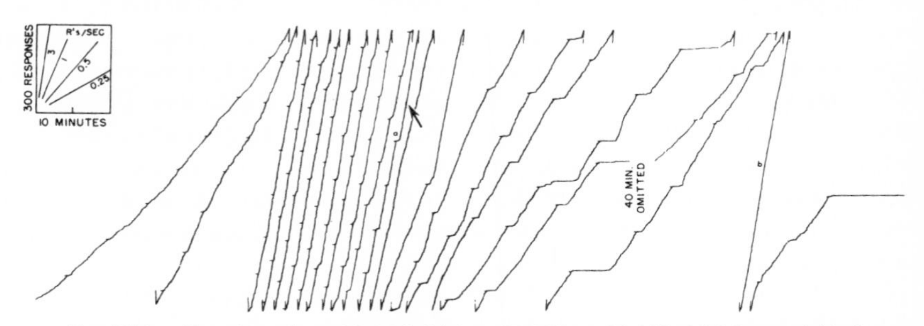

Fig. 596. Transition from FR 150 drl 6 to FR 150 to FR 150 drl 6 (second bird)

develops, reaching the maximum size at *a*. At the arrow the drl contingency was again introduced. The subsequent performance parallels the earlier transition from FR to FRdrl. At *b* a sustained period of fast responding occurs, although most fixed-ratio segments no longer show the effect of the previous FR reinforcement. The large number of responses per reinforcement now occurring as the result of the high rates developed under FR without drl now produces pausing after reinforcement, with one particularly extensive example.

The removal of the drl contingency in a third bird produced a much slower increase in rate. The entire 1st session on FR 150 without drl 6 showed no change over a prevailing FRdrl performance. The same performance continued on a 2nd session without the drl contingency, as the 1st segment in Fig. 597 shows. No change in performance has occurred, although the number of responses per reinforcement corresponds to the fixed ratio precisely. Higher rates of responding begin to emerge during the next 3 excursions of the recorder, particularly at *a* and *b*; and following the

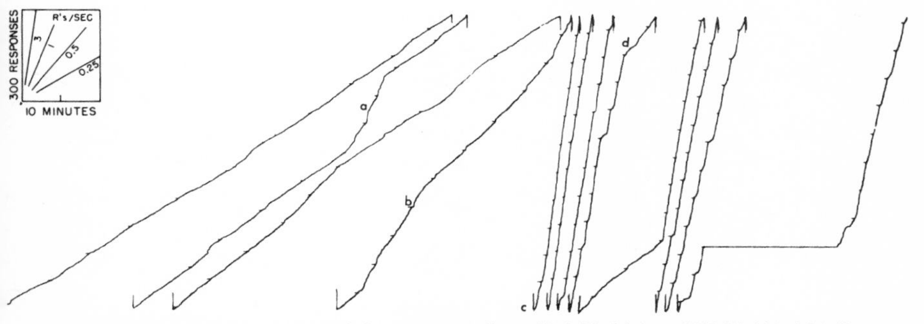

Fig. 597. Second session of the transition from FR 150 drl 6 to FR 150 (third bird)

reinforcement at *c*, a normal fixed-ratio performance emerges with an over-all rate of approximately 3 responses per second and local rates of 3.3 responses per second. The normal fixed-ratio performance is sustained for almost 4 excursions of the recorder, when there is a sudden shift to the previous drl performance, at *d*. This is continued through 4 fixed-ratio segments, and is followed by the normal fixed-ratio performance again. For the remainder of the session, the pause and period of acceleration following the reinforcement increase, leading to a pause of over 20 minutes. The entire session contains 120 reinforcements, approximately twice the normal number for the bird; and some of the pausing and lower rates of responding toward the end of the session may be due to satiation rather than a normal development of the schedule.

In the following session, shown in Fig. 598, the first 2 reinforcements on FR 150 without the drl continue the performance developed in the previous session. When drl 6 is again added at the arrow, the transition to a typical FRdrl performance is somewhat more rapid than the 1st transition described in Fig. 589.

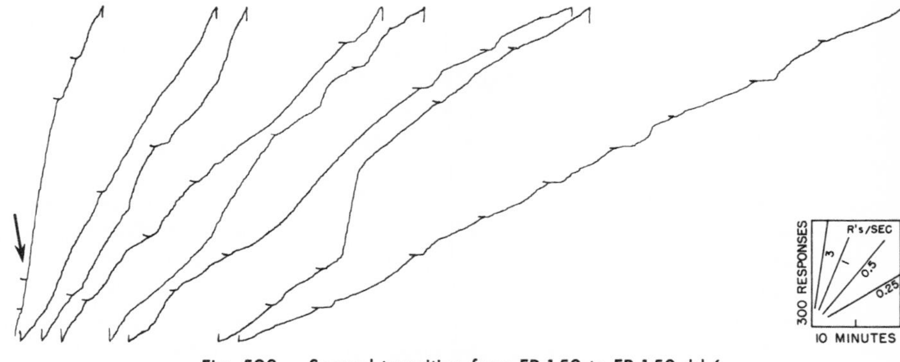

Fig. 598. Second transition from FR 150 to FR 150 drl 6

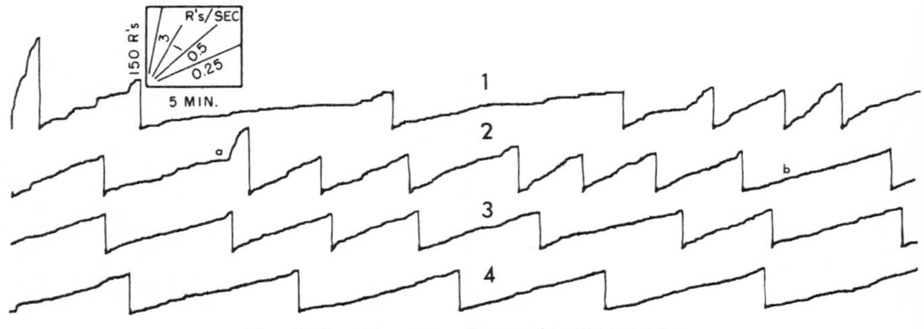

Fig. 599. Transition from crf to FR 60 drl 6

The development of an FRdrl performance immediately following crf

Figure 599 shows a 1st session on FR 60 drl 6 after crf. The pen resets at reinforcement. In the 1st segment in the graph, an extinction curve from the preceding crf, nearly 200 responses are emitted before the rate falls enough for a 6-second pause to occur. The rate continues to fall during the 2nd segment, and a 2nd reinforcement is received after approximately 80 responses, with the help of the rough grain. The next 2 ratios show very low rates, probably because of extinction at this value of FR. But a more stable performance soon emerges. A break-through of a higher rate occurs at *a*; but otherwise the drl contingency sustains a generally constant rate. Note in particular the segment at *b*. This rate prevails for the remainder of the session. All reinforcements are received within approximately 10 responses of the completion of the ratio. The figure resembles a normal transition from crf to FR except that the linear phase is prolonged.

The same schedule was continued for a total of 584 hours; Fig. 600 shows the further development of the performance. In Record A, after 13 hours, the rate tends to

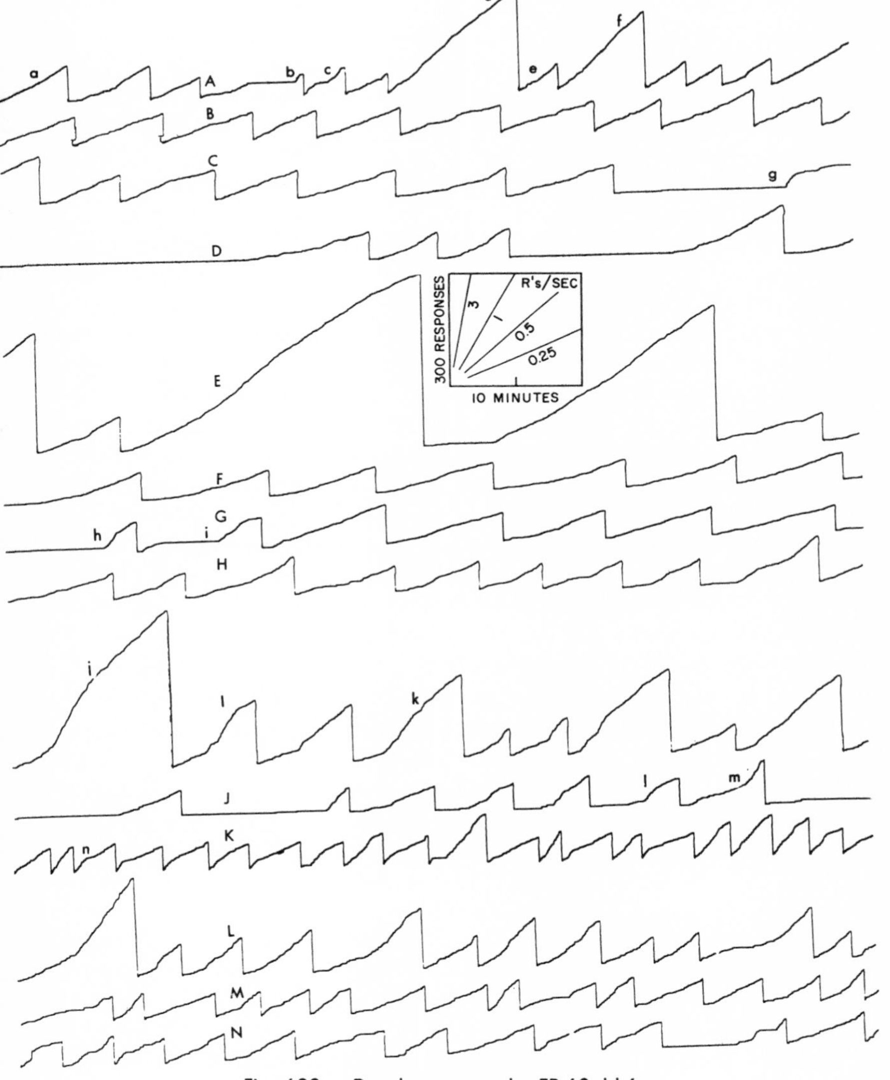

Fig. 600. Development under FR 60 drl 6

increase toward the end of the ratio segment. This increase is slight but continuous at *a* and *e,* but large following a pause, as at *b* and *c.* At *d* and *f* the rate postpones reinforcement far beyond the number of responses in the ratio. In Record B, after 26 hours, the rate of responding is roughly linear, with reinforcements occurring at roughly the fixed-ratio value. In Record C, after 54 hours, a pause and acceleration after many reinforcements appear. At *g* a high rate of responding immediately follows a long pause. In Record D, after 63 hours, the pause after the reinforcement becomes longer, and the rate increases throughout most of the fixed-ratio segment. In spite of the high rate, however, pauses occur frequently enough so that the number of responses does not greatly exceed the size of the fixed ratio. At 108 hours, in Record E, the pause and acceleration following the reinforcement have increased the rate beyond the point where the drl contingency is met, and segments with very large numbers of responses appear. In Record F, at 151 hours, the performance is temporarily stable, with a pause and slow increase in rate to a low value. In Record G high rates of responding emerge during the middle of the ratio segment, as at *h* and *i.* The segments in Record H, after 239 hours, show a general increase in rate through the ratio segment, with a tendency for a higher rate of responding to appear temporarily in the middle of the segment. In Record I, 10 hours later, the rate of responding is high when the ratio is completed, and most segments show a subsequent negative acceleration, with reinforcement occurring when the rate of responding has fallen. Two of the more conspicuous S-shaped curves appear at *j* and *k.* Record J, after 285 hours, shows a wide variety of rates, including a positively accelerated segment at *m* and a negatively accelerated segment at *l.* In spite of the generally increased local rate, the number of responses in the fixed ratio is not being greatly exceeded because of the grain. By Record K, after 329 hours, a different performance emerges which results in a much higher frequency of reinforcement. The rate tends to be highest immediately after reinforcement and falls off during the remainder of the segment, completing part of the fixed-ratio requirement at a high rate but slowing down subsequently when the drl requirement is operative. The fine grain of the record shows responses occurring in rapid bursts separated by discrete pauses, and a reinforcement such as

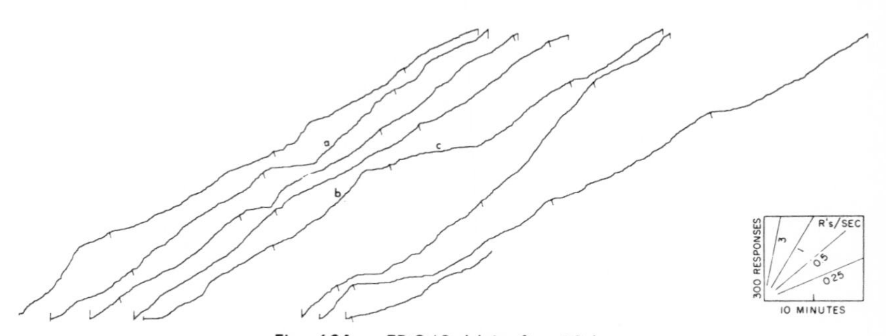

Fig. 601. FR 240 drl 6 after 60 hr

that at *n* occurs because of the grain rather than the low over-all rate. Marked examples of negatively accelerated segments are evident among linear performances. In spite of the high frequency of reinforcement under this type of performance, it is not stable. Records L, M, and N, taken after 362, 376, and 382 hours, respectively, show a variety of rate changes during the ratio segments, including nearly all of those already described. Responding still occurs in bursts, however. This is perhaps the only progressive development in the experiment.

The ratio was then increased to 240, the drl remaining at 6 seconds, for 60 hours. The result is shown in Fig. 601, where the pen does not reset after each reinforcement. The bird sustains a performance on this size of ratio showing a low over-all rate, rough grain, and wide oscillations in the over-all rate. The high rate at *a* appears to compensate for the preceding pause, but at *b* the increase precedes a period of slow responding (at *c*). Little or no curvature exists after reinforcement, despite the roughly fixed number of responses per reinforcement.

A second bird on FR 60 drl 6 satisfied the drl contingency by a rough grain much earlier in the experiment. The final performance showed many more instances of high rates at the beginning or in the middle of the ratio segment. The fixed ratio is thus counted out in a short time, and the subsequent retardation achieves reinforcement. Figure 602 summarizes the experiment. In the 2nd session (Record A) the over-all and local rates have already fallen; a typical performance consists of a pause and slow acceleration to a slightly S-shaped curve. A single instance occurs of the emergence of a high rate, possibly a residue of the previous crf. This rate could also result from the reinforcing effect of increasing the count. In Record B, after 15 hours, the over-all rate is low and almost linear, with reinforcements occurring approximately every 10 minutes. Record D shows two instances in which a high rate early in the ratio segment completes the ratio requirement and the rate then falls off to satisfy the drl before reinforcement.

Following the session represented by Record D, the drl was removed briefly, with the result shown in Fig. 603. The first four lines represent the 1st session of FR 60. The over-all and local rates increase steadily. The 1st segment on the 2nd session without drl (beginning at *a*) starts with a long pause and extended acceleration, as in the previous session; but in the 2nd fixed-ratio segment the rate increases beyond the final rate in the previous session. Toward the end of the session, beginning at *b*, the long pause following reinforcement reappears for 4 segments, after which a rapid acceleration leads to an even higher over-all rate. The portion of the record at the end of the 2nd session beginning at *c* shows 12 reinforcements occurring within 15 minutes. The 3rd session without the drl, beginning at *d*, shows no further increase in the rate of responding, although some local rates are considerably higher than any in the previous sessions—for example, at *e*, *f*, *g*, and *h*. In spite of the absence of the drl contingency, the grain continues to be rough. Responses occur in bursts separated by frequent pauses of the order of 6 seconds. At *i*, toward the end of the 3rd session, the drl contingency is reinstated, but it has little effect. The 3 sessions of reinforce-

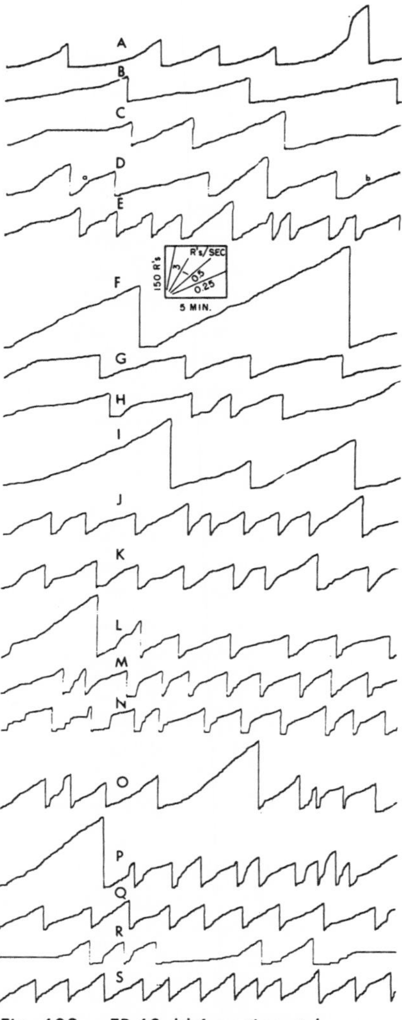

Fig. 602. FR 60 drl 6 portions taken over 335 hr (second bird)

ment without the drl had not been effective in progressing toward a fixed-ratio performance; in spite of the new schedule, the bird was still performing appropriately to drl. The final portion of the figure shows the start of the next session under FR 60 drl 6, in which the low over-all rate seen at the start of the figure returns.

With the drl again in effect, the fixed-ratio segments became roughly negatively ac-

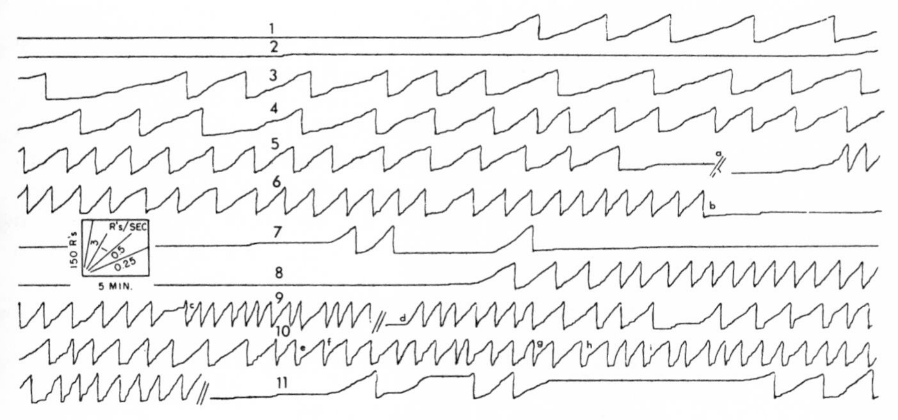

Fig. 603. Temporary omission of the drl 6

celerated, producing a higher frequency of reinforcement, as Record 602E, after 39 hours after crf, shows. In Record F, after 47 hours, a constant rate is too high to satisfy the drl requirement even with a rough grain, and several hundred responses per reinforcement are emitted. Record G, after 52 hours, contains smooth, negatively accelerated segments, which become more marked in Record H, after 68 hours. In Record I, after 71 hours, the curves are accelerated, and the rate is too high at the time the fixed ratio is counted out to satisfy the drl requirement, so that the segments contain a large number of responses. In Record J, at 160 hours, the performance consists of a high rate early in the fixed-ratio segment, followed by a shift to a lower rate which satisfies the drl contingency. Similar performances are shown in Record K, at 164 hours; Record L, at 168 hours; and Record M, at 173 hours. In one segment, for example, the whole fixed ratio is emitted at almost a normal fixed-ratio rate, and a sudden drop to a lower rate then satisfies the drl requirement. The remaining performances in the figure, Record N at 181 hours, Record O at 190 hours, Record P at 210 hours, Record Q at 211 hours, Record R at 299 hours, and Record S at 335 hours, show a variety of performances. Some segments show a high frequency of reinforcement because of high rates of responding early in the ratio followed by abrupt shifts to lower rates which satisfy the drl requirement. Others show intermediate rates where the grain of the record satisfies the drl requirement as soon as the fixed ratio is counted out. Still others show extended scallops which considerably exceed the number of responses in the ratio.

Figure 604 shows a complete session following Record P in Fig. 602. Reinforcements are received almost on schedule because many ratio segments begin at a high rate of responding and drop to a low rate. In any case, the rough grain is important.

Following the session represented in Record S in Fig. 602, we explored the size of the ratio at larger values after brief reinforcement on FR 30 and 40 with the drl contingency. Reinforcement was continued on larger fixed ratios with drl 6 for almost 200 hours, during which there was almost no progress toward a stable performance. Figure 605 contains segments from the various sessions. Record A on FR 220 drl 6, after a total of 390 hours on various FRdrl schedules, shows a moderate over-all rate of responding. The grain is rough and the segments negatively accelerated, either with smooth curvature, as at *b,* or a break-through to a higher rate early in the segment, as at *a.* Record B, at 401 hours, shows roughly linear segments except for the pause at *c* followed by a marked compensation in rate. Record C, after 402 hours, shows a

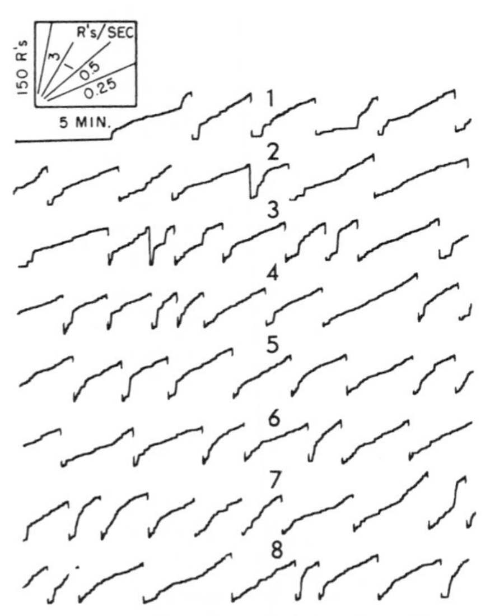

Fig. 604. FR 60 drl 6 showing negatively accelerated segments and rough grain

somewhat different type of pause and compensation, at *d.* Records D and E, at 420 and 424 hours, respectively, show instances of higher rates of responding immediately after the reinforcement, as at *e, f,* and *g.* Other segments, as at *h* and *i,* show marked local rate changes, rough grain, and short bursts of rapid responding. Record F, at 449 hours, shows marked oscillations in rate within the ratio segment. When the ratio was reduced, high initial rates with strong retardation, S-shaped curves, and a somewhat lower maximum rate appeared, as in Record G, after 534 hours. During the remaining 83 hours of the experiment the performance varied over a very wide range. In Record H, after 617 hours, a long pause follows reinforcement, and responding is slow and steady. This is a return to a performance which prevailed very early in the experiment.

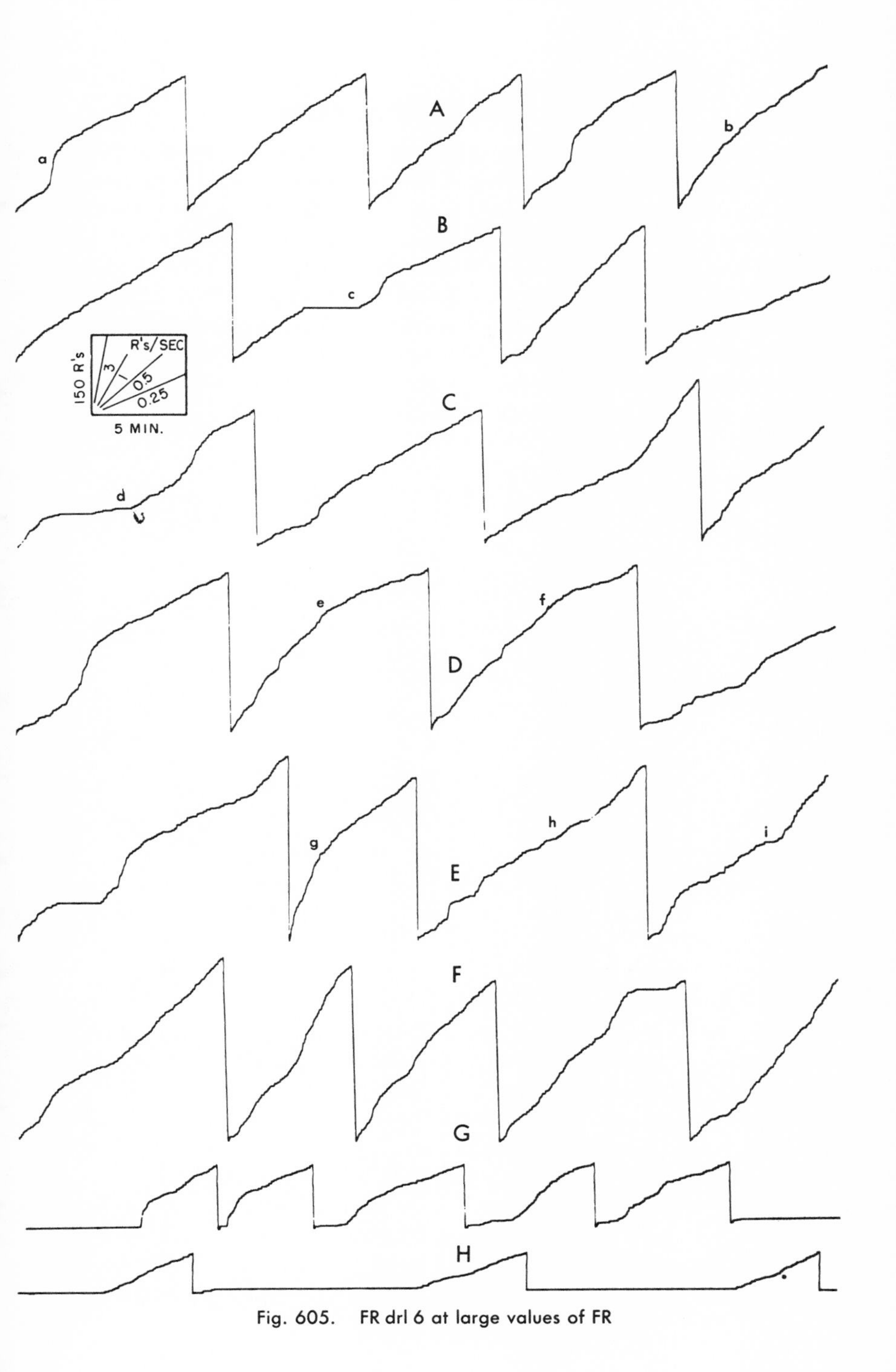

Fig. 605. FR drl 6 at large values of FR

In a similar experiment 2 birds with a very long history of FR were exposed to FRdrl, where the drl was first of the order of a few milliseconds. Such a drl contingency was easily satisfied by the existing rate and grain. No change in performance was observed during the period of exposure. The drl was then gradually increased and had very little effect until the drl reached 1 second, when negative curvature appeared. Reinforcements were received roughly on schedule. When the drl was then abruptly changed from 1 second to 2 seconds, the change in performance was more pronounced, as in Fig. 606. The drl contingency now postpones many reinforcements; and before the end of the session, marked changes in rate appear in the ratio segments. A rough grain emerges, doubtless because of the drl.

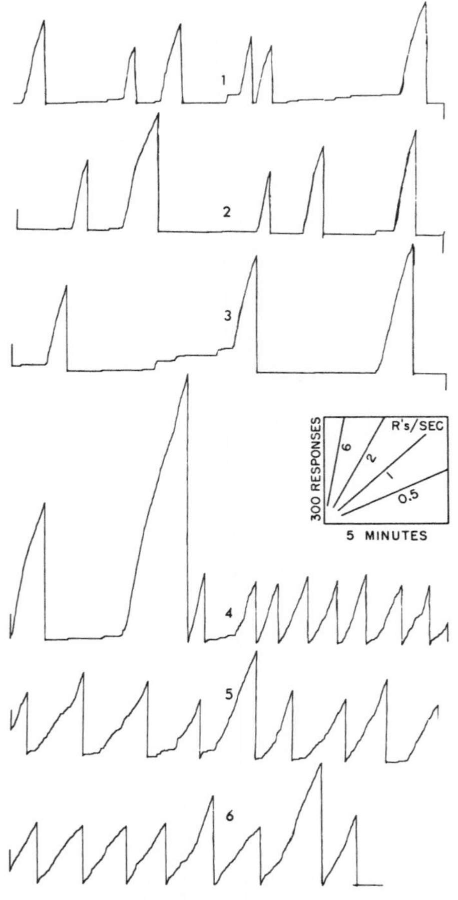

Fig. 606. FR 160 drl 2 after FR 160 drl 1

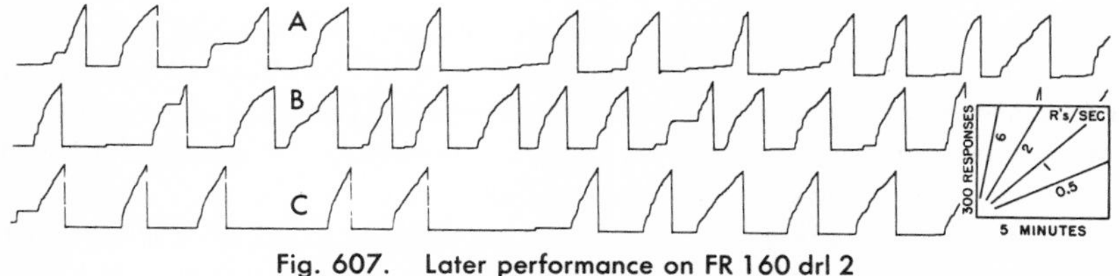

Fig. 607. Later performance on FR 160 drl 2

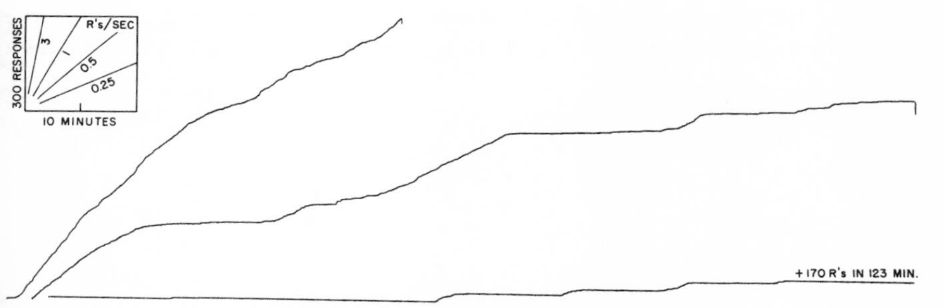

Fig. 608. Extinction after FR 60 drl 6

Figure 607 shows representative samples from the 6th, 8th, and 11th sessions on FR 160 drl 2. This is one of the most consistent and sustained examples we have observed of a performance under FRdrl in which part of the ratio is run off at a normal ratio rate, with negative acceleration then leading to a low terminal rate that, aided by grain, satisfies the drl contingency.

Extinction after FRdrl

Figure 608 contains an extinction curve after 36 hours of reinforcement on FR 60 drl 6 at the stage of development following Record B in Fig. 600, where the prevailing performance was a fairly stable low rate. The curve shows a fairly regular decline from an intermediate rate to a near-zero rate by the end of the figure. The high rates previously shown in FR and occasionally on FRdrl do not appear. The curve suggests extinction after VI rather than FR, and little evidence exists that the fixed-ratio component of the FRdrl schedule has any effect.

A similar extinction curve was taken just after Record M in Fig. 600, after more than 376 hours of reinforcement on FRdrl. The over-all rate declined continuously, no high rate appearing.

VARIABLE INTERVAL WITH DIFFERENTIAL REINFORCEMENT OF HIGH RATE

Figure 11, Chapter Three, has already described a device designed to close a circuit when a given number of responses has occurred within a designated time interval.

It has been used for the study of the differential reinforcement of high rates (drh).

The number of responses required at a given ratio and the rate could be varied more or less independently. In the following experiment several values were used. The conditions are relatively arbitrary and will not be specified in detail, since the following experiments are intended merely to demonstrate the general effects of drh rather than to make a parametric analysis of the variables.

In one experiment 2 birds were reinforced on an arithmetic 1-minute VI schedule after crf. Figure 609A shows a stable performance after 8 sessions on VI 1. We changed the schedule of reinforcement to VIdrh by withholding reinforcement on VI until the rate of responding was high enough to satisfy the drh contingency. To the extent that reinforcements are postponed because the rate of responding is not high

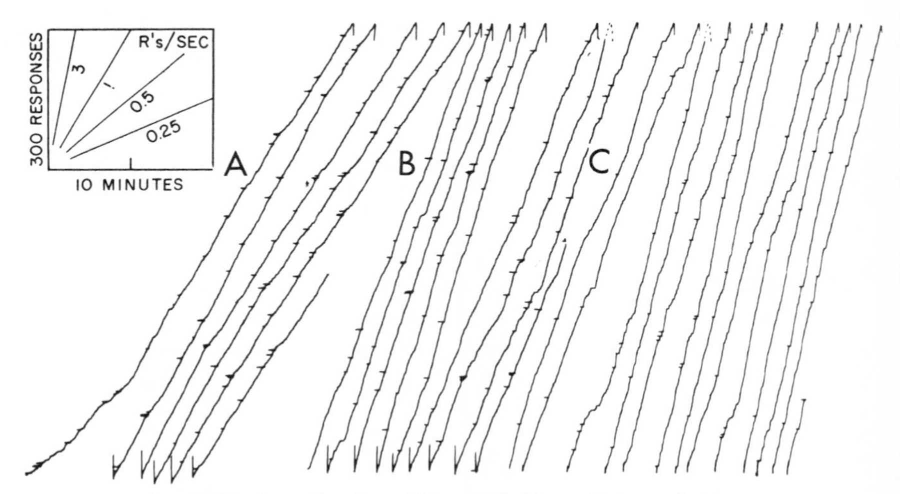

Fig. 609. Transition from VI 1 to VI 1 drh and later performance

enough when a reinforcement is set up, the mean interval (and particularly, the shortest interval) of the VI schedule is increased. By advancing the value of drh progressively in small steps, the rate on VI can be increased without an appreciable increase in the mean interval of reinforcement.

Record B in Fig. 609 shows a performance after 10 sessions on VIdrh. The overall rate is between 1.5 and 2 responses per second, compared with slightly less than 1 response per second in Record A without drh. Three sessions later, after the number of responses required for the drh is increased slightly, the rate increases to between 2.5 and 3 responses per second (Record C). The fine grain of the record shows short bursts at very high rates separated by pauses.

Figure 610 shows a later stage of exposure to VIdrh. In Record A, after 26 sessions, reinforcement is contingent on a higher rate, sustained for a larger number of responses than in Fig. 609. The over-all rate is between 3 and 3.5 responses per second. Fifteen sessions later, in Record B, the required rate has been increased, but the number of responses at that rate remains the same. The over-all rate reaches 4 responses per second, while local rates in sustained runs without pauses reach 5 responses per second. Note the absence of intermediate rates of responding. A second bird on the same procedure showed the same result.

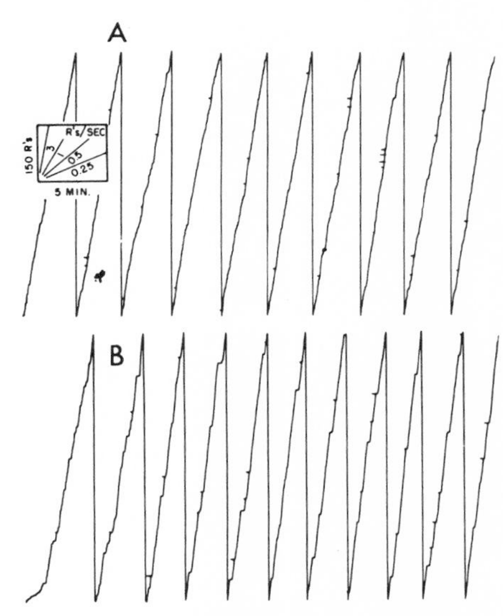

Fig. 610. Later performance on VI 1 drh

Extinction after VIdrh

Figure 611 shows an extinction curve after VIdrh under a value of drh which made it difficult for the bird to maintain a normal frequency of reinforcement. The 5 segments before the reinforcement at *a* show a performance which never satisfied the current drh. For some time after *a*, reinforcements are fairly often received at the same drh. The magazine was disconnected at the arrow and the session continued for 24 hours. At least 80% of the responses in the curve are emitted by the end of the second line in the figure, a little more than 1 hour after extinction began. Most of the later responses occur at a rate comparable with the high rate under drh. The intermediate over-all rates at *d*, *e*, and *f* are composed of short bursts at high rates separated by pauses. A sustained burst of responding at 4 responses per second occurs at the end of

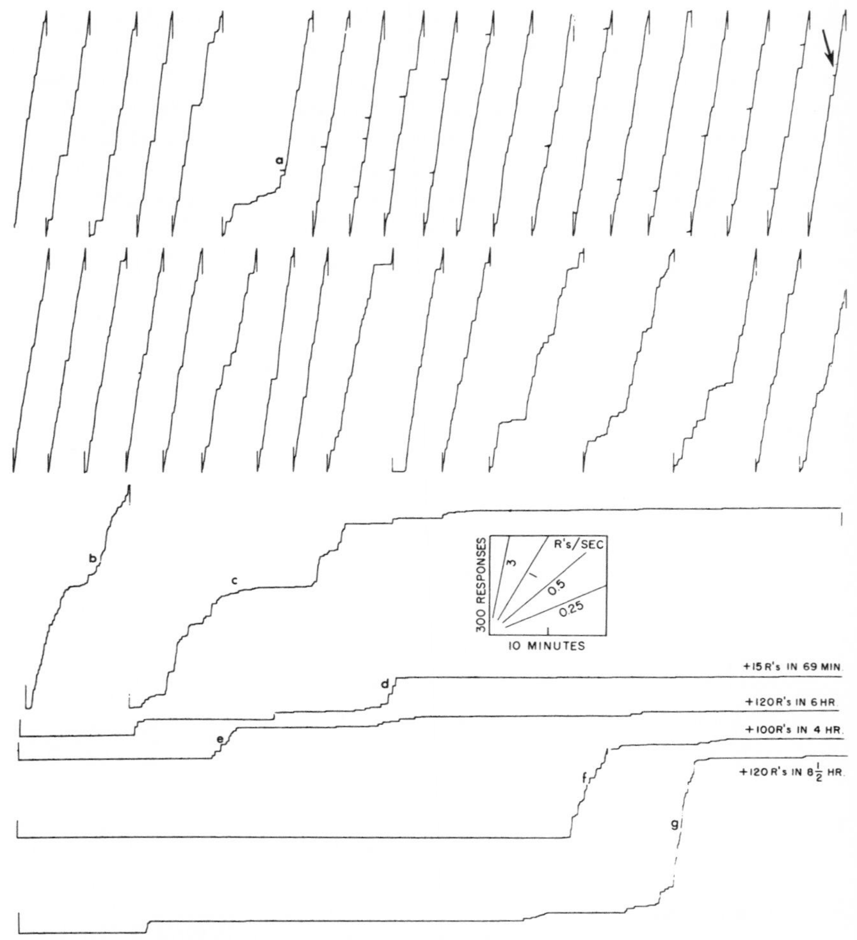

Fig. 611. Extinction after VI drh showing intermediate rates

the session, at *g*. Intermediate rates of responding occur at *b* and *c*, as the rate changes from the high over-all rate in the second line to the near-zero rate during the remainder of the session.

Figure 612 shows an extinction curve after 51 sessions of exposure to VI 1 drh when the bird maintains an almost normal rate of reinforcement. Extinction is begun at the arrow, and the resulting curve consists of periods of sustained responding at the normal VIdrh rate alternating with increasingly longer pauses. The periods of sustained responding grow shorter as extinction proceeds. The actual rates of responding, even at the end of the session, are the same as under VIdrh.

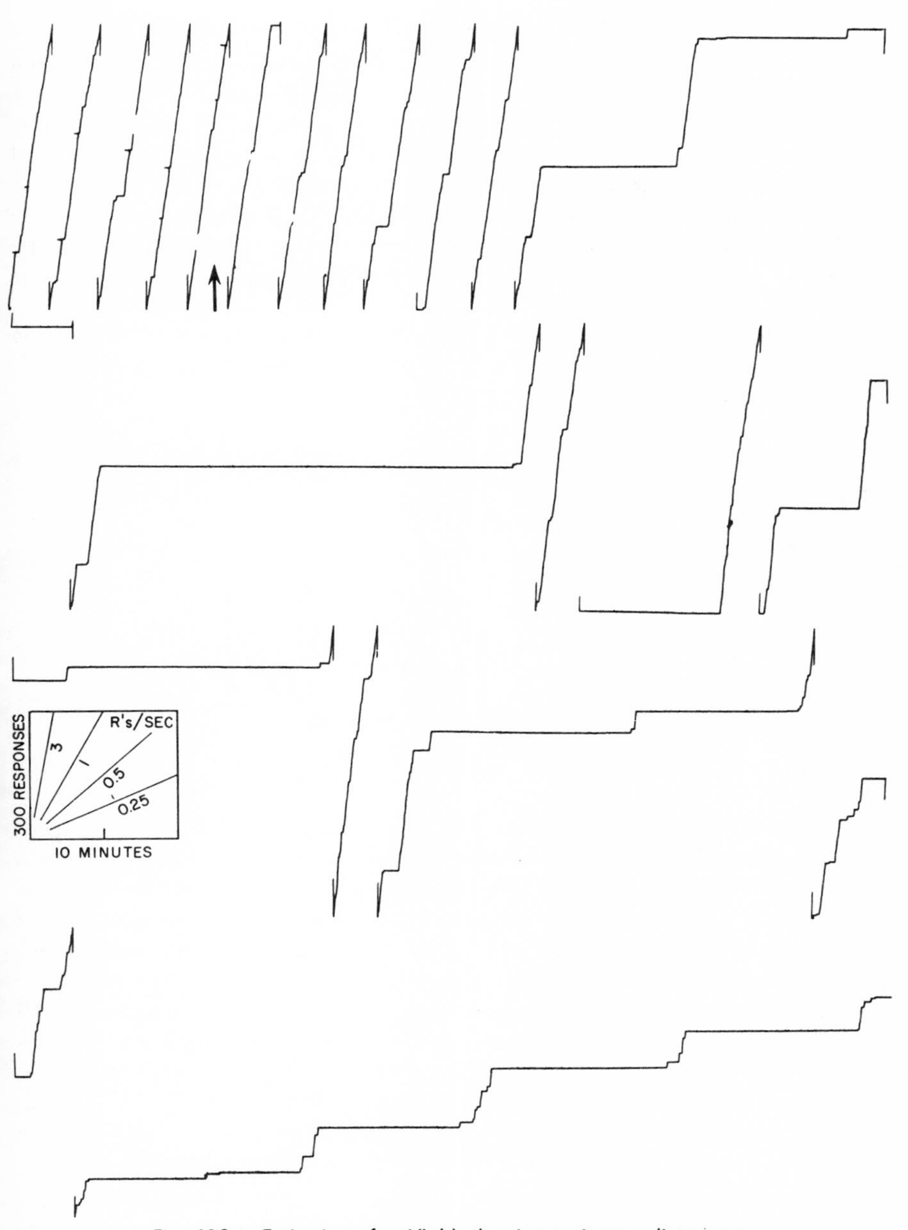

Fig. 612. Extinction after VI drh showing no intermediate rates

DIFFERENTIAL REINFORCEMENT OF RATES WITH PACING[1]

When the rate of responding required for reinforcement is specified by the setting of an upper limit, as with drl, the bird has only to respond at any rate lower than the requirement to be reinforced. When a lower limit is specified, as in drh, reinforcements occur whenever the rate is at any higher value. However, both an upper and a lower limit can be specified. A response must occur at least *x* seconds since the preceding response, and not more than *y* seconds since the preceding response. The number of successive responses which must meet these specifications may also be specified. The technique is called "pacing." A response meeting such requirements may be called a paced response.

A pacing contingency is made effective in the following way. The bird is put on VI 1 until its performance is stable. Every paced response is then reinforced until the rate almost always satisfies the contingency.

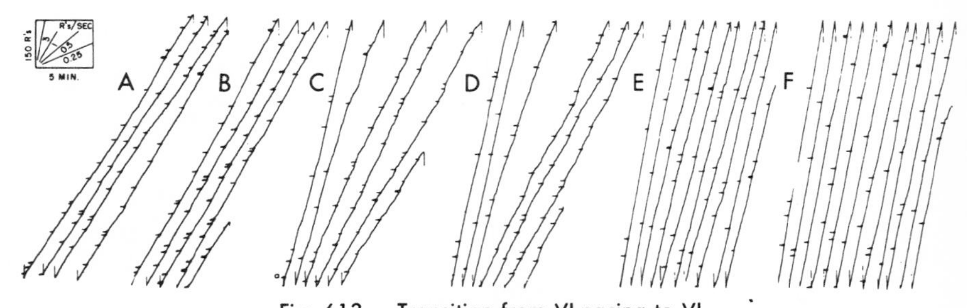

Fig. 613. Transition from VI pacing to VI

VI with pacing

The pacing contingency is added to VI 1 in a combined schedule. In one experiment the contingency required that 2 successive responses should not occur more than 2 seconds or less than 1.5 seconds after the preceding response. This schedule may be written: VI 1 pacing (2) 1.5-2.0. Figures 613A and B show stable samples of paced behavior under VI 1 pacing (3) 0.8-1.3. When the pacing contingency was removed, its effect survived for several sessions. Figure 613C shows the 1st session without the pacing contingency. A 1st response, reinforced at a high rate at *a*, reinstates the VI performance, which is now reinforced on schedule. Nevertheless, before the short session is over, the rate has fallen to the value prevailing under the pacing contingency. This general character prevailed for several sessions. Record D in Fig. 613 is for the 4th session, and some slight survival may be seen in the 6th session in Record E. The 7th session, in Record F, shows a stable high level of performance under VI 1.

[1] These experiments were carried out by Dr. Alfredo V. Lagmay, now of the University of the Philippines, to whom we are grateful for permission to report the results here.

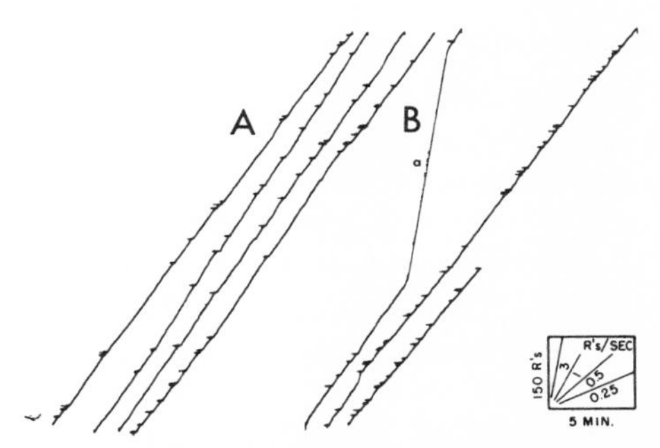

Fig. 614. Break-through of a high rate under pacing

The bird was then shaped to a lower pacing rate under conditions of VI 1 pacing (2) 1.5-2.0. Figure 614A shows a stable performance. As the rate is forced to lower values by pacing, high rates occasionally break through. An extreme instance of this is shown in Fig. 614B at *a*. The record is for the session immediately following Record A and shows, in general, the same type of performance. Nevertheless, at *a* several hundred responses are emitted at the rate appropriate to the original VI seen in Fig. 613F.

Figure 615A shows a still lower rate maintained under VI 1 pacing (3) 2.0-2.5. Reinforcements are characteristically followed by pausing, and occasional periods of a higher rate are apparent, as at *a, b,* and *c*. When the pacing contingency was removed, the return to a VI 1 performance was much slower than in Fig. 613. Figure 615B is for the session following that of Record A and shows a fairly smooth acceleration to a higher over-all rate. This curve is marked by conspicuous instances of the emergence of the VI rate, as at *d* and *e*. Records C and D are for the sessions immediately following Record B and show a progressive increase to a VI performance. The survival of the extreme pacing contingency is nevertheless clear in irregularities, as at *f*, and periods of slower responding, as at *g*.

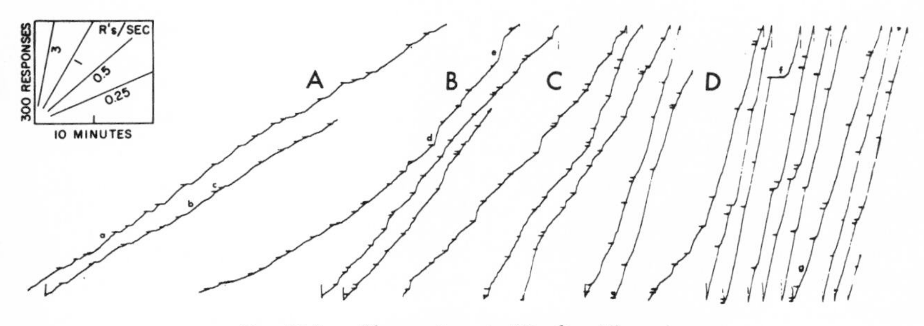

Fig. 615. Slow return to VI after VI pacing

Fixed ratio with pacing

A schedule of FR 35 pacing (2) 1.5-2.0 was substituted for a VIpacing schedule after a stable pacing performance had developed. The experiment was designed to see whether the constancy of number of responses at reinforcement and pacing conditions would begin to produce an effect by making the production of number reinforcing. Figure 616 shows the development from the 5th through the 50th session. Record A is for the 5th session on FR 35 pacing after VI 1 pacing. Responding is still mainly at the paced rate, although some pausing or slow responding appears after reinforcement. This condition prevails through the 6th session (Record B), although the terminal rate is increasing slightly. Eventually, this condition makes it difficult to satisfy the pacing contingency and the number of responses at reinforcement in-

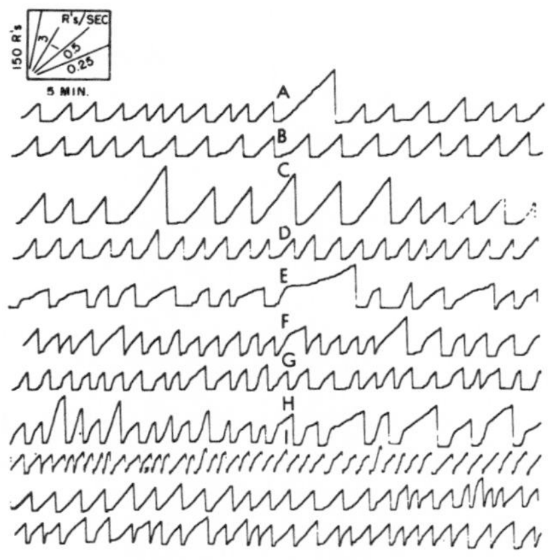

Fig. 616. Development of an FR performance with pacing

creases greatly, as Record C shows for the 8th session. The increase in the number of responses per reinforcement eventually produces a reduction in rate which corrects this change, and by the 15th session (Record D) the number of responses at reinforcement has again become fairly constant. Slight negative curvature appears toward the end of many of these ratio segments, however, and by the 17th session, in Record E, the curves have come to resemble strongly those under FRdrl. (See the earlier section in this chapter.) The remaining records in the figure are from the 19th, 23rd, 29th, 33rd, and 50th sessions. The performance shows the same kind of instability seen under FRdrl.

Figure 617 shows a complete session (considerably longer than usual) from the series shown in Fig. 616. The session occurred between Records D and E in the earlier figure. As with FRdrl, various types of performance come and go throughout the session.

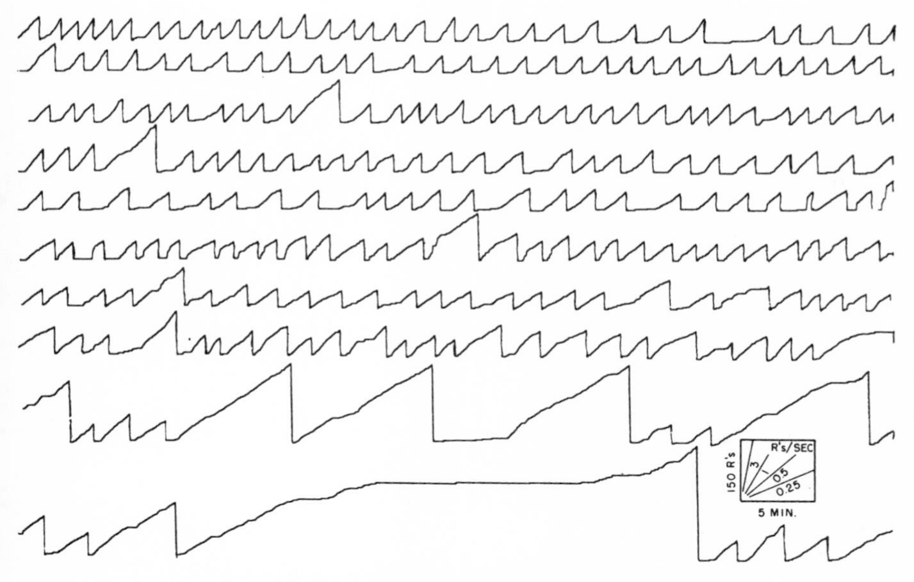

Fig. 617. Complete daily session on FR with pacing

When VIpacing was again restored after FRpacing, the earlier performance was, in general, recovered. However, the pausing after reinforcement persists and is usually to some extent compensated for by a short period of rapid responding. Figure 618 shows a 3rd session on VI 1 pacing (3) 1.1-1.4 after the long series of experiments just described on FRpacing. Note the consistent pause after reinforcement and, in most cases, a compensatory run which restores the curve to approximately the extrapolation of the preceding segments.

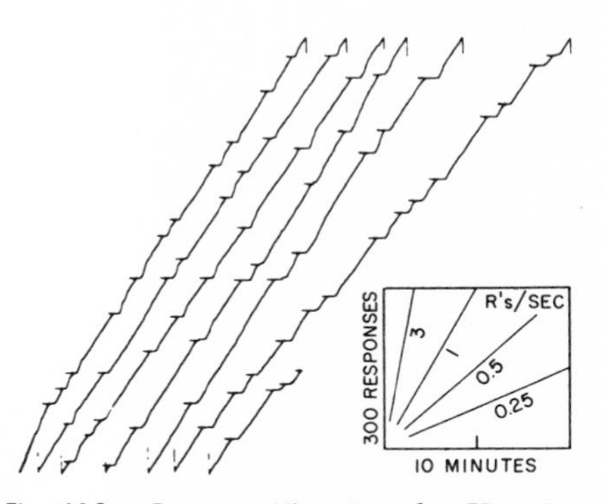

Fig. 618. Return to VI pacing after FR pacing

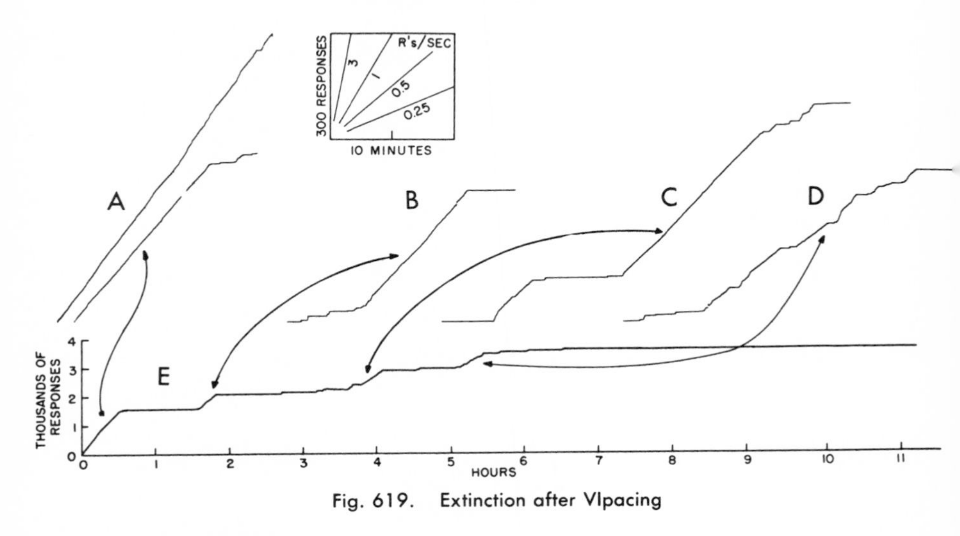

Fig. 619. Extinction after VIpacing

Extinction after VIpacing

Figure 619E shows an extinction curve taken after a well-established VIpacing. The principal portions of the curve during which the bird was responding are shown at A, B, C, and D for the segments indicated. If responding occurs at all, it occurs at approximately the earlier paced rates. Segment D contains some interesting pauses and compensatory increases in rate.

Chapter Ten

• • •

MULTIPLE SCHEDULES

INTRODUCTION

A MULTIPLE SCHEDULE consists of two or more alternating schedules of reinforcement with a different stimulus present during each. The schedules may alternate simply or at random. Schedule changes are usually made after reinforcement, although this condition is not essential. The fact that performances under the various components of the multiple schedule are usually quite independent of each other provides a technique for arranging control performances within a single subject and a single session. For example, a variable-interval schedule as one component of a multiple schedule may be used as an index of motivational conditions whose effect upon the other component schedules is being studied. As already noted, a multiple schedule is designated with the word "mult" followed by designations of the schedules—for example, mult FR 50 FI 10.

MULTIPLE FIXED-INTERVAL FIXED-RATIO SCHEDULES

Development and final performance

Development after crf. A multiple schedule may be put in force immediately after crf. Figure 620 shows parts of the first 3 sessions of the transition from crf to mult FI 10 FR 20. Record A shows the first 3 hours after crf. The first 2 reinforcements at *a* and *b* happened to occur on the FI 10 schedule while the key was red. They re-

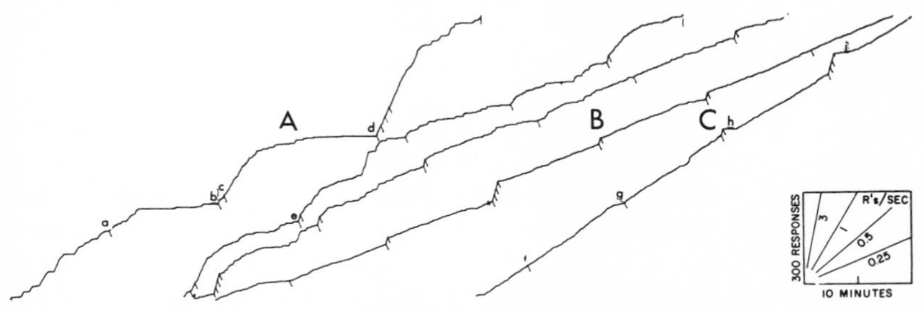

Fig. 620. First three sessions on mult FI 10 FR 20 after crf

flect mainly the extinction of the crf, during which the key had been both red and green from time to time. Immediately after the reinforcement at *b,* the key becomes green, and the 3rd reinforcement at *c* occurs on the FR 20 schedule. A negatively accelerated segment follows during the next FI 10 segment on a red key, and reinforcement occurs after a long pause at *d.* Three FR 20 segments then occur with the key green at the highest rate so far observed. The FI 10 segment which follows begins at the same high rate. Beginning with the fixed-ratio segment at *e,* all fixed ratios in the presence of the green key-color are run off at a fixed-ratio rate. Meanwhile, responding immediately after reinforcement declines progressively during the FI 10 segments, when the key is red. A corresponding increase in rate appears toward the end of the interval. By the end of Record A the rate in the presence of the red key is essentially constant throughout the interval segment.

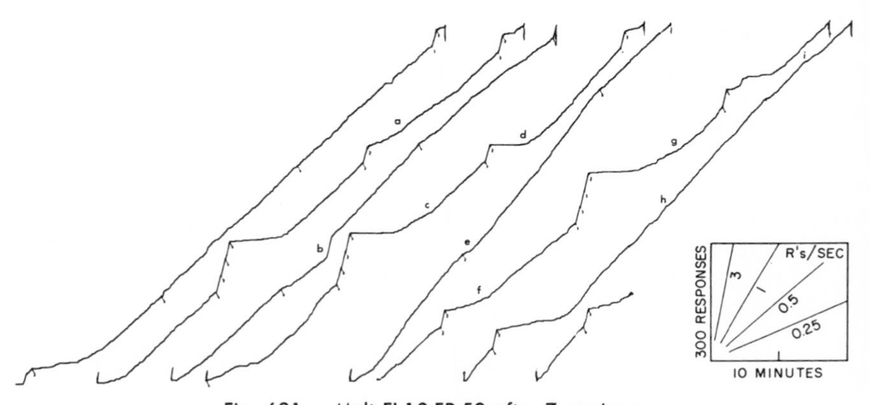

Fig. 621. Mult FI 10 FR 50 after 7 sessions

After $7\frac{1}{2}$ hours of exposure to the multiple schedule (Record B), a slightly lower rate immediately after reinforcement appears in the interval color. The rate is constant at 3 responses per second when the key is green. In Record C, after 13 hours of reinforcement on mult FI 10 FR 20, the 3rd session following crf, a marked pause follows reinforcement under the fixed-interval color if the preceding schedule was fixed-ratio, as at *h* and *i,* but not when 2 intervals follow each other, as at *f* and *g.* Note the small "bite" between *g* and *h.*

By the 7th session following crf, a fairly stable performance under the multiple schedule has developed, as seen in Fig. 621. The fixed ratio had been increased from 20 to 50 responses 7 hours earlier. A marked fixed-interval scallop occurs at *c* and *g,* where the preceding schedule is fixed-ratio; but at *e, h,* and *i,* where the preceding schedule is also fixed-interval, the rate is almost constant throughout the segment, except for a slight curvature immediately after reinforcement. In general, the extent of the fixed-interval scallop depends upon the number of preceding fixed-ratio segments. For example, the pause and scallop in the segments at *c* and *g*, following 3 ratios, are more pronounced than those at *a, d,* and *f*, following 1 ratio. At *b* a burst of approximately 50 responses occurs at the fixed-ratio rate in the middle of a fixed-interval segment in the presence of the fixed-interval color. Figure 622 shows a much later performance for the same bird on mult FI 10 FR 50, 131 hours, or 39 sessions, after crf. Pausing and acceleration now follow reinforcement, even in interval segments preceded by interval segments. Scalloping during fixed-interval segments preceded by fixed-ratio segments is more marked, however; and in many cases, as at *a* and *c*, a pause and low rate extend through most of the segment, the terminal rate not being reached. Scallops in fixed-interval segments preceded by fixed-interval segments vary from a brief period at a slightly lower rate of responding, as at *b*, to a substantial pause, as at *d*. The over-all rate has fallen progressively during the development of the performance on the multiple schedule, largely because of the increase in the extent of the pause and scallop in the fixed-interval segments.

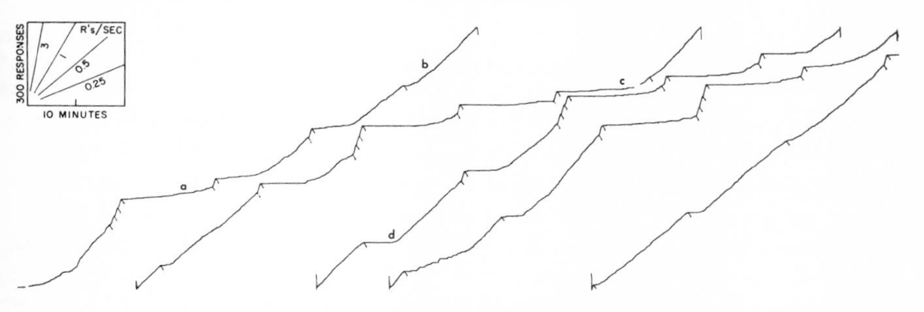

Fig. 622. Mult FI 10 FR 50, 39 sessions after crf

The second bird in the experiment showed a similar transition to a similar final performance.

Final performances were also recorded for 2 birds after an extended history on mult FI 5 FR at various fixed ratios. Figure 623A shows a segment after a history of 44 sessions of a multiple fixed-interval fixed-ratio, during which the performance was probed with TOs after reinforcement in several sessions; and in several other sessions the schedule was FI 5 only. In the performance in Fig. 623A the rate under the fixed-ratio schedule is 4.5 to 5 responses per second, compared with a terminal rate of 3.5 to 3.7 responses per second in the FI 5 schedule. The difference is by no means so great as in the previous experiment. Under the FI 5 reinforcement the pause and acceleration after reinforcement are the same whether or not the preceding segment was interval or ratio. Note the knee at *a*. The performance in Record B was recorded 21 sessions later, when the fixed ratio had been increased to 100. The extent of the pause under the FI 5 reinforcement has increased, and the terminal rate has fallen. The rate under the fixed-ratio schedules has increased. The FI rate at *b*, for example, is 2.5 responses per second, while the FR rate at *c* is 4.5 responses per sec-

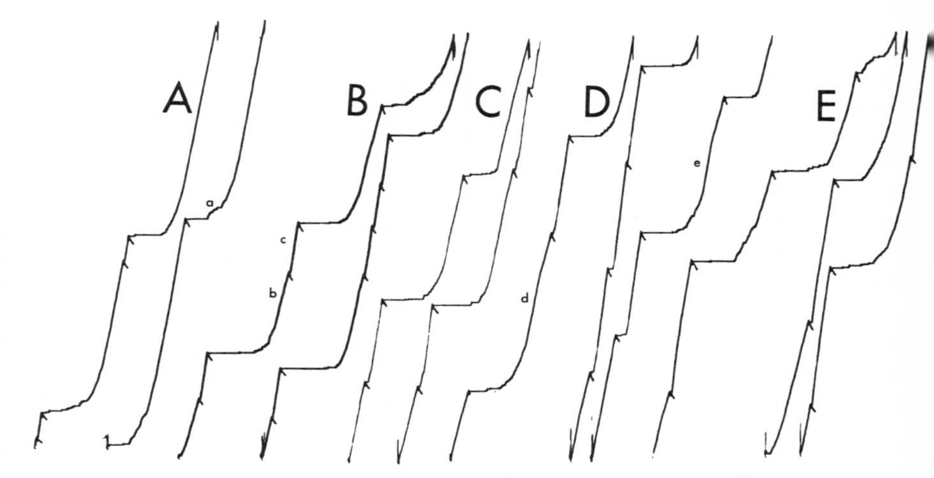

Fig. 623. Final mult FI 5 FR performance under various FRs

ond. In Record C, 7 sessions later, the fixed ratio has been increased to 175, and the performance remains similar to that in Record B. In Record D, 3 sessions later, the fixed ratio was increased to 225. Rates of approximately 4 responses per second now appear during the fixed-interval segments (at *d* and *e*), but fall to a lower value before reinforcement. A slight pause appears at the start of some ratio segments. In Record E, 5 sessions later, the fixed ratio has been increased to 300. The number of responses being emitted in the ratio and interval segments is roughly of the same order, although the rate of responding in the ratio segments is 4 responses per second, compared with rates of the order of 2.5 responses per second at the end of the interval segments. A second bird in the experiment showed a similar performance.

The fixed ratio was then increased to 375. Figure 624 shows the performance 15 sessions after Fig. 623. The multiple control is still being maintained, with terminal interval rates of approximately 2 per second and ratio rates of more than 4 per second.

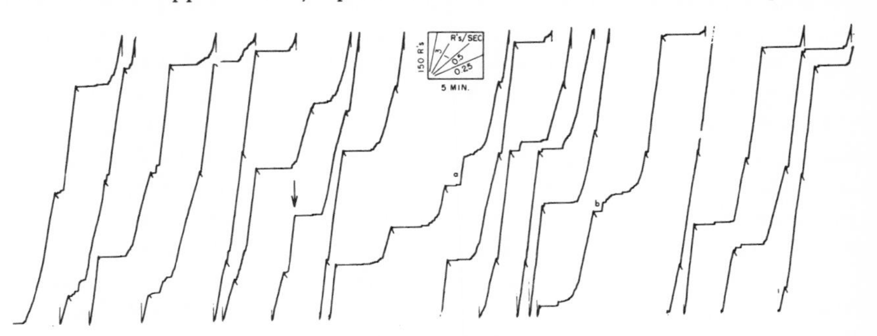

Fig. 624. Early performance on mult FI 5 FR 375

Nearly all fixed-interval segments begin with a substantial pause and contain some acceleration to the terminal rate. Pauses at the start of the fixed-ratio segments are either absent or only a few seconds long. Some signs of loss of multiple control begin to appear at *a* and *b*. Here, after a short pause in the *interval* color, responding begins at the *ratio* rate before it shifts to a lower rate and resumes the normal interval pattern. An extreme example is marked by the arrow. The transitions from the pause to the terminal rate in the interval segments are becoming more abrupt.

Reinforcement was continued on mult FR 375 FI 5 for slightly more than a month, after which the performance shown in Fig. 625 developed. The terminal rate in the interval has increased so that all segments are approximately of the same length and it is difficult to differentiate between ratio and interval segments. Ratio segments have been marked by dots at the *start* of the segment. Performances in some segments are

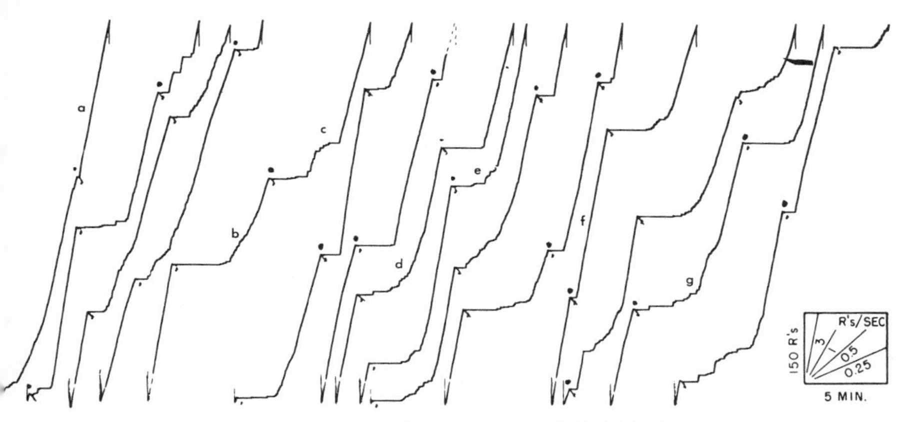

Fig. 625. Later performance on mult FI 5 FR 375

clear-cut, as, for example, the ratio segments at *a* and *f* and interval segments at *b* and *d*. On the other hand, some ratio segments, as at *c, e,* and *g,* resemble interval segments.

A very different set of values was chosen in another experiment in which the ratio was low (26) and the interval long (18 minutes). Figure 626 shows a final performance after a history of 357 hours of reinforcement on mult FIFR with occasional intramuscular injections of sodium pentobarbital. As in some of the previous figures, extended scallops occur in interval segments preceded by ratio performances (as at *a, c,* and *d*), while interval segments preceded by other fixed-interval performances, as at *e* and *f,* show a constant rate of responding throughout the 18-minute segment. The terminal ratio rate is of the order of 5 responses per second; the maximal rate during a fixed-interval segment is 0.73 response per second.

Development of mult FI 10 FR 70 after extended FR 70. Two birds, whose later performances will be described in Chapter Fourteen under "Multiple Chains," developed

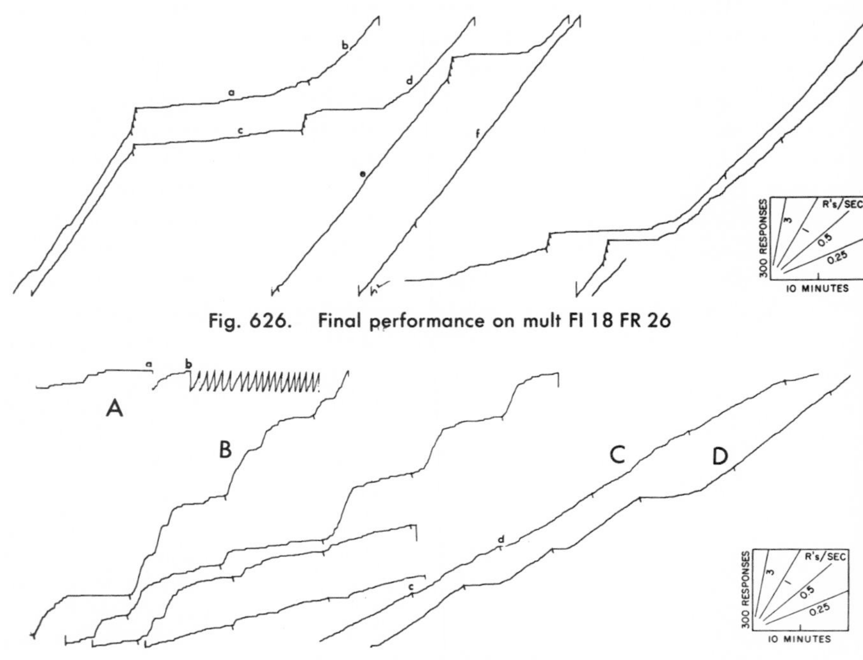

Fig. 626. Final performance on mult FI 18 FR 26

Fig. 627. Transition from FR 70 to mult FI 10 FR 70

a mult FI 10 FR 70 performance after previous exposure to FR 70 alone. Figure 627 shows the 1st transition to the multiple schedule. The 2 components of the multiple schedule are recorded separately. In Record A the session begins under the fixed ratio in the presence of a red key. The key-color is novel and has the usual effect. The bird takes more than 10 minutes to emit the ratio and receive the reinforcement at *a*. The 2nd ratio segment ending at *b* was programmed next. It is negatively accelerated. The key then changed to blue, the color present during the earlier FR 70, and the 1st interval segment in Record B was recorded. The 3rd ratio segment was then executed. It shows positive acceleration to a high terminal rate. The ratio rate increases progressively for the remainder of the session, with an improvement in grain. The fixed-interval performances interspersed with these fixed-ratio performances are negatively accelerated, as in a transition from crf, but the curves show some effect of the previous fixed-ratio reinforcement. Record C contains a segment of the interval performance from the 2nd session on the multiple schedule, after a total exposure of 6 hours. Here, the performance is roughly linear, with some slight pause or lower rate of responding following reinforcement, as at *c* and *d,* where fixed-ratio segments preceded. The 3rd session, exemplified by Record D, after 10 hours of reinforcement on the multiple schedule, shows marked scalloping whenever the interval segments are preceded by ratio segments, and an almost constant rate otherwise, as at the last 2 reinforcements in the record. Meanwhile, the fixed-ratio performances remain similar to the last segments of Record A.

Mult FIFR in the rat. Three rats gave a similar performance on mult FI 5 FR 20. The previous history included mult crf extinction FI 10 (to establish a stimulus for TO), FR 20, and FR 30. In the mult FIFR the component schedules appeared in simple alternation. The general illumination flashed on and off once per second when the schedule was fixed-ratio, and remained steady when the schedule was fixed-interval. The results must be qualified by the inadequate amount of food used as reinforcement. As already noted, it was discovered later that more stable performances would result with 0.1 to 1.5 grams per reinforcement instead of 0.05 gram, as in this experiment.

Figure 628 contains segments from the first 38 hours on mult FI 5 FR 20, during which a fairly stable performance developed. The first exposure to the multiple schedule (Record A) shows the effect of the preceding fixed-ratio reinforcement, which becomes less pronounced as the session progresses. By the end of this 1st session (Record B) clear evidence exists of multiple control. Responding begins immediately after reinforcement during the flashing light when the schedule is fixed-ratio; and a pause with subsequent acceleration after reinforcement occurs when illumination is steady and the schedule is fixed-interval. In Record C, after 11 hours of reinforcement on the multiple schedule, the multiple control is clear. The flashing light controls a rate of approximately 2.5 responses per second under the fixed-ratio schedule. The steady light produces a pause of up to 2.5 minutes at the beginning of the interval and a terminal rate of 1.5 responses per second under the interval schedule. Occasionally, as at *a*, the rat responds ratio-wise immediately after reinforcement, but then pauses and accelerates to the terminal interval rate. Toward the end of the session the rates during the fixed-interval stimulus fall to low values, as at *b* and *c*, where responding does not begin until near the end of the segment and the terminal rate is not reached. Record

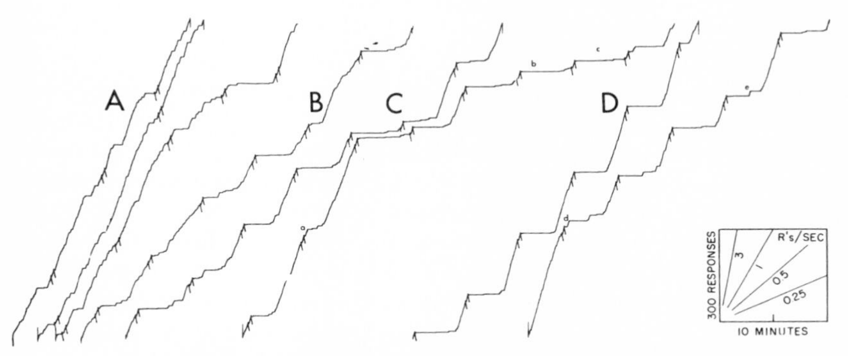

Fig. 628. Segments from the first 38 hr of mult FI 5 FR 20 (rat)

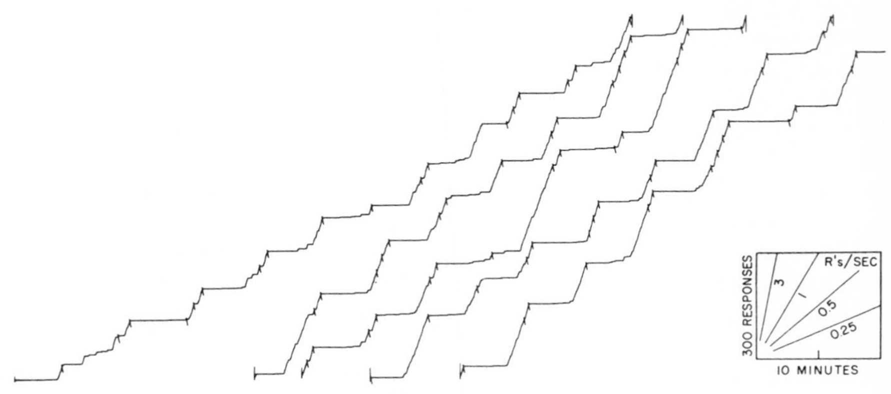

Fig. 629. Later performance on mult FI 5 FR 40 (rat)

D shows a late stage of development, after a total of 38 hours on the multiple schedule. The terminal rate under the fixed-ratio schedule is of the order of 3 responses per second, compared with slightly over 2 responses per second at the end of the interval segments. The fixed-interval performances are uniform from segment to segment with a substantial pause extending through most of the interval and a fairly abrupt shift to the terminal rate. Instances occur at *d* and *e* where the ratio rate appears briefly during the interval segment.

Figure 629 shows a much later performance for this rat on mult FI 5 FR 40.

A second rat in the experiment showed roughly the same development of a multiple performance except during the early stages, where the grain was very rough. A well-marked multiple control developed in the 2nd session, and later performances resemble that of Fig. 628C or 628D. A third rat showed a similar course of development except that a very low over-all rate developed after the 3rd session, and long pauses appeared, particularly at the beginning of the fixed-ratio segments. Later, the performance returned to the patterns of Fig. 628D.

Extinction after multiple FIFR

Extinction after a multiple schedule may be carried out in a number of ways, including the following:

(1) The response is extinguished in the presence of the stimulus correlated with one schedule, the multiple schedule is re-established, and the response is then extinguished in the presence of the stimulus correlated with the other schedule.

(2) The response is extinguished in the presence of the stimulus correlated with one schedule, and then in the presence of the stimulus correlated with the other schedule.

(3) Extinction is carried out while the stimuli are rotated. For example, after the number of responses in the fixed ratio has been emitted, the key changes to the color associated with the fixed-interval reinforcement; this key-color is present for the dura-

tion of the fixed interval. The color then changes if the next schedule would have been a ratio, or remains the same if another interval occurs in the series.

(4) Reinforcement may be continued on one schedule, while extinction is carried out in the presence of the second stimulus appropriate to the other schedule. The last alternative is technically either mult FRext or mult FIext.

Several types of extinction curves were produced by a bird which had had an extensive history on mixed fixed-interval fixed-ratio schedules (see Chapter Eleven), and then on a multiple schedule in which the stimulus correlated with the component schedules was either a steady or flashing green key-light. Figure 630 shows a stable performance on the mult FI 10 FR 125 schedule in the portion of the record before the arrow. Fixed-ratio and fixed-interval schedules alternate with each other, and the performance is stable and similar to those which have already been described. At the arrow the reinforcement was omitted, and the key-light was simply changed to the pattern correlated with the fixed-ratio schedule. For the remainder of the session the hatches indicate changes in key-light pattern rather than reinforcements. The fact that the resulting extinction curve shows marked deviations from the previous multiple FIFR performance indicates that the multiple performance depended upon the reinforcement as much as on the stimuli on the key. The rate does not drop to zero when the stimulus correlated with the fixed-interval schedule appears at *a*. Instead, the rate of responding from the previous fixed-ratio segment is maintained for a short time, after which it falls off abruptly to an intermediate value which then increases slightly as a normal fixed-interval segment. In the "interval" segment at *d* the rate continues to fall throughout. Fixed-ratio performances also change markedly with the omission of reinforcement. At *b* the segment begins at the rate from the previous interval performance, which is maintained for half of the ratio before the rate increases; and in the segment at *c* an intermediate rate is maintained throughout.

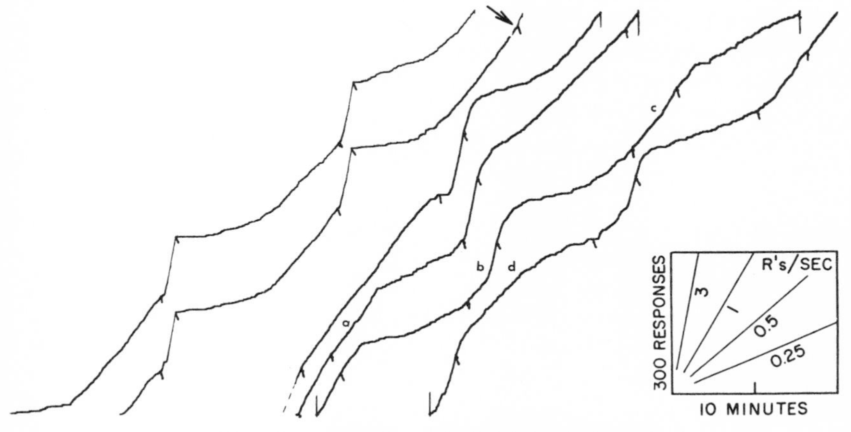

Fig. 630. Extinction after mult FIFR (stimuli alternating)

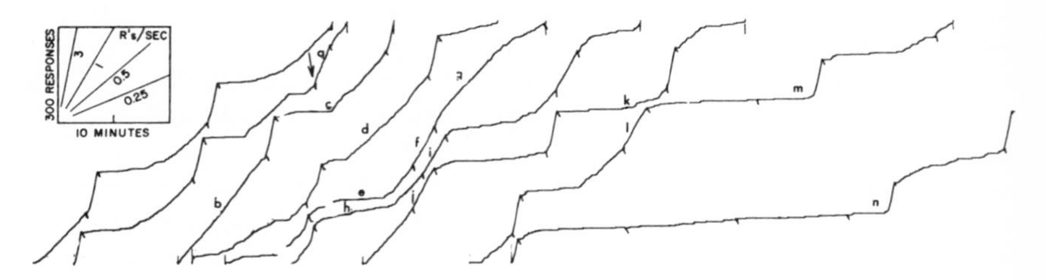

Fig. 631. Transition from mult FIFR to mult ext FR

Two sessions after the extinction curve in Fig. 630, we carried out partial extinction of the multiple schedule by discontinuing reinforcements in only the fixed-interval color. The session shown in Fig. 631 begins with a standard multiple performance as in Fig. 630. Beginning at the arrow, reinforcements at the end of the 10-minute, fixed-interval period were omitted. The stimulus simply changed to the other pattern at the end of the designated interval. Immediately, the rate during the 1st ratio segment (a) falls below the usual value, with rough grain. The next fixed-interval segment (*b*) shows a constant intermediate rate throughout. For the remainder of the session the over-all rate of responding in the presence of the fixed-interval stimulus falls off progressively. The normal character of the fixed-interval performance is sometimes preserved, as at *c, d,* and *k.* Examples are present, however, of the pattern disintegrating, as at *e* and *h,* where substantial responding follows reinforcement before the rate falls and accelerates to the terminal value. At *g* the initial rate in the interval segment is almost equal to the fixed-ratio rate, and it declines continuously during the remainder of the segment. The pattern of responding is disrupted, even though each performance is preceded by the reinforcement of the preceding fixed ratio. The fixed-ratio performance is also affected by the extinction. Segments at *f, i,* and *j* are run off at the terminal interval rate. Toward the end of the session, along with fixed ratios at lower rates of responding, as at *l,* a pause may follow reinforcement, as at *m* and *n,* although the terminal rate is normal. The ability to sustain a large ratio in a multiple schedule may be due in part to induction from the fixed-interval reinforcements. Although extinction in the presence of the fixed-interval stimulus is not complete by the

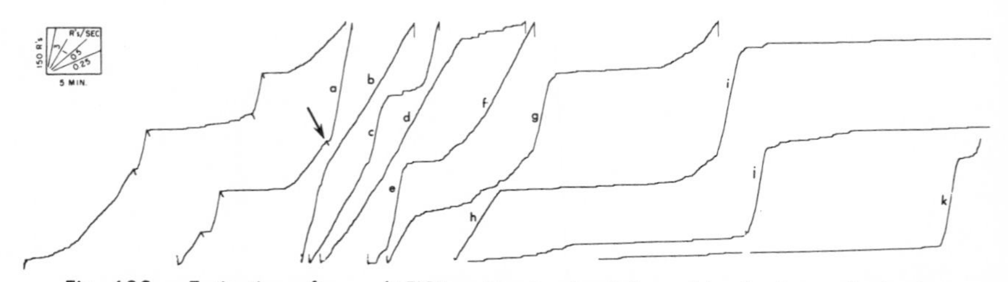

Fig. 632. Extinction after mult FIFR; ratio stimulus followed by the interval stimulus

end of the session, it has reached a low value and the ratio performance begins to "strain."

Five sessions after Fig. 631, a third type of extinction was carried out, in which the stimulus previously appropriate to the fixed-ratio schedule of reinforcement was present continuously for the first part of the session, and the stimulus previously correlated with the fixed-interval schedule of reinforcement was present during the latter part. This session, shown in Fig. 632 and 633, begins with multiple FI 10 FR 125; and the performance before the arrow in Fig. 632 is typical of this schedule. Following the reinforcement at the arrow, the magazine was disconnected. The stimulus remained appropriate to the ratio component. The extinction curve which follows shows responding at the earlier high fixed-ratio rate, as at *a, c, e, g, i, j,* and *k.* It also shows periods of zero or near-zero rate. Sustained responding also occurs, however, appro-

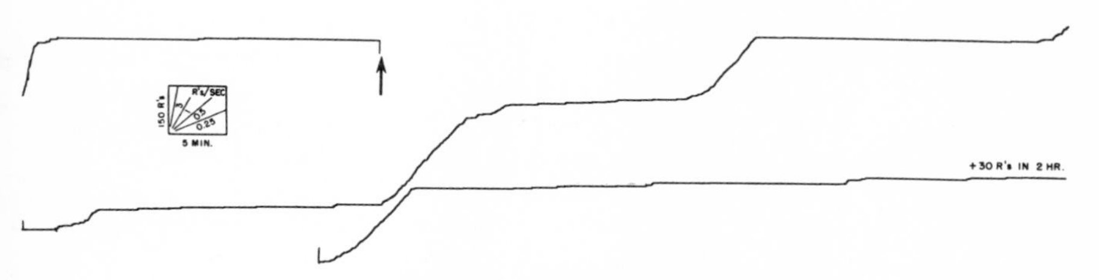

Fig. 633. Continuation of Fig. 632

priate to fixed-interval reinforcement, as at *b, d, f,* and *h.* The transitions from one rate to another are relatively abrupt.

The same extinction session is continued in Fig. 633. At the arrow the stimulus has changed to the pattern previously correlated with fixed-interval reinforcement. Extinction then continued for $4\frac{1}{2}$ more hours, during which approximately 1000 responses were emitted. The curve is composed largely of 3 wave-like changes, each resembling a fixed-interval segment with a fairly smooth acceleration to a terminal rate of approximately 0.75 response per second. The small number of responses in this part of the curve indicates that the previous extinction in the presence of the fixed-ratio stimulus has affected the amount of behavior available in the presence of the fixed-interval stimulus.

Another bird was extinguished in the presence of the fixed-ratio stimulus after prior extinction in the presence of the fixed-interval stimulus, after 191 sessions on various values of mult FI 5 FR. The complete extinction curve is presented in Fig. 634C in reduced form. The recorded form is shown at A and B. The extinction in the presence of the fixed-interval stimulus (Record B) shows various intermediate rates as well as 1 period of very rapid responding for more than 2000 responses. The 2nd extinction curve, in the presence of the fixed-ratio stimulus, consists of 2 large segments at the

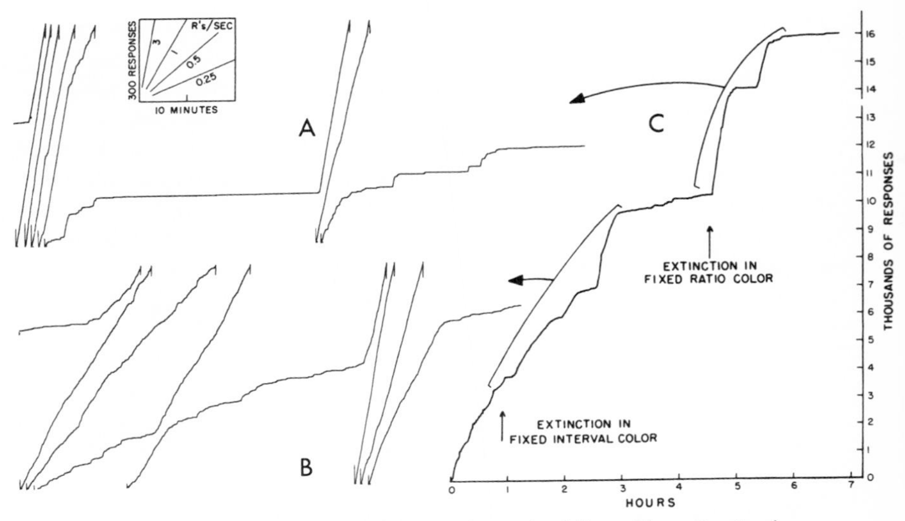

Fig. 634. Ext after mult FIFR: interval stimulus followed by ratio stimulus

ratio rate. The decline to a near-zero rate is not immediate, but the high fixed-ratio rate of responding is abruptly resumed.

Increasing the over-all rates on a multiple schedule by reinforcement of the separate members

Figure 635A shows a given stage in the development of a mult FI 10 FR 65 for a bird whose performance has already been considered (Fig. 620, 622, 631, 632, and 633). The scalloping in the interval is well-marked, except that intervals which follow intervals show no strong tendency to scallop (*b* and *c*). Meanwhile, the ratio performance is

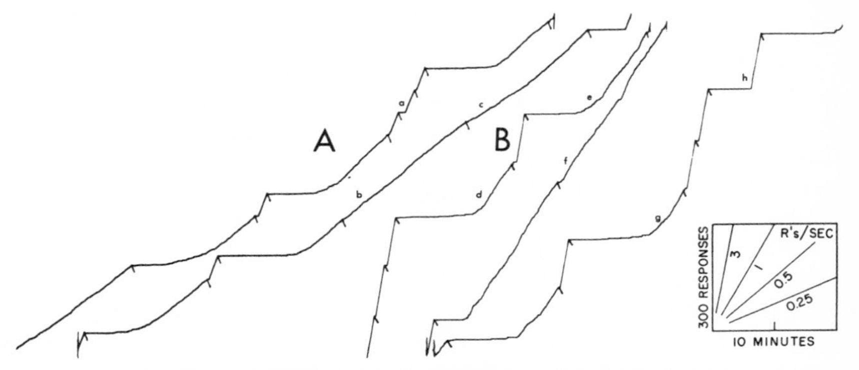

Fig. 635. Increase in over-all rate by separate reinforcement of components of a multiple schedule

irregular, showing pauses, as at *a,* and an unusually low terminal rate. We tried to raise the rate by reinforcing continuously on the ratio schedule and color (abandoning the multiple schedule temporarily) and by advancing the size of the ratio. Other sessions contained only reinforcement of the interval schedule under the appropriate color. This improved the multiple performance, as Record B in Fig. 635 shows. The ratio is now 130 rather than 65, and although some pausing occurs, as at *h,* the rate is now above 3 per second. The terminal rate in the interval is also higher than that in Record A. The pausing in the interval is also more pronounced, so that the scalloping is more marked (*d, e,* and *g*). The bird is still not scalloping, however, when one interval follows another (*f*).

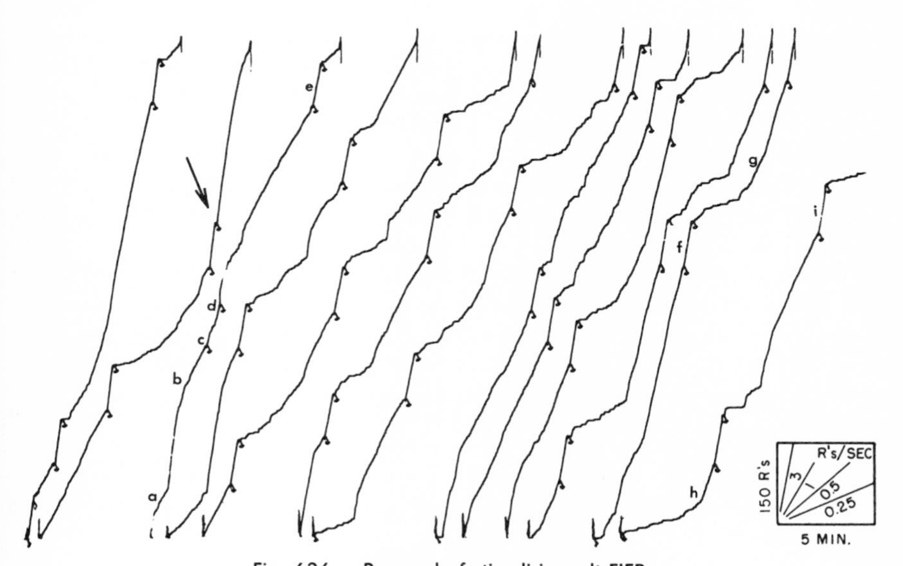

Fig. 636. Reversal of stimuli in mult FIFR

This "treatment" of a weak FR on a multiple schedule builds up the ratio in the absence of contrary inductive processes from a multiple FI schedule. There appears to be a reverse influence upon the interval performance in the form of a higher terminal rate after the ratio has been so built up.

Reversal of stimuli

The performance before the arrow in Fig. 636 developed after 12 hours of reinforcement on mult FI 5 FR 80, with a blue light on the key during the FI 5 reinforcement and a red light on the key during the FR 80 reinforcement. At the arrow the stimuli in the multiple schedule were reversed. The stimulus which had been present during the fixed-interval reinforcement now occurred when the schedule of reinforcement was fixed-ratio, and vice versa. The first stimulus after the reversal was the former fixed-ratio color, now present during fixed-interval reinforcement. The fixed-ratio rate of responding is maintained for several hundred responses, with a slight decline in the rate at *a* and again at *b* just before reinforcement. At the end of the FI segment at *c* the schedule of reinforcement was changed to fixed-ratio, and the stimulus on the key changed to the former interval color. The fixed-interval segment following the reinforcement at *d* began with a brief period of responding at the fixed-ratio rate, followed by a shift to an intermediate rate for the remainder of the interval. A high rate developed almost immediately in the presence of a new fixed-ratio stimulus (*e*); and responding decreased progressively during the fixed-interval segment, as the effect of the previous fixed-ratio reinforcement wore off and the fixed-interval reinforcements took effect. Before the end of the session the new stimuli were in almost complete control of the fixed-interval and fixed-ratio performances. The fixed ratios were now being run off at a constant high rate of responding, as at *f* and *i*, and many of the fixed-interval segments showed smooth and extended curvature, as at *g* and *h*.

Reinforcement was continued on the reversed multiple schedule for a total of 33 hours. Figure 637A shows a well-developed performance. A high rate prevails in the presence of the fixed-ratio stimulus, and a marked scallop and acceleration to a lower terminal rate in the presence of the fixed-interval stimulus. Scalloping is less marked in those intervals not preceded by fixed-ratio segments, as at *a*. At the start of the next session (Record B) the stimuli in the multiple schedule were again reversed, returning to conditions before the reversal in Fig. 636. The 2nd transition took place much more quickly and resulted in a more orderly performance. A normal fixed-ratio performance emerged beginning at *b*, with a rate of more than 5 responses per sec-

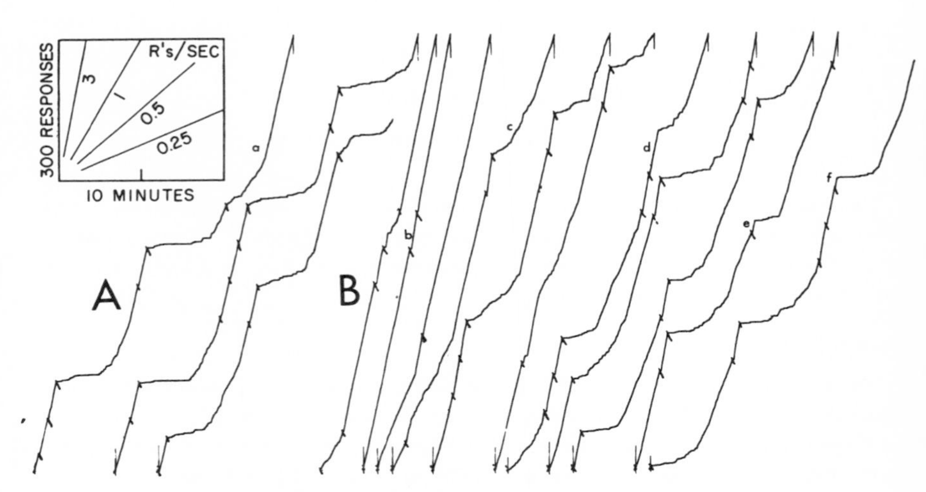

Fig. 637. Second reversal of stimuli in mult FIFR

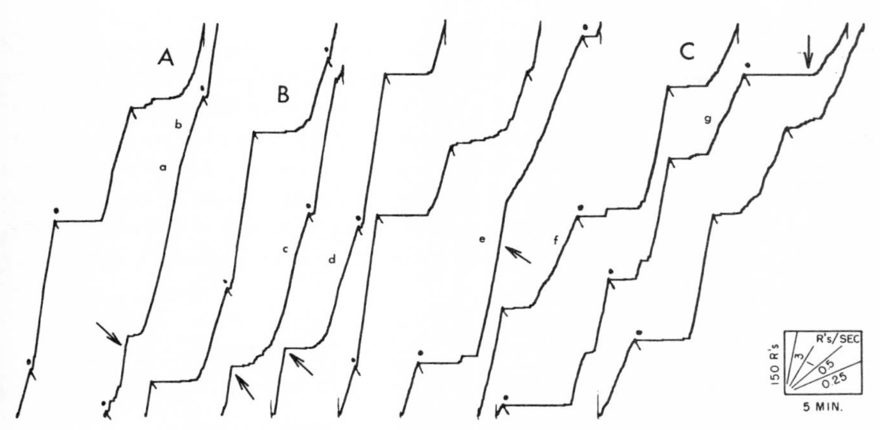

Fig. 638. Sudden changes from the ratio to interval stimulus

ond. The reversed fixed-ratio rate was maintained through most of the first 6 interval segments, but at *c* a scallop began to form. Curvature developed progressively toward the end of the session. An effect of the previous fixed-ratio reinforcement survived in several fixed-interval segments which began at the ratio rate at *d, e,* and *f.*

The effect of a sudden change of stimuli on a multiple FIFR schedule

In the experiments on multiple FIFR already described the *interval* color was occasionally changed when the bird was in the midst of a *ratio* performance. Figure 638 shows some typical effects. The beginning of the ratio segments are marked by the dot over the reinforcement. In Record A the arrow indicates the change to the FI color after a rather rough ratio performance. The ratio rate continues with only a slight drop for about 30 responses. The rate then breaks; after a pause, it accelerates to essentially the ratio value at *a.* It then drops to the interval rate at *b.* Two other examples in which the ratio rate holds for a short time in the interval color occurred in the 1st session of such probing.

Thereafter, the change was immediate, as Curve B shows. At each arrow a change was made from the ratio to the interval color. An interval scallop followed immediately. Both of these interval segments show a slight drop to a lower rate before the interval is completed (at *c* and *d*).

At *e* the ratio rate continues for almost 100 responses before the bird shifts abruptly to the interval rate, as at *f* and *g.* In Record C the interval color was also introduced at the arrow during the pause at the beginning of a ratio. Here, the interval terminal rate was not reached quite so abruptly.

Figure 639 shows a change for another bird together with an additional effect. The record begins with 2 interval performances (*a* and *b*) and then changes to a ratio. At the first arrow the key was changed to the interval color. There was an immediate drop to the terminal interval rate, which continued until the reinforcement at the second arrow. Thereafter, the ratio color prevailed and extinction was permitted to take place. This procedure gives some indication of what would have happened at the first arrow if the bird had been permitted to continue to respond under the ratio color but without reinforcement (as was the case because of the change to the interval performance). The figure also supplies an example of extinction in the ratio color of a multiple FIFR schedule. In general, the major responding occurs at the ratio terminal rate, although some negative curvature is present as that rate falls off at *c* and *e*, executed mainly by a roughness in grain. The return to the high rate is usually very abrupt, as at *d* and *f*.

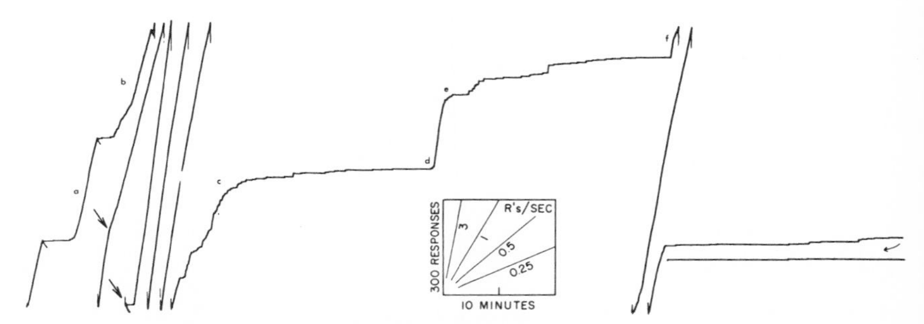

Fig. 639. Sudden stimulus changes followed by extinction

Change to the FI stimulus during a strained ratio on a multiple FIFR

A pigeon which had difficulty in holding mult FI 5 FR 275 provided an opportunity for testing the effectiveness of the interval color during prolonged pauses on the FR component. Figure 640 shows an example, together with the current state of the performance. Two interval segments with a moderate terminal rate are followed by a "strained" ratio at *a*. In the next interval only about a hundred responses occur before reinforcement. Another example of this occurs later at *d*. A second ratio shows a long pause at *b*. At the first arrow the color of the key was changed to that appropriate to the FI schedule. Responding began immediately and accelerated to the terminal interval rate. At the second arrow the ratio color was replaced. An exceptionally long pause followed before the ratio was completed at *c*. Evidently the conditions producing the pause at the beginning of the ratio segment clearly involve the stimuli appropriate to the ratio performance and are not due to emotional or motivational conditions or fatigue.

Figure 641 shows other examples of this effect. Record A is a single ratio period between reinforcements in which a momentary change was made to the interval color at one point. Responding began immediately and accelerated to the interval terminal rate. In a similar example at Record B acceleration was slower and produced only

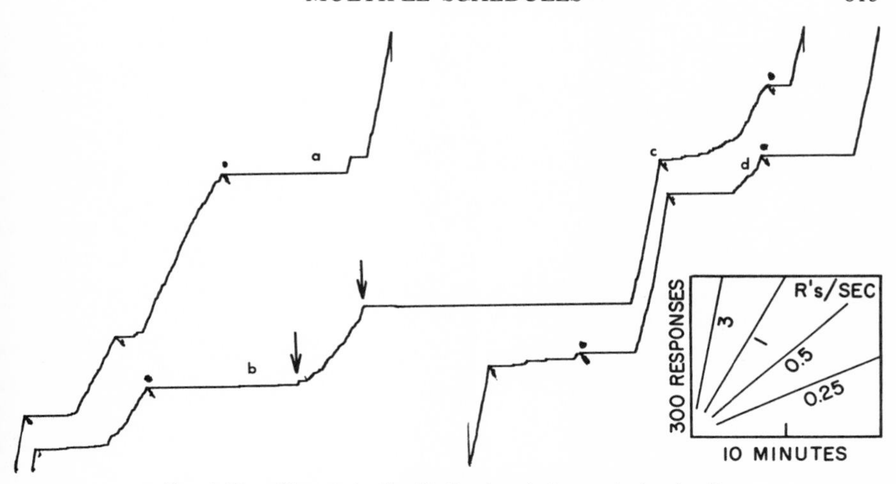

Fig. 640. Change to the FI stimulus during a strained ratio

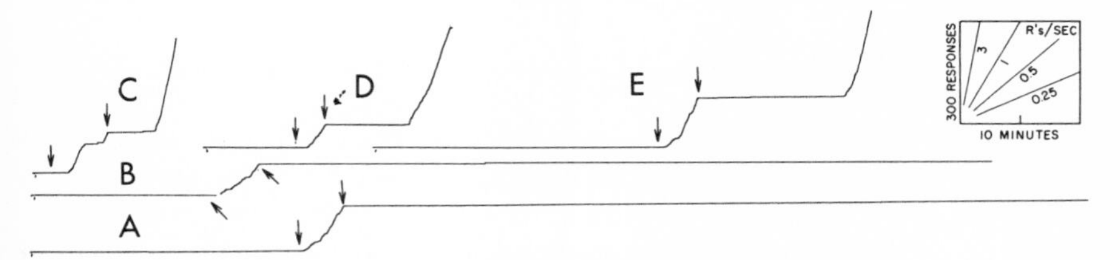

Fig. 641. Change to the FI stimulus during a strained ratio

a moderate rate before the ratio color was restored. Other examples of the same procedure appear in Records C, D, and E.

Figure 642 illustrates a further example of the effectiveness of the interval color while the ratio is badly strained. The session began (top line) with an interval performance, followed by the ratio color in which a few responses were emitted in the region of *a*. A pause of about 80 minutes then occurred. At *b* the interval color was introduced, and a fairly typical performance followed and was reinforced at *c*. A return to

Fig. 642. Change to the FI stimulus during a strained ratio

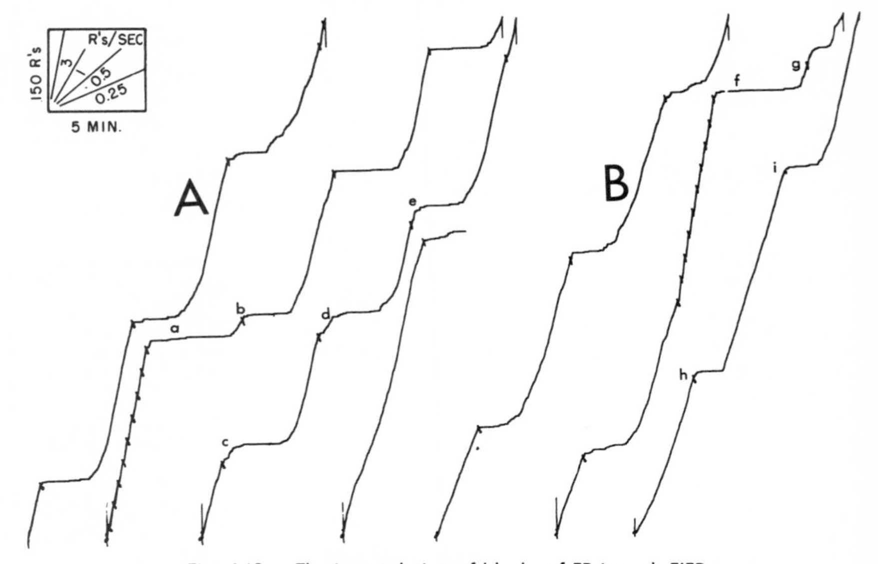

Fig. 643. The interpolation of blocks of FR in mult FIFR

the ratio color then produced another pause of about 40 minutes, at the end of which a change to the interval color again produced an interval performance with reinforcement at *d*. The color then remained appropriate to an interval, and another reinforcement was received at *e*. A change to the ratio color was then followed by a pause of about 45 minutes. As the session continued (not shown in the figure), the color was changed to that appropriate to the interval and the first response was reinforced. The response occurred within a few seconds of the change. A return to the ratio color produced an 83-minute pause; at the end of this the interval color was restored, and a response occurred in a few seconds and was reinforced. A return to the ratio color produced another pause of 35 minutes when the session was ended. The only responding in the ratio color throughout the entire session occurred at *a*. The interval color, however, produced appropriate interval performances whenever introduced.

The interpolation of blocks of FR in mult FIFR

The interrelation of component members in a mult FIFR is most obvious in the effect upon the interval scallop of a preceding ratio performance. We examined this relationship by inserting blocks of ratios in an FI schedule after a history of mult FIFR.

One bird with a substantial history of mult FI 5 FR 40 was put on FI 5, and a block of 9 FR reinforcements under the appropriate former FR color was interpolated in each of 2 sessions. Figure 643 shows the results. In both sessions the interval following the interpolation contains an exceptionally long pause (*a* and *f*) and, consequently, only a small number of responses before reinforcement (*b* and *g*). Although some evidence exists of running-through after reinforcement on this schedule before interpolation of FR under stimulus control, running-through becomes very conspicuous immediately after the interpolation. Slight running-through occurs at *a* and *b*, and 3 very marked examples occur later at *c*, *d*, and *e*. On the 2nd day, there is running-through at *f*, a conspicuous case at *g*, and slight examples at *h* and *i*. An acceleration to a brief run at the ratio rate appears following *g*.

FI after mult FIFR

One characteristic of mult FIFR is that the FI performances are smoothly scalloped and give the appearance of FI with TO after reinforcement. The FR components serve as TOs for the interval schedule. We checked on this by changing to straight FI 5 after mult FI 5 FR 40. Figure 644 shows the entire session. The previous history included 68 hours of mult FI 5 FR 40-80. In the figure the last ratio appears at *a*. At this time the interval shows smooth scalloping. Beginning at the arrow the schedule was FI 5. Fairly consistent scalloping continues for some time. At *b*, however, a small run-through appears, and at *c*, a well-marked knee. The pause at *d* is brief and is followed by a very long run at essentially the ratio rate at *e*. Later, an anomalous linear low rate occurs at *f*, and the following reinforcement is followed by a brief "priming" run of a ratio-like character at *g*. At *h* there is a complete running-through, characteristic of FI without TO. Running-through is common thereafter. Some instances are well-marked, as at *i*, and extremely rapid following short pauses, as at *j*. Another immediate assumption of the terminal rate occurs at *k*, a well-marked run-through followed by an interval curvature at *l*, and a run-through with negative acceleration marked by a bite, as at *m*. Some of this behavior may be attributed to the earlier ratio component of the multiple schedule, and suggests priming in a mixed FIFR. (See Chapter Eleven.) Other parts of it suggest FI without TO, and show clearly enough that the FR member of the multiple schedule was functioning as a TO with respect to the interval member.

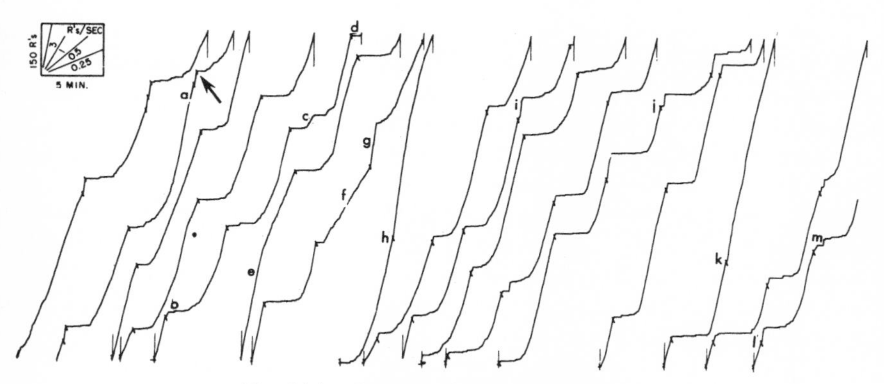

Fig. 644. Transition from mult FIFR to FI

The possibility should not be overlooked that the high rate and constant count at reinforcement in the ratio has an effect in suppressing the behavior at the beginning of the interval member. Thus, interval scalloping on mult FIFR is probably more pronounced and begins with longer pauses than on FITO.

The disruptive effect of TO 8 on FI 5. Record A of Fig. 645 shows a performance a few sessions after that in Fig. 644 on FI 5. A substantial scallop occurs after a pause or slight running-through after reinforcement. At the start of the following session (Record B), an 8-minute TO followed each reinforcement for the first time in the bird's history. The record begins with a TO at *e*, during which some responding occurs. A great deal of responding occurs during the TO following the next reinforcement at *f*. A high rate is maintained throughout the first 2 segments, and the pause and scallop following the reinforcement are completely eliminated. Slower responding follows the reinforcement at *g* for a short time, but this lower rate is not consistent until *h*. Thereafter, a lower rate of responding during the first part of the interval develops pro-

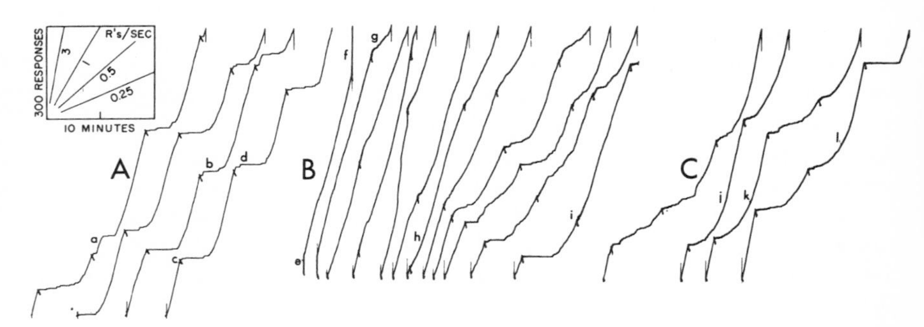

Fig. 645. The effect of TO after reinforcement on FI after mult FIFR

gressively except at *i*, where a high rate of responding is sustained throughout the interval. Record C presents a segment of the performance after 11 hours of further exposure to the TO. Occasional scallops are quite smooth, with some pause following the reinforcement, as at *j*, *k*, and *l*. But many cases of rough grain and irregular changes in rate are also present.

Figure 646 shows a performance after 37 hours with TO after reinforcement. A consistent scallop develops progressively during the session after a high rate was sustained throughout the 1st interval. The pause after reinforcement is brief or completely absent. In the following session the TO was removed. Record B shows the performance. The effect of removing the TO is not nearly so severe as the first introduction of the TO. Smooth FI performances soon occur, although responding may occur immediately after the reinforcement, and the terminal rate may be sustained throughout the interval.

The extreme disruption of the FI performance by TO appears to be due to the history of mult FIFR, since none of the other cases of FITO produced so radical a change.

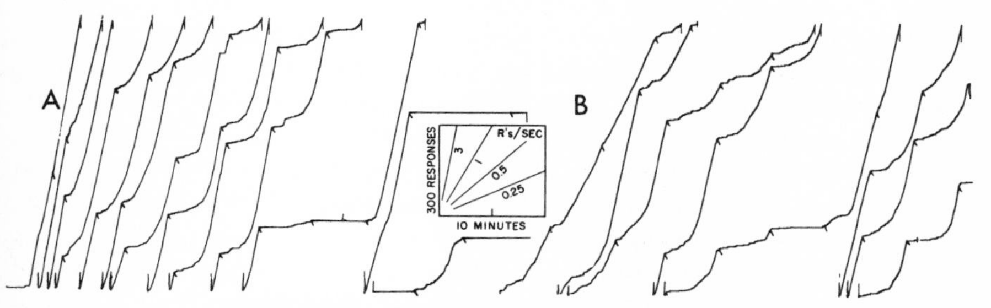

Fig. 646. Transition from FITO to FI after mult FIFR

This is further evidence for the view that the fixed-interval scallop in the multiple schedule is based on different factors from those in a pure fixed-interval schedule. Hence, although the performance in Fig. 645A resembles that on FI 5, the events which the bird uses in the fixed-interval performance are different from those in a simple FI 5 schedule.

The effect of an 8-minute TO on mult FIFR

Figure 647A shows a fairly stable performance on mult FI 10 FR 50. Very little scalloping occurs when one interval follows another (at *a* and *e*), except when the preceding interval has shown a low rate, as at *d*. The terminal ratio rate is higher than the interval rate, but some pausing or slow responding are evident at the start of the ratio, as at *b* and *c*. An 8-minute TO was then added immediately after each reinforcement in the following session (Record B). This began with a long interval at *f*, at the end of which a response was reinforced after only a small number of responses had been emitted. Some responding to the key occurs in the 8-minute TO

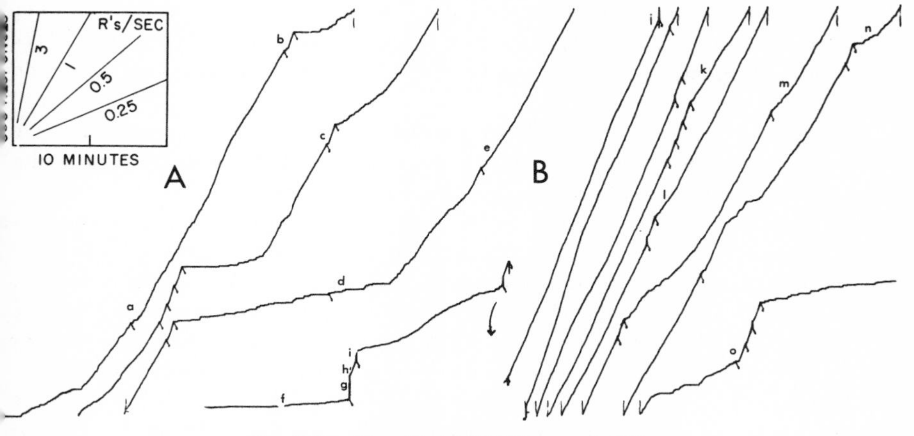

Fig. 647. The effect of TO 8 on mult FIFR

at *g,* and a ratio was then run off at slightly less than its usual rate at *h.* A further small amount of responding in the TO appears at *i,* after which an interval was run off at a fairly low rate. Further responding occurs in the TO at *j,* after which a ratio was executed at below the usual rate. The remainder of this record shows a high over-all rate at approximately the terminal value of the preceding interval, with only a slow development of scalloping after reinforcement toward the end of the session, as at *k, l, m,* and *n.* The ratio performances hold at approximately their preceding rate, with only an occasional example of pausing, as at *o.*

Eventually, a fairly adequate multiple-schedule performance is obtained in spite of the 8-minute TO after each reinforcement. Figure 648A shows the 8th session on the schedule. Here, the intervals are run off with smooth scallops which begin at a fairly high rate, suggesting the scallops in Fig. 646 for FITO. The ratios at this point,

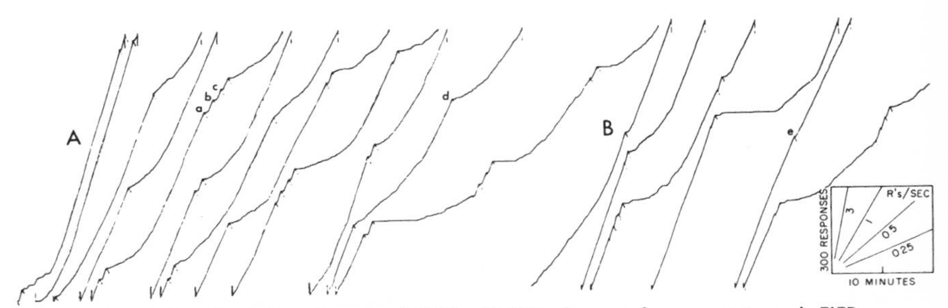

Fig. 648. Transition from mult FIFR with TO after reinforcement to mult FIFR

in general, show pausing before responding begins, as at *a, b,* and *c.* Note that the presence of the TO produces scalloping even when one interval follows another—for example, at *d.* On the following day (Record B) the TO was omitted, and the earlier performance was restored without difficulty. Note that now little or no scalloping appears when one interval follows another, as at *e.* In general, slight pausing still occurs before responding at the ratio color, but it is of the order prevailing before the appearance of the TO.

Another bird exposed to the same set of schedules showed a lower over-all rate. A low terminal rate is characteristic of the interval, as Fig. 649A shows. The ratio rate is high and shows little or no pausing. An 8-minute TO added to each reinforcement shows little effect (Record B). (The ratio at *a* contains too many responses, possibly because of responding in the TO, although the record does not clearly indicate it.) Little change occurs in the tendency to scallop when one interval follows another, although some such tendency already exists in Record A. On the following day the TO has a more disrupting effect, as Fig. 650A shows. The characteristic ratio rate appears at *a* in the 1st interval, although the usual terminal rate is reached before reinforcement at *b.* Again at *c* the ratio rate appears in the interval color, but again the rate drops to the interval rate before reinforcement. A similar break-through ap-

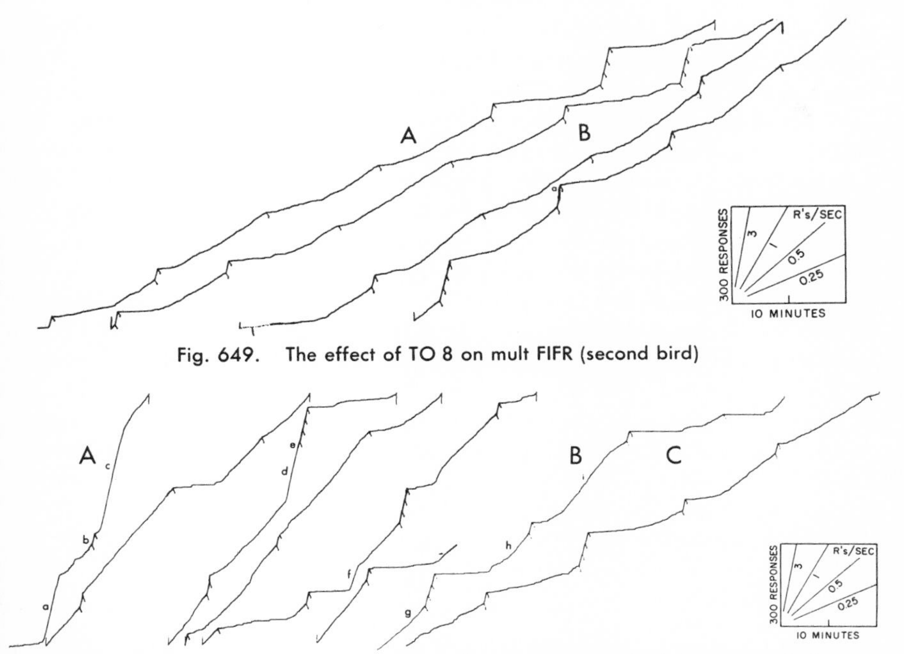

Fig. 649. The effect of TO 8 on mult FIFR (second bird)

Fig. 650. Second and eighth sessions on mult FIFRTO (second bird) and removal of TO

pears at *d*, and the high rate is maintained until reinforcement at *e*. A similar breakthrough of the ratio rate also occurs at *f*. By the end of the 9th session with the TO the performance is quite similar to that prevailing before TO was introduced. Record B shows a single excursion at this time. Except for the higher rates in the interval at *g*, *h*, and *i*, the general characteristics of Fig. 649A are apparent. Upon the removal of the TO, there was an almost immediate return to the earlier performance, as Record C indicates.

MULTIPLE SCHEDULES AND DISCRIMINATION

Multiple schedules involve the process of discrimination. The organism reacts in a different way to two stimuli when they are appropriate to schedules having different effects. The commonest techniques for discovering whether an organism can discriminate between two stimuli use multiple schedules of which one component is crf and the other ext. All "right" responses are reinforced; all "wrong" responses are extinguished. Standard techniques usually use a percentage of choices between alternative responses. The use of the rate of responding on an appropriate schedule to make the change in probability of the responses clearer has been reported by Dinsmoor (1952). The use of extinction as one component is not necessary, although it usually brings about a rapid change.

We used a variation on this technique in studying the visual acuity of the pigeon. A transparent key was mounted over Ronchi rulings illuminated from behind and presented in horizontal and vertical positions as the 2 stimuli in mult VIext. In the following experiment the rulings were 32 to the inch and clearly discriminable by the pigeon. The procedure consisted of changing stimuli on a variable-interval schedule; independent series of intervals were used to program the durations of presentation of the rulings in vertical and horizontal positions. In the horizontal position a response was reinforced at the end of the interval. In the vertical position the stimulus automatically changed to horizontal at the end of the interval.

The experiment begins as mult VIVI, with reinforcements on both stimuli until a stable performance is developed on both. These are separately recorded. When the discriminative procedure is then established, the rate declines under the stimulus no longer correlated with reinforcement. The process may be slow. Figure 651 shows the over-all change in rate on the nonreinforced component during 11 experimental sessions of various durations. The small circles represent mean rates for each session. The initial rate of the 1st session, lost in averaging, is indicated by the dashed line *a*. Besides the over-all negative curvature shown in Fig. 651, the daily session also shows a marked decline. The segment at A in Fig. 651 is shown in Fig. 652A, that

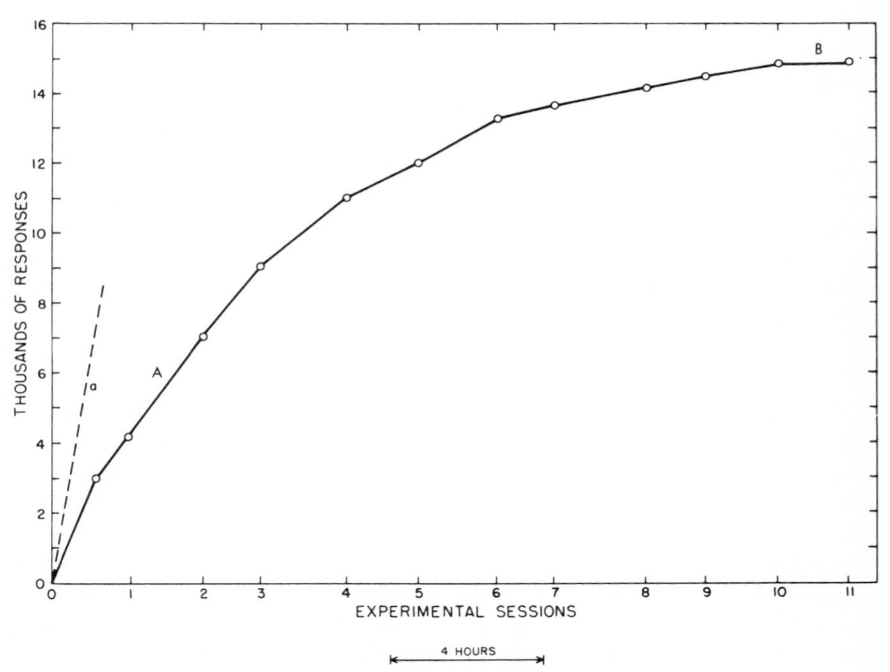

Fig. 651. Over-all rate change during the first 11 sessions of mult VIext

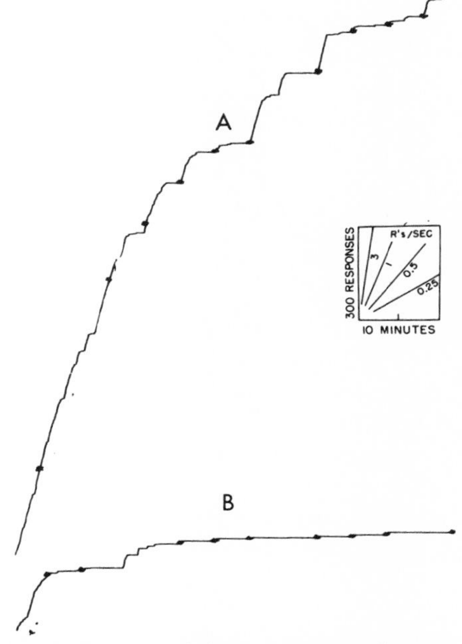

Fig. 652. Two daily sessions on mult VIext

at B, in Fig. 652B. Note the strong negative acceleration within most of the variable intervals of Fig. 652A.

A similar procedure was also used in an experiment in which the discriminative stimuli were white and red general illumination lights in the experimental box. Figure 653 shows the VI performance in the presence of the white "house-lights" 35 hours after the beginning of the discrimination. This is essentially the VI performance prevailing before the discrimination was undertaken. Figure 654 shows the corresponding performance under red illumination in the same session. Since the rate in the white light was slightly higher at the beginning of the experiment before a discrimination was possible, the difference between Fig. 653 and 654 is even less significant than it appears. In other words, the general illumination is not controlling substantially different rates, even after 35 hours of differential reinforcement on mult VIVI.

In an effort to check the importance of the location of a color discrimination, we used a new pair of stimuli and put the colors directly on the key, using colored bulbs behind the key as usual. The performance for the 1st session on the white key was similar to that in Fig. 653. The green key was a novel stimulus, however, and no responses were made to it during the entire 1st session. The same condition prevailed in the 2nd session. On the white key the performance shown in Segments 1, 2, and 3 in Fig. 655 was recorded. Meanwhile, no responses were being made to the green

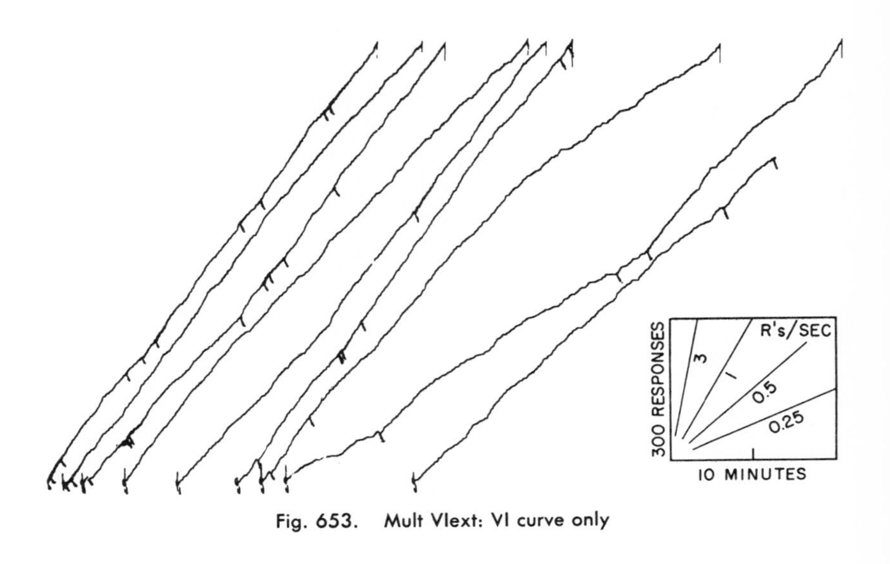

Fig. 653. Mult VIext: VI curve only

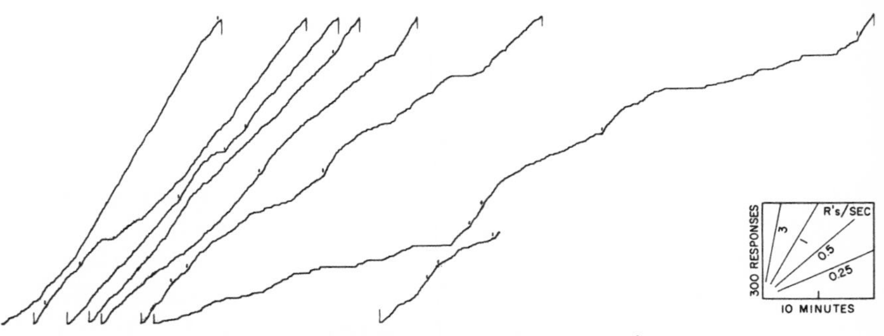

Fig. 654. Mult VIext: extinction curve only

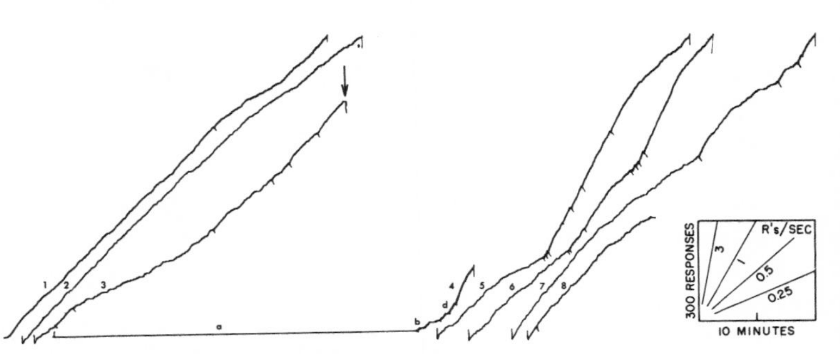

Fig. 655. Reversal of color discrimination: VI stimulus

key. Because of the effect of a novel stimulus, the discrimination was in effect complete, without differential reinforcement. The relation of the stimuli was then reversed: all responses to the white key were extinguished while reinforcements were set up in case responding should begin on the green key. Figures 655 and 656 show the result. The portion of the record at *a* in Fig. 656 is for the green key in its earlier position as correlated with extinction. At the arrow the key-color changes to white, and responding begins as a direct continuation of Segment 3 in Fig. 655. Throughout the rest of the session the response undergoes a fairly progressive extinction, as the rest of Fig. 656 shows. Meanwhile, in alternate periods, responses continue not to be made to the now green key at Segment *a* in Fig. 655. At *b*, however, a 1st response to the green key occurs and is reinforced. Two fairly short intervals at *c* and *d*, at the end of which the response is also reinforced on the green key, produce an acceleration; and

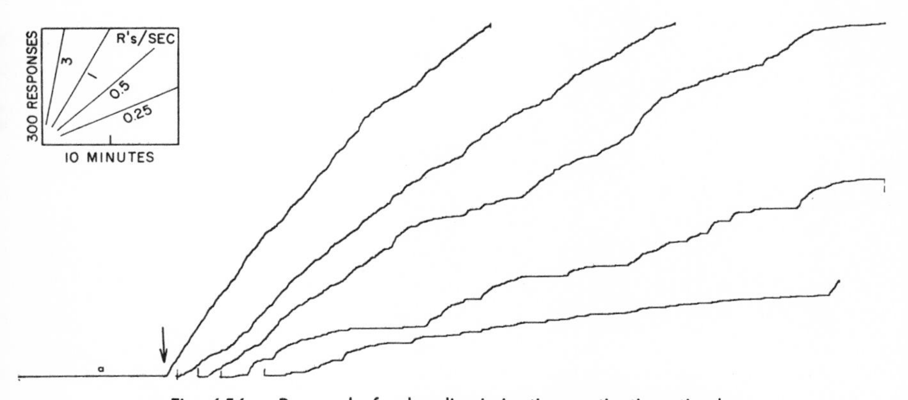

Fig. 656. Reversal of color discrimination: extinction stimulus

during the remainder of the session on the VI green component of the mult VIVI, substantial responding occurs, as the balance of Fig. 655 indicates. A tendency toward extinction occurs in the longer intervals, but the rate holds generally at about the over-all level of the performance on the white key in the first 3 segments of the session. We may conclude from the much more rapid process of extinction on the white key shown in Fig. 656 that the slowness of the process apparent in Fig. 653 and 654 must be due to the position of the lights. The color of the general illumination is evidently much less important to the pigeon than the color of the key which it strikes.

MULTIPLE VIVI

Two birds with a history of VI after crf were placed on mult VI 1 VI 3. The stimulus color was changed from blue to orange on an irregular schedule not correlated with reinforcement. A different mean value of VI prevailed on each. After 11 ses-

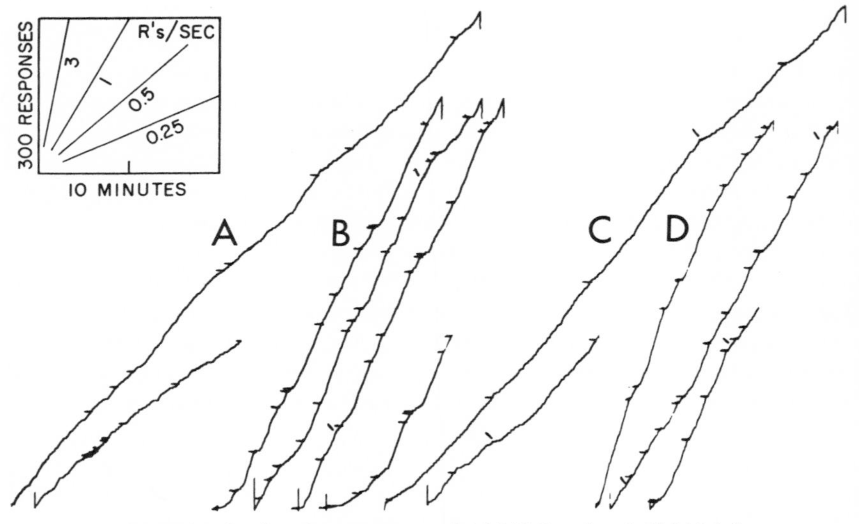

Fig. 657. Final performance on mult VI 1 VI 3 and mult VI 1 VI 6.5

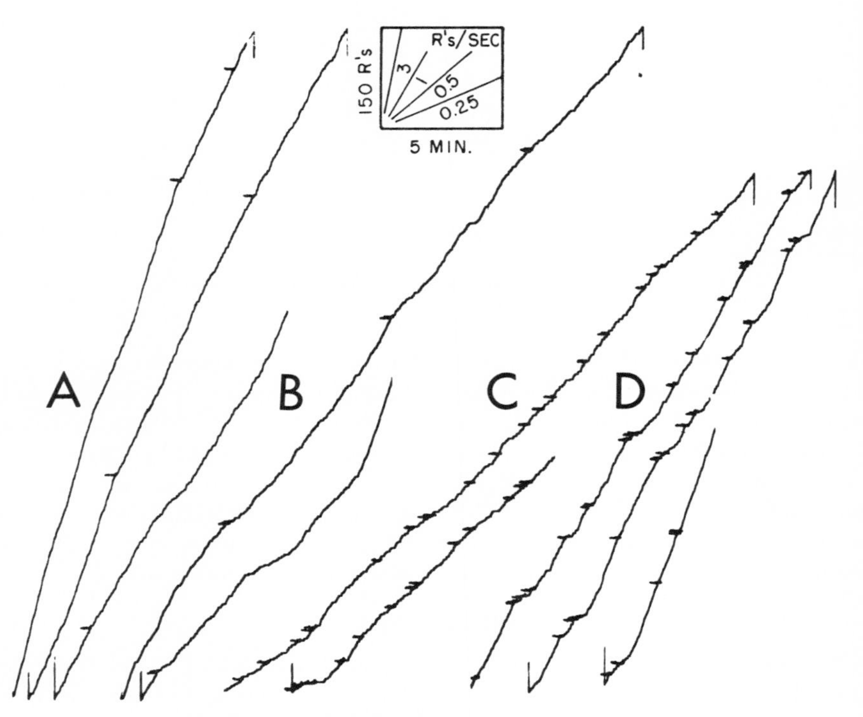

Fig 658. Reversal of discrimination on mult VI 1 VI 6.5

sions the rates on the two were clearly different and stable, as evident in Fig. 657A and B. Record A shows the whole session on a blue key on VI 3; Record B shows the whole session on an orange key on VI 1. Some points of changeover are marked in the record by diagonal dashes, although all were not clearly recorded. Later, the larger mean interval was increased to 6.5. After 25 sessions on this mult VI 1 VI 6.5, the performance was as shown in Records C and D. The VI 1 performance (Record D) remains essentially as in Record B, except that at the beginning of the session a higher rate is commonly observed for the first few minutes. The rate on the larger mean VI has held to essentially the same value as on the smaller at Record A, possibly because of some induction from the other schedules, although, as we have seen in Chapter Six, the VI rate generally resists change.

As a check on the extent of the stimulus control, the colors were reversed. The orange key, which had previously been correlated with VI 1, was now correlated with VI 6.5. Figure 658 shows the first and last sessions under the reversed conditions. Records A and B give the first and last performances on VI 6.5 under the control of the orange key. Some extinction of the higher rate previously associated with VI 1 occurs. Records C and D, which give the beginning and final sessions on the blue key under VI 1, show a steady adjustment to the new schedule.

MULT FR 60 FR 200 WITH PROLONGED EXPOSURE TO EACH CONTROLLING STIMULUS

The frequency with which stimuli are rotated in a multiple schedule is, of course, arbitrary. In the experiments already described stimuli were changed frequently, usually at reinforcement. We investigated the possibility of a multiple schedule where a single stimulus is in control for a full session on 3 birds with a long history on FR 60, on a blue key. At the beginning of a new session the key-color was orange, and the reinforcement was on FR 200. A fairly good ratio pause and curvature appropriate to FR 200 developed before the end of the session. On the following session the key was again blue and the schedule was FR 60. Very little disturbance from the preceding day on FR 200 was evident. On the 3rd day, the schedule was again FR 200 on the orange key; on the 4th, FR 60 with the blue key; and so on. The FR 200 schedule performance developed rapidly. The over-all rate declined somewhat throughout each session; this decline could be due to the intervening days on FR 60, but it is also fairly characteristic of ratio performances. Figure 659 shows the condition on the 12th and 13th days of this procedure. Record A gives the entire session with the blue key on FR 60; Record B gives the following entire session on FR 200 with the orange key. Note the decline in over-all rate during the session on FR 200. Note also the long pauses controlled, even at the beginning of the experiment, by the orange (FR 200) key.

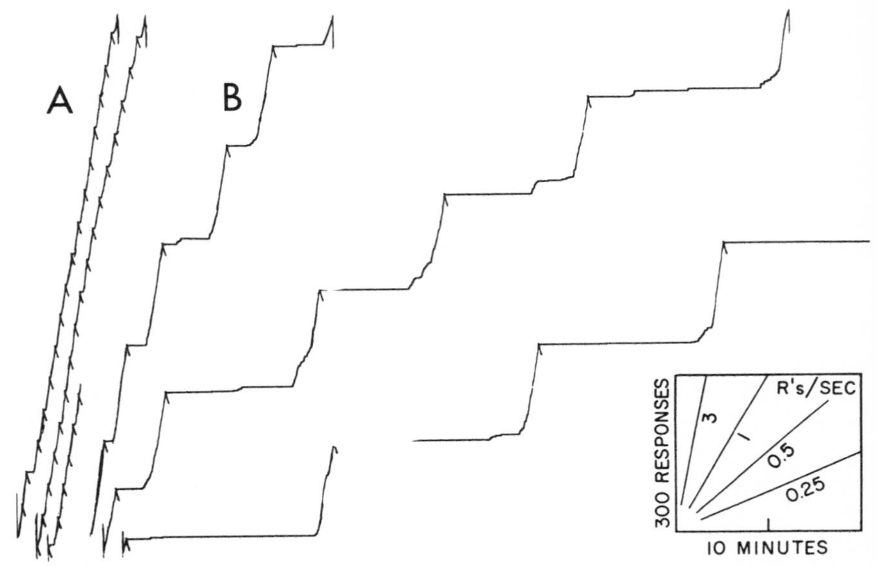

Fig. 659. Mult FR 60 FR 200 with prolonged exposure to each controlling stimulus

MULTIPLE FIFI

In many experiments on multiple schedules a mere change in stimuli seemed to have some effect apart from the correlation with schedules. We studied the effect of a simple change in a mult FI 15 FI 15. The key could be either green or white, changing color after reinforcement at random. A schedule of FI 15 was in force in each case. Figures 660 and 661 show a stable performance under these conditions for the white and green keys, respectively. The reinforcements after which the key changed color have been marked with dashes above the record. Although the characteristic is not inevitable, scalloping tends to occur when the color changes and running-through when it does not. In other words, a different key-color, even though on the same schedule, functions as a TO.

In a total of 6 sessions for this bird, the intervals following changeover began with one of three types of performance as follows: pause, 12%; run-through and pause, 74%; and complete run-through at terminal rate, 14%. In the same sessions the intervals not following changeovers showed: pauses, 32%; run-through and pause, 28%; complete run-through at terminal rate, 40%. A second bird showed a much more consistent picture. Intervals following changeovers showed: pause, 90%; run-through and pause, 7%; and complete run-through at terminal rate, 3%. Intervals following no changeover showed: pause, 0%; run-through and pause, 75%; and complete run-through at terminal rate, 25%.

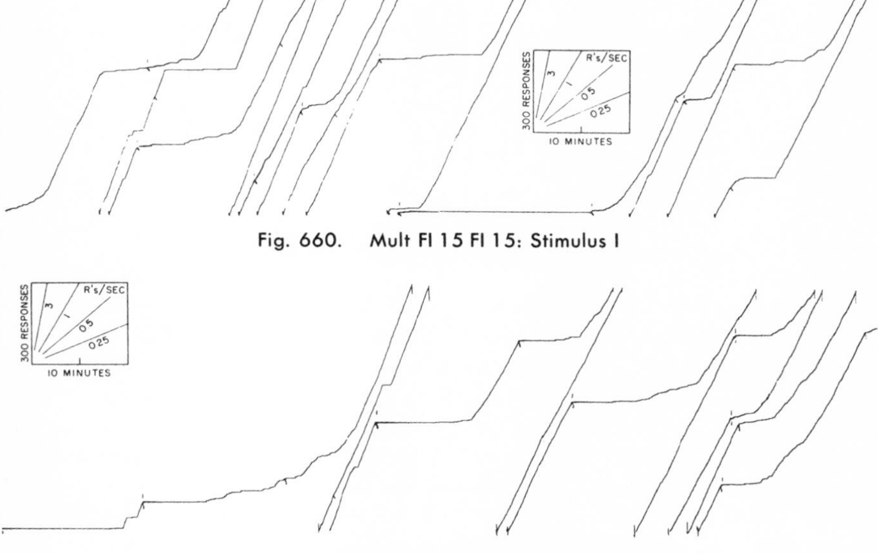

Fig. 660. Mult FI 15 FI 15: Stimulus I

Fig. 661. Mult FI 15 FI 15: Stimulus II

When the schedule was changed to FI 12 on one key-color, the performance changed only slightly. When one schedule was changed to FI 8 while the other remained at FI 15, however, a multiple control clearly developed, as Fig. 662 and 663 indicate. Figure 662 shows the FI 8 intervals, and Fig. 663, the FI 15. Pausing and slow responding are much less extended in the shorter interval. Figure 662 contains several instances of running-through when the color does not change. Figure 663, however, has no such case. On the contrary, a well-marked scallop in the subsequent interval often develops without a color change.

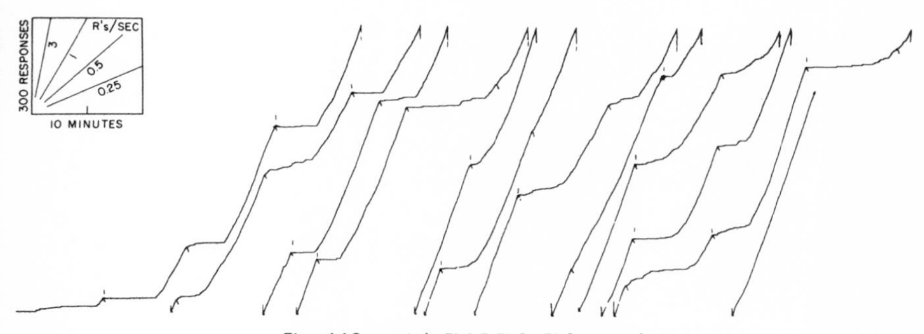

Fig. 662. Mult FI 15 FI 8: FI 8 record

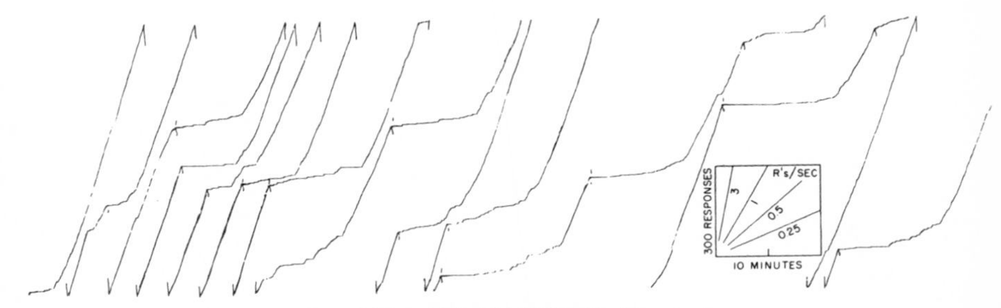

Fig. 663. Mult FI 15 FI 8: FI 15 record

The interval on the white key then was reduced to 4 minutes. Figure 664 shows the 1st, 9th, and 30th sessions for FI 4 and Fig. 665 the 30th session for FI 15. The FI 15 remains stable and is not affected by the reduction on the white key. The FI 4 (Fig. 664) immediately eliminates the pause after reinforcement, and only a slightly lowered rate for the first few responses after reinforcement remains. The development is slow. In Record B, after 9 sessions of the new multiple schedule, a short pause as well as a more extended curvature are apparent. However, consistent pausing and marked curvature appropriate to FI 4 do not develop until Record C, after 30 sessions of the multiple schedule.

At this size of interval little relation exists between what happens immediately after reinforcement and the condition of the preceding interval. This may be due to the size of the interval, or to the prolonged exposure to the multiple schedule.

The other bird showed a very similar performance, but with somewhat shallower scalloping and a tendency to develop a more prolonged linear terminal rate. The over-all rate was higher so that the number of responses per interval tended to be larger, but conditions of curvature were quite similar.

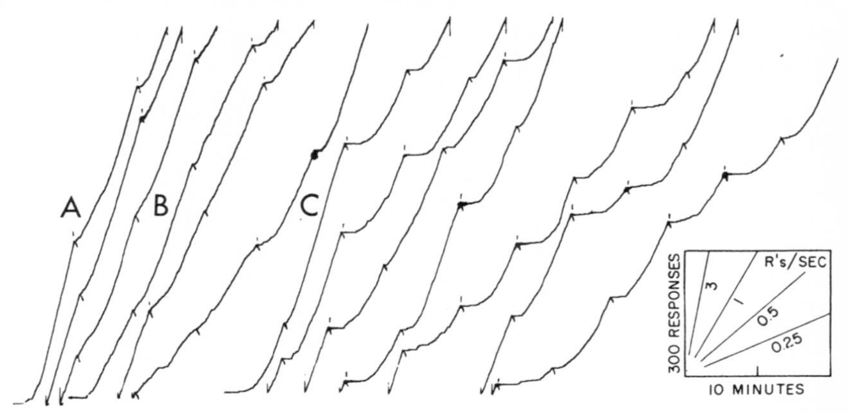

Fig. 664. Mult FI 15 FI 4: segments from the 1st, 9th, and 30th sessions of FI 4 record

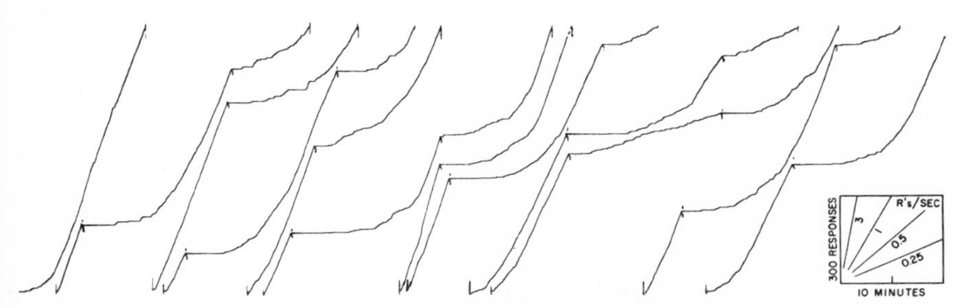

Fig. 665. Mult FI 15 FI 4: 30th session of FI 15 record

We then reduced the shorter interval further. After 11 sessions on mult FI 2 FI 15 the performances in Fig. 666 were recorded. Excellent scalloping occurs on the FI 15 (Record A) and on FI 2 (Record B). In general, the depth of the scalloping in FI 2 depends upon whether there has been a change from the other schedule. Changeovers tend to be followed by more marked scallops, although this is not always the case. The FI 15 performance shows persistent scalloping whether or not a change in key-color has just been made. In contrast, a simple FI 15 schedule shows occasional running-through unless a TO occurs after reinforcement.

We then tested the effect of a very large FI component in a mult FIFI. The FI 15 component on the green key remained unchanged. The interval on the white key was increased to 56 minutes. After 41 sessions a well-marked multiple performance had stabilized, as Fig. 667 and 668 indicate. These figures show the full session, except for the part of the last interval on FI 56 missing from Fig. 668. Well-marked FI 56 scallops appear in this figure, as at *a, b, d,* and *f.* Also, one instance of running-through at the terminal rate occurs at *c,* and one instance of slight running-through with a very shallow scallop at *e.* The intervals at *g* and *h,* not shown complete, have very few responses. The FI 15 scallops in Fig. 667 alternate with this performance at random.

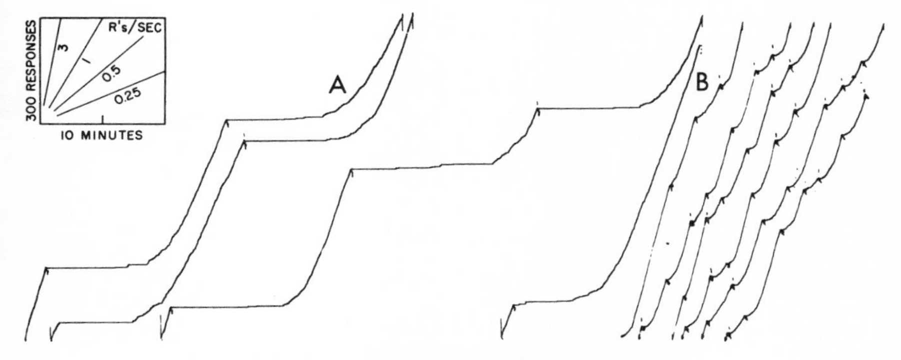

Fig. 666. Mult FI 15 FI 2 after 11 sessions

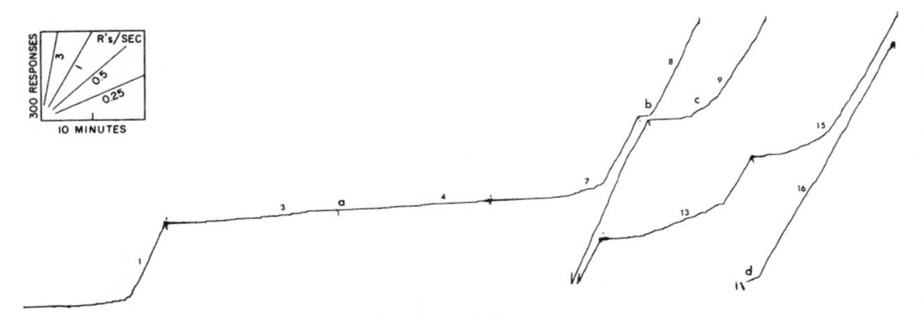

Fig. 667. Mult FI 15 FI 56: FI 15 record

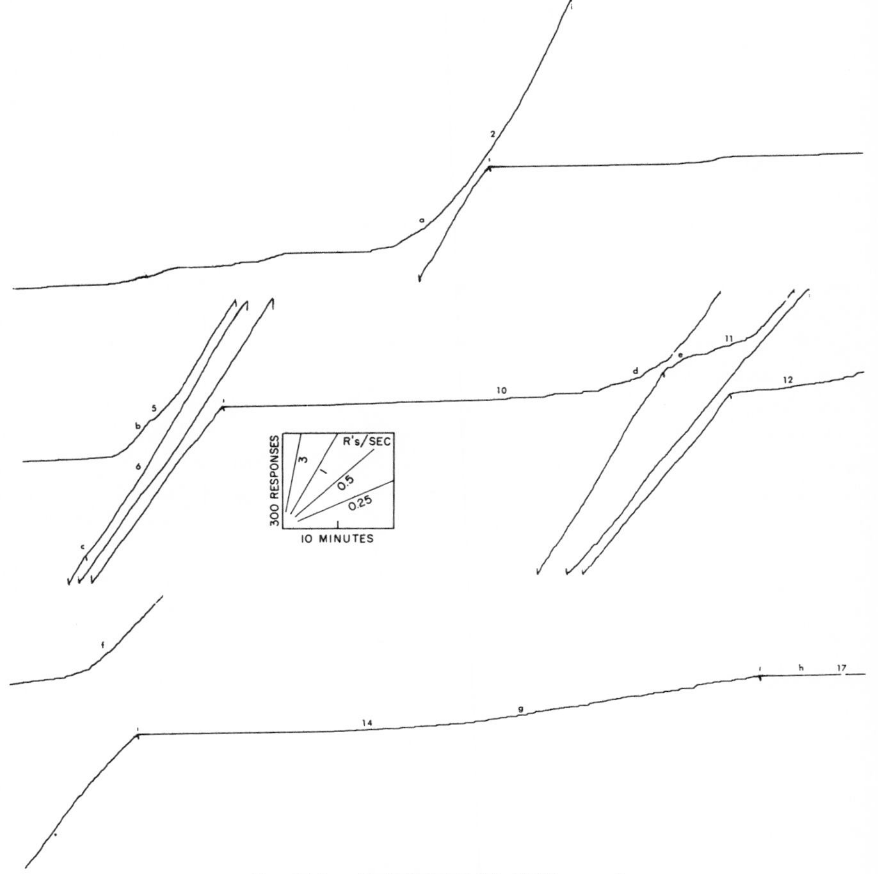

Fig. 668. Mult FI 15 FI 56: FI 56 record

These are generally typical of the interval, and some scalloping occurs whether or not the schedule has just changed. A marked instance of such scalloping appears at *c*, and minor instances at *b* and *d*. The low rate in the region of reinforcement at *a* is occasionally evident at this stage of the experiment. The actual order of occurrence of the segments in the 2 figures is shown by numbers.

Mult FI 2 FI 10

Performances were recorded on mult FI 2 FI 10 in an experiment in which the discriminability of the correlated stimuli was manipulated. Three birds were used which had a long history of FI with fading counter. (See Chapter Five.) Figure 669 is a record of the performance on both stimuli, the 2 stimuli being grossly different. The key was red during the 2-minute intervals, and blue during the 10-minute intervals. In the figure a fairly smooth scallop and a high terminal running rate are common in

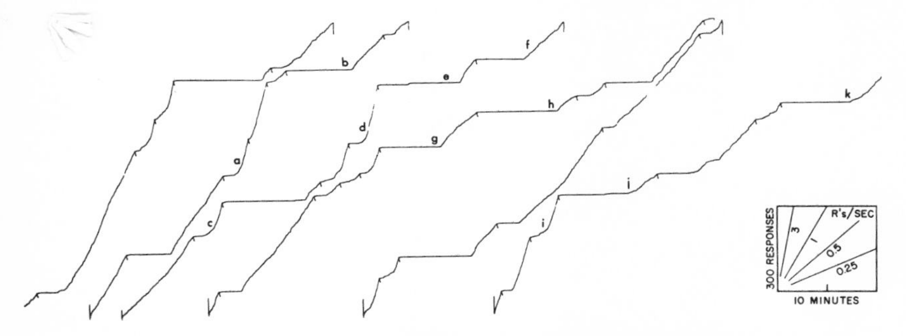

Fig. 669. Mult FI 2 FI 10 with maximum stimulus contrast after 47 sessions

the shorter intervals. Particularly good examples are at *a, c, d,* and *i*. The FI 10 component shows long pausing after reinforcement when the schedule has just changed, but also shows substantial pausing when the preceding interval was 10 minutes. Responding after a long pause is often triggered very late, and shows a rapid start with negative acceleration. Instances of this are at *b, e, g,* and *h*. Elsewhere, fairly smooth acceleration is observed, as at *j* and *k*.

When some red light was added to the blue, both over-all and terminal rates are lower. Late triggering and sharp negative acceleration in the longer intervals are occasionally quite marked, as at *a, b,* and *d* in Fig. 670, 20 sessions after Fig. 669. An instance of this in the short interval at *c* follows another short interval in which the reinforcement was delayed by an exceptionally long starting pause.

Figure 671 is the final picture on this schedule after many changes in color contrast. Here, pausing in the shorter interval is frequently quite long, as in *b, c,* and *d*. The longer intervals often show relatively late triggering, as at *a* and *e*. The terminal rates are now high again, however, and the longer intervals show a large number of responses before reinforcement. In general, the stimulus control is good.

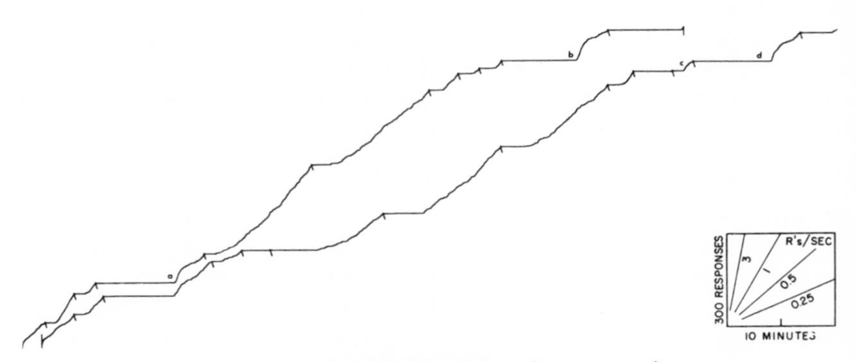

Fig. 670. Mult FI 2 FI 10 with less clear-cut stimuli

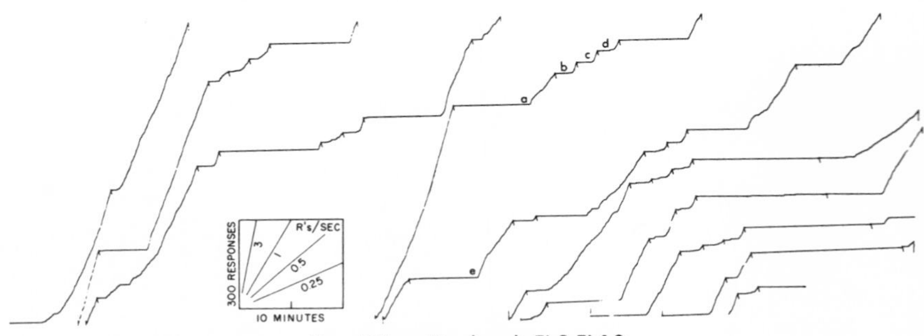

Fig. 671. Final mult FI 2 FI 10

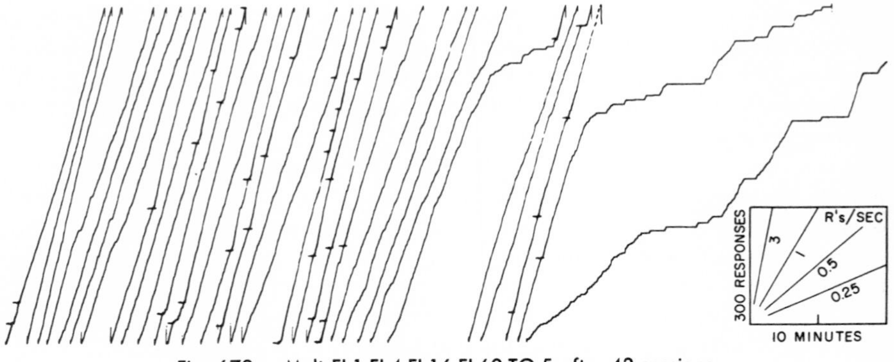

Fig. 672. Mult FI 1 FI 4 FI 16 FI 60 TO 5 after 42 sessions

Mult FI at four values

Two birds were exposed to a mult FI 1 FI 4 FI 16 FI 60 TO 5 for 42 sessions of 8 hours each immediately after crf. Little evidence of multiple control developed. The schedule had the general effect of a VI. The key-colors (red, blue, white, and green, respectively) appeared to have little effect. Figure 672 is a record of a complete session after 42 sessions. The two periods of low, irregular responding toward the end of the session are in the presence of the green key on FI 60. Slightly lower rates on this component appear earlier in the record.

During this part of the experiment a time out of 5 minutes followed each reinforcement. Later, the TO was abandoned and the shortest interval omitted. The schedule was then mult FI 4 FI 16 FI 60. Under these conditions some multiple control developed. Figure 673 shows the performance after 22 sessions or 176 hours of this

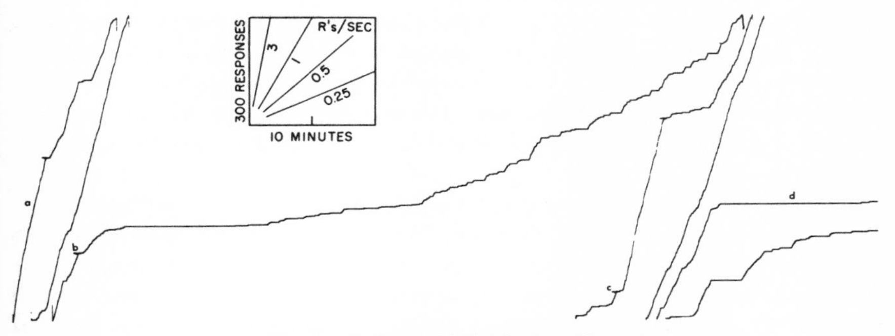

Fig. 673. Mult FI 4 FI 16 FI 60 after 22 sessions

schedule. The record begins with the end of a 4-minute interval on a blue key, during which the bird responds at a high rate (at *a*). The key then became white, and a 16-minute interval passed before the next reinforcement, at *b*. Although much of this performance is at a rate close to that of the 4-minute segment, some acceleration occurs at the beginning. A considerable irregularity is characteristic of the 16-minute intervals. The color was then changed to yellow and a 60-minute interval begun. Although some running-through occurs (after *b*), the bird breaks into a long and fairly smoothly accelerated scallop appropriate to the 60-minute schedule. Reinforcement occurs at *c*. The key was then blue again, and a high rate was maintained for most of a 4-minute interval. The color then changed to yellow; and after a slight scalloping, a fairly high rate developed in the following 60-minute interval. This high rate leads to some extinction and breaking at *d*. The net result is a not-very-smooth multiple performance; nevertheless, a distinguishable difference exists in the effects of the 3 key-colors.

The 16-minute interval was then dropped out (the schedule then being mult FI 4

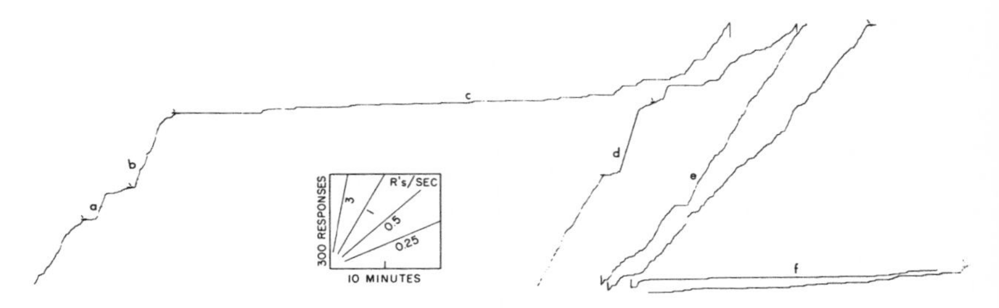

Fig. 674. Mult FI 4 FI 60 after 12 sessions

FI 60 on blue and yellow keys, respectively); Fig. 674 illustrates the performance after 12 sessions. This figure shows three 4-minute intervals at *a, b,* and *d,* and three 60-minute intervals at *c, e,* and *f.* An unusual feature of the short interval has developed. After a short pause, the rate assumes a value appropriate to the interval; but a sharp negative acceleration occurs before the interval is completed. This condition is particularly clear in *a* and *b* but can also be detected in *d.* It became practically inevitable on the short interval for both birds. We have no explanation for it. Note that, in general, possibly because of the complicated history of the experiment, the performance is quite erratic.

Later, the intervals were changed to 1.5 and 16 minutes, and the colors appropriate to them in the 1st schedule were reinstated. The schedule was now mult FI 1.5 FI 16 on red and white keys, respectively. Figure 675 shows the performance during the sixth 8-hour session. The scalloping appropriate to the 16-minute interval is much more uniform and characteristic of that schedule. Negative acceleration in the short interval still appears, particularly at *a, b, c, d,* and *e*; it is noticeable in every instance except at *f,* where the rate remains low throughout. The same characteristic was marked in the behavior of the other bird.

The experiment showed good multiple control on mult FIFI when the stimuli were sufficiently different, a poor but distinguishable multiple control on a mult FIFIFI, and failure to achieve multiple control on a 4-ply FI after a long exposure.

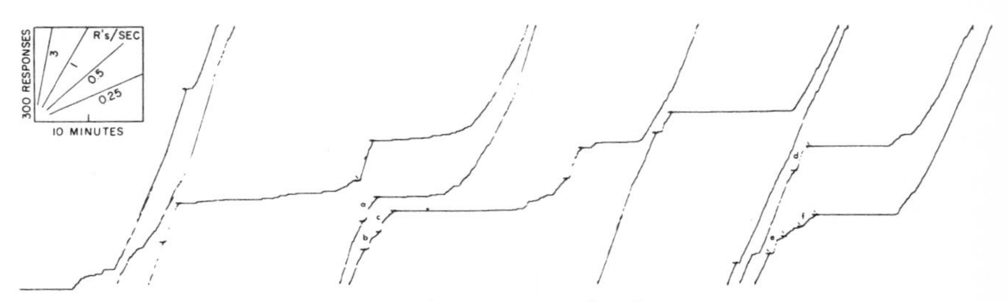

Fig. 675. Mult FI 1.5 FI 16 after 6 sessions

Multiple FI 30 tandem FI 30 FR 10

We attempted to bring the "ratio end effect" under stimulus control by placing 2 birds on a 30-minute FI with 2 key-colors changing at reinforcement at random and then by adding one tandem FR 10 under one color. The birds had had an extended history of mult FIFI with different values. At the beginning of the experiment both key-colors (green and white) were appropriate to FI 30. The tandem ratio was added to the FI 30 when the key was white. The tandem FR had little if any effect. The terminal rate was already about 2 per second at the start of the experiment, and remained at about that value throughout. However, the ratio grain developed conspic-

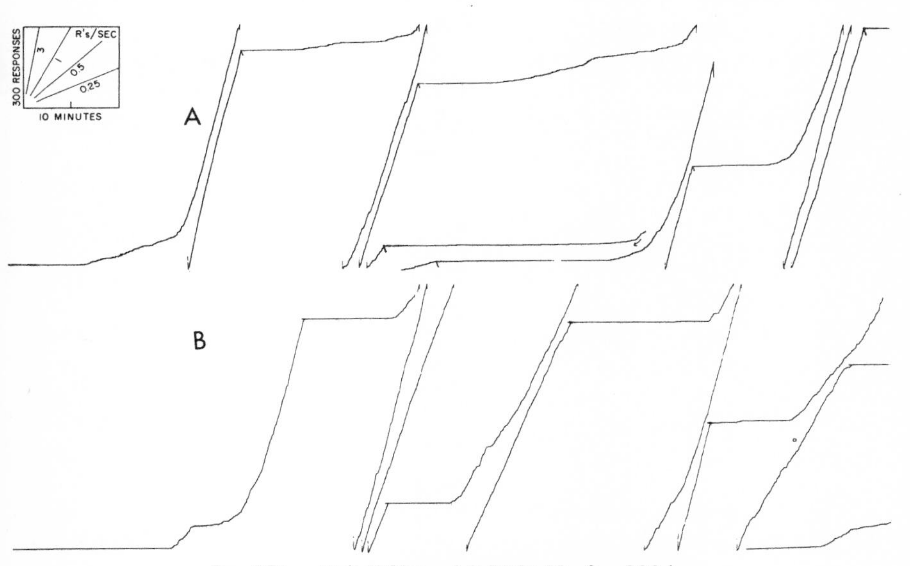

Fig. 676. Mult FI 30 tand FI 30 FR 10 after 200 hr

uously. Figure 676 shows a representative session after approximately 200 hours on mult FI 30 tand FI 30 FR 10. Record A is the performance on the green key on FI 30. Long pauses occur after reinforcement and also a steady, smooth acceleration to a terminal rate, which may be maintained for as many as 2000 responses before reinforcement. Record B is the alternating performance on tand FI 30 FR 10. Some rough grain, producing a lower terminal rate, appears, especially at *a*. Some indication exists that the curvature along which the bird approaches a terminal rate is sharper on the tandem color. The over-all performance is similar to that in Record A. Note that the performance is relatively free of running-through. The bird seldom had any tendency to begin responding immediately after a change of key-color. The second bird occasionally continued at the terminal rate for the succeeding interval when the color happened to remain the same. As in other multiple experiments, changing the color has some of the effect of a time out. In the present experiment the schedule is not actually changed with the change of key-color, since the tandem FR did not have a differential effect.

MULTIPLE VIVIdrh

The 3 birds previously studied on VIdrh were placed on a multiple schedule at the completion of that experiment. When the key was red, reinforcement was VI 1 drh; when the key was white, the schedule was VI 1 without drh. The change from one color to another occurred at reinforcement. Twenty-one sessions later all 3 birds had developed marked differences in rate on the 2 key-colors.

Figure 677 shows a complete session. The bars mark periods on the white key, where the rate is consistently lower. The white-key segment frequently begins with a short continuation of the rate prevailing on the red key, notably at *a, b,* and *c.*

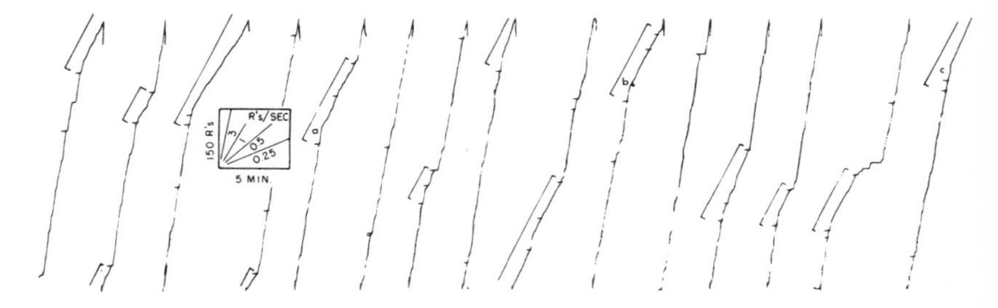

Fig. 677. Mult VI 1 drh VI 1 after 21 sessions

When the performance shown in Fig. 677 had been developed, a drl was added on the white key. The schedule was now mult VI 1 drh VI 1 drl. The drl contingency was easily satisfied in the early stages of the next session, which is illustrated in Fig. 678. (We lost part of a segment in recording the top record.) Fairly long runs occur on the white key, as at *a* and *c.* Nevertheless, before the end of the session, a substantially lower rate has developed under the drl contingency, as at *e, g,* and *h.* Note that the tendency to begin at a high rate on the white key, as at *b, d,* and *f,* still exists.

Eventually, the drl rate becomes quite similar to a simple VIdrl performance. Figure 679 shows the last session. The change to a low rate is now practically immediate, and the rate extremely stable. The drh meanwhile continues to show running-through. This performance is due to the gap-setting on the schedule which requires a certain minimal number of responses for reinforcement. There is no appreciable difference in the drh performance as the effect of the alternating performances on drl.

When the drl contingency was removed from the white key, the bird returned essentially to the performance which prevailed before it was added. Figure 680 shows

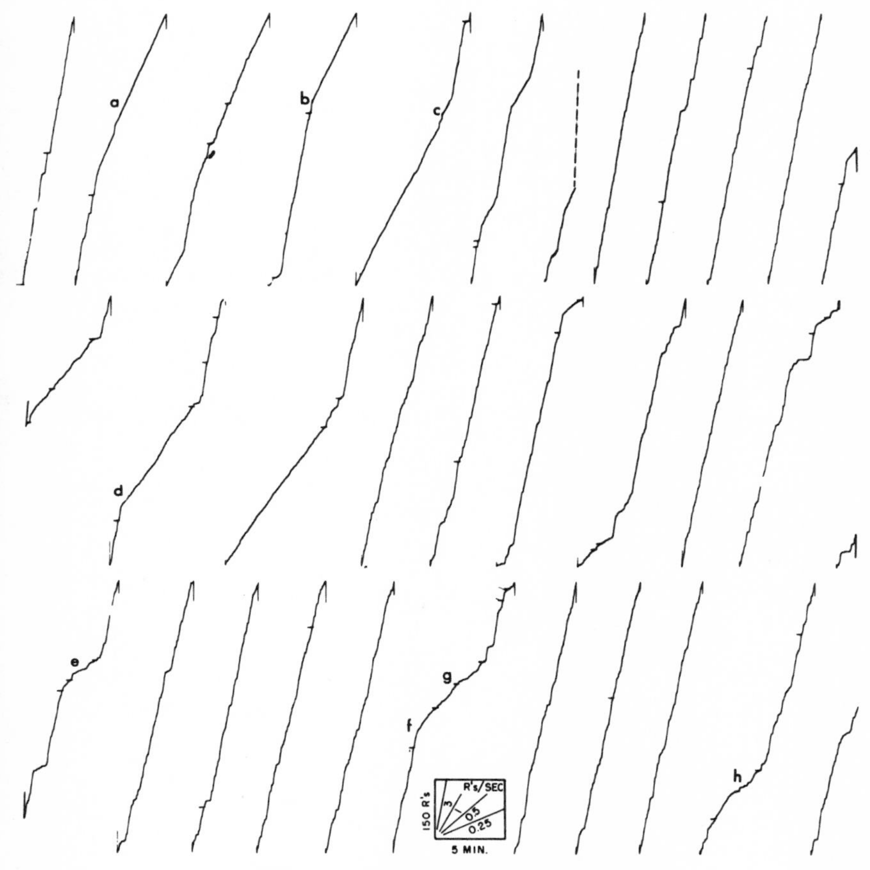

Fig. 678. First session on mult VI 1 drh VI 1 drl

the 3rd session after the removal of drl. Note that with the higher over-all rate on the white key (segments marked by bars), the tendency to begin responding at a high rate, as at *a, d,* and *e,* returns. At this stage a tendency also exists to return to this high rate (as at *b* and *c*) during the white phase, which now has no differential-rate reinforcement.

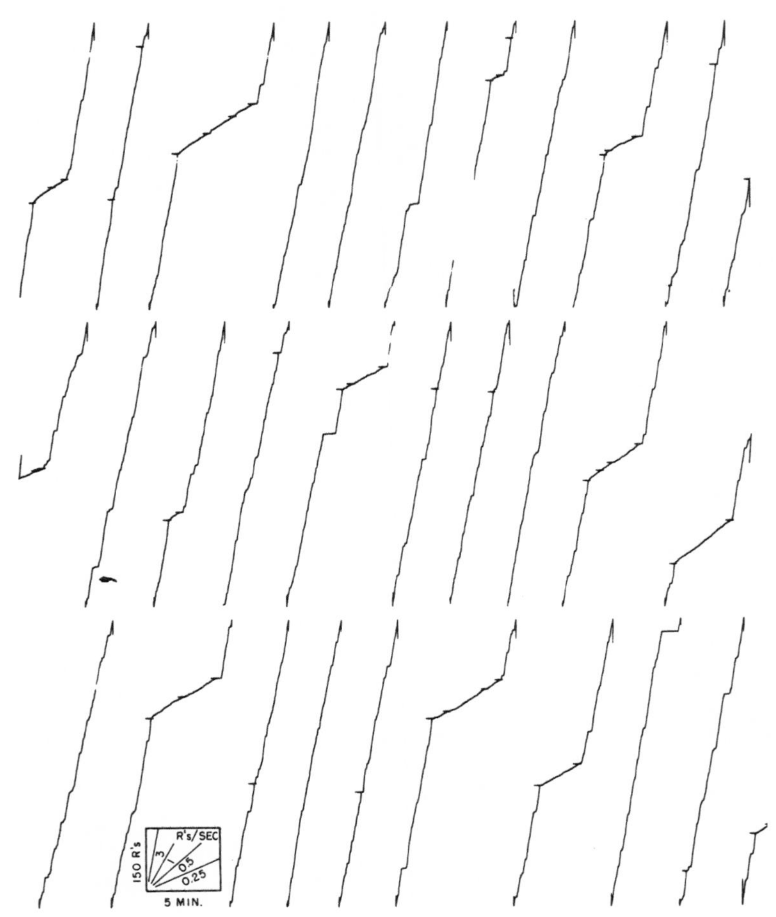

Fig. 679. Mult VI 1 drh VI 1 drl after 8 sessions

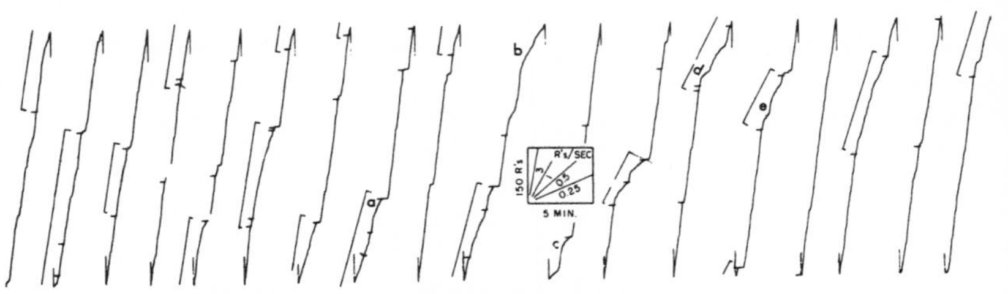

Fig. 680. Return to mult VIdrhVI after mult VIdrhVIdrl

MULTIPLE FIdrhFI

After reaching performances as in Fig. 680 on mult VIVIdrh, 2 of the birds in the preceding experiment were placed on mult FI 15 FI 15 drh. Figure 681 shows the changeover for 1 bird (not in the preceding figures). The first stimulus presented was the red key, which had previously been correlated with VIdrh and is now correlated with FIdrh. The rate begins at the usual VIdrh value at *a*. At *b*, after extinction due to the change to FI 15, the rate drops spontaneously to a lower value than that observed in the previous VI without differential reinforcement. This lower rate

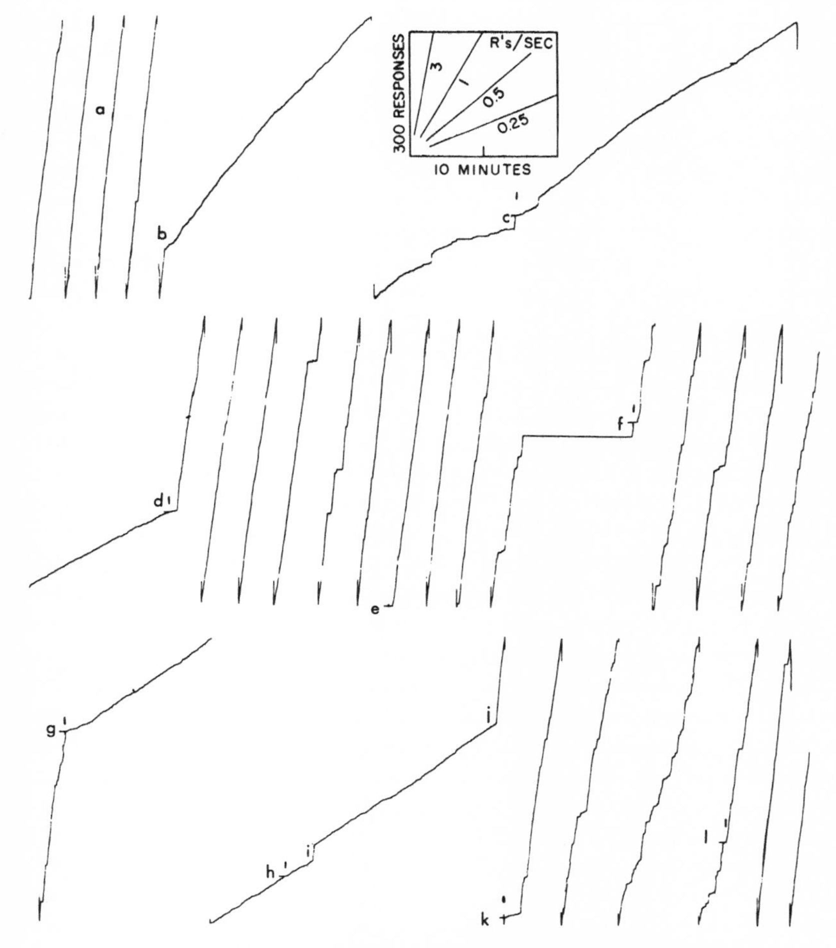

Fig. 681. First session on mult FI 15 FI 15 drh after mult VIVIdrh

suggests a return to a performance appropriate to the much earlier VIdrl. The short, rapid run at *c* satisfies the rate contingency, and reinforcement is received. The stimulus then changes to the color formerly appropriate to a VI without differential reinforcement, and a low rate is maintained for 2 intervals on the FI 15 schedule. Although no differential-rate contingency is in force, the rate actually shown is considerably below that previously prevailing on mult VIVIdrh and suggests again a return to the earlier condition of VIdrhVIdrl. After the key changes again to red at *d*, a high rate prevails until a reinforcement is received at that rate, at *e* and again at *f*. Just before *f* a substantial square break occurs which is common on drh when reinforcements are infrequent. The schedule continues on the red key at drh, and a 3rd reinforcement is received at *g*. When the key-color changes to white, the rate drops immediately to the lower value, and a reinforcement is received at *h*. Shortly thereafter, a momentary break-through occurs of the high rate at *i*, but a sharp return to the low value follows. When the higher rate again breaks through at *j*, a reinforcement is received at this rate at *k* (although there is no drh on the white key). This single reinforcement at a high rate suffices to bring out a full interval of responding on the white key very near the rate appropriate to the red key under the previous contingencies. However, some rough grain produces a general low over-all rate near the end of the interval when a reinforcement is received at *l*. The schedule then changes to FIdrh on the red key, and the experiment is terminated before the interval is complete.

The final state to which this schedule was carried is illustrated in Fig. 682, which shows a performance after 25 sessions for the same bird as in Fig. 681. The session starts with a pause appropriate to the FI schedule, and the 1st interval is concluded at *a* at a high terminal rate. Responding begins soon afterwards; it continues with only some decline in the terminal rate at the very end of the interval before reinforcement is received at *b*, where the drh contingency is satisfied by a short run. A long pause then intervenes before a return to a high terminal rate, and a break occurs at *c* before the interval had been completed. The drh contingency is not satisfied by the subsequent performance until the point of reinforcement at *d*. The key-color then changes to white, and a fairly smooth scallop follows on the nondifferentially reinforced schedule. A knee is evident at *e*, and the grain is not very smooth; however, the curvature is clear. Two other intervals follow on this key-color; and although the rate rises and the scalloping becomes less marked, some evidence of interval curvature still exists. At *f* the key-color changes to red, and a pause appropriate to the FI schedule follows. The high rate of responding which sets in is not maintained until the completion of the interval, and a long pause follows at *g*, the reinforcement being received only when the short run at *h* satisfies the drh contingency. The key-color then changes to white, and a rough scallop follows in which marked knees suggest the break-through of the drh rate. When the color changes to red at *i*, a long pause intervenes followed by a rapid rate which achieves reinforcement at the end of the first horizontal section of the figure. The schedule continues with the red key, and a good instance of what has

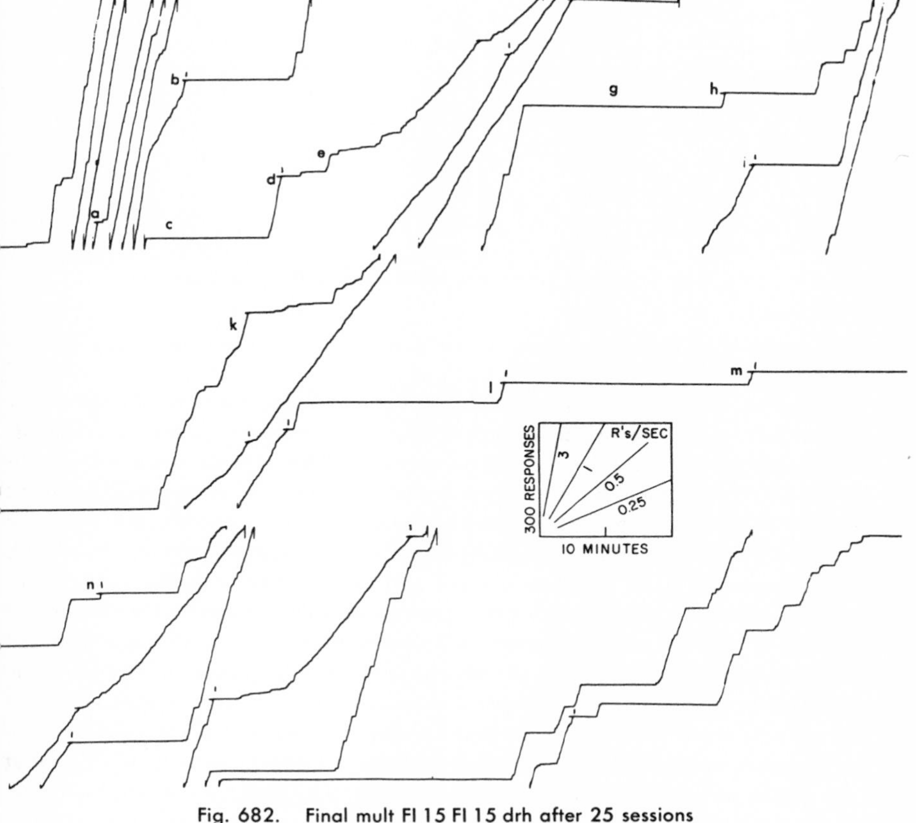

Fig. 682. Final mult FI 15 FI 15 drh after 25 sessions

developed by way of an interval scallop under the drh contingency follows at *j* to *k*. On the white key 2 shallow interval-scallops then follow; upon the return to the red key, reinforcements are received at *l*, *m*, and *n* after very little responding during the interval. Two intervals on the white key repeat the previous performance, and the subsequent behavior follows the same pattern. In general, the performance on the white key is not very different from a standard FI performance. Roughness exists, and induction from the other schedule appears to emphasize the knees which are occasionally seen on such a schedule. Pausing after reinforcement is only moderate. Each interval begins fairly soon at a substantial rate. The pauses on the drh contingency are extreme, however, and the over-all interval pattern, insofar as one develops, is a relatively square curve. The grain is rough and reflects the small "ratio" runs resulting from the gap-setting on the drh apparatus. (The actual interval on drh de-

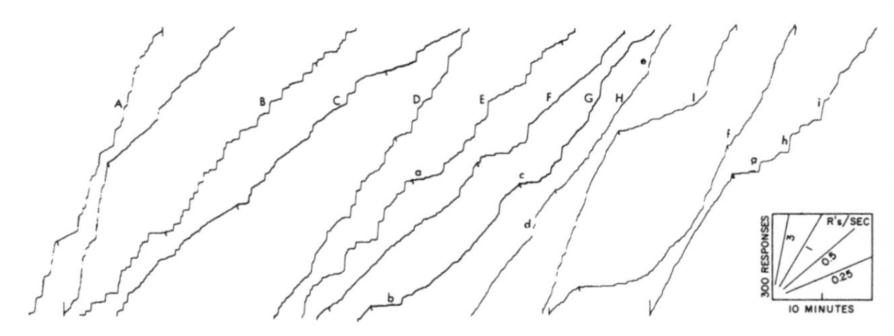

Fig. 683. Mult FI 15 FI 15 drh: development of FI 15 performance

pends upon when the bird satisfies the drh contingency, and may be longer than 15 minutes.)

Figure 683 shows the earlier development of the performance on the white key without drh on FI 15, separately recorded. Record A is for the 3rd session. Considerable induction from the drh component appears. This induction becomes particularly marked in Segment B from the 5th session. Segment C is from the 7th session; Segment D, from the 8th. Some pausing after reinforcement, particularly when the preceding interval has been on drh components, appears at *a* in Segment E for the 9th session, and again more conspicuously at *b* and *c* in Segment G for the 14th. (Segment F is from the 13th session.) A tendency not to scallop remains when the preceding interval has been on the white key, as a segment from the 16th session at Segment H demonstrates. The segment also shows, as do other segments in the figure, the occasional break-through of the higher rate, as at *d* and *e,* through induction from the drh contingency. Well-marked interval scallops appear by the 17th session when the preceding schedule has been drh, as 3 segments at I show. Note, however, that at *f* the reinforcement on the white key is followed by a quick return to a high rate. Rough grain still appears in the early stages in the scallop, as well as pauses followed by rapid runs, as at *g, h,* and *i.*

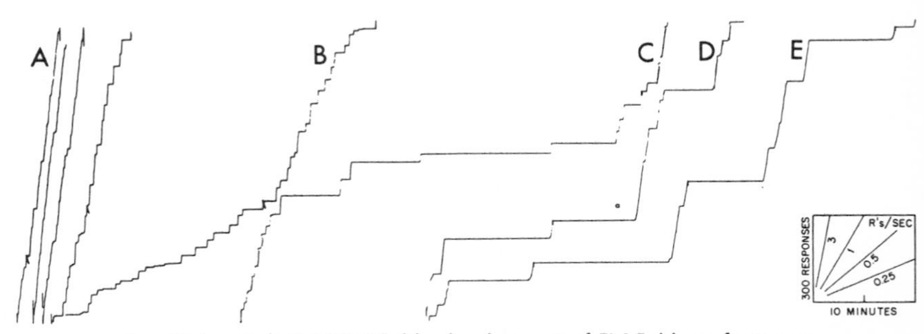

Fig. 684. Mult FI 15 FI 15 drh: development of FI 15 drh performance

Figure 684 shows the development of the FI performance with drh. The rate was originally quite high, as Record A for the 3rd session shows. Reinforcements are being received much less frequently than under the preceding VI 1 schedule, however, and the over-all rate falls. Short runs corresponding to the gap-setting in the drh apparatus appear. In the 4th session, shown at Record B, the rate is lower and has marked step-wise breaks. Record C is for the 10th session. The actual progress of the performance at this time depends upon arbitrary settings of the drh contingency, which were changed as the performance permitted. As the requirements were increased, the runs became longer. Some curvature appropriate to the interval begins to develop. This curvature is particularly evident at *a,* Record D (for the 16th session), and becomes fairly consistent by the 25th session at Record E, which immediately precedes Fig. 682.

The other bird on this schedule showed a much better development of an interval

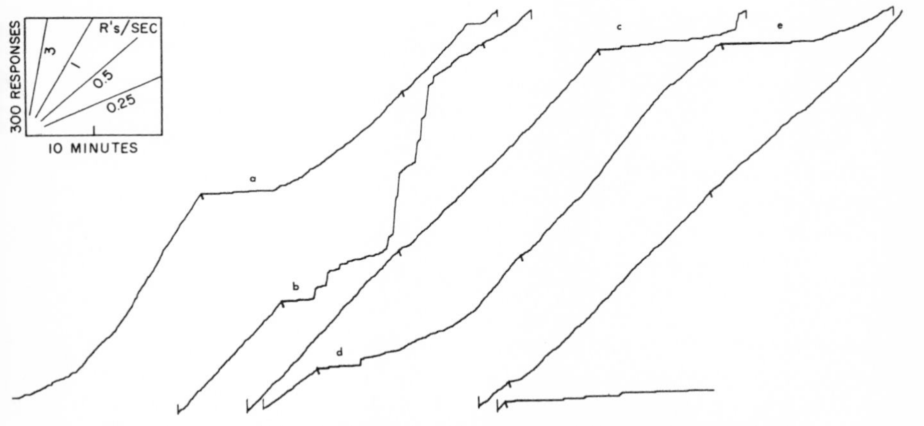

Fig. 685. Mult FI 15 FI 15 drh after 13 sessions: FI 15 record (second bird)

performance without differential reinforcement, but great straining on the drh contingency. A fairly early development on FI alone is shown for the 13th session in Fig. 685. Here again, after changing from the other color, the interval shows good scalloping, as at *a, b, c, d,* and *e*; but elsewhere, only slight evidence of scalloping exists. The only substantial example of a break-through of the drh rate is in the interval beginning at *b,* although slight examples appear in the intervals beginning at *c* and *d.* Later, these disappear, and a very satisfactory performance on FI alone is maintained, as Fig. 686 shows for the 19th session. After changing from the other color, the interval scallop is well-marked; otherwise, it is not. The rate is often so low following a change of schedule, as for example at *a, b,* and *c,* that the reinforcement occurs at a low count. This condition produces a second-order effect, if the schedule does

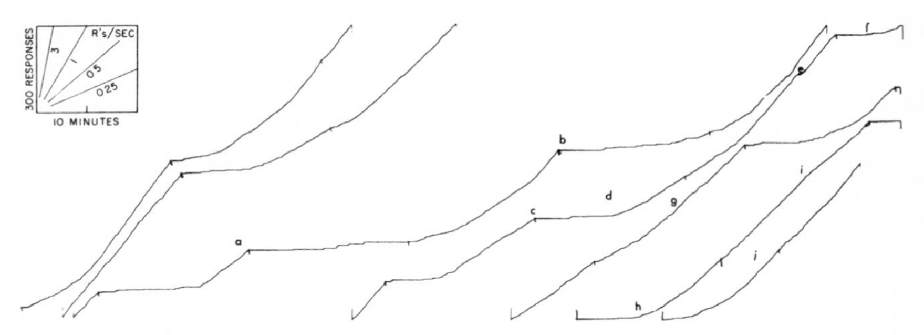

Fig. 686. Mult FI 15 FI 15 drh after 18 sessions: FI 15 record (second bird)

Fig. 687. Mult FI 15 FI 15 drh after 18 sessions: FI 15 drh record

not change, in which 2 successive intervals together describe a smooth scallop (as from *a* to *b, d* to *e,* and *h* to *i*).

Meanwhile, the performance on FI 15 drh under multiple control and alternating more or less at random with FI 15 shows rather long ratio runs corresponding to the gap-setting, but also prolonged periods of no responding. Figure 687 illustrates a fairly representative example, alternating with the performances on FI 15 alone in Fig. 686. Thus, Segments *d* and *e* in Fig. 686 occur at *b* in Fig. 687. The scallops at *f* and *g* in Fig. 686 occurred at point *d*; and the single-interval scallop at *j,* at point *e.* In general, after a change in schedule, a period of low responding follows which suggests a rather rough interval curve; but when no change has occurred from the preceding schedule, the interval curvature may be lacking as before *a* and *c.*

MULTIPLE VIVI PACING

Dr. Lagmay's experiments on pacing described in Chapter Nine included an exploratory attempt to establish multiple control. A bird which had been paced and then reinforced on VI 1 without pacing was again subjected to VI 1 pacing (3) 1.5-2.0. The key had previously been white, but it was red when the pacing contingency was re-introduced. Figure 688 shows the effect. Possibly because of the color change, a pacing performance is quickly recaptured.

Multiple control was then attempted, with the white key controlling VI alone during one-half of the session and the red key controlling paced VI during the other half. Figure 689 describes the first 2 sessions on this multiple schedule. The 1st session (Record A) begins with the key red. The bird occasionally breaks through with the VI rate alone, but a fairly adequate paced performance appears. When the key changes color at the arrow, a high rate on VI is observed. The following session (Record B) begins on the white key on VI, and a change to the red key at the arrow leads almost immediately to a paced performance. In later sessions the multiple control is less effective. A marked break-through of the unpaced rate is apparent in Fig. 690A at *a,* and 2 similar break-throughs, in a later session, in Record B at *b* and *c.* A second bird gave very similar performances.

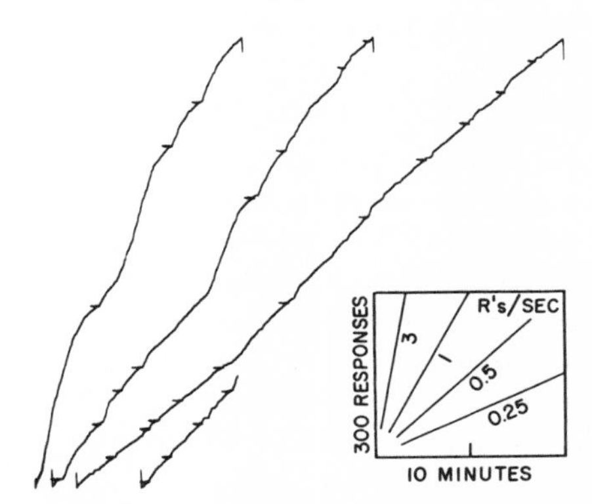

Fig. 688. Return to VI 1 pacing prior to mult VIVIpacing

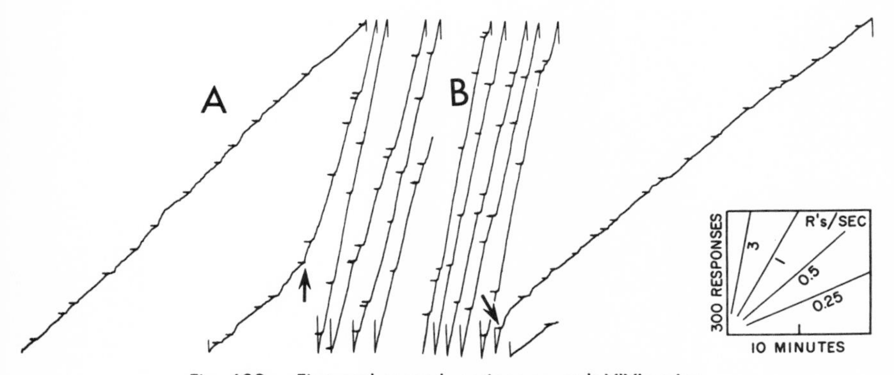

Fig. 689. First and second sessions on mult VIVIpacing

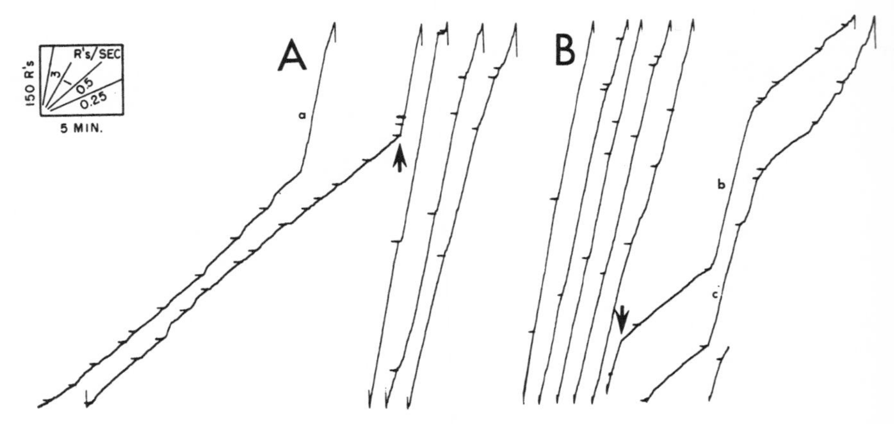

Fig. 690. Mult VIVIpacing showing loss of multiple control

MULTIPLE SCHEDULE WITH THREE COMPONENTS

As a demonstration of what may be done with multiple schedules containing more varied components, an apparatus was designed to reinforce on the following schedules: FI 6, VI 1 drl 4, and FR 50 or FR 70. The controlling key-colors were white, red, and blue, respectively. Figures 691 and 692 show stable performances of 2 birds. In Fig. 691 the session begins with a brief period of drl at *a*. Other periods of drl under the control of the red key appear at *c, f, i, l, o, r,* and elsewhere. The 2nd schedule in the session was FI 6 and the performance appears at *b*. Other interval performances are at *e, h, k, n, q,* and elsewhere. The first FR 50 appears at *d*; and other instances, at *g, j, m, p,* and elsewhere. The other bird showed a similar performance except that the over-all rate tended to be much higher in drl. The session in Fig. 692 begins at a fairly low over-all rate. Three interval segments, *a, b,* and *c,* contain very few re-

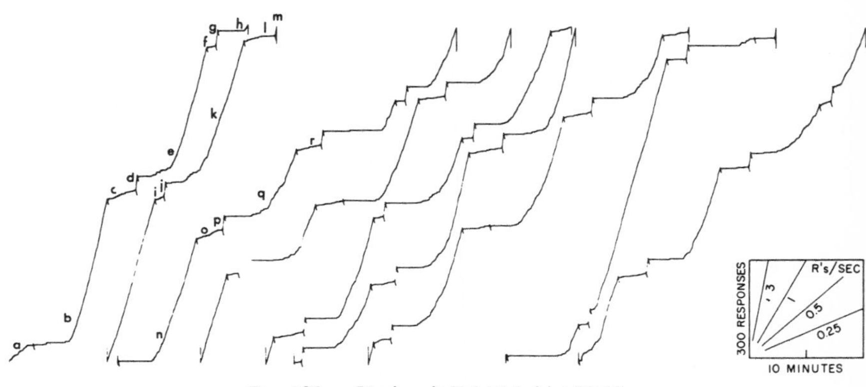

Fig. 691. Final mult FI 6 VI 1 drl 4 FR 50

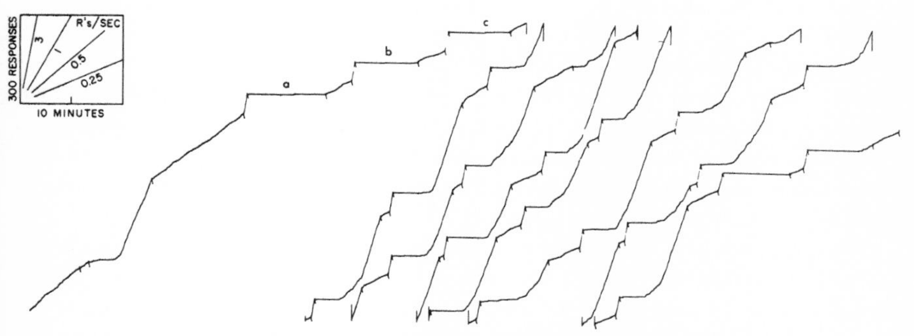

Fig. 692. Final mult FI 6 VI 1 drl 4 FR 70 (second bird)

sponses. Otherwise, the performance is similar to that of Fig. 691. The ratio here is FR 70.

A MULTIPLE SCHEDULE WITH FOUR COMPONENTS

Two birds with a long history of multiple and mixed schedules were placed on a mult FI 2 FI 11 FR 50 FR 250 schedule. In the early stages, blocks of reinforcements on each schedule were arranged under appropriate stimuli. The birds adjusted quickly. Figure 693, for example, shows appropriate performances on all 4 schedules as they were presented in blocks of 4 or 8. Large intervals are marked I; small intervals, i; large ratios, R; and small ratios, r. The 4 large intervals beginning at *a* show fairly good scalloping and an intermediate terminal rate. The rate oscillates a good deal at this time. At *b* a block of 4 short ratios is run off with only slight pausing and at a high rate. Four large ratios beginning at *c* show some pausing but a high terminal rate. Beginning at *d,* 4 short intervals show an essentially linear perform-

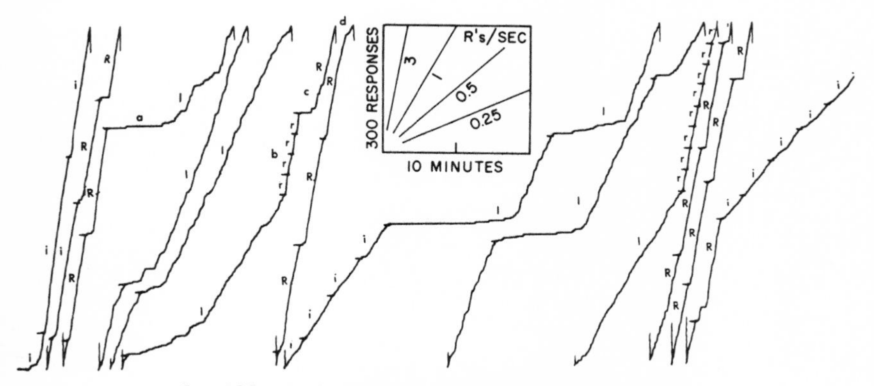

Fig. 693. Mult FI 2 FI 11 FR 50 FR 250 after 9 sessions

ance at the lowest of the 4 rates. The general pattern is continued throughout the session.

Later, the 4 schedules were presented at random. This procedure caused some trouble, and the larger ratio was temporarily omitted from the set. The other 3 schedules were distinguished accurately. Figure 694 shows the 7th session on mult FR 50 FI 2 FI 11, with long pauses and scalloping in many long intervals, only slight pausing and a high terminal rate in the small ratios, and a short pause or no pause and an intermediate terminal rate in the short intervals.

The large ratio was then added to the series and the birds exposed to mult FR 50 (red) FR 250 (white) FI 11 (blue) FI 2 (yellow) for a larger number of sessions. Figure 695 gives a performance after 32 sessions. The separate schedules now appear in short blocks of 2 or 3 each, and appropriate performances are exhibited. The other bird showed essentially the same result; Fig. 696 gives its final performance on the four-fold multiple schedule after 38 sessions. Here, the short ratio shows no pausing and a high terminal rate. The large ratio shows a high terminal rate, although a good deal of difficulty occurs at the beginning and later fairly smooth acceleration, as at *a* and *c*. The small interval now shows some pausing and occasional smooth curvature. The large interval shows a long initial pause, with very irregular development

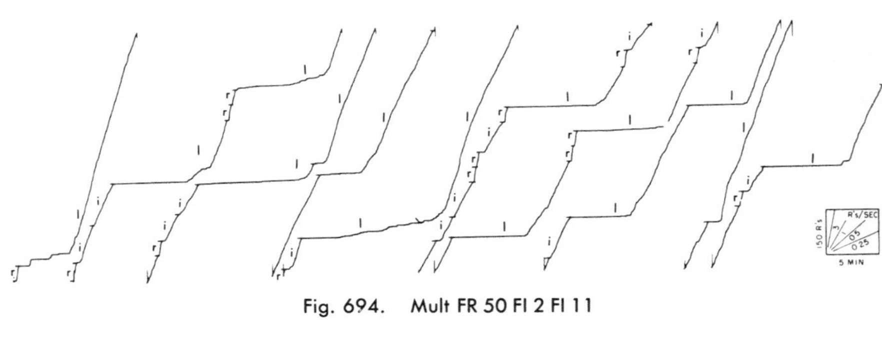

Fig. 694. Mult FR 50 FI 2 FI 11

Fig. 695. Mult FR 50 FR 250 FI 2 FI 11 after 32 sessions

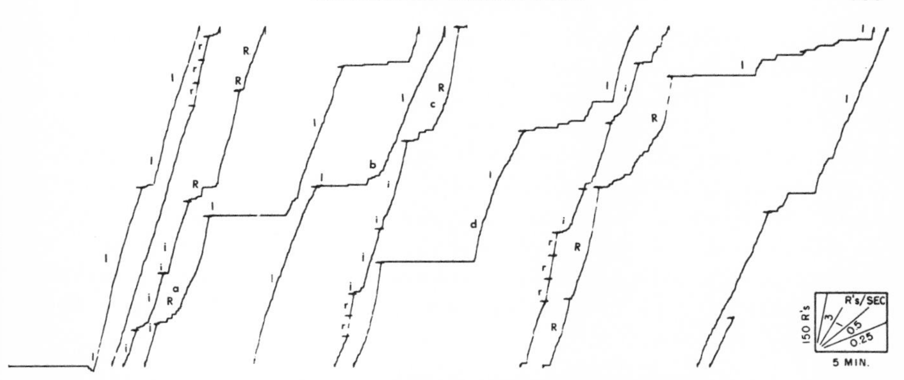

Fig. 696. Mult FR 50 FR 250 FI 2 FI 11 after 38 sessions (second bird)

of the terminal rate. Some instances show a fairly smooth positive acceleration, as at *b*; others show late triggering followed by negative curvature, as at *d*. In general, however, both birds demonstrate 4 separate performances under the control of 4 key-colors.

A MULTIPLE SCHEDULE INVOLVING EIGHT CONTROLLING STIMULI AND EIGHT SCHEDULES

In order to get a larger number of clearly discriminable patterns, small figures were projected on the key. A figure could be either a triangle or a circle, red or blue, and large or small. The 8 possible combinations of these properties were correlated with 8 separate schedules as follows. The figures were red on all interval schedules, green on all ratio schedules. They were triangles when there was no differential reinforcement of rate, and circles when the schedule included drl. They were large when the interval was 16 or the ratio 100; they were small when the interval was 2 or the ratio 40. This set of relations was designed to permit a separate control *via* color, shape, or size, apart from the specific schedule.

Four recorders were arranged to record separately: (1) large FRs, (2) small FRs, (3) large FIs, and (4) small FIs. Each recorder contained a marker indicating when the drl contingency and its appropriate stimulus were present. Thus, 8 different segments could be separately recorded, although each schedule with and without drl appeared in a single record. The order of presentation was semi-random, and only 1 recorder ran at a given time.

Three birds went directly from crf to VI 3, during which all 8 stimuli were presented in random order. Some color preference was observed. With one bird, for example, the blue pattern generated a rate of about 1.25 responses per second; the red pattern, about 0.95. When the VI performance had stabilized, except for this difference, the final eight-fold multiple schedule was introduced at once. The results for the bird showing the best stimulus control will be reported in terms of each of the component schedules.

First bird

Figure 697 shows the performance in the 5th session. (Each session contained 56 reinforcements.) At Record A the parts of the record under the bars were executed on a small blue circle and the schedule was FR 40 drl. The drl is effective relatively late, and a large number of responses is generally emitted. The short portions of the record not under bars were executed on small blue triangles and a simple FR 40. In Record B the bars mark portions of the record executed on large blue circles and are

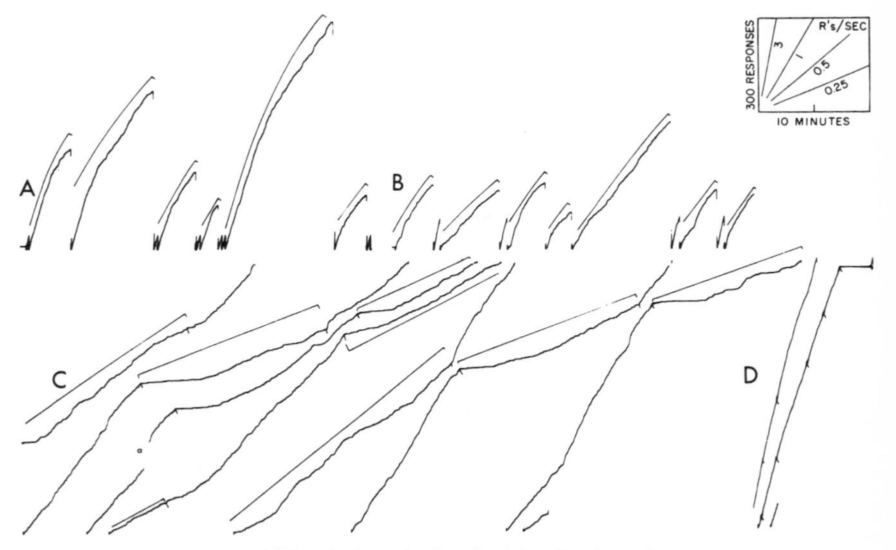

Fig. 697. 8-ply multiple schedule after 4 sessions

for FR 100 drl. The drl here is more effective than that in Record A; and despite the larger basic ratio, reinforcements are often received sooner. The portions of the record not under bars are for the standard FR 100 on a large blue triangle. Record C gives the performance on the larger interval, the bars marking the drl contingency. In general, a low rate with only slight curvature is generated by the large red circle correlated with FI 10 drl, while the standard FI 10 performance on the large red triangle does not yet show very marked interval curvature. Note the single instance of a break-through of a rate appropriate to a ratio performance at *a*. Record D contains the performance on the small red circle (not marked) under FI 1 drl or on the small red triangle on straight FI 1. The drl is almost ineffective. The stimulus condition during the horizontal segment cannot now be identified. At this stage the bird shows the effect of drl under the control of the size of the spot on all schedules except the smallest interval.

Three sessions later the stimulus control has developed further. Figure 698A shows the large ratio with and without drl, and the control is obvious. On the shorter ratio, however, in Record B, the drl contingency begins to be felt only after several times the number of responses have been emitted. This factor could be due to the higher rate of responding controlled by the appropriate stimulus (the size of the pattern) or to induction from Record A. In the shorter interval, at Record C, the drl contingency is now clearly effective, although not invariably so. For example, the higher rate

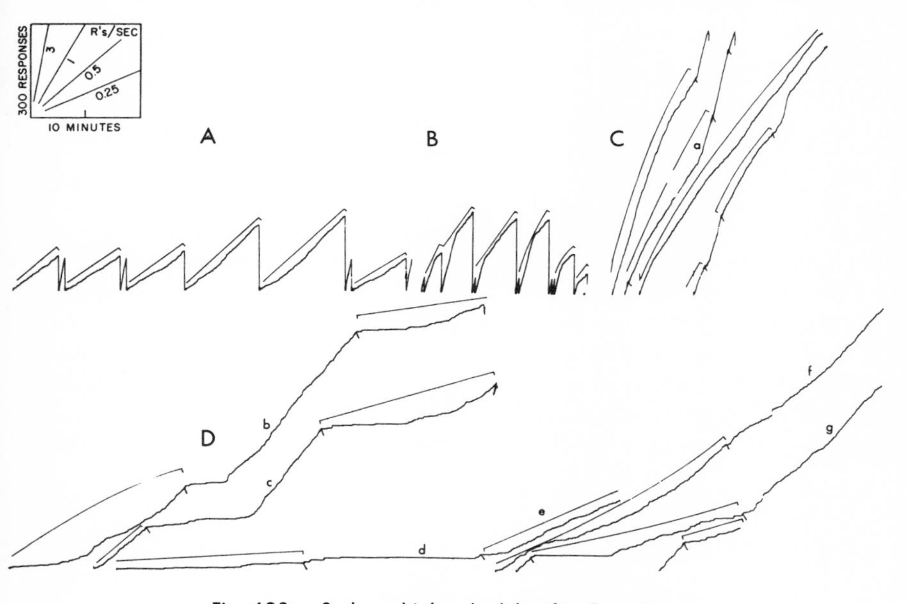

Fig. 698. 8-ply multiple schedule after 8 sessions

breaks out at *a* under the drl stimulus. On the large interval (Record D), scalloping is now marked both with and without the drl contingency. Fairly normal FI 16 scallops show at *b* and *c*, and the adjacent drl segments also show scalloping. At *d* an interval passes with very few responses, although the drl contingency is not in effect; and the following interval at *e* with the drl shows a higher rate. However, a still higher rate appears first at *f* and later at *g* when the drl contingency is not in effect.

Figure 699 gives the final performance after 44 sessions on the eight-fold multiple schedule. The stimulus patterns have been suggested, solid figures representing blue patterns. Record A illustrates the characteristic high rate under the drl until the smaller ratio has been counted out. At Record B the drl on the larger ratio produces an appropriate low rate from the very beginning of the ratio. At Record C on the

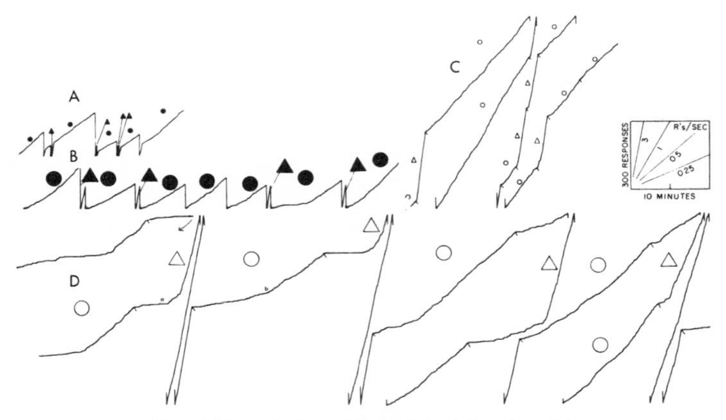

Fig. 699. 8-ply multiple schedule after 44 sessions

small interval the drl is now clearly effective; but the rates are still too high for immediate reinforcement, and intervals of perhaps 10 or 20 minutes often elapse before a reinforcement is received. At Record D the large interval shows scalloping under drl, as at *b*; but a higher terminal rate develops in the absence of the drl contingency, as at *a*. The fact that the drl is more effective on the larger than on the smaller interval indicates that with the small ratio and the small interval the more frequent reinforcement conflicts with the drl contingency.

In an attempt to check the cross-induction effects, we omitted the drl contingency

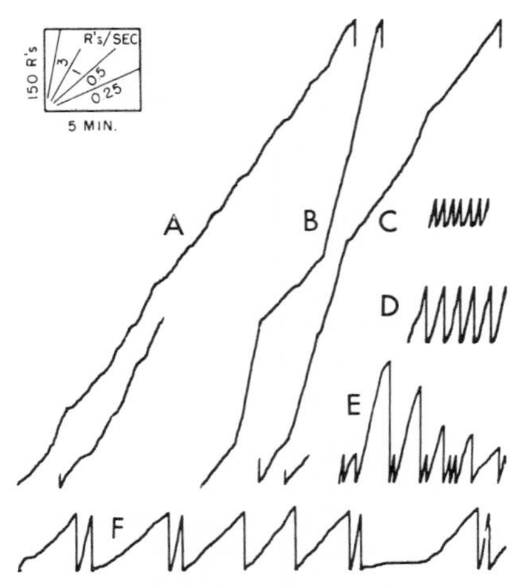

Fig. 700. 6-ply multiple schedule (omitting FR 40 drl and FR 100 drl)

from the smaller ratio. This omission left a seven-fold multiple schedule. After 6 days, the drl was omitted from the larger ratio, leaving a six-fold schedule. The effect of this omission, contrary to expectation, was to decrease the maximal rate on all schedules. The rate of the small ratio declines steadily throughout the 18 days in which the FR 40 drl was omitted, and a similar change occurs after omission of FR 100 drl in the FR 100 curve. The FI curves also show a loss of the high terminal rate. The final state 18 days after the usual eight-fold multiple performance is shown in Fig. 700 and 701, which should be compared with Fig. 699. Figure 700C shows the final condition of FR 40, which should be compared with the small samples of this schedule included in Fig. 699A. Record D is the final condition of the FR 100, which should be compared with the 4 samples of this in Fig. 699B. Record A is the final performance on both FI 1 and FI 1 drl; the curve lacks the high terminal rate in the ab-

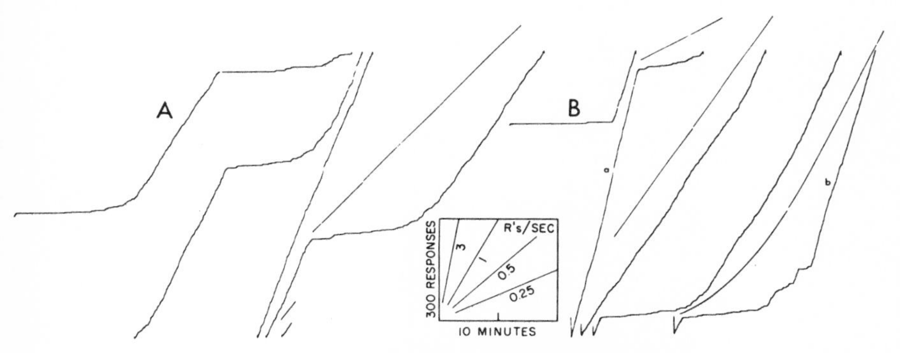

Fig. 701. A. 6-ply multiple schedule: FI 16 record
B. Return to 8-ply multiple schedule

sence of drl in Fig. 699C. Figure 701A shows the resulting performance under FI 16, to be compared with Fig. 699D, the high terminal rate of which has now disappeared.

The following day the original eight-fold schedule was reinstated with immediate effect. Thus, Fig. 700B represents the change in the performance at Record A as a result of re-introducing the drl on the 2 ratio schedules. The effect is to increase the terminal rate. Record E shows the effect on the smaller ratio, including the drl. The portions which are not drl are steeper than those in Record C. Record F is the combined performance for FR 100 and FR 100 drl, and the portions from FR 100 alone are steeper than those in Record D. The effect on the larger interval is not so complete in the 1st session; nevertheless, it shows an increase in the terminal rate, as *a* in Fig. 701B reveals. Note that at *b,* under the drl stimulus, a fairly high rate appears.

In a later stage of the experiment the eight-fold multiple was continued; and the larger ratio, both with and without drl, was increased to 185. This increase produced considerable straining and long positive accelerations in the ratio on drl, but scarcely any pausing on the non-drl member. Figure 702 shows an example of the perform-

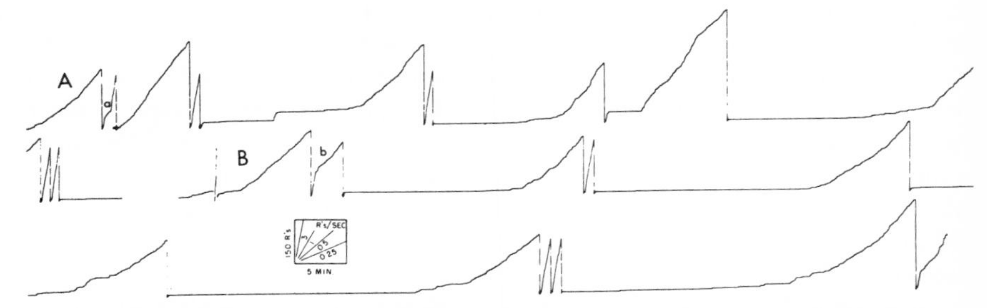

Fig. 702. 8-ply multiple schedule 14 sessions after large FR was increased to 185: FR and FRdrl record

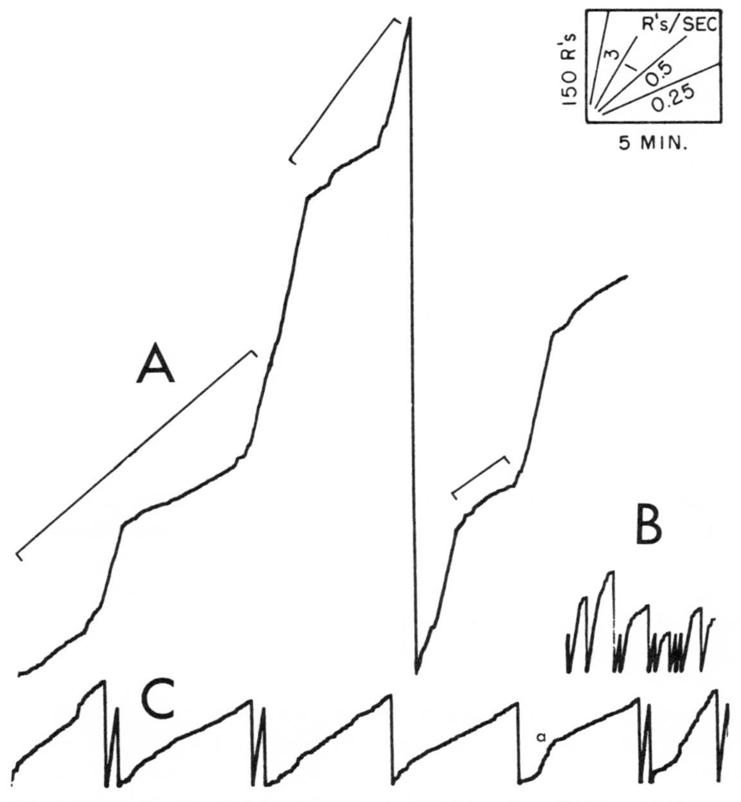

Fig. 703. 8-ply multiple schedule after 24 sessions: FR and small FI records (second bird)

ance on the large ratio with and without drl. The FRdrl "feels the strain" and the straight FR does not. This condition could be the indirect result of the larger mean ratio produced by the failure to satisfy the drl contingencies as soon as the scheduled ratio is counted out. Note the occasional imperfect multiple control at *a* and *b,* where the bird slows down under non-drl conditions. At this stage the drl on the small ratio continues to produce runs at a high rate of from one to three times the ratio; this high rate is followed by a sharp break to a low rate, which, however, often fails to receive reinforcement for several minutes. The rough grain satisfies the drl contingency in this case.

Second bird

A second bird on the same program developed multiple control to the extent seen in Fig. 703 and 704 after 24 sessions. Figure 704 reveals an almost complete suppression of rate by drl. On FI 16 alone, however, a high rate is assumed immediately upon presentation of the appropriate pattern, and continues without interruption for many thousands of responses. Two exceptions to this performance occur late in the session, at *a* and *b,* where the bird "primes" itself into a drl performance on the non-drl pattern. Meanwhile, on the short interval (Fig. 703A), a high rate appears in the absence of drl, and a somewhat more intermediate rate under drl. The small ratio (Fig. 703B) shows the same characteristic as in the performance by the other bird: the ratio is run off at a high rate, the count being somewhat exceeded before a low rate appears. In the larger ratio (Record C) the drl contingency is generally effective from the beginning, and the whole ratio is run off at a low rate, except for an occasional break-through, as at *a.*

When the drl contingencies were omitted on the ratios, the same effect was observed as with the other bird: the ratio rate fell to approximately half the preceding value. The high rate in the larger interval was also reduced. When the drl contingencies were added in the following session, high rates appeared at both intervals, and the ratio performances returned essentially to the condition of Fig. 703.

After returning to the eight-fold multiple schedule, this bird developed considerable "confusion": it appeared to respond more to its own behavior as a controlling stimulus than to the patterns on the key. Figures 705 and 706 show an intermediate state after 74 sessions on the multiple schedule, including the six-fold and seven-fold cases. The small-interval performance in Fig. 705A generally shows a single high rate in the absence of drl, as at *b, c, e, f,* and *g,* but a tendency to break into this rate while under drl, as at *a* and *d.* The small ratio with drl (Record B) continues to show a high rate before slow-down, although in some instances the ratio is not counted out before the drl becomes effective (for example, at *h* and *i*). The small ratio without drl remains high. On the larger ratio (Record C) the performance without drl is standard; but with the drl it shows considerable roughness of grain, which now evidently satisfies the drl contingency. The contingency makes itself felt, however, at the beginning of the exposure to this stimulus. Meanwhile, on the large interval (Fig. 706) much

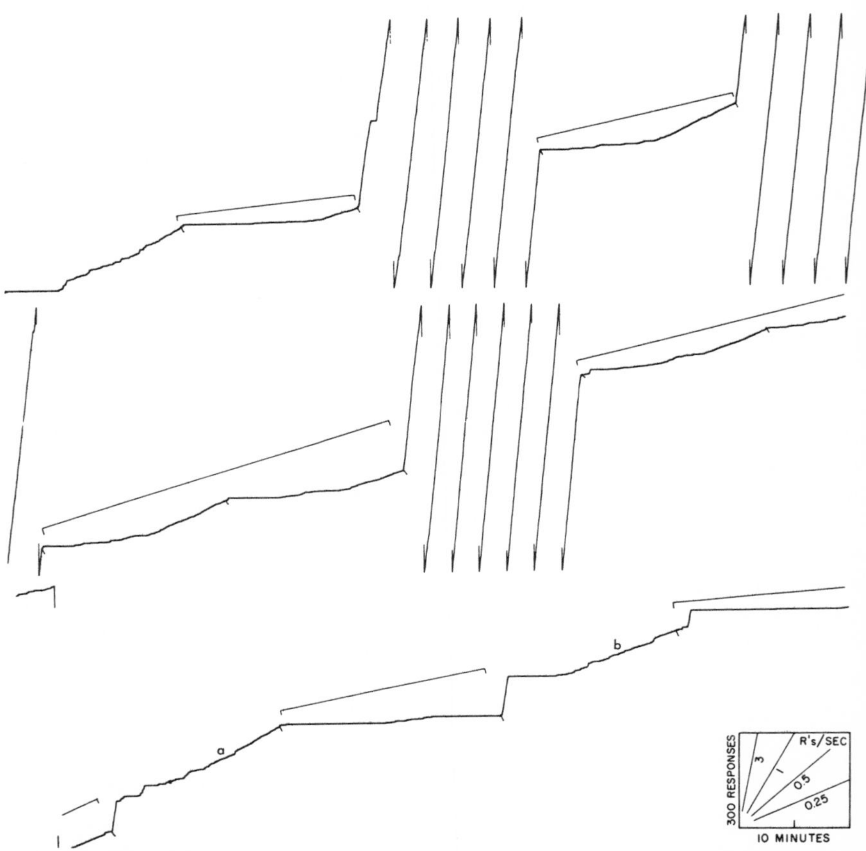

Fig. 704. 8-ply multiple schedule after 24 sessions: large FI record (second bird)

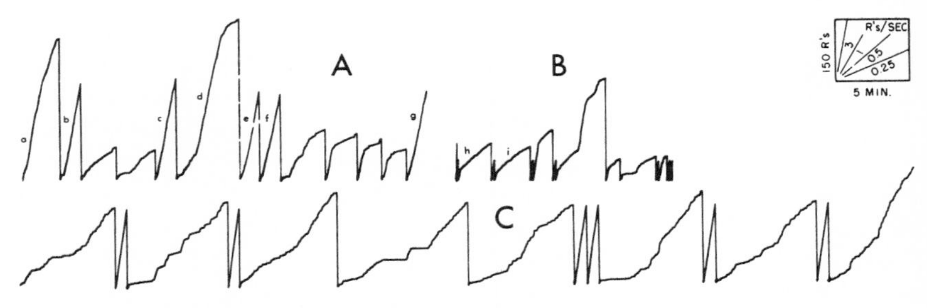

Fig. 705. 8-ply multiple schedule showing "confusion" after 74 sessions. FR and small FI records

Fig. 706. 8-ply multiple schedule showing "confusion" after 74 sessions: large FI record

of the performance shows a fine step-wise grain, which may appear either with or without the drl contingency. (The record resets at reinforcement.) Some periods of drl show the appearance of the high terminal rate characteristic of FI 16 alone, as at *a*, and *b*. In contrast, the FI 16 schedule without drl frequently shows periods of low responding, as at *c* and *d*.

The final picture for this bird shows an even more erratic stimulus control. In Fig. 707A the performance on the small ratio continues to show marked negative curvature. A high rate occasionally breaks through on the large ratio under drl (Record

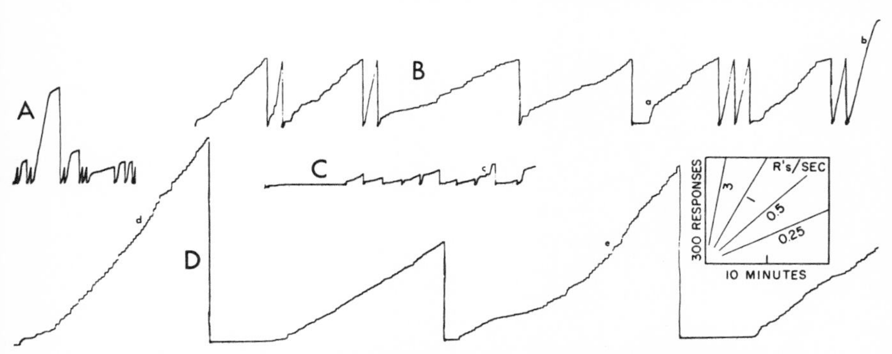

Fig. 707. 8-ply multiple schedule showing "confusion" after 86 sessions

B), either at the beginning of the segment, as at *a,* or throughout, as at *b.* The performance on the short interval (Record C) suggests strong induction from the ratio schedule. The interval characteristically begins with a few responses at a high rate; and regardless of drl conditions, is completed at a very low steady rate, occasionally with a break-through, as at *c.* On the large interval (Record D) some suppression occurs in the presence of drl; but the normal FI 16 is being run off at a much lower rate, and the performance is marked by strong step-wise grain, also suggesting induction from the ratio, as at *d* and *e.*

Third bird

The third bird never achieved good control under the eight-fold multiple schedule. High ratio rates developed early, and the large-ratio drl showed a tendency to begin at the normal ratio rate, contrary to the result for the other 2 birds. The large interval never developed a high terminal rate, although some control of the drl was evident. The smaller interval developed drl control, but the curves were irregular.

At one point a technical defect resulted in an extensive extinction curve in the stimulus appropriate to the large interval without drl. Figure 708 shows the entire session. The 1st interval under drl indicates fairly adequate control. Extinction begins at the arrow, at approximately the previous over-all rate for FI 16 alone. The record shows characteristic break-throughs of the ratio rate, which become more extensive as pauses develop later in the interval. Early break-throughs are seen at *a, b, c, d,* and elsewhere. Later, after long pauses, extensive break-throughs appear at *e, f, g, h,* and elsewhere. The rate during these break-throughs remains constant throughout the ex-

Fig. 708. Extinction with large FI stimulus after 8-ply multiple (third bird)

periment, although the over-all rate otherwise falls off fairly smoothly from a value such as that following *b* to that at *i*, and then *j* and *k*.

MULTIPLE FI 10 AVOIDANCE RS 30 SS 30

In the technique for studying avoidance behavior developed by Sidman (1953), an aversive stimulus (e.g., a shock) is presented periodically. The arbitrary time between shocks is the shock-shock interval (SS). A response, such as pressing a lever, postpones the next shock by another arbitrary interval (RS). We shall designate a given schedule in this way: avoid SS 30 RS 30, where the intervals are expressed in seconds.

Three rats with a history of FI and mix FRFR (see Chapter Eleven) were placed on avoid RS 30 SS 30 for 15 sessions. The schedule was then changed to mult FI 10 avoid RS 30 SS 30. The stimulus appropriate to the fixed interval was a buzzer; the stimulus appropriate to avoidance was the absence of the buzzer. A performance appropriate to the FI 10 developed very slowly in spite of the earlier history. Figure 709 shows performances under both stimuli after 260 hours (58 sessions). A well-marked scallop has developed in FI (Record A); a characteristic low stable rate appears in avoidance (Record B). The hatches mark shocks not avoided, and the dashes above the curve indicate the points at which the schedule changed.

The condition in Fig. 709 was not stable, however. The schedules were being changed in simple alternation, after reinforcement on FI and after 10 minutes on the avoidance schedule. Since responding was fairly constant at both points, many accidental contingencies between a response and change of schedule arose. To some extent the performance on the FI schedule was being punished by a change to the avoidance stimulus, while responses on the avoidance contingency were being reinforced by a change to the FI stimulus. Evidence of the effect of the spurious contingency under the avoidance stimulus appears in the form of bursts of responding which are not apparent on avoidance alone. Figure 710 shows two examples. Records A and B are for a single session recorded as in Fig. 709. Record C is an example from another session, where the avoidance schedule remained in force for a longer mean period. At *a* and *b* hundreds of responses emerge at a rate appropriate to the interval schedule. The accelerations are smooth.

In an effort to check the nature of these bursts of responses, we discontinued the

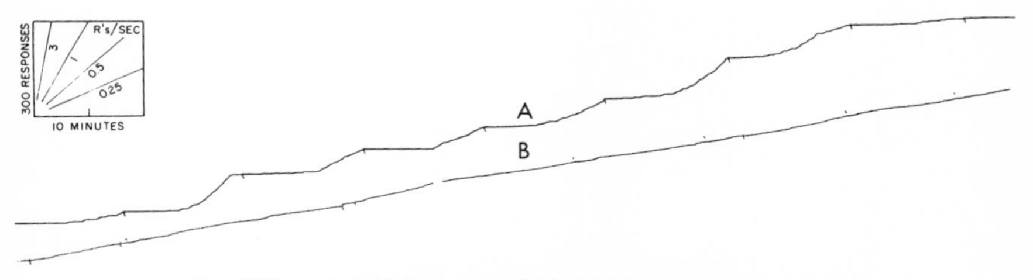

Fig. 709. Mult FI 10 avoid RS 30 SS 30 after 58 sessions

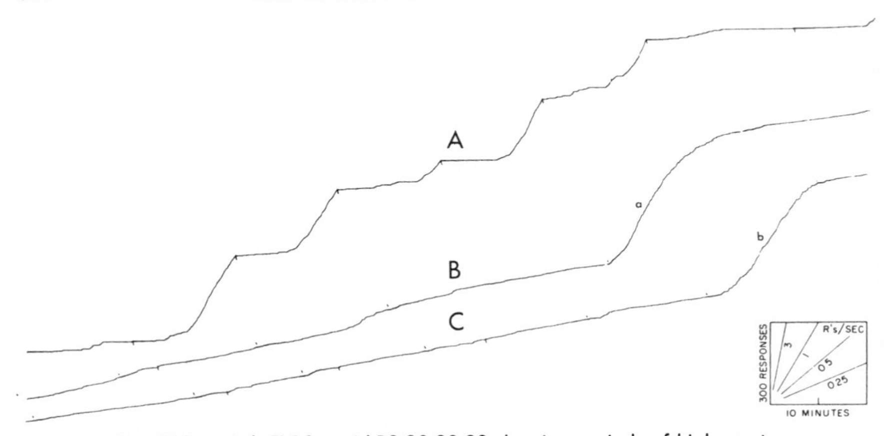

Fig. 710. Mult FI 10 avoid RS 30 SS 30 showing periods of higher rates in the avoidance stimulus

shock, and the avoidance behavior underwent extinction. This procedure had the effect of reducing the low sustained rate in Fig. 710B and C, but the bursts of responses remained in the performances of all three rats. Figure 711A shows a performance on the FI component for a second rat in the experiment. It does not show so smooth a scallop as the preceding figure; the rate is much higher and greater irregularity is evident. This rat also showed low sustained responding on the avoidance schedule, with break-throughs; but when the shock was discontinued (Record B), the sustained rate declined slowly, and responding ceased for long periods, as at *b, c, e, g,* and *h.* But the rate is by no means zero, and occasional bursts of responding at a rate appropriate to the interval schedule appear, as at *d* and *f.* An occasional period of slow responding with rough grain also occurs, as at *a.* This behavior could be the result of reinforcement by the change to the FI schedule stimulus, as experiments in Chapter Twelve will show. It could also be induction or the imperfect stimulus control seen in the 8-ply multiple performance.

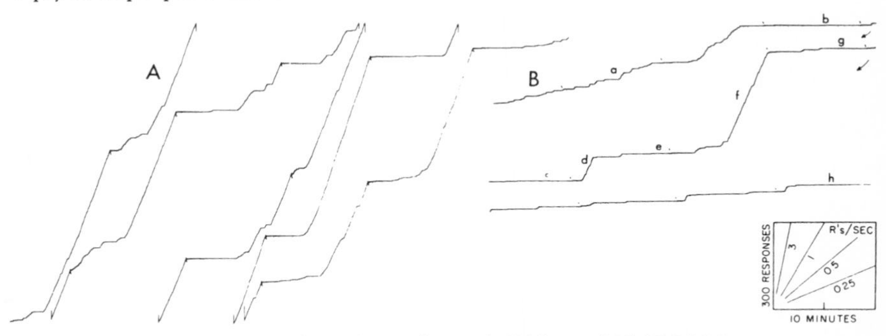

Fig. 711. Mult FI 10 ext after mult FI 10 avoid RS 30 SS 30

MULTIPLE DEPRIVATION

Two magazines were installed in one box; one contained food and the other delivered small amounts of water. (A dipper brought a small amount of water into position where the bird could drink it. The actual amount drunk depended upon the shape of the beak and the drinking behavior of the bird.) In a study of the interaction of types of deprivation, 2 multiple schedules were established. In one case a response was reinforced with water (on a schedule) whenever the key was green and with food whenever it was white. In a second case, 3 stimuli were correlated with (1) water, (2) food, and (3) either food or water at random. The color correlated with the third possibility was blue.

Three pigeons were placed on VI 3 with water reinforcement. During the early stages of the experiment, technical problems connected with the water reinforcement were solved, and the history during this period is rather complex. Later, records were

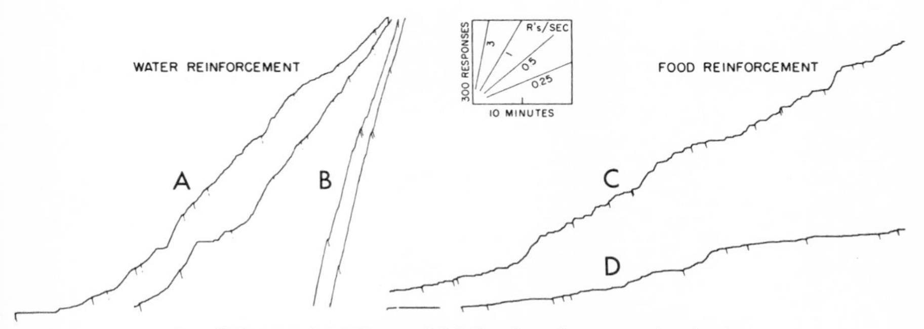

Fig. 712. Mult VI 3(water)VI 3(food) under water deprivation

obtained of the performance when the color was appropriate to water reinforcement (with no food deprivation). (In each case a large dish was available in the experimental box at all times containing the substance of which the pigeon was not at the moment deprived.) Some interaction exists among these two forms of deprivation. A thirsty bird cannot eat and is therefore presumably in some stage of food deprivation. The color on the key changed in simple alternation at this stage; later in the experiment the presentation was random. Changes occurred every 4 minutes without respect to reinforcement.

Figure 712 shows a relatively late stage of mult VI 3 (water) VI 3 (food) under water deprivation. Record A shows a performance with water reinforcement on VI 3 on a green key. Record D is the alternating performance on the white key, where a response is reinforced with food on VI 3; but the bird has undergone no food deprivation and is supplied with food in the box. Note that substantial responding is present in Record D, although the rate is low. Record B is the VI 3 performance for a second bird with water reinforcement; and Record C, the alternating performance under food reinforcement. Both rates are relatively higher than those of the first bird. Again, responding on the food-reinforced key is substantial. The points at which the schedule changed have not been marked.

Shortly after these curves were recorded, the birds were given free access to water and were deprived of food. Figure 713 shows a set of records comparable with those of Fig. 712 with the deprivations reversed. Record A is a VI 3 performance for food reinforcement; and Record D, the alternating performance under the color reinforced by the operation of the water magazine (water being freely available in the box). Record B is the performance of a second bird on food reinforcement, while Record C is the alternating performance under the color reinforced by the operation of the water magazine. (The presentations are now randomized rather than alternated.) A similar condition prevails: responding occurs on the color appropriate to the reinforcement for which no corresponding deprivation exists.

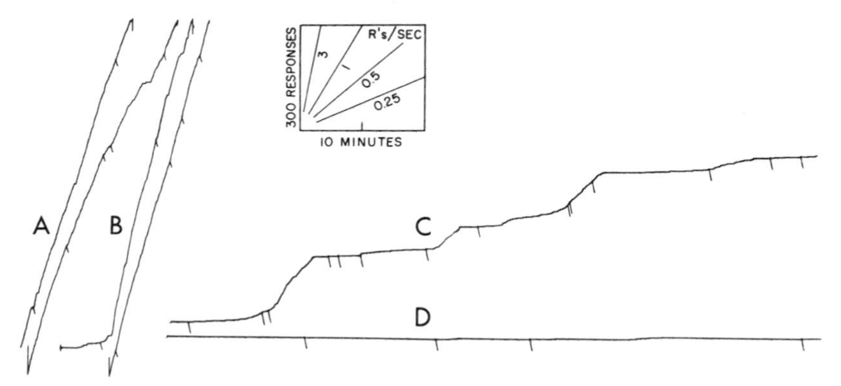

Fig. 713. Mult VI 3(water)VI 3(food) under food deprivation

The operation of the water magazine possibly has conditioned reinforcing properties from its similarity to the operation of the food magazine. When the water magazine was disconnected, responding to the green key dropped almost to zero, except for an initial burst of responses at the beginning of each session for one bird. A more plausible explanation is that the bird is still forming a discrimination between the key-colors. Much later in the experiment (see, for example, Fig. 718) near-zero rates are observed on the "satiated color."

We attempted to examine the performances under the 2 stimuli under simultaneous deprivation in both fields. The experiments are complicated, however, by the large amount of interaction between the 2 deprivations. The effect of a given degree of water deprivation will be affected by the extent of the food deprivation and vice versa.

In general, the VI curves obtained under this procedure are quite irregular. They show tendencies to drop to intermediate rates, and the curvature after reinforcement may be quite variable. Record A in Fig. 714 shows a performance on the key-color reinforced with water; and Record B, that on the key-color reinforced with food. The over-all rates are comparable. Note that Record B begins with about 20 minutes of responding at a low rate during several presentations of the "food" key-color. This performance frequently appears in food records, and may indicate that until the bird has received a few water reinforcements, it is unable to eat or at least eats at a rate suggesting a much lower level of food deprivation than that which actually prevails. After a small amount of water is received, the full effect of the food deprivation is evident. The rate which breaks through at *a* and again later in the record is much higher than the mean over-all rate.

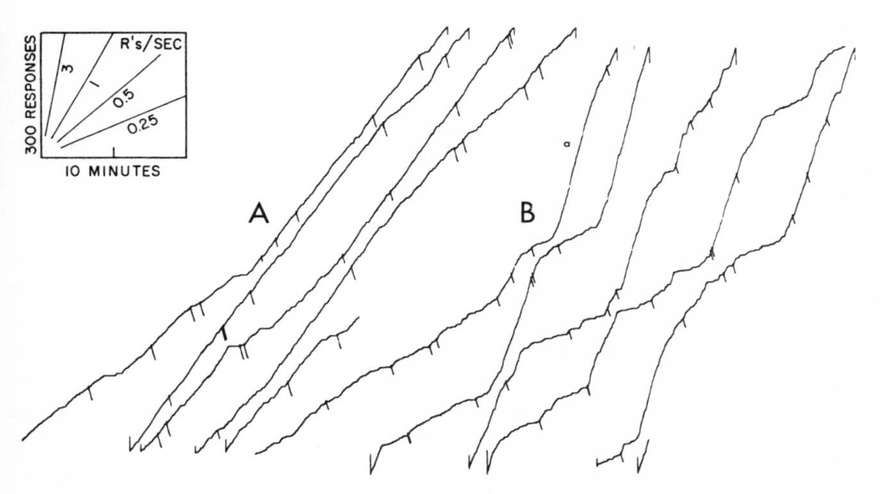

Fig. 714. Mult VI 3(water)VI 3(food) under food and water deprivation

In Fig. 715, for a second bird, a higher level of deprivation in both fields is suggested by the higher over-all rate. Record A shows responding on the green (water) key; and Record B, on the white (food) key. In this case, food behavior is not suppressed at the beginning of Record B as the result of concurrent water deprivation. Note that both records show an occasional drop to a lower rate, as at *a* and *b*.

In Fig. 716, for the third bird, the rate of the green (water) key at Record A shows a decline during the session that is probably due to satiation. The change in rate from *a* to *b* is of the order of 4:1. Concurrently, responding on the key-color reinforced with food shows a very low level of responding until the last few minutes of the session (Record B). The bird had been deprived of food as well as water, but the combined deprivation has here produced low rates on the food key.

An earlier stage with the same bird shows a decline in rate during the session on water reinforcement, while a high rate is maintained under food reinforcement (Fig. 717).

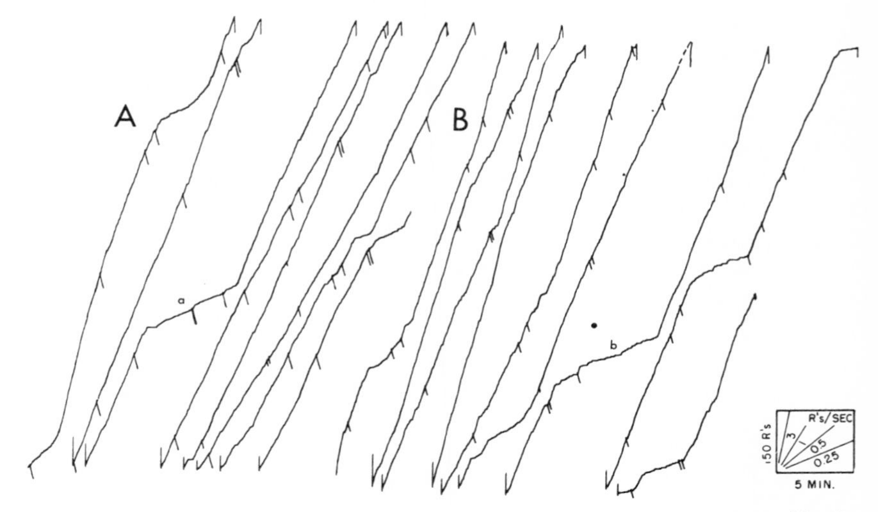

Fig. 715. Mult VI 3(water) VI 3(food) under food and water deprivation (second bird)

The over-all rate of responding on the water color changes from a high value at *a* to an extremely low value at *b*. A brief period of slow responding occurs on the food key at *c*, which may represent the suppressing effect of the concurrent water deprivation.

When both deprivations were substantial and fairly high rates were being shown on both key-colors, a glass of water was placed in the experimental box. Two birds consumed some time in drinking; this time appeared as breaks in responding on the food key, which happened to be present then. When the water color was on the key, how-

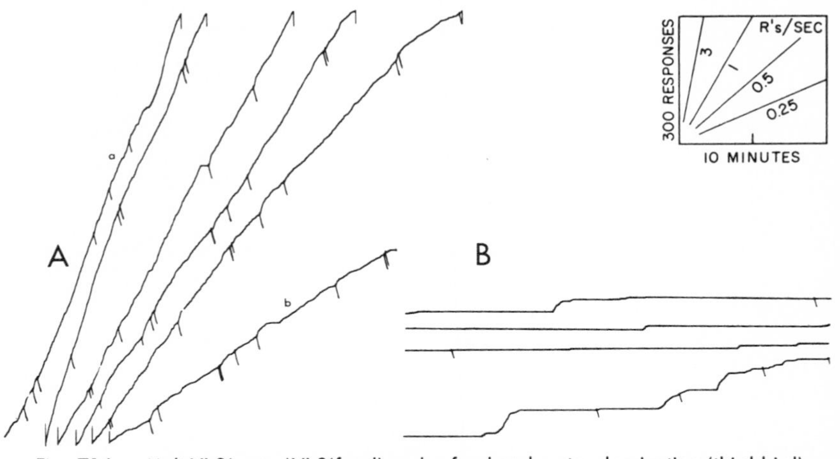

Fig. 716. Mult VI 3(water)VI 3(food) under food and water deprivation (third bird)

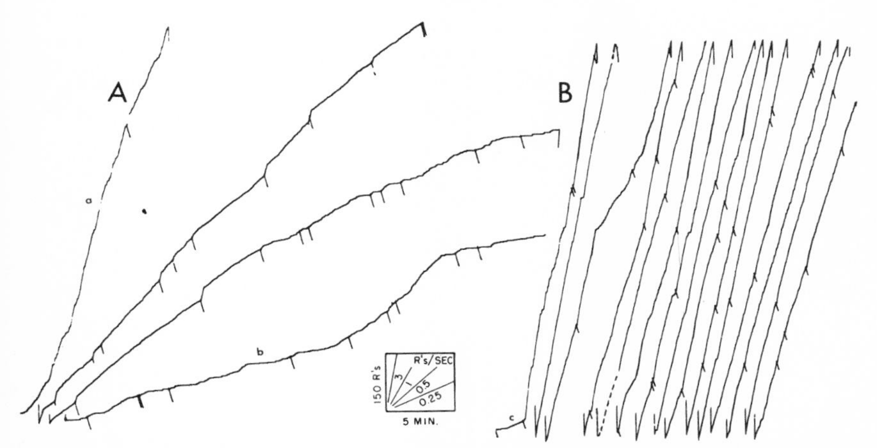

Fig. 717. Mult VI 3(water)VI 3(food) under food and water deprivation showing satiation under water reinforcement (third bird)

ever, little or no change occurred in the food rate. Figure 718 shows an example. Record A is for water; Record B, for food. The session begins with a high rate on the food color at *b,* and an intermediate rate with some irregularity on the water color at *a.* At the arrows, water was introduced. The remainder of the session shows very little responding on the water color. Responding on the food-reinforced key continues, although with increasing irregularity. The periods of low responding at *c, d,* and *e* all begin after a changeover from the water color. They possibly indicate some carryover of the low level of responding on the other component of the multiple schedule.

In a somewhat similar experiment the 3 birds were given free access to food and

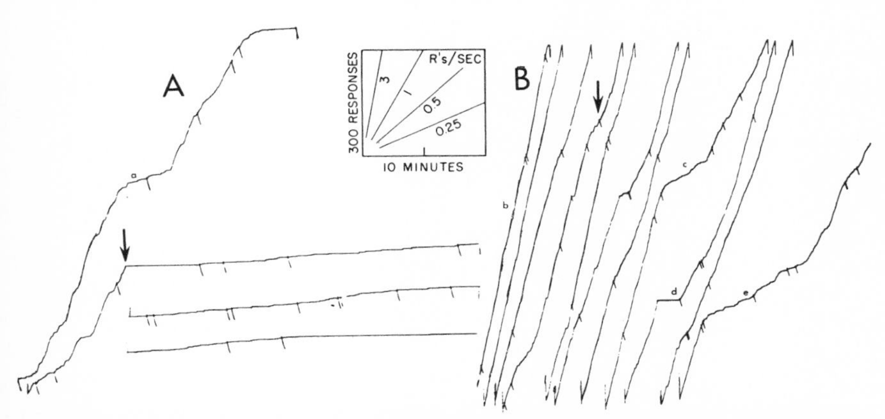

Fig. 718. Transition from mult VI 3(water)VI 3(food) under food and water deprivation to free access to water

deprived of water until stable rates were established. The deprivations were then reversed. Figure 719 shows the results. Each curve in the graph represents 2 daily experimental sessions. At the left of the vertical line the prevailing condition was water deprivation with full access to food. Performances after 24 hours of food deprivation are at the right of the line; but a supply of water is now available in the experimental box for the first time. Because of the water deprivation, the body-weights were near the 80% level in spite of the free access to food. Thus, a 24-hour food deprivation had severe effects.

Each bird is represented in Fig. 719 by 2 curves, the segments being recorded in random alternation in the same session. Consider, for example, the pair of curves E and F. Curve F shows responding on the water key on both days; E shows responding on the food key. On the day represented to the left of the vertical line a substantial, though irregular, rate is maintained on the water key (Curve F). Responding to the food key was negligible (Curve E). The following day, under reversed deprivations, the water record falls to almost zero, while the food record shows a fairly steep slope. (The first few minutes in all curves to the right of the line show low rates. The birds presumably spent most of this time in drinking and exploring the cup of water.)

The other 2 pairs of curves show a similar result, except that the key-colors are less effective controllers of rate. Substantial responding occurs on the water key after water satiation (right-hand portions of Curves A and B) and on the food key under access to food (left-hand portions of Curves C and D).

The interaction between food and water deprivation was illustrated by another experiment, reported in Fig. 720, where the numbers of responses to the food and water keys during the first hour of each daily session were plotted. The 2 points at the 1st session show a rate of approximately 6000 responses per hour on the food key and fewer than 100 responses per hour on the water key. At this time the conditions were free access to water and 80% body-weight controlled in the usual way. Immediately after this 1st session, water was withheld. An increase in the rate on the water key is not marked until the 4th session; and by this time the rate on the food key has dropped to below 2000 responses per hour. This lower rate may be due in part to additional food given because the body-weight is lowered by the dehydration; but it also probably reflects a lower level of activity under a given schedule of food deprivation as the result of the concurrent water deprivation. Later in the experiment, a substantial but varying rate appears on the water key. The only water now being received is that used for reinforcement during experimental sessions. In general, there is a rough tendency for the rate on the food key to rise when the rate on the water key falls. This tendency is particularly noticeable in the 13th session. When the bird is given free access to water again in the 14th and 15th sessions, the rate on the food key rises to roughly its original value.

In the later stage of the experiment, 3 key-colors were used. On one the response was reinforced with food; on another it was reinforced with water; and on the third, with either food or water in an unpredictable series and equally often. The perform-

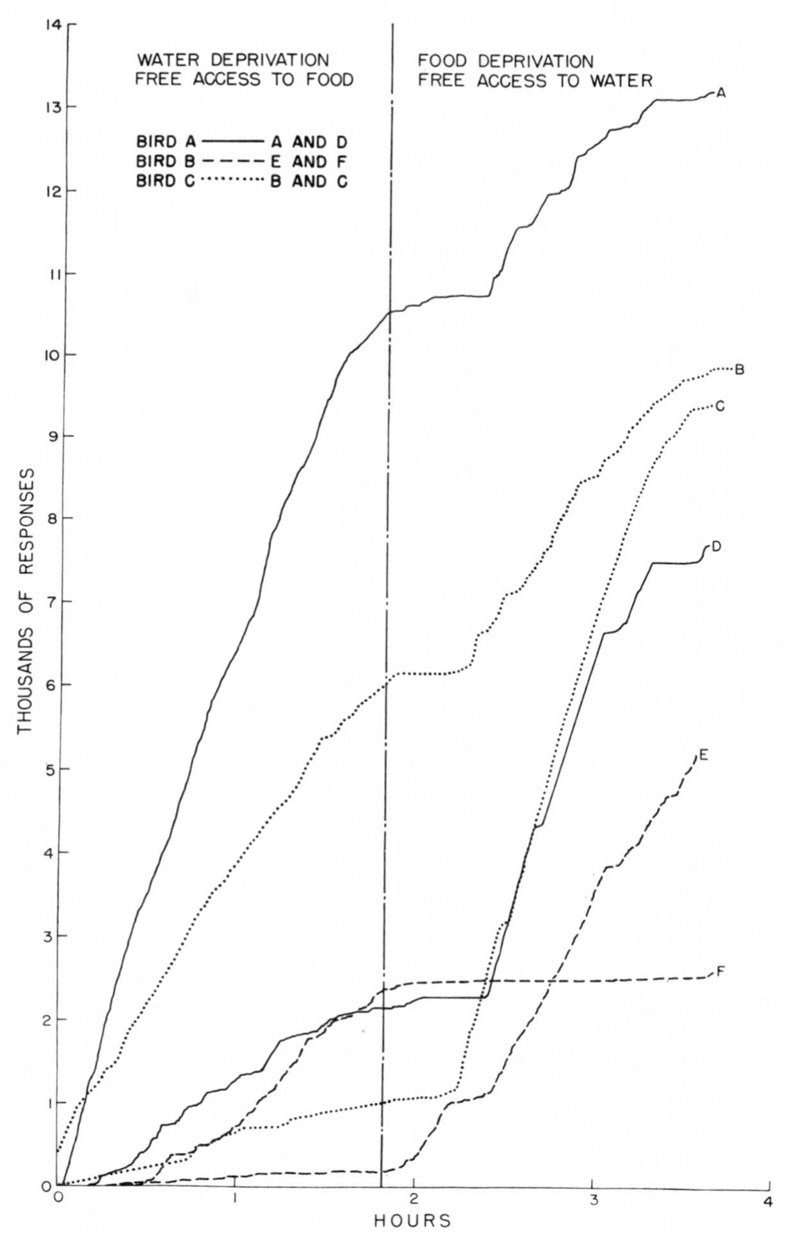

Fig. 719. Transition from water deprivation and free access to food to food deprivation and free access to water

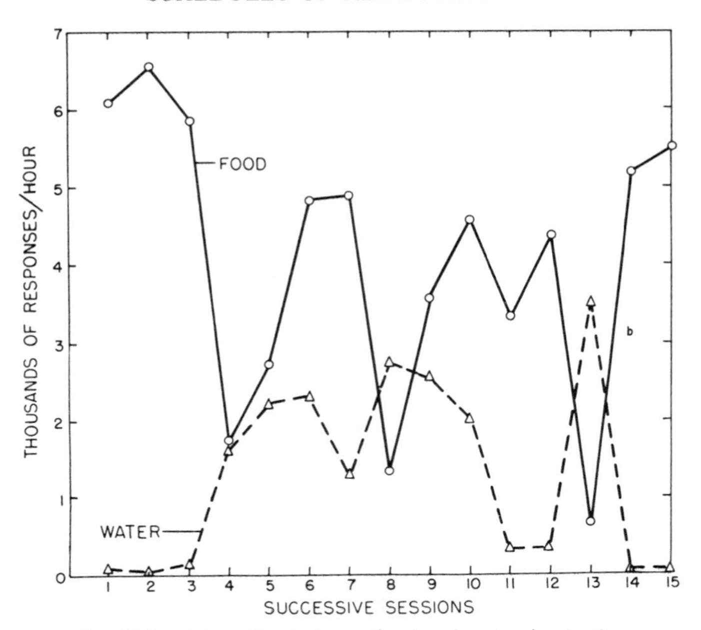

Fig. 720. Interaction between food and water deprivation

ance on the third key will depend to some extent upon the preceding history. If, for example, water deprivation has been moderate but food deprivation severe, performance on the mixed key is likely to show the effect of reinforcement with food and to retain signs of this effect later when conditions of deprivation are changed. Eventually, however, the key reinforced with both food and water assumed a value which appears to be a simple arithmetic mean of the rate on the other 2 keys under any given set of deprivations. Figure 721 shows an example in which both deprivations were substantial. Record A is the performance on the green key reinforced with water. Record B is the performance on the white key with the food reinforcement. An important feature is the tendency, already noticed, for the rate to decline temporarily following a changeover from the water key, as at *a*. Record A shows a moderate rate under water reinforcement and water deprivation. The slope of Record C on the key reinforced with both food and water is close to the mean for the other two over-all slopes.

Figure 722 shows an example of this averaging when one deprivation is slight. On the white key the bird is reinforced with food and responds rapidly (Record A); its rate tends to fall shortly after changeover from the water key, as at *a*. Record C illustrates a very low rate under water reinforcement under free access to water. The slope of Record B on the key reinforced with both food and water is again close to the mean of the other 2 curves.

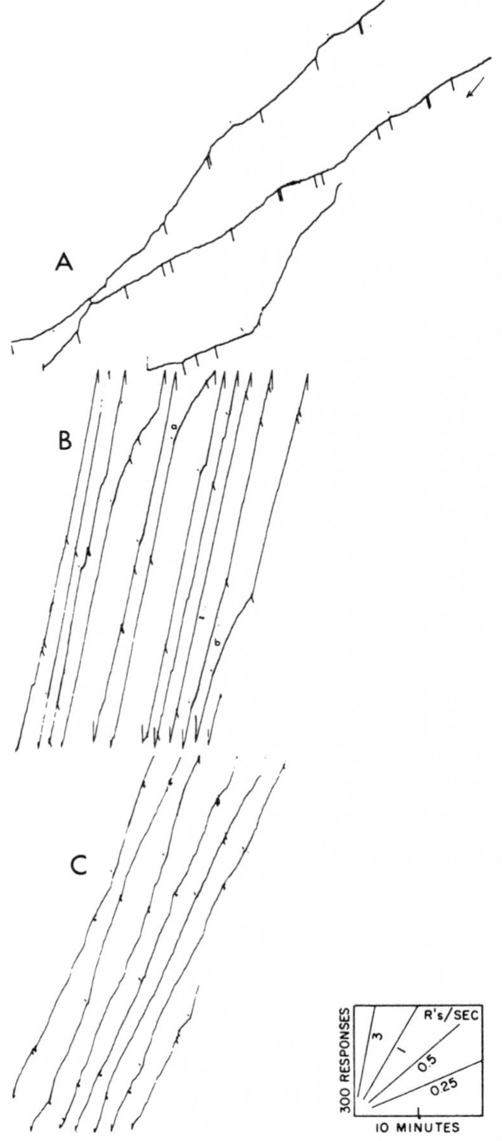

Fig. 721. Mult VI 3(food)VI 3(water)VI 3 (food or water) under food and water deprivation

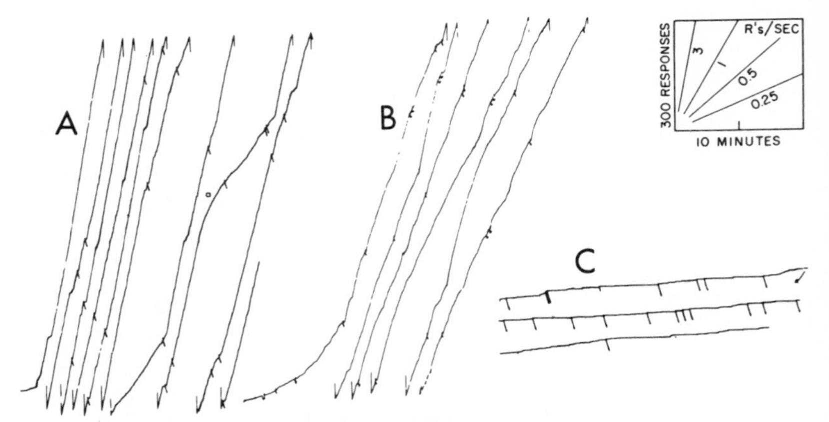

Fig. 722. Mult VI 3 (food)VI 3(water)VI 3(food or water) under food deprivation and free access to water

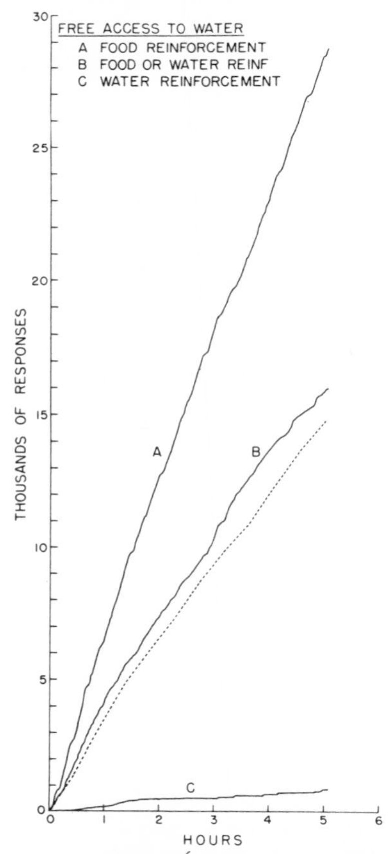

Fig. 723. Mult VI 3(food)VI 3(water)VI 3(food or water) under food deprivation and free access to water

Figure 723 shows continuous cumulative curves on the 3 key-colors, changing at random throughout the session and with free access to water. Curve A shows responding on the key-color reinforced with food under an existing high food deprivation. Curve C shows the performance on the key-color reinforced with water under free access to water. Any responding here is probably to be attributed to either conditioned reinforcing properties of the operation of the water magazine or an incomplete key-color discrimination. Curve B shows responding to the key-color reinforced with either food or water in a random order and an equal number of times. A dotted curve has been drawn to show the mean of Curves A and C. The curve on the color reinforced with both food and water is somewhat higher than the mean of the other curves, but is still a fair approximation.

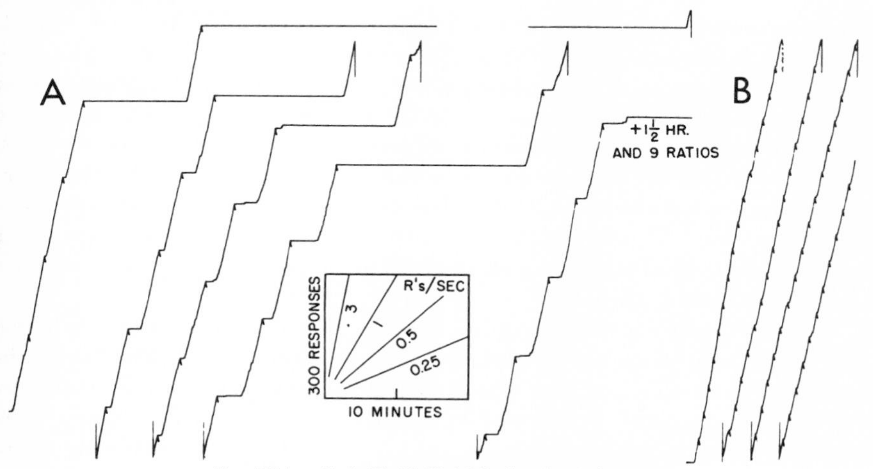

Fig. 724. Mult FR 55 FR 185 after brain lesion

PERFORMANCE ON MULTIPLE SCHEDULES AFTER BRAIN LESIONS

Figures 384 and 385 demonstrated a performance on FI 2 after brain lesion. The bird was subsequently reinforced on FR 60 for 18 sessions on a blue key, followed by 1 session on FR 185 on an orange key. Other sessions on FR 60 on the blue key followed. Figure 724B shows the performance in the next session when the schedule was FR 55 on the blue key. This performance was recorded 39 days after an operation, effects of which were determined post mortem as follows: "Autopsy revealed large cyst in the forward half of the brain, 9 mm in depth and 3 mm in diameter. This extended all the way from the top to the base of the right cerebral hemisphere. Microscopic examination shows marked loss of tissue on one side with evidence of trauma of half the remaining brain."

The performance is appropriate to the schedule, with an over-all rate of about 3 responses per second and pauses after reinforcement varying from zero to a few seconds.

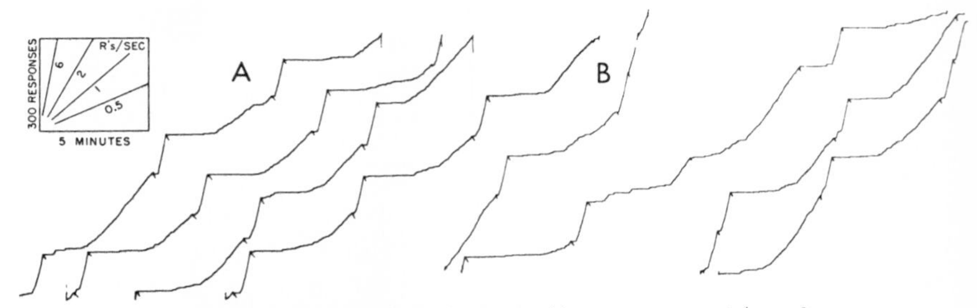

Fig. 725. Mult FIFR before brain lesion and best post-operative performance

In the following session, Record A, the key was orange and the schedule FR 185. Only brief pauses follow the first 3 reinforcements. The pauses increase, however, to the order often produced by a fixed ratio of this size. The rate at which the pauses develop and the lack of any disturbance of the usual fixed-ratio pattern suggest that a great deal of stimulus control has already developed from the two exposures to FR 185 on the orange key. We cannot say whether or not the fact that this bird's performance is adequate on mult FRFR and not on FI represents a lesion which produces a specific deficit. A return to the fixed-interval schedules of reinforcement under which atypical curves were previously recorded would possibly now produce behavior appropriate to these schedules.

In a control experiment on the effect of brain lesions on performances on multiple schedules, 2 birds which had established stable performances on mult FRFI were subjected to a sham operation where the surface of the brain was exposed. Post-mortem microscopic examination showed slight injury to the brain surface. One of these birds had shown the performance in Fig. 725A on mult FI 10 FR 125 before the operation. This rather large ratio is run off with only slight pausing or curvature at the beginning, and the interval scallops are quite consistent and usually smoothly accelerated. The first experimental session 13 days after an operation begins as in Fig. 726. The multiple control is present from the start. Ratio performances show some curvature, as at *a,* although instances appear in which the entire ratio is run off at a high rate,

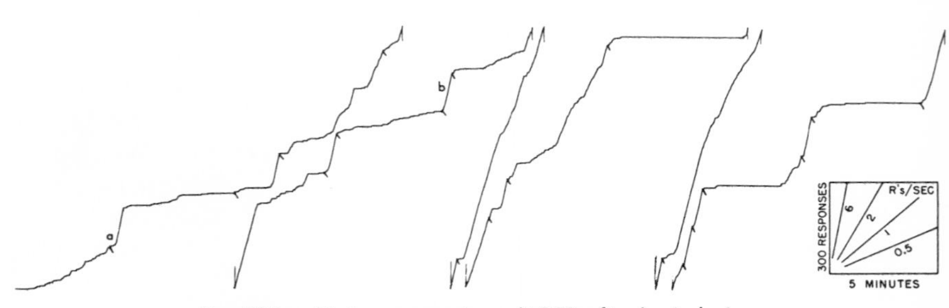

Fig. 726. First exposure to mult FIFR after brain lesion

as at *b*. The interval curvature has been disturbed by an interruption of the experiment and by the incidental effects of the operation. Toward the end of this 1st session, not shown, a fairly clear interval curvature begins to emerge. The performance on the multiple schedule 3 sessions later is shown in Fig. 725B, which may be compared with Record A in the same figure, taken before the sham operation. In comparing the 2 records, note that interval and ratio were alternating in Fig. 725A and changing more or less at random in Fig. 725B. A simple alternation generally gives a more stable performance. The other bird in the experiment also showed recovery of a multiple performance in the 4th session following operation.

Chapter Eleven

• • •

MIXED SCHEDULES

Mixed schedules are the same as multiple schedules except that no stimuli are correlated with the component schedules. Thus, under mix FR 100 FI 10 the organism is reinforced either on FR 100 or on FI 10, and there is no difference in the stimuli present in the two cases. Every multiple schedule has a corresponding mixed schedule. The methods of programming the components are also the same in the two cases.

MIXED FIXED-RATIO FIXED-RATIO SCHEDULES

Experiment I

In an early experiment 2 birds were reinforced on mix FR 190 FR 30 after crf. One of them developed the intermediate performance shown in Fig. 727A. The small ratio shows a slight pause and a high, well-sustained terminal rate. The large ratio shows inflections corresponding roughly to the size of the smaller ratio. Each long segment contains at least one, the distance of which from the preceding reinforcement is slightly more than the small ratio. Segments with two inflections are common. As many as three may occur, as at *a*. Evidently, the bird's own behavior serves as a stimulus in the manner of a multiple schedule. The emission of approximately the number of responses in the smaller ratio "primes" a pause appropriate to the larger ratio. This performance becomes clear later, when the larger ratio is divided into 4 almost equal parts. Later on, after 56 hours, the bird develops an almost perfect priming from the short to the long ratios shown in Record B. After almost every reinforcement, there is a run slightly in excess of the smaller ratio, followed by a marked break which may extend to as much as 5 minutes, followed in turn by a single sustained run for the balance of the long ratio. Almost no pausing occurs in the short ratio, but a characteristic long pause develops in the long ratio when the controlling stimulus has been set up by the initial run.

Figures 728 and 729 show various stages in the performance of the other bird. Figure 728A shows multiple inflections similar to those in Fig. 727A except that the

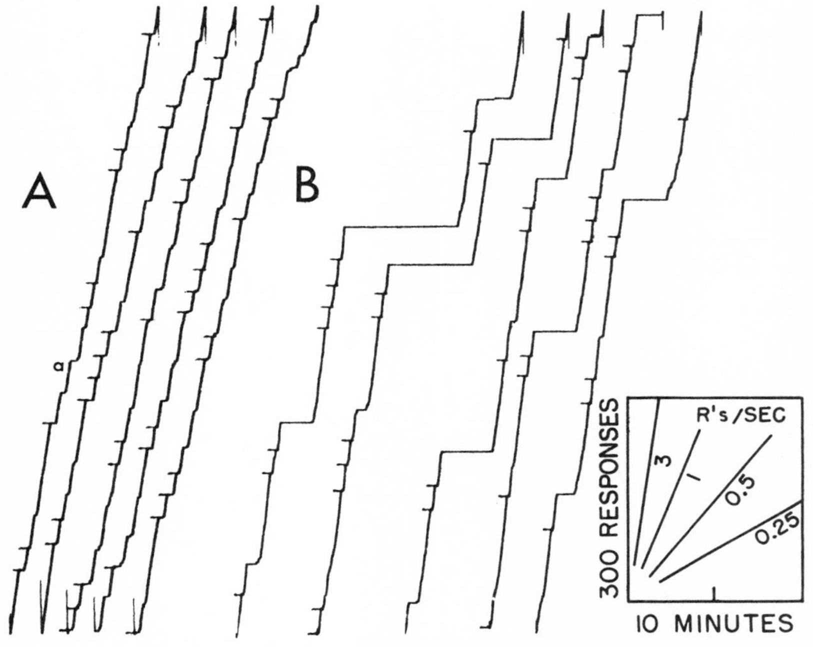

Fig. 727. Early and later performances on mix FR 30 FR 190

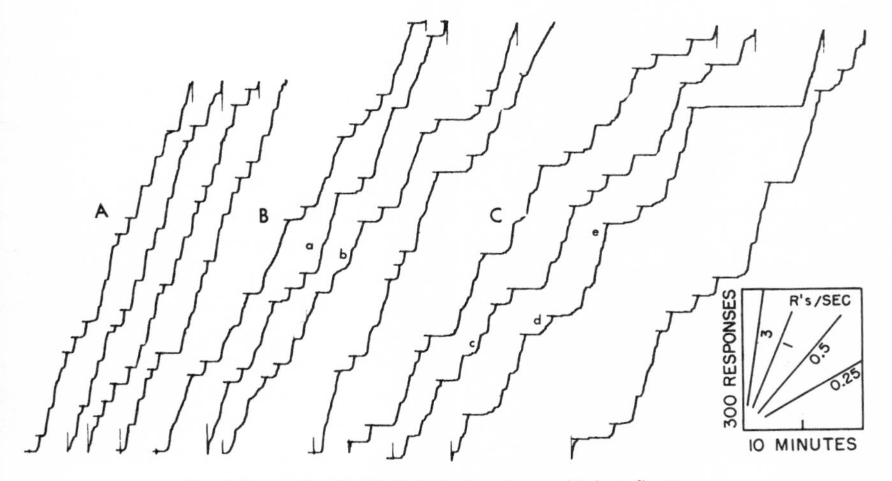

Fig. 728. Mix FR 30 FR 180 showing multiple inflections

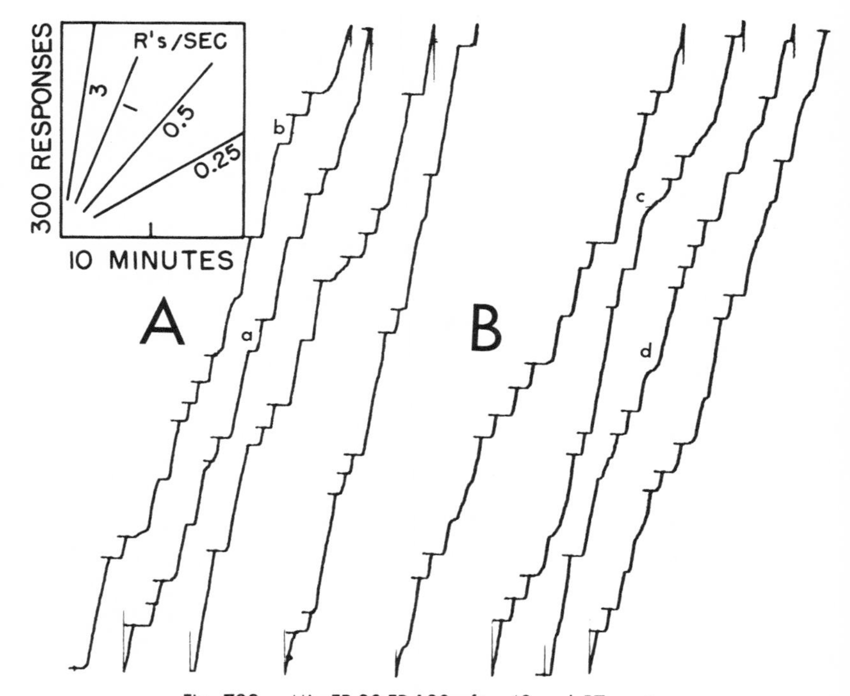

Fig. 729. Mix FR 30 FR 180 after 48 and 57 sessions

changes in rate are not so abrupt. Figure 728B reveals better priming into the balance of the longer ratio. The first break and inflection are more marked than any occurring later in the long ratio, for example, at *a* or *b*. The wavelike character of the record is damped during the long ratio. The factors responsible for the oscillation are to some extent controlled by the distance the bird has progressed from the preceding reinforcement. This damping is still evident in Record C, as for example at *c*, and the first pause after reinforcement is much more marked than any break occurring later in the long ratio. Here, for example, from *d* to *e*, is an over-all curvature appropriate to the longer ratio. In Fig. 729A the first break occurs later. This record was taken 48 sessions after the first exposure to the mixed schedule. In many instances, as for example, at *a* and *b*, the break occurs after the middle of the longer ratio has been passed. Record B shows another temporary phase containing S-shaped curves, as at *c* and *d*. The curve at *d*, for example, is similar to those in Fig. 727B except that in place of sharp breaks, the curvature is fairly smooth.

Note the consistent difference between the 2 birds in Fig. 727 and 729. The first effect after reinforcement is appropriate to the small ratio in Fig. 727 (no pause) and to the larger ratio in Fig. 729 (relatively long pause after reinforcement).

Experiment II

Two other birds with a history on multiple and mixed schedules were reinforced on mix FRFR, where the size of the larger ratio varied from FR 250 to FR 380; the smaller remained at FR 60 throughout, and the proportion of large to small ratios was varied. A 2.5-minute TO followed every reinforcement throughout the experiment. The ratios were recorded separately, the pen resetting at reinforcement.

Figure 730 shows 4 stages in the development of a mix FRFR performance. Record A shows the larger ratio after 20 sessions on mix FR 60 FR 360; Record B, after 22 sessions; Record C, after 29 sessions; and Record D, on FR 60 FR 380, after 30 sessions. Note the multiple breaks at *a* and *b*. The terminal rate in the large ratio has fallen by Record D (at *c, d, e,* etc.).

The other bird developed a much more consistent performance on mix FR 60 FR 360. Figure 731 shows the large ratios in the 23rd session. The paper speed has doubled. Note the sharp breaks and the relative consistency of pausing after a number of responses greater than the short ratio has been run off. There is no pausing after reinforcement, except at *d* and *f*. The breaks occur considerably beyond the FR 50 point; the break at *c,* for example, occurs after about 150 rather than 50 responses after reinforcement. The remaining "ratio" after this pause is of the same order as that before it. The bird is, in essence, being reinforced 50% of the time on a ratio varying around 150, in addition to the reinforcement on FR 60. In 3 cases, at *a, b,* and *e,* the larger ratio is run off without a noticeable change in rate.

The sharp break in Fig. 731 is only a temporary phase. Under various manipulations of the percentage of large and small ratios, the performance shown in Fig. 732 was generated. Here, the over-all rate has declined, and multiple breaks frequently appear, as at *b, c,* and *d,* within a single long ratio. Occasional sharp breaks still occur, as at *f*; but even so, the balance of the interval shows some negative acceleration and poor grain. Note at this stage the occasional appearance of pauses after reinforcement, as at *a* and *e*. Little relation exists between the rate and a preceding reinforcement or pause. The performance suggests a variable ratio with rather rough grain, as one might expect from the information in the preceding paragraph.

Figure 733 shows a still later performance for the same bird on mix FR 60 FR 360 after an intervening history of straight FR 360. The separate long-interval segments have been "stacked" for better reproduction. The break occurs much closer to an actual reading of 60 responses, but the acceleration at the break is seldom abrupt. The balance of the long interval begins to show an over-all slight positive curvature. Here, responding on the short ratio "primes" the bird into a performance appropriate to the long ratio. These curves should be compared with those under mix FRFI with respect to the accuracy of the priming runs.

Figure 734 presents a complete session for this bird at this stage, in which both ratios are shown in the order in which they occurred. The shorter ratios are generally completed without pausing, and only an occasional pause occurs after reinforcement, as

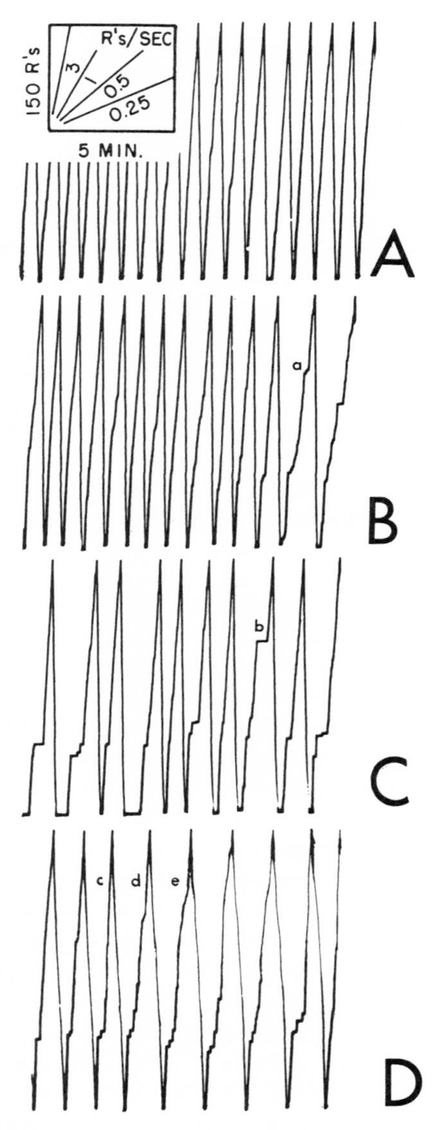

Fig. 730. Four stages in the development of mix FRFR

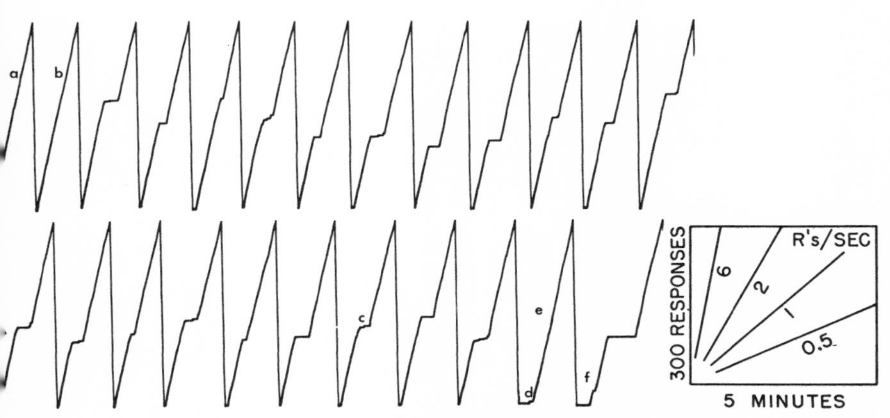

Fig. 731. Early development of mix FRFR performance (second bird)

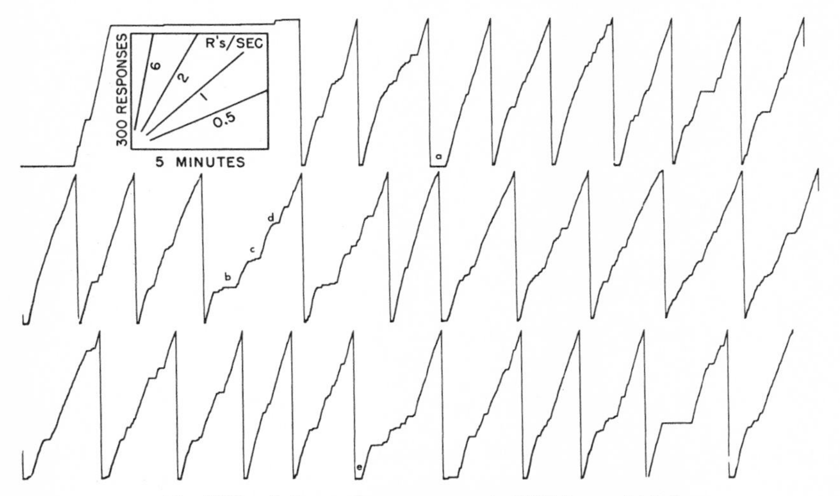

Fig. 732. Later performance on mix FRFR (second bird)

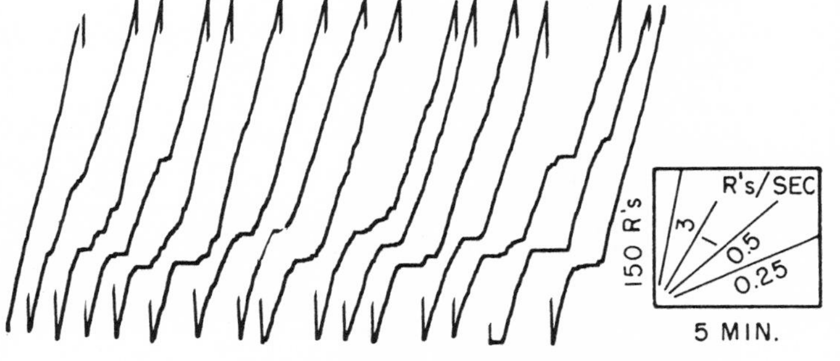

Fig. 733. Mix FRFR showing good priming (second bird)

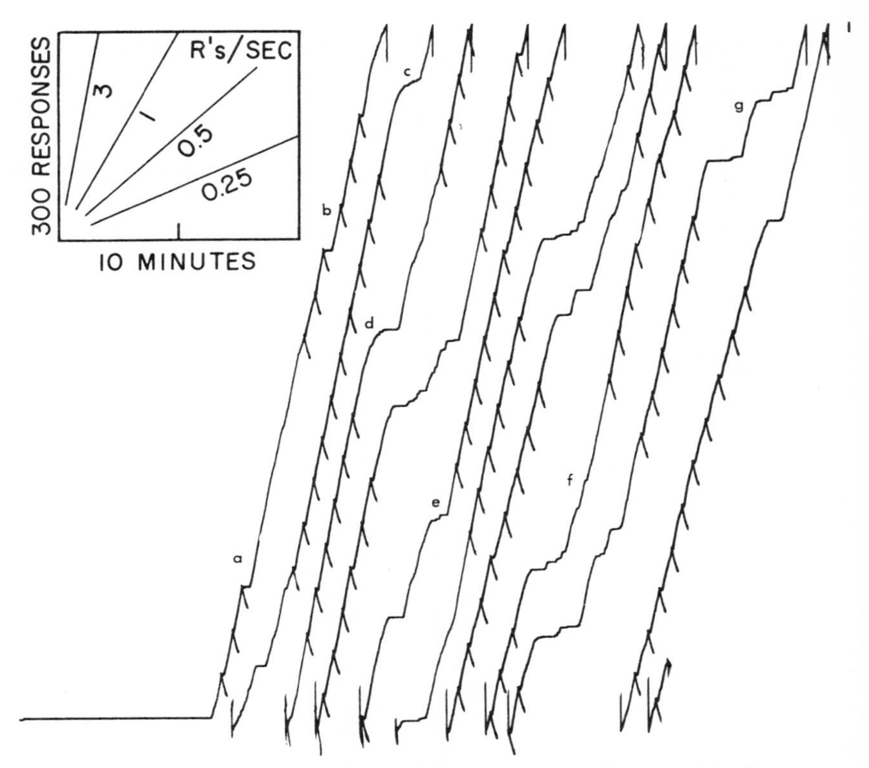

Fig. 734. Complete daily session on mix FR 60 FR 380 (second bird)

at *a* and *b*. A well-marked negative break occurs at a number of responses equal to approximately twice the smaller ratio, as at *c*, *d*, and elsewhere. The balance of the long ratio may contain additional breaks, as at *e* and *g*; or it may continue with a fairly good acceleration suggesting a larger ratio performance, as at *f*.

Figure 735 contains a single session for the first bird, recorded in two ways, at approximately the stage shown in Fig. 730C. Record C gives the whole performance. (One hour has been omitted in which only a response or two was emitted.) Pausing due to the shorter ratio is fairly well-developed, and multiple breaks occur, as at *m* and *n*. The segments on the larger ratio have been separately recorded at Record A, and on the smaller, at Record B. Corresponding performances are as follows: *a* and *i*, *b* and *j*, *c* and *o*, *d* and *p*, *e* and *h*, *f* and *k*, and *g* and *l*.

Whether pauses will appear on mix FRFR depends upon the mean of the ratios, and hence varies with the sizes of the ratios and the proportions in which they are used in the program. A bird which had been on a mix FR 60 FR 360 for many sessions, in which only two instances of the longer ratio occurred for every ten of the shorter (yielding a low mean ratio), paused little or not at all, as Fig. 736A shows. When the proportions of the two ratios were changed so that three of the larger ratios occurred for every nine of the smaller, pausing quickly appeared after reinforcement, as at *e, f, g,* and *h,* and soon after the short ratio had been completed, as at *a, b, c, d,* and elsewhere.

Figure 737 reports a full daily session, in which for the first time the schedule is merely 360 after a history of mix FRFR. Although some trace of a break exists after the shorter ratio, particularly late in the session, as at *c, g,* and *h,* the main change is the development of pauses after reinforcement, as at *a, b, d, e, f,* and elsewhere. Here, the discontinuation of the reinforcement on the shorter ratio permits extinctions

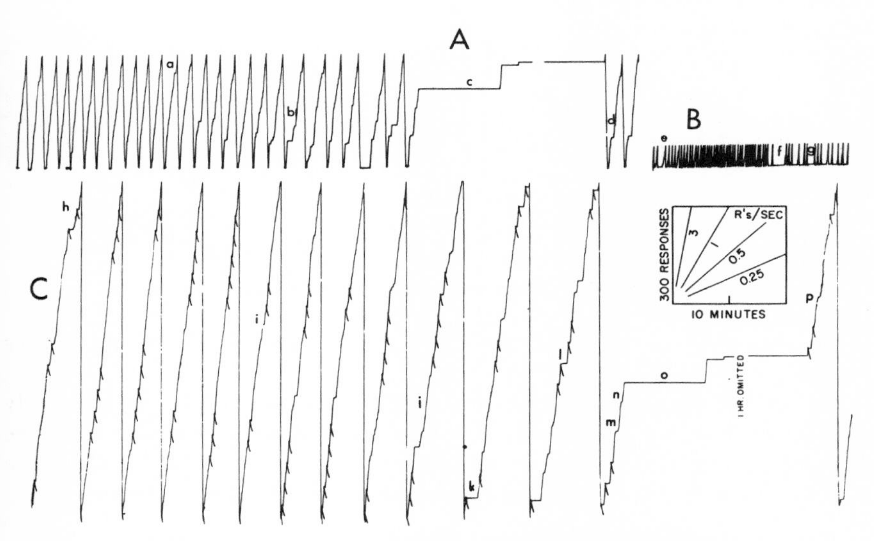

Fig. 735. Complete daily session on mix FR 360 FR 60 (first bird)

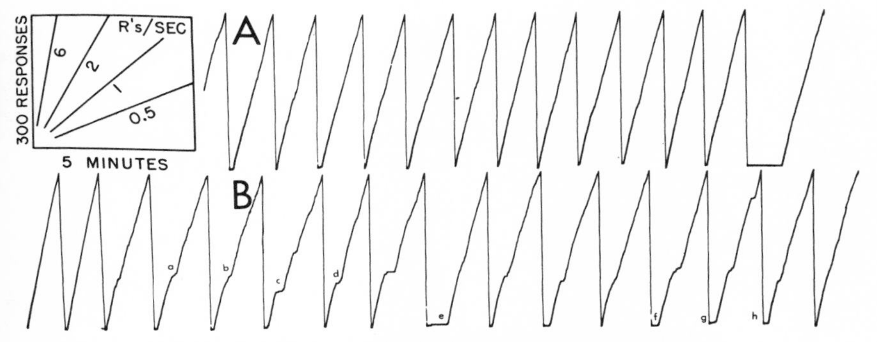

Fig. 736. Change in the proportion of short to long ratios

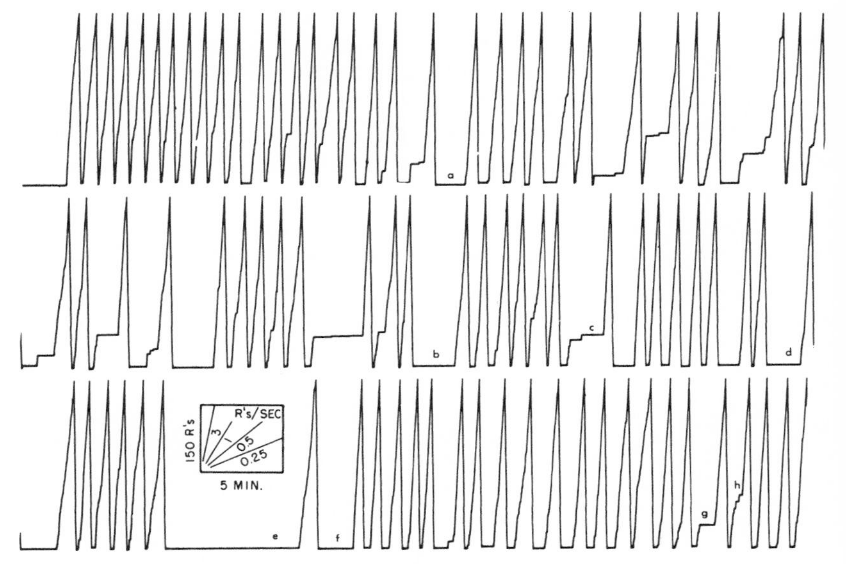

Fig. 737. First session on FR 360 after mix FR 60 FR 360

of the priming break; but it generates a higher mean, which produces pausing after reinforcement.

Another bird, changing for the first time to FR 360 after mix FRFR, gave an entirely different result, as Fig. 738 shows. Here, the increased mean ratio produces severe straining and emphasizes the breaks surviving from the earlier schedule. After more than 3 hours on FR 360, breaks occur frequently at the appropriate points, at *g, h, i,*

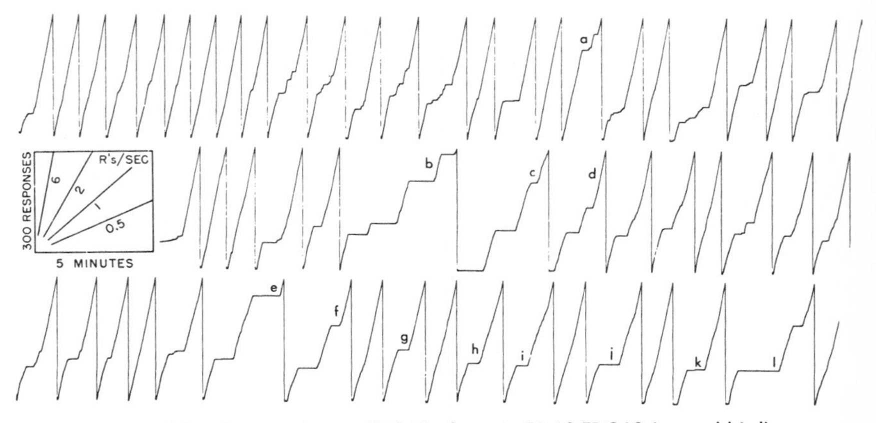

Fig. 738. First session on FR 360 after mix FR 60 FR 360 (second bird)

j, k, l, and elsewhere, although reinforcement has not been received at the short ratio since the preceding session. Note that frequent reinforcements have been received after short ratios *after pausing,* as for example at *a, b, c, d, e,* and *f.* Because this bird was showing multiple inflections, the removal of the short ratio from the schedule did not greatly alter the conditions of reinforcement except to change the mean ratio, or at best the percentage reinforcement on short ratios. Under these conditions the pausing becomes more, rather than less, pronounced.

This bird still showed breaks at the end of the fourth session on FR 360, as Fig. 739A illustrates. The performance is still characteristic of FR 360 in Fig. 738, and breaks still occur, as at *a, b, c,* and *d.* When short ratios were again added to the schedule to compose a mix FR 60 FR 360 on the following session, the performance in Fig. 739B

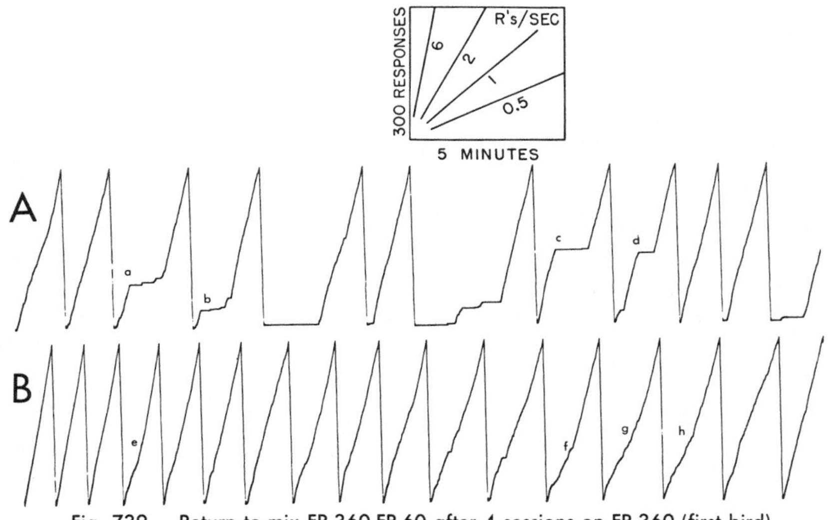

Fig. 739. Return to mix FR 360 FR 60 after 4 sessions on FR 360 (first bird)

began. The reduction in the mean ratio reduces the pausing after reinforcement, and any substantial pausing after the completion of the short ratio also disappears. An S-shaped character appears in many segments, clearly seen at *e, f, g,* and *h,* which points to surviving stimulus control.

The change in Fig. 739 is from breaks to fewer or less sharp breaks. The other bird showed the opposite effect. After 4 sessions on FR 360, almost no indication exists of the shorter ratio, as Fig. 740A shows. Considerable straining occurs, but the ratios are generally run off quickly once they have begun. On the following day, the schedule was changed to mix FR 360 FR 60 in which 1 long ratio occurred for every 4 short ratios. This procedure tended to reduce the mean ratio and led to an almost complete absence of pausing after reinforcement, as Record B shows. The reinforce-

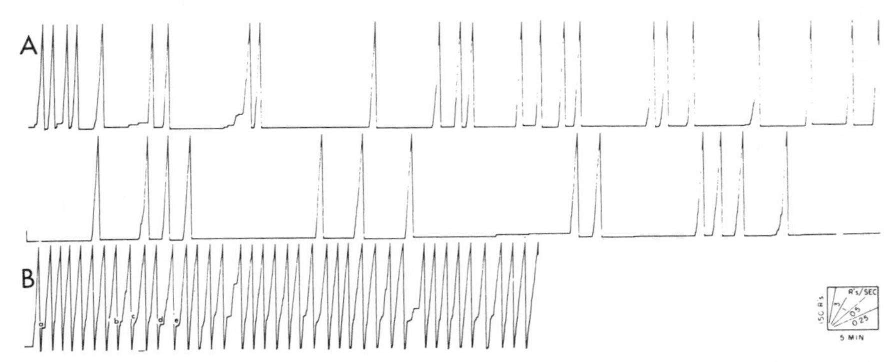

Fig. 740. Return to mix FR 360 FR 60 after 4 sessions on FR 360 (second bird)

ment on the shorter ratio, however, began to reinstate the break after the short ratio run in the long ratio, as is evident at *a, b, c, d, e,* and elsewhere.

Mixed fixed-ratio fixed-ratio schedules in rats

Two rats with a history of 62 hours on FI 10 were reinforced on mix FR 160 FR 20. A slight indication of self-generated stimulus control appears in the 1st session, shown complete in Fig. 741A. Brief pausing after reinforcement becomes consistent, and pauses after a number of responses approximately equal to the shorter ratio begin to appear, as at *a, b,* and *c.* By the end of the next (3-hour) session, multiple breaks are common, as Record B shows. The terminal rate has increased, and the pauses during the long intervals are larger than those after reinforcement.

Later stages are shown in Fig. 742. Record A is from the 15th hour; Record B, from the 24th; Record C, the 33rd; and Record D, the 43rd hour. "Damped" multiple breaks appear in Record C, and a fairly consistent single early break appears in Record D.

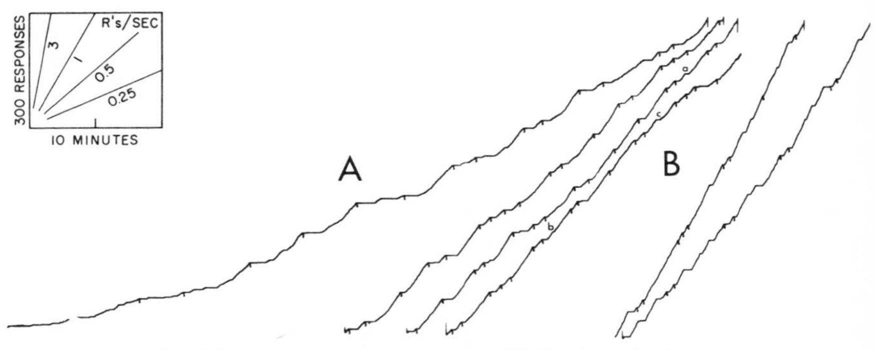

Fig. 741. Early development of mix FR 20 FR 160 in the rat

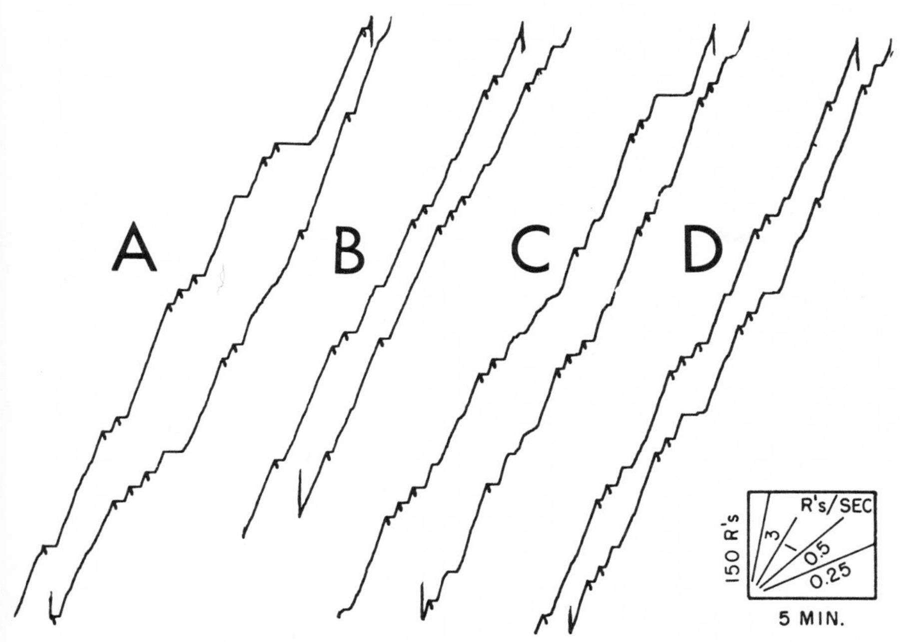

Fig. 742. Later development of mix FR 20 FR 160 in the rat

A comparison of mult FRFR and mix FRFR

Two birds with a long history at different values of FI were reinforced on FR 50 for 3 sessions. Except for slightly different rates, their performances were quite comparable. In the following session, one bird was placed directly on mult FR 300 FR 50, while the other was placed on mix FR 300 FR 50. The latter held the mixed schedule well, and several values of ratio were interchanged in successive sessions without producing long pauses at any point in the bird's performance. On the multiple schedule, however, the other bird broke almost immediately and had great difficulty in holding the schedule, even though the longer ratio was later considerably reduced. Figure 743A illustrates a fairly stable performance on mult FR 120 FR 50. This is the end of the 7th session at several values of the larger FR. During this part of the experiment, the key-colors were red and white on the shorter and longer ratios, respectively. In the following session the key-color was blue (the color used on the earlier straight FR), and the ratios remained 50 and 120. Record B shows that this mixed schedule is held well, although considerable pausing occurred on the multiple schedule with the same values of ratio. Mixed and multiple schedules were then alternated in a nonsystematic fashion, and the values of the larger ratio were changed. Thirteen sessions after Fig. 743B the bird was sustaining a performance on a mix FR 50 FR 240, as Fig. 744A shows. At the beginning of the following session, the same ratio

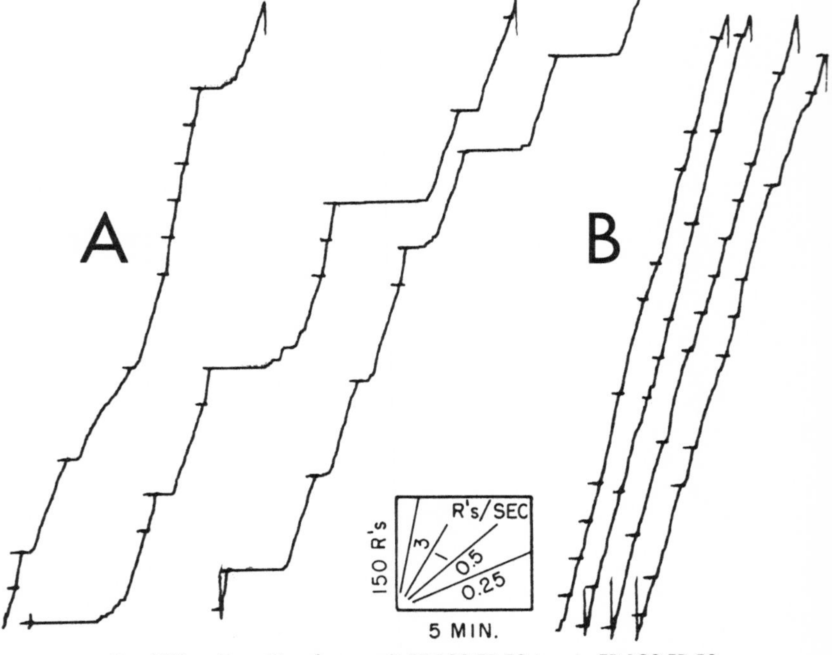

Fig. 743. Transition from mult FR 120 FR 50 to mix FR 120 FR 50

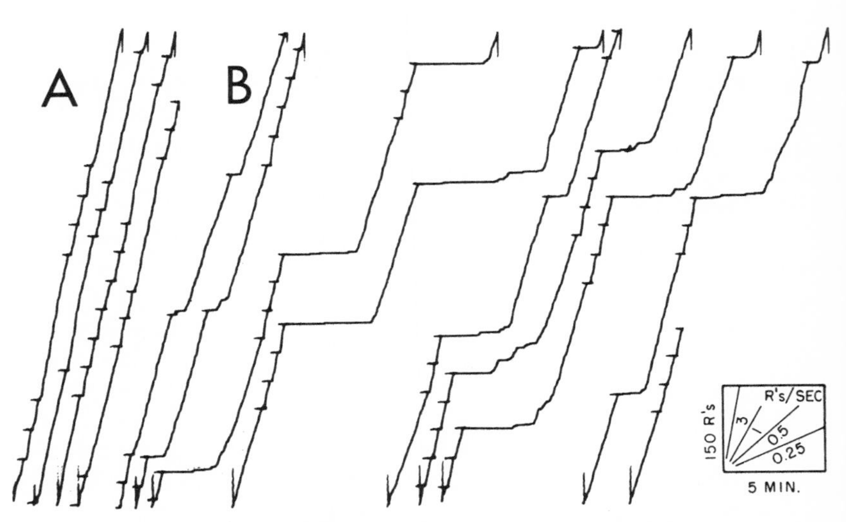

Fig. 744. Transition from mix FR 50 FR 240 to mult FR 50 FR 240

values prevailed, but the red and white keys had replaced the blue to provide a multiple schedule. The quick development of severe pausing after reinforcement in the longer ratios is clear in Record B. This bird showed, without exception, an ability to hold mix FRFR, although long pauses appeared in the larger ratios on the comparable multiple schedule.

The other bird, which went directly to the mixed schedule after 3 sessions on FR 50, held the performance well, both at the original settings of 50 and 300, and at various other values of the larger ratio during the next few sessions. When the key-color was changed from blue to red and white for purposes of a multiple schedule, the bird also held a variety of values of the larger FR fairly well, although some pausing and curvature usually occurred on the longer ratios. Figure 745A displays a final performance on mult FR 50 FR 240. Here, most of the longer ratios, under stimulus control, show a marked break and some scalloping. In the following session, the key-color was blue, and the schedule was therefore mixed. The over-all rate rises immediately, and the pausing after reinforcement is considerably reduced.

Multiple and mixed schedules were compared similarly in another experiment on 2 birds with a history of approximately 2000 reinforcements on mix FRFR and 900 reinforcements (in 10 sessions) on mult FRFR. Figure 746A illustrates a performance

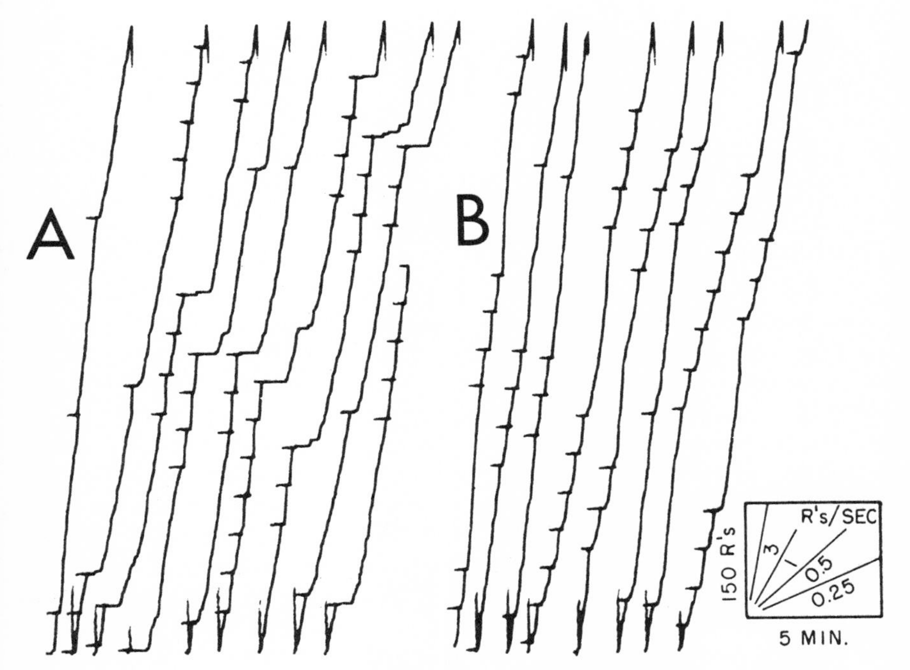

Fig. 745. Transition from mult FR 50 FR 240 to mix FR 50 FR 240 (second bird)

on mult FR 275 FR 50 with a TO 8 after each reinforcement. The larger ratio is breaking severely. At one point, 50 minutes of no responding have been omitted from the graph. Meanwhile, the short ratio is run off with no pausing. After 2 more sessions of this multiple schedule, the controlling stimuli were removed, and a mixed schedule was then in force. The bird adjusted to this quickly; Record B shows the performance after 4 sessions. The curves are essentially linear, except for a slight indication of a break in the long ratio shortly after the completion of a portion equal to the short ratio, as at *a* and *b*.

All 3 birds show that mixed schedules are more easily held than multiple with comparable values. The mixed schedule has the effect appropriate to the *mean* ratio. The multiple case shows breaking on the longer ratio, comparable with that in straight fixed-ratio performances.

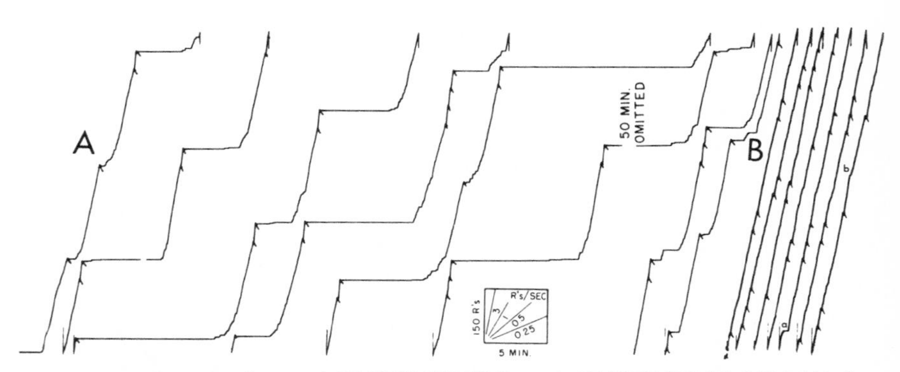

Fig. 746. Transition from mult FR 50 FR 275 TO 8 to mix FR 50 FR 275 TO 8 (third bird)

Extinction after mix FRFR

We obtained examples of extinction during the experiments just described, usually through failure of the apparatus. As a rule, the performance in extinction depends upon the performance which has been reached on the mix FRFR schedule. In Fig. 747, for example, the performance prevailing at the moment appears at the left of the record, from *a* to *b*. Small single breaks occur after reinforcement in the long ratio. (The ratios were not being counted accurately at this time. The apparatus eventually broke down at *d*, and no further reinforcements were forthcoming.) The performance in extinction shows, in general, runs approximately equal to the short ratio, as at *c, e, f, g, h, i, j*, and elsewhere. Occasionally, a more sustained performance occurred characteristic of that prevailing in the balance of the long ratio, as at *d* and *l*. The beginning of the run, from *k* to *l*, shows the damping of the oscillation which this bird had already exhibited in Fig. 728.

Figure 748 gives an extinction curve for the bird shown in Fig. 734 and elsewhere. Record A shows the first 3 excursions on the previous day under mix FR 360 FR 60 TO

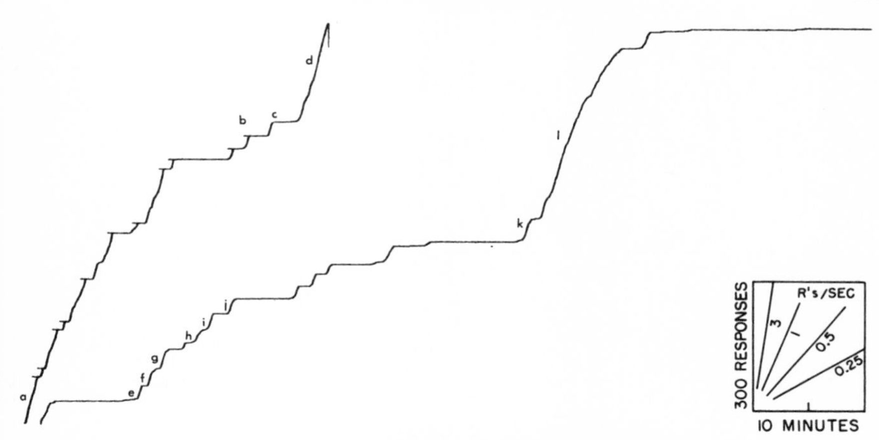

Fig. 747. Extinction after mix FR 50 FR 250

2.5. They show the characteristic break after a number of responses roughly equal to two of the smaller ratios. This break is occasionally omitted, as at *a*. At the beginning of the session (shown in Record B), the apparatus was not reinforcing, and an extinction curve was recorded. It begins with a run of over 2000 responses, ending at *b*. A run of more than 1000 responses occurs later, ending at *c*. The rest of the record is composed of runs roughly of the order of one or two times the smaller ratio, and these tend to show the same negative curvature as the breaks appearing after reinforcement in the performance in Record A.

Figures 749 and 750 present similar extinction curves for two other birds. Each figure contains a sample of a performance on mix FR 275 FR 50 at Record A, and an extinction beginning at the start of the following session at Record B. The grain of each extinction curve resembles that of the longer ratios on mix FRFR.

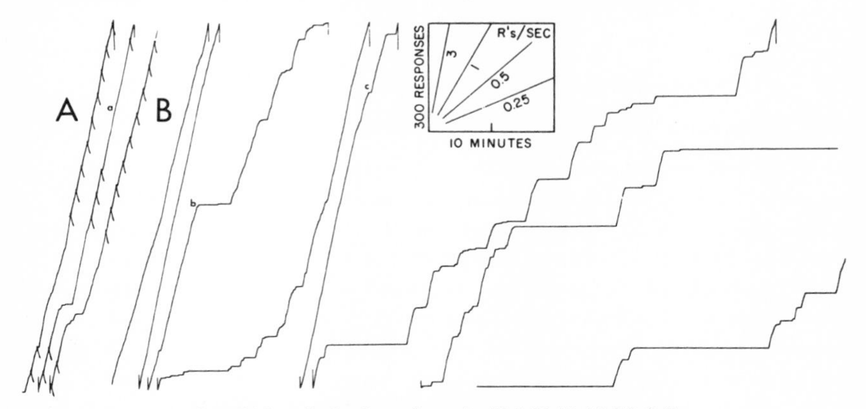

Fig. 748. Extinction after mix FR 360 FR 60 TO 2.5

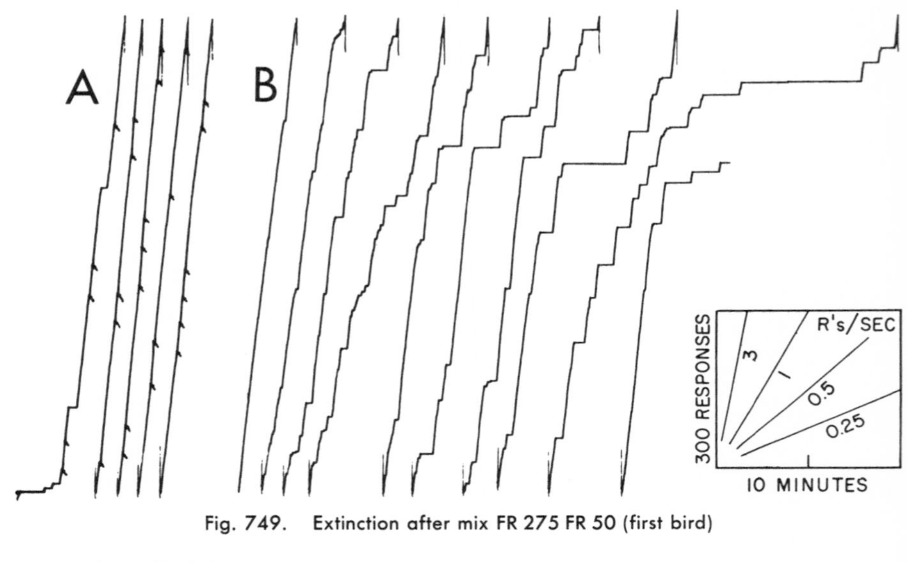

Fig. 749. Extinction after mix FR 275 FR 50 (first bird)

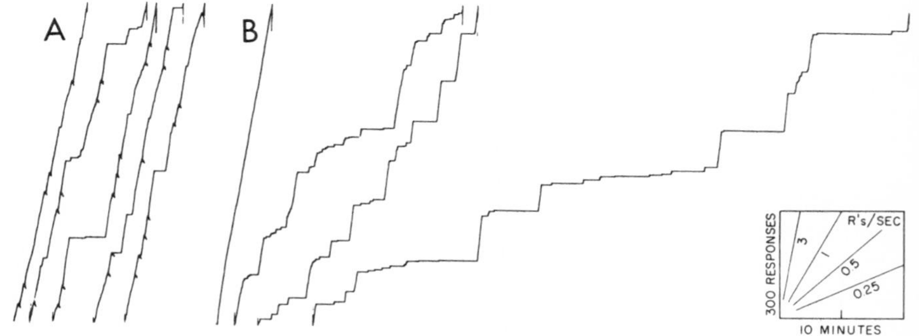

Fig. 750. Extinction after mix FR 275 FR 50 (second bird)

Effect of pentobarbital on mix FR 60 FR 360

A bird whose performance at another stage in the experiment described in Fig. 737 later showed a consistent slight break in the long intervals on mix FR 360 FR 60. An example appears at *a* in Fig. 751. After the reinforcement at *b*, 5 milligrams of pentobarbital ("Nembutal") was injected intramuscularly. This dosage was sufficient to abolish practically all responding for more than 1 hour. A response occurs at *c*, and a few at *d* and *e*. The first post-injection ratio is eventually completed at *f*, nearly 2 hours after injection. This reinforcement is followed again by a long pause of nearly 20 minutes, after which a full ratio is run off characteristically at *g*. The next reinforcement is also followed by a short pause at *h*, but a fairly typical ratio performance

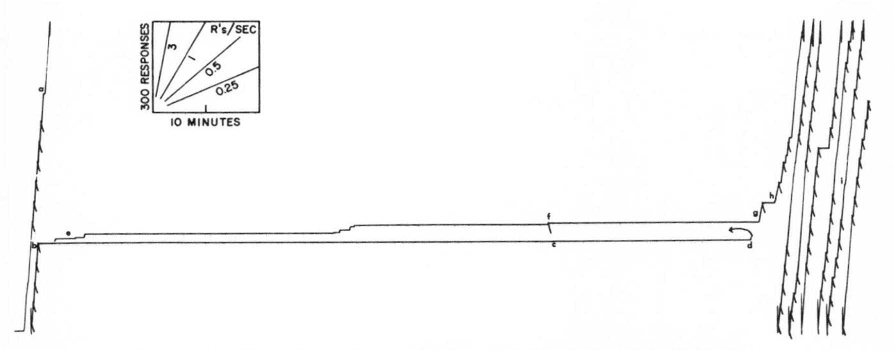

Fig. 751. The effect of pentobarbital on mix FR 360 FR 60 TO 2.5

then ensues. The 7 long ratios in the rest of the schedule during this session do not show the priming breaks which are characteristic of the sessions both before and after the session shown. This factor may be due to the hyper-excitable phase which pentobarbital generally produces at approximately this stage. A slight break occurs at *i*.

MIXED FIXED-INTERVAL FIXED-INTERVAL SCHEDULES

Mix FI 330 sec FI 30 sec[1]

The variable schedules discussed in Chapters Six and Seven are essentially extreme examples of mixed schedules. Some of the series used, especially those with many short intervals or ratios, produce a priming run after reinforcement similar to the mix FRFR above. Conversely, a mixed schedule may have the effect of a variable schedule until the behavior can stabilize in a performance appropriate to the component members.

A pigeon which had produced fairly good scalloping on FI 2.5 (Fig. 127) returned to a linear performance suggesting a VI schedule when first exposed to mix FI 30 sec FI 4. Figure 752A shows a sample from the end of the first 2-hour session. In the following session, greater irregularity began to appear; the rate has some tendency to be slightly higher following reinforcement, but drops after perhaps 100 responses, as at *a* and *b* in Record B (from the middle of the second session, or after 3 hours of exposure to the schedule). The effect is already more pronounced by the 4th hour at the end of this 2nd session shown in Record C. By the end of the 3rd session, or after 6 hours of exposure to the schedule, occasional instances occur of a much more marked priming, as at *c* in Record D. Records E, F, and G show the final segments of the next 3 sessions, after 8, 10, and 12 hours, respectively, of exposure to the schedule. By the time Record G is reached, a well-marked priming exists after the shorter interval, and a falling-off into a curvature appropriate to a longer interval, as at *d, e,* and *f.*

[1] The experiments on mix FIFI represented by Fig. 752 to 754 were carried out by Mr. George Victor, to whom the authors express their indebtedness.

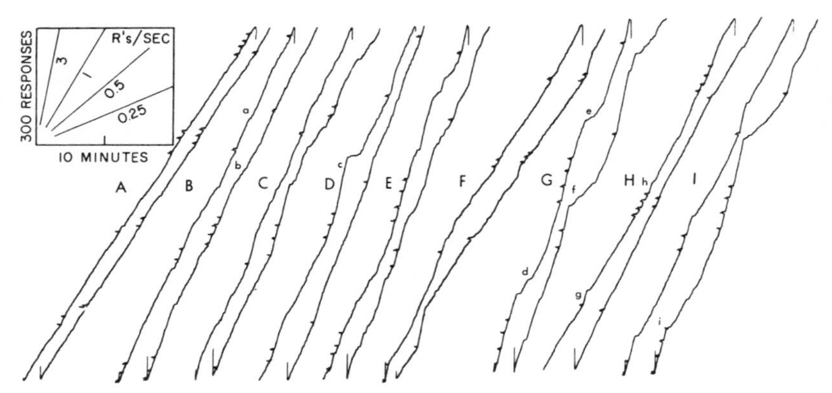

Fig. 752. Development of mix FI 330 sec FI 30 sec during the first 7 sessions

The segments at Record G are from the middle of the session. By the end of the session, a rather linear performance has temporarily reappeared, with only slight priming, as at *g* and *h* in Record H. Record I presents the performance at the end of the next session, after 14 hours of exposure to the mix FIFI. Here, fairly consistent priming occurs after reinforcement, except at *i*, and a sustained lower but slightly accelerating rate in the longer intervals.

The bird was further exposed to the same mix FIFI schedule; and after a total of 36 sessions, or 72 hours, the performance was of the character shown in Fig. 753.

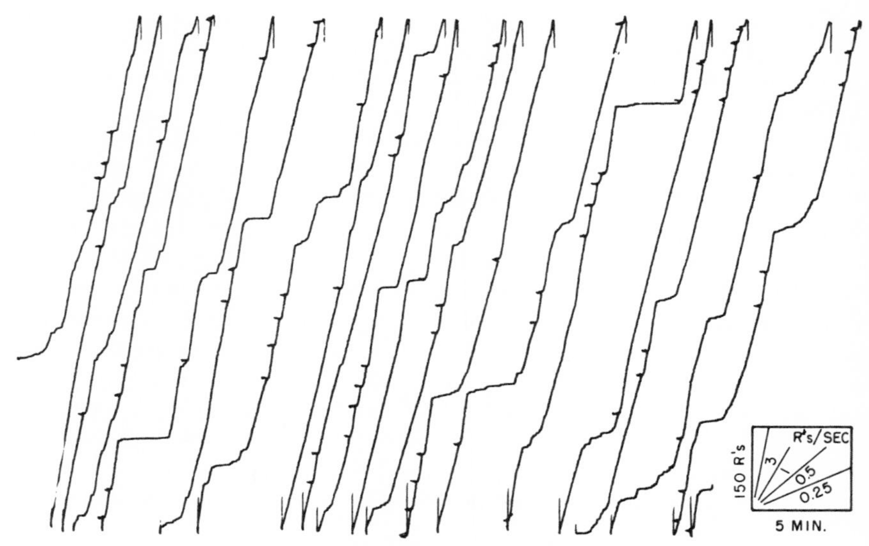

Fig. 753. Mix FI 330 sec FI 30 sec after 36 sessions

Here, very well-marked priming runs occur, long pauses after the priming runs, and, in general, a much higher over-all rate. A TO 2 was then added after every reinforcement. The first effect was to eliminate the performance characteristic of the mix FIFI. Figure 754A, which follows directly upon the session in Fig. 753, is for the end of the 1st session. A slightly higher rate after reinforcement begins to reappear. In the 3rd session (Record B), after 6 hours, a performance characteristic of the mixed schedule is more marked. Record D shows the final state observed in this experiment at the end of 14 hours. Here, double priming runs begin to appear in greater num-

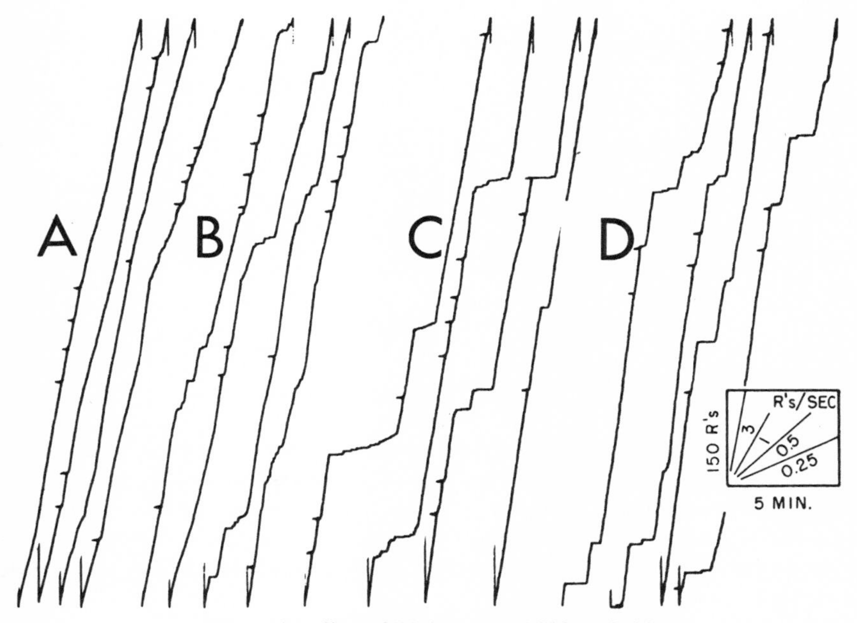

Fig. 754. The effect of TO 2 on mix FI 330 sec FI 30 sec

bers. The TO appears to eliminate whatever stimuli the bird had been using for the performance on the mixed schedule, but the new schedule with TO has a rapid effect and is eventually similar to that without TO.

Mix FI 5 FI 1

First bird. In another experiment 4 birds were exposed to mix FI 5 FI 1 for nearly 400 hours each. We have shown the development of an appropriate performance by selecting pairs of excursions from representative points throughout the process in Fig. 755. The pairs of excursions in the figure are samples from the following hours of exposure: A, 11; B, 27; C, 28; D, 66; E, 72; F, 114; G, 131; H, 153; I, 170; J, 224; K, 254; L, 289; M, 312; N, 373; O, 378; P, 388; Q, 391; and R, 395. In Record

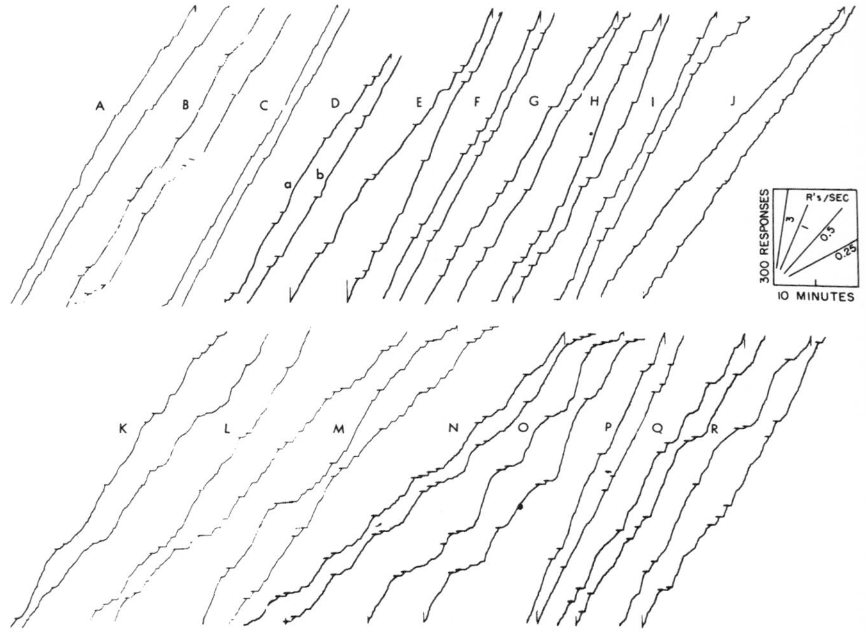

Fig. 755. Mix FI 5 FI 1: development during 400 hr

B pausing is already developing after reinforcement, but the over-all rate is quite irregular. In some instances the curvature appears to correspond with the interval, but this effect is not uniform. In Record C the performance is nearly linear, except for pauses after reinforcement. The effect of the mix FIFI begins to be felt in Record D, where the evidence is clear, as at *a* and *b*, of a decline in rate following an increase in rate after reinforcement.

Record H begins to show a sustained performance in the long segments after the priming run, though multiple breaks are evident, particularly at a later stage, in Records L and M. Fairly well-described scallops appropriate to the longer interval are clear in Record N and become pronounced at Record O, where a considerable negative curvature appears before the longer intervals are completed. The bird recovers from this performance to almost a linear type of record at Record P, which is not very different from, for example, the performance in Record D many hundreds of hours earlier. The curvature in Record O has generated a new set of contingencies which destroy this performance. Later, negative curvature returns, as in Record Q, in which the decline during the longer intervals is occasionally fairly abrupt. Abrupt changes in rate are more clear in Record R. Multiple inflections during the long intervals still occasionally occur at this stage.

Figure 756 shows a complete daily session for this bird early in the development of the performance under mix FIFI between Records C and D of Fig. 755.

Second bird. A second bird showed many of the same effects. Figure 757 gives sample records showing some of the principal features of the performance of this bird. At Record A a slight pause follows reinforcement, but responding is otherwise fairly linear. At Record B some inflection in the long intervals suggests a priming from the

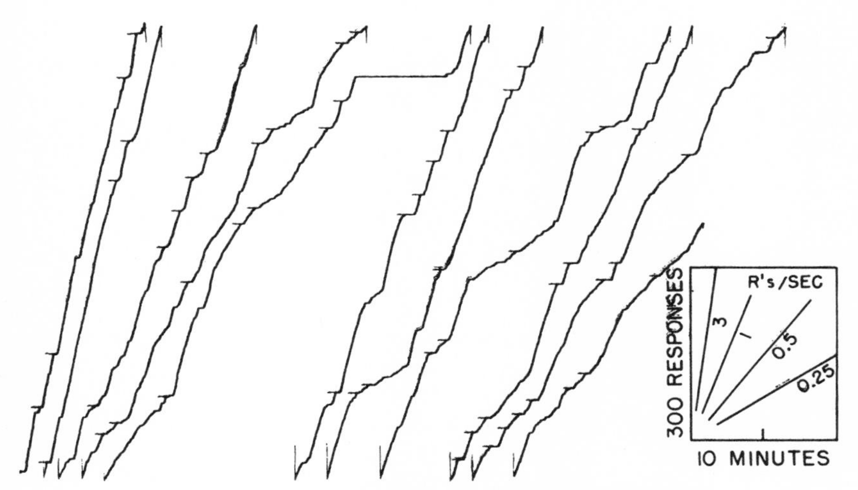

Fig. 756. Mix FI 5 FI 1: a complete daily session after 48 hr

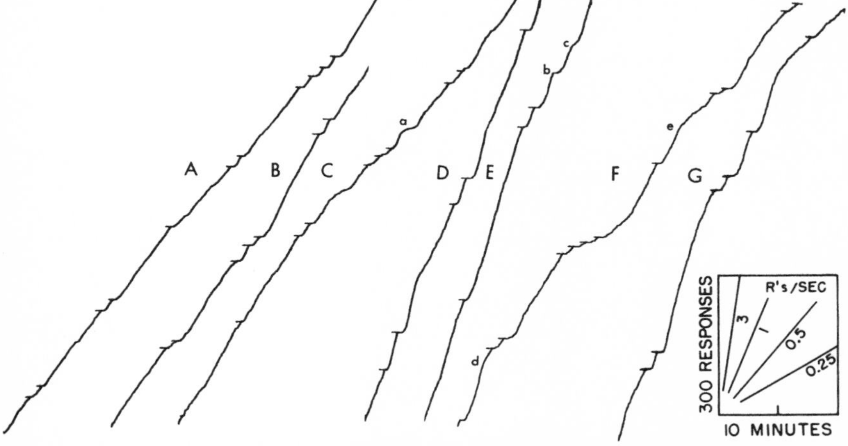

Fig. 757. Principal features during the development of mix FI 5 FI 1 (second bird)

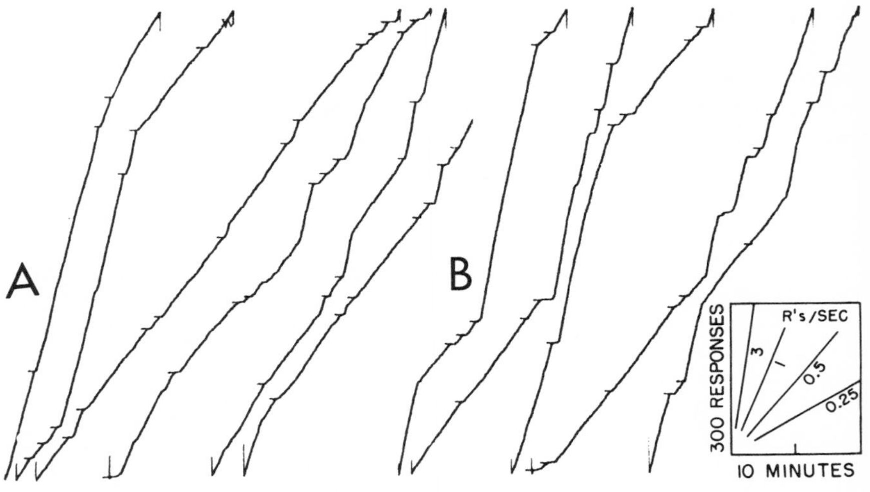

Fig. 758. Sudden shifts to a second stable over-all rate

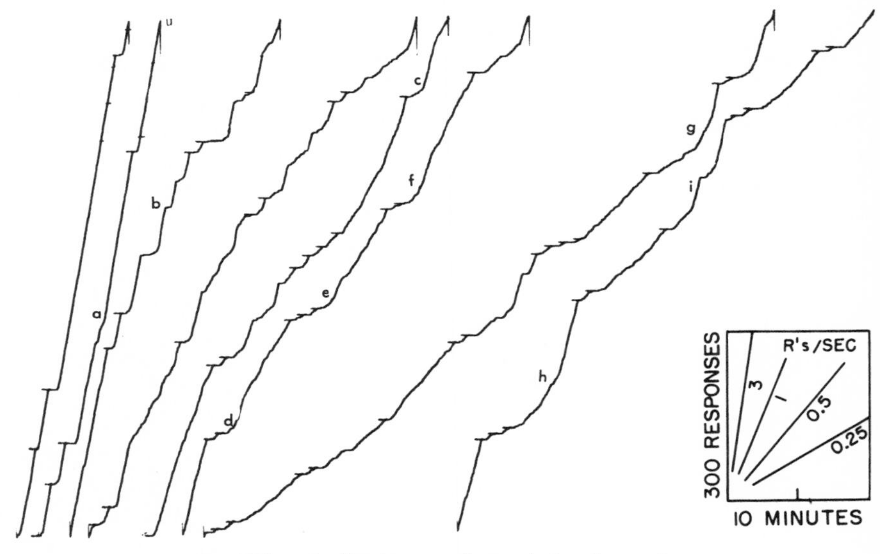

Fig. 759. Decline in over-all rate during the session

short-interval performance. At Record C marked inflections begin to appear, as at *a*. In Record D the priming run leads directly to a single linear rate rather than through a second period of acceleration. Record E shows an example of 2 inflections in a single long interval, at *b* and *c*. Record F shows negative acceleration in the long interval, particularly at *d* and *e*, which is less abrupt in Record G. This second bird showed another effect, however—a sudden shift to a second stable over-all rate. This shift appeared for the first time in the session shown in Fig. 758A. The high rate with which the session begins is much above the normal running rate of the bird. It reappears elsewhere during the session; changes from one rate to another are usually fairly abrupt and not too clearly related to reinforcement. This was later found to be a consistent pattern. The two rates are in a ratio of about 3:1. Record B shows a sample from the end of the 4th session after Fig. 758A. It was found upon direct observation that during the recording of the lower rate the bird was pecking on the panel at the side of the key in such a way that the response was not recorded. These double rates have been observed elsewhere, and are usually associated with mixed schedules.

Figure 759 shows a general tendency for this bird to start fast and to retard throughout the session. This was a useful fact, because the characteristic features of mix FIFI are in part a function of the over-all rate. A single daily session shows several of them, with other conditions remaining fairly constant. When the rate is highest, the only important feature is the pause after reinforcement, the rate otherwise being high and uniform. A slight break first appears at *a*, but the over-all high rate is maintained consistently until reinforcement. The first long interval at a slightly lower rate (*b*) shows 4 scallops, as the effect of the mix FIFI. Multiple inflections persist for several intervals following. Later, the first scallop and recovery in the interval become the only important deviations, as at *c*, *d*, *e*, and *f*. As the rate continues to fall, the terminal rate declines, although a high rate may break through occasionally in a well-marked interval scallop, as at *g* and *h*. At this stage the rate in the shorter interval has become low, and together with the pause, produces only a small number of responses at reinforcement in the shorter intervals. Perhaps this is why not much priming occurs in the longer interval. A conspicuous exception occurs at *i*; this performance resembles the occasional double scallop seen in the FI performance when responding is interrupted by some incidental disturbances and the pause suffices to set off a second scallop.

Third bird. A third bird went directly to mix FI 5 FI 1 from crf. The first 3-hour session showed a slow positive acceleration, in which 700 responses were emitted. On the 2nd day, no responses were emitted during the first $1\frac{1}{4}$ hours. The first response was reinforced, as indicated at *a* in Fig. 760. A smooth acceleration then followed for the second time. Note the runs at *b* and *c*.

The third bird also showed a temporary phase in which 2 rates prevailed. Figure 761 describes the effect at its height. In general, however, the bird gave an extremely uniform linear performance throughout the experiment. Practically all of the characteristic features already noted appear in Fig. 761. An anomalous drop to the lower

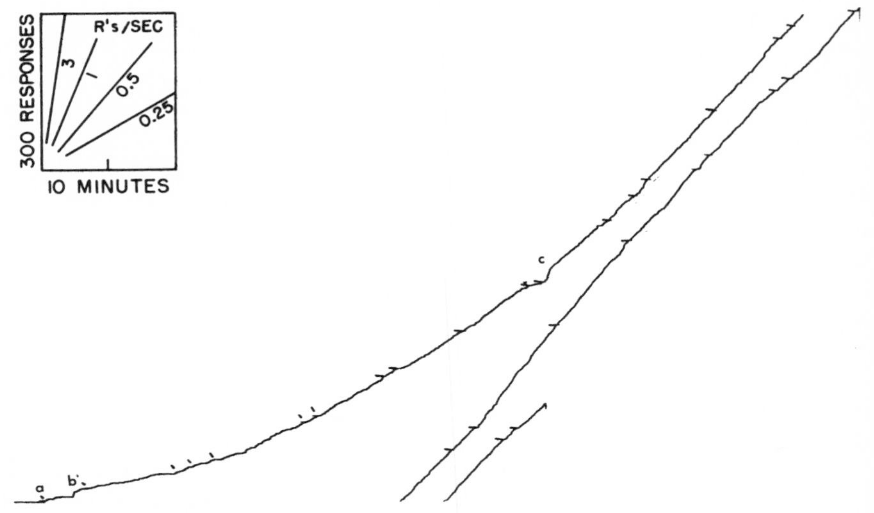

Fig. 760. Second session on mix FI 5 FI 1 after crf (third bird)

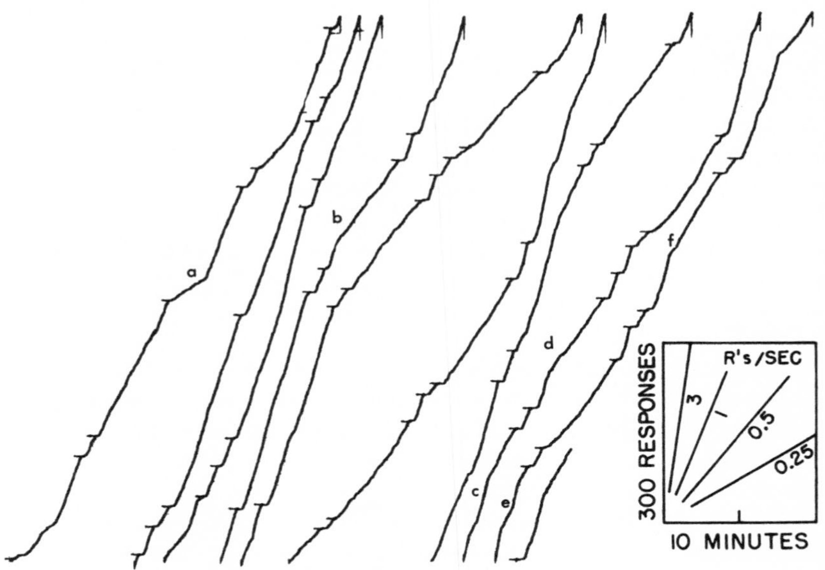

Fig. 761. Sudden shifts to a second stable over-all rate (third bird)

rate following reinforcement at *a* is also reflected in the general tendency for the over-all rate to assume one of two values throughout the session. Several instances occur in which approximately the usual number of responses in the shorter interval are emitted after reinforcement; then, a drop occurs to a low rate for the balance of the interval, as at *b, d,* and *f.* Multiple inflections during the long interval appear at *c* and *e.* This record is for an entire daily session, 120 hours after the transition from crf to mix FIFI.

Fourth bird. The fourth bird showed the same characteristic features. Figure 762 shows a daily session 118 hours after crf. The record is an especially good demonstration of sharp priming runs after reinforcements; they contain somewhat more responses than those usually emitted in the short interval, followed by sharp declines—often to zero before the intervals are completed. Many instances of multiple inflection occur

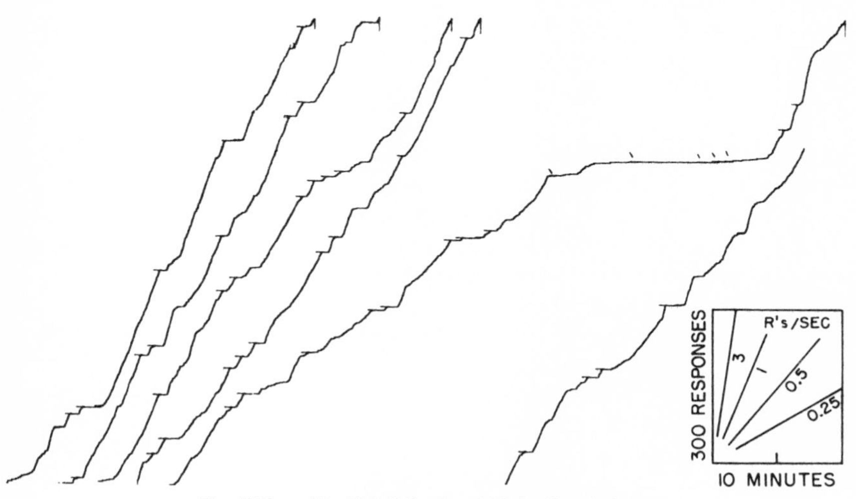

Fig. 762. Mix FI 5 FI 1 after 118 hr (fourth bird)

in a single long interval. The decline during the experimental session is characteristic of this bird and resembles that of Fig. 759.

Mix FIFI with counter or clock

Mix FRFR and mix FIFI are possibly the best evidence we have that the emission of a given number of responses at a given rate may serve as an effective stimulus in controlling subsequent behavior. The addition of a clock or counter to such a schedule therefore should have obvious significance.

Two birds were reinforced on mix FI 10 FI 2 with the added counter already described. The spot of light on the key grew to 1 inch as its maximal length as 300 responses were emitted. Figure 763 shows a final performance after 44 hours. In general, the long interval shows a break after emission of a number of responses that is

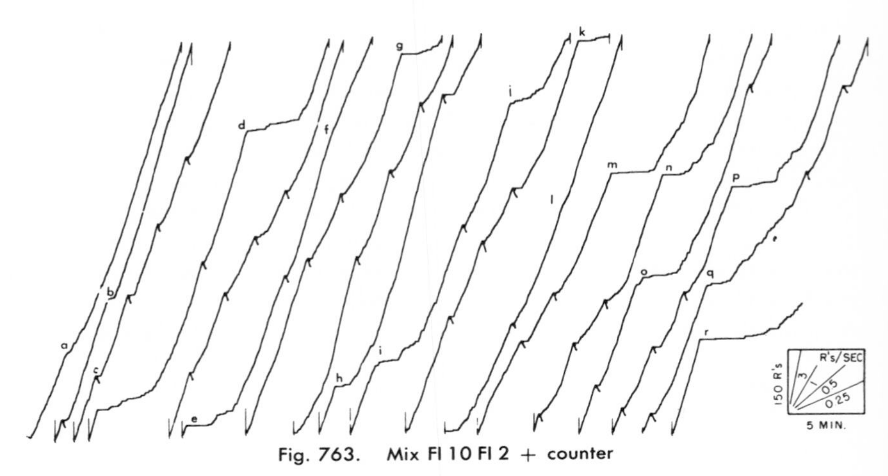

Fig. 763. Mix FI 10 FI 2 + counter

roughly twice the number shown in the shorter interval. Instances of these breaks are at *d, e, g, h, i, j, k, m, n, o, p, q,* and *r*. Only slight breaks appear early in the record, at *a* and *b*; and the rate changes only slightly, in lieu of a break in the long intervals, at *f* and *l*. The extent of the scalloping in the balance of the interval is fairly uniform, and much greater than the slight scalloping after reinforcement in the shorter intervals or at the beginning of longer intervals.

Unfortunately, we have no performance on mix FI 10 FI 2 without counter for this bird for purposes of comparison.

When the counter added to mix FI 10 FI 2 showed "fading" (see Chapter Five), it had a less consistent effect. Figure 764 illustrates the final performance of the bird shown in Fig. 763 with the fading counter. One fairly stable rate is considerably above the value required to increase the size of the spot, another is clearly below, and one is at approximately the value needed. The second part of the longer interval tends to

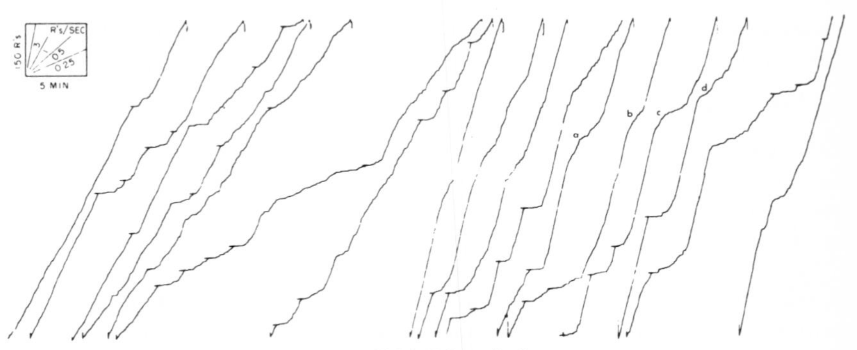

Fig. 764. Mix FI 10 FI 2 + fading counter

take on positive curvature. The drop in rate is no longer abrupt where the number of responses characteristic of the shorter interval has been emitted in the longer interval. For example, the curves at *a, b, c,* and *d* are fairly smoothly S-shaped.

We showed that the added counter is responsible for these features of the performance, rather than prolonged exposure to mix FIFI alone, in a later session when the counter was removed. Figure 765A shows the first 3 and the last 3 excursions from the 1st session. The first effect of removing the counter is to remove practically all changes in rate. By the end of the session, some slight pausing occurs after reinforcement, although pauses are usually preceded by a few responses, as at *a* and *b*. Nothing like a standard mix FIFI performance appears at any time during the session.

On the following day, a counter was introduced in which the spot reached full size only after 1200 responses. A break in the long interval is immediately reinstated (at *c* in Record B), but it is greatly postponed because of the slow change in the counter "reading." The second long interval contains a similar postponed break at *d*. La-

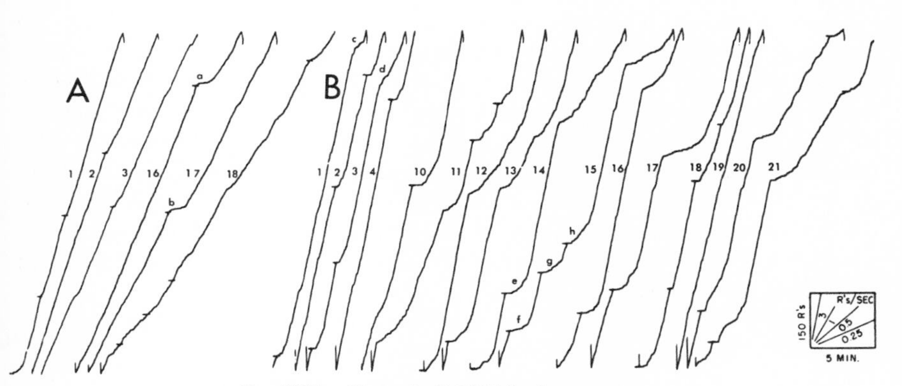

Fig. 765. First mix FI 10 FI 2 after counter

ter in the session, well-marked breaks occur; but, in general, they appear relatively late in the long intervals. The scalloping after reinforcement is at times quite marked and smooth, as at *e, f, g,* and *h,* although later instances occur in which the pause is much shorter and the return to a high rate much more rapid.

When the counter was replaced by a *clock,* which reached its maximum value in 10 minutes (equal to the longest interval), the behavior showed some change. Figure 766 shows a session beginning after 41 hours on this schedule. A somewhat more consistent pausing after reinforcement and a somewhat more abrupt assumption of a higher rate are characteristic, although many instances also occur in the longer intervals in which a fairly smooth scallop begins after the break in rate, as in *a, b, c, d, e,* and *f.* Note that these scallops appropriate to the longer interval are much more marked than in the shorter interval. They are not, however, characteristic of a clock performance. At the beginning of the excursion marked *g* the apparatus was changed

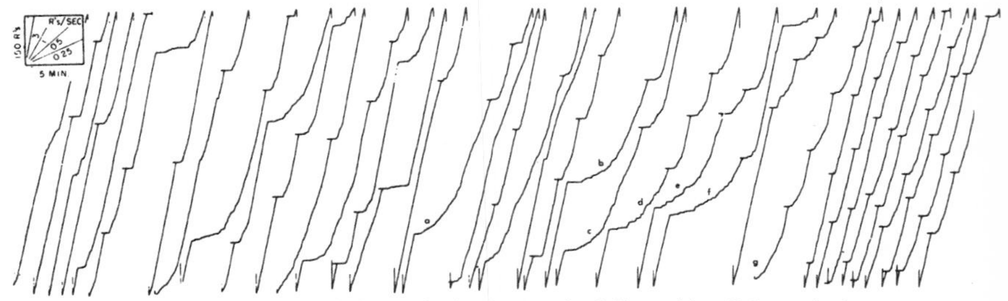

Fig. 766. FI 10 FI 2 + clock after 41 hr followed by FI 2 + clock

during a brief period of TO, so that all subsequent intervals would be 2 minutes with clock. The first 5 intervals following show curvature comparable with that which has been prevailing up to this point under mix FIFI; but the performance quickly goes over to a fairly good short FI with clock.

Figure 767 shows a final performance with mix FIFI with clock after 95 hours. In the session in Record A most of the breaks in the long intervals are abrupt and are followed by smooth curvature, for example, at *a* and *b*. On the following day, with a slightly higher over-all rate but otherwise under the same conditions, the longer intervals tend to show a smooth decline in rate to a somewhat more intermediate value (Record B). Instances are at *c, d, e,* and *f*. A single instance of an abrupt break appears at *g*. This difference from one day to the next is unexplained. It suggests that the factors responsible for the sharp break are subtle and depend upon some temporary condition produced by the schedule. Figure 354 shows the same clock on FI 10, without the effect of the mixed FI 2.

In another experiment the schedule was changed from FI 10 with clock to mix FI 2 FI 10 with clock. A good performance appropriate to the new schedule developed. This is apparent after 31 hours of exposure to the schedule in Fig. 768. An acceleration occurs at *a* appropriate to the frequent reinforcement at this reading of the clock. The bird then drops to a lower rate at *b* appropriate to the reading on the clock

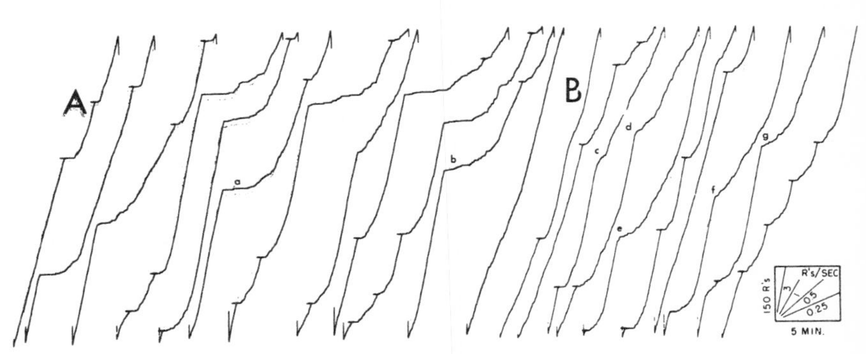

Fig. 767. Mix FI 10 FI 2 + clock after 95 hr

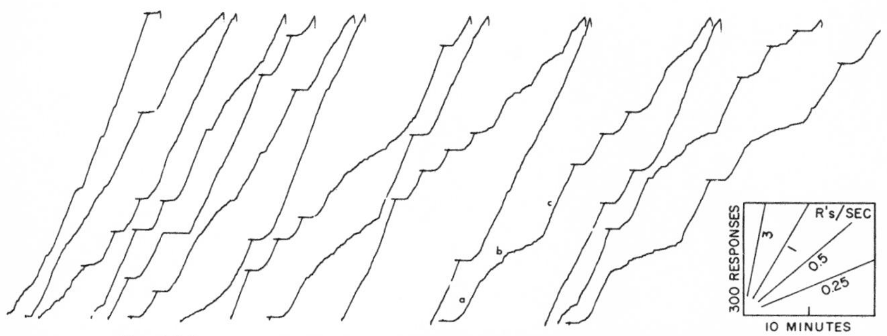

Fig. 768. Mix FI 10 FI 2 + clock after FI 10 + clock (second bird)

lying between the 2 intervals. It assumes a high rate again at *c* appropriate to the clock reading at the end of the longer interval. In general, all the long intervals follow this pattern. There appears to be too much induction from the two readings associated with reinforcement to permit the bird to stop responding altogether between the 2- and 10-minute intervals.

In an earlier stage of the experiment this bird had been on a simple FI 10 with clock (after a history of mix FIFI with counter). The resulting clock performance aids in the interpretation of Fig. 768. Figure 769 illustrates the development of FI 10 with clock. The records shown are the terminal portions of 4 successive days, Records A, B, C, and D, and the middle portion of the 5th day at Record E. The terminal excursions on the 5th day are above the usual rate and not typical of the day's performance. The figure shows the development of a fairly representative performance on FI 10 with clock. The pause and terminal rates are much greater than those in Fig. 768, and the curvature is much sharper.

Another bird with an unusually ragged performance on FI 10 with clock was transferred to mix FI 10 FI 2 with the same clock. Figure 770A shows a sample of the terminal performance on the simple FI. Record B gives the first part of the session on mix FI 10 FI 2 with clock. Scarcely any responding occurs in the 2-minute interval because of the earlier clock history. Reinforcements, as at *a, b, c, d,* and *e,* are fol-

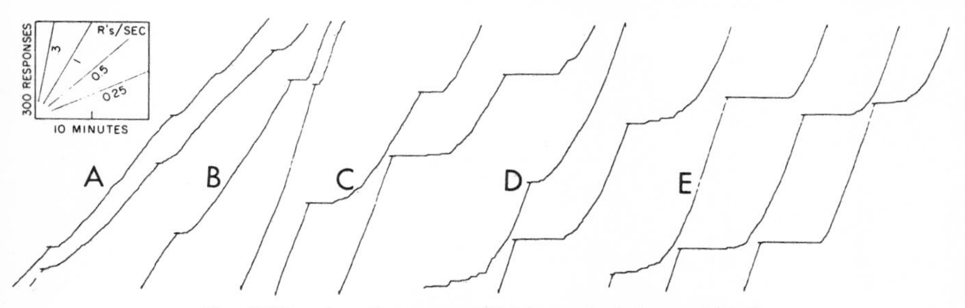

Fig. 769. Development of FI 10 + clock (second bird)

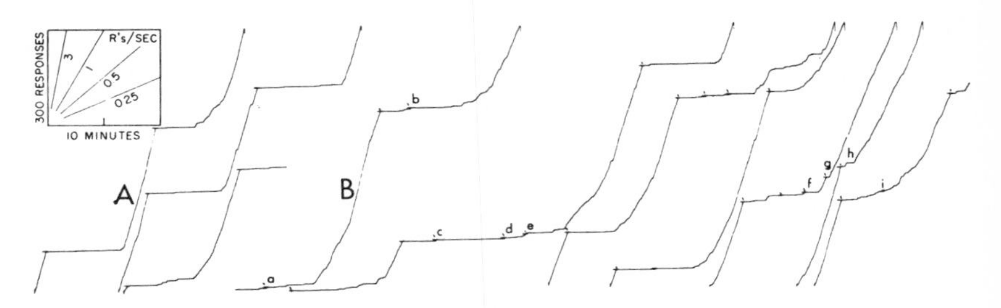

Fig. 770. Transition from FI 10 + clock to mix FI 10 FI 2 + clock

lowed by further periods at a low rate as a function of the previous clock. The first result is that responses are reinforced at low readings on the clock. This effect shows at *f, g,* and *h,* where the rate at the low reading has increased substantially. Even the scallop at *i* no longer begins with the prolonged pause characteristic of the clock performance.

As with the bird in Fig. 768, although this bird is capable of a good clock performance on FI 10, it is unable to develop an appropriate performance on mix FI 2 FI 10 with clock. Figure 771 shows the terminal performance in this experiment. The bird tends to respond rapidly, for example, at *l,* when the clock reading is that which characteristically prevails in the shorter interval. The rate drops again to zero when the clock reading exceeds this value, and is therefore appropriate to no reinforcement (at *m*). It returns to a higher value of responding when the clock again reaches a reading associated with reinforcement at *n*. However, the terminal rate at *n* is not characteristic of a good clock performance; and it will be apparent that although this over-all pattern is roughly followed in many long intervals (*g, h, i, j, k,* and *o*), it is by no means inevitable. At *a* the bird fails to stop responding when the clock is midway between readings for the shorter and longer intervals. At *b* it retards at that time, but fails to develop a good terminal run appropriate to a final reading of the clock. Seg-

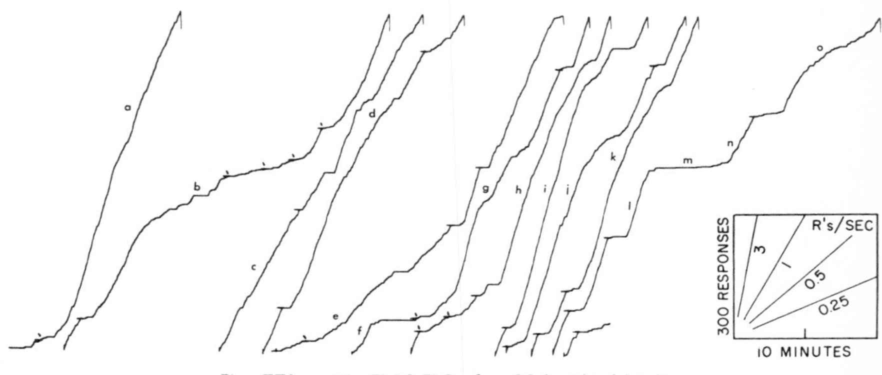

Fig. 771. Mix FI 10 FI 2 after 20 hr (third bird)

ment *c* resembles Segment *d,* and the emergence of a good terminal rate with benefit of clock at *d* is missing. In the interval at *e,* no high rate develops at any time. At *f* a terminal rate is reached, but it is not maintained until reinforcement. Much longer exposure to the schedule with this clock might have produced a sharper performance similar to that at *l, m,* and *n,* with better terminal rates.

Figure 772A shows a performance under mix FI 5 FI 2 with nonfading counter. Little departure exists here from a single running rate; but the longer intervals contain breaks which appear after the number of responses emitted is equal to two or more times the number characteristically occurring in the shorter interval. Good examples are at *a, b, c,* and *d.* The other long intervals show few significant rate changes. The

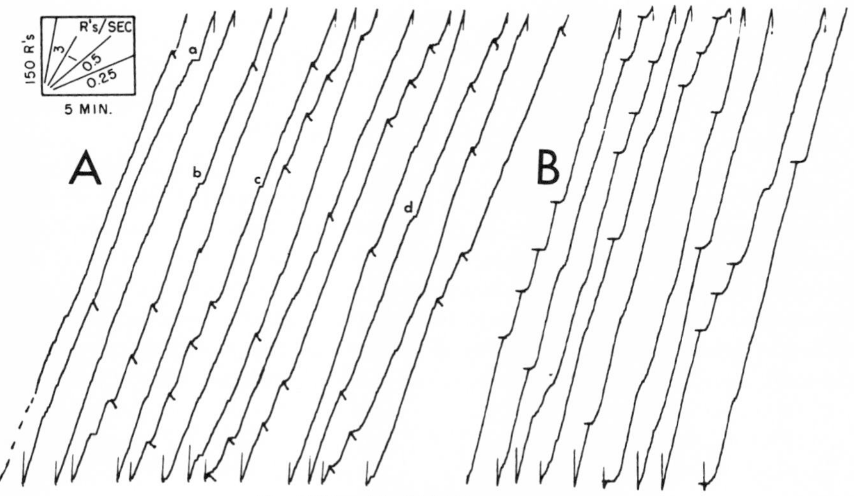

Fig. 772. Mix FI 5 FI 2 + counter

record was taken 23 hours after exposure to the mix FIFI schedule with counter.

Much later in the experiment, after a fading counter had been used, the bird was again placed on a mix FI 5 FI 2 with counter. The performance is quite stable and smoothly accelerated. Record B gives a sample. Here, the long intervals characteristically show an inflection at a point corresponding to about twice the number of responses in the shorter intervals.

Figure 773 shows a fairly consistent performance on mix FI 5 FI 1 with fading counter. The history of the bird includes extended reinforcement on mix FI 5 FI 1 with nonfading counter. A period of relatively rapid responding follows each reinforcement and, in the longer intervals, a sharp break to essentially a constant intermediate terminal rate then occurs. The "compensatory" effect of the runs after reinforcement may be seen by sighting along these curves.

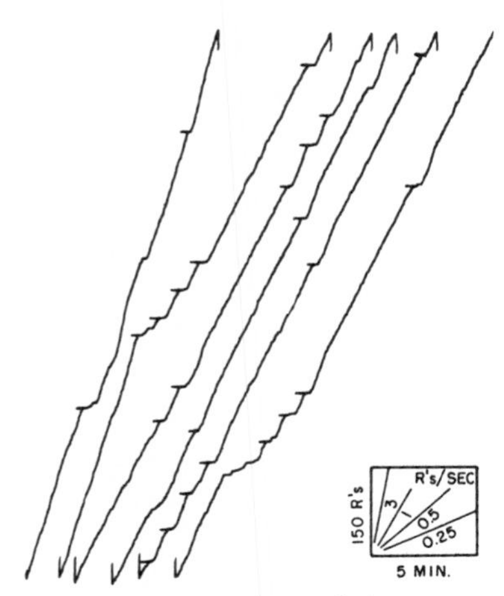

Fig. 773. Mix FI 5 FI 1 + fading counter

The effect upon the rate is clearest when the change is made from a fading to a nonfading counter and vice versa. Figure 774 shows one instance of each. In Record A, 4 excursions with a fading counter show the general characteristics of Fig. 773. At the arrow the fading element was removed, and the rate immediately rises to a much higher value. The character of the performance is not greatly changed, although the difference between the rate just after reinforcement and the running rate is somewhat reduced. In Record B, on the following day, the first 4 excursions show the perform-

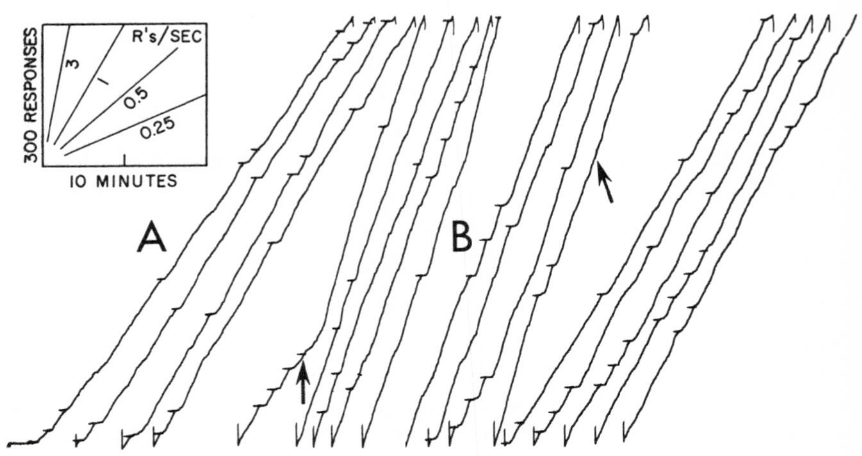

Fig. 774. Mix FI 5 FI 1: A. Transition from fading to nonfading counter
B. Transition from nonfading to fading counter

ance with the simple counter. At the arrow the fading element was introduced. At this point the spot was presumably large, and at the current rate, must have remained large until reinforcement. Thereafter, however, the fading element is important, and an immediate drop to a lower rate for the balance of the figure is clear. With this drop the well-marked character of Fig. 773 with the fading counter also returns.

When the values on the mixed schedule were changed to 10 and 2 rather than 5 and 1 minutes, respectively, the transition followed a predictable course. A performance appropriate to the new schedule emerged fairly clearly. Figure 775 shows the 13th hour. Some of the longer intervals reveal marked breaks, followed by scalloping for the balance of the interval, as at *a, c, d, f, h,* and *i.* Occasionally, a knee occurs in the early part of the interval, notably at *b, e,* and *g,* which may be to some extent due to the earlier shorter intervals. The ease with which the bird develops a break appropriate to mix FI 2 FI 10, where no very marked break had appeared

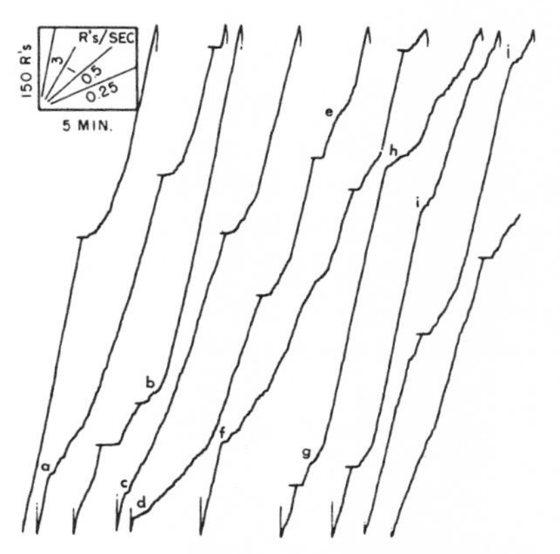

Fig. 775. Mix FI 10 FI 2 + counter 13 hr after mix FI 5 FI 1 + counter

after prolonged exposure to FI 1 FI 5, suggests either that the size of the interval or the extent of the change of the spot on the key is important.

Larger values in mixed fixed-interval schedules

A pigeon with an extended history of FI and VI was exposed to mix FI 20 FI 4. Figure 776 shows a fairly early performance. The long interval at *b* is convenient for inspection. It begins with a scallop similar to that in the short interval, as at *a,* and the rate then drops off smoothly to a stable value, suggesting a VI rather than a large FI performance. Two other birds on this schedule passed through phases of this sort. (Note that in this figure the movement of the pen indicating reinforcement was not always completed. The return movement was accomplished during perhaps the next

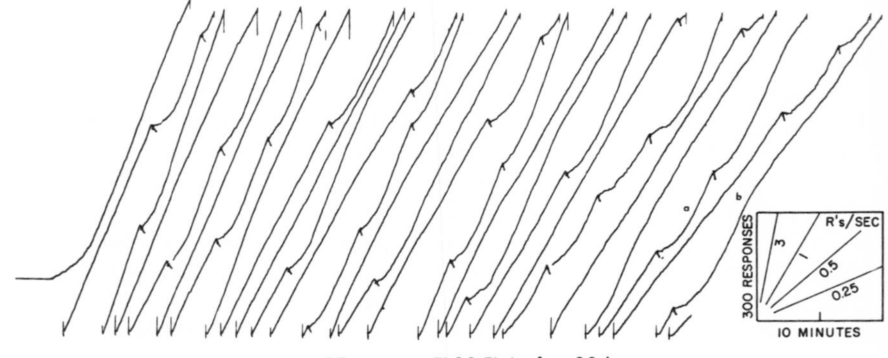

Fig. 776. Mix FI 20 FI 4 after 32 hr

10 or 20 responses. The over-all curvature, of course, is not affected by this defect of recording.)

When a counter was added, the increase in over-all rate obscured the effect of the mixed schedule. A sample performance appears at the beginning of Fig. 777. When a fading element is added to the counter, at the arrow, the rate is currently so high that the counter reading is not affected. Immediately after reinforcement at *a,* where the counter reading is minimal, the fading element is effective for the first time. The result is to reduce responding for some time to a value below that at which the spot will grow. The bird emerges from this low rate in a smooth acceleration, and reaches

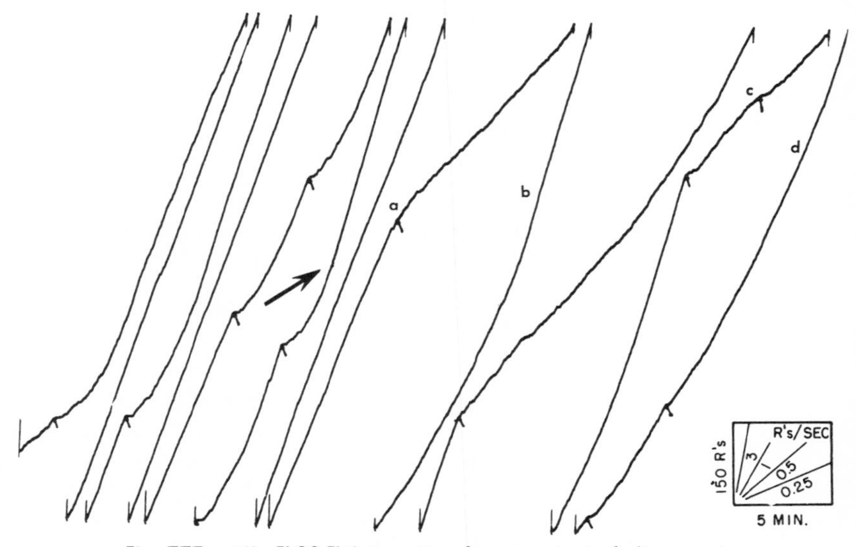

Fig. 777. Mix FI 20 FI 4: transition from counter to fading counter

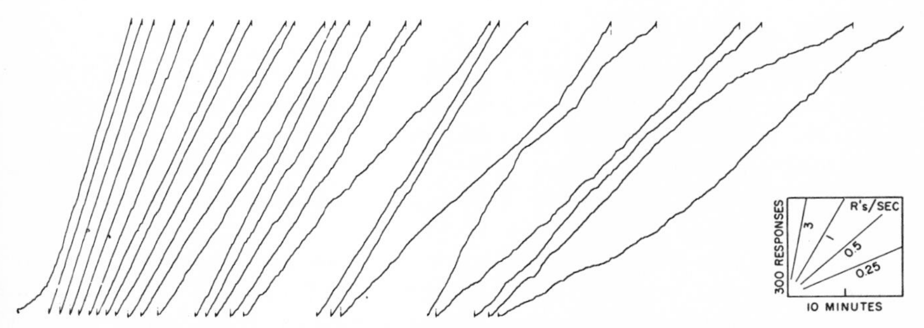

Fig. 778. First extinction curve after mix FI 20 FI 4 + counter

a terminal rate at *b*, at which the spot grows at a fairly rapid rate and reaches its maximal size before reinforcement. A very similar performance follows for the next interval, which is also long. A short interval is then terminated at *c*. During a series of 3 short intervals, some acceleration occurs; and the long interval which follows begins at a higher-than-usual rate and shows the same smooth acceleration to the high terminal rate at *d*. At no point in these long, smooth accelerations does the beginning movement of the spot seem to have any effect.

Extinction after mix FIFI

During a series of experiments in which the prevailing schedule was mix FI 20 FI 4 with counter, 1 bird was extinguished for at least 7 hours on 2 occasions with the slit at large. Figures 778 and 779 show the resulting curves. In both figures, some slight shifting from one rate to another appears, often quite abrupt; but the general linear character of the behavior through these long experimental periods is clear, as is also the tendency for the over-all rate to fall off fairly smoothly during the period. Both records show a slight warm-up, a quick acceleration to the terminal rate in Fig. 778, and a gradual increase occupying the 1st excursion of Fig. 779. The curves suggest extinction after VI. As has already been pointed out, VI is a species of mixed schedule.

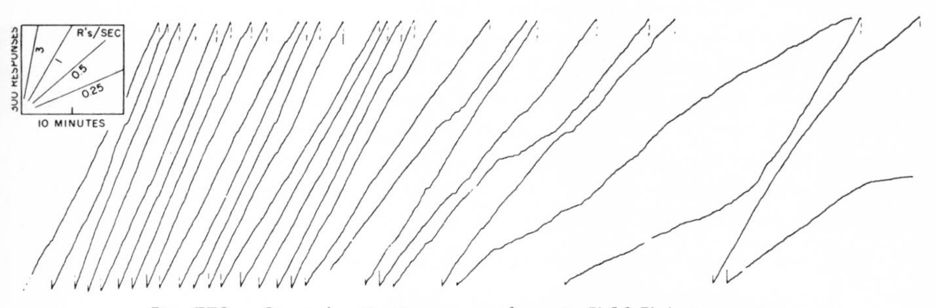

Fig. 779. Second extinction curve after mix FI 20 FI 4 + counter

Mixed FR extinction[1]

Two birds with a history of crf and 20 sessions of FR 50 were placed on a schedule in which, in random order, a response was reinforced after 50 responses, or no response was reinforced during an extinction period of 20 minutes. Later, the extinction period was extended to 1 hour. The component schedules were separated by a 10-second TO; some such period is necessary to distinguish between the end of extinction and the beginning of the ratio.

Figures 780, 781, and 782 give the development of an appropriate performance for 1 bird. (Some inaccuracy of timing may be detected in these figures. The points under consideration should not have been affected, however.) In Fig. 780 the segments occur in the order shown from top to bottom. The 1st reinforcement was received after a ratio at *a*. During the next 20 minutes the extinction shown at *b* followed. Thereupon, after a 10-second TO, a second extinction curve followed at *c*, also for 20 minutes. After a 10-second TO at *d*, a ratio was run off at *e*. This was of course not controlled by a current stimulus, so that *b*, *c*, and *e* are essentially consecutive stages of extinction after the single ratio reinforcement at *a*. Now, however, a reinforcement occurs at *f* upon completion of the ratio. The extinction curve shown at *g* then follows, after which 2 ratios were reinforced at *h* and *i*. These were followed in turn by a somewhat more smoothly declining extinction curve at *j*, which, in turn, was followed by a ratio at *k* and another extinction curve at *l*. Then, after a 10-second TO, a further period of extinction followed at *m*. This was not preceded by a ratio reinforcement and essentially continues the curve at *l*.

Figure 781 shows the later development in the same session and for 3 sessions following. The first segment in Record A follows immediately upon the last segment in Fig. 780. Thereafter, every 4th extinction period is represented. The experimental session lasted 14 hours; Curve 6 shows the last extinction that day. The following session (Record B) shows a marked change in the extinction curvature. The ratio performances at this stage, though omitted from the figure, are suggested by the priming runs at the beginnings of the extinction curves shown. The 1st extinction in the 2nd session contains a large number of responses, but Segments 2, 3, 4, and 5 show a rapid drop in the amount of extinction responding. Some responding returns on the following day in Record C, but again it rapidly drops to a fairly low over-all rate during the extinction period. In the 4th session, Record D, the period of extinction was somewhat shorter, because of difficulties in timing; but it shows a drop to essentially the final performance under such a schedule. Segment D2 shows an excellent example of a priming run followed by an almost complete absence of responding for the balance of the extinction period. Instances of multiple priming runs develop before the end of the period with some consistency; excellent examples are apparent in Seg-

[1] We are indebted to Dr. Merle Moskowitz for carrying out this experiment.

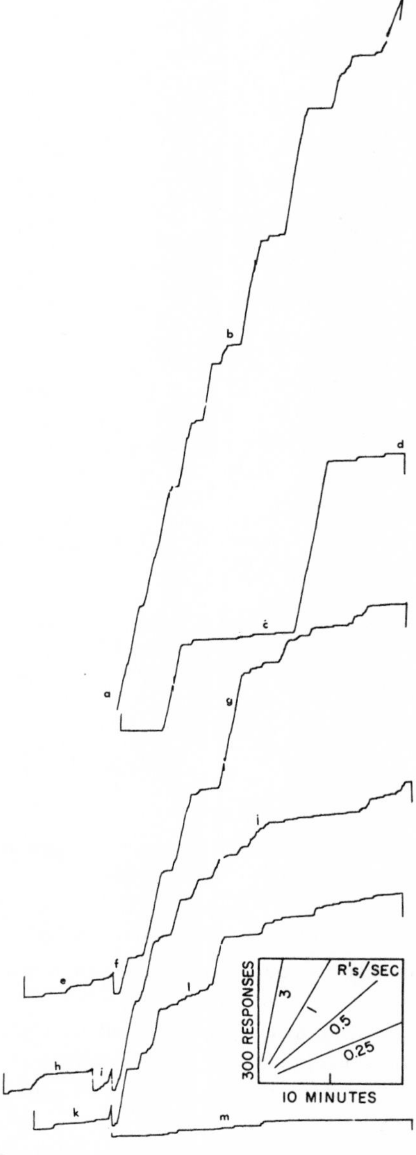

Fig. 780. First session of mix FR 50 ext 20 after FR 50

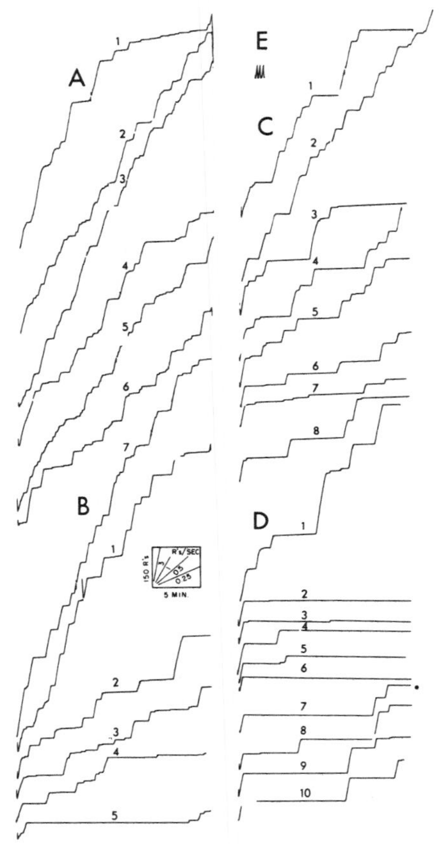

Fig. 781. Segments from the first 4 sessions on mix FR 50 ext 20

ment 10. The performance when the schedule is FR 50 is represented by the segments in Record E, which are, of course, similar to the first responses during the extinction period.

The further development of mix FR 50 ext is followed in Fig. 782A. The figure shows part of a session about 3 hours long on a single recording. In the next session (Record B) the multiple breaks become rare. Several sessions later, with the extinction period now lengthened to 30 minutes, the performance shown in Record C was recorded.

A second bird confirms all the features of the preceding experiment, with about the same speed of development. Figure 783A shows the performance of this bird with

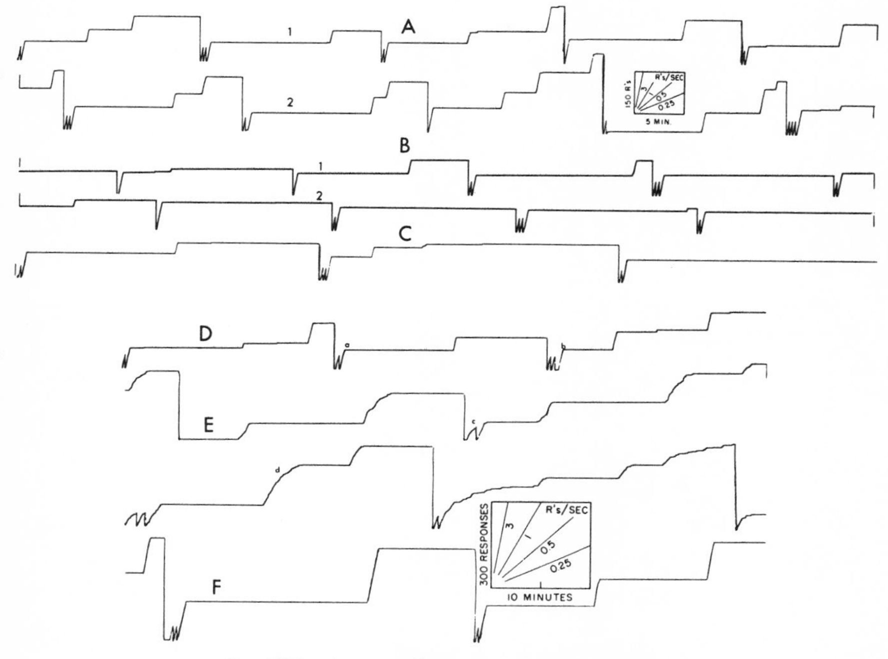

Fig. 782. Later performance on mix FR 50 ext

an even longer period of extinction. (The TO is now 6 minutes rather than 10 seconds.) This is the second session in which the extinction periods have been 1 hour long, following an earlier period with 30-minute extinction. Multiple breaks are clear. "Square" changes in rate are obvious, and the priming runs at the beginning of the session have, in general, the same magnitude. Later, in a somewhat shorter extinction period (50 minutes), the multiple runs tended to disappear. A stage in which this phase is well-developed appears for 4 consecutive extinction periods from a later session in Record B. This bird showed a temporary exceptional phase in the earlier development of the performance. Figure 782D represents a period in the exposure to the schedule comparable with that in Fig. 782A. The running rate is not quite so high and shows some slight curvature in the priming runs, as at *a* and *b,* becoming much more marked on the following day, shown in Record E. The ratio rate has fallen abruptly, as at *c,* and marked curvature appears. This is best seen in the breaks occurring later during extinction, particularly at *d.* The bird recovered from this characteristic within a few days, and eventually showed the performance seen in Record F, which is for 2 sessions preceding Fig. 783.

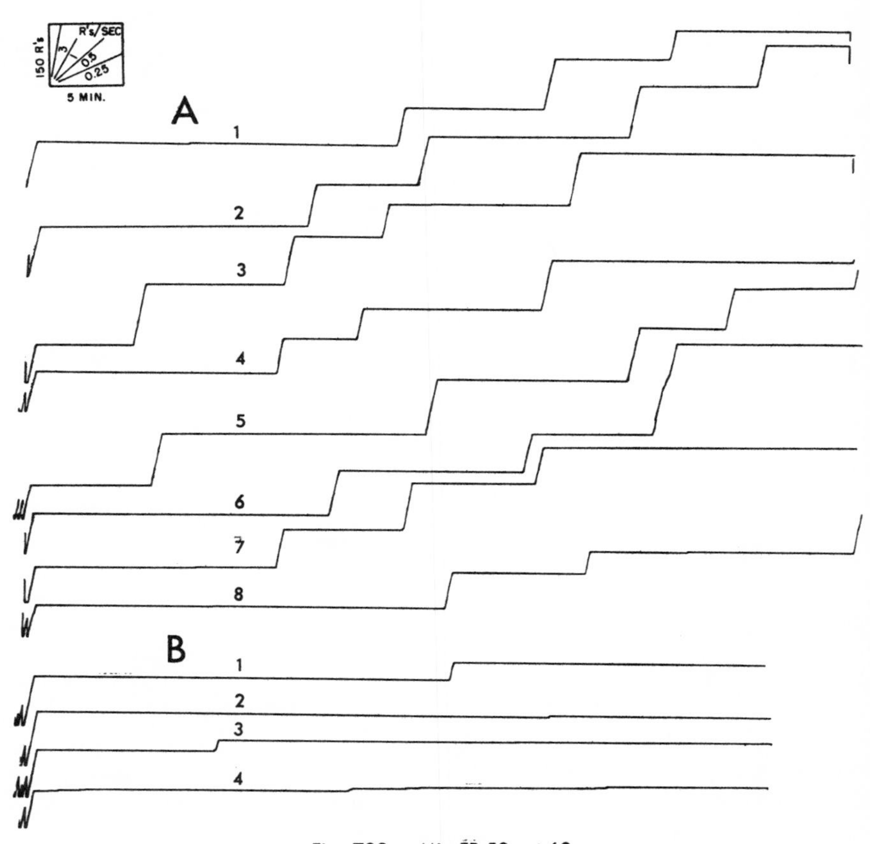

Fig. 783. Mix FR 50 ext 60

MIXED FIXED-INTERVAL FIXED-RATIO SCHEDULES

Mix FI 10 FR 40[1]

A pigeon which had been reinforced on mult FR 40 FI 10 with green and white key-colors, respectively, for 63 hours had developed the performance shown in Fig. 784A. The ratios are all run off at high rates. But the intervals show considerable variation in curvature and in the number of responses; and an interval following an interval usually begins at the preceding terminal rate with no scalloping. At the beginning of the following session, the key-light was changed to orange and the schedule was then mix FR 40 FI 10. The new color has some disturbing effect, as the first part of Record B shows, but a reinforcement is quickly reached at *a,* and a second is received at *b,*

[1] We are indebted to Dr. Clare Marshall for carrying out this experiment.

also on the ratio component. The key-light remains unchanged, of course, but the schedule now shifts to FI 10; and during this interval nearly 1000 responses are emitted. The rate varies somewhat. A second interval follows, and a reduction to a lower rate ensues. The next reinforcement is on the ratio schedule, which causes a return to a fairly sustained performance during the 3rd interval ending with a reinforcement at *c*. By this time reinforcements have occurred on the orange key on essentially a VI or VR basis, and a sustained performance completes the experimental session. Record C follows Record B in the same session after 3 hours, the intervening performance having been omitted from the figure. Some tendency to pause or to reduce the rate after reinforcement begins to develop. This is apparent, for example, at *e*, and is accumulated into an over-all lower rate in the region of *d* and *f*.

The first well-marked effect of mix FRFI appears only after 3 sessions on the schedule. The whole 4th session is shown in Fig. 785, which begins 13 hours after the introduction of the mix FRFI. The early part of the session shows essentially a linear performance. But beginning at *a* reinforcement tends to be followed by an increase in rate, and this rate tends to fall off after something of the order of two or three hundred responses have been emitted. The decline in rate takes various forms, as at *b*, *c*, and *d*, and it becomes progressively sharper later in the session, as at *e*, *f*, and *g*. Curvature begins to occur after the priming run.

In a session 26 hours after the introduction of the mixed schedule, the priming after reinforcement is practically inevitable; and at this stage the break into the curve appropriate to the interval member of the schedule is either smoothly curved or fairly sharp. Figure 786 shows the performance. Priming runs curve off smoothly, as at *a*, or *b*, or fairly sharply, as at *c* or *d*.

Eventually, a sharp break always occurs, and at a count approximately twice that of the ratio. Figure 787 gives a daily session, which begins after 95 hours on mix FIFR (the ratio having been reduced to 27). Here, almost all breaks are sharp, and the performance during the balance of the interval tends to show an initial long pause, followed by a smooth acceleration.

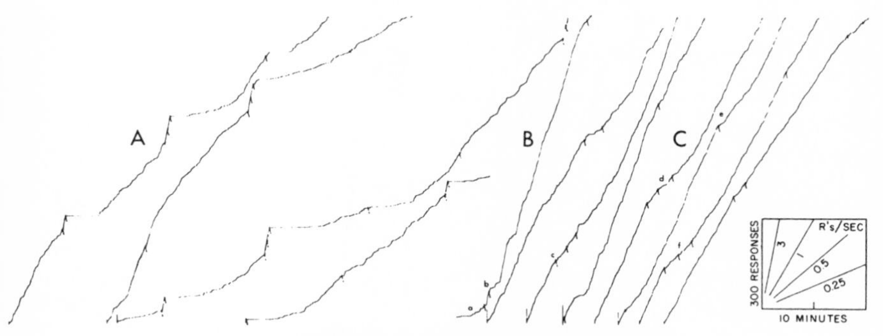

Fig. 784. Transition from mult FI 10 FR 40 to mix FI 10 FR 40

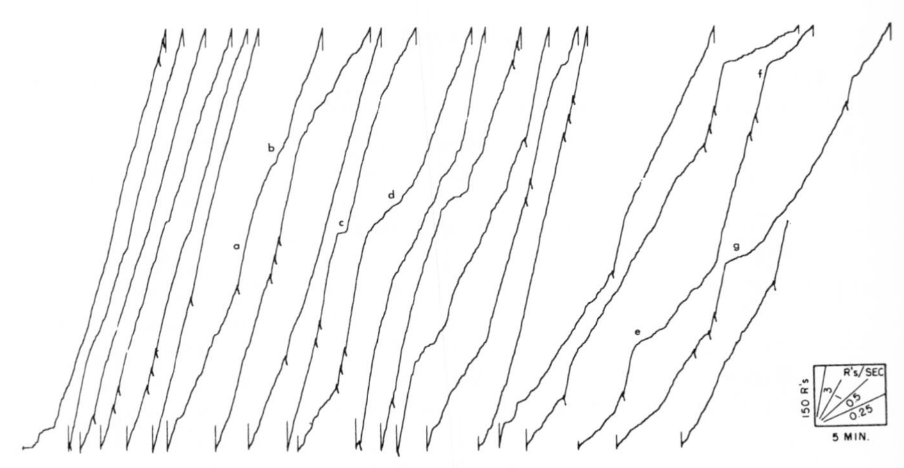

Fig. 785. Mix FI 10 FR 40 after 3 sessions

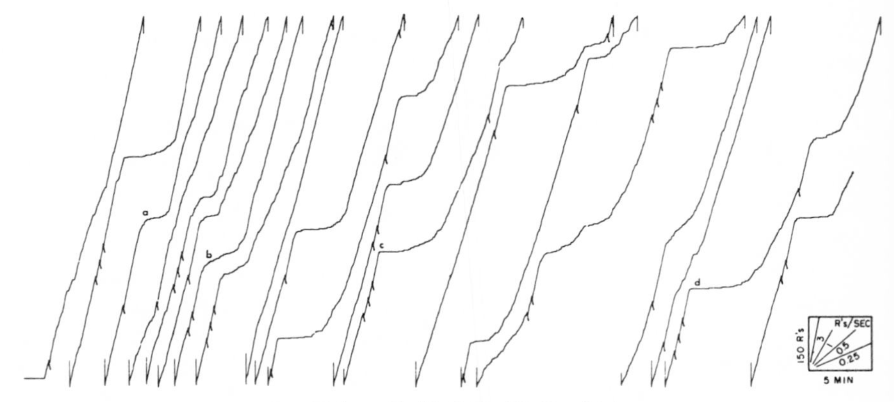

Fig. 786. Mix FI 10 FR 40 after 26 hr

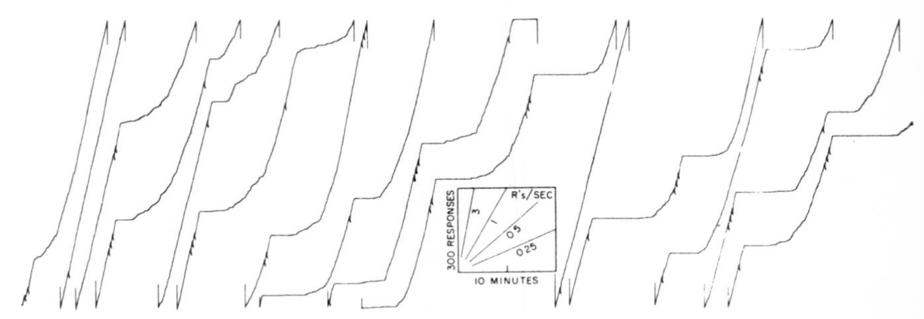

Fig. 787. Mix FI 10 FR 27 after 95 hr

A second bird only partially confirms the records just described. The earlier multiple performance was somewhat sharper, leading to a higher terminal rate than that in Fig. 784. The general character of Fig. 784C was confirmed in the 1st session. The performance in Fig. 786 was also duplicated in its essentials after about the same number of hours on the schedule. Shortly thereafter, however, the second bird began to show repeated breaks in the interval segment. This anomalous performance can be explained by the fact that the bird was occasionally pausing after reinforcement. Pausing after the first break in the interval segment can therefore trip off a second ratio run. Examples of pauses after reinforcement appear in Fig. 788 at *a, c, d, f,* and elsewhere. When the bird primes itself into a pause, as, for example, at *b, e,* and *g,* it may emerge from this pause with a ratio performance rather than an interval performance. The actual priming effect shows damping, just as in some of the mix FIFR experiments described elsewhere. Good examples of damping are at *e* and *g*. The tendency to pause and to respond with another ratio performance declines as the interval progresses. Although some segments show a priming run leading into a fairly smooth sustained rate until reinforcement (at *h*), the occasional instances in which pauses are followed by reinforcement after roughly 40 responses make it impossible for the bird to develop the pause and smooth acceleration characteristic of the FI segments in Fig. 786.

Figure 789 shows a segment of the following session. Here, reinforcements are occasionally followed by pauses, as at *a* and *b,* and these pauses are invariably followed by

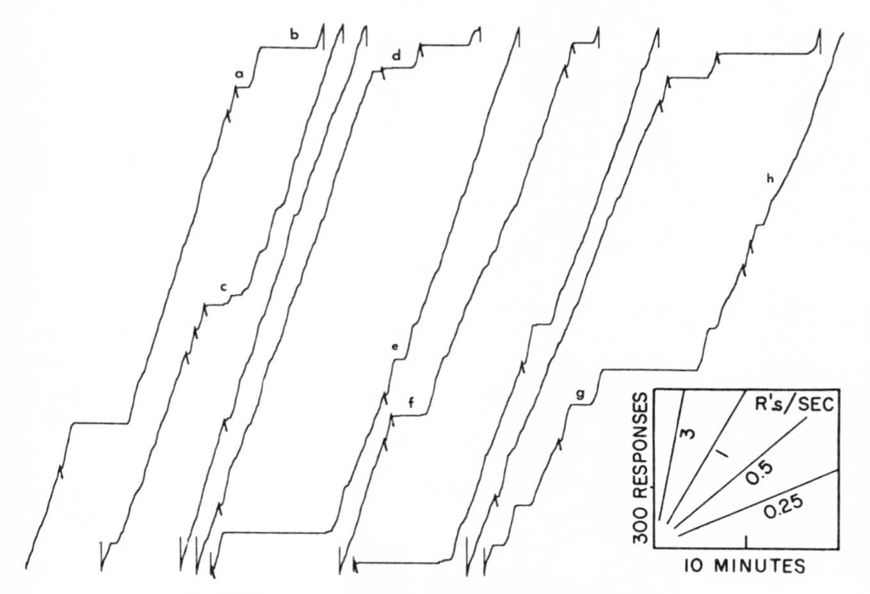

Fig. 788. Mix FI 10 FR 40 after 40 hr (second bird)

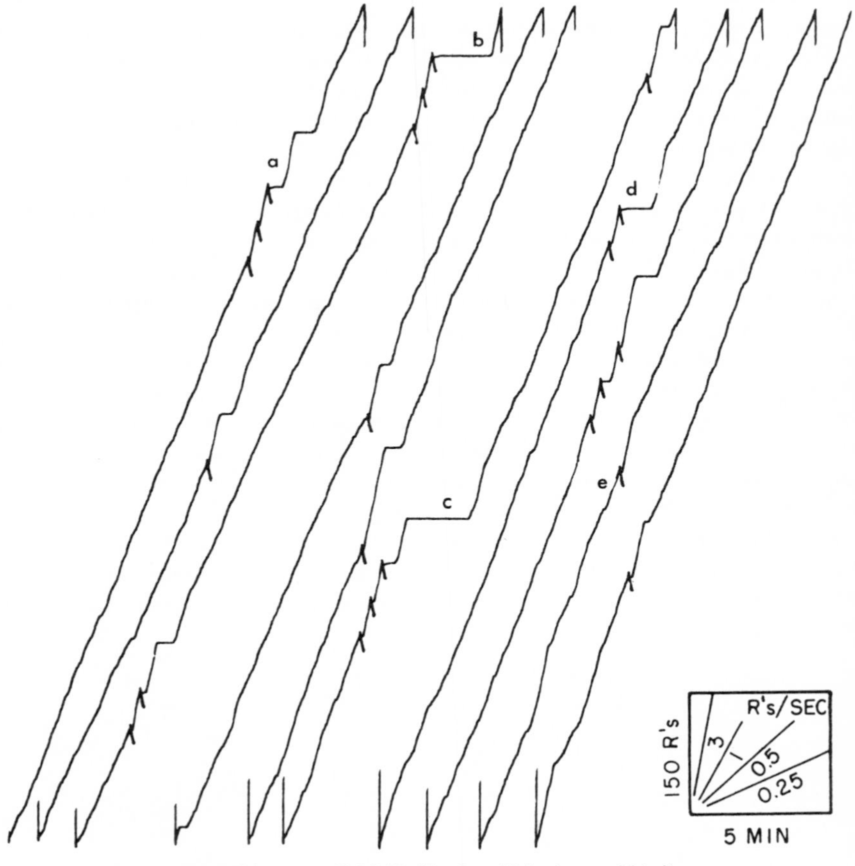

Fig. 789. Mix FI 10 FR 40 after 45 hr (second bird)

ratio runs containing a number somewhat greater than 40. The run may be followed by another pause followed by another run, as at *c*, or it may prime directly into an intermediate constant rate, as at *d*. Occasionally, the run after reinforcement primes directly into an intermediate rate, as at *e*. The figure shows a variety of combinations of these effects.

Figure 790 illustrates the best mix FIFR performance for this bird. It contains several examples of pauses immediately after reinforcement, after which a ratio of 40 responses is reinforced, as at *a*, *e*, *i*, *j*, and *n*. It also has examples in which such pauses occurring immediately after reinforcement are followed by an interval reinforcement, as at *h*, *m*, and *o*. Particularly toward the end of the session, instances still occur, as in Fig. 789, in which a run after reinforcement is followed by a pause which then primes into an intermediate terminal rate, as at *p* and *q*. However, the figure shows several instances in which there is a priming run of approximately twice the ratio of 40 followed by a marked pause followed by a run for the balance of an interval. There is often fairly good positive curvature, suggesting the mix FIFR performance of Fig. 787, as at *b, c, d, f, g, k,* and *l*. To this extent the bird is responding in a fashion which we might call appropriate to mix FIFR. (This schedule had an equal number of interval and ratio reinforcements, and it might have been possible to bring the performance around by increasing the number of ratios. This increase would presumably have reduced the pausing after reinforcement and would have reduced the variation in the actual contingencies.)

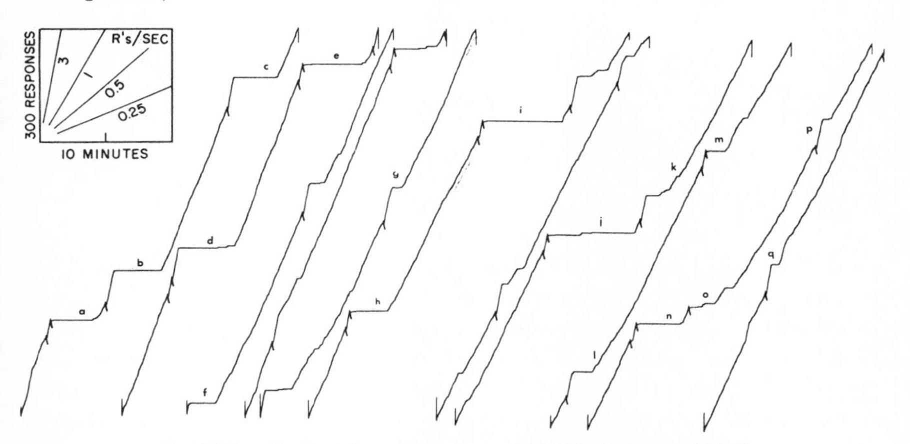

Fig. 790. Final performance on mix FI 10 FR 40 (second bird)

Mix FIFR in rats

Three rats studied on mult FRFI, as already described, were transferred to mix FRFI at the same values in the presence of the stimulus previously correlated with the ratio. In the earlier multiple-schedule experiment, the 2 stimuli had been alternated rather than presented at random; this was also done in the mixed case. Figures 791 and 792 contain 2 examples of first sessions on mix FR 20 FI 5. The result shows clearly that the interval performance on the preceding multiple schedule was due as much to the priming stimulus from the preceding ratio performance as to the external stimulus control, because all 3 rats very quickly developed an essentially multiple performance on the mixed schedule. In other words, they quickly dropped any priming run at the beginning of the interval, and showed good interval scalloping in alternation with good ratio segments. No priming run is needed, since the ratio performance which invariably precedes serves as the necessary stimulus. (The third rat developed the performance a little more slowly, but reached the same final patterns.)

We easily checked this interpretation by changing the schedule to a *random* presen-

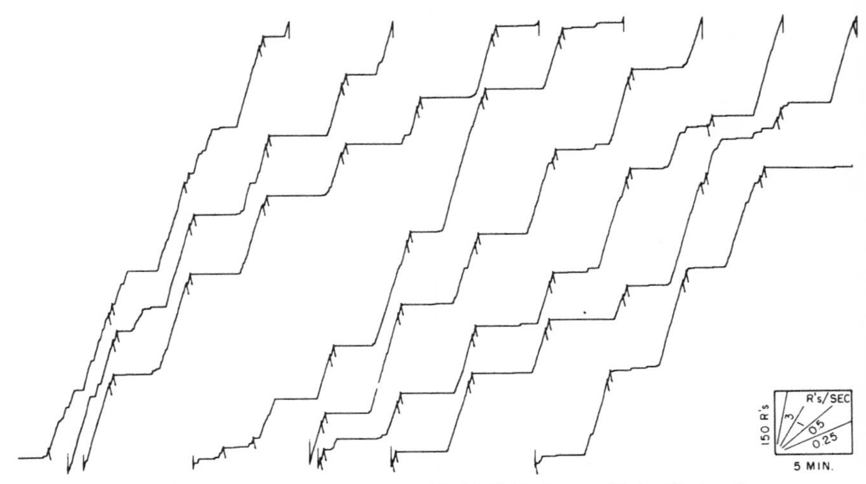

Fig. 791. First session on mix FR 20 FI 5 after multiple (first rat)

tation of ratios and intervals. All 3 rats were unable to maintain a "multiple-type" performance on the random mixed schedule. Figure 793 shows an example for the rat in Fig. 791. The preceding performance on an alternating mixed schedule was essentially like that shown in Fig. 791. Here, however, we see repeated runs of approximately $1\frac{1}{2}$ to 2 times the number of responses in the ratio occurring thoughout the interval, with an occasional longer run appropriate to the end of an interval. Before

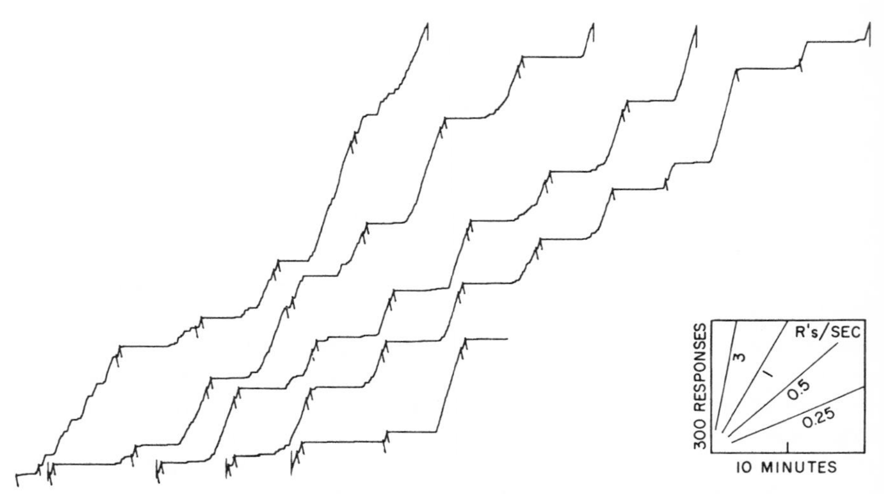

Fig. 792. First session on mix FR 20 FI 5 after multiple (second rat)

the end of the session, a fair approximation to a mix FRFI develops. This is interspersed with instances in which the ratio begins with a long pause at *e* and *h*, and hence where an interval is occasionally run off with a fair performance without priming, as at *a*. The commoner case, however, is that in which a primed run leads to a completion of the interval segment in a manner typical of a mixed schedule, as at *b, c, d, f,* and *g*. (This schedule was in force for only a short time; the reinforcement was inadequate to produce a normal ratio performance. Since pauses frequently occur before the ratio run, a percentage reinforcement of ratio-sized runs follows. This maintains the strained condition originating under the inadequate reinforcement.)

Upon returning to a multiple schedule, however, the rat is able to develop and hold a reasonably good performance. Figure 629 has illustrated an example. In order to get this type of performance, however, it was necessary to amplify the controlling

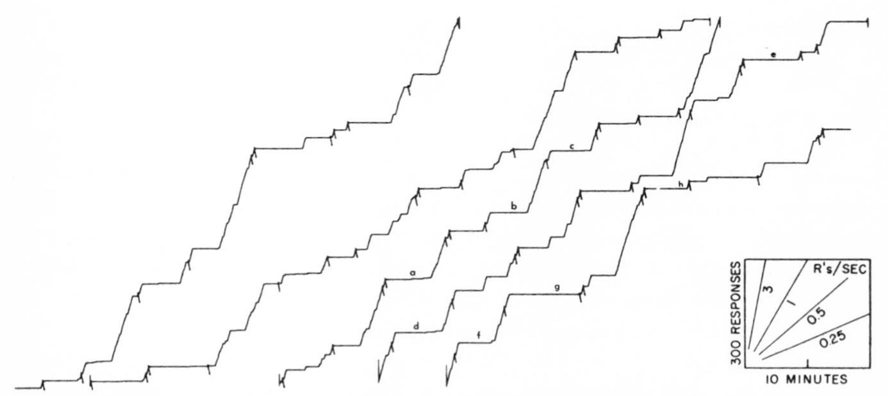

Fig 793. Random sequence of schedules in mix FIFR for the first time after simple alternation

stimuli by adding to the flashing or steady light a buzzing sound or the absence of such a sound. Under this combined stimulus, a fairly good multiple performance developed; but it is marked by an occasional priming run and a rather rough grain in the interval scallop. The terminal rate observed in the interval is not markedly higher than that of the final running rate in the ratio.

The effect of drugs on mix FIFR [1]

The effect of sodium pentobarbital on mix FRFI. A bird had developed a stable performance on mix FR 40 FI 10. It was injected intramuscularly with 2 milligrams of sodium pentobarbital at the arrow in Fig. 794. The immediate effect was a disturbance of the 1st interval performance and the elimination of priming into an interval scallop

[1] This experiment was conducted in collaboration with Drs. P. B. Dews and Clare Marshall. We are indebted to Dr. Marshall for carrying out the experiment.

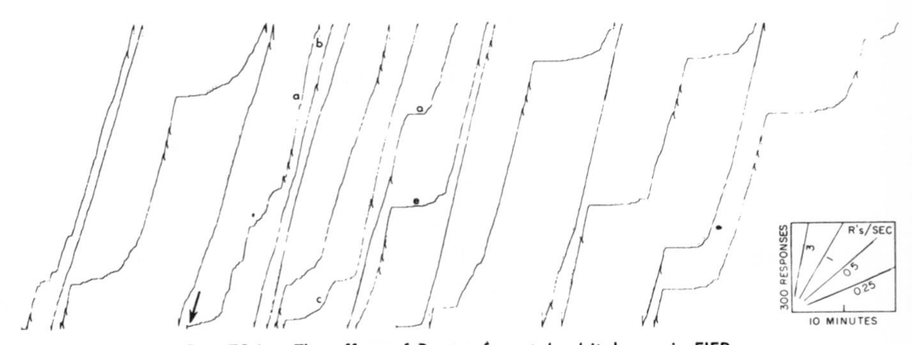

Fig. 794. The effect of 2 mg of pentobarbital on mix FIFR

in the 2nd and 3rd intervals, at *a*. Some evidence exists of a disturbance of the interval curvature at *b, c,* and *d,* but the performance is probably normal at *e*. The horizontal portion of the interval curve toward the end of the session seems to be prolonged, suggesting a later depressive effect.

In a similar experiment a few sessions later, 4 milligrams of sodium pentobarbital were injected intramuscularly at the arrow in Fig. 795. The effect appears almost immediately as a very rough grain and a tendency to depart from the usual terminal rate. Ratio performances were also disturbed (at *a* and *b*). Little evidence is present of any priming into the scallop appropriate to the interval for a number of reinforcements; the first evidence appears at *c,* while at *d* a fairly normal case exists. The bird recovers toward the end of the experimental period, giving good performances toward the latter half of the session. Here, the excitatory effect of the drug can be detected for about 2 hours. It takes the form of a disturbance in grain and the removal of the pausing appropriate to the interval schedule, with some disturbance even immediately after reinforcement, where the ratio is usually run off rapidly. Again, a later possible depressive effect appears in the long pauses at *e* and *f*.

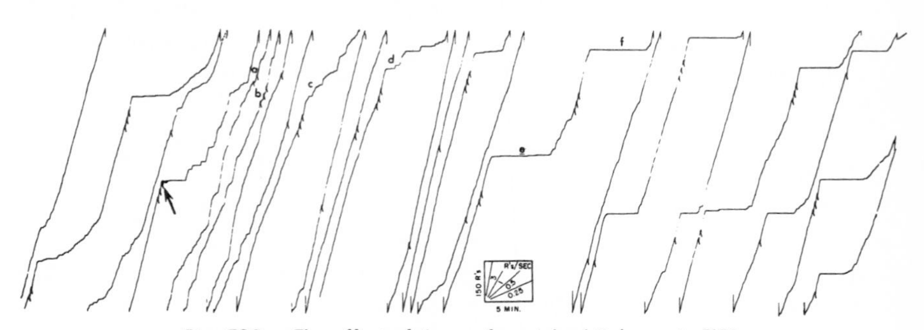

Fig. 795. The effect of 4 mg of pentobarbital on mix FIFR

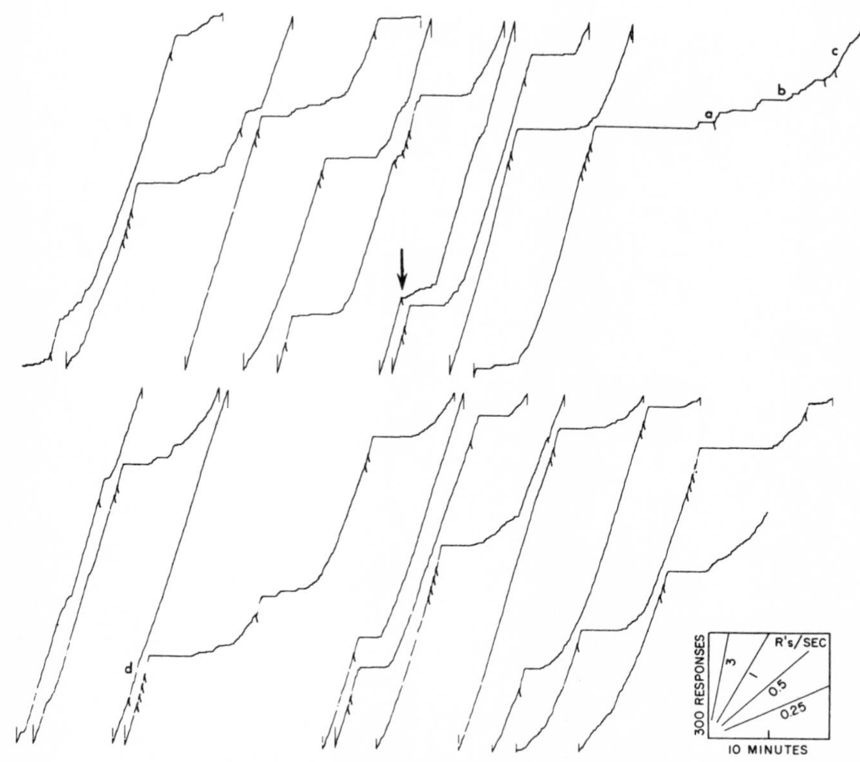

Fig. 796. The effect of 10 mg of synhexyl on mix FIFR

The effect of synhexyl on mix FRFI. In the session following Fig. 787 the bird was injected with 10 milligrams of synhexyl (a marijuana-like drug) at the arrow in Fig. 796. The lack of a priming run immediately after the injection can be attributed to the handling of the animal. No reliable effect appears until the intervals at *a, b, c,* and *d.* The effect of this amount of the drug is not great. Three days later, however, the same bird was injected with 20 milligrams of synhexyl, as shown in Fig. 797. The interval immediately following injection is disturbed; but this disturbance may be due to handling. A much more profound effect occurs later; the disturbance of the grain of the record begins at *a* but becomes marked at *b* and *c.* This pattern resembles the effect of sodium pentobarbital evident in Fig. 795 at *a* and *b.* A later marked depressant effect of synhexyl appears in the intervals *d, e, g,* and *h,* in spite of the fairly normal ratio runs at *f.* Some disturbance in grain occurs throughout the balance of the session.

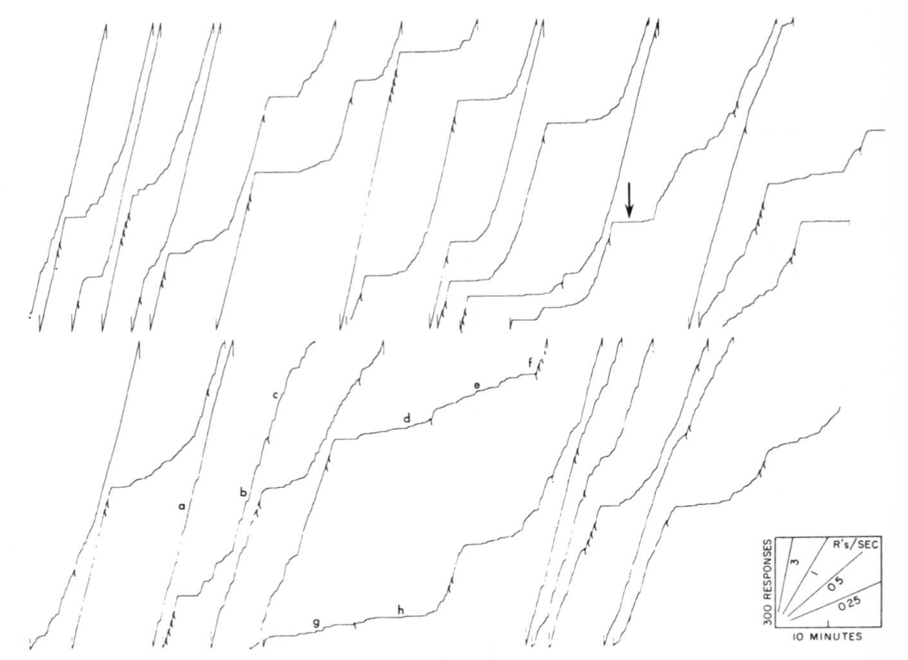

Fig. 797. The effect of 20 mg synhexyl on mix FIFR

MULTIPLE FR PRIMED FI

First bird

The transition from multiple to mixed schedules was first studied with the use of a priming stimulus. On mult FR 120 FI 10, for example, the key-color was green during the fixed ratio and flashing green (1 flash per second) for 2 minutes at the beginning of the interval. During the remainder of the interval the light was steady green. Under these circumstances no actual stimulus difference exists between the ratio and the last 8 minutes of the interval. But the flashing light during the first 2 minutes of the interval primes the organism into an interval scallop, and this stimulus may control the bird for the balance of the interval.

A bird had been on mult FR 120 FI 10 in which the stimulus for the ratio was a green key, and the stimulus for an interval was a red key. In the present experiment the ratio stimulus remained green, but the interval, as already noted, was also green with a flashing prime for the first 2 minutes. Figure 798 shows the first session in which this schedule was first presented after the earlier mult FRFI. The first 3 ratios are run off in the former ratio color. (Some slight delay at the start of the experiment was not unusual.) The first appearance of the flashing light follows reinforcement at *a*. It suppresses the behavior; but the bird recovers fairly quickly and assumes the ratio rate, which is maintained except for slight pausing until the 1st-interval reinforcement at *b*. Except for a short period of flashing light, following *a*, all of this occurs on the green key. The following reinforcement occurs after FR 125. The segment shows some disturbance resulting from the previous long run at the ratio rate and in the ratio color. Responding begins before the light stops flashing; and during the interval which follows more than 1000 responses occur at the ratio rate, but with some rough grain. Reinforcement is then received at *c*. More marked breaking begins to occur at *d, e, f,* and *g,* and the interval which follows the reinforcement at *h* shows a very rough grain. The first development under the priming schedule appears to be the suppression of behavior during the flashing light, arising from the fact that no reinforcement occurs while the light is flashing or for some time thereafter. The reinforcements at *i, j,* and *k* are followed by intervals and hence by flashing lights for a 2-minute period. The bird slows down during the flashing light and returns immediately to a high rate when the flashing terminates. Figure 799A shows the 3rd hour of the 1st session. Responding in the flashing light is further reduced, instances occurring here at

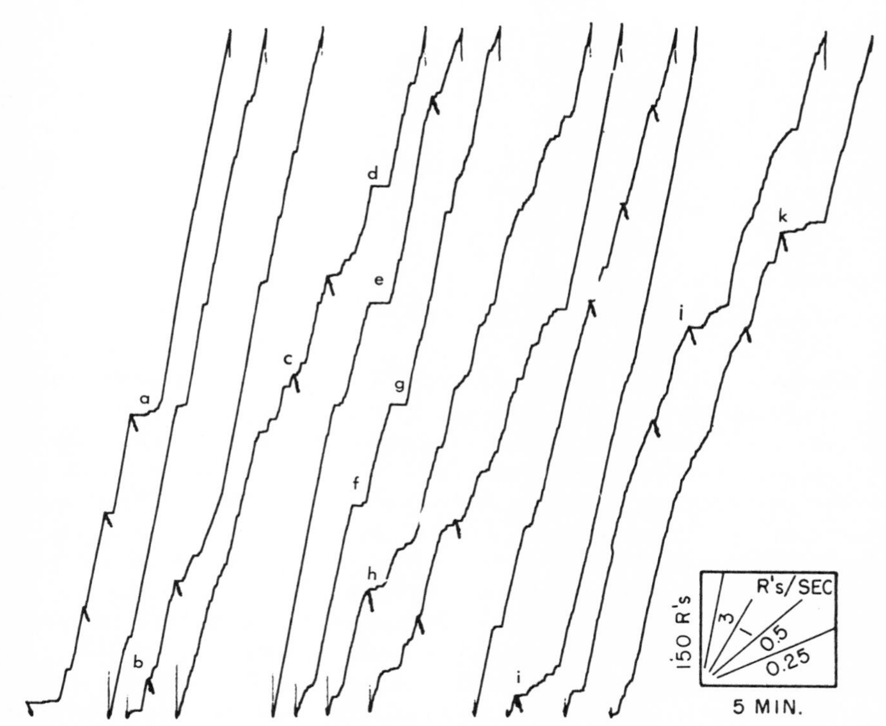

Fig. 798. First session on mult FR 120 primed 120 sec FI 10 after mult FI 10 FR 120

b, d, f, and *i.* These are followed by a high rate reminiscent of the earlier multiple performance, at *c, e, g,* and *j.* Reinforcement never occurs at these high rates, because they tend to appear just after the cessation of the flashing light in the interval. Note, however, the appearance of a second over-all rate, notably at *a, h,* and *k,* at an intermediate value resulting from the unusual schedule of reinforcement now prevailing on the steady green light.

Further developments appear by the 2nd hour of the 3rd session, 9 hours after the beginning of the schedule, as Record B shows. Here, the rate is reduced during the flashing light, as at *m*; and these low rates generally are followed by the highest rates observed in the session, as at *n, o,* and *p.* Most of the time, however, the bird is running at an intermediate rate (for example, at *q*) resulting from the residual schedule on the steady green light.

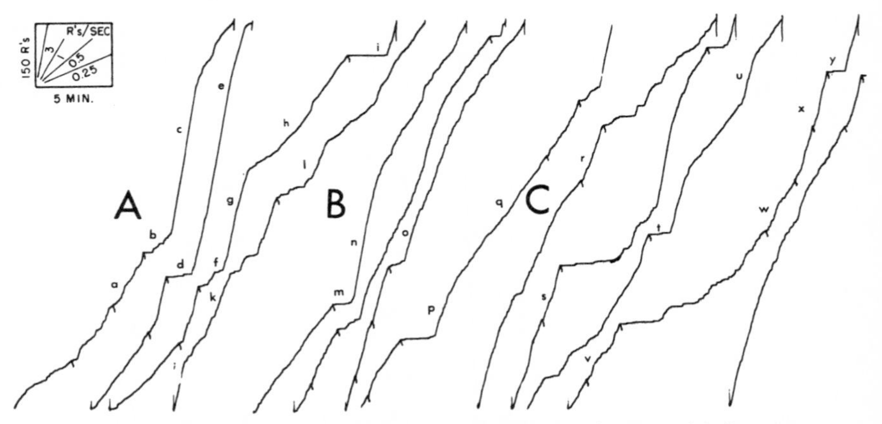

Fig. 799. Development of mult FR primed FI 10 during the 1st and 3rd sessions

These general features persist for some time. Then, another change occurs in that a "steady green light following reinforcement" begins to control the higher rate appropriate to the inevitable fixed ratio existing under these conditions. In Record C, which is for a later part of the 3rd session at the 12th hour on the schedule, the performance under the steady green light following reinforcement is strengthened—at *r, s,* and *x.* Lower rates nevertheless occasionally appear, as at *v* and *w.* Scattered responses sometimes appear during the flashing light, as, for example, at *t* and *y*; but the low rate here is usually followed by a high rate reminiscent of the earlier ratio. In the figure exceptional instances show the emergence of this high rate later in the interval after the intermediate rate has already been assumed (as at *u*).

Figure 800A shows the beginning of the 2nd hour in the following session. The steady green key just after reinforcement now controls a high constant rate. A low rate appears during the flashing green light which primes an FI scallop at *b.* The fixed-ratio rate breaks through following the termination of the flashing light at *a.*

A high rate of responding is now being reinforced in the steady green light because of the restoration of the ratio performance. This factor leads to frequent breaking-through of the ratio rate later in the balance of the *interval.* Breaking-through had not appeared earlier because high rates were not being reinforced. Record B gives examples of this stage from the 17th hour in the 5th session. Reinforcements of FR 125 on a steady green light at high rates occur at *f, h, i,* and *j.* The flashing green light produces a lowered rate at *c, e, g,* and *k.* A somewhat lesser reduction in rate appears during the flashing light at *l* and *m.* These suppressions are generally followed by periods of very rapid responding, with exceptions at *k* and *n.* But rapid runs drop to the intermediate rate shown more characteristically in Record A. After the rate has fallen to this intermediate value, however, periods of rapid running may now break through, as at *d.* This is the by-product of the improved contingencies of reinforcement on the steady green light under the ratio schedule, which is now being primed in its turn by "reinforcement not followed by a flashing light."

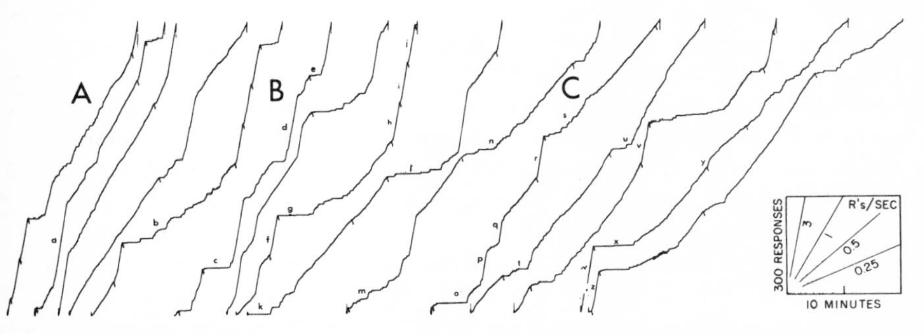

Fig. 800. Segments from the fourth, fifth, and sixth sessions on mult FR 120 primed FI 10

A further development of all these characteristics appears in Record C, where a reinforcement followed by a steady green light uniformly produces a typical ratio performance, as at *r, v, w, z,* and elsewhere. The flashing light after reinforcement now produces some running-through and a rather ragged low rate rather than a complete suppression. Examples are at *o, s, t, u,* and elsewhere. These periods of slow responding are now not inevitably followed by ratio runs. In fact, such runs are becoming exceptional, as at *p* and *q.* We have here the beginning of a performance in the interval appropriate to the interval schedule. A good example is the interval between *x* and *y,* where the preceding ratio was run off at a high rate at *w.*

A performance on the mult FR 125 primed FI 10 appears in Fig. 801, showing the 7th, 8th, and 9th sessions in Records A, B, and C, respectively. Record A contains appropriate performances in both interval and ratio. "Slips" occasionally occur, as at *a,* where in an interval segment the bird reverts back to the interval rate after emitting approximately 50 responses at the ratio rate; at *b,* where the termination of the

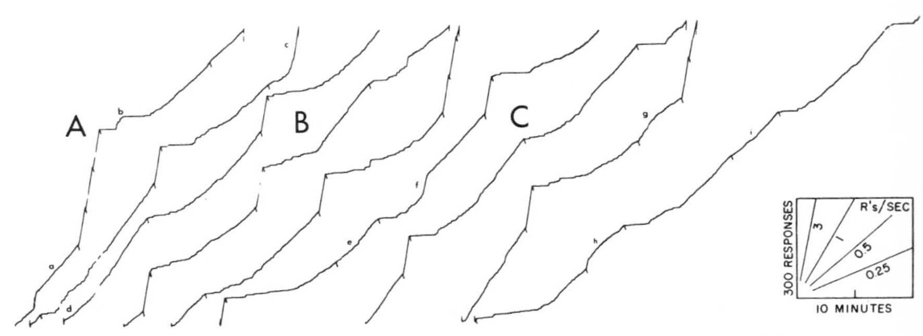

Fig. 801. Seventh, eighth, and ninth sessions on mult FR 125 primed FI 10

flashing light produces a small burst of responses, after which the bird assumes the "proper" interval performance; and at *c*, where a burst of about 150 responses is emitted at the ratio rate upon termination of the flashing light, after which the bird reverts back to the interval rate. Record B shows 2 examples of slips. At *e* a whole ratio is run off at the interval rate; and at *f* an interval begins normally, but accelerates into the fixed-ratio rate for approximately 50 responses before returning to the interval rate for the remainder of the interval. More marked deviations occur in Record C, where 3 ratios, at *g, h,* and *i,* occur at the fixed-interval rate. In general, however, the multiple control at this stage of the schedule is good. Note that the intervals have curvature whether the preceding schedule was fixed-interval or fixed-ratio. This is unlike the performance under a straight multiple schedule.

At this stage of the experiment, TO probes were used to examine the process, and further development was to some extent disturbed. Later, however, this bird showed the much more advanced state of the primed mult FRFI evident in Fig. 802. This is the 25th session on the priming schedule. (An anomalous reinforcement hatch appears in the record at *b*; this may have been merely an erroneous marking operation or an actual reinforcement.) Although some exceptions occur, the typical performance at this stage may be described as follows. When a reinforcement is followed by the steady green light, a standard ratio performance is run off, as at *f, g,* and *h.* When a

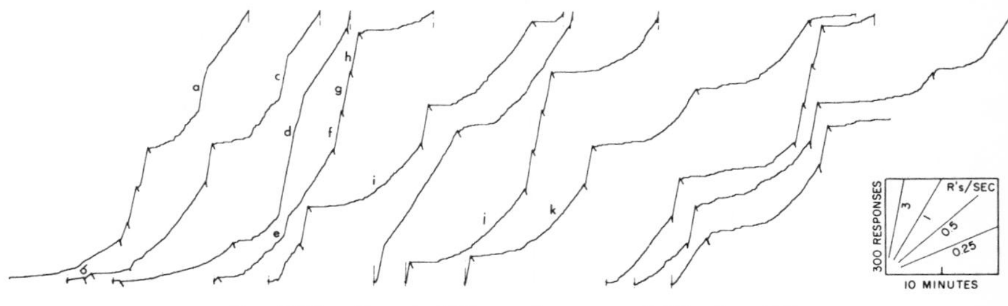

Fig. 802. Twenty-fifth session on mult FR primed FI

reinforcement is first followed by 2 minutes of flashing light and then by the steady green light, however, a curve appropriate to an interval performance appears, as at *i*, *j*, and *k*. The exceptions are exemplified at *a*, *c*, *d*, and *e* in the early part of the session. Here, the acceleration developing during the interval sets off too high a rate; since the key-color is appropriate to FR, this rate continues for approximately the number of responses in the ratio, only to prime the bird into the linear performance prevailing under the earlier, rather anomalous schedule under the steady green light. It should be noted that "mistakes" on the steady green light take the form of running at a rate appropriate to a ratio performance on an interval schedule rather than at an interval rate during a ratio. It should be noted, too, that the runs which break through in the interval schedules are usually as long as the ratio or longer. An exception appears at *e*.

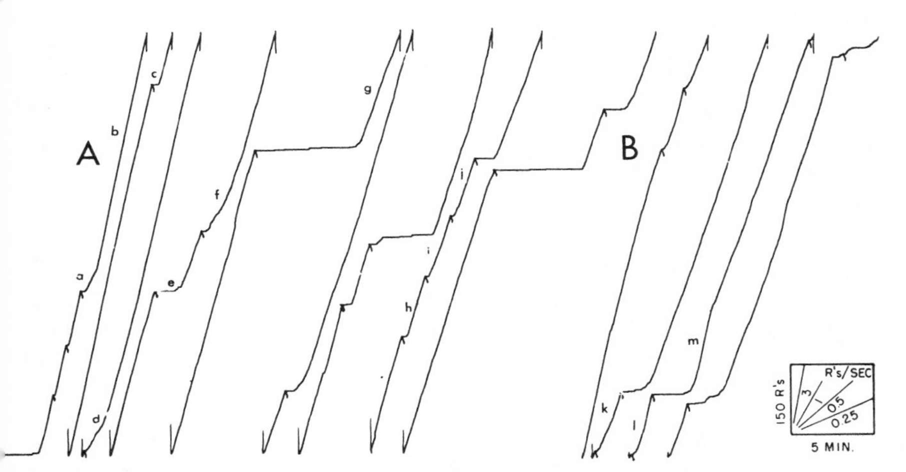

Fig. 803. Second and third sessions on mult FR 125 primed FI 10 (second bird)

Second bird

The 1st day on which a second bird changed from a mult FRFI with green and red lights to mult FR primed FI with a green light flashing for 2 minutes as a prime was lost in recording; but the 2nd day revealed a rather different performance, as Fig. 803A shows. The over-all rate is higher, and the bird was not disturbed by the flashing light to the same extent. At *a*, for example, some responding occurs during the flashing period at almost exactly the rate of flashing. In other experiments a degree of stimulus control has been observed under which responses can be evoked or stopped within a fraction of a second by manipulation of the light. Here, the bird was evidently striking the green light each time it appeared in the 1-per-second flashing. As soon as the steady green is present without flashing, the ratio rate is maintained for more than 1000 responses at *b*, and the following ratio begins with a slight pause, at *c*. The following interval at *d* shows responding during the flashing light and an almost immediate return to the ratio rate for the balance of the interval. The effect upon the following ratio was to lengthen the pause at *e*. The following interval at *f* shows responding during the flashing light and again a return to the ratio rate; but the following interval shows a prolonged pause, possibly because of the straining due to the maintenance of the ratio rate during the entire interval. This pause is followed, however, by a return to the ratio rate near the end of the interval at *g*. The following interval again contains more than 1000 responses. The pauses at the beginnings of intervals now usually exceed those in ratios—for example, at *h*, *i*, and *j*. The flashing light begins to function not only for its duration but for some time following. When responding resumes, however, the rate tends to be at the ratio value.

On the following day the performance is repeated, as Fig. 803B shows, except that the ratios now show some strain, as at *k* and *l*, while the intervals start with pauses or low rates. In one case (*m*) the initial pause is followed by a high rate. Where the performance in Fig. 802 showed a ratio performance breaking into the interval performance, the reverse is true here. The terminal interval performance on "steady green" carries into the ratio.

Figure 804 shows a late performance for this second bird under a priming sched-

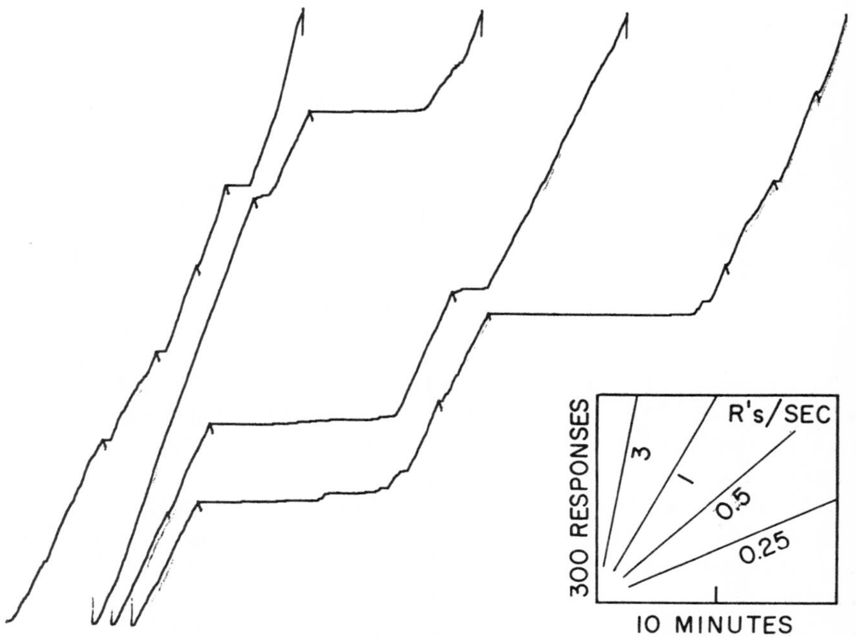

Fig. 804. Mult FR 125 primed FI 10 after 23 hr (second bird)

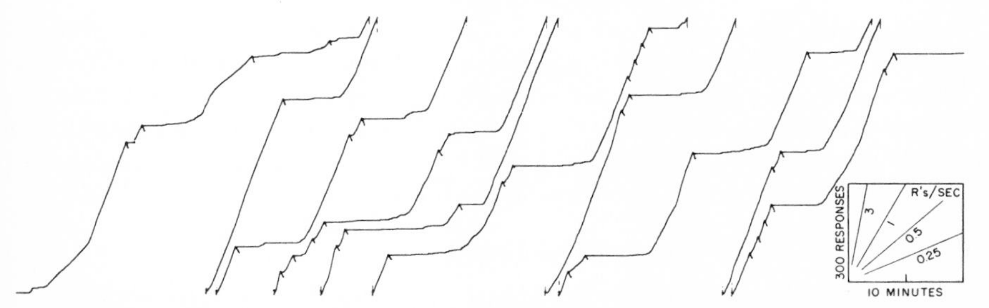

Fig. 805. The effect of reducing the size of the FR component on mult FR primed FI

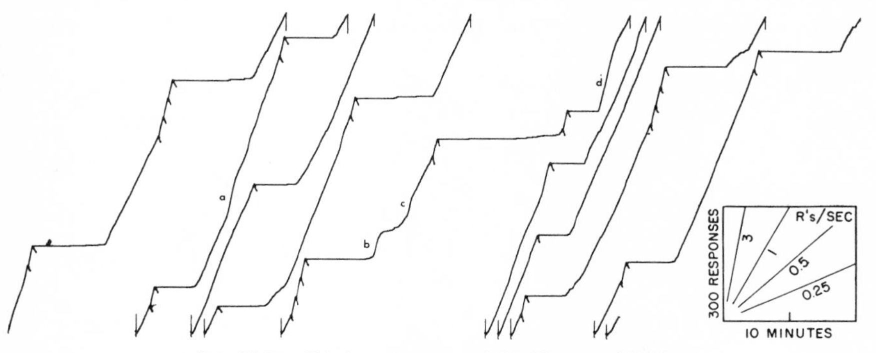

Fig. 806. Third session on mult FR 50 primed FI 10

ule before probes were introduced. The ratio performance has suffered badly. Ratios are frequently run off at the terminal interval rate. The intervals tend to be marked by prolonged pauses but, in general, to end at the same rate as the ratios.

Figure 808 shows a still later performance for this bird. At this time, TO probes were being used, and the performance may show the effect. Here, however, a much better curvature has developed in the interval, although the ratios are still being run off at much below a normal ratio rate, with rough grain and curvature.

In an effort to improve the ratio performance for this bird, the ratio was reduced from 125 to 50. Figure 805 shows the 1st day on the new schedule. At the beginning of this session, difficulty still exists in starting the ratio; but before the end of the session, the ratio rate is accelerating to a value higher than the terminal rate in the interval. On the following day the 2 rates clearly break apart; and by the 3rd session with the shorter ratio, a reasonably satisfactory performance under mult FR primed FI appeared, as evident in Fig. 806. Here, the "steady green" directly after reinforcement controls a high rate appropriate to a ratio. "Flashing green" after reinforcement sets up a pause which primes the bird into a prolonged period of very slow responding, or no responding at all, which leads to a fairly standard interval segment. Exceptions appear at *a*, *b*, and *c*, where the ratio rate breaks through in the interval; and at *d*, where the pause following reinforcement leads immediately to the ratio rate, at which a number of responses equal to approximately 3 times the ratio is emitted before the rate drops to the value characteristic of the terminal-interval performance. This figure confirms Fig. 802 in the essentials of the primed FRFI performance and indicates that the earlier failure to conform was due to the selection of too high an FR.

Effect of a TO probe on multiple primed FI

The first bird in the experiment just described showed at one stage on mult FR 125 primed FI 10 (between the stages represented by Fig. 801 and 802) a performance in which ratio segments contained either a high rate, a smooth intermediate rate, or, occasionally, a high rate dropping to a lower value during the ratio. The interval per-

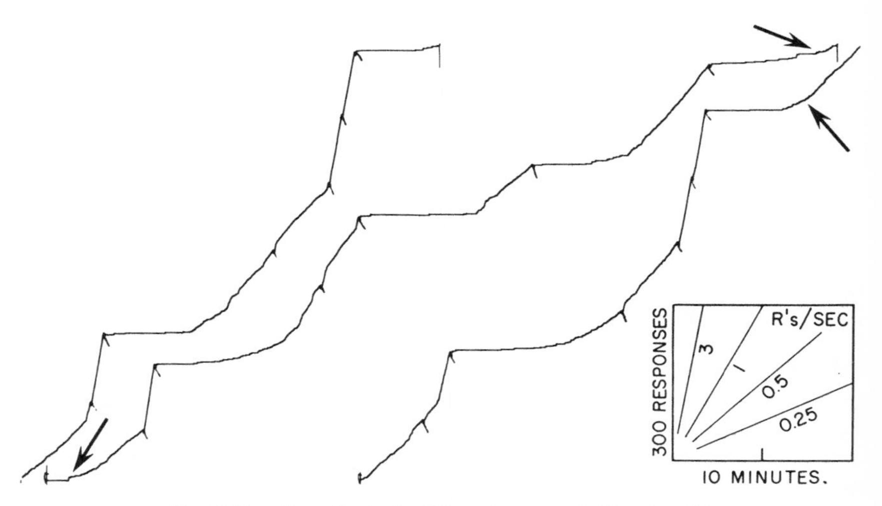

Fig. 807. The effect of a TO probe on mult FR primed FI

formances at this time generally show prolonged periods of low rate after reinforcement, with fairly smooth acceleration to an intermediate linear terminal rate. Probes consisting of 1-minute TOs were introduced approximately midway in an occasional interval. The first two showed some disturbance; shortly after the probe terminated, responding accelerated to the ratio rate for a brief period. Thereafter, however, the probe had remarkably little effect. The duration of the TO was later increased to 5 minutes. Three instances at 5 minutes appear in Fig. 807, where, at the arrows, a TO 5 was introduced midway in the interval curvature or shortly after acceleration had begun. In each case the interval was completed in essentially a normal fashion in spite of this period of TO.

As we have just seen, the second bird had developed a performance under mult

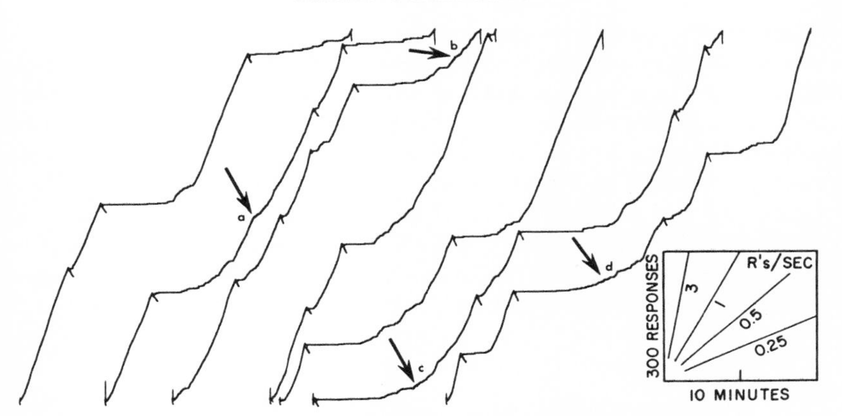

Fig. 808. The effect of TO probes on mult FR primed FI (second bird)

FR primed FI which showed difficulty in the ratio. Nevertheless, when 5-minute TO probes were introduced into the fairly well-developed interval scallop, as shown at the arrows in Fig. 808, little effect upon the over-all curvature was noticeable. At *a* the rate is somewhat reduced, but it quickly returns to a value characteristic of the interval. At *b, c,* and *d* little or no change is evident in over-all rate. Whatever is responsible for the interval curvature at this point survives a 5-minute period of TO.

Effect of reducing priming stimuli on mult FR primed FI

First bird. We continued the experiment on priming the interval in a multiple schedule by reducing the duration of the flashing green light, which served as the priming stimulus. Shortening the flashing green light from 2 minutes to 1 minute had no effect on the performance. After several days, the duration of flashing was dropped to 30 seconds. Figure 809 shows the effect. The record begins with a well-developed primed multiple performance with smooth curvature in each interval (except for an anomalous break at *a*). Beginning at the arrow, the priming stimulus lasted only 30 seconds. The 1st interval at *b* is run off normally. The 2nd, however, shows break-

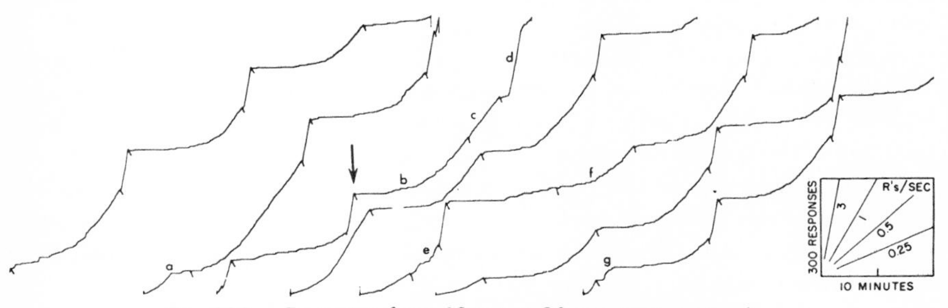

Fig. 809. Transition from 60-sec to 30-sec priming stimulus

through of the ratio rate (at *d*). Note that the bird occasionally ran the ratio at the terminal interval rate (as at *c*). Some disturbance of the priming stimulus is also evident at *g*. The main result appears to be in the ratio performance. The ratio at *e*, which is run off in the steady green light, immediately following the reinforcement, shows very rough and irregular grain. The ratio at *f* is run off with curvature appropriate to the terminal parts of an interval. Such curvature may be due to the fact that the rate did not accelerate to the normal terminal rate in the preceding interval.

The priming stimulus was later reduced to 15 seconds, and the schedule was continued for 3 sessions. Figure 810 shows a final performance with a 15-second prime in the interval. Except for the slight burst of responses at *b* and the low rate at *a*, this

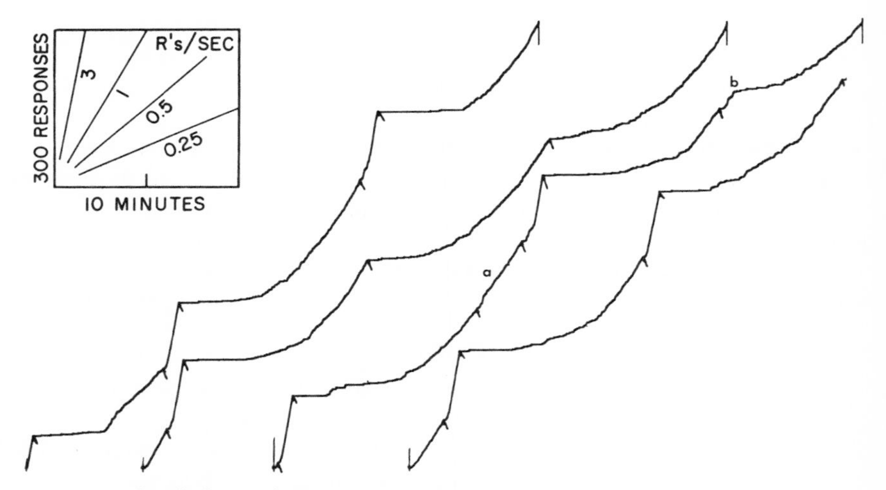

Fig. 810. Mult FR 125 primed FI 10 with 15-sec priming stimulus

duration of the priming stimulus is effective in producing a multiple performance. These segments were taken from the middle of the session, however. At the beginning of the session the bird showed irregularities. Similar irregularities early in the session appeared the next day.

At the beginning of Fig. 811 the prime is still 15 seconds long, and it fails to bring the bird into the interval performance at *a* and *b*. The bird primes itself by running off from 1 to 2 ratios at the ratio rate. At *c*, however, the performance typical of the later part of Fig. 810 has developed.

At the arrow in Fig. 811 the priming stimulus was reduced to 5 seconds. The immediate effect was a failure to produce the pausing appropriate to the beginning of the interval, which runs off at almost the terminal rate throughout the interval, as at *d*. Another interval (at *e*) is smoothly executed, and a ratio is then run off

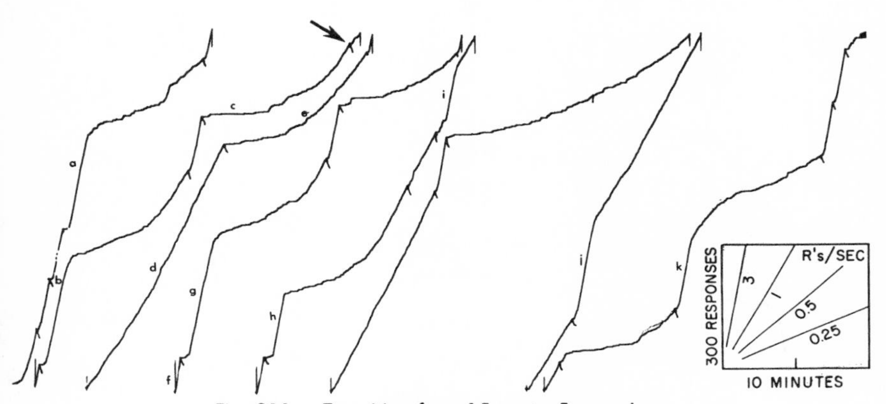

Fig. 811. Transition from 15-sec to 5-sec prime

normally at *f*. The priming stimulus fails to work in the interval which follows at *g*, and the bird primes itself into an interval scallop by running at the ratio rate for about 300 responses. The 5-second prime also fails to operate effectively at *h*, *i*, *j*, and *k*.

When the priming stimulus was removed altogether, the schedule became, of course, a mix FIFR. This occurred 2 sessions after the ineffective 5-second prime of Fig. 811, and is shown in Fig. 812. Here, all intervals begin with self-priming runs, after which the bird decelerates fairly quickly but smoothly to a pause followed by a fairly typical interval scallop. Meanwhile, the ratios are for the most part run off at standard rates, although exceptions appear at *a* and *b*. (For a further development of this

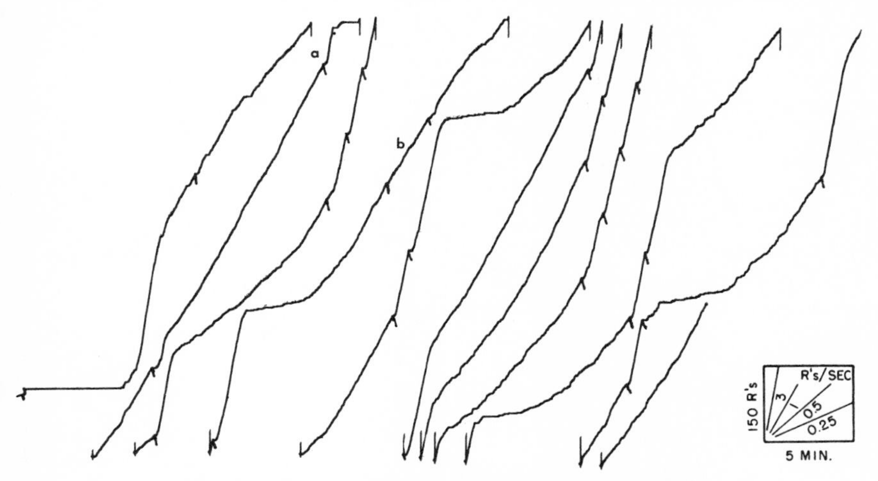

Fig. 812. Transition from mult FR primed FI to mix FIFR

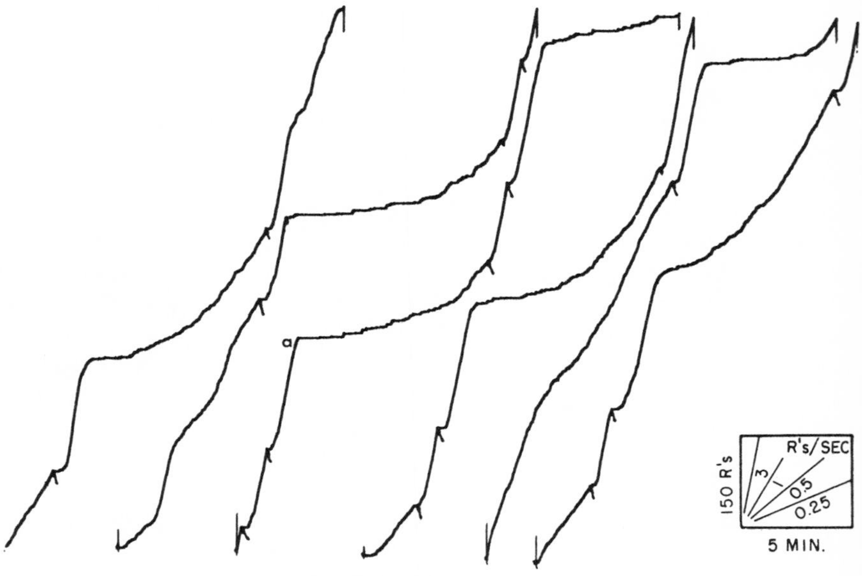

Fig. 813. Mix FI 10 FR 125 after 6 sessions

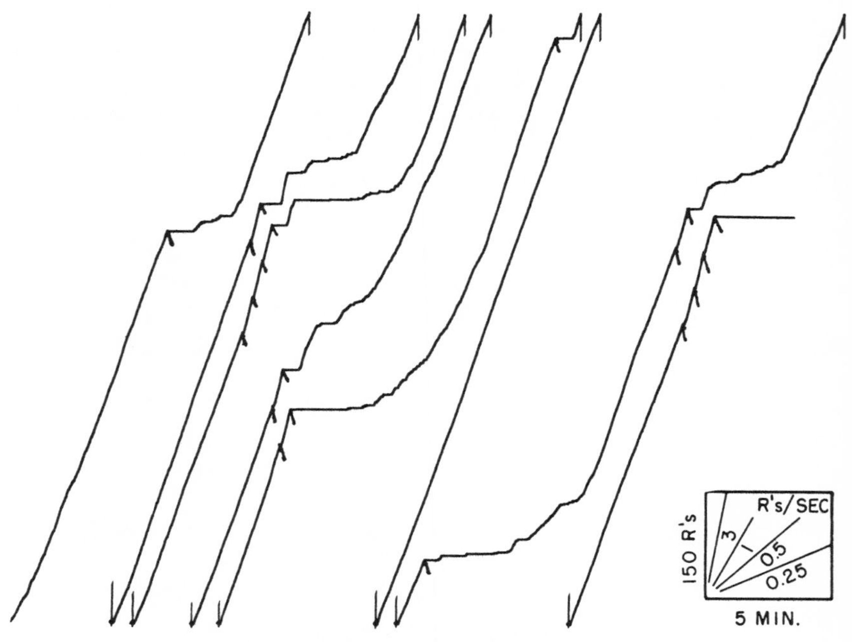

Fig. 814. Mult FR 50 primed FI 10 (second bird)

type of schedule, see the section on mix FIFR.) Figure 813 describes the final performance for this bird 6 sessions after Fig. 812. The run after reinforcement before the break to a low rate has become somewhat shorter, and the break is occasionally more abrupt, as at *a*.

Second bird. When the second bird was tested on a shorter flashing light as a prime, trouble appeared at the 30-second duration. The bird showed a tendency to resume the ratio rate at the termination of the flashing light, or shortly thereafter, and to count out approximately one ratio before priming itself into the interval scallop. Occasionally, a standard interval performance was observed; and, occasionally, the short run after the pause led into the terminal-interval rate for the balance of the segment. Some of these effects appear in Fig. 814. Later, the bird showed difficulty with a prime of 15 seconds and almost complete disruption with one of 5 seconds, when the performance was essentially the same as on a mixed schedule.

INTERPOLATED SCHEDULES

A relatively short period on 1 schedule may be inserted into a background schedule which occupies the main part of the session. When no stimuli exist appropriate to the 2 schedules, we shall speak of the short schedule as "interpolated." It has the effect of a complex probe. In a case to be described the schedule during most of the session was FI 15, but for a certain number of successive reinforcements it was FI 1. How often a brief period on another schedule is interpolated per session depends upon the length of a session and the time needed to observe the consequences. Points to be noted include the changeover to the interpolated schedule, the development of a performance on the interpolated schedule, and the changeover to the background schedule. Of course, many combinations are possible. We have studied the interpolation of a short FI in a long FI, a small FR in a long FI, and a small FR in VI.

FI inter FI

Two birds previously used in an experiment on matching colors were placed on FI 15 for 30 hours. The daily sessions were 8 hours long for one, and 6 hours long for the other. A brief period of FI 1 was then interpolated in the middle of each session. In general, this period contained 16 reinforcements, although the number varied somewhat.

For one bird the results of the early interpolations were lost through poor recording and apparatus defects. By the 15th interpolation, however, a well-marked pattern had emerged, which appears in Fig. 815A, showing the important middle section of the day's run. The bird is not pausing substantially after reinforcement on FI 15 in spite of rather long exposure to that schedule.

The interpolated schedule as a whole seems to have some of the effect of a VI. The early reinforcements on FI 1, the interpolation beginning at *a,* also show only slight pausing, but a marked scallop develops during the interpolation. Upon the return to

Fig. 815. FI 15 inter FI 1 after 14 interpolations

the 15-minute interval, the scallop appropriate to the short interval begins, at *b*. It runs past the usual point of reinforcement and then breaks sharply into a compensatory pause, at *c*. The compensatory nature of this pause is suggested by the fact that it is not characteristic of the longer-interval performance. The bird then returns to an essentially linear FI 15 performance. Some variation in the over-all rate is evident, however. A marked deviation appears in the record at *d*, where after a progressive decline in rate, the curve accelerates to the terminal slope in the short interval. Record B shows roughly the same performance for the following session. Scallops develop on the interpolated schedule, and a run appropriate to FI 1 appears at the beginning of the first subsequent FI 15 at *e*. This run is followed by the same break and, in this case, what appears to be a late compensatory increase at *f*. The performance is repeated in the following session (Record C), but Record D does not show any marked inflection upon entering the subsequent 15-minute interval at *g*. The earlier performance reappears in the following session, however, as Record E shows. In this case

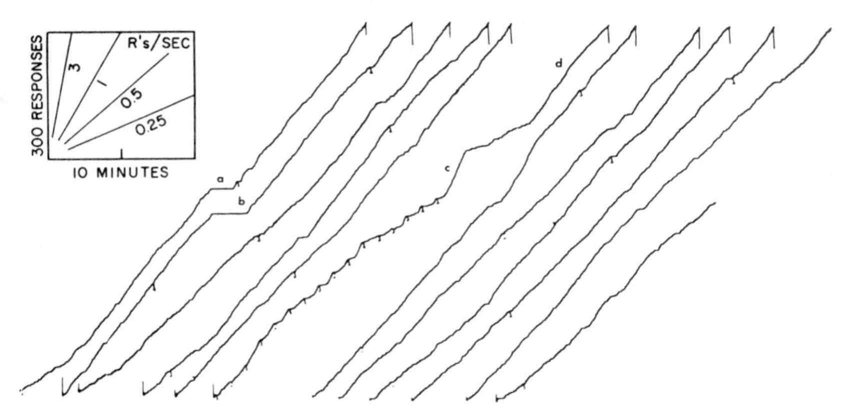

Fig. 816. FI 15 inter FI 1: entire session after 20 interpolations

a period of rather slow responding follows before the return to the normal FI 15 rate at *h*. Figure 816 shows the entire following session. The early breaks at *a* and *b* are exceptional for this bird; and, in general, no marked, consistent pausing appears after reinforcement. Scallops developed during the interpolated period; these are followed by a substantial run at *c*, which is followed again by a break to a low rate before the return to the standard FI 15 rate at *d*. The record clearly shows that pausing occurs on FI 1 but not on FI 15, and that the pause on FI 1 must be reinstated each day through reconditioning in the early part of the interpolated schedule. Figure 816 occurred beween Records E and F of Fig. 815. An over-all disturbance is evident as a result of the interpolation. Later interpolations are shown in Fig. 815 F, G, H, and I.

The latter effect was much more marked for the other pigeon. This bird had already developed some tendency to pause after reinforcement, and the first effect of the interpolated FI 1 was to summate this tendency to produce a decline in rate. Figure 817A shows the first interpolation. The short intervals beginning at *a* to *c* and *d* reduced the rate very nearly to zero, from which a much higher rate recovers, although without consistent scalloping. (Some negative acceleration actually appears, particularly in the 11th and 12th intervals.) When the schedule returns to 15 minutes, the normal performance is quickly reinstated, at *e*. The tendency to respond at a higher rate after reinforcement, noted particularly at *f*, disappears in the later part of the session not shown, when there is a slight tendency to pause after reinforcement. The 2nd interpolation, in the following session, has the same effect. It leads to a pause of about 5 minutes, at *g*. This pause is followed again by a recovery on the short-interval schedule and a return to a stable rate on FI 15 thereafter. In subsequent sessions, pauses and scallops after reinforcement on FI 15 develop further; but a performance appropriate to FI 1 also emerges, although scalloping develops late. The 7th interpolation is illustrated in Fig. 817C; and the 11th, at Record D. By this time

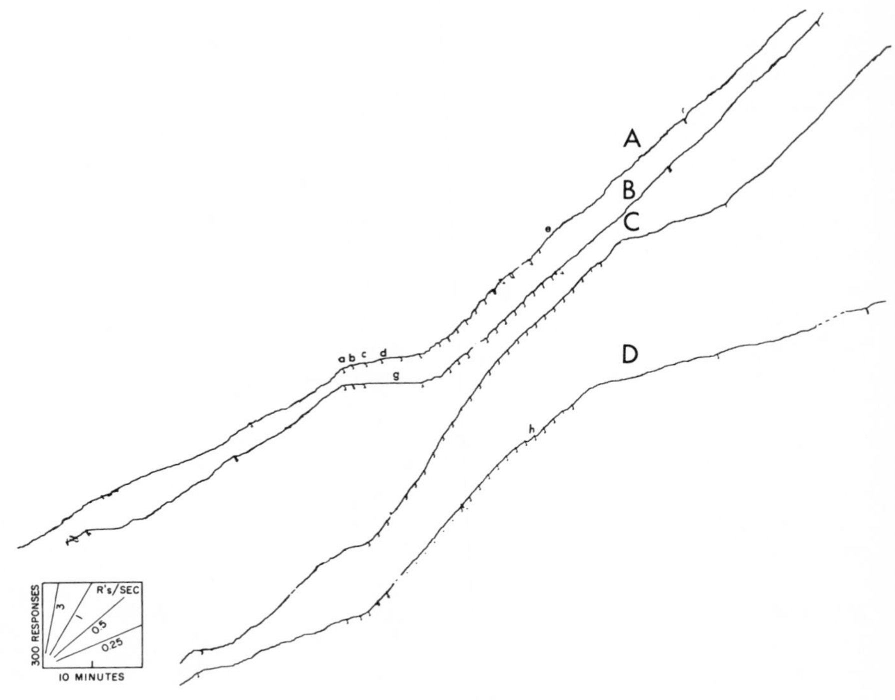

Fig. 817. FI 15 inter FI 1 (second bird)

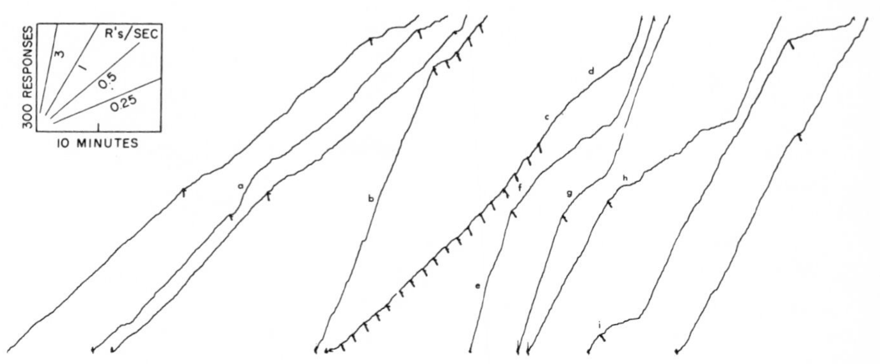

Fig. 818. FI 15 inter FI 1: 3rd interpolation (second bird)

there is a considerable tendency to pause on the 15-minute schedule; but such pauses appear only occasionally on FI 1, notably at *h* in this record.

Where an interpolated FI 1 against FI 15 had something of the effect of VI for the first bird, the performance of the present bird suggests a mix FIFI. As early as the 3rd interpolation, long runs occur at 2 rates on FI 15 alone which are common in mix FIFI. A substantial portion of the 3rd session is shown in Fig. 818 (follows Fig. 817B). Note the sudden emergence of a much higher rate of responding on FI 15, at *a* and *b*. The effect is more common after the block of reinforcements on FI 1. The interpolated performance is followed by a somewhat higher rate at *c*, which then declines to the lower of the 2 running rates at *d*. The rate rises again to the exceptionally high value at *e*. The negative curvature after reinforcement shown at *f, g, h,* and *i* resembles that at *c* and appears to be the result of the interpolated FI 1.

Later, the performance resembles that of the first bird more closely, except that the rate following the interpolation is generally lower for some time. The 20th, 21st, 22nd, and 23rd interpolations for the second bird are given in Fig. 819, which may

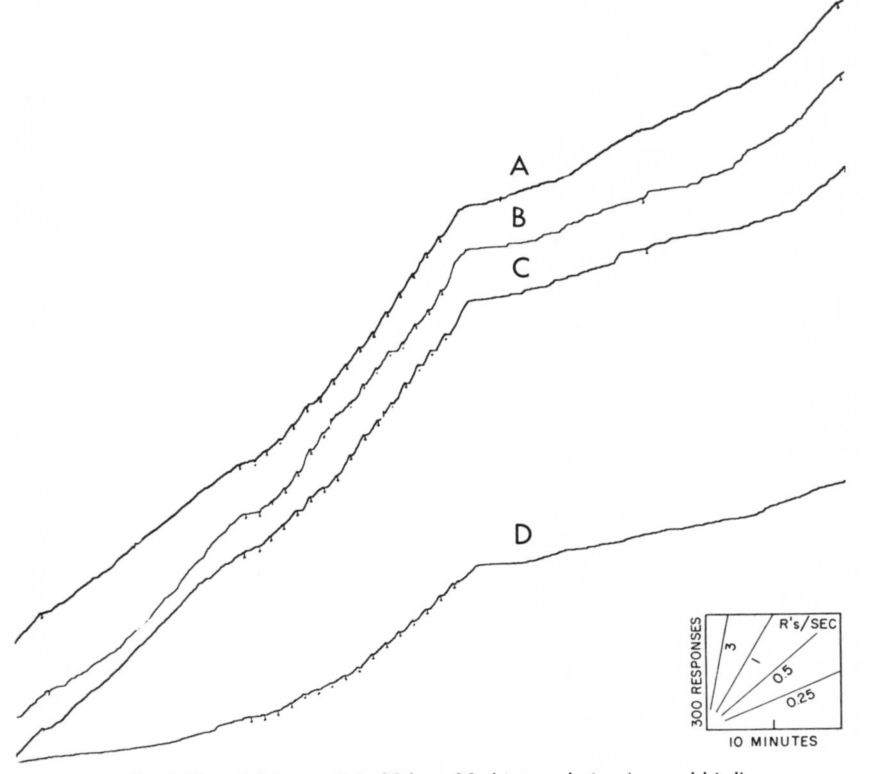

Fig. 819. FI 15 inter FI 1: 20th to 23rd interpolation (second bird)

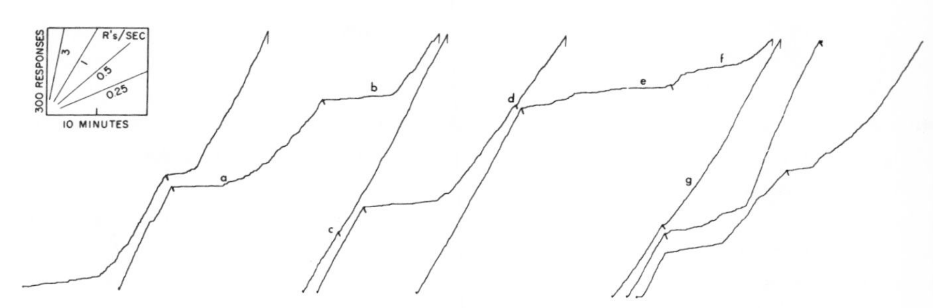

Fig. 820. FI 15 inter FI 1: 31st interpolation (second bird)

be compared with Fig. 816. Well-marked scallops appear on FI 1, but they develop again each day. The FI 15 curve now shows a longer and more marked scalloping. The first long interval after interpolation shows priming, but the ensuing rate for 15 to 30 minutes following the interpolation is consistently lower. At this time the bird still shows wide variations in running rate.

Figures 820 and 821 show the final performance for the second bird, at the 31st interpolation. At the beginning of Fig. 820, well-marked FI 15 scallops appear, as at *a* and *b*, with an occasional example of running-through without scalloping, as at *c* and *d*. Some evidence of a second-order effect appears in 3 intervals, *e*, *f*, and *g*. The performance is no longer characterized by frequent, sharp shifts from one running rate to another. Toward the end of the session, however, the performance is somewhat more variable. Figure 821 shows this performance, together with the interpolation. An early run appears, suggesting the FI 1 performance, at *a*. This run may be due

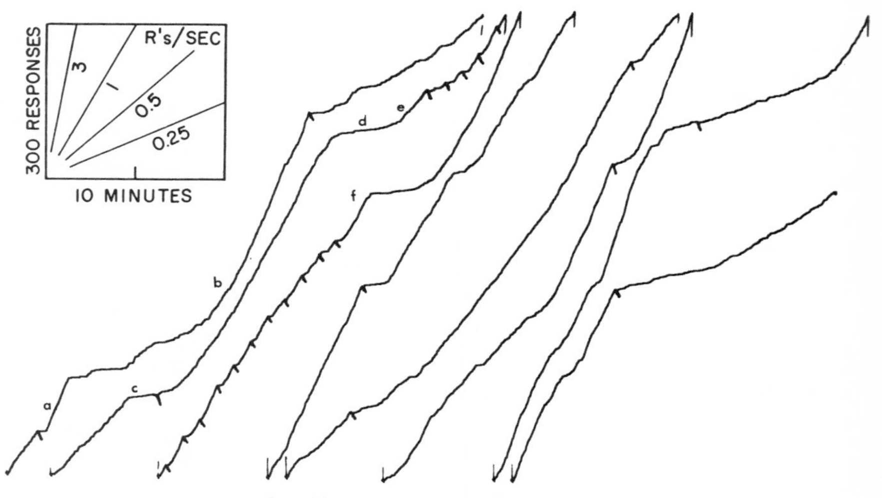

Fig. 821. Continuation of Fig. 820

to a temporal discrimination of the time in the session when the FI 1 interpolation is normally made. It is followed, however, by a normal interval scallop at *b*. The next interval shows an abrupt break before reinforcement at *c*, which would suggest some straining at this point. The following interval begins early; the rate again declines to a low value at *d*, but exhibits some recovery at *e* before reinforcement. Under the FI 1 interpolation, scallops develop quickly. The usual FI 1 performance occurs at the beginning of the first 15-minute interval at *f*, which primes the bird with an interval sequence. The balance of the session shows oscillation in rate, rather rough grain, and frequent breaking to lower rates.

FI inter FR

In a continuation of the experiment just described, the interpolated member was changed from FI 1 to FR 30 and, later, FR 50. Under FR inter FI the slight scalloped characteristic of the early short interval disappears. The small ratio is held eas-

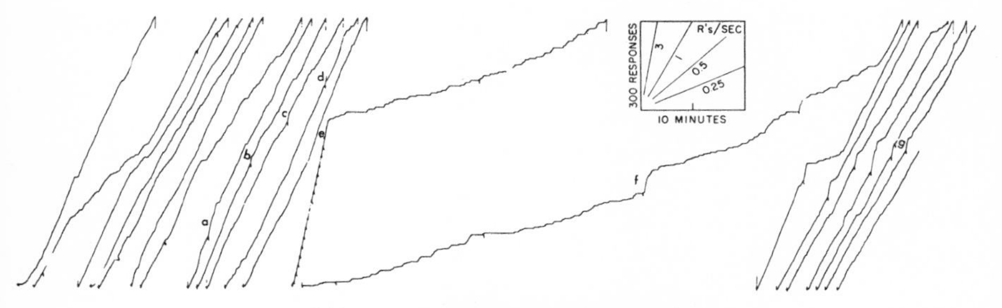

Fig. 822. FI 15 inter FR 30: 4th interpolation

ily. Some priming begins to appear elsewhere in the daily session. The 4th interpolation of FR 30 is given in Fig. 822, containing the whole session. Figure 816 showed the last performance for this bird with an interpolated FI. In the present figure the over-all rate has risen sharply. The background performance is roughly linear. Little evidence appears of any change in rate after reinforcement until the reinforcement at *a*. This lack of change may show some control exercised by the position in the daily session, because the following 3 reinforcements at *b, c,* and *d* also show brief, short priming runs. The interpolation runs off rapidly at a much higher over-all rate than the background FI 15, and at the end a short priming occurs into a very low rate at *e*. A very much depressed rate for several intervals following the interpolation is characteristic of this bird. The extent of this depression, which in this pigeon is more than an hour long, appears to be reduced as the experiment progresses. For the first 3 reinforcements after interpolation, no consistent priming run occurs; but later in the session, every reinforcement is followed by a substantial rur at the interpolated ratio rate. These begin at *f* and extend through the last reinforcement at *g*.

When FR 50 inter FI 15 was continued after the stage shown in Fig. 822, well-marked interval scallops began to appear in the background schedule. Many of these begin with short ratio runs suggesting a mixed schedule. Figure 823 shows the 11th session following Fig. 822. The first reinforcements, *a* through *d*, are followed by fairly standard interval scallops. The fifth, at *e*, is followed by a priming run, however, which is repeated again at *f* and *g* with only slight pausing. The following reinforcement at *h* is followed by a substantial pause, as is also the next at *i*; and the intervals show the terminal rate throughout. The reinforcement at *j* produces an irregular run, and the interpolated high rate follows throughout the segment. All of this may be a temporal discrimination in "anticipating" the interpolated schedule. The ratio reinforcements are run off at a still higher rate, however. A priming break into the interval pause appears at *k*, and the next 4 intervals show a low sustained rate. A brief priming run follows the reinforcement at *l*. At *m*, evidence of a priming effect is unclear, because the bird runs through the subsequent interval. At *n*, however, a well-marked priming run occurs followed by a smooth scallop. Inconsistency of the performance is shown at *o*, where a similar scallop follows without a priming run.

Although some progressive change in the size of the first priming run is to be ex-

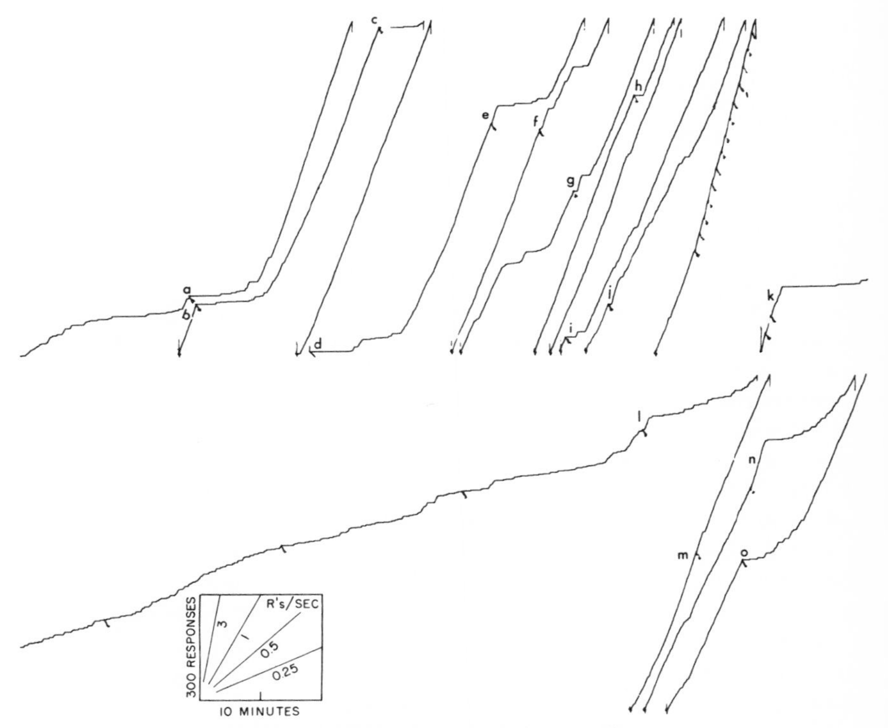

Fig. 823. FI 15 inter FR 50: 15th interpolation

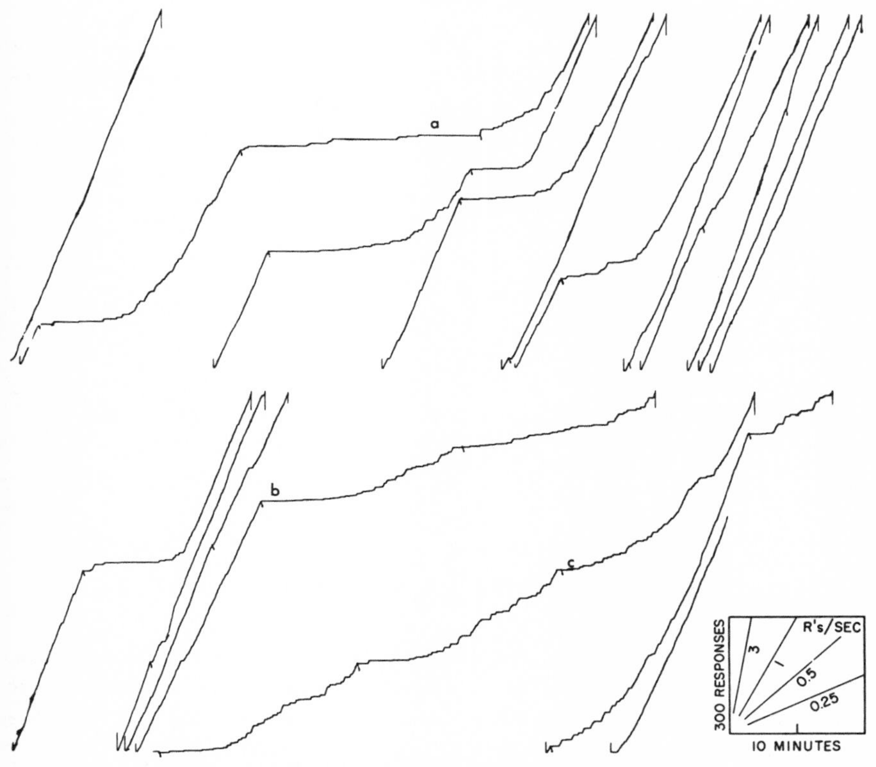

Fig. 824. FI 15, three sessions after the last interpolated FR

pected as the result of prolonging exposure to the interpolated schedule, it is probably significant that the bird shows the clearest break at the end of the priming run at an average of 77 responses on FR 30 as against 123 responses on FR 50. The run is roughly equal to $2\frac{1}{2}$ ratios.

The prolonged period of slow responding which follows the interpolated FR for this bird persisted at approximately the same point in the session when the interpolation was withheld. A lower rate was evident for 4 sessions after the block was withdrawn. Figure 824 shows the 3rd session. A single interval reveals a low rate early in the session, at *a,* but the performance then continues in a fairly standard fashion for FI 15 until *b,* after which 5 intervals show a low rate from which the bird gradually recovers. The 6th interval, beginning at *c,* is fairly normal. The over-all curvature of the session strongly resembles that of sessions with interpolated FR.

Figure 825 indicates the performance of the second bird in the 5th session with an interpolated FR 30. The part of the session shown in the figure was followed by 6 hours during which an over-all rate of about 1200 responses per hour was maintained.

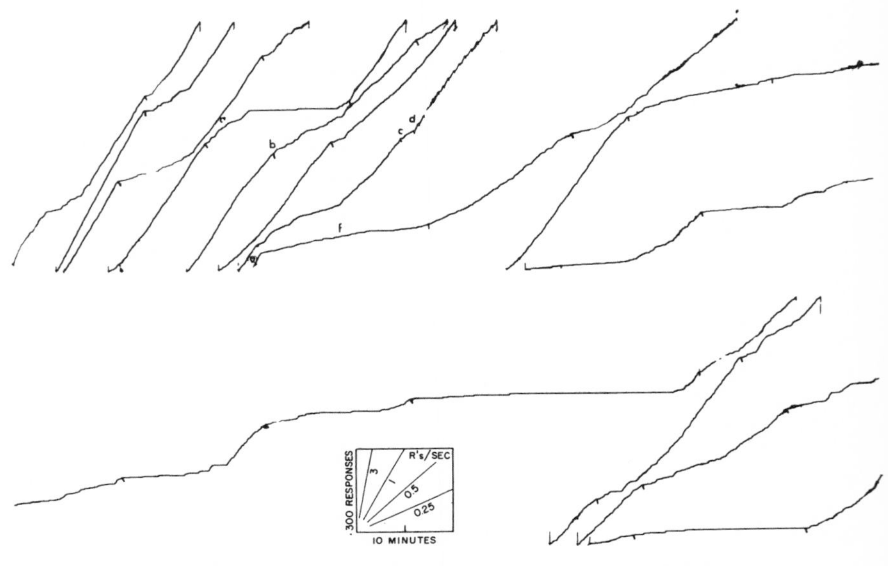

Fig. 825. FI 15 inter FR 30: 5th interpolation (second bird)

The figure shows some early instances of negative curvature following reinforcement, as at *a* and *b*, although priming is not pronounced in the early part of the session. The interpolated block begins with a fairly characteristic depressed rate after reinforcement at *c*; but the 1st reinforcement on the ratio (at *d*) produces a fairly characteristic ratio performance except for rather marked scalloping. The last reinforcement on the ratio schedule at *e* is followed by a priming run, leading into a very low rate at *f*. The following 2 intervals show an unusually low number of responses at reinforcement. In this respect the record confirms the performance of the first bird; however, little evidence exists of sharp priming after reinforcement at this stage. The next 2

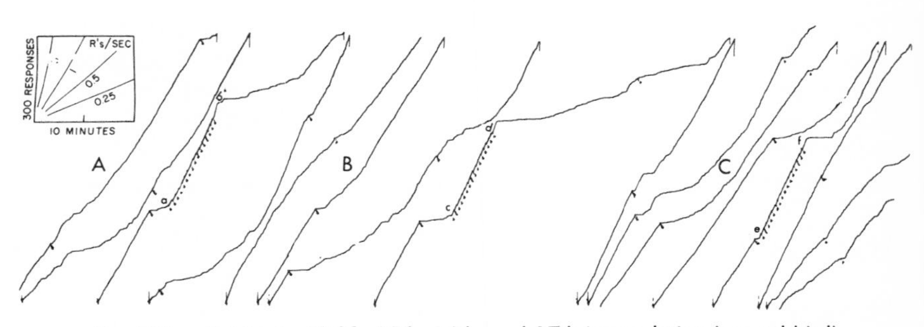

Fig. 826. FI 15 inter FR 30: 15th, 16th, and 17th interpolation (second bird)

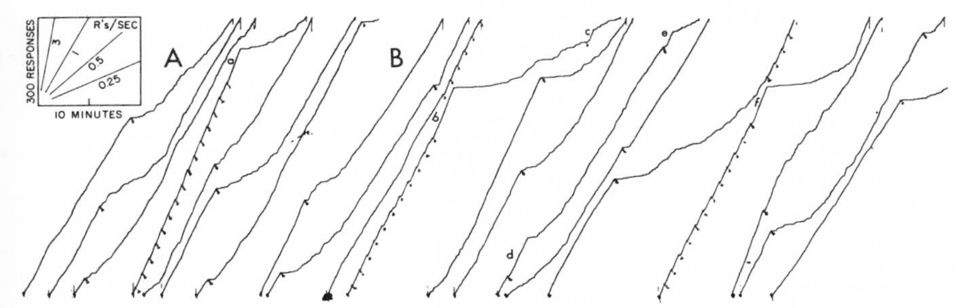

Fig. 827. Transition to FI 15 inter FR 50 after FI 15 inter FR 30 (second bird)

or 3 hours show in general a low over-all rate, with wide oscillations in running rate.

Figure 826 shows the 15th, 16th, and 17th sessions under interpolated FR for the second bird. Portions of each session before and after the interpolated block of ratios are shown. More evidence of random priming exists, but it is by no means so clear as in Fig. 822, for example. The first reinforcements on the ratio schedule, at *a, c,* and *e,* suffice to generate typical ratio performances during the interpolations; and in each case a priming run follows the last such reinforcement, at *b, d,* and *f,* leading into a good interval scallop. Occasional priming runs follow other reinforcements after the interpolation. Especially in Record B, a general depression in rate follows the interpolated block.

When the interpolated ratio was increased to 50, the effect upon the following priming run was immediate. Figure 827 illustrates the 3 sessions following Fig. 826C. Portions before and after the interpolated ratios are again shown. Here, the priming runs at *a, b,* and *f* are much more extended than those at *b, d,* and *f* in Fig. 826. If the run at *a* is significantly larger than hitherto, it must be wholly because of the preceding block of ratios, since this was the first day the ratio had been increased. Priming runs elsewhere in the session are at *c, d,* and *e.*

When the interpolation was withheld, the bird being reinforced simply on FI 15 for

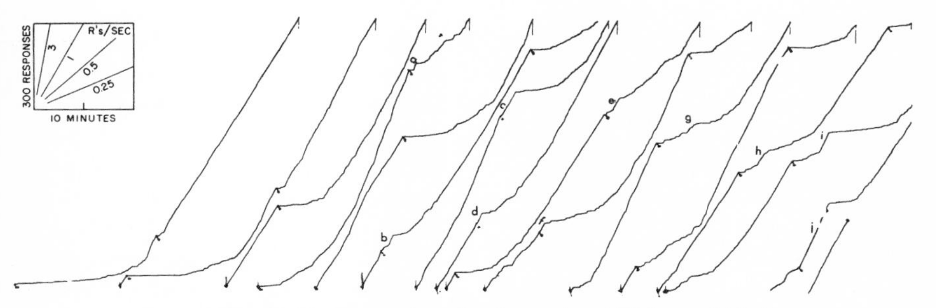

Fig. 828. FI 15: first session after interpolated FR (second bird)

the whole session, many instances of priming continued to be seen. Figure 828 reports the whole session following Fig. 827C. Well-marked primes appear at *a, b, c, d, e, f,* and *j*. What may be regarded as late primes appear at *g, h,* and *i,* although these resemble the knees often seen in interval scallops. Note that this bird does not show a pattern for the whole session suggesting the temporal effects of Fig. 823.

By the 5th session after the removal of the interpolated block of ratios, only occasional evidence of a priming run appears. Figure 829 reports the first part of this session; it shows possible primes at *a* and *b,* but elsewhere merely a rather ragged FI 15 performance. The session was continued for $1\frac{1}{4}$ hours beyond the part shown in the graph. About 4000 responses were emitted during this time, with the same general character shown.

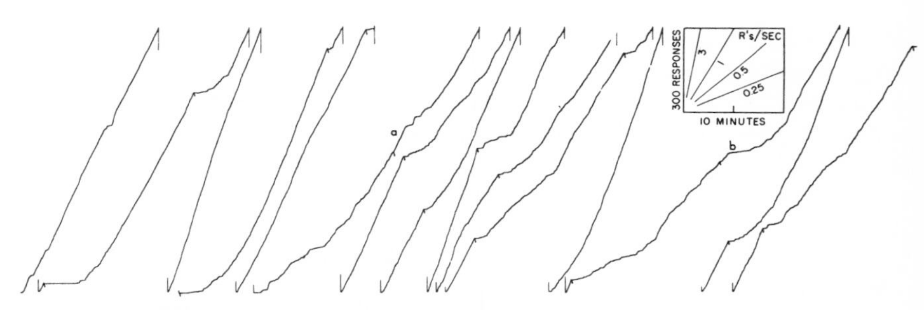

Fig. 829. FI 15: fifth session after interpolated FR (second bird)

VI 5 inter FR 50

The background schedule was then changed to VI 5 ("geometric") for a few sessions. A block of 10 reinforcements on FR 50 was interpolated every hour. The programming equipment developed trouble, and some of the material was lost; but Fig. 830 shows 1-hour segments of all recorded instances of the interpolated FR, with some of the behavior before and after in the order of its occurrence. The FR rate is not much higher than that of the VI, and not quite typical of an FR 50 performance. After each interpolation, a priming run of from 80 to 180 responses occurs, followed, in general, by a rather sharper break than that appearing elsewhere in the VI performance. Instances of priming occur elsewhere, particularly in the later segments. Note the rather poor VI performance, apparently due to the frequent interpolation of the block of FRs.

The second bird was reinforced on this schedule for only 4 interpolations during 2 sessions. Two of these show marked priming runs followed by substantial decreases, more marked than in Fig. 830. The other 2 interpolations were followed by only slight increases in rate, which were not clearly priming runs. The experiment was not continued long enough to see whether this other bird would, in general, give the performance of the present figure.

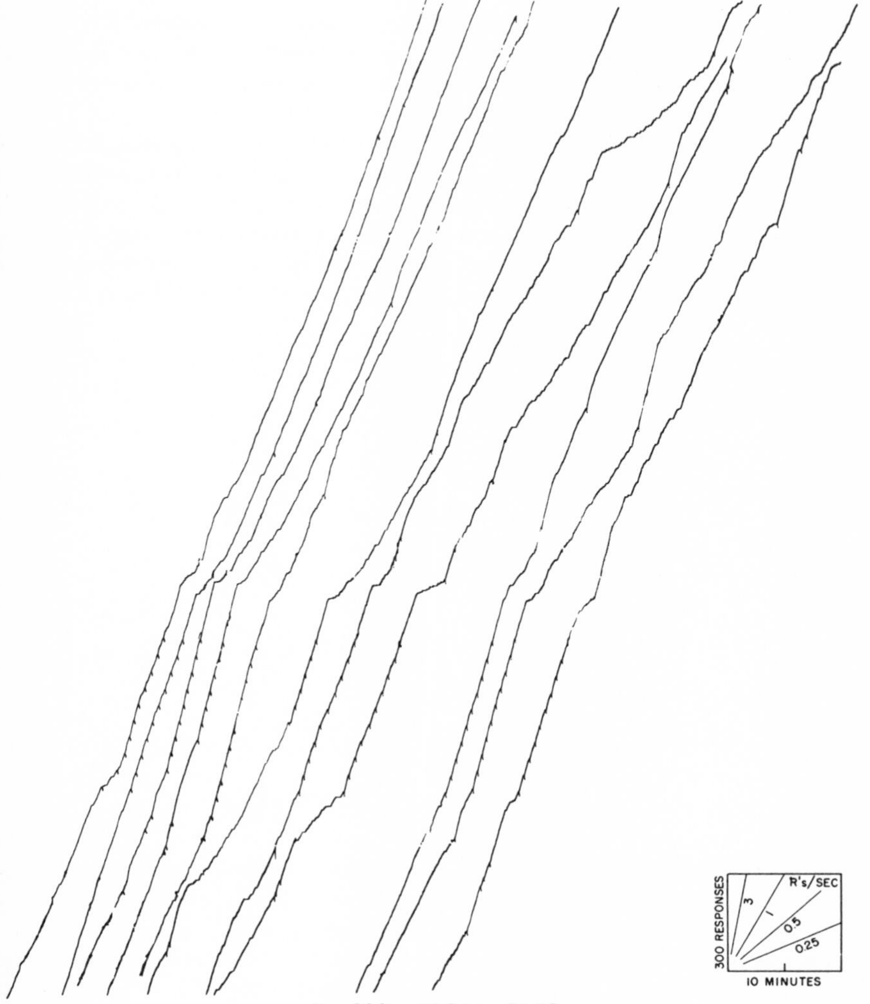

Fig. 830. VI 5 inter FR 50

A MULTIPLE SCHEDULE COMPOSED OF FR AND MIX FRFR

We attempted to bring the characteristic priming effect of the mix FRFR under stimulus control, and to separate it from a straight FR under the control of a different stimulus. Three rats which had been on FI for 62 hours following crf were then reinforced on mix FR 20 FR 80. Figure 173 showed later performances on FI. The beginning of the new procedure was given in Fig. 741, where traces of the earlier FI performance were evident. Record B showed examples from the end of the 2nd session after 6 hours of exposure to the multiple schedule. The larger FR had been increased to 160. Figure 742 showed the subsequent development over a period of 13 sessions. The position of the first break in the long ratio became more stable and occurred after fewer responses as the experiment progressed.

This mix FRFR schedule was continued under the control of one stimulus (steady light in the experimental box); and, at the same time, under a different stimulus (a flashing light), responses were reinforced on the longer ratio alone. If successful, this procedure should yield a performance on the longer ratio showing step-wise breaks under the steady light and no breaks under the flashing. In the time devoted to the experiment (100 hours) this was achieved with only 1 of the 3 rats. Figure 831 illus-

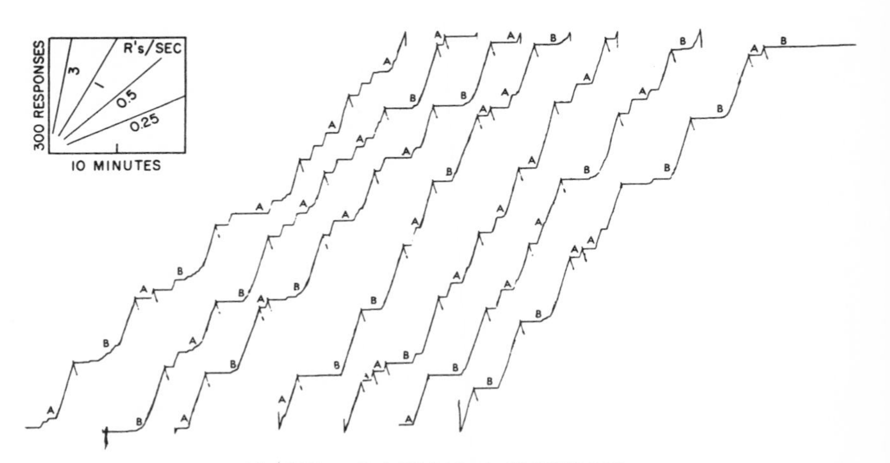

Fig. 831. Mult FR 160 mix FR 20 FR 160

trates the next-to-last session for this rat. A indicates the steady light under which reinforcement occurred after 20 responses or after 160. B indicates the flashing light under which reinforcement occurred at the end of 160 responses. A fairly good multiple control is evident. In general, under Stimulus B the ratio is run off with some curvature and possibly a somewhat lower terminal rate, but without gross breaks. Under Stimulus A the longer intervals usually show breaks. There are four exceptions. Two breaks occur under Stimulus B which are not justified by the schedule; they could be regarded as induction from the other stimulus. In three cases a break fails to occur under Stimulus A. (A slight break appears in one of these.) Out of 42 long intervals the curvature is appropriate to a mix FRFR or to a single FR in 37 or 38 instances. On the following day, when a similar performance prevailed, 7 "mistakes" occurred under Stimulus B, and 2 or possibly 3 under Stimulus A. When the large ratio was dropped to 135, and the experiment continued otherwise in the same way, the effect disappeared.

One of the other rats may have failed to show this effect because the over-all rate was high and little breaking occurred as the result of the mix FRFR. Note that on this schedule very few short ratios are reinforced.

Chapter Twelve

• • •

CHAINED SCHEDULES

Two or more responses may be "chained" together if the first produces a stimulus in the presence of which the second is reinforced. Chains of any length are conceivable. The separate members of the chain are usually identified by the topography of the response. Another type of chain is possible in which a response of a single topography is emitted under different stimuli, these stimuli being produced in succession by repeated reponses. In this case the members of the chain are identified by the reinforcements which they produce or the stimuli under which they occur. It has been assumed that some such chaining occurs in a well-developed fixed ratio in which each response made in the presence of a given number of responses already emitted is reinforced by an increase in that number.

In a chained schedule the bird first responds in the presence of one stimulus on a given schedule and is reinforced by a change in the stimulus under which it is then reinforced, usually on a different schedule. Ultimately, a response is reinforced with food. Chained schedules are under the stimulus control characteristic of multiple schedules, but they differ in that only one eventual reinforcement with food maintains all members of a schedule.

A demonstration of a chained schedule is produced by beginning with a multiple schedule containing extinction as one component, for example, mult FI 1 ext, in which the periods of extinction are terminated according to a random distribution of intervals similar to that used in VI 1. The reappearance of the stimulus controlling FI 1 is not correlated with a response, nor does a TO occur. However, the stimulus appropriate to the FI 1 schedule can now be made to appear when the organism occasionally responds during the extinction stimulus. Any increase in responding during the part of the schedule previously showing extinction demonstrates the reinforcing effect of the stimulus appropriate to FI.

CHAIN VIFI

Chain VI 1 FI 1[1]

In one such experiment a bird was reinforced on mult FI 1 ext. The final performance generated is shown in Fig. 832A and C; but, for the moment, only the 1st seg-

[1] See Ferster (1953) for an application of chained schedules to the problem of delayed reinforcement.

ment of Record A is relevant. These performances show a high rate of responding on FI 1 (C), with little or no pausing after reinforcement, and a very low rate on extinction (A), both under appropriate stimuli. At *a*, in Record A, the production of the stimulus appropriate to FI 1 was made contingent upon a response under the other stimulus on a VI 1 schedule. The first such contingency was felt immediately after point *a*. A single 1-minute interval was then executed. The burst of responding after *a* was recorded in the return to the stimulus formerly appropriate to extinction. The second production of the FI 1 stimulus occurred at *b*. Subsequently in the session, the rate in Record A increased steadily. In general, responding tended to be most rapid after the return from the FI 1 schedule.

Records B and D show the 2nd session. A fairly stable VI performance (B) has been generated by reinforcement consisting simply of a change to the key-color on

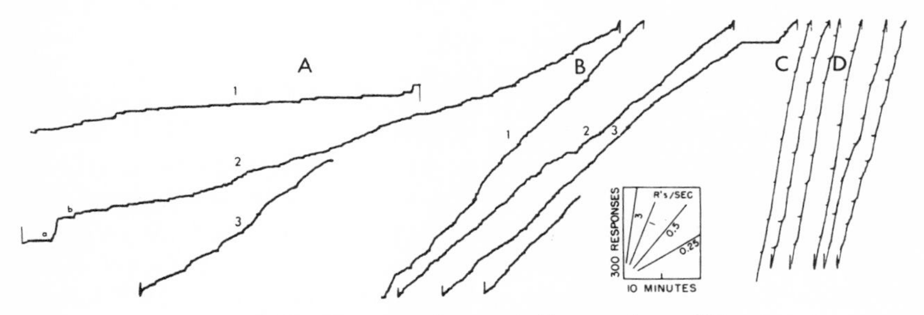

Fig. 832. Transition from mult Flext to chain VI FI

which reinforcements are scheduled in FI 1. Record D gives the performance on the latter key.

Two other birds were reinforced for 40 sessions on mult FI (blue key) extinction (red key). The 2 schedules were then chained together by letting responses on the red key produce the blue key in VI 1, a response on the red key being reinforced a fixed interval later. A substantial rate on the red key developed very slowly. Several values of FI (1, 1.5, 2, and 5 minutes) were introduced into this general pattern. Stable conditions on some of these are shown in the following figures.

Figure 833A shows a performance on VI 1 (red key) reinforced, as marked, by the production of the blue key. Record B shows the alternating performance on FI 1 on the blue key. The over-all rate is higher in Record A, the first member of the chain, than in Record B, the member reinforced with food. While pauses occur in both cases, they are more marked in B, and the transition to a terminal rate is more abrupt. Figure 833C shows a performance on VI 1 (red key) reinforced, as marked, by the production of the blue key. Record D shows the alternating performance on FI 1.5 on the blue key. The over-all rates are here of the same order of magnitude. The VI schedule in Record C is generally successful in eliminating the pause after reinforcement, while the segments in Record D usually show FI pauses.

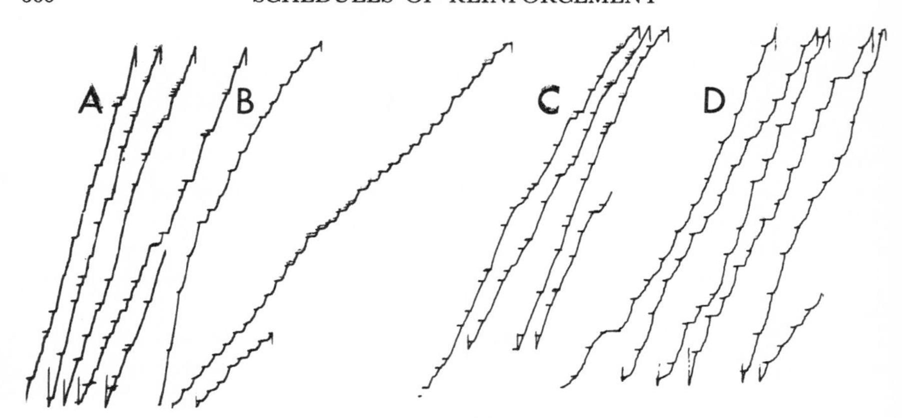

Fig. 833. Final performance on chain VI 1 FI 1.5

Increases in the size of the FI component on chain VI 3 FI

In another experiment, 2 birds were stabilized on chain VI 3 FI in which the value of FI was somewhat systematically varied. These birds had had an extensive history of mult VI 3 FI 3 in which, of course, both components were separately reinforced with food. A transition to the chained schedule was made by the omission of the food reinforcement on VI 3; instead, the production of the stimulus appropriate to FI 3 was substituted. Both birds substantially lost the performance on VI 3 formerly prevailing under the multiple schedule before developing a chained performance.

First bird. After 60 sessions on the chained schedule, the fairly stable conditions in Fig. 834 prevailed. Record A is the VI 3 performance, where each reinforcement, as marked, consists of the production of the stimulus under which FI 2 was reinforced

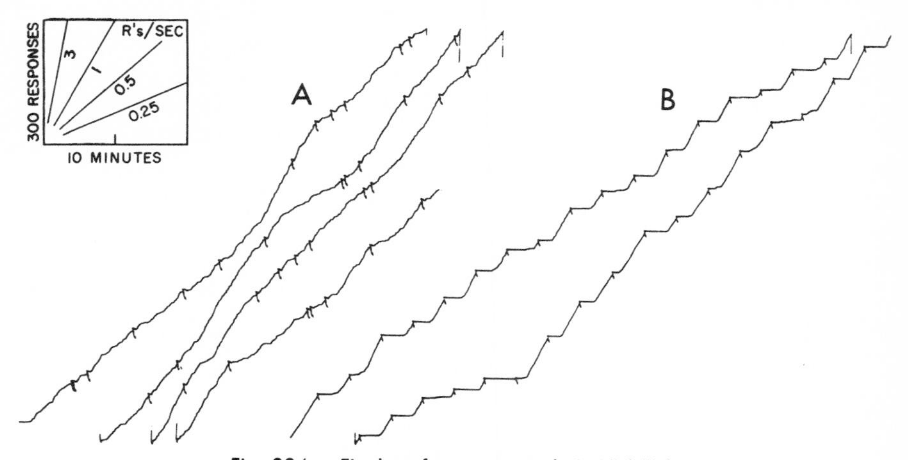

Fig. 834. Final performance on chain VI 3 FI 2

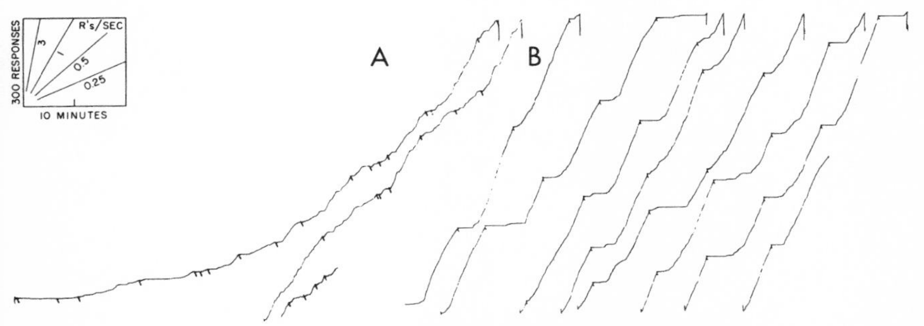

Fig. 835. Final performance on chain VI 3 FI 5.5

with food. Record B gives the alternating FI 2 performance. The over-all rates are perhaps a little lower for the second member of the chain. The VI performance showed considerable irregularity and a tendency to adopt the terminal FI rate. Slight pausing occurs after reinforcement on the VI schedule, and substantial pausing and curvature, on the FI schedule.

Chain VI 3 FI 5.5

The FI interval was then extended to 5.5 minutes and the new schedule remained in force for 48 sessions. Figure 835 shows the final performance. The FI 5.5 exhibits some negative curvature, giving an S-shaped character to many interval segments. Pauses follow most reinforcements, but instances also appear of running-through and of fairly rapid acceleration to the terminal rate. The performance under the first schedule in the chain shown at Record A is very irregular, with a low over-all rate leading to marked acceleration during the session.

The FI interval was then further increased to 7.5. Figure 836 describes the performance 15 sessions later. The FI schedule shows a lower terminal rate than that in Fig. 835 and considerable irregularity in the interval curvature. Marked instances of negative acceleration before reinforcement appear. (The somewhat variable per-

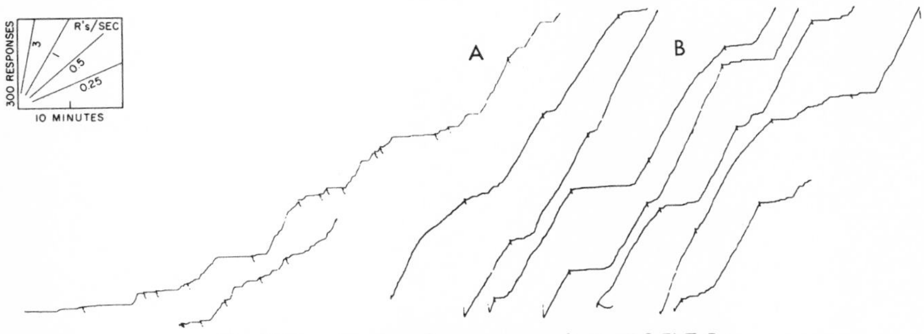

Fig. 836. Final performance on chain VI 3 FI 7.5

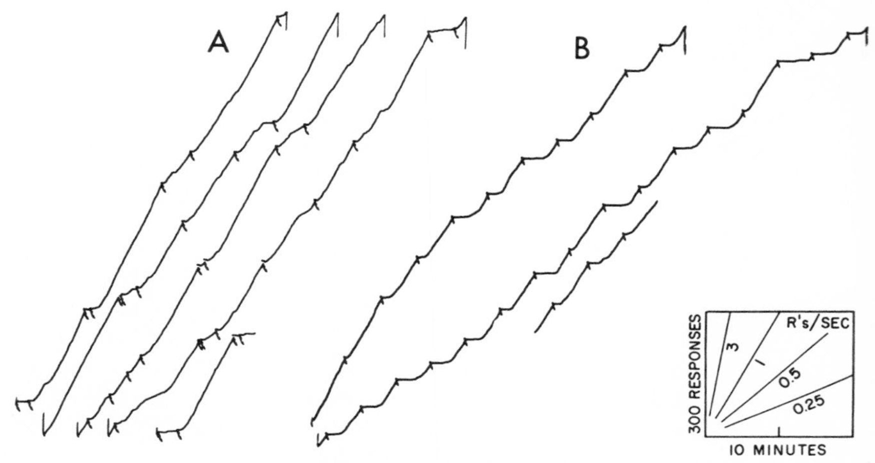

Fig. 837. Return to chain VI 3 FI 2

formance on FI in such a schedule may be due in part to the irregular VI schedule of the preceding member. The interval performance is sometimes executed after a very short interval on the first component of the chain and in other instances after a long interval.) In the session following Fig. 836 the original value of FI (2 minutes) was restored. The rate on the first member of the schedule increased immediately. Ten sessions later, the stable performance apparent in Fig. 837 was recorded. The FI 2 performance at Record B is very similar to that in Fig. 834B, while the VI performance

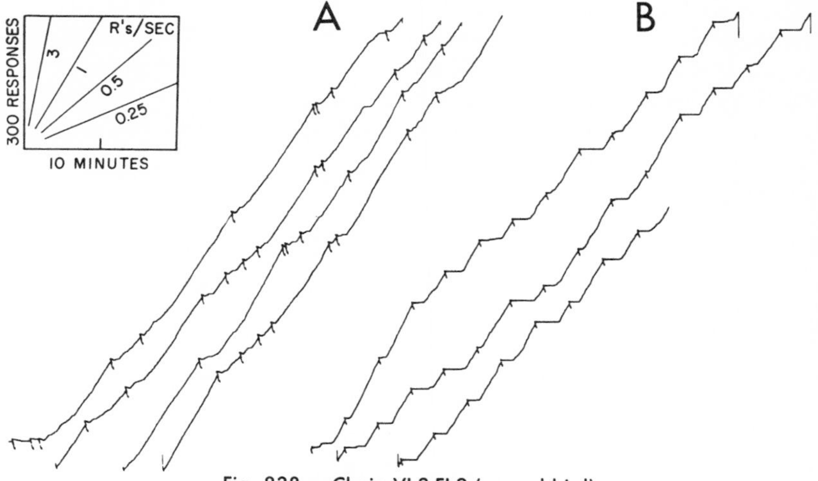

Fig. 838. Chain VI 3 FI 2 (second bird)

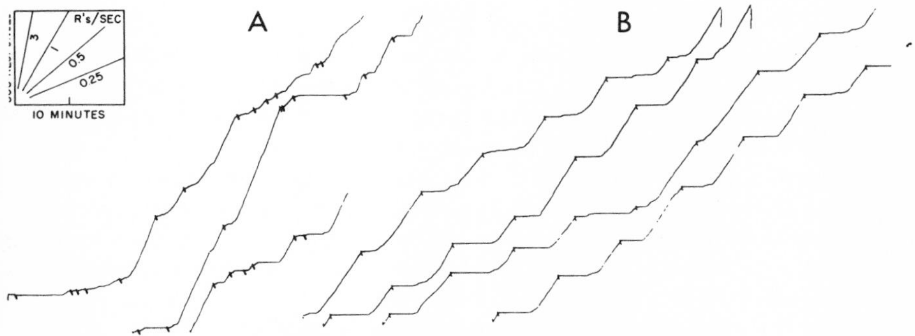

Fig. 839. Chain VI 3 FI 5.5 (second bird)

on the first member shows a much smoother grain. Some pausing after reinforcement is now developing on the VI schedule. The over-all rate on the first member of the chain is now slightly higher than that on the second and is higher than the rate previously observed on VI 3 in Fig. 834.

Second bird. We studied a second bird on the same program; in general, its performance confirms the essential points just made for the first bird. Figure 838 shows the stable performance on the original chained VI 3 FI 2. Figure 839 illustrates a performance 50 sessions after the change to VI 3 FI 5.5. The VI performance on the first member of the chain is somewhat more severely disrupted than it was for the other bird. The over-all rate is still of the same order as that occurring under FI 2 in the second member of the chain. Figure 840 shows the final performance on chained VI 3 FI 7.5. The FI performance closely resembles that of the other bird. The VI performance shows a tendency to approach the terminal FI rate and much more prolonged pauses than those in Fig. 836. No effort was made to recover the original performance with this bird.

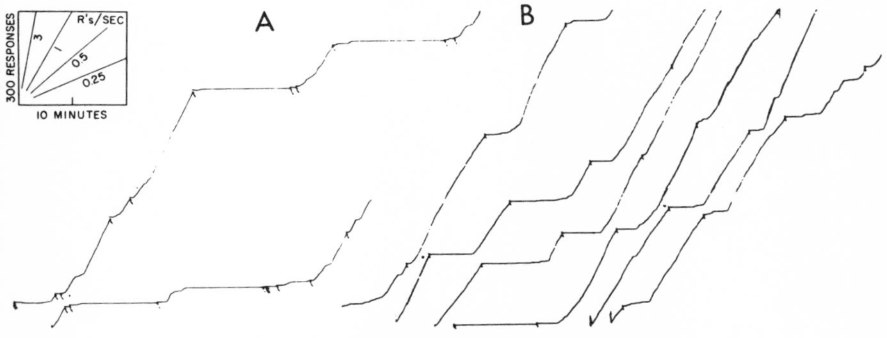

Fig. 840. Chain VI 3 FI 7.5 (second bird)

CHAIN FIVI

In another experiment on 2 birds the 2nd (food-reinforced) component of a chained schedule was maintained at VI 3 throughout the experiment. The 1st member reinforced by the change to the stimulus appropriate to the 2nd member began at FI 2, but subsequently became FI 3.5, FI 5, and then FI 7.5. Figures 841, 842, 843, and 844 illustrate the results. Figure 841 shows a stable chained performance after 60 sessions on chained FI 2 VI 3. The FI 2 shows pauses after reinforcement, smooth curvature in many instances, and some second-order effects in spite of the intervening periods on the other key. The VI performance shows no consistent relation between local rate and recent reinforcements and a well-sustained high over-all rate. On FI 3.5 in Fig. 842 the interval performance shows increasingly longer pauses and some variation in the terminal rate. The figure was recorded after 16 sessions on chained FI 3.5 VI 3. The performance on the VI component remains essentially as in Fig. 841. On FI 5.5 (Fig. 843), pauses after reinforcement on the interval schedule are often exceptionally long, with an abrupt change to a terminal rate, as at *a* and *c*, or an abrupt increase followed by negative curvature leading again into the terminal rate, as at *b*. An occasional interval is run off at an intermediate rate, as at *d*. The over-all rate has not changed its order of magnitude. The performance on the 2nd component on VI 3 (B) remains unchanged. When the 1st component is FI 7.5 (Fig. 844), the FI performance shows increasingly longer pauses and frequently abrupt changes to a rate near the terminal rate. The terminal rates throughout the experiment tend to be lower than normal on an FI schedule alone, and the pauses are also longer than normal throughout. The VI schedule on the 2nd component in Fig. 844 remains essentially unchanged (B).

A repetition of the experiment with the second bird produced stable performances

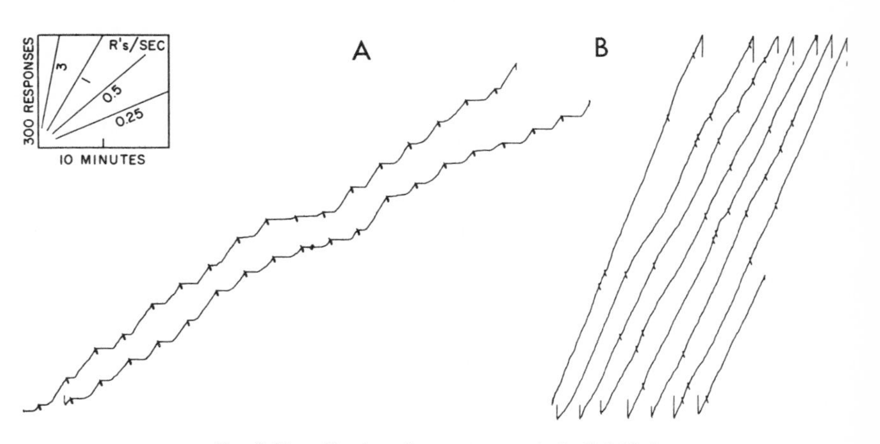

Fig. 841. Final performance on chain FI 2 VI 3

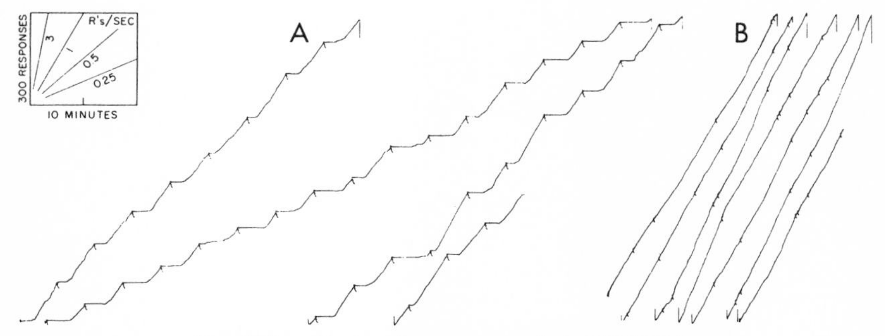

Fig. 842. Final performance on chain FI 3.5 VI 3

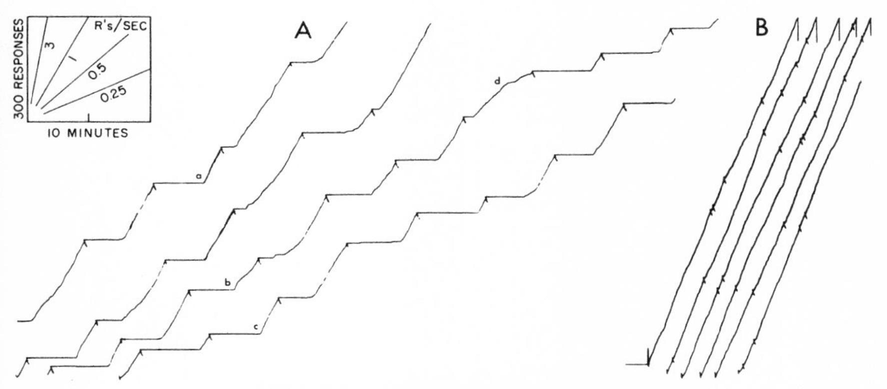

Fig. 843. Final performance on chain FI 5.5 VI 3

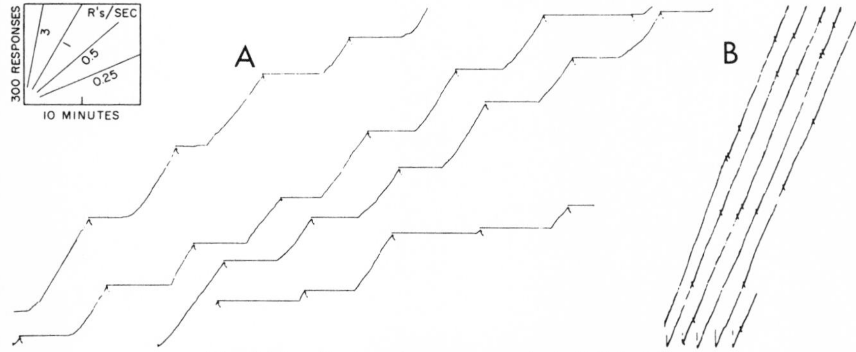

Fig. 844. Final performance on chain FI 7.5 VI 3

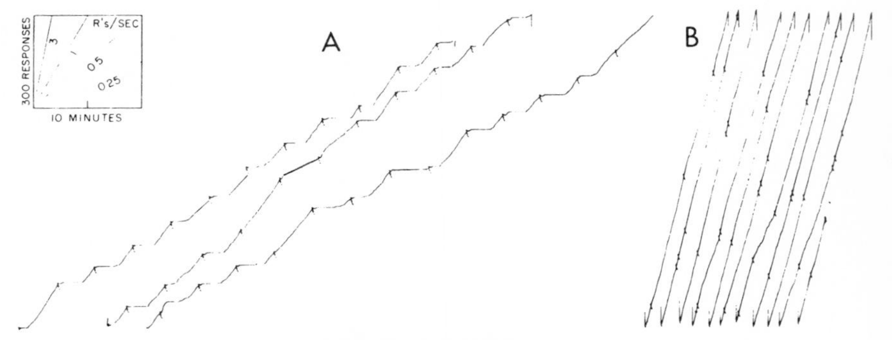

Fig. 845. Chain FI 3.5 VI 3

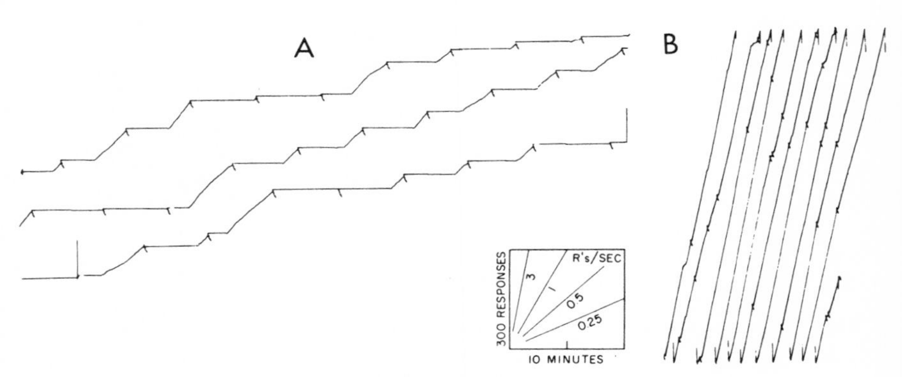

Fig. 846. Chain FI 5.5 VI 3

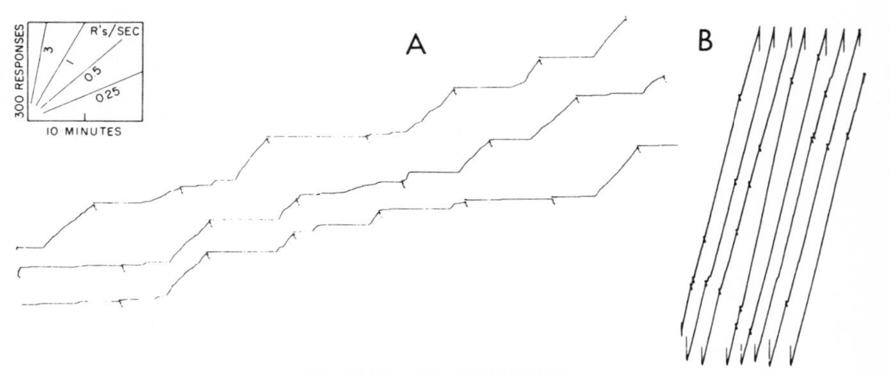

Fig. 847. Chain FI 7.5 VI 3

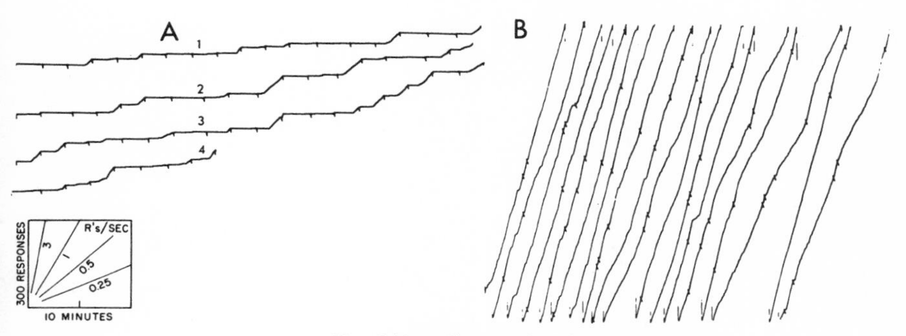

Fig. 848. Chain FI 2 VI 3

when the first component was FI 3.5, FI 5.5, and FI 7.5, shown in Fig. 845, 846, and 847, respectively. The second VI component (B) remains unchanged throughout, but the over-all rate declines severely on the initial chained component on FI. Pauses become extremely long, often delaying reinforcement, and the terminal rates are exceptionally low. When we tried to recover a higher rate on the 1st component of the chained schedule by reducing the interval to FI 2 after Fig. 847, a higher over-all rate failed to return. Figure 848A illustrates the performance. In Fig. 848B, the VI performance also begins to show irregularity.

CHAIN VIFR

Four birds were reinforced on FR after crf until the performance stabilized. A discrimination was then set up. Responses continued to be reinforced on FR 50 on a blue key; but the key changed at reinforcement to orange, and no responses were reinforced until, after an interval which varied in the manner of a VI 1 schedule, the key again changed to blue. Discriminations developed quickly in 3 of the 4 birds. Figure 849 gives the 1st session for 1 bird. The performance on the blue key appears at Record A (a defect in the recorder was responsible for the break at *a*); and the declining rate on the orange key at Record B. The latter shows considerable irregularity and a frequent return to the ratio rate, as at *b* and *c*, during extinction. The performance on the blue key on the 6th day appears at Record C; and on the orange key, at Record D. The rate is essentially zero on the orange key. This performance continued for the rest of the 10 sessions devoted to discrimination. (The exceptional bird continued to respond at the beginning of the experimental session on the orange key, but its rate dropped to essentially zero before the end of each session.)

On the 11th day the change from orange to blue was made contingent upon a response to the orange key. This contingency produced a chain in which a response on the orange key was reinforced on VI 1 by a change to a blue key. The 50th response on the blue key was then reinforced with food. Figure 850 shows the immediate effect for 2 birds. Both birds developed substantial rates of responding on the orange key

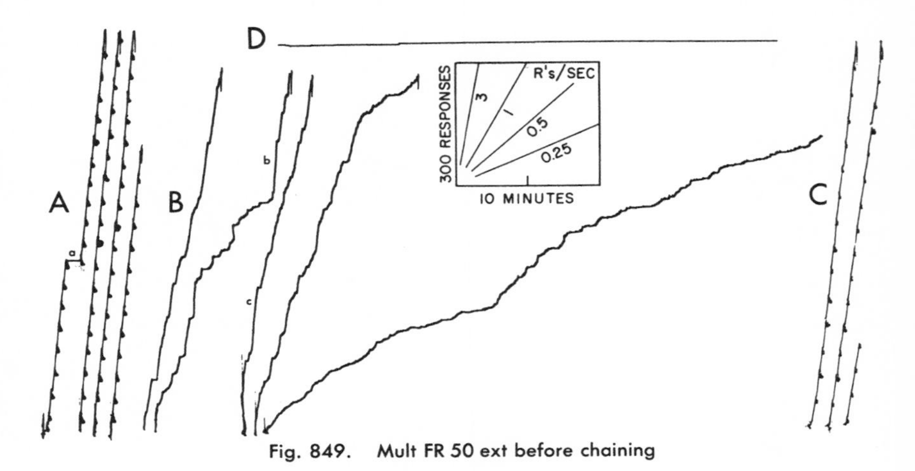

Fig. 849. Mult FR 50 ext before chaining

during the session, as in Records B and D. Responding on the blue key continued, as Records A and C show.

The exceptional bird already noted showed the performance of Records A and B in Fig. 851 on the last day of the discrimination. Record A gives the responding on the blue key; Record B shows the characteristic high rate on the orange key at the beginning of the session, leading to a near-zero rate before the end of the session. When the responses on the orange key were reinforced by the production of the blue key, an increase in rate occurred as in the other birds; but this increase was added to the original high rate in the session. Records C and D show the resulting performance.

All birds were maintained on the chained VI 1 FR for 60 sessions, during which the values of the ratio were changed. In 2 birds the ratio was advanced to the point at which severe breaking after reinforcement was observed. Figures 852 and 853 show

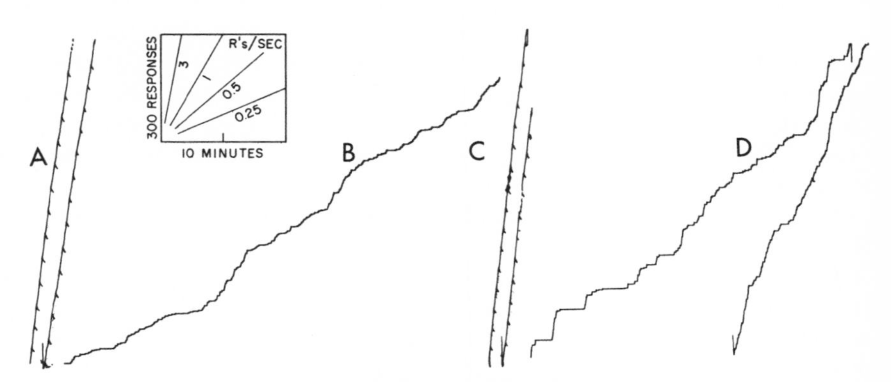

Fig. 850. First session on chain VIFR after mult FRext

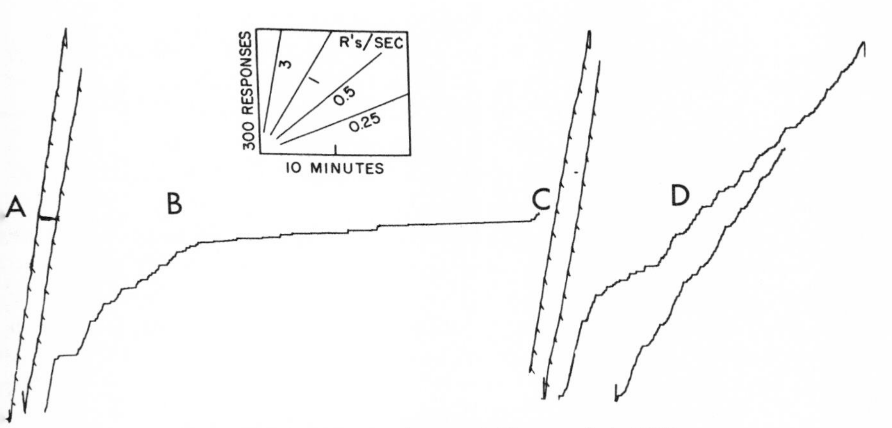

Fig. 851. Transition from mult FRext to chain VIFR

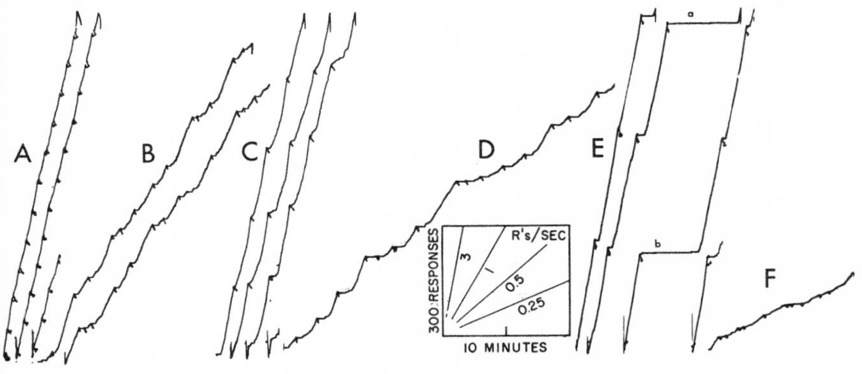

Fig. 852. Chain VI 1 FR (75-300)

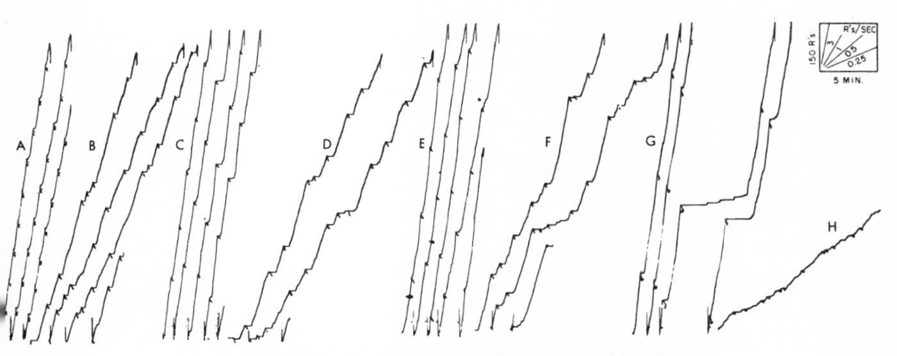

Fig. 853. Chain VI 1 FR (90-300) (second bird)

the course of the performance. In Figure 852, Records A, C, and E show performances on the blue key on ratios beginning at 75 at Record A and ending with 300 at Record E. These are for the 10th, 35th, and 60th sessions in the experiment. The development of marked breaks on FR 300 is evident, as at *a* and *b*. The performance on the orange key which is reinforced by a change to these ratio schedules varies significantly. At Record B, in the performance on the same day as Record A, pauses appear after reinforcement; some rough grain and positive acceleration occur during the longer intervals. Pausing and scalloping become more marked by the 35th session at Record D (corresponding to the performance at Record C on the ratio); and in the 60th session, when the ratio has been extended to 300 and is breaking, the rate on the first member of the chain becomes very low, as Record F reveals.

Essentially the same points may be made for the second bird in Fig. 853. Pausing

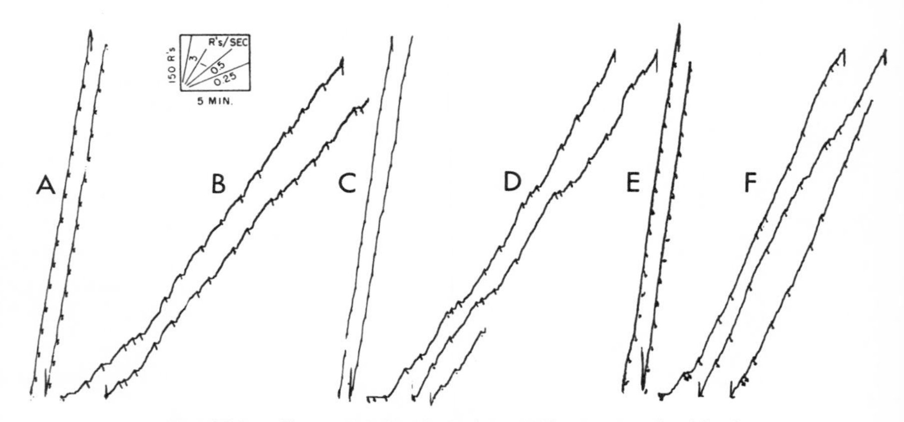

Fig. 854. Chain VI 1 FR 50: 15th to 57th sessions (third bird)

after reinforcement in the VI schedule appears at Records B, D, and F. The corresponding ratio performances are at Records A, C, and E, respectively; these are for the 10th, 23rd, and 34th sessions. Record F begins to show marked pausing and scalloping similar to that in Fig. 852D. When, as at Record G in the 54th session, the ratio of 300 produces marked breaks, the VI performance is reduced to the low level at Record H.

The ratio was not advanced for the other 2 birds, so that straining did not develop. However, pauses after reinforcement in the VI member of the chain appear. Records A and B in Fig. 854 are for the 15th session. The performance on the ratio (Record A) shows no pausing at the start of the ratio run; but the performance on VI 1 at Record B shows pauses after reinforcement, generally followed by some compensatory increase in rate, with negative acceleration during the longer intervals. Records C and D are for the 20th session. Record D still shows irregularities in the VI perform-

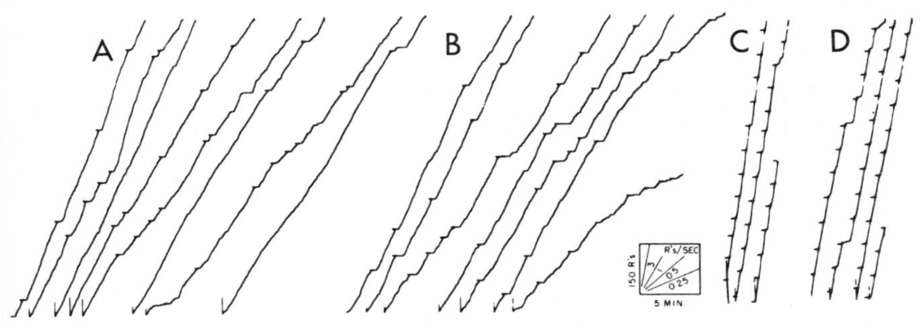

Fig. 855. Chain VI 3 FR 70

ance, with pausing after reinforcement. By the 57th session, however, shown at Records E and F, the ratio and VI performance are normal.

The results with the fourth bird in the series were similar.

In another experiment, 2 birds with a history of mult FRFI and mult FIext were reinforced on chain VI 3 FR for 57 sessions during which the FR was increased from 20 to 70. Figure 855 shows performances from the first part of the last session. Records A and B give the first components of the chain on VI 3. Some pausing after reinforcement occurs, and some variation in rate. In general, however, a fairly good VI performance is evident. Records C and D show the corresponding performances on the second component under FR 70. Except for an occasional pause after reinforcement, the ratio is well-sustained.

CHAIN FRVIdrl

In an experiment to be described later (see Fig. 862) a stable performance was established on chain VI 2.5 VI 2.5 drl 4. The 1st member of the chain on a blue key was then made FR 35, while the schedule on the 2nd component on the orange key remained VI 2.5 drl 4. The sequence of events was as follows. On the blue key the 35th response was reinforced by a change to an orange key on which a response was reinforced on VI drl 4. Figures 856A and B show a final performance after 27 sessions of this schedule. The ratio performance at Record B is quite irregular, and the ratio rates are exceptionally low for so small a value. This condition developed gradually during the 27 sessions. Initially, the ratio performance was practically normal; a sample from the 2nd session appears at Record C. Meanwhile, the responding on the orange key shows some slight decline during the session, and the over-all rate is much lower than that in Fig. 862. The ratio grain could be due to interference from mediating behavior on the drl schedule. Nevertheless, the performance illustrates the possibly unexpected condition that responding may occur at a fairly high rate when the reinforcement consists of the production of a stimulus controlling a low rate.

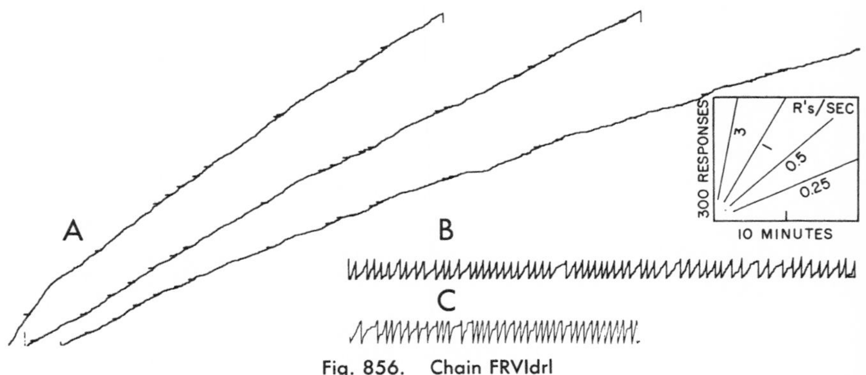

Fig. 856. Chain FRVIdrl

The same result was obtained with a second bird, also on chain VI 2.5 VI 2.5 drl 4. When the schedule on the first (blue) key was changed to FR 50, a very high rate immediately appeared on the ratio (a high rate existed on that key under VI 2.5). The performance for the 1st session on the ratio member appears in Fig. 857C, which corresponds to Fig. 856C. Later, the ratio performance showed trouble, as it did with the other bird. Figure 857B shows the performance on the ratio after 28 sessions of the chain FRVIdrl, during which the value of the ratio was dropped to 35. Record A contains the very low stable performance under VI 2.5 drl 4. The rate has fallen below that of Fig. 868, partly because of the prolonged exposure to the drl contingency.

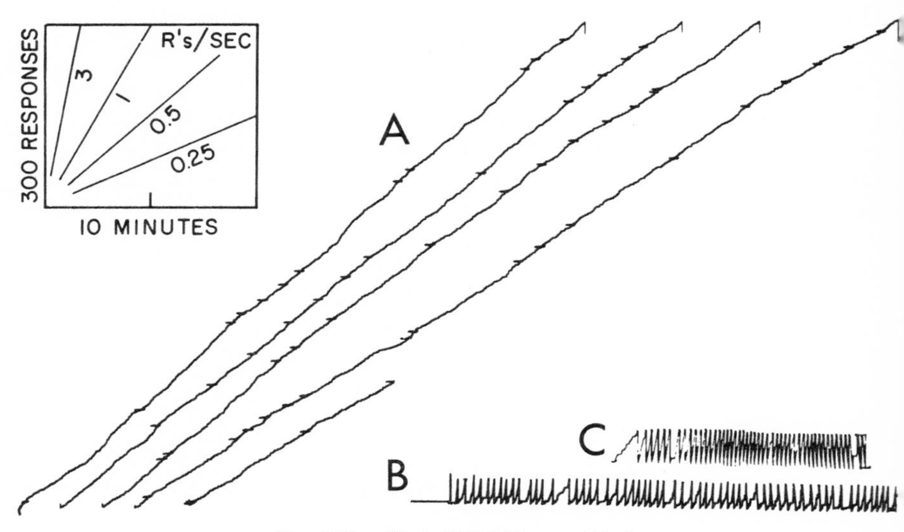

Fig. 857. Chain FRVIdrl (second bird)

CHAIN VIVI

Two birds were reinforced on mult VI 1 (blue key) ext (orange key) until a near-zero rate developed on the orange key. The periods of extinction varied in the manner of a VI 1. The 2 schedules were then linked together in a chain; responses on the orange key produced the blue key on a VI 1 schedule. Figure 858 illustrates the development of a substantial rate on the orange key in 3 sessions. Record A shows the performance on the orange key on the last day of the multiple schedule. Only a few responses occur during the session. Record B is for the 1st day upon which a response to the orange key produces the blue key on a VI 1 schedule. The rate increases slightly during the session, and at least one sustained run occurs (at *a*) at essentially the rate on the blue key. Records C and D show the 2nd and 3rd days of the chained schedule; a higher rate develops and in Record D conspicuous pausing occurs after reinforcement, as at *b, c, d,* and elsewhere. Record E shows the VI 1 performance on the blue key on the last day of the multiple schedule corresponding to Record A. Record F shows the performance on the blue key (2nd component) on the 3rd day of the chained schedule corresponding to Record D. The appearance of an unusually high rate at *e* dropping at *f* to the normal rate was exceptional and was not evident in the other bird or on the omitted days between Records E and F for the present bird.

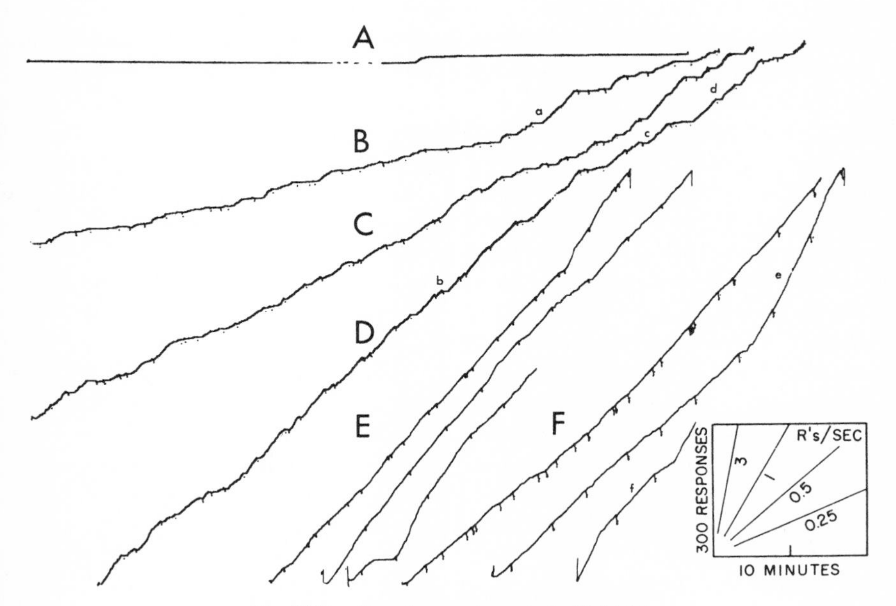

Fig. 858. Development on chain VI 1 VI 1

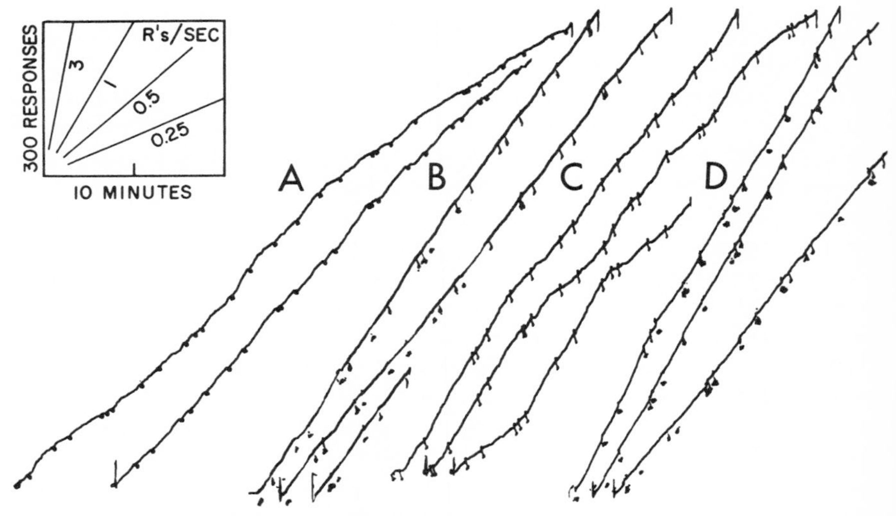

Fig. 859. Chain VI 1 VI 1 at two levels of deprivation

The level of deprivation was then varied. Figure 859 shows performances at 2 levels for the same bird as above. The weight during the sessions shown at Records A and B was approximately 10% above that in Records C and D. Record A shows responding on the orange key; and Record B, on the blue key 35 sessions after the beginning of the chained schedule. Note the slightly higher rate on the blue key, the absence of any pausing on the blue key, and a rather rougher grain on the orange key, much of this contributed by slight pausing after reinforcement. Records C and D show the performance at the lower body-weight, 43 sessions after Fig. 858. The rates here are substantially higher than those at Records A and B. Responding to the orange key is more irregular, and the rate is again relatively high for the blue key.

Figure 860 shows a similar set of records for a second bird. Records A and B illustrate performances on the orange and blue keys, respectively, at a relatively high body-weight. The performance on the orange key is somewhat more orderly, but the same over-all rates prevail. Records C and D indicate the performances at a 12% lower body-weight. The performance on the blue key at Record D is now quite linear and shows almost no pausing after reinforcement. Responding to the orange key at Record C shows some pausing and compensatory running after reinforcement; but, in general, a high rate is also maintained. The rates are of the same order on both the orange and blue keys. A lower body-weight produces a higher rate on *both* members of the chain.

Two birds were reinforced on mult VI 3 (blue) VI 3 (orange) immediately after crf. The VI rates showed a color preference for blue in the ratio of 9:7. We then set up a chain by reinforcing responses to the blue key with a change to the orange key instead

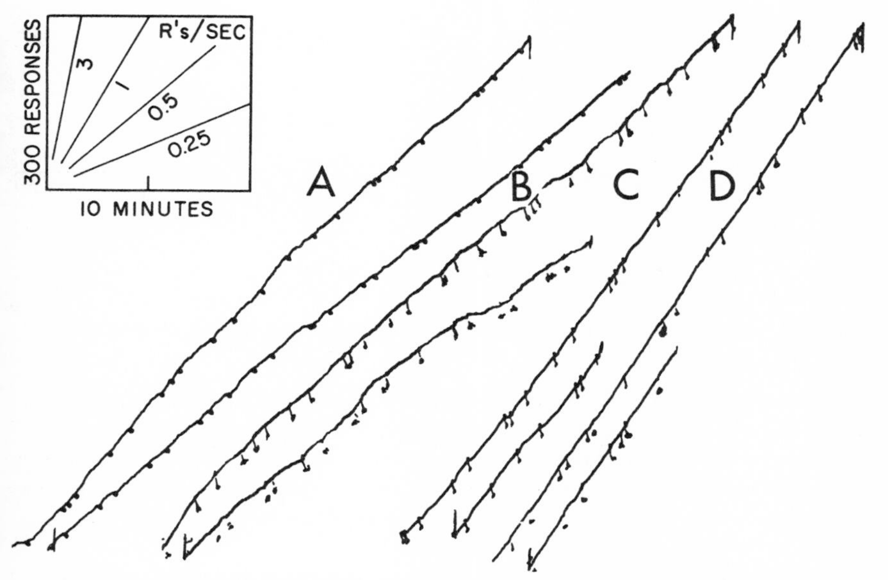

Fig. 860. Chain VI 1 VI 1 at two levels of deprivation (second bird)

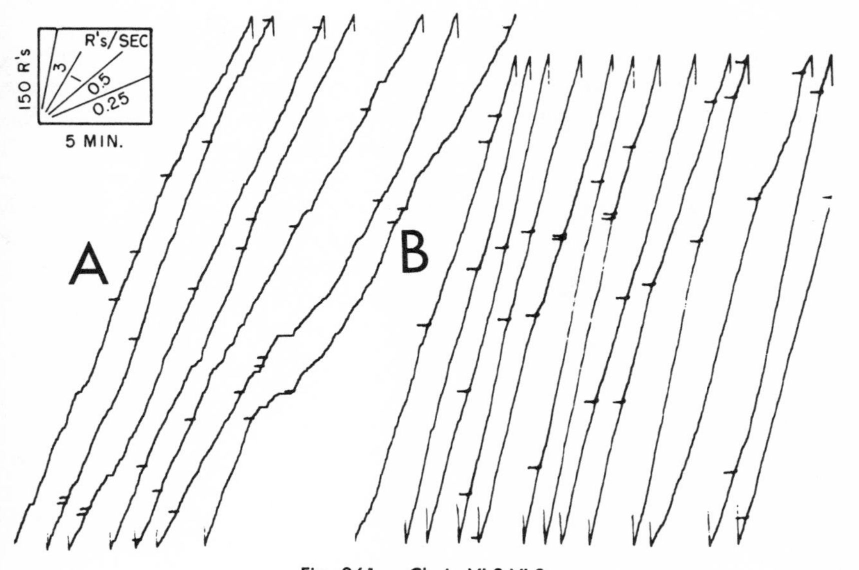

Fig. 861. Chain VI 3 VI 3

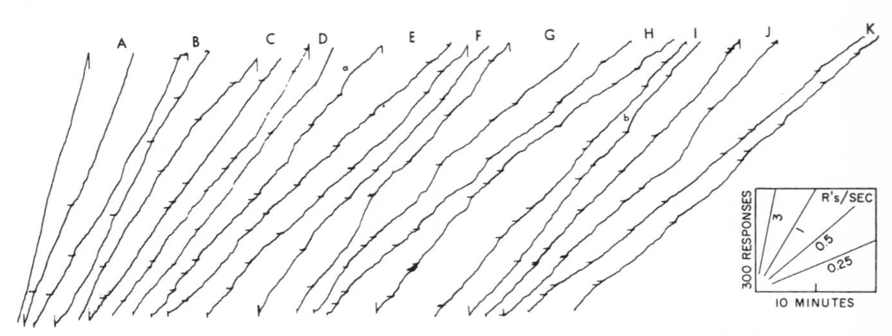

Fig. 862. Chain VI 3 VI 3 drl: segments from the VI 3 drl record showing the development of a low rate

of with food. This procedure reduced the total density of food reinforcement by one-half. The rate on the blue key declined during 20 subsequent sessions. Figure 861 illustrates the entire 21st session. Record A shows responding on the previously preferred blue key, reinforced, as marked, by changes to the orange key. Record B shows responding on the orange key, reinforced, as marked, with food. After food reinforcement, the key became blue immediately.

A drl 4 was then added to the VI schedule on the orange key. This was the 2nd component of the chain reinforced with food. The decline in rate is extremely slow for this contingency. Figure 862 gives the 2nd and 3rd excursions from each of 11 successive sessions. Responding tends to be especially rapid immediately after the changeover from the blue key, as at *a* and *b*. More marked examples of this tendency occur elsewhere, and possibly represent an emotional effect arising from the release from the blue key and the presentation of the orange key, responses to which are reinforced with food. One result is the lengthening of the shorter intervals in the VI schedule. Concurrently with the decline shown in Fig. 862, the performance on the blue key becomes very irregular. Long periods of no responding occur, the grain is very rough, and the over-all rate declines. At times, however, a fairly high rate emerges. Figure 863 shows the entire performance on the blue key corresponding to the session from which the segments at K in Fig. 862 were taken. Very rough grain,

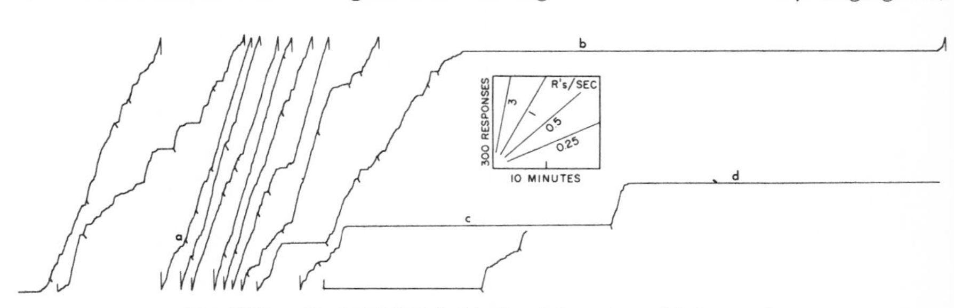

Fig. 863. Chain VI 3 VI 3 drl after 10 sessions: VI 3 record

periods of quite rapid responding, as at *a,* and exceptionally long pauses, as at *b, c,* and *d,* occur. The rapid responding at *a* is as extreme as it ever becomes, and the pauses at *b, c,* and *d* are exceptionally long. However, the record correctly reports the general disturbance in the 1st member of the chain in which the 2nd member contains a drl. This disturbance could in part reflect interference from mediating behavior.

When the drl contingency was then removed, the bird returned very rapidly to the high uniform rate of the orange key. Figure 864 shows the 2nd session of the new schedule. The performance on the 1st member of the chain (Record A) has already recovered much of the character of Fig. 861A.

When the drl was again added to the contingency on the orange key, the rate fell off within a single session to the level reached only after 11 sessions in Fig. 862. Figure 865 shows the entire 1st session. Note the marked instances of a high rate immediately

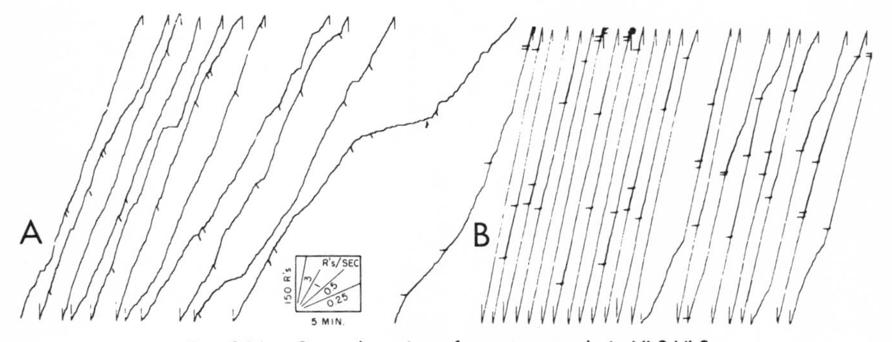

Fig. 864. Second session after return to chain VI 3 VI 3

following the change from the blue key, as at *a* and *b.* Following this change in the contingency on the orange key, the responding on the blue key began to show the extreme ranges in rate already seen in Fig. 863.

A second bird showed initially a higher rate on both key-colors under mult VI 3 VI 3, with a preference of approximately 9 to 8 in favor of the blue key. This performance is in the same direction as that of the other bird, but the relation does not change during the chaining. During 21 sessions in which the 2 schedules were chained, the bird did not show a decline in rate on the blue key. Figure 866 shows the 22nd session complete. When drl was added, the performance showed the usual decline; but the rate on the blue key remained essentially unchanged. The performance was similar to a later session under these conditions, shown in Fig. 868. When the drl is again omitted, the rate rises quickly, as Fig. 867 shows. Record A illustrates the performance on the blue key reinforced by the appearance of the orange key; and Record B, the performance on the orange key reinforced with food. Note the beginning of a tendency to respond more rapidly on the blue key after the preceding reinforcement on

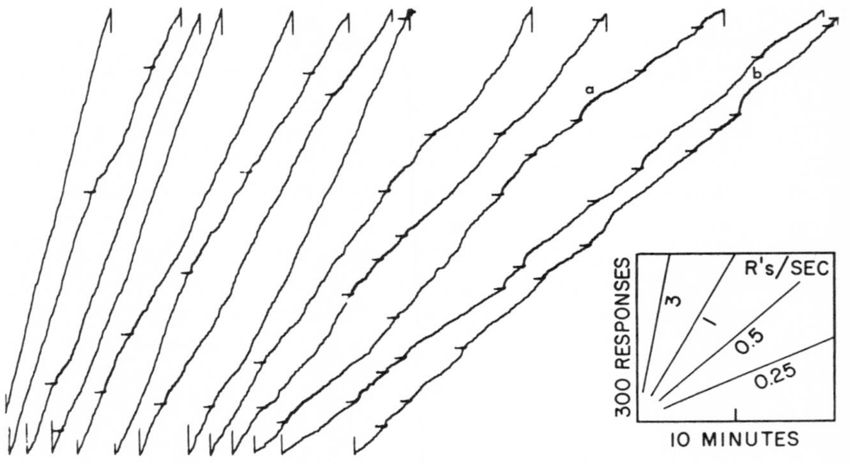

Fig. 865. Return to chain VI 3 VI 3 drl: VI 3 drl record

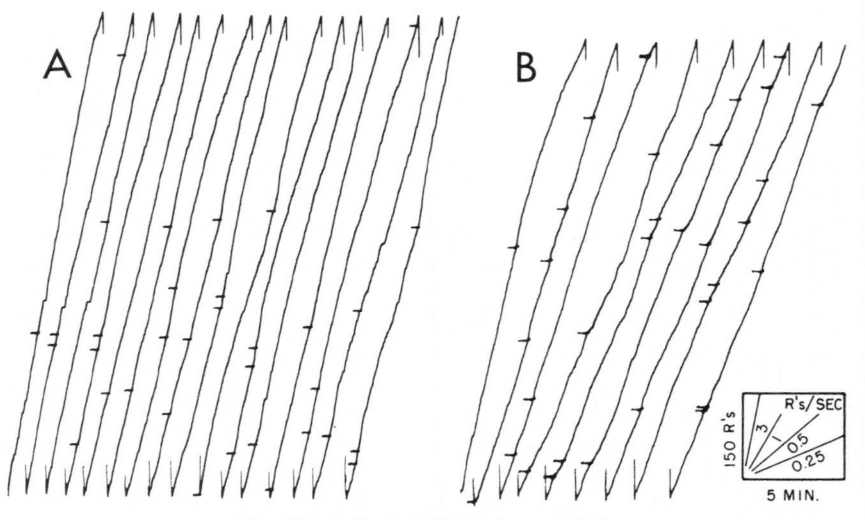

Fig. 866. Chain VI 3 VI 3 (second bird)

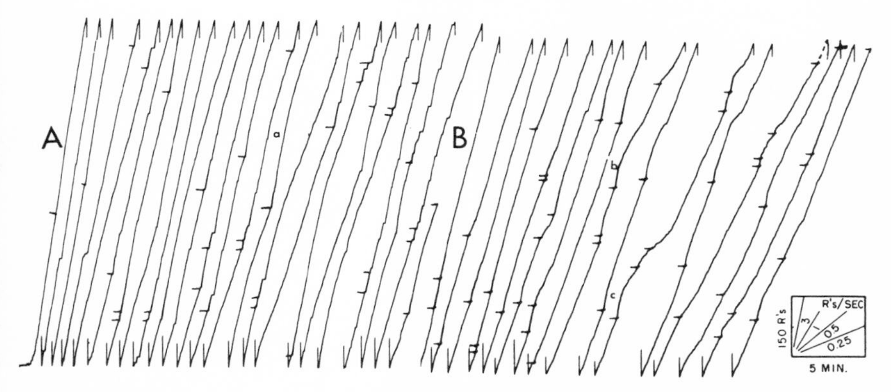

Fig. 867. Return to chain VI 3 VI 3 after chain VI 3 VI 3 drl (second bird)

the orange key, as at *a*. Note also, in Record B, a much sharper curvature resulting from a high rate immediately after changeover from the blue key, as at *b* and *c*.

When the drl was again added to the 2nd member, the rate fell. Figure 868 gives a sample of the performance. On the blue key (Record B) a higher rate immediately after reinforcement is now common. It is followed by a fairly smooth decline whenever the interval permits, as at *a, b, c,* and *d*. Note that this performance represents a tendency for the rate not to rise above that previously seen on this key (Fig. 867A) but to decline to a lower over-all value.

Two of the birds in the preceding experiment, one on chain VI 3 VI 3, the other on chain VI 1 VI 1, were left in the experimental box for 18¾ and 16 hours, respectively. During these long sessions, substantial amounts of food were received in reinforcements. Figure 869 gives the result. Record A is the cumulative performance of the bird on the 2nd component of chain VI 3 VI 3 and Record B is for the 1st component. Satiation reduces the rate in the 1st component, but has little effect on the second. The same result held for the other bird. Record C shows the 2nd-component rate holding well, while the 1st component (Record D) declines in satiation.

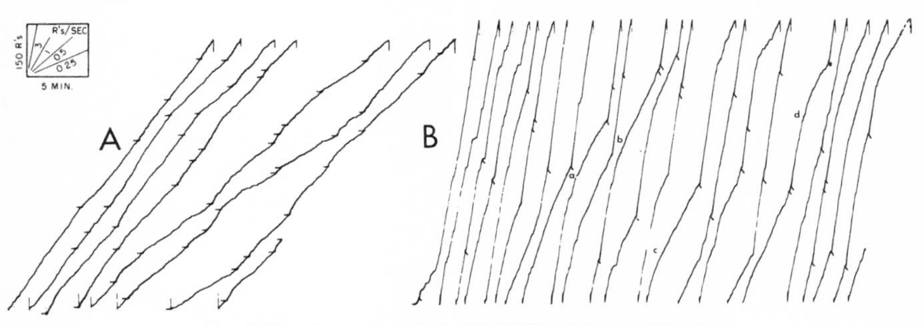

Fig. 868. Return to chain VI 3 VI 3 drl (second bird)

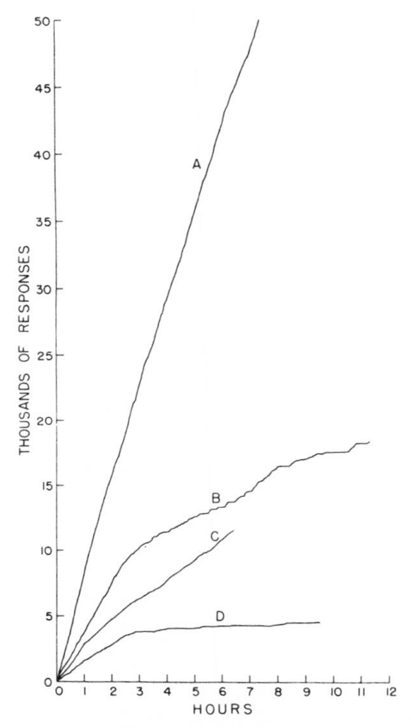

Fig. 869. Satiation curves on chain VIVI

CHAINED FIFR

Two birds with a history of mult FIFR were reinforced for 30 sessions on chain FI 4 FR, where the FR varied between 20 and 60. Figure 870 illustrates the last observed performances on chain FI 4 FR 60. Record A shows the performance on the

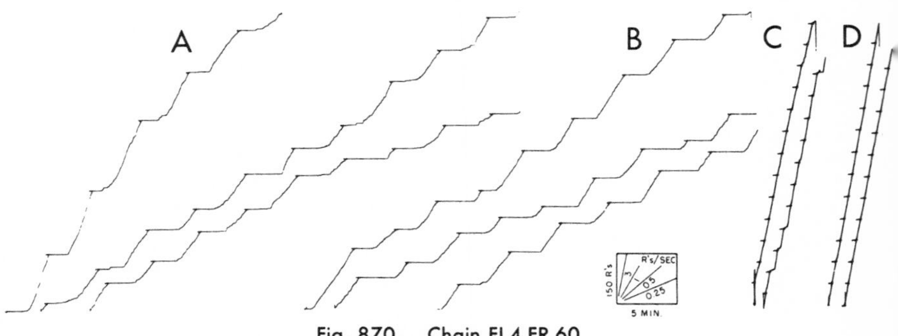

Fig. 870. Chain FI 4 FR 60

1st-component FI for one bird. Pauses are longer than those for a simple FI 4. Record C shows the corresponding 2nd-component performance on FR 60. A few small breaks occur, but the terminal rate and general pattern are typical of a ratio performance. Records B and D show the comparable records for the second bird.

CHAIN FRFI

Two birds, also with a history of mult FIFR, were reinforced for 58 sessions on chain FRFI, during which the FI values ranged from 1 to 3 minutes while the ratios varied from 20 to 50. Figure 871 shows performances for 1 bird under 2 sets of interval values. Records A and C illustrate 1st and 2nd components, respectively, on chain FR 20 FI 1. The ratio performance at Record A shows consistent pausing after reinforcement, which would be quite unusual on a simple ratio of this size. The over-

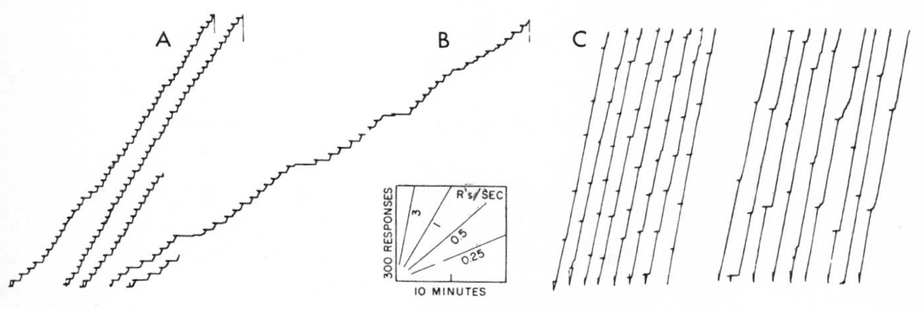

Fig. 871. Chain FR 20 FI 1 and 2

all and terminal rates in Record A are lower than those in Record C. Meanwhile, the FI 1 performance in Record C lacks the pausing after reinforcement and scalloping evident in a simple FI 1. This performance is related to the position of the FI 1 on the 2nd component of the chained schedule. Records B and D show performances on the 1st and 2nd components of chain FR 20 FI 2. The FI performance shows some slight scalloping. But the main effect of the larger interval is felt in the preceding ratio performance (Record D), where both over-all and terminal rates are low. The second bird showed similar results, except that it could not maintain a comparable over-all rate on the ratio component.

CHAIN FRFR

The "block" counter described in Chapter Four is in reality a species of chain FR FRFR. . . . In the experiment described there the color of the key could change at prescribed points during the emission of the ratio. The last response in the ratio on the last color of the counter was, of course, reinforced with food. But the production of this last color thereupon became reinforcing for the last response in the preceding color, and so on. In some of these experiments the arrangements of colors during the ratio produced effects which are relevant here.

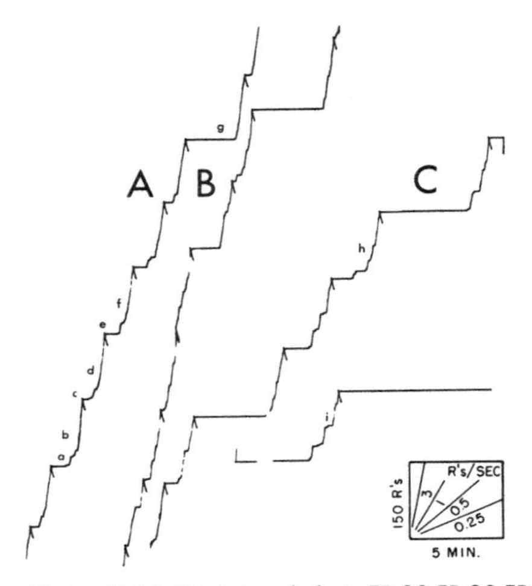

Fig. 872. Chain FR 15 FR 105 and chain FR 30 FR 30 FR 30 FR 30

In one case a color change occurred at only one point during the ratio. Fifteen responses on a red key produced a blue key, and 100 responses on the blue key produced food. The performance (Fig. 872) shows a pause after each reinforcement, as at *a, c,* and *e.* After the change from red to blue at 15 responses, a 2nd but smaller pause frequently occurs, as at *b, d,* and *f.* At *g,* where the pause after reinforcement is exceptionally long, practically no pausing occurs on the change in color. Separate recording of the 2 parts of the ratio would show a low over-all rate, with strong pausing on the red key and a high over-all rate with only slight pausing on the blue.

Record B shows an example of a performance in which the key-color changes after each block of 30 responses. Three breaks can usually be detected in each segment. However, the pausing after reinforcement is often longer than any pause upon change of color. Later changes frequently have some tendency to show less pausing. This tendency is particularly clear in Record C, which is for the session following Record B. Record C shows the damping of the oscillation already mentioned in Chapter Four, as at *h.* Note that damping tends to occur when a ratio is straining. (The 7 reinforcements shown in the figure are the only ones received in a session of more than 1 hour.) The damped ratio at *i* followed a long pause not shown in the graph.

The usual effect of changing from one set of colors to another is to break up the control of the chain and to produce a temporarily high rate. In Fig. 873A the bird is showing a badly strained performance on the 4-color schedule just described. At the beginning of the following session the ratio of 120 responses was subdivided in successive blocks of 60, 30, and 30 responses. The 1st color was that under which responses had formerly been magazine-reinforced. A change to a previously nonreinforced color occurs after 60 responses. After 30 responses, the previously reinforced color returns for the last 30 responses. Record B shows the effect. The pausing after reinforcement characteristic of the earlier 4-color performance is immediately lost, since the presence of the previously reinforced color leads to an immediate start, as at *a*, *c*, and *e*. The change to the previously nonreinforced key after 60 responses produces a slight pause, however, as at *b*, *d*, and *f*. The pause at 60 responses continues for some time, and is present in the last 2 ratios in the session at *j* and *k*, although these now also show pauses after reinforcement. At one stage the whole sequence functions as a single color ratio, as at *g*, *h*, and *i*. On the following day, the break at 60 responses

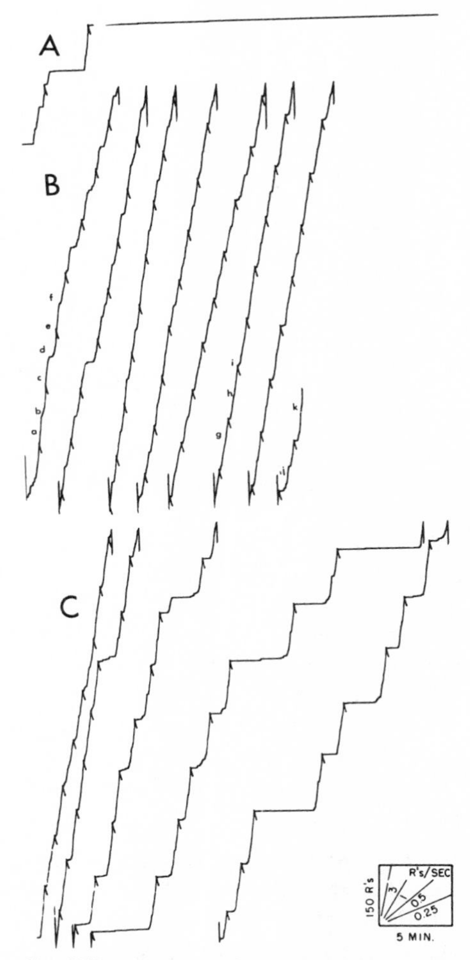

Fig. 873. Transition from chain FR 30 FR 30 FR 30 FR 30 to chain FR 60 FR 30 FR 30

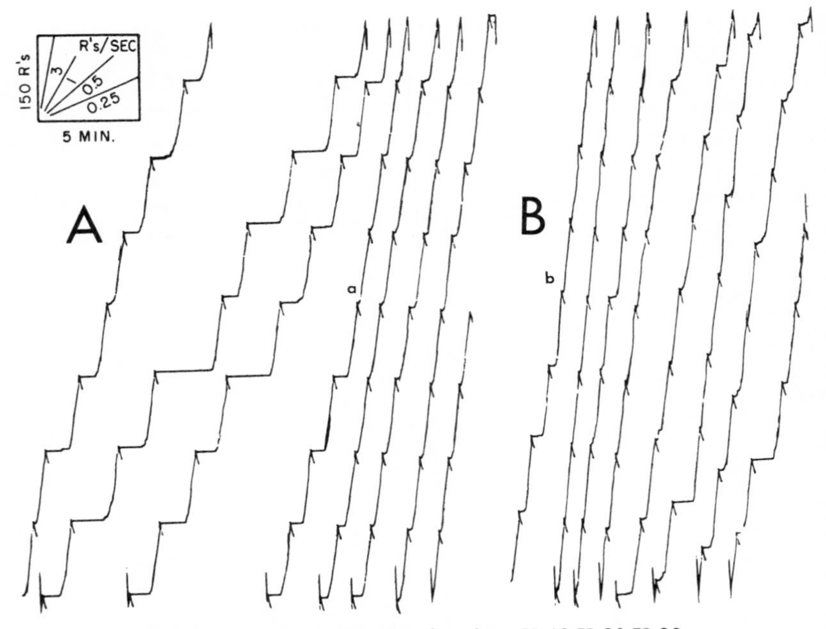

Fig. 874. Return to FR 120 after chain FR 60 FR 30 FR 30

continues to appear from time to time. The increase in the over-all rate produced by the new color schedule begins to disappear. This increased rate is characteristic of changing a color sequence in submultiples of FR. Record C continues the same procedure as Record B.

After pausing had developed in Fig. 873C, the color previously present at reinforcement was maintained throughout the ratio. Figure 874 shows 2 resulting performances. In Record A a bird not previously discussed requires more than 25 reinforcements to recover from the breaking on the "chained" ratio and to reach the high over-all rate at *a*. The pausing occurring before that point was characteristic of the preceding chained ratio, similar to that of Fig. 873C. The other bird (Record B), from the preceding figure, also requires a few reinforcements to reach the new high rate shown at *b*. Here, the rate declines, and the pauses emerge on the new schedule with the single key-color within the session. A comparable decline for the other bird appeared in a later session.

CHAIN VI UNCORRELATED FI (Reinforcement by the Production of a Stimulus in the Presence of Which Food Is Presented Unrelated to a Response at the End of 1 Minute)

We tried to show whether the performance on the 2nd component on a chained schedule or the reinforcement received under the stimulus appropriate to that perform-

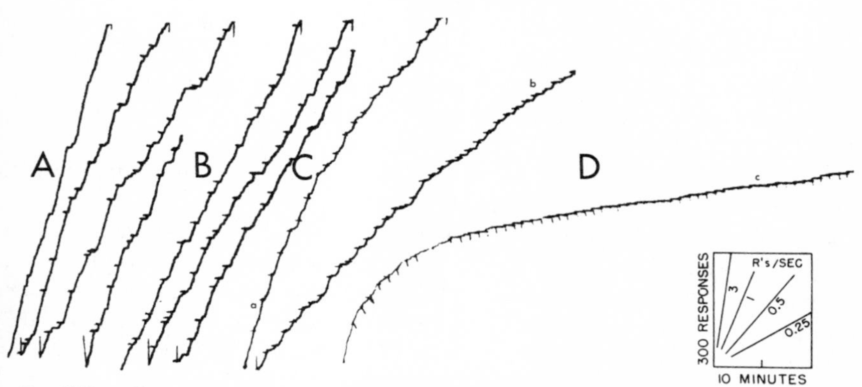

Fig. 875. Chain VIFI with the magazine opening independently of a response at the end of the FI

ance maintains behavior on the 1st component. To this end we changed the procedure on a chained VIFI so that food was presented on an FI 1 schedule but without correlation with a response. At the end of 1 minute under the 2nd-component color, the magazine operated whether or not the bird was responding. Under such conditions the rate fell off, but usually not entirely to zero because of accidental reinforcements. Figure 875A and C shows the 1st day under the change. The FI 1 rate on the preceding bona fide chain was similar to the starting rate in Record C at *a*. During the session the rate declines markedly to the value at *b*. Record A shows the responding in the 1st component on this day.

Within a few sessions the rate during the last half of each session dropped to a low value. Each new session, however, began at a high rate followed by marked negative acceleration. Record D shows a performance, after 17 sessions, with this initial negative acceleration. Meanwhile, the 1st component of the chain, where the VI 1 reinforcement is the production of the 2nd color, undergoes no substantial decline, as seen in Record B. The last excursion in Record B corresponds to the period of very slow responding in the 2nd color at *c*. Therefore, the conditioned reinforcement of the property of the stimulus for the 2nd schedule would appear to derive from its correlation with food, regardless of whether or not the bird is responding in the presence of the 2nd stimulus. It should be noted, however, that the rate in the 2nd color does not reach zero in this experiment. We could have forced the rate to zero by withholding reinforcement every time a response occurred.

CHAIN FRcrfdrl

Experiment I

Two birds with a long history of FRdrl described elsewhere were changed to a schedule in which the drl phase was accompanied by a stimulus change on the key. As soon as the bird "counted out" the ratio, the key-color changed from red to purple, and the first response which followed a 6-second pause on the purple key was reinforced. The added stimulus should eventually permit the bird to run the ratio at a high rate, and then to slow down immediately for reinforcement. The resulting schedule is chain FR(red)crfdrl(purple).

With both birds the result suggests some release from the previous drl effect because of the appearance of the purple light.

One bird in particular tended to respond faster when the purple light appeared. The over-all rate for the 1st day for this bird, in a session lasting 6 hours, did not increase as a result of the added stimulus. Toward the end of this period, in fact, pauses as long as 30 minutes appear. The performance under the ratio is essentially that which prevailed under FRdrl. Segment 1 of Fig. 876 shows the performance during

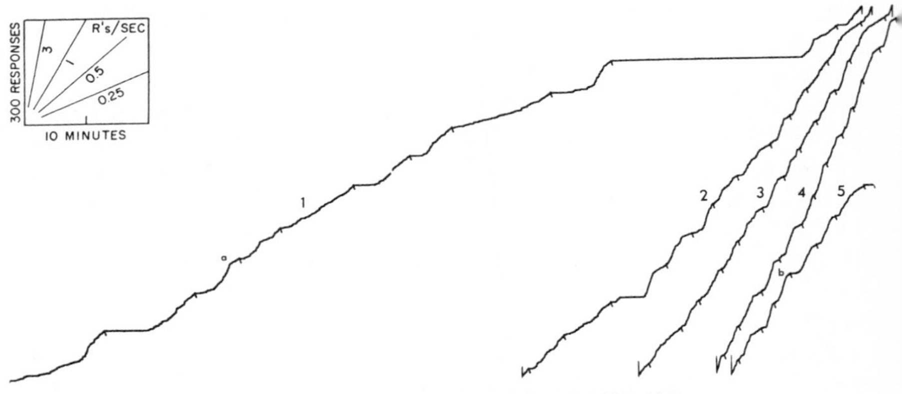

Fig. 876. Second session on chain FR 20 crfdrl

approximately the 1st hour of the next session. The tendency to run fast when the purple key appears is no longer evident. Early in the segment signs appear of some stimulus control for the drl phase in the fairly sharp breaks, as at *a*. In the 2nd segment, acceleration is noticeable; and before the end of the period the performance is fairly appropriate to the new schedule. At *b* in Segment 5, for example, the ratio is completed at a fairly high rate after only a brief starting pause, and the rate then drops abruptly almost to a value sufficient to achieve immediate reinforcement. The other bird reached this stage much more rapidly. The 1st session on the new schedule shows signs of the tendency to accelerate upon the appearance of the new stimulus, as at *a*, *b*, *c*, and *d* in Fig. 877, although the effect is much less marked than with the other bird. Otherwise, the performance for the first 2 segments resembles that under the former FRdrl. The acceleration to a new performance takes place relatively rapidly in the region between *e* and *f*. An excellent approximation of the final performance under this schedule appears at *g* and elsewhere.

Figure 878A illustrates the most advanced development of a performance on chain FRcrfdrl in the experiment. Here, with very few exceptions, the ratio is counted out, the light changes color, and in a few responses at a low rate the reinforcement is received. The ratio performance is now standard for this value of the ratio (95) at this level of deprivation. The first bird (Record B) did not develop quite so high a rate on the ratio, and obviously occasionally had trouble in slowing down after the change in color. Nevertheless, the over-all rate is much higher than in the case where there was no stimulus change at the completion of the ratio.

An intermediate stage of the development of this performance for the bird in Fig. 876 and 878B may help reveal the nature of the scalloping occasionally observed on FRs. In Fig. 879, 4 segments are reproduced from the end of the 10th session on

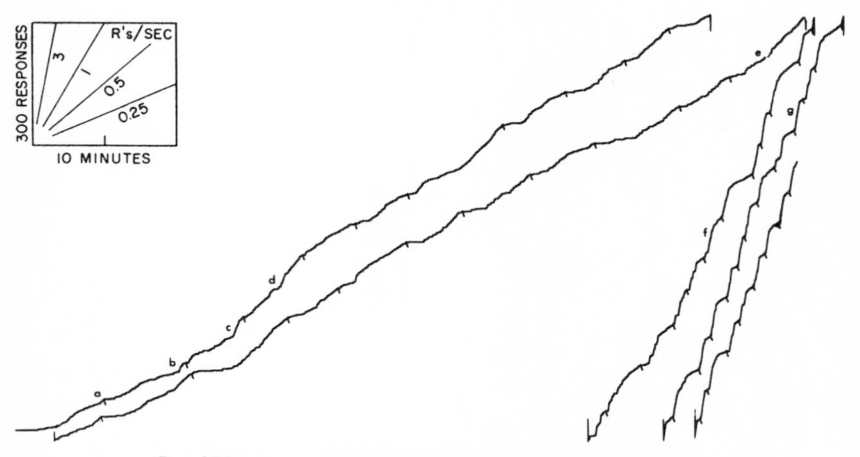

Fig. 877. First session on chain FR 70 crfdrl (second bird)

chain FR 120 crfdrl 6. Straining in the ratio is evident in the long horizontal segments of the record; there is also difficulty in adjusting to the drl contingency, for example, at *a, b, c,* and *d.* Nevertheless, fairly good stimulus control is clear in the sharpness with which the rate falls after the ratio has been counted out. The curious feature of the performance is that all the ratio runs are smoothly scalloped, suggesting FI rather than FR curves. This factor may be due to some inductive effect from the low rate under the drl contingency. Under this schedule the ratio is reinforced by the change of color at its completion. The rate at the moment of reinforcement with food is very low. The ratio performance is therefore usually reinforced when the bird is responding rapidly, but the major reinforcement occurs at a low rate. The temporary effect is to produce a fairly smooth curvature during the ratio; this may be related to the chain FRFI schedules reported earlier, where the FR component in the 1st member showed low rates and curvature. The second bird never exhibited this degree of

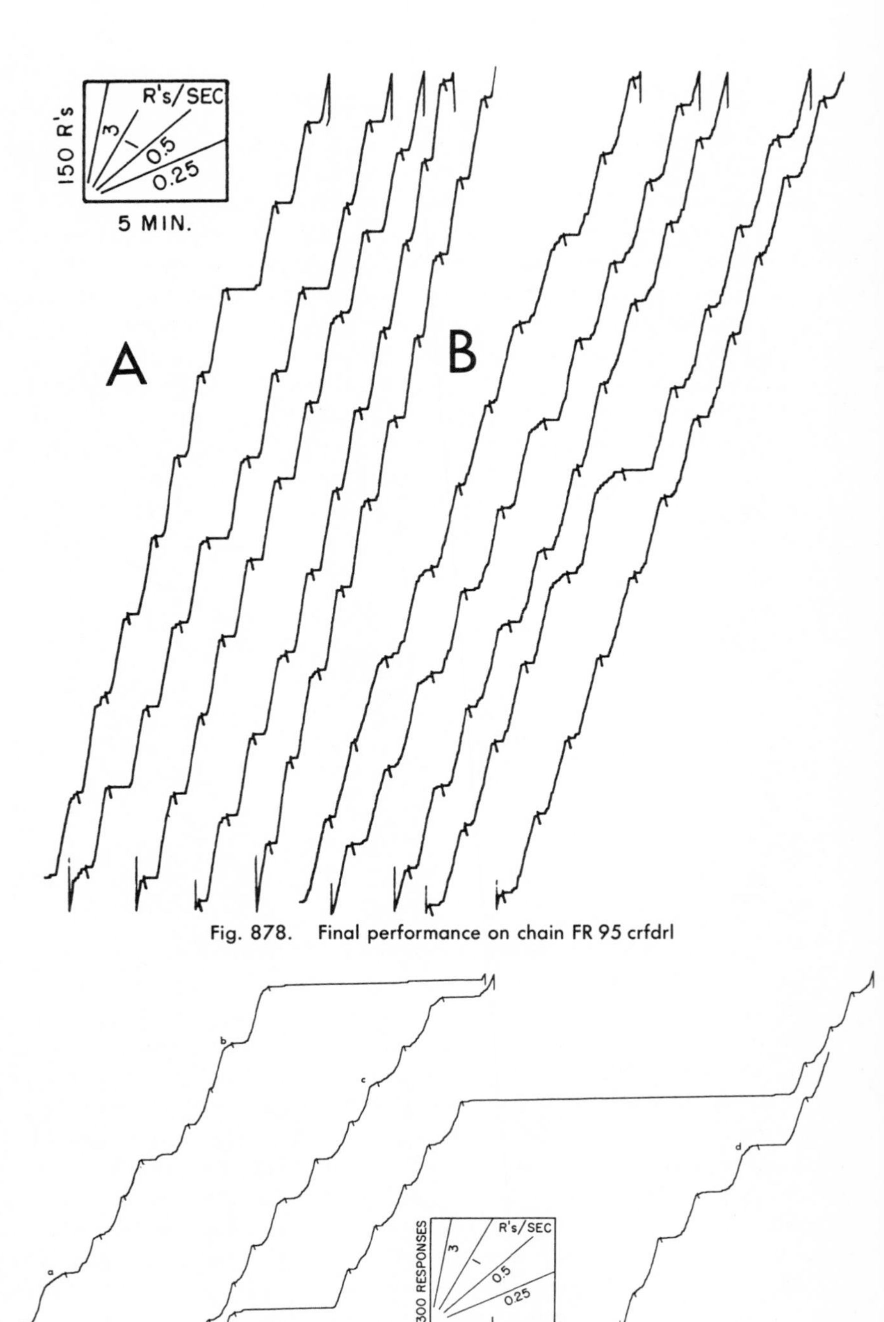

Fig. 878. Final performance on chain FR 95 crfdrl

Fig. 879. Chain FR 95 crfdrl developing after reinforcement (first bird)

straining and, in general, showed no positive curvature during the ratio performance at any stage in the development of the earlier performance under FRdrl.

Experiment II

Another bird with a long history (442 hours) on FRdrl, with occasional tests of the effect of removing the drl contingency, was then reinforced on a chain FR 80 crfdrl 6. The development of the performance is illustrated in Fig. 880, where pairs of segments have been chosen from various sessions throughout the process. Record A shows the performance prevailing at the end of the experiment on FRdrl. Beginning with the next session the "house lights" were dimmed upon completion of the ratio. Dimming thus served as a stimulus which could control the drl contingency and reinforce the FR. Some effect begins to appear in the 3rd hour of the session (Record B), where the breaks at *a, b, c, d,* and elsewhere mark the stimulus change. The condition

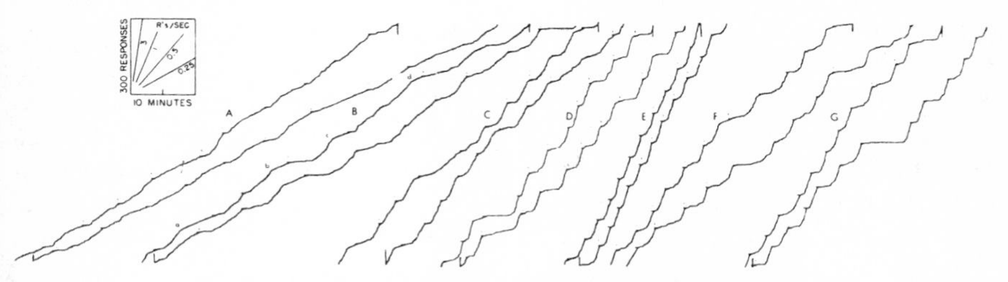

Fig. 880. Transition from FRdrl to chain FRcrfdrl

at the 9th hour in a later session is given at Record C; at the 18th hour, at Record D; at the 22nd hour, at Record E; at the 33rd hour, at Record F; and at the 37th hour, at Record G. The control exercised by the dimming of the light develops progressively. The rate falls immediately after the ratio is completed. However, a good ratio fails to emerge under these conditions in the later stages of the experiment, and the over-all rate has some tendency to fall off, with signs of straining on the FR 80 schedule.

Figure 881 shows the final condition under the chained schedule for the complete last session. Considerable breaking occurs on the ratio, although the terminal rate is fairly high. As usual, rather marked curvature in the scallop on the ratio is correlated with breaking. The stimulus control is generally adequate although occasionally, as at *a, b,* and *c,* several responses are made before the criterion of a 6-second delay is satisfied. A comparison of this figure with Fig. 878 reveals that the ratio rate is here relatively low. This condition may be due in part to the fact that the dimming of the house lights is not very clear-cut as a stimulus; but it may be a general property of chains in which the 1st member is an FR. (See *Chain FRFI.*)

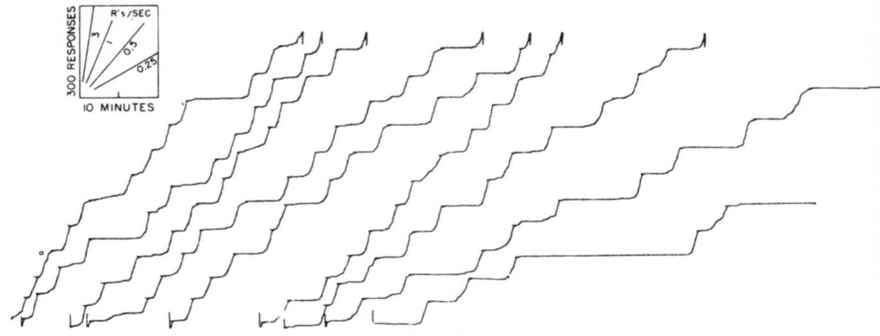

Fig. 881. Final performance on chain FR 80 crfdrl 16

The effect of reducing the difference in stimuli on a chain FR 120 crfdrl 6

The stimuli used in an experiment on chain FRcrfdrl already described consisted of a red 6-watt, 115-volt bulb behind a translucent key. A blue light of the same wattage was added to this key to produce the purple key specified in that experiment. We tested the importance of the difference between the 2 stimuli by adding a potentiometer to the circuit of the blue light so that its relative intensity could be reduced. We made no effort to analyze these lights accurately. Differences will be expressed in terms of the resistance of the potentiometer. In one series, daily runs were made with the potentiometer set at 800, 1100, 1150, 1250, 1450, and 1500 ohms, respectively.

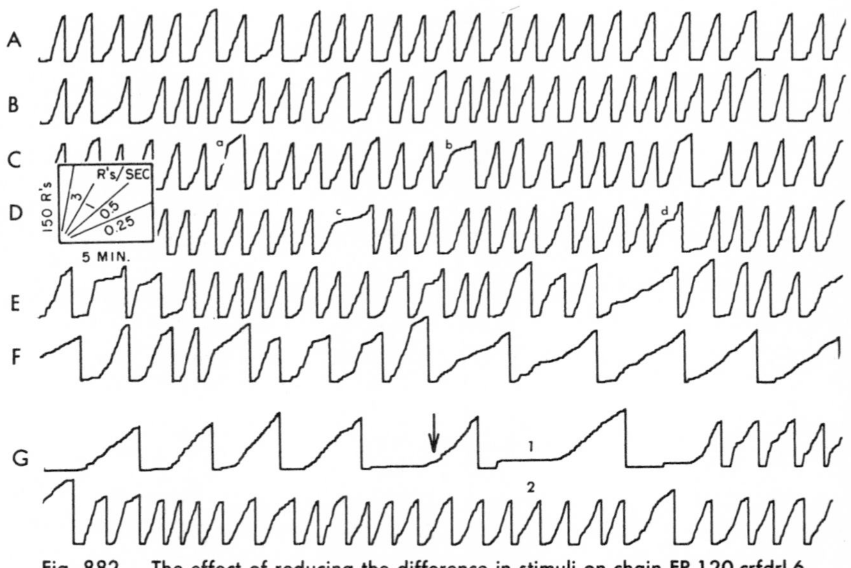

Fig. 882. The effect of reducing the difference in stimuli on chain FR 120 crfdrl 6

The change in voltage at the blue lamp undoubtedly changed the hue as well as the brightness. Ohm readings were used mainly to reproduce a given value from day to day.

The "threshold" proved to be near the 1500-ohm setting. Figure 882A is for the last part of a session with the setting at 800 ohms. The performance is not essentially different from that in the earlier experiment with the full addition of the blue light, shown in Fig. 878B. Only by the 3rd day, at 1150 ohms, from which Record C represents the terminal part of the session, is there evidence of a premature break at *b* and of a tendency not to respect the drl contingency sufficiently, as at *a*. In the following session, with a setting of 1250 ohms, 2 premature breaks occur at *c* and *d*; this difficulty becomes more serious in Record E in the following session with a setting of 1400 ohms. At 1500 ohms on the following day (Record F), some ratios are run off at a

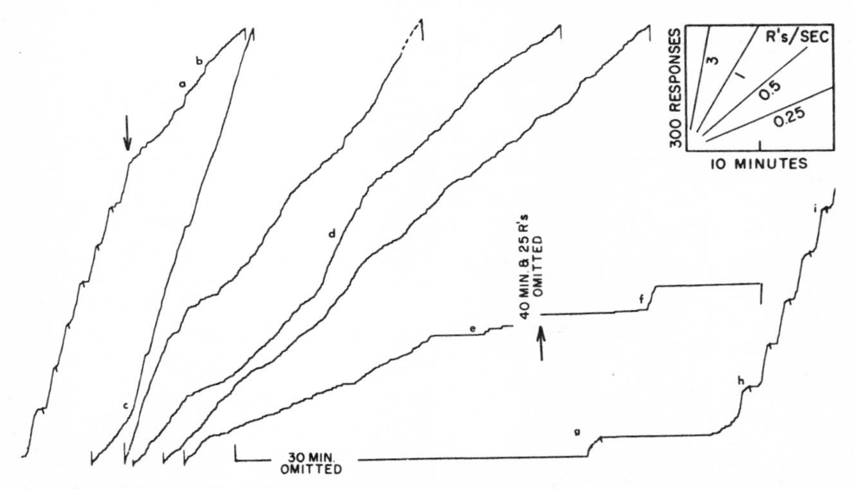

Fig. 883. Extinction after chain FRcrfdrl in crfdrl color followed by the FR color

fairly constant rate, suggesting the performance for this bird on FR 70 drl 6 given in Fig. 600. The maximum stimulus differences were reinstated at the arrow in Record G, and the original performance was quickly recovered. This effect might provide a useful technique for the investigation of threshold values, where the actual properties of the stimuli were carefully controlled and measured, and the mean time taken to reach the ratio used as a measure of the effectiveness of the stimulus.

Extinction after chain FRcrfdrl

After the development of the terminal performance seen in Fig. 878B, the magazine was disconnected and the stimulus allowed to remain purple after the completion of the first ratio. Figure 883 shows the resulting extinction, following a short period on the earlier schedule before the first arrow. Except for short break-throughs of the ratio rate at *a* and *b*, the rate remains appropriate to drl until *c*. After *c* a rate appropriate to the ratio emerges for about 1000 responses, declining again to a rough grain at the lower rate. Some instances exist of a return to a higher rate, notably at *d*, but the over-all decline brings the rate to practically zero by point *e*. A 40-minute period on the same procedure has been omitted, following *e*, during which only 25 responses were emitted. The purple light was then changed to red (second arrow). The short burst at *f* does not complete the ratio. At *g*, however, whereupon the light changed to purple and a reinforcement was received after a short period of responding at the drl rate. This reinforcement reinstated the original schedule quickly, as *h* and *i* show.

The original FRdrl was maintained for 2 full sessions, and in the following session extinction was permitted to take place in the "ratio" color (red) alone. Figure 884

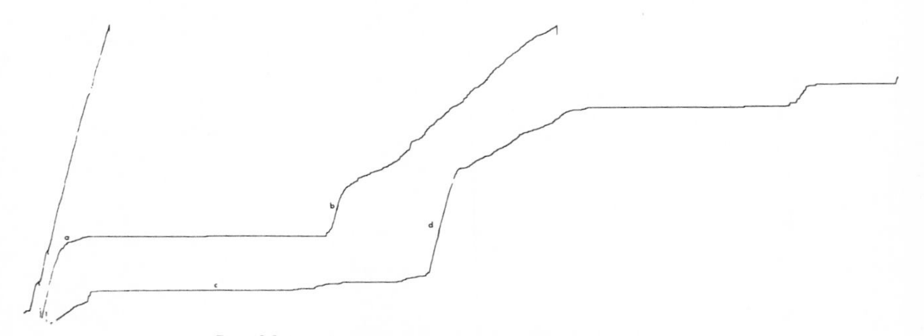

Fig. 884. Extinction after chain FRcrfdrl in FR color

illustrates the result. About 1000 responses are emitted under the red stimuli before the rate falls off, smoothly but rather rapidly, at *a*. After a substantial pause, a brief period of responding at the ratio rate appears at *b*; again it falls off fairly smoothly, this time to a rate roughly appropriate to drl, although the stimulus previously correlated with drl is not present. After another period of no responding at *c*, the bird returns again to the ratio rate at *d*, and the over-all rate ultimately falls off to a low value. Here, then, is an instance in which the rate appropriate to the drl contingency emerges for some time under the stimulus previously correlated with the ratio part. After completion of the performance shown in Fig. 884, the stimulus was changed to purple, the stimulus appropriate to the drl contingency. This change reinstates some responding, in Fig. 885, which follows continuously after Fig. 884. In this figure responding at the ratio rate emerges briefly, although the color is now appropriate to the drl contingency. The rate falls off to a very low value before the end of the period.

A second bird with the same history was also extinguished in the color appropriate to drl. Extinction begins after the completion of a ratio at *a* in Fig. 886. Although the ratio rate appears briefly as at *b*, *c*, and *d*, the over-all rate falls off strongly to

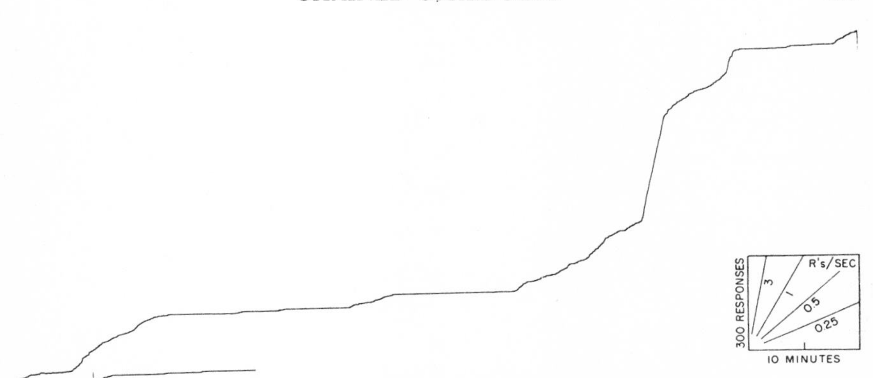

Fig. 885. Continuation of Fig. 884 showing extinction in the drl color

practically zero before the point *e* is reached. At this point the color was changed to red (the color previously correlated with the ratio component of the chained schedule). The ratio rate appears for approximately 700 responses and then breaks sharply (at *f*) to a low value. Responding reappears briefly at a higher rate, as at *g*. Beginning abruptly at *h*, a fairly substantial rate declines smoothly to a low rate at which only 280 responses were emitted in the final 8.5 hours of the session (not shown in the figure).

In another experiment with the same birds, a well-developed performance on chain FR 120 crfdrl 6 was extinguished while the stimuli were changed every 5 minutes. Figure 887 shows 2 curves for one bird. In the actual experiment the segment of Record A marked *a* occurred first. At the dot above the record at *a* the stimulus changed. Segment *b* in Record B was then recorded, the recorder in Record A stopping. At the small dot above the record beyond *b* the stimulus was changed to the first setting again, and Segment *c* in Record A was recorded. This portion was followed in turn by Segment *d* in B, Segment *e* in A, Segment *f* in B, and so on. The 2 curves are appropriate to the respective parts of the chained schedule. Record A is a good ex-

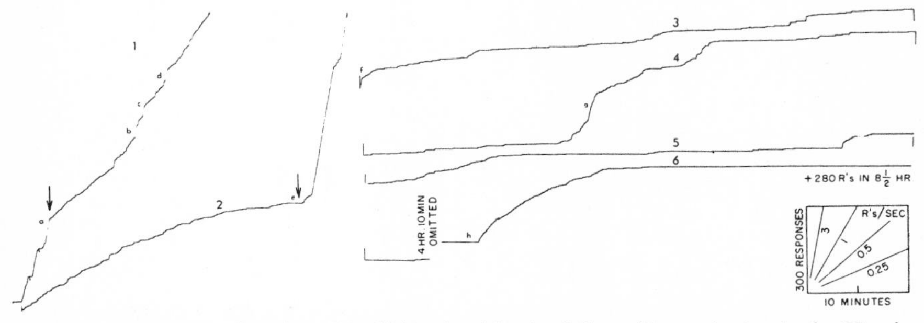

Fig. 886. Extinction after chain FRcrfdrl in the drl color followed by extinction in the FR color

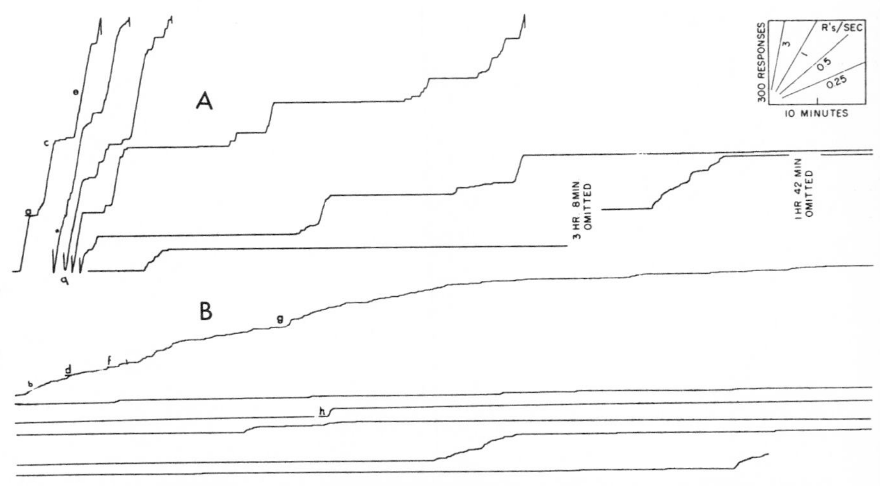

Fig. 887. Extinction after chain FR 120 crfdrl 6 with stimuli alternating every 5 min

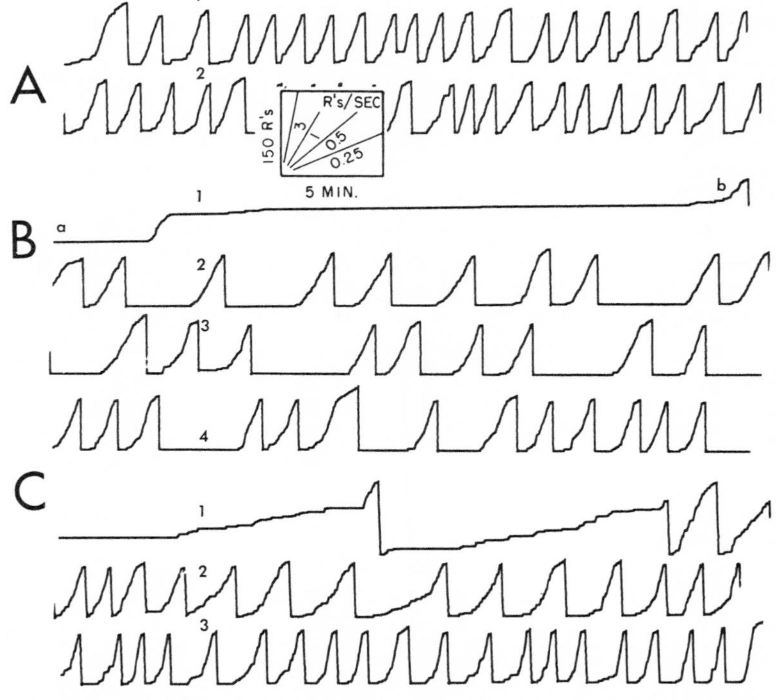

Fig. 888. The effect of Hystadyl and lysergic acid (diethylamide) on chain FRdrl

ample of extinction after FR, when responding occurs at the ratio value or in very rough grain at lower values. (Several hours of little or no responding have been omitted from this figure as marked.) The complete curve in Record B, meanwhile, shows only the rate appropriate to the drl contingency, except possibly for very brief runs at *g* and *h*.

The effect of certain drugs upon chain FRdrl

Two birds showing a fairly stable performance on chain FR 120 crf drl 6, as shown in Fig. 888A, were given 0.06 milligram of lysergic acid (LSD) in 3 cubic centimeters of water, administered orally at the start of the following session (Record B). A considerable depressive effect appears immediately following the administration of the drug at *a*. The 1st ratio is completed after about 35 minutes (at *b*). Pauses following reinforcement are conspicuously longer throughout the experimental session. (This was the only substantial effect of LSD found in a series of 4 administrations of 0.005, 0.005, 0.01, and 0.06 milligram, respectively.) Pauses are extended and the running rate is reduced, but the stimulus control of the drl contingency and the behavior under that contingency remain essentially unchanged. Three doses of lysergic acid, from 0.005 to 0.02 milligram, showed no marked effect in the second bird.

Record C illustrates another example of the effect of a suppressing drug on the performance. After 2 ratios had been run off normally (not shown), the bird was given a large dose of Histadyl (an antihistamine) by mouth in water solution, and was replaced in the apparatus. During the next 130 minutes only 30 responses were emitted. These occurred in groups of 2 or 3 and indicated that the bird could respond at this time. The performance shown in Record C then followed. The first 2 ratios require about 15 minutes each, but the drl contingency is soon effective. Some disturbance is evident in the rather marked positive curvature in many of the ratios in Segment 2, but the performance is essentially normal before the end of Segment 3 at the end of the experimental session. (These experiments are offered simply to show the character of records influenced by depressant drugs.)

A COMPARISON OF THE REINFORCING EFFECTS OF THE TWO CONTROLLING STIMULI IN A MULT FIFR

In this experiment the apparatus contained 2 keys, and 2 colors were possible on each. On the right key, red and blue eventually control the ratio and interval members of a mult FIFR, respectively. On the left key the schedule was always VI 3. Reinforcement on this key was the presentation of one or the other of the colors on the right key. When the left key was white, the right key was unlighted; but a response to the left key would produce, on VI 3, a red key on the right. In the presence of the red key a response was then reinforced on FR. The left key was unlighted during the FR. When the left key was green, a response would, on VI 3, change the right key from unlighted to blue, a response on the blue key being reinforced on FI 10. Four performances were recorded: (1) the VI performance on the white left key; (2) the VI performance on the green left key; (3) the FI performance on the blue right key; and (4) the ratio performance on the red right key. For convenience, the latter 2 performances were recorded in succession on a single recorder.

The two VI rates on the different colors on the left key provide direct measures of the reinforcing effects of the 2 controlling stimuli on the right key. In an actual experiment, for example, the session might begin with the left key white and the right key unlighted. Responding on the white key would, after a variable interval, produce a red light on the right key (the white light immediately going out on the left key). The bird would then respond on the red key and be reinforced upon completing 50 responses. The red light would then go out, and the left key would be either white or green. In either case the bird would respond on the left key. If the key had become green, a response would, after a variable interval, produce blue on the right, and the green key

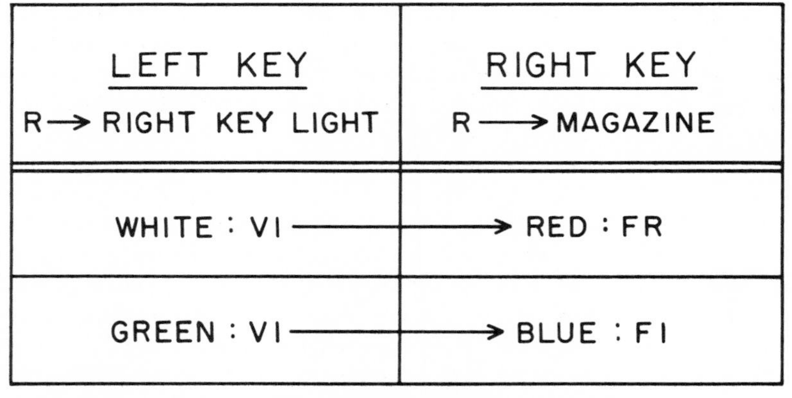

Fig. 889. Table of contingencies

would darken. Responding on the blue key would, after 10 minutes, produce a reinforcement. Figure 889 represents the contingencies.

The 2 birds used had had an extended history on FR. They were placed in a new apparatus and put directly on mult FRFI. The novel stimulus in the apparatus had, as usual, an effect similar to a performance immediately after crf. This was shown for 1 bird in Fig. 627, in Chapter Ten, where the development of the multiple schedule was described.

The 2 birds were then changed to the multiple-chain procedure just described. Figure 890 describes a well-developed performance under the mult chain VI 3 FI 10 chain VI 3 FR 50 after 12 sessions. Record A shows the multiple performance on the right key with interval scallops at *a, c,* and elsewhere and ratio performances at *b, d,* and elsewhere. Note that this is a relatively low ratio rate. The figure also shows vertical portions of the curve, as at *e, f,* and elsewhere, which represent responding to the unlighted right key. Since these always precede ratio performances, they indicate that when the bird is responding on the white key at the left (one such response producing the red key on the right), there is some tendency to move to the right key before the red light appears. This tendency corresponds with the higher rate shown on VI 3 on the white left key. Record B shows the separately recorded VI 3 performance on the white key, reinforced by the appearance of the red key on the right. A substantial slope is maintained throughout the session. Record C, on the other hand, shows the very low rate on the left green key when a response is reinforced occasionally by the production of the blue key on the right. (This part of the multiple schedule is unduly long, because the periods of no responding lengthened the intervals appearing on the VI 3 schedule. The segment shown is only part of the session, and was followed by 2 hours in which only 70 responses occurred.) Roughly speaking, the bird responds

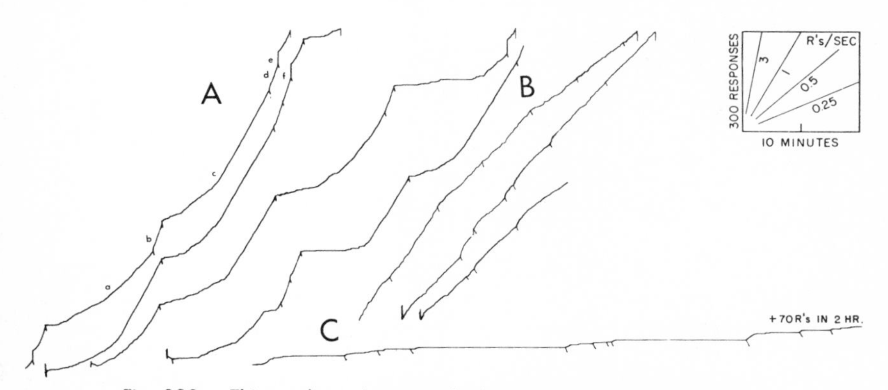

Fig. 890. Thirteenth session on mult chain VI 3 FI 10 chain VI 3 FR 50

at a substantial rate to produce the stimulus for FR 50 but scarcely at all to produce the stimulus for FI 10.

We tested the result by reversing the reinforcing conditions on the left key: where a response to the white key had previously produced a red key on the right, it now produces a blue; and where a response to the green key on the left had previously produced blue on the right, it now produces red. The effect is almost immediate.

Figure 891A, before the arrow, shows responding on the green (pre-FI) key. Record B, before the arrow, indicates responding on the white (pre-FR) key. The rates before the arrows are of the same order as those in Fig. 890B and C. Record C gives the performance on the right key of both colors on one record. At the arrows the contingencies were reversed. Record A now shows responding on the white key and Record B on the green. The performance at Record A declines slowly during the session; that on Record B rises somewhat more rapidly to a substantial rate. A complete reversal of rate on the colors on the left key resulted from the reversal of the reinforcing contingencies in the change to the right key. (The numbers set opposite the segments indicate the actual recorder order of segments in Fig. 891 for part of the session.)

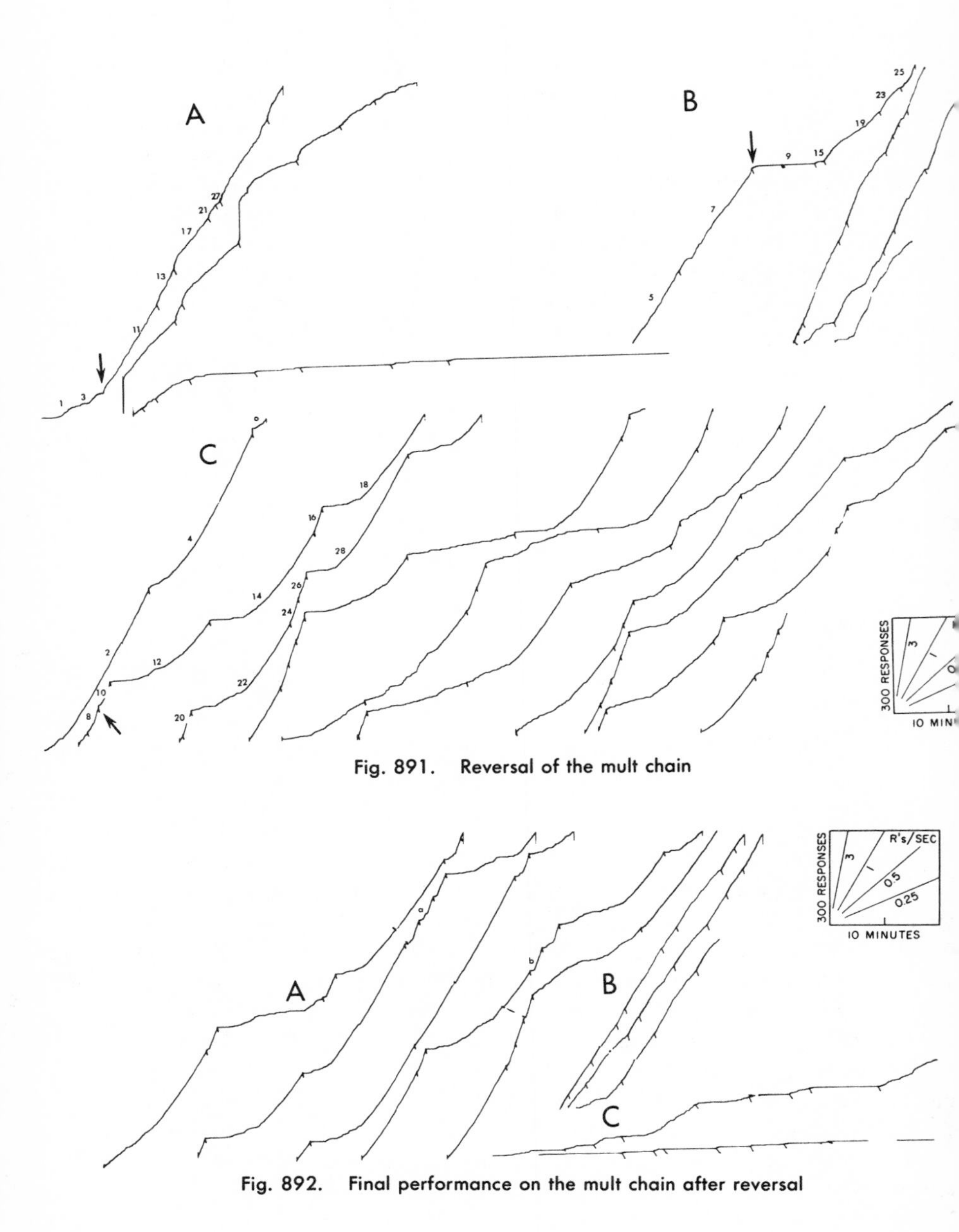

Fig. 891. Reversal of the mult chain

Fig. 892. Final performance on the mult chain after reversal

On the following day the rate on the white key showed some recovery from the low value at the end of Record A, but declined quickly to a very low value for the balance of the session. The rate on the other color of the left key remained substantial.

Figure 892 shows a final performance on the reversed contingencies. The VI performance on the pre-ratio color appears at Record B, and on the pre-interval color at Record C. Record A contains part of the multiple performance on the right key. The rates on all colors are slightly higher than those in Fig. 890. The ratios in Record A now show pauses, as at *b*, and small knees, as at *a*. Note that the tendency to respond on the right key before it is lighted has disappeared.

The length of the fixed interval was then reduced slowly until the performance was practically indistinguishable from that on fixed ratio on the multiple schedule. The actual reinforcement with food is, of course, brought closer to the presentation of the appropriate controlling key-color. The rate immediately after the appearance

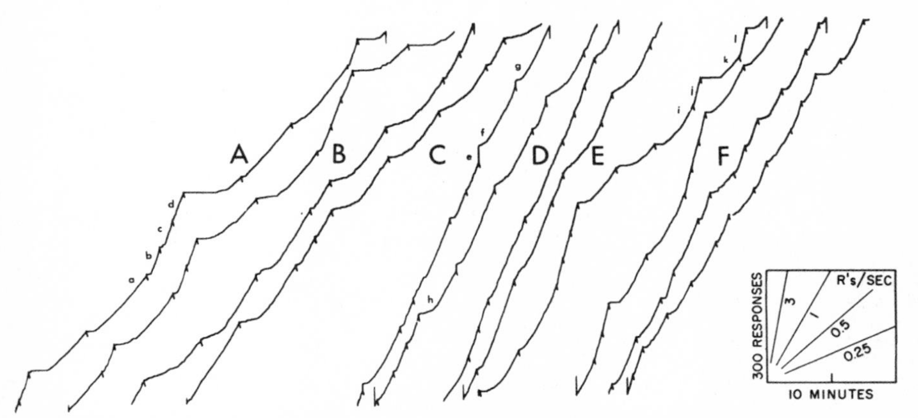

Fig. 893. Changes in the size of the FI in the mult FIFR component of the multiple chain

of the key-color is also changed. Both of these may affect the reinforcing properties of the colors. Figures 893 and 894 indicate changes in the multiple schedule through these various stages. Figure 893A shows a sample of the performance when the fixed interval is first reduced to 5 minutes, the fixed ratio remaining 50. (The session follows immediately after Fig. 892.) Partly because of the first shortening of the interval, the terminal rate in the interval is reduced relative to the ratio rate in the multiple schedule; the rate at *a,* for example, is much lower than that at *b, c,* and *d.* In Record B (3 sessions later) the interval was further shortened to 4 minutes. This change appears to have an effect upon the ratio, and the terminal rates are scarcely different at this stage. The interval was then further decreased to 3 minutes; Record C shows a sample of a performance from the 3rd session. Here, for the most part, the terminal rates are indistinguishable from the ratio rates, although some pausing and curvature are still present in the early part of the interval—for example, at *f, g,* and *h.* The vertical line at *e* and elsewhere shows responding on the key before the appearance of the color controlling the *interval* schedule. The earlier instances of this when the FI was 10 minutes had preceded the ratio schedule only. Record D, part of the 5th session on FI 3, shows somewhat more marked scalloping. Record E gives the 6th session on FI 3. Here, marked scallops have returned, and the terminal rates are again lower than the ratio rates. (The rates at *i* and *k* are substantially lower than those at *j* and *l*.) When the FI is reduced to 2 minutes, distinguishing between the interval and ratio performances becomes difficult, as Record F shows.

Records A and B in Fig. 894 show the 1st and 17th sessions on FI 1. In Record B no distinguishable difference exists between interval and ratio performances. Later

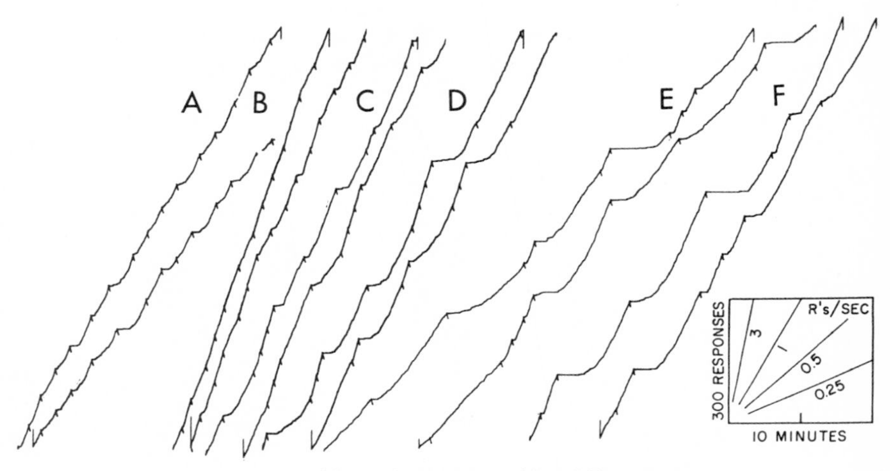

Fig. 894. Continuation of Fig. 893

sets of values were as follows: FI 2 FR 50 (Record C); FI 3 FR 50 (Record D); FI 5 FR 50 (Record E); and FI 5 FR 100 (Record F). The FR 100 performance in Record F is inferior to that at FR 75 in Fig. 893A.

Concurrently with these changes in the multiple schedule, changes appeared in the performances on the left key. Figure 895 shows various conditions on the color on which a response was reinforced with the presentation of the blue (interval) key on the right side. (Note that these are not in the order in which they occurred.) Record I is a sample of an early performance shortly after Record C in Fig. 892. The reinforcing schedule is FI 5. Record H shows an example of a break-through (at *a*) at a high rate produced by the drop to FI 3 as the reinforcing schedule. In Records G and F the reinforcing schedule was FI 5. When the interval was reduced to 1 minute, the rate on the pre-interval key rose. Record E shows the performances shortly after Fig. 894A, with the FI set at 1. A later performance under FI 1 appears at A at a stage between Curves A and B in Fig. 894. Record C, with FI 3, follows Record D,

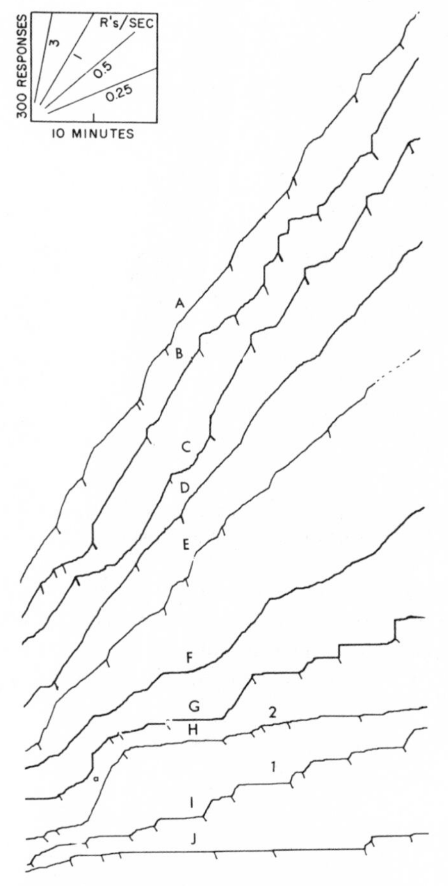

Fig. 895. Pre-interval VI performance during changes in the size of the FI in the mult FIFR component of the multiple chain

with FI 1. Record B shows a somewhat later stage, falling between Records D and E in Fig. 894.

The performance on the pre-ratio key, meanwhile, also undergoes some change because of the modification in the size of the interval. Figure 896 gives samples. Record A corresponds to Fig. 892B. Record B shows the first sign of a disturbance on FI 2; this is for the 2nd session before F in Fig. 893. Further irregularity develops in Record C on FI 1 for the 4th session of this schedule; and in Record D, for the 8th. Early performances recover when the interval is again increased to 3 minutes, as Record E indicates. Later, with the interval holding at 5, a slight tendency for nega-

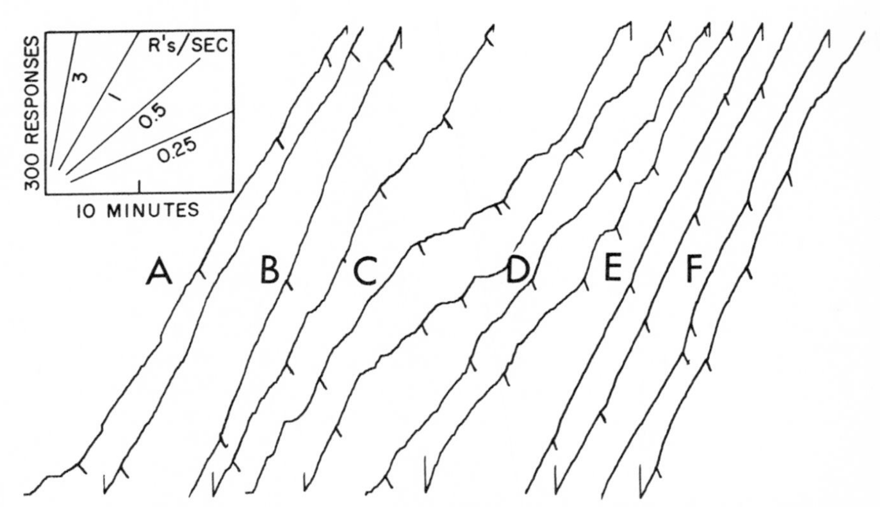

Fig. 896. Pre-ratio performance during changes in the size of the FI in the mult FIFR component of the multiple chain

tive curvature develops and becomes characteristic of the performance. This appears in Record F for the 15th session after the return to FI 5, and corresponds to a session a few days before F in Fig. 894. This curvature may be nothing more than the eventual adjustment to the VI schedule on the pre-ratio (and pre-interval) key.

Chapter Thirteen

• • •

CONCURRENT SCHEDULES

CONCURRENT SCHEDULES ON TWO KEYS

IN SOME EXPLORATORY EXPERIMENTS, pigeons were given access to 2 keys on the same wall of the experimental chamber. A single magazine could be arranged to reinforce a response to either key. A separate programming circuit controlled reinforcement on each key. Separate recorders were also used.

Concurrent VIVI

Figure 897 shows concurrent performances under VI on the 2 keys. A slightly higher rate appears on the right key in Record B. Otherwise, the performances are characteristic of VI.

Figure 898A and B show similar performances for the same bird on a later session. A slightly higher rate now appears on the left key at Record A. Record C shows a

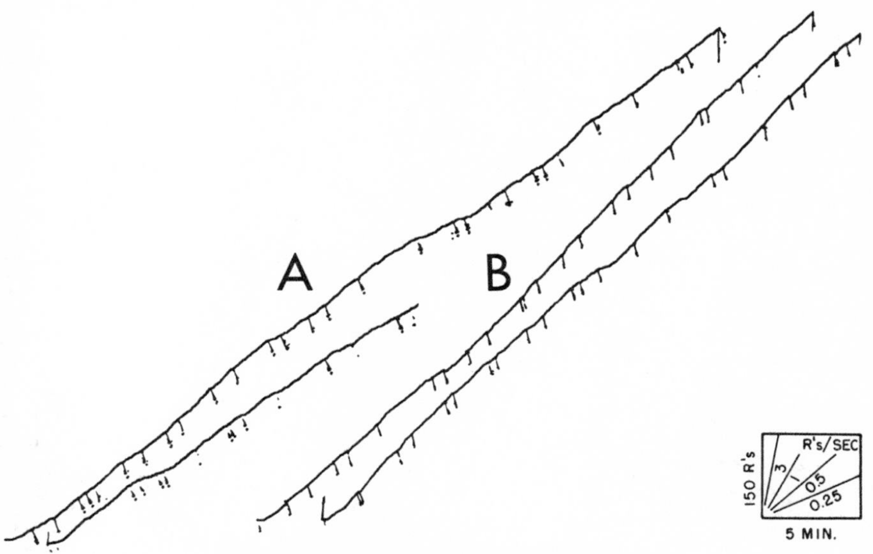

Fig. 897. Concurrent VI 1 VI 1 after 9 sessions

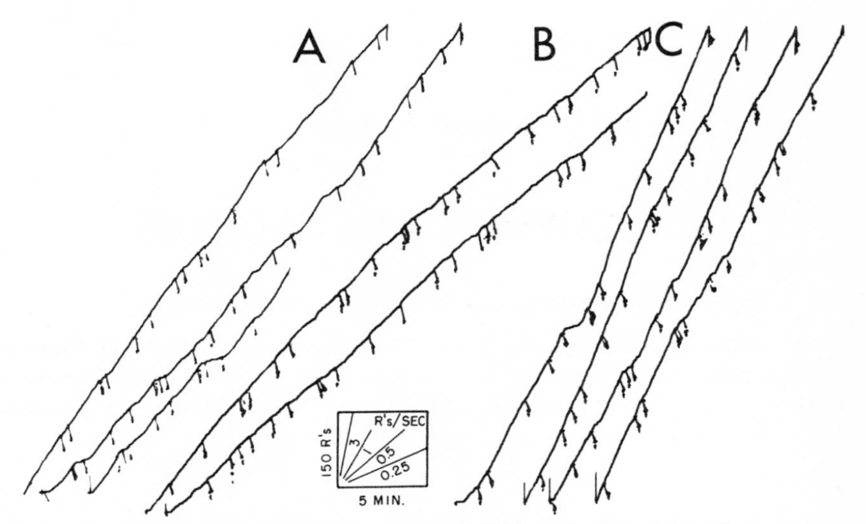

Fig. 898. Transition from concurrent VI 1 VI 1 to VI 1 on a single key

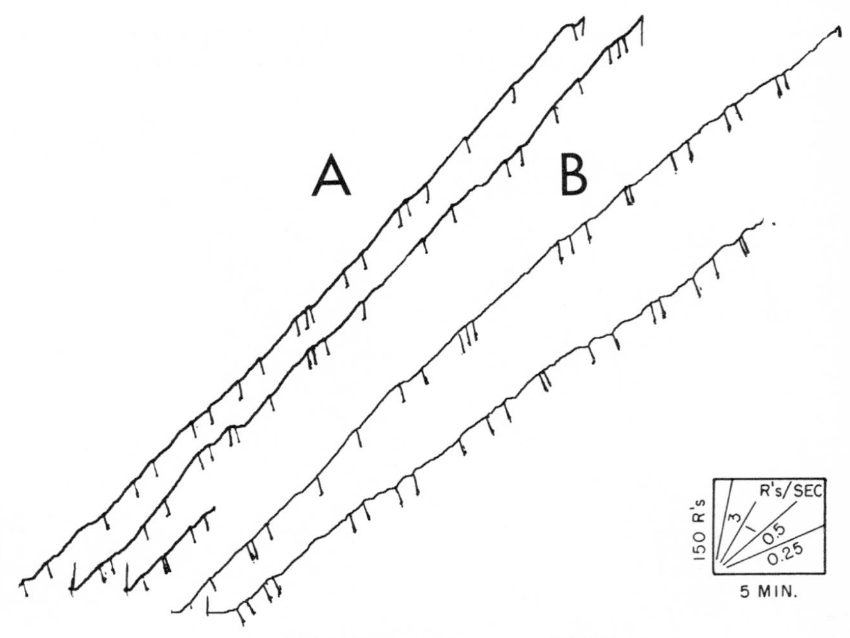

Fig. 899. Return to concurrent VI 1 VI 1

record taken on the following session, when the left key was covered. Reinforcements on the right key were programmed by both circuits and, of course, occurred twice as frequently as before. The over-all rate is approximately twice that shown on the same key for the preceding session.

The experiment was continued with the left key covered for 11 sessions. When both keys were made available at the same time, the performances in Fig. 898A and B were recovered, as Fig. 899 shows.

Concurrent FRFI

In a further experiment the schedule on the left key was FR 125; and that on the right, an FI 8. The value of the ratio was varied from time to time. Figure 900 gives an early example of the performances. Record A shows the performance on FI 8 on

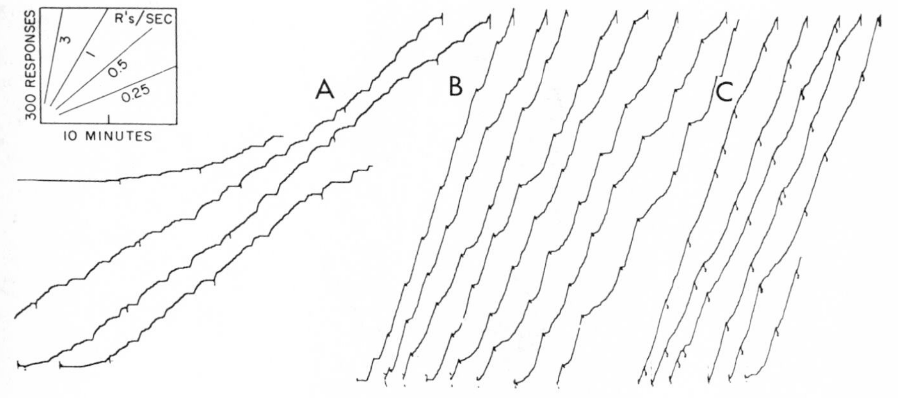

Fig. 900. Concurrent FR 125 FI 8

the right key, Record B the performance on FR 125 on the left key. No scalloping appropriate to the interval has developed, and the negative bursts of responding are the inversion of the positive accelerations on the ratio performance in Record B. In other words, the bird responds on the interval schedule when not responding on the ratio, but the ratio performance dominates. This performance is generally successful in achieving the interval reinforcement when it is due, and permits the animal to execute a fair FR 125 performance. The competing key seems to emphasize a smooth acceleration on the ratio schedule. Record C shows a recording of all responses on both keys. Only the last 5 excursions in the session are shown. The over-all performance is roughly linear, with occasional scalloping, much of which is probably due to the ratio schedule. If so, responding on the interval key does not make up for the pauses after reinforcement on the fixed-ratio schedule. Complete compensation is not to be expected, however, since the interval contingency on the right key causes a lower rate than the ratio on the left.

Figure 901 illustrates a much later performance with reduced FR on the left key against FI 5 on the right key. Here, the ratio performance at B is much squarer, and shows a fairly normal tendency to fluctuate between periods of long and short pauses. The behavior on the interval key varies inversely. When the rate in Record B is high in the region of *a*, it is low in Record A at *b*. When the rate is low at *c* in Record B, it is high at *d* in Record A. After the rate returns again to a reasonably high value in Record B at *e*, it falls sharply to a low over-all value at *f*. The performance is successful in achieving the interval reinforcements almost as soon as scheduled, and permits the ratio performance to proceed practically without interference from the other

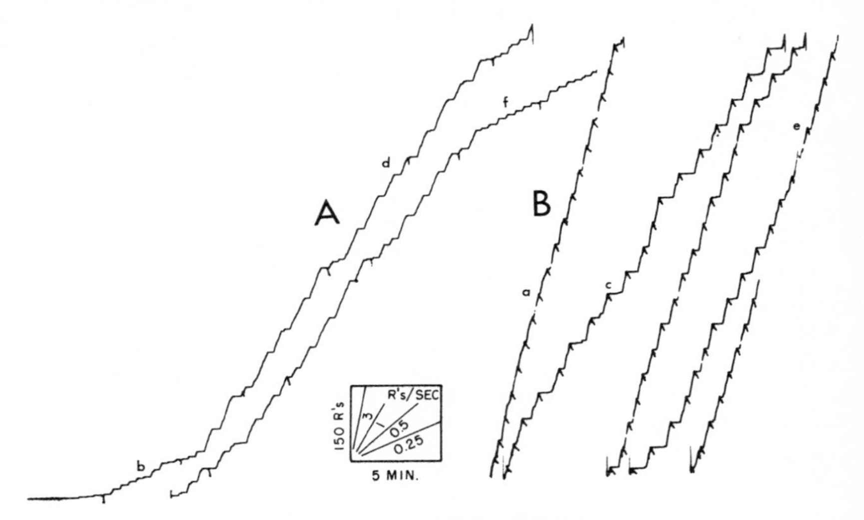

Fig. 901. Concurrent FR 50 FI 5

key. The finer step-wise grain of the interval curve is also, of course, the reciprocal of the ratio curve.

A second bird in the transition to FR on one key and FI on the other showed, at an early date, a good ratio performance on FR 35 and an FI performance illustrating the general features of Fig. 900 and 901. Responding on the interval occurred during pauses on the more powerful ratio schedule.

When the ratio was strained, however, by greatly increasing it, the ratio character of the record was lost. Some evidence exists of a temporary tendency for the FI schedule to affect both keys, with higher rates on the FR key during the scallop of the short FI. In Fig. 902A, for example (for the second bird in the experiment), fairly good scalloping occurs on the FI 2 key. Responding on the larger ratio key (the ratio is now 245) occurs when the bird is not responding on the interval key (B). When the ratio was increased to 385 (Fig. 902C and D), the record occasionally showed examples of smooth, inverted interval curves on the ratio key, as at *a*. (In this figure the left-

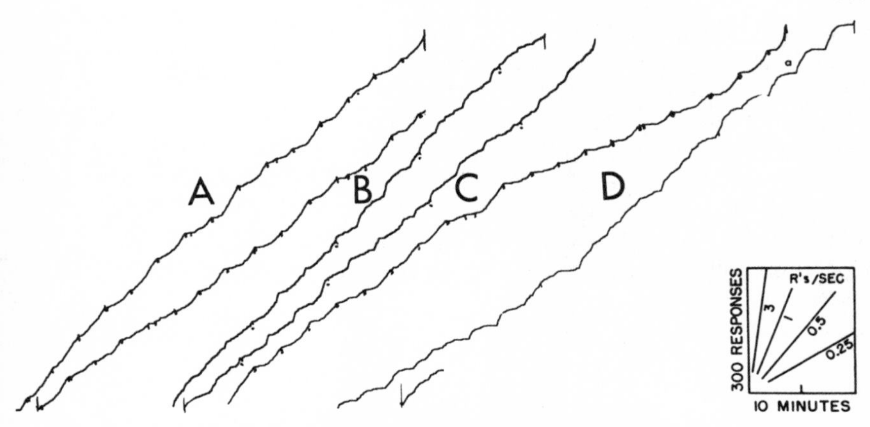

Fig. 902. Concurrent FI 2 FR 245 and FI 2 FR 385 (second bird)

key reinforcements are also marked on the right key. The interval schedule permits the reader to determine which of these were on the right key.)

This bird never showed a good step-wise break on a 5-minute FI except when the ratio on the other key was quite small—FR 25. By varying the size of the ratio, however, we could reproduce these general features of concurrent FIFR on 2 keys. Figure 903 shows a late performance with FR 75 on the left key and FI 5 on the right key. The record on the left key has been ıeproduced in the recorded form. The 3 segments are continuous. Segments from the record on the right key have been cut and reassembled in order to indicate simultaneous features of the 2 curves. At *a* a reinforcement on the interval schedule produces a pause on both keys, at *b* and *c*. In general, the pauses in the FI curves correspond to periods of high rates of responding on the ratio curve. Instances have been marked by the connecting lines. Thus, re-

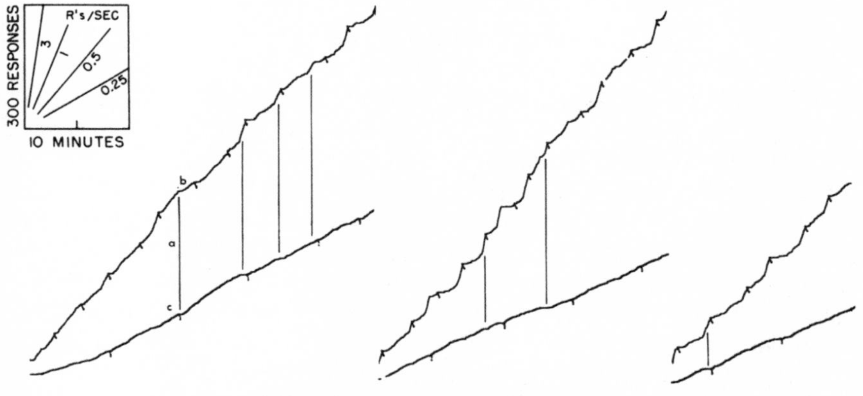

Fig. 903. Concurrent FI 5 FR 75 (second bird)

sponding on the interval curve (the lower curve) occurs when the ratio schedule is not commanding a high rate. Note that this bird does not develop a good terminal rate on the ratio comparable with that in Fig. 901.

Figure 904 illustrates a late development of a similar performance for a third bird in the same experiment, with FR 150 on the left key and FI 5 on the right. The FR performance in the upper curve is almost standard for that schedule. Substantial responding occurs on the other key during the pauses in the ratio performance, particularly as these develop during the experimental session. Where the FR schedule shows only small pauses at *a*, the rate on the FI key is negligible, as at *b*. Where the

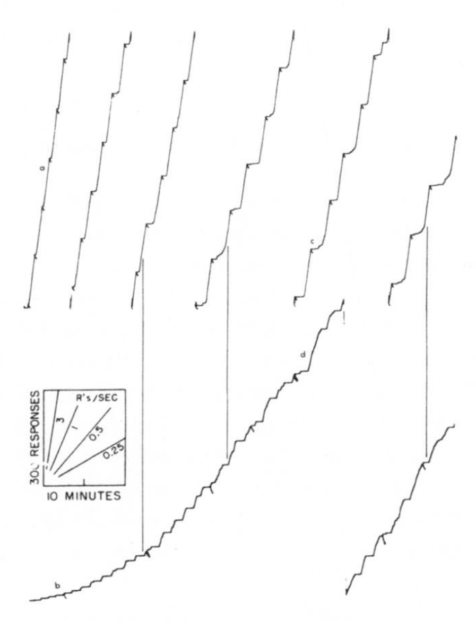

Fig. 904. Concurrent FI 5 FR 150

pauses have become prolonged in the latter part of the period, as at *c*, very sustained responding occurs on the FI key at *d*. A few corresponding runs and pauses in the 2 records have been connected by lines.

A fourth bird was put on several settings of the 2 schedules without much effect. It tended to show performances appropriate to a variable schedule on both keys. Late in the experiment, however, well-marked scallops began to appear on FI 2. On a large ratio of 385 concurrently on the other schedule, the performance closely resembled a VI schedule except that some trace of the curvature in the 2-minute scalloping appeared in the reverse form, as in Fig. 902C and D. Later sessions showed an even greater similarity to the behavior of the pigeon in the figure.

Concurrent VIFI

Two birds on a concurrent VI 2 FI 2 on 2 keys failed to develop marked scalloping on FI, although some traces are clear in the later stages in the experiment. In Fig. 905, sighting along the lower record will reveal scalloping on FI 2. However, scallops in progress are often broken shortly before reinforcement by the occurrence of reinforcements on the other key. Lines have been drawn to connect 4 such instances. Under these conditions the bird responds approximately with a pattern of 2 responses on the left key, 1 on the right, 2 on the left, 1 on the right, etc. In general, pauses oc-

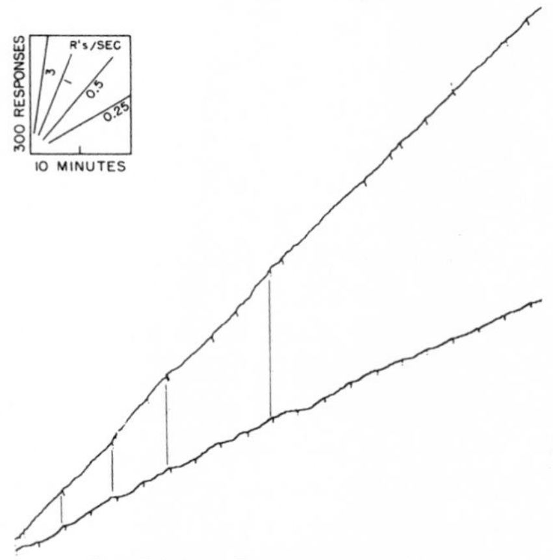

Fig. 905. Concurrent VI 2 FI 2

cur on the right key (lower record) after reinforcement, but slight runs occur on the VI key after reinforcement. The performance on the VI key is essentially linear, and the bird does not appear to show any tendency to respond more often on that key while periods of low responding are in effect on the FI key. This performance is therefore not comparable with that in which an FR schedule on one key controls rapid responding on the other during the FI pause.

CONCURRENT SCHEDULES ON A SINGLE KEY

Concurrent FI 10 avoid RS 30 SS 30

Three rats with a history of mult FI 10 avoid RS 30 SS 30 were reinforced on FI 10 alone for 34 hours. The general type of performance seen in Fig. 906 for a full session appeared. In general, good pausing and a high terminal rate occur. An occasional period of low responding appears and, at *a,* one instance of postponing the reinforcement beyond the 10-minute interval. The avoidance (RS 30 SS 30) schedule was reinstated in the following session, the stimulus situation being the same as on FI 10

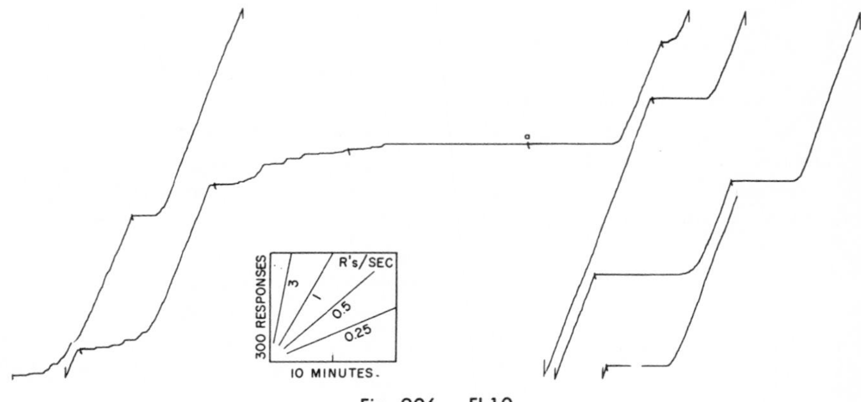

Fig. 906. FI 10

in the earlier history. Two schedules—FI 10 and avoidance—were now being programmed at the same time but independently. Figure 907 gives an early stage of the resulting performance. A few shocks reinstate the fully developed avoidance behavior, although they now occur in the presence of the stimulus which had appeared only during the FI 10. A typical avoidance performance at a fairly low, steady rate replaces the long pauses at the beginning of the intervals. This rate may have some effect in encouraging "knees," but these tended to disappear in later sessions. The terminal rate remains approximately the same. A final performance under this schedule appears in Fig. 908, where the whole session is represented. The sustained avoidance rate is pointed up in a few cases by lines which estimate the slope. Occasional knees appear. During the session the rat received 22 shocks for failing to maintain the basic avoidance rate.

When the avoidance contingency was changed to RS 15 SS 15, the slope in the early part of the interval remained about the same. Figure 909 gives the 3rd session on this schedule. The interval performance is somewhat more orderly, possibly because

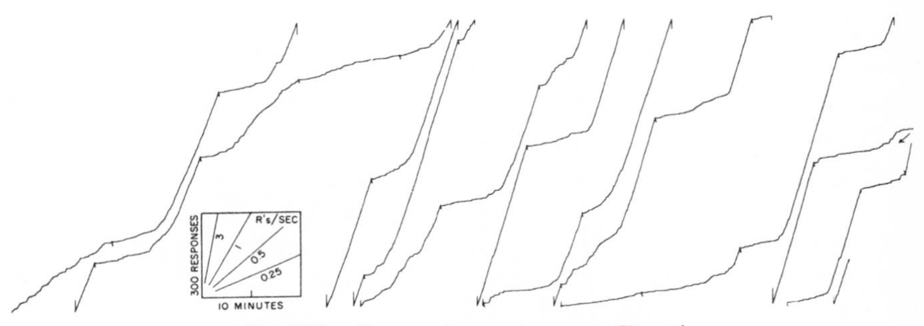

Fig. 907. First session on concurrent Flavoid

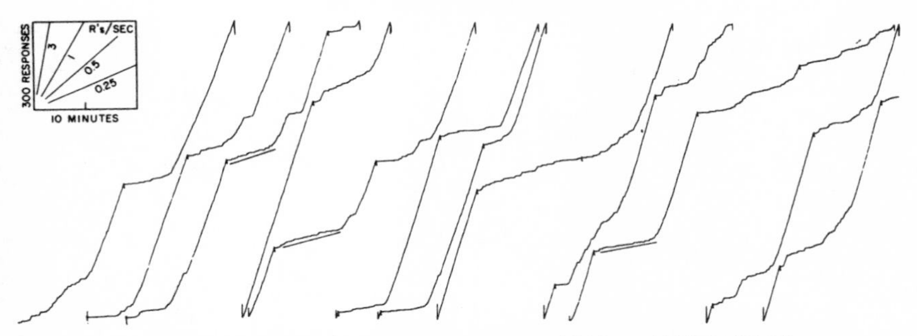

Fig. 908. Final performance on concurrent FI 10 avoid RS 30 SS 30

of longer exposure to the schedule, as well as adaptation of emotional responses to shock. In the session following Fig. 909 the avoidance contingency was changed further to RS 7 SS 7. This change was made at the arrow in Fig. 910. The effect is an immediate increase in the basic avoidance rate at *a* which is maintained throughout the interval. Acceleration to the usual terminal rate is missing. Acceleration is lacking elsewhere, as at *b* and *c*. At *d* a late acceleration reaches something less than the usual terminal rate before reinforcement. This development may be due to conditioned suppressing properties of the shocks, which are now received more frequently. Later, interval segments uniformly ended at a high terminal rate. (See the beginning of Fig. 912.)

The suppression of the scallop appropriate to FI 10 noticeable in some instances in Fig. 910 was much more marked in a second rat. Here, the change to the RS 7 SS 7 contingency was made at the beginning of the experimental session. Figure 911A gives a sample of the preceding day on RS 15 SS 15. The complete following session at RS 7 SS 7 appears in Record B. An acceleration does not break through until very late in the session, at *a,* near the end of the interval. Thereafter, some suggestion of interval curvature appears for the balance of the period. This rat showed similar

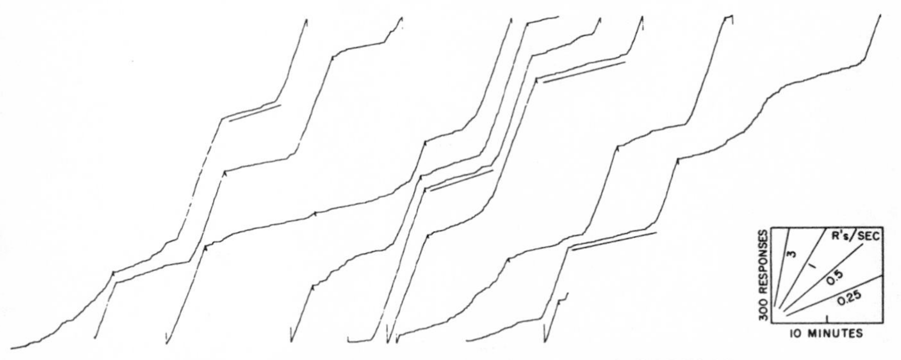

Fig. 909. Third session on concurrent FI 10 avoid RS 15 SS 15

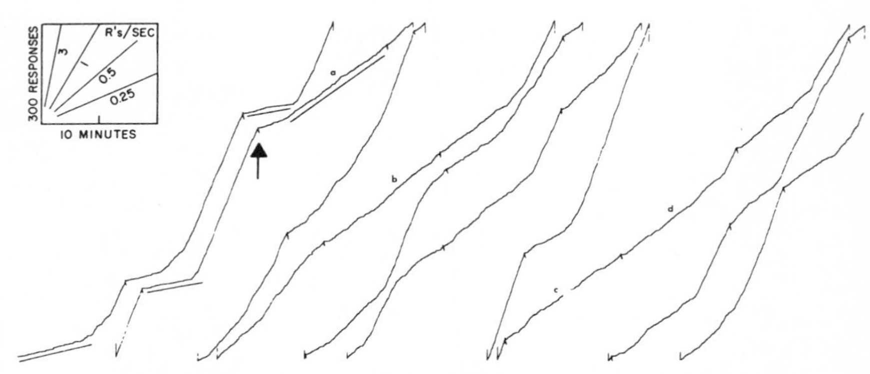

Fig. 910. Transition from concurrent FI 10 avoid RS 15 SS 15 to concurrent FI 10 avoid RS 7 SS 7

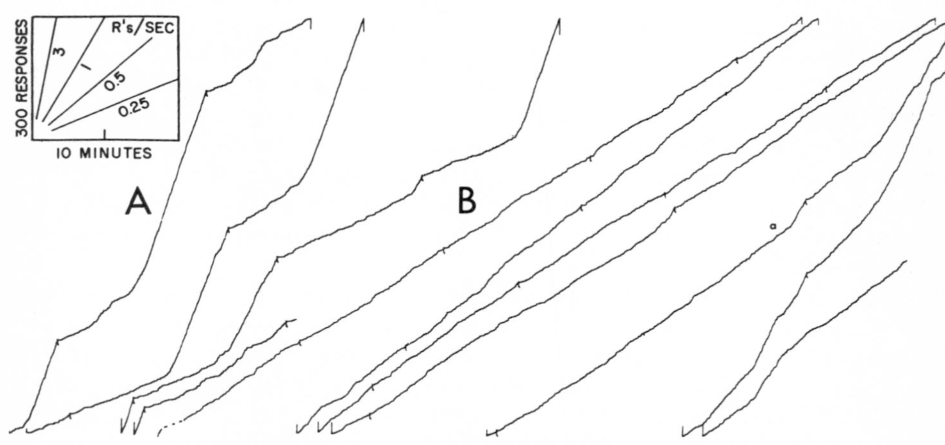

Fig. 911. Transition from concurrent FI 10 avoid RS 15 SS 15 to concurrent FI 10 avoid RS 7 SS 7

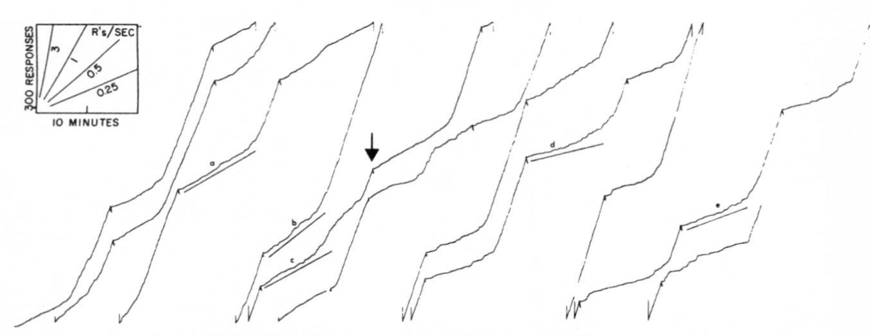

Fig. 912. Transition from concurrent FI 10 avoid RS 7 SS 7 to concurrent FI 10 avoid RS 15 SS 15

periods of no positive curvature during the 3 other sessions at which the schedule remained at RS 7 SS 7. Upon returning to RS 15 SS 15, the FI 10 behavior was restored.

When the contingency was restored from RS 7 SS 7 to RS 15 SS 15 for the first rat discussed above, at the arrow in Fig. 912, the rate decreased almost immediately in the early part of the interval. Compare the slopes at *a, b,* and *c,* for example, with those at *d* and *e.*

A multiple schedule in which one stimulus controls a concurrent FI 10 avoid RS 30 SS 30 while another controls FI 10

Two rats from the experiment on concurrent FI 10 avoid RS 30 SS 30 continued to be reinforced on that schedule in the presence of a buzzer. In the absence of the buzzer

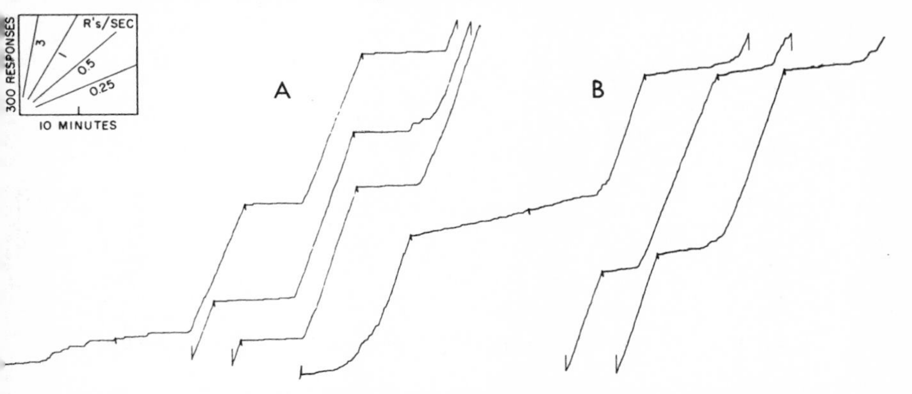

Fig. 913. Early performance on mult FI 10 concurrent FI 10 avoid RS 30 SS 30

they were reinforced on FI 10 alone. Figure 913, for the 5th session, shows an early development of a multiple performance. Responses on FI 10 appear in Record A, while those in the absence of the buzzer on the concurrent schedule appear in Record B. The 2 stimuli alternated in a simple fashion at reinforcement. The concurrent FI 10 avoid RS 30 SS 30 evidently has the effect previously described (at Record B), while the multiple controlled FI 10 is fairly normal (at Record A).

These characteristics are even more marked in the later development evident in Fig. 914, after 31 sessions on the multiple schedule. (A block of 5 of these sessions was devoted to extinction.) Note the long pauses under the stimulus controlling the FI performance in Record A, and the parallel segments of the typical avoidance rate at the beginning of each interval in Record B. In Fig. 913 the terminal rate in the concurrent case was somewhat higher than that in the straight FI. This difference is even more apparent in Fig. 914, where the terminal rates in Record A are of the order of 1.5 responses per second, but in Record B they are of the order of 2 responses per second. A slight roughness of grain is evident in the terminal straight FI rate.

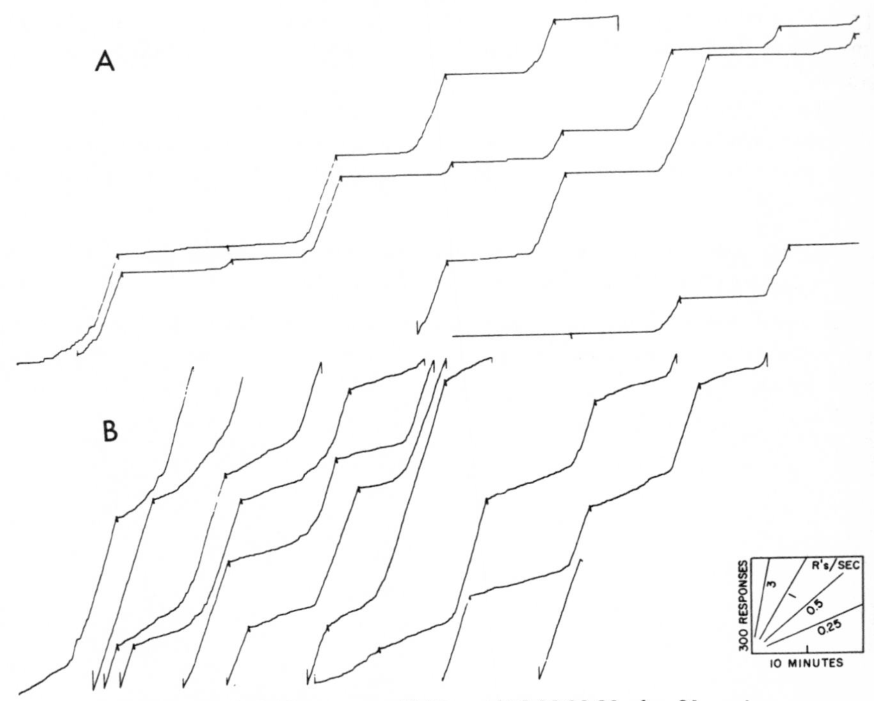

Fig. 914. Mult FI 10 concurrent FI 10 avoid RS 30 SS 30 after 31 sessions

When the avoidance contingency was omitted so that both stimuli now controlled FI 10, the avoidance performance slowly disappeared under the stimulus previously controlling it. Figure 915 gives a sample of the performance after 15 sessions. Record A shows the performance with the buzzer off on FI 10; and Record B, the alternating performance with the buzzer on at FI 10. If the shocking circuit had been connected during Record B, the rat would have received approximately 194 shocks during the period. In the earlier performance on the former multiple schedule, including the avoidance schedule, the rat received only 2 shocks in maintaining the basic avoidance rate during the session shown in Fig. 914.

Figure 916 is a plot of the shocks actually received by both rats on the concurrent schedule together with the number of shocks they would have received after the avoidance contingency was removed. The first 7 points show the number of shocks actually received on the concurrent schedule. The points thereafter are the number that would have been received if the shocking circuit had been connected. The number rises significantly only after 4 sessions, but eventually reaches a high value at the stage shown in Fig. 915. When the shock was again introduced, there was a very rapid return to the previous concurrent schedule, as the 3 last points in Fig. 916 indicate.

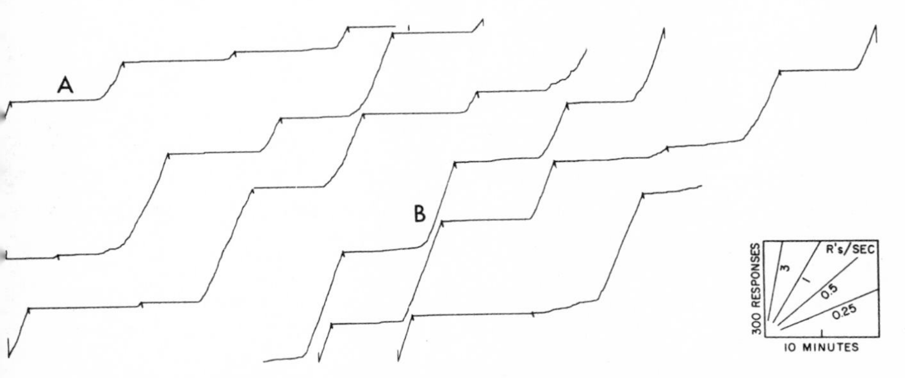

Fig. 915. Mult FI 10 FI 10 after mult FI 10 concurrent FI 10 avoid RS 30 SS 30

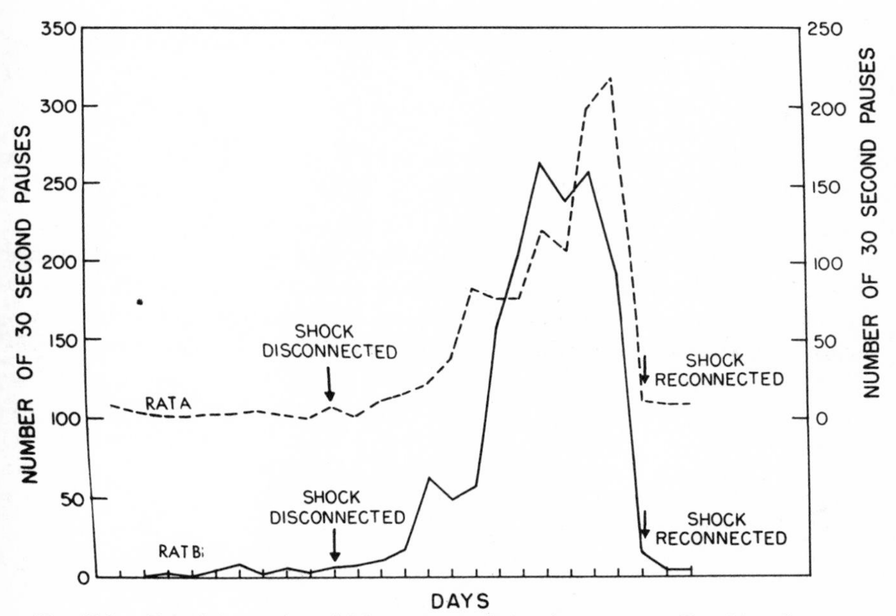

Fig. 916. Plot of the number of 30-sec pauses during the concurrent Flavoid performance

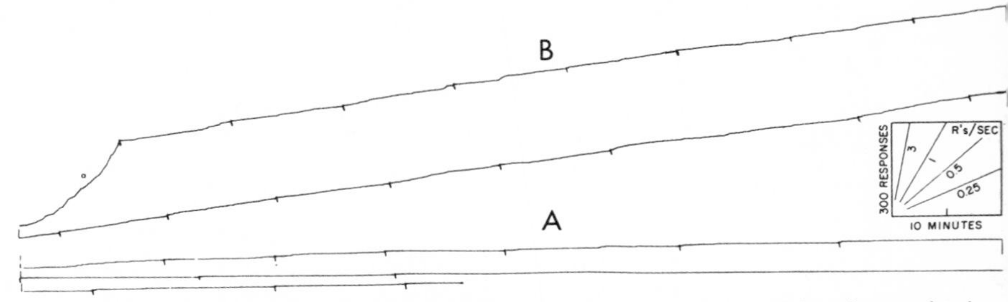

Fig. 917. Extinction after mult FI 10 concurrent FI 10 avoid RS 30 SS 30 by allowing food magazine to operate empty

The extinction mentioned in connection with Fig. 914 occurred about halfway between that figure and Fig. 913. The magazine feed jammed, so that the magazine operated with the usual auditory stimuli, but no food appeared in the trough. The result was the extinction of the interval scallop and the survival of the avoidance performance. Figure 917 gives the performance in the 5th session. One FI scallop at *a* in Record B shows surviving control of stimuli at start of the session.

The balance of Record B is characteristic of the avoidance schedule. Alternating segments under the other stimulus (Record A no buzzer) show only a low trickle of responses, and many "reinforcements" are postponed.

The effect of chlorpromazine and pentobarbital on a multiple FI 10 (concurrent FI 10 avoid RS 30 SS 30)

Figure 918A illustrates the performance of one rat on the schedule just described under the stimulus controlling the concurrent part of the schedule. (The performance on FI alone is not shown.) This is a performance similar to that of Fig. 914. In the following session 2.5 milligrams of chlorpromazine was injected intraperitoneally immediately before the start of the session. The effect was the elimination of prac-

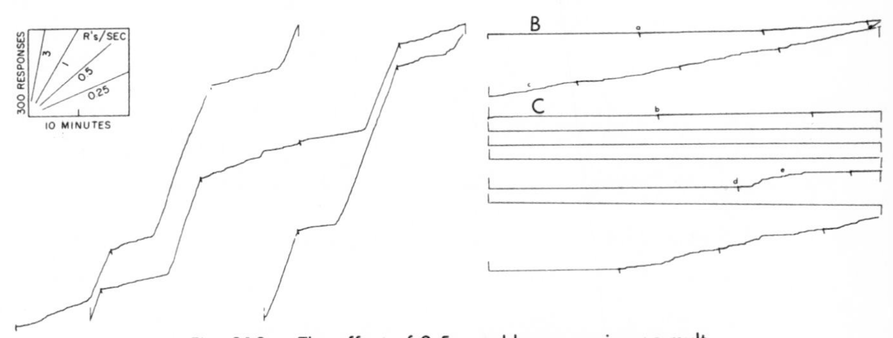

Fig. 918. The effect of 2.5 mg chlorpromazine on mult FI 10 concurrent FI 10 avoid RS 30 SS 30

tically all activity for some time. Record B in the figure is for responding under the concurrent FI avoidance component. Record C is the alternating performance on FI alone. Responses occur on both components sufficiently often to produce the alternation of stimuli. For example, at *a* in Record B a first response changes the stimulus to that appropriate to FI alone; and a somewhat delayed response at *b* in Record C changes the stimulus back. In each case a pellet is produced as reinforcement. The pellets were eaten, and observation of the rat showed it to be awake, alert, but otherwise not responding. Shocks were received on the avoidance schedule whenever the rate fell below the critical point, as it clearly did during the first 2 segments of Record B. After 3 or 4 intervals, however (that is, after about 40 minutes), the avoidance performance is reinstated, beginning in marked form approximately at *c*. This continues for the balance of the session. Very little responding occurred on the FI alone, however, one interval being extended to almost 2½ hours, when a response occurred (at *d*) and the schedule was reversed. When the schedule later returned to the FI stimulus, a trickle of responses emerged in the interval at *e*. In no case does the ordinary FI 10 performance under food reinforcement appear. Evidently the drug greatly

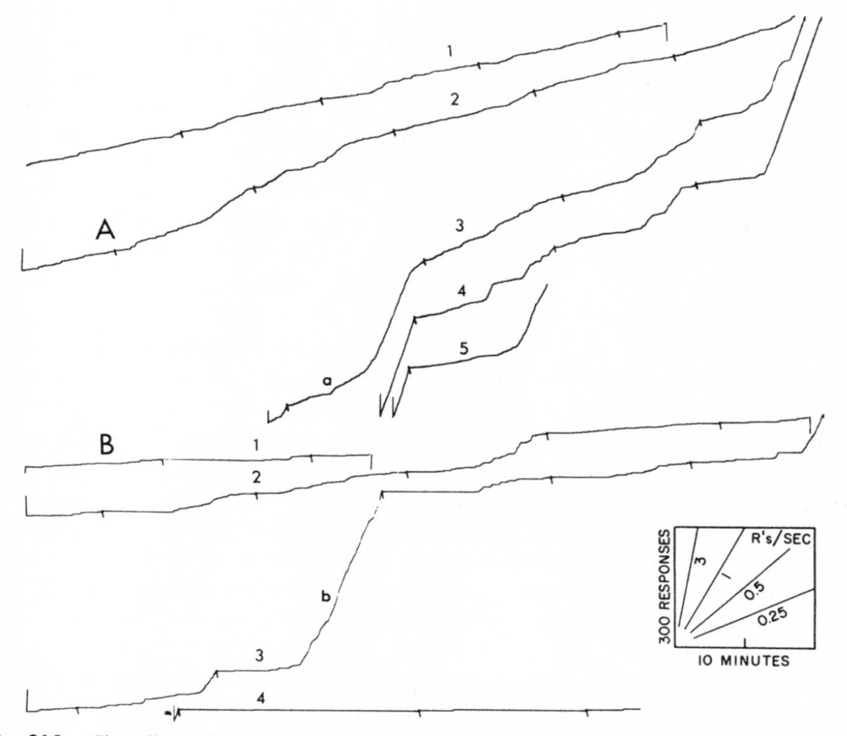

Fig. 919. The effect of 1 mg chlorpromazine on mult FI 10 concurrent FI 10 avoid RS 30 SS 30

reduces the behavior appropriate to the food reinforcement while leaving a substantial level of avoidance behavior.

In another session a smaller dose of the drug was given 1 milligram at the start of the session. Figure 919 illustrates the performances. In Record A, 10 intervals under the concurrent schedule show a rather rough avoidance performance with no evidence of the interval scallop; and at Record B the initial 9 intervals on FI 10 only also show a very low rate. A fairly standard interval scallop appears beginning with the avoidance rate at *a,* although the rate falls before reinforcement is received. On the FI 10 schedule a single interval performance emerges at *b.* The rate then declines. Several examples of interval curvature appear, however, in combination with the avoidance contingency on the concurrent schedule. In this experiment the avoidance contingency is essentially untouched, although the performance becomes somewhat more irregular under the administration of the drug. The FI 10 performance is removed for the first 3 or 4 hours, and makes only an irregular appearance thereafter. A 1-milligram dose of the same drug given to the second rat produced an even more marked effect in removing the interval scallop from the concurrent schedule. Only 3 slight traces of this appear in the 3rd hour of the session. On the concurrent FI schedule alone, 2 fairly substantial interval scallops appear near the end of the 3rd hour.

Later, an injection of 2 milligrams of sodium pentobarbital had little effect upon the performance on either component schedule. The rat had shown a sustained, fairly stable condition on FI 10, in which pauses were occurring after all reinforcements, but the durations of the pauses covered a fairly narrow range.

AN UNCLASSIFIED SCHEDULE

An unclassified complex schedule had the following properties. The bird was reinforced upon the completion of 160 responses as in a simple FR. If the bird paused for 2 minutes after reinforcement, however, it was reinforced for the 1st response. Thus, if the ratio began to strain so that the pause after reinforcement exceeded 2 minutes, reinforcement occurred on FR 1. This has the same effect as an adjusting schedule: straining is corrected by additional reinforcements, not (as in the adjusting case) by a slight change in the size of the ratio but by a temporary change in the mean ratio. Figure 920 illustrates a series of sessions on such a schedule. Record A shows a sample of the behavior under FR 160. The effect of reinforcing the 1st response if the pause after reinforcement exceeds 2 minutes is almost immediate. Unfortunately, the 1st record was poorly made, but Record B gives a part of the 2nd session. By this time a few reinforcements on crf drl 2 min have led to responding soon after reinforcement, although the terminal rate in the ratio is not immediately reached. A single instance in which a response is reinforced because of a pause greater than 2 minutes appears at *a.*

One of the first effects seems to be to encourage a low rate after reinforcement. Record C shows a segment of a session following Record B in which the final completion of the ratio is much disturbed, and a scattering of responses fills in the interval.

Note that a response after a pause of 2 minutes is reinforced only when it is the first response after reinforcement. Several responses after pauses greater than 2 minutes occur in Record C, but no single response is reinforced at this time. The following session shows recovery; and the spacing of reinforcements is now of the same order as those in Record A, but the pauses are now filled with a few scattered responses. Record E, a sample for the session following Record D, shows rather prolonged responding at a low rate. The first response after reinforcement always occurs too soon

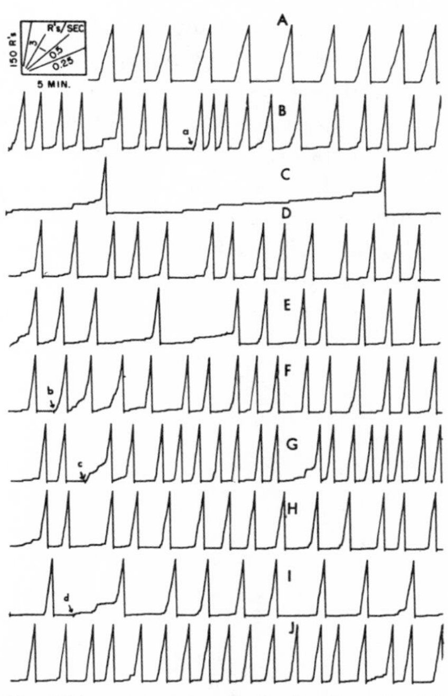

Fig. 920. An unclassified adjusting schedule

to be reinforced on FR 1. Record F is for the session following Record E. Another example of a single response reinforced after a pause of 2 minutes appears at *b*.

Another example, shown in Record G at *c*, is followed by a momentary intermediate rate of responding before the terminal rate is reached. In Record H, for the session following Record G, the ratios are spaced out roughly as in Record A, but no pause after reinforcement exceeds 2 minutes. Record I at *g* gives a single example of a reinforcement of 1 response. All other cases have at least one other response within the required 2 minutes. The final performance observed in this experiment is represented by Record J, where the ratios are separated by substantial periods of time, but a few responses follow soon after reinforcement and fail to satisfy the drl contingency included in the schedule.

This experiment appears to demonstrate a successful technique for preventing the ultimate extinction of a ratio which cannot otherwise be sustained by the organism.

ADJUSTING SCHEDULES

Organisms differ in the maximal size of ratio under which they will reach and sustain an FR performance. These differences are a function of deprivation level, health, general reactivity of the organism, and so on. The history through which the maximal ratio is approached is also important. We investigated a technique for determining the maximal ratio a bird can hold by approaching the maximal ratio with the benefit of a natural correction of the advance in size of ratio. This procedure was designed to avoid the usual arbitrary program of advancing the ratio. The resulting program is an example of a general type which may be called an "adjusting" schedule.

The special case of an "adjusting" ratio may be defined as a schedule in which the number of responses emitted before reinforcement is changed progressively in terms of some characteristic of the behavior of the bird during the preceding ratio. The bird begins at some value of the ratio which it can conveniently hold. This value is then

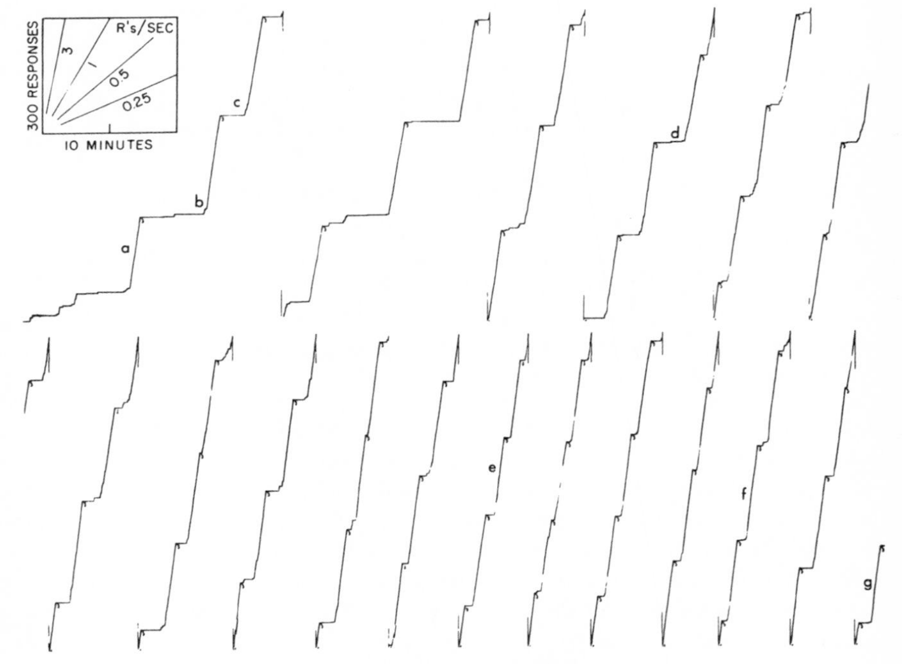

Fig. 921. Adjusting FR

increased in terms of its performance. If any straining appears, the size of the ratio is automatically decreased. The basis for increasing or decreasing the ratio chosen in this experiment was the pause after the reinforcement. Other possibilities might be the total time to complete the ratio, the time to complete some fraction of the ratio, and so on. Note that the adjustment made at each ratio affects other ratios. Adjusting schedules therefore differ sharply from interlocking schedules, when a standard set of specifications is resumed at each reinforcement.

We studied 3 birds on an adjusting ratio schedule for approximately 300 sessions; during this time, we varied the criteria for changing the ratio and made incidental variations in the deprivation level. With the techniques and criteria employed, the maximal ratios sustained by the 3 birds without prolonged pausing were approximately 445, 600, and 650.

Figure 921 gives an example of the way in which the adjustment permits the bird to set its own ratio appropriate to the conditions of deprivation, health, etc. The criterion of adjustment at this time is a 25-second pause. During the first 25 seconds the ratio decreases slowly unless the bird responds. If a response occurs, the ratio is increased by 5 responses. The session begins with a badly strained ratio at *a*, *b*, and *c*. At this point the ratio has the value of approximately 270. Because of the breaking, the ratio was automatically adjusted downward, and reaches about 235 at *d*. Still some pausing and curvature appear at this point, and a further reduction follows. By *e* the ratio has been reduced to about 200, and here the performance holds well. A slight pause results; but it is fairly stable, and the adjustment is therefore reversed. The ratio is now slightly increased at each reinforcement. By *f* it has returned to about 250 responses, some pausing reappears, and the ratio is adjusted downward, approaching 200 by *g*.

The experiment shows that the pause after reinforcement is an important property of the ratio performance. The adjusting ratio has some of the self-corrective characteristics of interval schedules, although the criteria must lie within a fairly narrow range to protect the behavior from extreme strain.

Glossary

A

accidental chaining A process in which a response which frequently precedes a reinforced response shares in the effect of the reinforcement in such a way that the whole sequence becomes a stable part of the organism's behavior. A form of superstitious behavior (q.v.).

accidental (or incidental or spurious) reinforcement A coincidence of a response and a reinforcing event (e.g., in certain programs designed to establish a discrimination, the appearance of the discriminative stimulus may coincide with a response in its absence).

adaptation (1) As operation: exposing an organism to a stimulus. (2) As process: a change in the extent of the reaction of the organism to the stimulus.

ad lib body-weight The weight approached or reached by a mature organism under continuous access to food.

ad-lib feeding Providing continuous access to food.

adjusting schedule A form of schedule in which a value (e.g., of interval or ratio) is changed in some systematic way from reinforcement to reinforcement as a function of the performance (e.g., a fixed ratio is adjusted after each reinforcement according to some measured aspect of the performance in the preceding session, such as the length of the pause before the first response).

alternative schedule A response is reinforced on whichever of two schedules is satisfied first (e.g., after 5 minutes has elapsed or after 50 responses have been emitted, whichever occurs first).

anxiety, see **conditioned suppression.**

aversive stimulus A stimulus, the removal of which is reinforcing, or which may produce a low rate in the presence of a stimulus which frequently precedes it (which thus becomes a conditioned aversive stimulus).

avoidance behavior Behavior which postpones an aversive event and thus provides escape from conditioned aversive stimuli. Avoid RS 10 SS 10 is a notation for the avoidance procedure of Sidman; the numbers represent seconds, and RS stands for response-shock interval (q.v.) and SS for shock-shock interval (q.v.).

B

behavior (1) Broadly speaking, any activity of the organism; more particularly any activity which changes the position of the organism or any part thereof in space. The dependent variable in a science of behavior. (2) The events of (1) as they affect the organism as stimuli (traditionally called proprioceptive stimulation, feed-back, etc.)

bite A deviation from a smooth curve consisting of a period of relatively slow responding followed more or less abruptly by a compensatory increase in rate which restores the curve to the extrapolation of the earlier portion. Opposite of knee (q.v.).

body-weight (in control of level of deprivation) A collateral effect of a schedule of food deprivation used as a check upon such a schedule. It is usually measured immediately prior to an experimental session. A fairly constant body-weight from day to day is maintained by feeding the organism up to a given weight after completion of each session.

box An experimental chamber containing the organism during an experiment, usually with some degree of sound and light shielding, and containing one or more manipulanda, reinforcing devices, and manipulable stimuli.

C

chained responses A sequence of responses in which one response produces conditions essential to the next, as in making the next response possible or more likely to be reinforced. Successive responses may or may not have the same topography.

chained schedules (chain) A schedule in which responding under one stimulus on a given schedule is reinforced by the production of a second stimulus in the presence of which a response is reinforced on a second schedule with food, water, etc. Resembles a multiple schedule (q.v.) except that the reinforcement of the first component is simply the production of the stimulus of the second component.

clock A stimulus some dimension of which varies systematically with time, usually measured from the preceding reinforcement but possibly from some other point. (E.g., FI + clock means reinforcement on FI in the presence of such a stimulus.)

compensatory rate A higher than normal rate following one lower than normal, tending to restore the over-all rate to an earlier value. Also a lower than normal rate following a higher than normal one with the same result.

concurrent operants Two or more responses, of different topography at least with respect to locus, capable of being executed with little mutual interference at the same time or in rapid alternation, under the control of separate programming devices (e.g., responses to two keys present at the same time under separate schedules).

concurrent schedules (conc) Two or more schedules independently arranged but operating at the same time, reinforcements being set up by both.

conditioned reinforcer A stimulus having the effect of a reinforcer because of its relation to a stimulus already having that effect.

conditioned suppression A reduction in rate in the presence of a previously neutral stimulus which has characteristically preceded an aversive event. (E.g., a 3-minute tone which has repeatedly been followed by a strong shock will eventually suppress operant behavior in progress when the tone is introduced.) Sometimes equated with anxiety.

conditioned stimulus A stimulus which evokes a response or alters some other condition of behavior only because of a history in which it has been paired with a stimulus (often unconditioned) having the same effect.

conditioning, see **operant conditioning, respondent conditioning.**

conjunctive schedule (conj) A schedule in which two contingencies must be met to achieve reinforcement. (E.g., a response is reinforced after 10 minutes and after 100 responses have been emitted, both of these since the preceding reinforcement.)

contingency (of reinforcement or punishment) (1) In operant conditioning: the temporal, intensive, and topographical conditions under which a response is followed by a positive or negative reinforcing stimulus or the removal of either of these. (2) In respondent conditioning: the conditions under which unconditioned and conditioned stimuli are paired.

continuous reinforcement (crf) Reinforcement of every response. Sometimes called "regular" reinforcement. Nonintermittent reinforcement.

control (as in "stimulus control") An observed tendency for a probability or rate of responding to vary with the presence and absence of a variable (e.g. a stimulus).

counter A stimulus some dimension of which varies systematically with number of responses emitted, counted from the preceding reinforcement or some other marking event. FR + counter means reinforcement on FR in the presence of such a stimulus.

crf Continuous reinforcement (q.v.).

cumulative curve A curve showing the number of responses emitted plotted against time. Such a curve is conveniently recorded while the behavior is in progress. Rate of responding can be read from it as the slope at any given point, and compensatory changes in rate can be estimated from inspection.

curve Used in referring to the present figures to refer to any unitary portion of a recorded performance.

D

deprivation (1) As operation: withholding food, water, sexual contact, etc. Any given program establishes a "level" of deprivation. (2) As process: resulting changes in behavior, usually spoken of as an increasing "state" of deprivation. Any given state is a given "level" of deprivation.

deviation A change in rate against an established baseline, often followed by compensatory changes in rate.

differential rate reinforcement Reinforcement (continuous or intermittent) which depends upon the immediately preceding rate of responding. This may be the reciprocal either of the time elapsing between the reinforced response and the immediately preceding response or of the time required to execute three or more responses. See also **differential reinforcement** of high rates, of low rates, and of paced responses.

differential reinforcement Reinforcement which is contingent upon (1) the presence of a given property of a stimulus, in which case the resulting process is discrimination, (2) the presence of a given intensive, durational, or topographical property of a response, in which case the resulting process is differentiation, or (3) a given rate of responding. See also **differential rate reinforcement.**

differential reinforcement of high rates (drh) Reinforcements occur only when the rate is above some specified value.

differential reinforcement of low rates (drl) Reinforcements occur only when the rate is below some specified value.

differential reinforcement of paced responses (drp) or (pacing) Time between the preceding response and the reinforced response is specified more narrowly as falling between certain limits.

differentiation (1) As operation: the differential reinforcement of responses which satisfy some formal specification with respect to intensity, duration, or topography. (2) As process: the resulting change in relative frequency of responses showing the specified property.

disk, see **key.**

discrimination

operant discrimination: (1) As operation: the differential reinforcement of a response with respect to a property of a stimulus (e.g., responses to a red key are reinforced; responses to a green key are not). (2) As process: the resulting change in rate as a function of the properties of the stimuli, observed either concurrently or under later conditions. The organism "shows a discrimination" by responding more rapidly in the presence of the property correlated with reinforcement.

respondent discrimination: (1) As operation: arranging a third stimulus in respondent conditioning in the presence of which unconditioned and conditioned stimuli are paired and in the absence of which they are not (e.g., a tone is paired with food in the presence of a flashing light). (2) As process: the resulting change in behavior by virtue of which a conditioned response is elicited (by the tone) only in the presence of the stimulus present when the conditioned stimulus is paired with a reinforcing stimulus (the flashing light).

discriminative stimulus (1) In operant discrimination: a stimulus in the presence of which a response is reinforced and in the absence of which it goes unreinforced. (2) In respondent discrimination: a stimulus in the presence of which an unconditioned and conditioned stimulus are paired, and in the absence of which they go unpaired.

drh Differential reinforcement of high rates (q.v.).

drl Differential reinforcement of low rates (q.v.).

drp Differential reinforcement of paced responses (q.v.).

E

escape behavior (esc) Behavior which terminates an aversive stimulus (e.g., esc RS 10 designating a contingency in which a response produces the cessation of an aversive stimulus for 10 seconds).

excursion On the standard recorder the pen returns to a baseline after cumulating a block of responses of the order of 800 to 1000. In describing the present figures, each crossing of the paper by the pen is referred to as an excursion.

extinction (ext)

operant extinction: (1) As operation: the withholding of a reinforcement previously contingent upon a response. (2) As process: the resulting decrease in probability or rate.

respondent extinction: (1) As operation: the presentation of a conditioned stimulus occurring without the unconditioned or reinforcing stimulus. (2) As process: the resulting reduction in the magnitude or other dimension of the response elicited by the conditioned stimulus.

F

fine grain, see **grain.**

fixed-interval schedule (FI) A schedule of intermittent reinforcement in which the first response occurring after a given interval of time, measured from the preceding reinforcement, is reinforced. A given interval schedule is designated by adding a number to the letters FI to indicate minutes. (E.g., FI 5 is a schedule in which the first response which occurs 5 minutes or more after the preceding reinforcement is reinforced.)

fixed-ratio schedule (FR) A schedule of intermittent reinforcement in which a response is reinforced upon completion of a fixed number of responses counted from the preceding reinforcement. ("Ratio" refers to the ratio: responses/reinforcement.) A given ratio schedule is designated by adding a number to the letters FR to indicate the ratio. (E.g., FR 100 is a schedule of reinforcement in which the 100th response after the preceding reinforcement is reinforced.)

food magazine, see **magazine.**

FR Fixed-ratio schedule (q.v.).

G

generalization, see **induction.**

gradient A related set of values on a dimension of a stimulus, along which stimulus induction has occurred.

grain The character of the recorded cumulative curve arising from variability in inter-response times. Responses equally spaced regardless of rate produce a "smooth" grain. Groups of responses with pauses interspersed produce a "rough" grain.

The term usually applies to short intervals of time and hence only to local rates (q.v.). Occasionally, it is useful to distinguish between "fine" grain and somewhat larger irregularities in the local rates.

H

hold, see **limited hold.**

"house light" The light responsible for the general illumination in the experimental box.

I

incidental reinforcement, see **accidental reinforcement.**

induction Often called generalization.

stimulus induction: a process through which a stimulus acquires or loses the capacity to elicit a response, control a discriminative response, set up an emotional "state," etc., because of its similarity to a stimulus which has acquired or lost such a capacity through other means. (E.g., if a red light is established as a discriminative stimulus, an orange [or even yellow] light may be found to share the same function, though in lesser degree.)

response induction: a process through which a response changes its probability or rate because it shares properties with another response which has changed its probability or rate through other means.

interlocking schedule (interlock) A schedule of intermittent reinforcement in which the reinforcement is determined by two schedules, where the setting of one schedule is altered by progress made in the other. (E.g., in the schedule interlock FI 5 FR 250, the organism is reinforced at a ratio which is slowly reduced from 250 to 1 during 5 minutes. If responding is rapid, reinforcement occurs only after a large ratio has been completed; if responding is slow, reinforcement occurs at a much lower ratio; if no response occurs within 5 minutes, the first response is reinforced.)

intermittent reinforcement Noncontinuous reinforcement. A schedule according to which not every response is reinforced. See entries listed under schedules of reinforcement.

intermittent schedule A schedule involving intermittent reinforcement as contrasted with continuous reinforcement and extinction.

interpolated schedule (inter) A single block of reinforcements in one schedule is interpolated into a sustained period of responding on a different background schedule. (E.g., on FR 20 inter FI 15, several reinforcements on FR 20 are inserted into an experimental period in which the organism is otherwise reinforced on FI 15.)

inter-response time (IRT) Time elapsing between two successive responses; response-response interval.

interval schedules Schedules of intermittent reinforcement in which reinforcements are programmed by a clock.

IRT Inter-response time (q.v.).

K

key Any manipulable object, the movement of which closes or breaks an electrical circuit. In experiments with pigeons, a useful key is a translucent disk at a convenient height on the wall of the experimental box. When the pigeon pecks this disk, a circuit is made or broken. In experiments with rats, a horizontal bar parallel to and approximately ½ inch from the wall of the experimental box can be pressed downward against a light spring to close or break the circuit.

key light Light projected upon the translucent pigeon key, used as a stimulus. Lights of different colors or patterns may be used.

knee A deviation from a smooth curve, often seen in the early acceleration of interval or ratio segments, consisting of a brief period of rapid responding followed more or less abruptly by a compensatory low rate which restores the curve to an extrapolation of the earlier portion. Opposite of bite (q.v.).

L

lever, see **key.**

limited hold A short period during which a reinforcement arranged by an interval schedule is held available. At the end of the limited hold, a response will not be reinforced until another reinforcement has been set up.

local rate Rate of responding in some small region of a curve. Contrasted with overall rate. See also **rate of responding.**

M

magazine A mechanical device which makes food, water, etc., available to the organism, usually in reinforcement.

manipulandum Any movable object serving as a key (q.v.).

mean rate, see **rate of responding.**

mediating behavior Behavior occurring between two instances of the response being studied (or between some other event and such an instance) which is used by the organism as a controlling stimulus in subsequent behavior. (E.g., under drl the necessary delay in responding is often produced by the incidental reinforcement of mediating behavior which might be called "marking time.") A time out between response and reinforcement may not greatly reduce the effectiveness of the reinforcement if mediating behavior has been acquired during the TO.

mixed schedule (mix) Reinforcement is programmed by two or more schedules alternating usually at random. No stimuli are correlated with the schedules as in multiple schedules. (E.g., mix FI 5 FR 50 represents a schedule in which a reinforcement sometimes occurs after an interval of 5 minutes and sometimes after a "ratio" of 50 responses, the possibilities occurring either at random or according to a program in any determined proportion.)

multiple schedule (mult) Reinforcement is programmed by two or more schedules alternating, usually at random, each schedule being accompanied by an appropriate stimulus as long as the schedule is in force. Differs from mixed schedule simply in the presence of controlling stimuli. (E.g., mult FI 5 FR 100 represents a schedule under which reinforcement sometimes occurs after an interval of 5 minutes and sometimes after 100 responses, the possibilities occurring either at random or according to some program in any determined proportion, when an appropriate stimulus accompanies each schedule.)

N

negative reinforcer An aversive stimulus (q.v.).

nonintermittent schedules Continuous reinforcement (q.v.) and extinction (q.v.).

number Brief expression for the number of responses already emitted in executing a ratio, especially as these function as stimuli. (E.g., "number as reinforcer" means that the accumulation of a number of responses in the course of emitting a ratio acts as a conditioned reinforcer.) See **counter.**

O

operant A unit of behavior defined by a contingency of reinforcement. Pecking a key is an operant if instances are reinforced in a given situation. A class of responses, all members of which are equally effective in achieving reinforcement under a given set of conditions.

operant conditioning (1) As operation: arranging the reinforcement of a response possessing specified properties, or, more specifically, arranging that a given reinforcer follow the emission of a given response. (2) As process: the resulting increase in the rate of occurrence of responses possessing these properties.

operation Arranging or altering some condition in an experiment (e.g., arranging or changing a contingency of reinforcement, depriving an animal of food on a given schedule, introducing a conditioned aversive stimulus, etc.).

over-all rate The mean rate over a fairly large segment of behavior. See also **rate of responding.**

P

pacing, see **differential reinforcement of paced responses.**

pause A period of no responding, contrasted with neighboring fairly high rates, as after reinforcement at high ratios.

Pavlovian conditioning Respondent conditioning (q.v.).

percentage reinforcement On any given schedule, including continuous reinforcement, a certain percentage of reinforcements may be replaced by some other event, such as a time out. [E.g., FR 50 (20%) means that the completion of 50 responses is followed by reinforcement 20% of the time and by a time out 80% of the time, the order of reinforcements and time outs being random.]

performance Behavior characteristically observed under a given schedule (e.g., FI performance).

pre-aversive stimulus A stimulus repeatedly preceding an aversive stimulus.

prime In a multiple schedule a stimulus under which the organism embarks upon a schedule performance which usually sustains itself when the prime is withdrawn.

probability of response The probability that a response will be emitted within a specified interval, inferred from its observed frequency under comparable conditions.

probe A change in conditions at some arbitrary point in an experiment made to evaluate or test for the conditions currently in control. May be a TO, a discriminative stimulus, a schedule-controlling stimulus, etc.

process Any change in rate of responding, specifically as the result of an experimental operation.

programming Arranging a set of reinforcing contingencies, including schedules, stimuli, etc.

punishment An operation in which an aversive or conditioned aversive stimulus is made contingent upon a response.

R

rate of responding Responses per unit time, usually responses per second. It is convenient to distinguish between different rates according to the interval of time covered.

fine grain: rate distribution among smallest clusters of responses.

local rate: rate measured over a short time. The tangent of the cumulative curve at any given point, ignoring the fine grain.

mean rate: responses per unit time calculated for an interval during which changes in local rate have occurred.

over-all rate: mean rate for a still longer period of time (of the order of minutes or hours). Frequently applying to the rate between successive reinforcements, without respect to segment curvature.

running rate: sustained constant rate, often the only important single rate except for zero observed under a given schedule (as in some ratio performances).

terminal rate: the rate reached on the fixed schedules at the moment of reinforcement.

ratio schedules Reinforcements are programmed according to the number of responses emitted by the organism.

record Used in speaking of the present figures to identify a portion representing either a whole session or part of a session, as distinct from curve, excursion, etc.

recorder Device for obtaining a cumulative record of the responses of an organism.

"regular" reinforcement, see **continuous reinforcement**.

reinforcement

operant reinforcement: presenting a reinforcing stimulus when a response occurs, or arranging such presentation.

respondent reinforcement: presenting a conditioned and an unconditioned stimulus at approximately the same time. See also **contingency, differential reinforcement, intermittent reinforcement, continuous reinforcement,** etc.

reinforcer Any event which, when used in the temporal relations specified in reinforcement, is found to produce the process of conditioning.

respondent An unconditioned or conditioned reflex in the sense of a response elicited by a particular stimulus.

respondent conditioning The establishment and strengthening of a conditioned reflex through the roughly simultaneous presentation of unconditioned and conditioned stimuli, as in the Pavlovian experiment.

response (1) An instance of an identifiable part of behavior. (2) A class of such instances. In this sense, response is equivalent to operant (q.v.). See also **behavior.**

response-response interval (RR) Time elapsing between two successive responses. Inter-response time.

response-shock interval (RS) In avoidance conditioning the time elapsing between the last response and occurrence of a shock as an aversive stimulus. See also **avoidance behavior.**

RR Response-response interval (q.v.).

RS Response-shock interval (q.v.).

running rate, see **rate of responding.**

running weight A given body-weight, selected as an indicator of a schedule of deprivation, at which the organism is held from day to day during an experiment.

S

satiation (1) As operation: making food available to an organism, possibly until it stops eating. (2) As process: the resulting change in rate, usually a reduction.

schedules of reinforcement, see the following:

nonintermittent: continuous reinforcement (crf)
extinction (ext)

intermittent: fixed ratio (FR)
variable ratio (VR)
fixed interval (FI)
variable interval (VI)
tandem (tand)
multiple (mult)
mixed (mix)
interlocking (interlock)
alternative (alt)
concurrent (conc)
conjunctive (conj)
interpolated (inter)
chained (chain)

scallop Positively accelerated portion of the cumulative record, usually used in speaking of interval or ratio segments.

second-order effect A relation between the numbers of responses emitted in two or more successive intervals on FI in which an over-all acceleration may be observed throughout all segments.

segment Part of the record of an experimental session, usually between two reinforcements.

session Experimental period. The period during which an organism is exposed to experimental conditions, usually once per day.

set up To set up a reinforcement is to close a circuit so that the next response will be reinforced whenever it occurs.

shock-shock interval (SS) In avoidance conditioning, the interval between successive shocks when no response has been made in that interval. See also **avoidance behavior.**

speedometer A stimulus, some dimension of which changes as a function of the rate of responding measured over some arbitrary period of time.

spontaneous recovery A temporarily higher rate sometimes observed at the beginning of an experimental session, following a session in which the rate has declined (e.g., in extinction). This traditional term suggests that the earlier rate has "recovered" during the intervening time. A more plausible explanation is that stimuli closely associated with the beginning of the session control a higher rate because of earlier conditions of reinforcement and because there has not yet been an opportunity for this effect to be changed by the experimental changes made during the bulk of the preceding session.

spurious reinforcement, see **accidental reinforcement.**

SS Shock-shock interval (q.v.)

stimulus Any physical event or condition in the experimental situation including the organism's own behavior. Not to be confined, as it sometimes is, to those events or conditions which have a demonstrable effect on the organism. See particular names for different types of stimuli.

strength of response Sometimes used to designate probability or rate of responding.

superstitious behavior Behavior strengthened through reinforcing contingencies not explicitly arranged and possibly not frequent or permanent, but nevertheless effective in increasing the strength of the operant.

T

tandem schedule (tand) A schedule of intermittent reinforcement in which a single reinforcement is programmed by two schedules acting in succession without correlated stimuli. E.g., in tand FI 10 FR 5 a reinforcement occurs when 5 responses have been executed after a 10-minute interval has elapsed. In tand FRFI, a (usually) short interval must elapse after the completion of a ratio before a response is reinforced. It is often important to specify which of the two schedules composes the more substantial part of the schedule. This can be done by italicizing the important member.

terminal rate, see **rate of responding.**

time out (TO) Time (in minutes unless otherwise specified) during which the organism characteristically does not engage in the behavior being studied. With pigeons a convenient TO is arranged by turning off all lights in the apparatus. In the rat, a TO can be achieved through the use of a previously developed discriminative stimulus. TOs are used as probes, markers in a series of events, a method of eliminating effects of earlier behavior, etc.

U

unconditioned stimulus A stimulus, the capacity of which to elicit a response does not depend upon its having been paired with another stimulus possessing this capacity.

V

variable Any condition in an experiment, whether manipulable or merely observed, which can be changed or changes.

dependent variable: in these experiments, the behavior of the organism, or more specifically the rate of emission of a given type of response.

independent variable: any condition which is varied systematically in studying a change in the dependent variable (e.g., a stimulus, a reinforcing contingency, a schedule of deprivation, etc.).

variable-interval schedule (VI) A schedule of intermittent reinforcement in which reinforcements are programmed according to a random series of intervals having a given mean and lying between arbitrary extreme values.

variable-ratio schedule (VR) A schedule of intermittent reinforcement under which reinforcements are programmed according to a random series of ratios having a given mean and lying between arbitrary extreme values.

W

warm up Acceleration at the start of a session leading to the level of performance characteristic of the bulk of the session.

water magazine, see **magazine.**

weight The body-weight of the animal usually measured at the beginning of the daily experiment.

Y

yoked boxes A system of controlling frequency of reinforcement as one variable in experiments on ratio schedules. Reinforcements are "set up" in one box whenever they are set up in another on a ratio schedule. The schedule in the yoked box is, nevertheless, an interval schedule.

References

DEWS, P. B. Studies on Behavior. I. Differential sensitivity to pentobarbital of pecking performance in pigeons depending on the schedule of reward. *J. pharmacol. exptl. therap.*, 1955, *113*, 393-401

DINSMOOR, J. A. The effect of periodic reinforcement of bar-pressing in the presence of a discriminative stimulus. *J. comp. physiol. Psychol.*, 1952, *44*, 354-361

ESTES, W. K. and SKINNER, B. F. Some quantitative properties of anxiety. *J. exp. Psychol.*, 1941, *29*, 390-400

FERSTER, C. B. Sustained behavior under delayed reinforcement. *J. exp. Psychol.*, 1953, *45*, 218-224

GUTTMAN, N. and KALISH, H. I. Discriminability and stimulus generalization. *J. exp. Psychol.*, 1956, *51*, 79-88

HERRNSTEIN, R. J. Behavioral consequences of the removal of a discriminative stimulus associated with variable-interval reinforcement. *Unpublished doctoral dissertation*, Harvard University, 1955

HERRNSTEIN, R. J. and MORSE, W. H. Selective action of sodium pentobarbital on component behaviors of a single reinforcement schedule. *Science*, 1956, *124*, 367-368

MORSE, W. H. and HERRNSTEIN, R. J. The effects of drugs on characteristics of behavior maintained by complex schedules of intermittent reinforcement. *Proc. N. Y. Acad. of Sci.*, 1956, *65*, 247-356

SIDMAN, M. Avoidance conditioning with brief shock and no exteroceptive warning signal. *Science*, 1953, *118*, 157-158

SIDMAN, M. and STEBBINS, W. C. Satiation effects under fixed-ratio schedules of reinforcement. *J. comp. physiol. Psychol.*, 1954, *47*, 114-116

SKINNER, B. F. The abolishment of a discrimination. *Proc. Nat. Acad. Sci.*, 1933, *8*, 114-129

SKINNER, B. F. *The behavior of organisms.* New York, Appleton-Century-Crofts, 1938

SKINNER, B. F. *Science and human behavior.* New York, Macmillan, 1953

Index

(Terms marked *"G"* are defined in the Glossary)